ELEMENTARY
ALGEBRA

ELEMENTARY
ALGEBRA

Harold R. Jacobs

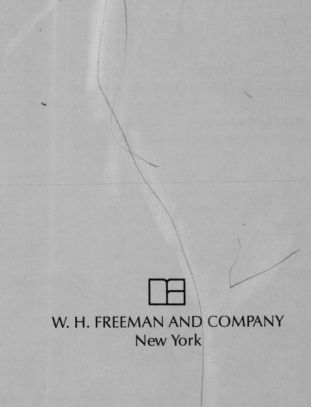

W. H. FREEMAN AND COMPANY
New York

The cover illustration is a periodic drawing by Maurits Escher. Reproduced with the permission of the Escher Foundation, Haags Gemeentemuseum, The Hague.

Library of Congress Cataloging in Publication Data

Jacobs, Harold R
    Elementary algebra.

    Includes index.
    1. Algebra. I. Title.
QA152.2.J33   512.9′042   78-10744
ISBN 0-7167-1047-1

Printed in the United States of America

Tenth printing, 2000

Contents

Contents **vii**

A Letter to the Student

The English philosopher and scientist Roger Bacon once wrote: "Mathematics is the gate and key of the sciences. . . . Neglect of mathematics works injury to all knowledge, since he who is ignorant of it cannot know the other sciences or the things of this world."

In turn, algebra is the gate and key of mathematics. For this reason, colleges and universities require mastery of algebra in preparation for studying not only the sciences, but also such subjects as engineering, medicine, architecture, philosophy, psychology, and law.

Although many problems that can be solved by algebra can also be worked out by common sense, their translation into algebraic form generally makes them easier to deal with. Because of this, algebra has become the language of science. The goal of this course is to learn how to use this language.

Success in algebra depends on a combination of talent and effort. A few people are so gifted in mathematics that they can succeed with very little effort. For most people, however, diligent practice is the key to success. Like developing ability in a sport, becoming good at algebra takes practice. It is my hope that this book will help you both to enjoy the subject and to be successful in your studies.

Harold R. Jacobs

ELEMENTARY
ALGEBRA

INTRODUCTION
A Number Trick

Think of a number from one to ten. Add seven to it. Multiply the result by two. Subtract four. Divide by two. Subtract the number that you first thought of. Is your answer five?

Number tricks such as this have long been popular. That the final result can be known by someone who doesn't know which number was originally chosen is surprising.

How does the trick work? If we make a table (like the one at the top of the next page) showing what happens when it is done with each number from one to ten, some patterns appear.

Would these patterns continue if the table were extended to include other numbers? If we began by thinking of eleven, would the answer at the end still be five? What if we began with one hundred? Would we get five at the end if we began with zero? Do you think it is correct to assume that the trick will work for *any* number you might think of?

Even though you may feel that the answer to every one of these questions is yes, *how* the trick works is still not clear. Merely doing arithmetic with a series

1

of different numbers cannot reveal the secret of why they all lead to the same result.

The number thought of:	1	2	3	4	5	6	7	8	9	10
Add seven:	8	9	10	11	12	13	14	15	16	17
Multiply by two:	16	18	20	22	24	26	28	30	32	34
Subtract four:	12	14	16	18	20	22	24	26	28	30
Divide by two:	6	7	8	9	10	11	12	13	14	15
Subtract the number first thought of:	5	5	5	5	5	5	5	5	5	5

There is a simple way, however, to discover the secret. Instead of writing down a specific number at the start, we will use a symbol to represent whatever number might be chosen. We will begin with a box.

Throughout the trick this box will represent the number originally chosen.

The next step in the trick is to add seven. We will represent numbers we know with sets of circles, and so seven will look like this:

To show the result of adding seven to the number, we draw seven circles beside the box.

If we illustrate the entire trick in this way, it looks like this:

The number thought of:

Add seven:

Multiply by two:

Subtract four:

Divide by two:

Subtract the number first
thought of:

The pictures make it easy to see why, no matter what number we start with, the answer at the end of the trick is always five. The box representing the original number disappears in the last step, leaving five circles.

Doing arithmetic with symbols rather than specific numbers is the basis of algebra. The explanation with the boxes and circles of what is happening throughout the number trick is an example of this. One of our goals in learning algebra will be to learn how to set up and solve problems using symbols such as these.

Exercises

1. Here are directions for another number trick and part of a table to show what happens when the trick is done with each number from one to five.

Think of a number:	1	2	3	4	5
Double it:	2	4	▥	▥	▥
Add six:	8	▥	▥	▥	▥
Divide by two:	4	▥	▥	▥	▥
Subtract the number that you first thought of:	3	▥	▥	▥	▥

a) Copy and complete the table.
b) Does your table prove that the trick will work for *any* number?
c) Show how the trick works by illustrating the steps with boxes and circles. The first two steps are shown below.

Think of a number: ☐

Double it: ☐☐

d) Do your drawings prove that the trick will work for *any* number?

2. The pictures below illustrate the steps of another number trick. Tell what is happening in each step in words.

Step 1. ☐

Step 2. ☐☐☐☐

Step 3. ☐☐☐☐ ○○○○○○○○

Step 4. ☐ ○○

Step 5. ☐ ○○○○○

Step 6. ○○○○○

3. In the next number trick, we will study the effect of changing some of the directions.

Step 1.   Think of a number.
Step 2.   Add four.
Step 3.   Multiply by two.
Step 4.   Subtract four.
Step 5.   Divide by two.
Step 6.   Subtract the number that you first thought of.

a) What is the result at the end of this trick?

b) Suppose that the second step were changed as shown below.

Step 1.   Think of a number.
Step 2.   Add six.
Step 3.   Multiply by two.
Step 4.   Subtract four.
Step 5.   Divide by two.
Step 6.   Subtract the number that you first thought of.

The trick will still work, even though the result at the end is changed. How is it changed?

c) Suppose instead that the fourth step were changed as shown below.

Step 1.   Think of a number.
Step 2.   Add four.
Step 3.   Multiply by two.
Step 4.   Subtract six.
Step 5.   Divide by two.
Step 6.   Subtract the number that you first thought of.

What effect does this have on the trick?

d) Suppose instead that the third step were changed as shown below.

Step 1.   Think of a number.
Step 2.   Add four.
Step 3.   Multiply by four.
Step 4.   Subtract four.
Step 5.   Divide by two.
Step 6.   Subtract the number that you first thought of.

What effect does this have on the trick?

4. Here is the beginning of a number trick. Can you make up more steps so that it will give the same answer for any number a person might choose?

Think of a number.
Triple it.
Add twelve.

Chapter 1

FUNDAMENTAL OPERATIONS

"Can you do Addition?" the White Queen asked.
"What's one and one and one and one and
one and one and one and one and one and one?"
    "I don't know," said Alice. "I lost count."

LEWIS CARROLL, *Through The Looking Glass*

LESSON 1
Addition

Soon after a child is able to count, he learns how to add. The two operations are closely connected, as anyone who has ever added by counting on his fingers knows. Consider the problem of adding the numbers represented by these two sets of circles:

o o o o o   o o

At first a child finds the answer by counting all of the circles. Then he learns the fact that $5 + 2 = 7$.

Another way to picture addition is by lengths along a line. This figure also illustrates the fact that $5 + 2 = 7$.

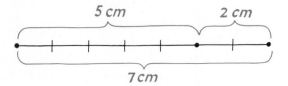

The result of adding two or more numbers, called their **sum**, does not depend on either the order of the numbers or the order in which they are added. To find the number of circles in the pattern above, for example, we could add the numbers of circles in the four rows from top to bottom:

$$1 + 2 + 3 + 4$$

or from bottom to top:

$$4 + 3 + 2 + 1$$

Either way, we get the same number: 10.

In algebra, it is often necessary to indicate the sum of two or more numbers without actually being able to add them. For example, in illustrating the number trick that appears in the introduction to this book, we used a box to represent the original number and a set of circles to represent the number seven:

Original number   Seven

To represent their sum, we drew the seven circles beside the box:

The sum of the original
number and seven

Instead of bothering to draw pictures like this, it is easier to represent the original number with a letter, such as x, and simply write

$$x + 7$$

The expression $x + 7$ means "the sum of x and 7." If we replace x with 1, $x + 7 = 1 + 7 = 8$. If we replace x with 2, $x + 7 = 2 + 7 = 9$, and so forth. Because x can be replaced by various numbers, it is called a **variable**.

If we know both numbers being added, such as 4 and 5, we can write their sum as a number, 9. If we know only one number or neither one, the best that we can do is to write an expression such as $x + 2$ or $x + y$. The length of the line segment below, for example, is the sum of the lengths of the three marked segments.

To indicate this sum, we can write $3 + x + 1$ or, more briefly, $x + 4$. Without knowing the length labeled x, we cannot simplify this answer any further.

Exercises

TEACHER: Haven't you finished adding up those numbers yet?
STUDENT: Oh, yes. I've added them up ten times already.
TEACHER: Excellent! I like a student who is thorough.
STUDENT: Thank you. Here are the ten answers.*

Set I

Find each of the following sums.

1. $1000 + 700 + 70 + 6$
2. $999 + 99 + 9$
3. $1 + 0.9 + 0.08 + 0.004$
4. $20 + 0.2 + 0.002$
5. $1 + 12 + 123 + 1234$
6. $1111 + 222 + 33 + 4$
7. $1 + 1.2 + 1.23 + 1.234$
8. $1.111 + 2.22 + 3.3 + 4$
9. $0.7 + 0.70 + 0.700 + 0.7000$
10. $0.5 + 0.55 + 0.555$

Set II

11. Write a number or expression for each of the following.

a) The sum of 10 and 7.
b) The sum of x and 7.
c) The sum of 10 and y.
d) The sum of x and y.
e) Four added to 8.
f) Four added to z.
g) The sum of 2, 5, and 1.
h) The sum of x, 5, and 1.
i) The sum of 2, y, and 1.
j) The sum of x, y, and 1.

*Alan Wayne, in *Mathematical Circles Revisited* by Howard W. Eves. © Copyright Prindle, Weber & Schmidt, Inc. 1971.

12. In the figures below, the box represents any number and the sets of circles represent specific numbers.

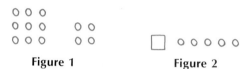

Figure 1   Figure 2

a) What addition problem is illustrated by Figure 1?
b) What is the answer to the problem?
c) Write an algebraic expression to represent the addition problem illustrated by Figure 2.
d) What is the answer to the problem if the box represents 2?
e) What is the answer to the problem if the box represents 4?

13. What is the length marked with a question mark in each of these figures?

a)
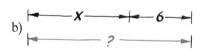

b)
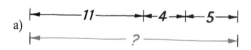

c)
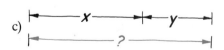

d)
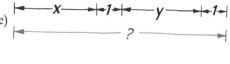

e)
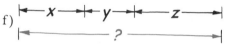

f)

14. The figure below can be used to show that $3 + 7$ and $7 + 3$ are the same number, depending on whether the figure is read from left to right or from right to left.

○ ○ ○   ○ ○ ○ ○ ○ ○ ○

Draw boxes and circles to show that
a) $x + 6$ and $6 + x$ mean the same thing.
b) $2 + x + 5$ and $x + 7$ mean the same thing.
c) $x + 4 + x$ and $4 + x + x$ mean the same thing.

15. The expression $x + y + 2$ represents the sum of x, y, and 2. If x is 1, it can be written as $1 + y + 2$ or $y + 3$. How can $x + y + 2$ be written if
a) x is 8?
b) x is 9?
c) y is 3?
d) y is 0?
e) x is 6 and y is 2?

16. Mr. Benny is 39 years old.
a) How old will he be in 5 years?
b) How old will he be in x years?
c) How old will he be 6 years after that?
Mrs. Benny is x years old.
d) How old will she be in 5 years?
e) How old will she be in y years?
f) How old will she be z years after that?

Set III

17. Write a number or expression for each of the following.
    a) The sum of 3 and 11.
    b) The sum of 3 and x.
    c) The sum of y and 11.
    d) The sum of y and x.
    e) Seven increased by 2.
    f) Seven increased by x.
    g) The sum of 9, 1, and 4.
    h) The sum of x, 1, and 4.
    i) The sum of 9, y, and 4.
    j) The sum of x, y, and 4.

18. In the figures below, the box represents any number and the sets of circles represent specific numbers.

Figure 1

Figure 2

    a) What addition problem is illustrated by Figure 1?
    b) What is the answer to the problem?
    c) Write an algebraic expression to represent the addition problem illustrated by Figure 2.
    d) What is the answer to the problem if the box represents 1?
    e) What is the answer to the problem if the box represents 5?

19. The *perimeter* of a figure is the sum of the lengths of its sides. What is the perimeter of each of these figures?

20. The figure below can be used to show that $4 + 5$ and $5 + 4$ are the same number, depending on whether the figure is read from left to right or from right to left.

    ○ ○ ○ ○   ○ ○ ○ ○ ○

    Draw boxes and circles to show that
    a) $2 + x$ and $x + 2$ mean the same thing.
    b) $8 + x + 1$ and $x + 9$ mean the same thing.
    c) $x + x + 3$ and $x + 3 + x$ mean the same thing.

21. The expression $x + 1 + y$ represents the sum of x, 1, and y. If x is 4, it can be written as $4 + 1 + y$ or $5 + y$. How can $x + 1 + y$ be written if
    a) x is 2?
    b) x is 0?
    c) y is 6?
    d) y is 9?
    e) x is 3 and y is 7?

22. Each week, Dashing Dan jogs one mile farther than he did the week before.
    a) If he jogs 18 miles this week, how far will he jog next week?
    b) If he jogs x miles this week, how far will he jog next week?
    c) If he jogged y miles three weeks ago, how far will he jog this week?
    d) If he jogged y miles z weeks ago, how far will he jog this week?

a)

b)

c)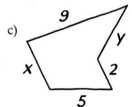

d)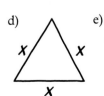

e)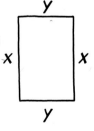

Set IV A Number Puzzle

Numbers have been written in four spaces in this tic-tac-toe design. If we add across the rows and down the columns, we get the sums shown in the second figure. If we now add across the bottom row and down the last column, the answers are the same number:

$$6 + 10 = 16 \quad \text{and} \quad 4 + 12 = 16$$

Is this just a coincidence or would it happen if we started with *any* set of four numbers?

Draw a tic-tac-toe design and, in the same spaces as those in the example above, write four numbers of your own choosing. Add the rows and columns and see what happens. Can you explain why?

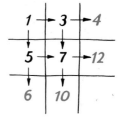

LESSON 2
Subtraction

The month of September 1752 was one of the strangest months in history. The day following September 1 was September 13!

This was done to bring the calendar back into line with the seasons. The calendar established by Julius Caesar in 45 B.C. had as its basis a standard year of 365 days with every fourth year, "leap year," having 366. This resulted in the average length of a year being 365.25 days, whereas the earth in fact travels once around the sun in about 365.24 days. For a short period, this error didn't amount to much, but after many centuries it became so great that it had to be corrected.

The number of days left in the month of September 1752 can be found by subtraction: $30 - 11 = 19$. Subtraction is the opposite of addition because we are "taking away" rather than "adding to." The two operations are closely related, however, because to every subtraction problem there corresponds an addition problem: $30 - 11 = 19$ because $19 + 11 = 30$.

To represent a subtraction problem such as $7 - 2$ by means of circles, we might draw seven circles from which two have been "taken away" by being crossed out.

⊘ ⊘ ○ ○ ○ ○ ○

Subtraction can also be pictured by lengths along a line. The figure below is another way of showing that $7 - 2 = 5$.

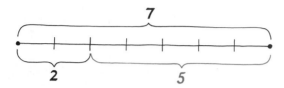

Although addition and subtraction are closely related, there is an important difference between the two operations. The sum of two numbers does not depend on the order of the numbers. The length marked with a question mark in the figure at the left below can be written either as $4 + 1$ or $1 + 4$.

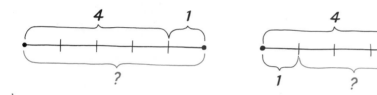

The result of subtracting one number from another, called their **difference,** does depend on the order of the numbers. The length marked with a question mark in the figure at the right is $4 - 1$, not $1 - 4$. When we refer to the difference between two numbers, we mean the number that results from subtracting the second number from the first.

Exercises

Set I

Find each of the following differences.

1. $22222 - 2000$
2. $666 - 77$
3. $1000 - 123$
4. $4.321 - 1$
5. $4.321 - 0.1$
6. $3.1416 - 3.1416$
7. $1 - 0.9$
8. $1 - 0.99$
9. $1812 - 18.12$
10. $181.2 - 1.812$

Set II

11. Write a number or expression for each of the following.
    a) The difference between 10 and 7.
    b) Six decreased by x.
    c) Six taken away from x.
    d) Three less than 11.
    e) One less than x.
    f) The difference between x and y.
    g) The result of subtracting x from 4.
    h) Four subtracted from x.

12. What is the length marked with a question mark in each of these figures?

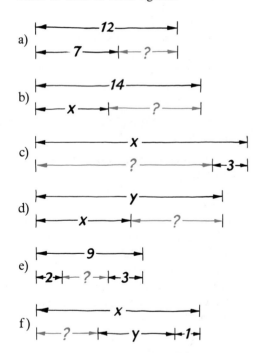

13. Find the value of each of the following expressions for the numbers given.
    a) $x - 4$ if x is 6.
    b) $x - 4$ if x is 7.
    c) $x - 4$ if x is 14.
    d) What happens to the value of $x - 4$ as x gets larger?

e) $15 - x$ if x is 3.
f) $15 - x$ if x is 4.
g) $15 - x$ if x is 10.
h) What happens to the value of $15 - x$ as x gets larger?

14. Find the value of each of the following for the numbers given.
    The sum of x and $y - 3$
    a) if x is 7 and y is 4.
    b) if x is 2 and y is 11.
    The difference between $x + y$ and 3
    c) if x is 7 and y is 4.
    d) if x is 2 and y is 11.
    e) Can you explain why the answers to parts c and d are the same as those to parts a and b?

15. The sum of the numbers on any two opposite faces of a die is 7. Suppose that a die is thrown.
    a) If the number showing on the top of it is 3, what is the number on the bottom?
    b) If the number showing on the top of it is x, what is the number on the bottom?
    Suppose that two dice are thrown.
    c) If the sum of the two numbers showing on top is 8, what is the sum of the two numbers on the bottom?
    d) If the sum of the two numbers showing on top is y, what is the sum of the two numbers on the bottom?

16. Babar weighs 7,000 pounds.
    a) If he loses x pounds, how much will he weigh?
    b) If he gains y pounds, how much will he weigh?

17. The amount of profit that Shirley Feeney makes selling sandwiches depends on how much they cost her and how much she sells them for.

a) If peanut butter sandwiches cost her 21 cents each and she sells them for 45 cents, how much profit does she make on each one?

b) If jelly sandwiches cost her x cents each and she sells them for y cents, how much profit does she make on each one?

c) If egg sandwiches cost her x cents each and she wants to make a profit of 30 cents, how much should she sell them for?

d) If she sells ham sandwiches for 95 cents each and makes a profit of y cents on each one, how much do they cost her?

Set III

18. Write a number or expression for each of the following.
    a) The difference between 9 and 3.
    b) Five taken away from x.
    c) Five decreased by x.
    d) Eight less than 20.
    e) Two less than x.
    f) The difference between y and x.
    g) The result of subtracting x from 7.
    h) Seven subtracted from x.

19. The perimeter of a figure is the sum of the lengths of its sides.

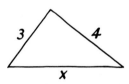

    a) How long is the side marked x in this triangle if the triangle's perimeter is 12?
    b) How long is it if the triangle's perimeter is y?

    Use the perimeters given below each of the following figures to tell the length of the side marked x.

20. Find the value of each of the following expressions for the numbers given.
    a) $x - 2$ if x is 9.
    b) $x - 2$ if x is 10.
    c) $x - 2$ if x is 20.
    d) What happens to the value of $x - 2$ as x gets larger?
    e) $8 - x$ if x is 1.
    f) $8 - x$ if x is 2.
    g) $8 - x$ if x is 8.
    h) What happens to the value of $8 - x$ as x gets larger?

21. Find the value of each of the following for the numbers given.
    The sum of x and $7 - y$
    a) if x is 5 and y is 1.
    b) if x is 13 and y is 6.
    The difference between $x + 7$ and y
    c) if x is 5 and y is 1.
    d) if x is 13 and y is 6.
    e) Can you explain why the answers to parts c and d are the same as those to parts a and b?

c)

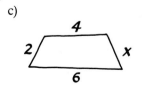

Perimeter is 15

d)

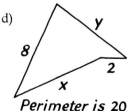

Perimeter is 20

e)

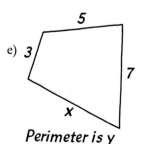

Perimeter is y

f)

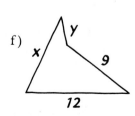

Perimeter is z

22. A log is cut into two pieces.
   a) If the log was 12 feet long and one piece is x feet long, how long is the other piece?
   b) If the two pieces are x feet and y feet long, how long was the log?
   c) If the log was x feet long and one piece is y feet long, how long is the other piece?

23. Laverne DeFazio has 2 dollars in her checking account.
   a) If she writes a check for x dollars, how much money will remain in her account?
   b) If she makes a deposit of y dollars, how much money will she have in her account?
   c) If her account increases to z dollars, how much money has she added to it?

24. The Swinging Singles Tennis Club has 100 members.
   a) If x of them are men, how many are women?
   b) If y people join the club, how many members will it have?
   c) If y people join the club and z people resign, how many members will it have?

Set IV

"Forty-eight, forty-nine, fifty, seventy-five, nine, ten, twenty."

This seems like a strange way to count and yet clerks in stores do it all the time. What is going on? Can you tell what problem is being solved? Is the problem being solved by addition or subtraction?

LESSON 3
Multiplication

**"Six times six is 54! Don't they teach you anything
at that school?"**

Learning the multiplication table is not an easy task. When you first learned how to multiply, you did it by adding. For example, the problem 3×5 can be illustrated by three sets of circles with five circles in each set.

o o o o o   o o o o o   o o o o o

The circles can also be arranged in three rows to form a rectangle.

o o o o o
o o o o o
o o o o o

Both patterns show that $3 \times 5 = 5 + 5 + 5 = 15$. In learning the multiplication table, you memorized the answers to problems such as this so that pictures and adding became unnecessary.

The result of multiplying two or more numbers is called their **product.** Another way to picture a product is by means of area. The rectangle at the right, for example, is divided into 4 rows of squares with 7 squares in each row: it contains

$$4 \times 7$$

squares in all. The area of the rectangle, 28, is the product of its dimensions, 4 and 7.

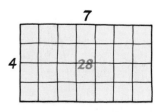

17

Something that helps in learning the multiplication table is the fact that if

$$4 \times 7 = 28$$

then it is also true that

$$7 \times 4 = 28$$

The product of two or more numbers, like their sum, does not depend on either their order or the order in which they are multiplied.

Each of the number tricks that we considered in the introductory lesson included a step consisting of multiplication. For example, if we are told to think of a number and multiply it by four, the result might be illustrated by a set of four boxes:

If we use a letter, such as x, to represent the number thought of, we might write:

$$4 \times x$$

Because the symbol for multiplication used in arithmetic looks so much like the letter x, however, it is not ordinarily used in algebra. Instead, we simply write $4x$ with the understanding that this means "4 times x." We can't indicate the product of two numbers such as 3 and 5 this way because 35 means "thirty-five," not "three times five." To indicate that the 3 and 5 are two separate numbers, we can either enclose them in parentheses, (3)(5), or insert a raised dot between them, $3 \cdot 5$.

In this lesson we have observed that the product of two numbers, such as $4x$, can be interpreted either as repeated addition,

$$x + x + x + x$$

or as the area of a rectangle whose dimensions are 4 and x.

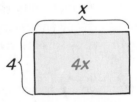

In the next lesson, we will see how these ideas can be applied to division.

Exercises

Set I

Find each of the following products.

1. $100 \cdot 360$
2. $(5)(142857)$
3. $271 \cdot 287$
4. $(0.05)(20)$

5. $(1.5)(8.23)$
6. $(8.23)(1.5)$
7. $(0.7)(1.1)(1.3)$
8. $(7)(1.1)(1.3)$

9. $(7)(11)(1.3)$
10. $2 \cdot 2 \cdot 2 \cdot 2 \cdot 5 \cdot 5 \cdot 5 \cdot 5$

Set II

11. Draw figures as indicated.
    a) A figure with circles to show that $4 \cdot 3$ and $3 \cdot 4$ are the same number.
    b) A figure with boxes to illustrate $5x$ if each box represents x.
    c) A rectangle divided into squares to illustrate $2 \cdot 7$.

12. Write a number or expression for each of the following.
    a) The product of 5 and 6.
    b) The sum of 5 and 6.
    c) The product of 5 and x.
    d) The sum of 5 and x.
    e) The product of x and y.
    f) The sum of x and y.
    g) The product of x and x.
    h) Eight multiplied by x.
    i) Eight subtracted from x.
    j) The sum of 2, 7, and x.
    k) The product of 2, 7, and x.
    l) The sum of 10, y, and 3.
    m) The product of 10, y, and 3.
    n) The sum of 4, x, and y.
    o) The product of 4, x, and y.

13. The multiplication problem $4 \cdot 3$ and the addition problem $3 + 3 + 3 + 3$ are equivalent. Write a multiplication problem equivalent to each of the following addition problems.

    a) $2 + 2 + 2 + 2 + 2 + 2$
    b) $6 + 6$
    c) $x + x + x + x + x$
    d) $\underbrace{7 + 7 + \cdots + 7}_{\text{11 of them}}$
    e) $\underbrace{7 + 7 + \cdots + 7}_{x \text{ of them}}$
    f) $\underbrace{y + y + \cdots + y}_{x \text{ of them}}$

    Write an addition problem equivalent to each of the following multiplication problems.

    g) $3 \cdot 17$
    h) $4x$
    i) $y \cdot 2$
    j) yz

14. The area of a rectangle is the product of its length and width. What is the area of each of these rectangles?

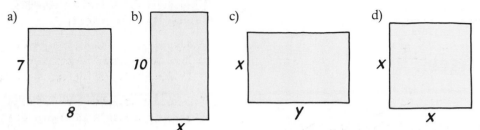

a) 7 / 8

b) 10 / x

c) x / y

d) x / x

15. Although their name suggests that they have 100 legs, some centipedes have only 28 legs whereas others have as many as 354.
    a) How many legs do 5 centipedes have altogether if each one has 28 legs?
    b) How many legs do x centipedes have altogether if each one has 354 legs?

16. Because there are 60 minutes in an hour, there are $60x$ minutes in x hours.
    a) How many days are there in x weeks?
    b) How many hours are there in x days?
    c) How many minutes are there in one day?

d) How many minutes are there in x days?
e) How many minutes are there in x weeks?
f) How many years are there in x centuries?
g) How many months are there in x centuries?

17. Miss Haversham's Hupmobile gets about 11 miles per gallon.
    a) Approximately how many miles should she be able to travel on a full tank of 15 gallons?
    b) Approximately how many miles can she travel on x gallons of gas?

Set III

18. Draw figures as indicated.
    a) A figure with circles to show that $2 \cdot 6$ and $6 \cdot 2$ are the same number.
    b) A figure with boxes to illustrate $3x$ if each box represents x.
    c) A rectangle divided into squares to illustrate $4 \cdot 5$.

19. Write a number or expression for each of the following.
    a) The product of 7 and 3.
    b) The sum of 7 and 3.
    c) The product of 7 and x.
    d) The sum of 7 and x.
    e) The product of y and x.
    f) The sum of y and x.
    g) The product of y and y.

h) Five multiplied by x.
i) Five subtracted from x.
j) The sum of 4, 6, and x.
k) The product of 4, 6, and x.
l) The sum of 5, y, and 12.
m) The product of 5, y, and 12.
n) The sum of x, y, and 2.
o) The product of x, y, and 2.

20. The multiplication problem $5 \cdot 8$ and the addition problem $8 + 8 + 8 + 8 + 8$ are equivalent. Write a multiplication problem equivalent to each of the following addition problems.

    a) $10 + 10 + 10$
    b) $3 + 3 + 3 + 3 + 3 + 3 + 3 + 3 + 3 + 3$

c) $x + x + x$

d) $\underbrace{4 + 4 + \cdots + 4}_{\text{15 of them}}$

e) $\underbrace{4 + 4 + \cdots + 4}_{x \text{ of them}}$

f) $\underbrace{x + x + \cdots + x}_{y \text{ of them}}$

Write an addition problem equivalent to each of the following multiplication problems.

g) $2 \cdot 19$

h) $5x$

i) $y \cdot 3$

j) $x \cdot x$

21. The volume of a rectangular box is the product of its length, width, and height. The volume of the box shown here, for example, is $5 \cdot 3 \cdot 2 = 30$ because it contains 30 cubes.

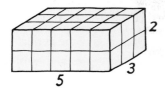

What is the volume of each of these boxes?

a)

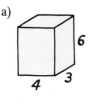
6
4 3

b)

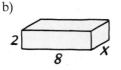
2
8 x

c)

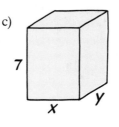
7
x y

d)

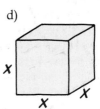
x
x x

22. At Mr. Kitzel's Bakery, one donut costs 9 cents and there is no discount for buying donuts in quantity.

a) How much would a dozen donuts cost?
b) How much would x donuts cost?
c) How much would y dozen donuts cost?

23. The trees in an orchard are arranged in rows with an equal number of trees in each row.

a) If there are x rows and each row contains 20 trees, how many trees are there in all?
b) If there are x rows and each row contains x trees, how many trees are there in all?

24. Because there are 100 centimeters in one meter, there are $100x$ centimeters in x meters.

a) How many millimeters are there in one meter?
b) How many millimeters are there in x meters?
c) How many meters are there in three kilometers?
d) How many meters are there in y kilometers?
e) How many centimeters are there in one kilometer?
f) How many centimeters are there in y kilometers?
g) How many millimeters are there in y kilometers?

Set IV

An old-fashioned method for multiplying two numbers is illustrated in the drawings shown here.

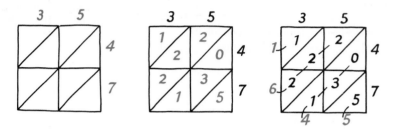

The numbers to be multiplied, 35 and 47, are written above and to the right of the figure as shown in the first drawing. Each digit of one number is multiplied by each digit of the other, $3 \cdot 4 = 12$, $5 \cdot 4 = 20$, $3 \cdot 7 = 21$, $5 \cdot 7 = 35$, and the answers written in the boxes as shown in the second drawing. The digits in each slanting column are added and their sums written below and to the left as shown in the third drawing. The answer is found by reading these digits in order from the upper left: $35 \cdot 47 = 1645$.

Try this method on the following problems. Does it give the correct answer in each case?

1. $52 \cdot 76$
2. $83 \cdot 29$

LESSON 4
Division

Problem: $\dfrac{12}{2} = ?$

Calculation:

$$
\begin{array}{rl}
12 & \\
-\ 2 & 1 \\
\hline
10 & \\
-\ 2 & 2 \\
\hline
8 & \\
-\ 2 & 3 \\
\hline
6 & \\
-\ 2 & 4 \\
\hline
4 & \\
-\ 2 & 5 \\
\hline
2 & \\
-\ 2 & 6 \text{ times} \\
\hline
0 &
\end{array}
$$

Answer: $\dfrac{12}{2} = 6$

If two people share a dozen clams so that each one gets the same number, how many will each one get? This is such an easy division problem that we know the answer immediately. One way to illustrate it is shown below.

$$
\circ\ \circ\ |\ \circ\ \circ\ |\ \circ\ \circ\ |\ \circ\ \circ\ |\ \circ\ \circ\ |\ \circ\ \circ
$$

Twelve circles have been separated into groups of two ("one clam for you and one for me, one for you and one for me," and so forth); the answer, six, can be found by counting the number of groups.

The method that some mechanical calculators use to divide is illustrated at the right. The calculator subtracts 2 from 12, 2 from the result, 2 from that result, and so on until it arrives at 0. The number of times 2 has been subtracted is the answer.

Although this may seem like a peculiar way to divide, it is related to the way that we have been picturing multiplication as repeated addition. The calculator is doing division by repeated subtraction.

Division can also be interpreted in terms of multiplication. The answer to the problem of dividing 12 by 2 is the number that must be multiplied by 2 to give 12. This interpretation can also be pictured by means of the relationship of the area of a rectangle to its dimensions, as the figure at the right illustrates.

The result of dividing one number by another is called their **quotient.** By the

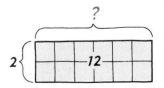

23

quotient of the numbers x and y, we mean the result of dividing x by y and write it as $\frac{x}{y}$. The quotient of two numbers, like their difference, depends on the order of the numbers. The quotient of 3 and 6, $\frac{3}{6}$, for example, is not the same number as the quotient of 6 and 3, $\frac{6}{3}$. When we refer to the quotient of two numbers, we mean the number that results from dividing the first number by the second.

Although quotients are found by *division*, they can be checked by *multiplication*. To check that $\frac{80}{5} = 16$, for example, we multiply 5 by 16 to see if the result is 80. In general, the quotient $\frac{x}{y}$ is the number that must be multiplied by y to give x.

Exercises

Set I

Find each of the following quotients.

1. $\frac{500}{10}$ 3. $\frac{7404}{6}$ 5. $\frac{4}{2.5}$ 7. $\frac{2.5}{4}$

2. $\frac{10}{500}$ 4. $\frac{111111}{37}$ 6. $\frac{40}{2.5}$ 8. $\frac{2.5}{40}$

Set II

9. Write a number or expression for each of the following.
   a) The quotient of 12 and 3.
   b) The difference between 12 and 3.
   c) Seven divided by x.
   d) Seven divided into x.
   e) The quotient of x and 2.
   f) The product of x and 2.
   g) The result of dividing 10 by x.
   h) The result of subtracting x from 10.

   i) The quotient of x and y.
   j) The product of x and y.

10. The figure below illustrates two division problems: $\frac{6}{2} = 3$ and $\frac{6}{3} = 2$.

What division problems are illustrated by the following figures?

a)

○ ○ ○ | ○ ○ ○ | ○ ○ ○ | ○ ○ ○

b)
○ ○ | ○ ○
○ ○ | ○ ○
○ ○ | ○ ○
○ ○ | ○ ○

c)
○ ○ ○ | ○ ○ ○ | ○ ○ ○
○ ○ ○ | ○ ○ ○ | ○ ○ ○

11. A common way in which to check a division is to multiply the answer by the dividing number to see if the result is equal to the number divided. For example, if $\frac{12}{6} = 2$, then $6 \cdot 2 = 12$. Write the multiplication problem that "checks" each of these division problems.

a) $\frac{15}{5} = 3$ e) $\frac{x}{10} = 7$

b) $\frac{92}{23} = 4$ f) $\frac{36}{x} = 12$

c) $\frac{0}{12} = 0$ g) $\frac{20}{4} = x$

d) $\frac{7.5}{7.5} = 1$ h) $\frac{x}{y} = 2$

12. Find the value of each of the following expressions for the numbers given.
a) $9x$ if x is 5.
b) $9x$ if x is 7.
c) $9x$ if x is 12.
d) What happens to the value of $9x$ as x gets larger?

e) $\frac{x}{4}$ if x is 4. g) $\frac{x}{4}$ if x is 100.

f) $\frac{x}{4}$ if x is 20.

h) What happens to the value of $\frac{x}{4}$ as x gets larger?

i) $\frac{30}{x}$ if x is 2. j) $\frac{30}{x}$ if x is 5.

k) $\frac{30}{x}$ if x is 60.

l) What happens to the value of $\frac{30}{x}$ as x gets larger?

13. A band of pirates has 300 bottles of beer, which the pirates plan to share equally.
a) If there are 15 pirates in the band, how many bottles will each one get?
b) If there are x pirates in the band, how many bottles will each one get?

14. Mr. Vanderbilt bought some gold at $170 per ounce.
a) How much did he pay if he bought x ounces?
b) How many ounces did he get if he paid $102,000?
c) How many ounces would he get for x dollars?

15. The common flea is capable of covering 12 inches in one jump.
a) How far can a flea travel if it makes x jumps?
b) How many jumps would a flea have to make in order to cover 600 inches?
c) How many jumps would it have to make in order to cover x inches?

16. Miss Haversham drove her Hupmobile 159 miles.
a) If it used 15 gallons of gas, how many miles per gallon did she get?
b) If it used x gallons of gas, how many miles per gallon did she get?

Set III

17. Write a number or expression for each of the following.

    a) The quotient of 8 and 2.
    b) The difference between 8 and 2.
    c) Five divided into x.
    d) Five divided by x.
    e) The quotient of 3 and x.
    f) The product of 3 and x.
    g) The result of dividing x by 12.
    h) The result of subtracting 12 from x.
    i) The quotient of y and x.
    j) The product of y and x.

18. Find the value of each of the following expressions for the numbers given.

    a) $7x$ if x is 3.
    b) $7x$ if x is 6.
    c) $7x$ if x is 11.
    d) What happens to the value of $7x$ as x gets larger?

    e) $\frac{x}{3}$ if x is 0.

    f) $\frac{x}{3}$ if x is 12.

    g) $\frac{x}{3}$ if x is 51.

    h) What happens to the value of $\frac{x}{3}$ as x gets larger?

    i) $\frac{18}{x}$ if x is 9.

    j) $\frac{18}{x}$ if x is 10.

    k) $\frac{18}{x}$ if x is 45.

    l) What happens to the value of $\frac{18}{x}$ as x gets larger?

19. Find the missing dimension for each of these rectangles. (The numbers inside represent their areas.)

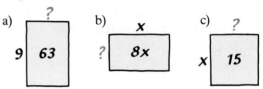

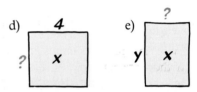

20. Most people learn how to do long division without knowing why it works. The method by which it is done is based on repeated subtraction. For example, compare the two methods below:

Long division	Repeated subtraction	
12		
$15)\overline{180}$	180	
-150	-150	10 fifteens
30	30	
-30	-30	2 more fifteens
0	0	12 fifteens subtracted

    a) Write this long division problem as a repeated subtraction problem.

$$\begin{array}{r} 46 \\ 21)\overline{966} \\ -840 \\ \hline 126 \\ -126 \\ \hline 0 \end{array}$$

b) Divide 875 by 7 using both long division and repeated subtraction.

21. Suppose that a dogcart were to travel at a steady rate for 2,000 meters.
    a) How long would it take if the dogcart traveled 100 meters each minute?
    b) How long would it take if the dogcart traveled x meters each minute?

22. The members of the River City School band are marching in a rectangular array of rows and columns.
    a) If there are x rows and y columns, how many people are in the band?
    b) If there are 80 people in the band and x rows, how many columns are there?
    c) If there are x people in the band and 8 columns, how many rows are there?

23. A cube has six faces.
    a) If each face of a cube has a surface area of 25 square inches, what is its total surface area?
    b) If each face of a cube has a surface area of x square inches, what is its total surface area?
    c) If the total surface area of a cube is 96 square inches, what is the surface area of one face?
    d) If the total surface area of a cube is y square inches, what is the surface area of one face?

24. On Monday, Mr. Kitzel made 10 dozen donuts.
    a) If it cost him $10.80, what was his cost per dozen donuts?
    b) If it cost him $10.80, what was his cost per donut?
    c) If it cost him x dollars, what was his cost per dozen donuts?
    d) If it cost him x dollars, what was his cost per donut?

Set IV   The Pilgrims and the Loaves of Bread

Two pilgrims stopped by the side of the road to eat. One had seven loaves of bread and the other had five loaves. A third traveler arrived before they had begun their meal and asked them to share their food with him. They agreed and the three shared the bread equally.

After they had finished, the third traveler got up, thanked the two pilgrims for the bread, and left twelve silver pieces in payment for his meal. The pilgrim who originally had seven loaves of bread thought that he should receive seven of the coins and his fellow pilgrim should receive five, in the same numbers as their original loaves of bread. The other pilgrim, however, thought that the coins should be split six and six, because the bread had been shared equally.

They could not agree, and so they asked a local wise man what to do. The wise man decided that the pilgrim who originally had seven loaves of bread should receive nine silver pieces and the one who originally had five loaves should receive only three.*

Can you explain why this is fair?

---

* Puzzles of this sort date from Roman times. An example similar to the one given here can be found in *Mathematical Recreations*, second edition, by Maurice Kraitchik (Dover, 1953).

LESSON 5
Raising to a Power

It is an amusing Speculation to look back, and compute what Numbers of Men and Women among the Ancients, clubb'd their Endeavours to the Production of a Single Modern.

BENJAMIN FRANKLIN, *Poor Richard's Almanack,* 1751

If you traced your family tree through ten generations, how many ancestors would there be in the tenth generation back? Because you are descended from two parents, each of whom had two parents, each of whom had two parents, and so on, the numbers in each generation back are:

2 parents,
2 · 2 grandparents,
2 · 2 · 2 great grandparents,
2 · 2 · 2 · 2 great great grandparents,
2 · 2 · 2 · 2 · 2 great great great grandparents,
and so on.

It can be seen from this pattern that a person has $2 \cdot 2 \cdot 2 \cdot 2 \cdot 2 \cdot 2 \cdot 2 \cdot 2 \cdot 2 \cdot 2$ ancestors in the tenth generation preceding his or her own.

A simpler way to write this number is 2^{10}. The 10, called an *exponent,* indicates that 10 twos are to be multiplied:

$$2^{10} = 2 \cdot 2 \cdot 2 \cdot 2 \cdot 2 \cdot 2 \cdot 2 \cdot 2 \cdot 2 \cdot 2 = 1,024$$

In the tenth generation back, you may have more than a thousand ancestors!

The solution of this problem requires repeated multiplication by the same number. Such an operation is called *raising to a power*. The number of grandparents that a person has is 2^2, or "2 raised to the second power." The number of a person's great grandparents is 2^3, or "2 raised to the third power." In each of these cases, we can represent the number by a geometric pattern of circles. Because 2^2 can be pictured as a square, it is also referred to as "2 squared." Because 2^3 can be pictured as a cube, it is also referred to as "2 cubed."

2^2 2^3

Space as we know it is limited to three dimensions, and so powers higher than the third do not have special names.

It is important to understand the difference in meaning between *multiplication* and *raising to a power* and in the symbols used to represent each of these operations. The following examples should make this difference clear.

$3 \cdot 4$ means "3 times 4" or $4 + 4 + 4$

4^3 means "4 to the third power" or $4 \cdot 4 \cdot 4$

$3x$ means "3 times x" or $x + x + x$

x^3 means "x to the third power" or $x \cdot x \cdot x$

nx means "n times x" or the sum of n x's

x^n means "x to the nth power" or the product of n x's

Exercises

Set I

Find each of the following powers.

1. 5^2
2. 2^5
3. 10^3
4. 10^7
5. 1^3
6. 1^7
7. $(1.3)^2$
8. $(3.1)^2$
9. $(0.4)^3$
10. $(0.4)^6$

Set II

11. The expression x^2 can be named in more than one way.
    a) Write two different names for it.
    b) What is the 2 called?

12. What numbers or expressions do these figures represent? Express each as a power.

a)   b)   c)

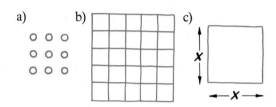

d)   e)

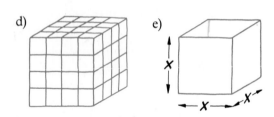

13. Write each of the following in symbols.
    a) Seven squared.
    b) Two raised to the sixth power.
    c) The number x cubed.
    d) The eighth power of x.
    e) Three raised to the xth power.
    f) The yth power of x.

14. The raising-to-a-power problem 3^5 and the multiplication problem $3 \cdot 3 \cdot 3 \cdot 3 \cdot 3$ are equivalent. Write a power problem equivalent to each of the following.
    a) $7 \cdot 7 \cdot 7 \cdot 7$
    b) $4 \cdot 4 \cdot 4 \cdot 4 \cdot 4 \cdot 4 \cdot 4$
    c) $x \cdot x \cdot x \cdot x \cdot x \cdot x$
    d) $\underbrace{2 \cdot 2 \cdot \ldots \cdot 2}_{\text{12 of them}}$

e) $\underbrace{2 \cdot 2 \cdot \ldots \cdot 2}_{x \text{ of them}}$

f) $\underbrace{x \cdot x \cdot \ldots \cdot x}_{y \text{ of them}}$

Write a multiplication problem equivalent to each of the following.

g) 8^5 i) 3^x
h) x^3 j) y^x

15. The figure below contains 7^4 dots.

a) How many dots is that?
b) Because the dots are arranged in a square pattern, their number can also be written as a certain number squared. What is it?

16. The number 243 can be written as a power of 3. To find out what power it is, we can make a list of powers of 3 until we come to 243:

$$3^2 = 9, \ 3^3 = 27, \ 3^4 = 81, \ 3^5 = 243$$

Express each of the following numbers as a power of the number given.
a) 729 as a power of 3.
b) 64 as a power of 2.
c) 64 as a power of 4.

d) 64 as a power of 8.

e) 10,000 as a power of 10.

f) 1,000,000,000 as a power of 10.

g) It is impossible to express 5 as a power of 1. Explain why.

17. If we know that 4^5 is 1,024, we can find 4^6 by observing that

$$4^6 = 4 \cdot \underbrace{4 \cdot 4 \cdot 4 \cdot 4 \cdot 4}_{4^5} \quad \text{so} \quad 4^6 = 4 \cdot 1,024 = 4,096$$

Use a similar method to find each of the following.

a) 2^9 if $2^8 = 256$.

b) 11^4 if $11^3 = 1,331$.

c) 3^7 if $3^5 = 243$.

d) 5^8 if $5^5 = 3,125$.

e) What would you have to multiply x^6 by in order to get x^7?

f) What would you have to multiply x^{10} by in order to get x^{12}?

Set III

18. The expression x^3 can be named in more than one way.

a) Write two different names for it.

b) What is the 3 called?

19. What numbers or expressions do these figures represent? Express each as a power.

a)

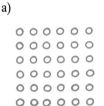

b)

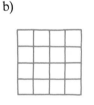

c)

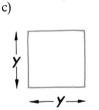

d)

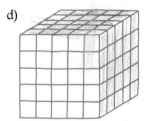

e)

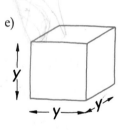

20. Write each of the following in symbols.

a) Two cubed.

b) Five raised to the tenth power.

c) The number x squared.

d) The fourth power of x.

e) Nine raised to the yth power.

f) The xth power of y.

21. The raising-to-a-power problem x^4 and the multiplication problem $x \cdot x \cdot x \cdot x$ are equivalent. Write a power problem equivalent to each of the following.

a) $6 \cdot 6 \cdot 6 \cdot 6 \cdot 6$

b) $11 \cdot 11 \cdot 11$

c) $y \cdot y \cdot y \cdot y \cdot y \cdot y \cdot y$

d) $\underbrace{3 \cdot 3 \cdot \ldots \cdot 3}_{10 \text{ of them}}$

e) $\underbrace{3 \cdot 3 \cdot \ldots \cdot 3}_{x \text{ of them}}$

f) $\underbrace{x \cdot x \cdot \ldots \cdot x}_{y \text{ of them}}$

Write a multiplication problem equivalent to each of the following.

g) 1^4

h) x^5

i) 5^x

j) x^y

22. The number 625 can be written as a power of 5. To find out what power it is, we can make a list of powers of 5 until we come to 625:

$$5^2 = 25, 5^3 = 125, 5^4 = 625$$

Express each of the following numbers as a power of the number given.

a) 343 as a power of 7.

b) 6,561 as a power of 81.

c) 6,561 as a power of 9.

d) 6,561 as a power of 3.

e) 1,000 as a power of 10.

f) 10,000,000 as a power of 10.

g) It is impossible to express 10 as a power of 1. Explain why.

23. This table shows the values of the second through sixth powers of 6.

$$6^2 = 36$$
$$6^3 = 216$$
$$6^4 = 1,296$$
$$6^5 = 7,776$$
$$6^6 = 46,656$$

a) Can you guess what one of the digits of 6^{100} might be?

b) Make a table showing the values of the second through sixth powers of 5.

c) Can you guess what any of the digits of 5^{100} might be?

This table shows the values of the second through sixth powers of 9.

$$9^2 = 81$$
$$9^3 = 729$$
$$9^4 = 6,561$$
$$9^5 = 59,049$$
$$9^6 = 531,441$$

d) Can you guess what any of the digits of 9^{100} might be?

Set IV

After fooling around all summer, Obtuse Ollie didn't want to work very hard in the first few weeks of school. He decided to study algebra one minute the first week, two minutes the second week, four minutes the third week, and so on, doubling the amount of time each succeeding week.

If he sticks to this plan and the semester contains twenty weeks, how many minutes will Ollie study algebra in the last week?

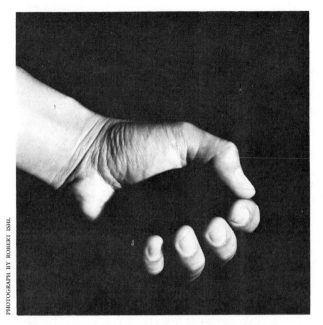

A visual paradox: How many objects is the hand holding?

Zero is the first of ten symbols—the digits—with which we are able to represent any of an infinitude of numbers. Zero is also the first of the numbers which we must represent. Yet zero, first of the digits, was the last to be invented; and zero, first of the numbers, was the last to be discovered.

CONSTANCE REID, *From Zero to Infinity*

LESSON **6**
Zero and One

Although the Alexandrian astronomer Ptolemy used the symbol o, an abbreviation of a word meaning "nothing," as a digit in his work, it was not until many centuries later that the idea of zero as a number was accepted. Because numbers originated with counting and it doesn't seem natural to count with zero, it was not considered to be a number. The *counting numbers,* also called the *natural numbers,* begin with one. Although zero is never used in counting, it is sometimes used to answer the same question that the counting numbers answer, the question of how many.

The behavior of the number zero in calculations differs from that of all other numbers in several basic ways. It is the only number that can be added to or subtracted from another number without changing that number.

▶ For every number x, $x + 0 = x$ (also, $0 + x = x$)   and $x - 0 = x$.

It is the only number that, regardless of what number it is multiplied by, always gives the same result: zero.

▶ For every number x, $x \cdot 0 = 0$.

If x is a counting number, such as 5, it is easy to see why: $5 \cdot 0 = 0 + 0 + 0 + 0 + 0 = 0$. Assuming that the product of two numbers does not depend on the order in which they are multiplied, it is also true that

$$0x = 0$$

Strange as it may seem, it is easy to divide zero by another number, *yet dividing a number by zero makes no sense at all!* Remember that the quotient of two numbers x and y, $\dfrac{x}{y}$, is the number that must be multiplied by y to give x. For example, $\dfrac{6}{2} = 3$ because $3 \cdot 2 = 6$. Now dividing zero by another number is okay: $\dfrac{0}{x} = 0$, because, as we have observed above, $x \cdot 0 = 0$.

On the other hand, dividing a number by zero leads to trouble. If we tried dividing 3 by 0, for example, the number $\dfrac{3}{0}$ would be the number that must be multiplied by 0 to give 3. But there is *no such number;* every number multiplied by 0 gives *zero* as the result.

Dividing zero by itself leads to trouble of a different sort. Suppose that $\dfrac{0}{0}$ is equal to some number x: if $\dfrac{0}{0} = x$, then it must be true that $x \cdot 0 = 0$. But this is true for *every number x*. Hence $\dfrac{0}{0} = 0, \dfrac{0}{0} = 1, \dfrac{0}{0} = 2$, and so forth. Because $\dfrac{0}{0}$ can mean anything, it is meaningless.

The number one plays the same role in multiplication and division that the number zero plays in addition and subtraction: it does not change the number that it is multiplying or dividing.

▶ For every number x, $\quad x \cdot 1 = x$ (also, $1x = x$) $\quad$ and $\quad \dfrac{x}{1} = x$.

Exercises

Beginning with this lesson, the exercises in Set I will review ideas from earlier lessons.

Set I

1. Show how the following number trick works by drawing boxes and circles to illustrate the steps.
   a) Think of a number.
   b) Add four.
   c) Multiply by three.
   d) Subtract nine.
   e) Divide by three.
   f) Subtract the number that you first thought of.
   The result is one.

2. Write a number for each of the following:
   a) The product of 3 and x.
   b) The sum of 3 and x.
   c) The difference between 3 and x.
   d) The quotient of 3 and x.
   e) The third power of x.
   f) The xth power of 3.

3. Write another expression equivalent to each of the following.
   a) $a + a$
   b) $5b$
   c) $c \cdot c \cdot c$
   d) d^4
   e) $\underbrace{e + e + \cdots + e}_{x \text{ of them}}$
   f) $\underbrace{f \cdot f \cdot \cdots \cdot f}_{y \text{ of them}}$

Set II

4. What do you know about the following?
   a) The sum of any number and zero.
   b) The difference between any number and zero.
   c) The product of any number and zero.
   d) The product of any number and one.
   e) The quotient of zero and any number.
   f) The quotient of any number and zero.
   g) The quotient of any number and one.

5. Sometimes it is easier to multiply than to add. Figure out each of the following:
   a) $0 + 1 + 2 + 3 + 4 + 5 + 6 + 7 + 8 + 9$
   b) $0 \cdot 1 \cdot 2 \cdot 3 \cdot 4 \cdot 5 \cdot 6 \cdot 7 \cdot 8 \cdot 9$
   c) $1 + 1 + 1 + 1 + 1 + 1 + 1 + 1 + 1$ $+ 1 + 1$
   d) $1 \cdot 1 \cdot 1 \cdot 1 \cdot 1 \cdot 1 \cdot 1 \cdot 1 \cdot 1 \cdot 1 \cdot 1$

6. The following questions are about powers of one.
   a) What is the value of 1^2? Why?
   b) What is the value of 1^7?
   c) What is the value of 1^x, in which x is a counting number larger than one?

7. The following remark appeared in a French arithmetic book published in 1485:

   "The digits are no more than ten different figures, of which nine have value and the tenth is worth nothing in itself but gives a higher value to the others."

   a) What digit "is worth nothing in itself"?
   b) Give an example of how it "gives a higher value" to another digit.

8. If possible, simplify each of the following.
   a) $1x$
   b) $0x$
   c) $x + 0$
   d) $x + 1$
   e) $x - 0$
   f) $\dfrac{0}{x}$
   g) $\dfrac{x}{0}$
   h) $\dfrac{x}{1}$

9. Each of the following expressions contains two unknown numbers, x and y. Simplify each expression as much as you can. You may assume that neither x nor y is zero.
   a) $1x + 1y$
   b) $1x - 0y$
   c) $0x + 1y$
   d) $0x + 0y$
   e) $\dfrac{x}{1} + \dfrac{y}{1}$

f) $\dfrac{0}{x} + \dfrac{0}{y}$

g) $\dfrac{x}{1} - \dfrac{0}{y}$

h) $\dfrac{0}{x} + \dfrac{y}{1}$

10. When an *even* number is divided by two, the remainder is zero. For example,

$$\begin{array}{r} 6 \\ 2\overline{)12} \\ -12 \\ \hline 0 \end{array}$$

   a) What is the remainder when an *odd* number is divided by two?
   b) What is the remainder when *zero* is divided by two?
   c) Is zero *even* or *odd*?

Set III

11. If possible, tell what number should replace ▓ in each of the following equations to make it true.
    a) $x \cdot ▓ = x$
    b) $x + ▓ = x$
    c) $\dfrac{x}{▓} = x$
    d) $\dfrac{x}{x} = ▓$ (Assume that x is not zero.)
    e) $\dfrac{0}{0} = ▓$
    f) $x - ▓ = x$
    g) $x - ▓ = 0$
    h) $x \cdot ▓ = 0$

12. One way to picture the product $3 \cdot 5$ is shown below.

    ○ ○ ○ ○ ○
    ○ ○ ○ ○ ○
    ○ ○ ○ ○ ○

    a) How could this figure be changed to picture the product $2 \cdot 5$?
    b) How could it be changed to picture the product $1 \cdot 5$?
    c) Could it be changed to picture the product $0 \cdot 5$?

13. The following questions are about powers of zero.
    a) What is the value of 0^2? Why?
    b) What is the value of 0^5?
    c) What is the value of 0^x, in which x is a counting number larger than one?

14. Each of the following expressions contains two unknown numbers, x and y. Simplify each expression as much as you can. You may assume that neither x nor y is zero.
    a) $1x + 0y$
    b) $1y - 1x$
    c) $1y - 0x$
    d) $0x - 0y$
    e) $\dfrac{y}{1} + \dfrac{x}{1}$
    f) $\dfrac{0}{x} - \dfrac{0}{y}$
    g) $\dfrac{y}{1} - \dfrac{0}{x}$
    h) $\dfrac{0}{y} + \dfrac{x}{1}$

15. In the ninth century, an Arab mathematician wrote:

    "When nothing remains in subtraction, put down a small circle so that the place be not empty, but the circle must occupy it, so that the number of places will not be diminished when the place is empty and the second be mistaken for the first."

    a) Use the problem below to show what he meant.

$$\begin{array}{r} 45 \\ -\ 5 \\ \hline \end{array}$$

    b) What does the zero in the answer to this problem mean?

16. The following questions are about the counting numbers.
    a) If x represents a counting number, what is the next larger counting number?
    b) If x represents a counting number larger than one, what is the next smaller counting number?
    c) What is the smallest counting number?

17. Obtuse Ollie says that, if you divide a number by zero, the answer is zero.
    a) Explain why $\dfrac{7}{0}$ is not equal to 0.
    b) Does it make sense to say that $\dfrac{7}{0}$ is equal to 7?

    Acute Alice says that, if you divide zero by a number, the answer is zero.
    c) Explain why $\dfrac{0}{7}$ is equal to 0.
    d) Does it make sense to say that $\dfrac{0}{0}$ is equal to 0?

18. Some automatic calculators do division by repeated subtraction, subtracting the dividing number over and over until the result is zero.
    a) If you tried to divide 12 by 0 on such a calculator, would it eventually arrive at zero? Explain.
    b) What do you suppose would happen if you tried dividing 0 by 0 on such a calculator?

Set IV

We have been using numbers larger than 1 as exponents to indicate repeated multiplication.

$$x^2 \text{ means } x \cdot x,$$
$$x^3 \text{ means } x \cdot x \cdot x,$$
$$x^4 \text{ means } x \cdot x \cdot x \cdot x,$$
and so forth.

What would 1 or 0 mean if we used them as exponents? It seems rather obvious from the pattern above that

$$x^1 \text{ means } x.$$

What do you think x^0 should mean? Rather than just making a guess, make a conclusion from the information in the table below.

x	x^4	x^3	x^2	x^1	x^0
4	$256 \rightarrow$	$64 \rightarrow$	$16 \rightarrow$	$4 \rightarrow$	?
3	$81 \rightarrow$	$27 \rightarrow$	$9 \rightarrow$	$3 \rightarrow$	?
2	$16 \rightarrow$	$8 \rightarrow$	$4 \rightarrow$	$2 \rightarrow$	?
1	$1 \rightarrow$	$1 \rightarrow$	$1 \rightarrow$	$1 \rightarrow$	?
0	$0 \rightarrow$	$0 \rightarrow$	$0 \rightarrow$	$0 \rightarrow$	?

"It's mighty good eating
    for the pennies it costs."

KOREN

LESSON 7
Several
Operations

What do you think is the correct value for the following expression?

$$2 \times 12 + 3 \times 10$$

It all depends on what you are trying to find. For example, suppose that Mrs. Naugatuck wants to buy 2 pounds of porcupine at 12 cents a pound and 3 pounds of iguana at 10 cents a pound. How much will the order cost?

To answer this question, we have to find

$$2 \times 12 + 3 \times 10$$

It is obvious from the situation that both multiplications should be done before the addition:

$$2 \times 12 + 3 \times 10 =$$
$$24 \quad + \quad 30 \quad =$$
$$54$$

The order will cost 54 cents.

Now consider this problem. Mrs. Naugatuck wants to buy 2 dozen duck eggs and 3 buffalo sausages. If they cost 10 cents each, how much will she have to spend?

The answer to this question is also

$$2 \times 12 + 3 \times 10$$

In this case, however, the operations are done in a different order. Multiplying 12 by 2, adding 3, and multiplying the result by 10, we get

$$\begin{aligned}
2 \times 12 + 3 \times 10 &= \\
24 \quad + 3 \times 10 &= \\
27 \quad \times 10 &= \\
270
\end{aligned}$$

She will have to spend $2.70.

The fact that the answer to a problem that requires several operations can depend on the order in which they are done has led mathematicians to make rules for dealing with such problems. The rules are:

First, figure out the powers if there are any.
Then do the multiplications and divisions in order from left to right.
Finally, do the additions and subtractions in order from left to right.

According to these rules, the answer to the problem written as

$$2 \times 12 + 3 \times 10$$

is 54. If we want to change the order of operations, as in the second problem, we use parentheses. It would be written as

$$(2 \times 12 + 3) \times 10$$

We will learn in the next lesson how to use parentheses to change the order of operations.

Examples of how the rules for order of operations are used are given on the next page.

EXAMPLE 1

Find the value of $5^2 - 2 \cdot 3^2 + 4$.

SOLUTION

Figuring out the powers first, we get

$$25 - 2 \cdot 9 + 4$$

(Notice that the 3 is squared before it is multiplied by 2 in the next step.) Doing the multiplication next, we get

$$25 - \quad 18 + 4$$

Finally, doing the addition and subtraction in order from left to right, we get

$$7 \quad + 4 =$$
$$11$$

EXAMPLE 2

Find the value of $3 \cdot 4^3 + 7 \cdot 5 - 11^2$.

SOLUTION

$$3 \cdot 4^3 + 7 \cdot 5 - 11^2 =$$
$$3 \cdot 64 + 7 \cdot 5 - 121 =$$
$$192 + 35 - 121 =$$
$$227 \quad - 121 =$$
$$106$$

EXAMPLE 3

Find the value of $\dfrac{28}{4} - \dfrac{6^2}{12} - \dfrac{32}{2^3}$.

SOLUTION

$$\frac{28}{4} - \frac{6^2}{12} - \frac{32}{2^3} =$$

$$\frac{28}{4} - \frac{36}{12} - \frac{32}{8} =$$

$$7 - 3 - 4 =$$

$$4 \quad - 4 =$$

$$0$$

Exercises

Set I

1. If possible, express each of the following numbers as a power of the number given.
   a) 125 as a power of 5.
   b) 10 as a power of 0.
   c) 64 as a power of 2.

2. A parking meter will take nickels or dimes.
   a) If it contains x coins and someone puts in a dime, how many coins does it contain in all?
   b) If it contains 17 coins of which x are nickels, how many dimes does it contain?
   c) If it contains x nickels and 24 dimes and someone puts in 2 more nickels, how many coins does it contain in all?

3. Mr. Webster is trying to improve his vocabulary.
   a) If he learns x new words each day, how many words will he learn in a week?
   b) If he learns x new words each day, how long will it take him to learn 1,000 new words?
   c) If he knows 15,000 words now and learns 10 new words each day, how many words will he know in x days?

Set II

4. The figure shown here illustrates the expression $2^3 + 2^2$.

Which figure below illustrates each of the expressions in parts a through g?
(The crossed-out circles indicate subtraction.)

Figure 1

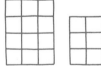

Figure 2

Figure 3

Figure 4

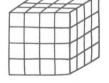

Figure 5

Figure 6

a) $4^2 + 4^2$
b) $4^3 + 3^2$
c) $4 \cdot 3 + 3 \cdot 2$
d) $6^2 - 3 \cdot 2$
e) $6 \cdot 2 - 3 \cdot 2$
f) $2 \cdot 4^2$
g) $4 \cdot 2^2$

5. Find the value of each of the following expressions.
   a) $2 \cdot 5 + 4 \cdot 10$
   b) $2 + 5 \cdot 4 + 10$
   c) $3 \cdot 2^4$
   d) $3 + 2^4$
   e) $5^2 - 4^2$
   f) $5^2 \cdot 4^2$
   g) $6 \cdot 7 - 12 + 3^3$
   h) $6 \cdot 7 + 3^3 - 12$
   i) $6 \cdot 7 - 3^3 + 12$
   j) $6 \cdot 7 + 12 - 3^3$
   k) $\dfrac{4^2}{8} + \dfrac{8^2}{4}$
   l) $\dfrac{4^2}{4} + \dfrac{8^2}{8}$
   m) $\dfrac{8^2}{4} + \dfrac{4^2}{8}$
   n) $11 - 2 \cdot 3 + 7 \cdot 2$
   o) $11 - 2^3 + 7^2$
   p) $11 \cdot 2^3 - 7^2$
   q) $11 \cdot 7^2 - 2^3$

6. Write an expression for each of the following.
   a) The sum of the squares of x and y.
   b) Ten decreased by the product of x and 5.
   c) The quotient of x and 5, decreased by 10.
   d) The product of 8 and the cube of x.
   e) The difference between the fourth power of y and y.
   f) Two more than the quotient of 12 and x.
   g) The sum of x and the product of x and y.

7. The value of the expression $x^2 + 3x - 2$ depends on the number with which we replace x. For example, if x is 5,

$$
\begin{aligned}
x^2 + 3x - 2 &= \\
5^2 + 3 \cdot 5 - 2 &= \\
25 + 15 - 2 &= \\
40 \qquad - 2 &= \\
38 &
\end{aligned}
$$

Find the value of $x^2 + 3x - 2$ if
   a) x is 1.
   b) x is 4.
   c) x is 10.
   d) x is 20.

8. Find the value of each of the following expressions for the numbers given.
   a) $2x + 7$ if x is 6.
   b) $15 - 3x$ if x is 2.
   c) $1 + 4x^2$ if x is 5.
   d) $x^3 - x^2$ if x is 10.
   e) $x^4 + x$ if x is 3.
   f) $5x^2 - x + 6$ if x is 4.

9. At Frankenfurter's Delicatessen, salami costs 80 cents a pound and liverwurst costs 95 cents a pound.
   a) How much would an order of 7 pounds of salami and 3 pounds of liverwurst cost?
   b) How much would an order of x pounds of salami and y pounds of liverwurst cost?

Set III

10. The figure at the right illustrates the expression $5^2 - 2 \cdot 3$. Which figure below illustrates each of the expressions in parts a through g?

Figure 1

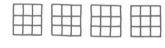

Figure 2

Figure 3

Figure 4

Figure 5

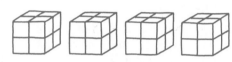

Figure 6

a) $3 \cdot 2 + 4 \cdot 2$

b) $2^3 + 2^3 + 2^3 + 2^3$

c) $3^2 + 4^2$

d) $4^2 - 3^2$

e) $4 \cdot 2 - 3 \cdot 2$

f) $2^3 \cdot 4$

g) $3^2 \cdot 4$

11. Find the value of each of the following expressions.

a) $20 - 6 + 3$

b) $20 - 6 \cdot 3$

c) $20 \cdot 6 - 3$

d) $5 \cdot 9 - 4 \cdot 7$

e) $5 + 9 \cdot 4 - 7$

f) $2 + 4^3$

g) $2 \cdot 4^3$

h) $2^4 \cdot 3$

i) $26 + 3 \cdot 8 - 5^2$

j) $26 - 5^2 + 3 \cdot 8$

k) $26 - 3 \cdot 8 + 5^2$

l) $26 + 5^2 - 3 \cdot 8$

m) $\dfrac{4^3}{2} - \dfrac{6^2}{3}$

n) $\dfrac{2^3}{4} - \dfrac{3^2}{6}$

o) $3 \cdot 3 \cdot 4 - 3 \cdot 2 \cdot 5$

p) $3 \cdot 3^4 - 3 \cdot 2^5$

q) $3^4 \cdot 3 - 2^5 \cdot 3$

12. Write an expression for each of the following.

a) One more than the product of x and 7.

b) The difference between the cubes of x and y.

c) Three times x, decreased by three times y.

d) Twelve increased by the quotient of x and 6.

e) The product of 5 and the square of x.

f) The sum of x and the fifth power of y.

g) The quotient of 1 and the product of x and y.

13. The value of the expression $x^3 - 2x + 4$ depends on the number with which we replace x. For example, if x is 3,

$$x^3 - 2x + 4 =$$
$$3^3 - 2 \cdot 3 + 4 =$$
$$27 - 2 \cdot 3 + 4 =$$
$$27 - 6 + 4 =$$
$$21 + 4 =$$
$$25$$

Find the value of $x^3 - 2x + 4$ if
a) x is 0.
b) x is 2.
c) x is 5.
d) x is 7.

14. Find the value of each of the following expressions for the numbers given.
a) $5x + 4$ if x is 8.
b) $17 - 2x$ if x is 3.
c) $1 + 3x^2$ if x is 4.
d) $x^2 + x^3$ if x is 10.
e) $x^4 - x$ if x is 5.
f) $6x^2 + x - 2$ if x is 1.

15. Acute Alice put a square snapshot of her aunt Edna in a square frame.

FROM THE MOTION PICTURE "ONLY YESTERDAY."
COURTESY OF UNIVERSAL PICTURES.

Find the area of the frame (the green region in the figure above) if
a) each side of the snapshot is 7 centimeters long and the outer sides of the frame are each 10 centimeters long.
b) each side of the snapshot is x centimeters long and the outer sides of the frame are each y centimeters long.

Set IV

Because very few people enjoy doing arithmetic, pocket calculators have become very popular. Although they are easy to use, getting the correct answer to a problem that requires more than one operation is not as simple as it might seem.

Consider the problem

$$12 \cdot 5 - \frac{8}{4} + 7 \cdot 2$$

for example. If you push the keys for these numbers and operations in the order shown here,

a calculator will give the wrong answer.

1. What is the correct answer to the problem?
2. What answer do you think the calculator might give instead?
3. Why would it give that answer?
4. What would you do if you wanted to use such a calculator to get the correct answer to the problem?

$$\boxed{1}\ \boxed{2}\ \boxed{\times}\ \boxed{5}\ \boxed{-}\ \boxed{8}\ \boxed{\div}\ \boxed{4}\ \boxed{+}\ \boxed{7}\ \boxed{\times}\ \boxed{2}\ \boxed{=}$$

DRAWING BY DEDINI; © 1974
THE NEW YORKER MAGAZINE, INC.

LESSON 8
Parentheses

Parentheses are among the most frequently used symbols in algebra. One way in which parentheses are used is to change the usual order of operations. For example, suppose that the sum of 3 and 5 is to be subtracted from 10. Because $3 + 5 = 8$ and $10 - 8 = 2$, the answer to this problem is 2. If we tried writing the problem as

$$10 - 3 + 5$$

however, we would get the wrong answer because, according to our rules of operation, additions and subtractions are done from left to right:

$$
\begin{aligned}
10 - 3 + 5 &= \\
7 \quad + 5 &= \\
12
\end{aligned}
$$

In order to show that we want to add 3 and 5 before subtracting the result from 10, we write

$$10 - (3 + 5)$$

The parentheses indicate that the operation inside them is to be done first:

$$10 - (3 + 5) =$$
$$10 - \quad 8 \quad =$$
$$2$$

▶ In an expression containing parentheses, the parentheses indicate that the operations enclosed within them are to be done before anything else.

Division is usually indicated in algebra by a fraction bar. To show, for example, that the sum of 9 and 3 is to be divided by the difference of 5 and 1, we write

$$\frac{9 + 3}{5 - 1}$$

The fraction bar here means not only to divide, but also to *add and subtract before dividing.*

$$\frac{9 + 3}{5 - 1} = \frac{12}{4} = 3$$

Because the usual procedure is to divide (and multiply) before adding and subtracting, the fraction bar acts here as a parentheses symbol.

Here are more examples of how the value of an expression containing parentheses is found.

EXAMPLE 1
Find the value of $(7 + 4)(7 - 4)$.

SOLUTION

$$(7 + 4)(7 - 4) =$$
$$11 \quad \cdot \quad 3 \quad =$$
$$33$$

EXAMPLE 2

Find the value of $4 + (11 - 2)^2$.

SOLUTION

$$
\begin{aligned}
4 + (11 - 2)^2 &= \\
4 + \quad 9^2 &= \\
4 + \quad 81 &= \\
85
\end{aligned}
$$

EXAMPLE 3

Find the value of $\dfrac{10}{6 - 5} + \dfrac{6 \cdot 5}{10}$.

SOLUTION

$$
\begin{aligned}
\frac{10}{6 - 5} + \frac{6 \cdot 5}{10} &= \\
\frac{10}{1} \quad + \quad \frac{30}{10} &= \\
10 \quad + \quad 3 &= \\
13
\end{aligned}
$$

Exercises

Set I

1. If possible, find the value of each of the following.

   a) $0 \cdot 100$ c) $\dfrac{0}{100}$ e) $\dfrac{1}{100}$

   b) $1 \cdot 100$ d) $\dfrac{100}{0}$ f) $\dfrac{100}{1}$

2. Find the missing dimension for each of these rectangles. (Some of your answers will be in terms of the letters.)

a)

5
? | 35

b)

?
x | 6x

c)

?
x | x^2

d)

x
? | 20

e)
?
1 | x

f)

x
? | y

3. This animal, a native of Madagascar called the tenrec, is capable of giving birth only ten weeks after it itself is born.

   a) How many generations of descendants of one of these animals could be born in 50 weeks?
   b) How many generations of descendants could be born in x weeks if x is a multiple of 10?

Set II

4. Tell whether or not the expressions in each
   of the following pairs are equal.
   a) $(11 + 5) + 2$ and $11 + (5 + 2)$
   b) $(11 - 5) - 2$ and $11 - (5 - 2)$
   c) $(11 + 5) - 2$ and $11 + (5 - 2)$
   d) $(11 - 5) + 2$ and $11 - (5 + 2)$

   e) $12 \cdot 6 \cdot 3$ and $12 \cdot (6 \cdot 3)$
   f) $12 + 6 \cdot 3$ and $12 + (6 \cdot 3)$
   g) $12 + 6 \cdot 3$ and $(12 + 6) \cdot 3$
   h) $\dfrac{12 + 6}{3}$ and $\dfrac{(12 + 6)}{3}$

5. Find the value of each of these expressions.
   a) $7 \cdot 3^2$
   b) $(7 \cdot 3)^2$
   c) $4 + 2 \cdot 3 + 5$
   d) $(4 + 2) \cdot 3 + 5$
   e) $4 + 2 \cdot (3 + 5)$
   f) $(4 + 2) \cdot (3 + 5)$

   g) $15 - 3 \cdot 4 - 2$
   h) $(15 - 3) \cdot (4 - 2)$
   i) $15 - (3 \cdot 4 - 2)$
   j) $15 - 3 \cdot (4 - 2)$
   k) $\dfrac{30}{10} + \dfrac{6}{2}$

   l) $\dfrac{30 + 6}{10 + 2}$
   m) $\dfrac{30}{10} \cdot \dfrac{6}{2}$
   n) $\dfrac{30 \cdot 6}{10 \cdot 2}$

   o) $5^2 - 5 \cdot 2^2$
   p) $(5^2 - 5) \cdot 2^2$
   q) $(5^2 - 5 \cdot 2)^2$

6. The figure shown here illustrates the
   expression $(2 + 3)^2$.

   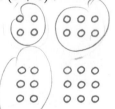

   Which figure below illustrates each of the
   expressions in parts a through h?

   a) $4^2 + 1^2$
   b) $(4 + 1)^2$
   c) $4(4 + 1)$
   d) $(4 - 1)^2$

   e) $4^2 - 1^2$
   f) $4(4 - 1)$
   g) $4^2 + 4$
   h) $4^2 - 4$

   Figure 1   **Figure 2**   **Figure 3**   **Figure 4**   **Figure 5**   **Figure 6**

7. To show that someone is to add x and 3 and then square the result, we write $(x + 3)^2$. Write an expression for each of the following sets of operations.
   a) Subtract 5 from x and then cube the result.
   b) Multiply x by 6 and then add y.
   c) Add y to 6 and then multiply by x.
   d) Divide 10 by x and then subtract y.
   e) Subtract y from 10 and then divide by x.
   f) Multiply the sum of x and 2 by the sum of x and 7.
   g) Divide the difference of x and y by twice x.
   h) Square the product of 3 and x and subtract the result from 11.
   i) Subtract the product of 3 and x from 11 and square the result.
   j) Add the cubes of x and y and multiply the result by 8.

8. Find the values of the following expressions for the numbers given.

   $$x^2 + 2x - 15$$
   a) if x is 3
   b) if x is 4
   c) if x is 10
   d) if x is 50

   $$(x - 3)(x + 5)$$
   e) if x is 3
   f) if x is 4
   g) if x is 10
   h) if x is 50

Set III

9. Tell whether or not the expressions in each of the following pairs are equal.
   a) $(14 + 6) + 1$ and $14 + (6 + 1)$
   b) $(14 + 6) - 1$ and $14 + (6 - 1)$
   c) $(14 - 6) - 1$ and $14 - (6 - 1)$
   d) $(14 - 6) + 1$ and $14 - (6 + 1)$
   e) $(10 \cdot 2) \cdot 4$ and $10 \cdot (2 \cdot 4)$
   f) $10 + 2 \cdot 4$ and $(10 + 2) \cdot 4$
   g) $10 + 2 \cdot 4$ and $10 + (2 \cdot 4)$
   h) $\dfrac{10 - 2}{4}$ and $\dfrac{(10 - 2)}{4}$

10. Find the value of each of these expressions.
   a) $3 \cdot 4^2$
   b) $(3 \cdot 4)^2$
   c) $8 + 3 \cdot 8 - 3$
   d) $(8 + 3) \cdot 8 - 3$

   e) $8 + 3 \cdot (8 - 3)$
   f) $(8 + 3) \cdot (8 - 3)$
   g) $12 - 2 \cdot 5 - 1$
   h) $(12 - 2) \cdot (5 - 1)$
   i) $12 - (2 \cdot 5 - 1)$
   j) $12 - 2 \cdot (5 - 1)$
   k) $\dfrac{32}{4} - \dfrac{4}{2}$
   l) $\dfrac{32 - 4}{4 - 2}$
   m) $\dfrac{32}{4} \cdot \dfrac{4}{2}$
   n) $\dfrac{32 \cdot 4}{4 \cdot 2}$
   o) $7^2 - 9 \cdot 2^2$
   p) $(7^2 - 9) \cdot 2^2$
   q) $(7^2 - 9 \cdot 2)^2$

11. The figure shown here illustrates the
    expression $(5 - 1)^2$.

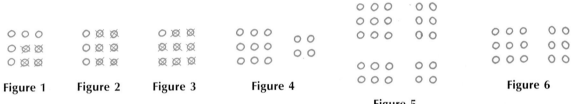

a) $3^2 - 2^2$ e) $3^2 + 2^2$
b) $(3 - 2)^2$ f) $3(3 + 2)$
c) $3(3 - 2)$ g) $3^2 - 3 \cdot 2$
d) $(3 + 2)^2$ h) $3^2 + 3 \cdot 2$

Which figure below illustrates each of the
expressions in parts a through h?

Figure 1   **Figure 2**   **Figure 3**   **Figure 4**   **Figure 6**

Figure 5

12. To show that someone is to subtract 2 from
    x and then multiply the result by 3, we
    write $3(x - 2)$ or $(x - 2)3$. Write an
    expression for each of the following sets of
    operations.
    a) Add 11 to x and multiply by y.
    b) Multiply 11 by y and then add x.
    c) Divide x by 3 and then subtract 1.
    d) Subtract 1 from x and then divide by 3.
    e) Add x and y and square the result.
    f) Add the squares of x and y.
    g) Multiply the difference between x and y
       by x.
    h) Cube the product of 2 and x and subtract
       the result from 9.
    i) Subtract the product of 2 and x from 9
       and cube the result.
    j) Divide the sum of x and y by 5 times y.

13. Find the values of the following expressions
    for the numbers given.

    $$x^2 + 4x - 12$$
    a) if x is 2
    b) if x is 4
    c) if x is 10
    d) if x is 15

    $$(x + 6)(x - 2)$$
    e) if x is 2
    f) if x is 4
    g) if x is 10
    h) if x is 15

Set IV

The value of the expression 1 ▓ 2 ▓ 3 ▓ 4 depends on the symbols of operation with which we replace the blanks. Examples are shown in the picture at the right.

1. Can you figure out which of the following symbols of operation, $+$, $-$, $\cdot$, and $\div$, should be used to replace the blanks in the expression

1 ▓ 2 ▓ 3 ▓ 4 ▓ 5 ▓ 6 ▓ 7 ▓ 8 ▓ 9 ▓ 10

   in order to make it as large a number as possible?

2. What is the value of the number?

3. Suppose that, in addition to replacing the blanks with symbols of operation, you may add parentheses wherever you wish. What would you do to make the expression as large a number as possible?

4. What is the value of the number?

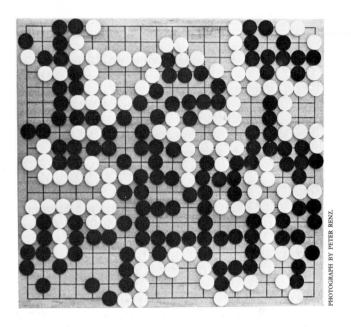

PHOTOGRAPH BY PETER RENZ.

LESSON 9
The Distributive Rule

The oldest game in the world may be the game of Go. It originated in China and is thought to have been played as long ago as the twenty-fourth century B.C.

Go is played with black and white stones on a square board. The object is to capture more territory than the other player while losing as few stones as possible in doing so. The photograph above shows how the board might look at the end of a game.

Although the way in which the stones are arranged on the board makes them difficult to count, the stones in the pattern below are easy to count. Two ways to count them illustrate a simple but very useful pattern called the *distributive rule.* One way is to multiply the sum of the numbers of black and white stones in one row, $6 + 4$, by the number of rows, 8:

$$8(6 + 4) = 8(10) = 80$$

The other way is to multiply each of these numbers, 6 and 4, by 8 and add the results:

$$8(6) + 8(4) = 48 + 32 = 80$$

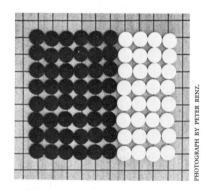

PHOTOGRAPH BY PETER RENZ.

Comparing the first way with the second, we see that

$$8(6 + 4) = 8(6) + 8(4)$$

This pattern is true for *any* set of three numbers.

▶ **The Distributive Rule (Addition)**

For any three numbers a, b, and c, $a(b + c) = ab + ac$.

Notice that this rule is about a relationship between multiplication and addition. Sometimes it is stated by simply saying that "multiplication distributes over addition."

There is a similar rule relating multiplication and subtraction.

▶ **The Distributive Rule (Subtraction)**

For any three numbers a, b, and c, $a(b - c) = ab - ac$.

Because the product of two numbers does not depend on the order of the numbers, the two distributive rules can also be written with the numbers in each product interchanged; that is, because

$$a(b + c) = ab + ac \quad \text{and} \quad a(b - c) = ab - ac$$

it is also true that

$$(b + c)a = ba + ca \quad \text{and} \quad (b - c)a = ba - ca$$

The distributive rules are among the most fundamental patterns of algebra. Here are examples of how they are used.

EXAMPLE 1

Use the distributive rule to write the product $10(x + 2)$ as a sum.

SOLUTION

$10(x + 2) = 10(x) + 10(2) = 10x + 20$

EXAMPLE 2

Use the distributive rule to write the product $(5 + x)y$ as a sum.

SOLUTION

$(5 + x)y = 5y + xy$

EXAMPLE 3

Use the distributive rule to write the product $x(x - 1)$ as a difference.

SOLUTION

$x(x - 1) = x(x) - x(1) = x^2 - x$

Exercises

Set I

1. Simplify each of the following expressions.
   a) $a + a + a + a + a$
   b) bbb
   c) $\dfrac{a}{1} + \dfrac{b}{1}$
   d) $0(a + b)$

2. Mr. Hunt can type 20 words per minute and Miss Peck can type x words per minute.
   a) If they type at the same time, how many words can they type in a minute?
   b) How many words can Miss Peck type in 5 minutes?
   c) How long would it take Mr. Hunt to type y words?

3. The largest pizza ever baked weighed 1,000 pounds.
   a) If it contained x pounds of cheese, how much did the other ingredients weigh?
   b) If the pizza were cut into y equal pieces, how much would each piece weigh?
   c) If 10 people ate z pounds each, how much would be left?

Set II

4. The figure below illustrates the pattern $2(4 + 3) = 2(4) + 2(3)$.

   Write a pattern illustrated by each of the following figures.

   a)

   b)

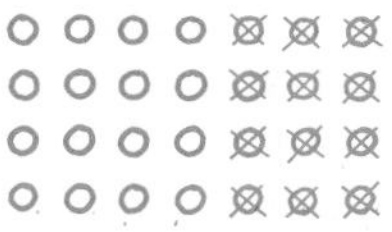

   c)

   d)

5. The multiplication problem $5x^2$ and the addition problem $x^2 + x^2 + x^2 + x^2 + x^2$ are equivalent. Write a multiplication problem equivalent to each of the following addition problems.
   a) $x^3 + x^3 + x^3 + x^3$
   b) $2x + 2x + 2x + 2x + 2x + 2x + 2x$
   c) $(x + 1) + (x + 1) + (x + 1)$
   d) $\underbrace{(x + y) + (x + y) + \cdots + (x + y)}_{\text{9 of them}}$

   Write an addition problem equivalent to each of the following multiplication problems.
   e) $2x^4$
   f) $5(3x)$
   g) $4(x + 7)$

6. According to the distributive rule,
   $4(x + 2) = 4x + 8$. One way to prove this
   is by writing $4(x + 2)$ as a repeated addition
   problem and rearranging the numbers being
   added:

$$4(x + 2) = (x + 2) + (x + 2) + (x + 2) + (x + 2)$$
$$= x + x + x + x + 2 + 2 + 2 + 2$$
$$= 4x + 8$$

Use the same method to prove that
a) $3(x + 5) = 3x + 15$
b) $2(x + y) = 2x + 2y$
c) $4(x^2 + 1) = 4x^2 + 4$

7. Use the distributive rule to write each of the
   following as a sum or difference.
   a) $8(x + 3)$ f) $(4 + x)y$
   b) $5(y - 2)$ g) $(y - x)7$
   c) $x(x + 1)$ h) $(x - 6)x$
   d) $y(x - y)$ i) $10(x^2 + 4)$
   e) $(x + 9)2$ j) $x(x^3 - 1)$

8. The way in which you learned to multiply
   numbers in arithmetic has as its basis the
   distributive rule. For example, to multiply
   51 by 32 we write

$$
\begin{array}{r}
51 \\
\times\ \ 32 \\
\hline
102 \\
+\,1530 \\
\hline
1632
\end{array}
$$

To see how the distributive rule applies, consider the fact that $32 = 2 + 30$ so that

$$32 \cdot 51 = (2 + 30)51$$
$$= 2 \cdot 51 + 30 \cdot 51$$
$$= 102 + 1530$$
$$= 1632$$

a) Do the following multiplication problem.

$$
\begin{array}{r}
72 \\
\times 43 \\
\hline
\end{array}
$$

b) Show, by using the distributive rule, why
   what you have done is correct.
c) Now do this mutiplication problem.

$$
\begin{array}{r}
43 \\
\times 72 \\
\hline
\end{array}
$$

d) Explain your method by using the
   distributive rule.

9. Write the total area of each of these
   rectangles in two different ways.

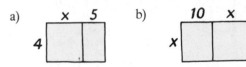

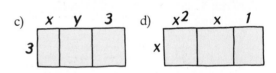

10. Buster Brown owns x pairs of tennis shoes
    and y pairs of loafers.
    a) If he has no shoes other than these, how
       many pairs of shoes does he own in all?
    b) Write the number of shoes that he owns
       altogether as a product.
    c) How many individual tennis shoes does
       he own?
    d) How many individual loafers does
       he own?
    e) Write the number of shoes that he owns
       altogether as a sum.

Set III

11. The figure below illustrates the pattern
    $3(6 - 2) = 3(6) - 3(2)$.

    ```
    o o o o ⊠ ⊠
    o o o o ⊠ ⊠
    o o o o ⊠ ⊠
    ```

    Write a pattern illustrated by each of the following figures.

    a)
    ```
    o o o o o      o o o o
    o o o o o      o o o o
    ```

    b)
    ```
    o o o    o
    o o o    o
    o o o    o
    o o o    o
    o o o    o
    o o o    o
    o o o    o
    ```

    c)
    ```
    o o o ⊠ ⊠ ⊠ ⊠ ⊠
    o o o ⊠ ⊠ ⊠ ⊠ ⊠
    o o o ⊠ ⊠ ⊠ ⊠ ⊠
    o o o ⊠ ⊠ ⊠ ⊠ ⊠
    ```

    d)
    ```
    o o o ⊠
    o o o ⊠
    o o o ⊠
    o o o ⊠
    o o o ⊠
    ```

12. The multiplication problem $3x^4$ and the
    addition problem $x^4 + x^4 + x^4$ are
    equivalent. Write a multiplication problem
    equivalent to each of the following addition
    problems.
    a) $x^2 + x^2 + x^2 + x^2 + x^2 + x^2 + x^2$
    b) $5x + 5x + 5x$
    c) $(x + 7) + (x + 7)$
    d) $\underbrace{(x + y) + (x + y) + \cdots + (x + y)}_{\text{10 of them}}$

    Write an addition problem equivalent to
    each of the following multiplication
    problems.
    e) $4x^3$
    f) $2(7x)$
    g) $3(x + 8)$

13. According to the distributive rule,
    $5(x + 1) = 5x + 5$. One way to prove this
    is by writing $5(x + 1)$ as a repeated addition
    problem and rearranging the numbers being
    added:

    $5(x + 1) = (x + 1) + (x + 1) + (x + 1) + (x + 1) + (x + 1)$
    $= x + x + x + x + x + 1 + 1 + 1 + 1 + 1$
    $= 5x + 5$

    Use the same method to prove that
    a) $2(x + 6) = 2x + 12$
    b) $4(x + y) = 4x + 4y$
    c) $3(x^2 + 2) = 3x^2 + 6$

14. Use the distributive rule to write each of the
    following as a sum or difference.
    a) $2(x + 5)$ f) $(6 + x)x$
    b) $4(y - 7)$ g) $(y - 4)5$
    c) $x(3 + x)$ h) $(x - y)y$
    d) $y(y - 1)$ i) $3(x^2 + 9)$
    e) $(x + 8)10$ j) $x^2(x - 2)$

15. The way in which you learned to multiply
    numbers in arithmetic has as its basis the
    distributive rule. For example, to multiply
    62 by 14 we write

    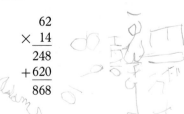

    ```
        62
    ×   14
       248
    + 620
       868
    ```

To see how the distributive rule applies, consider the fact that $14 = 4 + 10$ so that

$$14 \cdot 62 = (4 + 10)62 = 4 \cdot 62 + 10 \cdot 62$$
$$= 248 + 620 = 868$$

a) Do the following multiplication problem.

$$\begin{array}{r} 84 \\ \times 21 \\ \hline \end{array}$$

b) Show, by using the distributive rule, why what you have done is correct.

c) Now do this multiplication problem.

$$\begin{array}{r} 21 \\ \times 84 \\ \hline \end{array}$$

d) Explain your method by using the distributive rule.

17. Alice's Restaurant sells espresso coffee for 40 cents a cup. Suppose that one week it sells x cups of the coffee and the next week it sells y cups.
a) How many cups does the restaurant sell in all?
b) Write the total amount charged for the coffee during the two weeks as a product.
c) How much did the restaurant charge for the coffee during the first week?
d) How much did the restaurant charge during the second week?
e) Write the total amount charged for the coffee during the two weeks as a sum.

16. Write the total area of each of these rectangles in two different ways.

a)
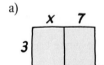

b)
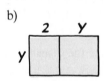

c)
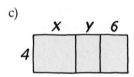

d)
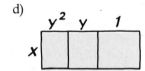

Set IV

You know from the distributive rule of multiplication over addition that, for all values of x and y,

$$2(x + y) = 2x + 2y$$

Is it also true that

$$(x + y)^2 = x^2 + y^2?$$

1. To find out, find the values of the following expressions for the numbers given.

$(x + y)^2$
a) if x is 2 and y is 0.
b) if x is 0 and y is 6.
c) if x is 3 and y is 4.
d) if x is 9 and y is 1.

$x^2 + y^2$
e) if x is 2 and y is 0.
f) if x is 0 and y is 6.
g) if x is 3 and y is 4.
h) if x is 9 and y is 1.

2. What do you conclude about $(x + y)^2$ and $x^2 + y^2$ on the basis of your results?

CHICAGO TRIBUNE—NEW YORK NEWS SYNDICATE, INC.

Summary and Review

In this chapter, we have reviewed the fundamental operations and their relationships.

Addition (*Lesson 1*) The result of adding two numbers, say a and b, is called their *sum* and is written as $a + b$. It does not depend on the order of the numbers, and so $a + b = b + a$.

Subtraction (*Lesson 2*) The result of subtracting one number from another, say b from a, is called their *difference* and is written as $a - b$. It may be understood to mean either "b taken away from a" or "the number that must be added to b to give a."

Multiplication (*Lesson 3*) The result of multiplying two numbers, say a and b, is called their *product* and is written as ab. As in addition, it does not depend on the order of the numbers, and so $ab = ba$. Multiplication can be understood as repeated addition; for example, $3a$ means $a + a + a$.

Division (*Lesson 4*) The result of dividing one number by another, say a by b, is called their *quotient* and is written as $\frac{a}{b}$. It is the number that must be multiplied by b to give a.

59

Raising to a Power (*Lesson 5*) To raise a number to a power means to multiply the number by itself one or more times; for example, a^4 is read as "*a* to the fourth power" and means $a \cdot a \cdot a \cdot a$. The 4 is called an *exponent*. The second and third powers of a number such as *a* are called "*a* squared" and "*a* cubed."

Zero and One (*Lesson 6*) Zero is the only number that can be added to or subtracted from another number without changing it. For every number *a*, $a + 0 = a$ and $a - 0 = a$.

Whenever any number is multiplied by zero, the result is zero. For every number *a*, $a \cdot 0 = 0$.

Although zero may be divided by another number, giving zero as the result, we never divide a number by zero. For every number *a* (except 0), $\frac{0}{a} = 0$;. $\frac{a}{0}$ and $\frac{0}{0}$ are meaningless.

One is the only number that can be multiplied by or divided into another number without changing it. For every number *a*, $a \cdot 1 = a$ and $\frac{a}{1} = a$.

Several Operations (*Lesson 7*) In performing a series of operations, we work from left to right, first raising to powers, then multiplying and dividing, and finally adding and subtracting.

Parentheses (*Lesson 8*) Parentheses are often used to change the usual order of operations by indicating that the operation inside them is to be done first. The fraction bar used to indicate division acts as a parentheses symbol.

The Distributive Rule (*Lesson 9*) The distributive rule relates multiplication and addition. It says that for any three numbers *a*, *b*, and *c*,

$$a(b + c) = ab + ac \quad \text{and} \quad (b + c)a = ba + ca$$

A similar rule relates multiplication and subtraction. For any three numbers *a*, *b*, and *c*,

$$a(b - c) = ab - ac \quad \text{and} \quad (b - c)a = ba - ca$$

Exercises

Set I

1. Write another expression equivalent to each of the following.

   a) $7 + 7 + 7 + 7$
   b) $7 \cdot 7 \cdot 7 \cdot 7$
   c) $2x$
   d) y^6

2. Write a number for each of the following:
   a) The number w squared.
   b) The product of 3 and x.
   c) The number y taken away from 17.
   d) The fifth power of z.

3. Here are directions for a number trick.

Step 1.	Think of a number.
Step 2.	Multiply by five.
Step 3.	Add eight.
Step 4.	Subtract three.
Step 5.	Divide by five.
Step 6.	Subtract the number that you first thought of.

   a) Show how the trick works by drawing boxes and circles to illustrate the steps.
   b) What is the result at the end of the trick?
   c) Two steps in the trick could be combined into one without changing the end result. Which are they?
   d) What would the step replacing them be?

4. Which figure below and at the right illustrates each of the following expressions?
   a) $4 + 3^2$ b) $4 \cdot 3^2$ c) $(4 + 3)^2$

5. This problem is about the powers of 4.
   a) Make a table showing the values of the second through sixth powers of 4.
   b) Can you guess what any of the digits of 4^{100} might be?

6. A chessboard contains 2^6 small squares.
   a) How many squares is that?
   b) Can you write 2^6 as a number squared?

7. The perimeter of a rectangle is the sum of the lengths of its sides. The area of a rectangle is the product of its length and width. What are the perimeter and area of each of these rectangles?

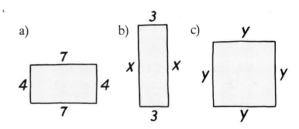

8. Find the value of each of these expressions.
   a) $30 - 9 - 7$
   b) $30 - (9 - 7)$
   c) $1 + 4^3$
   d) $(1 + 4)^3$

Figure 1

```
o o o    o o o    o o o    o o o
o o o    o o o    o o o    o o o
o o o    o o o    o o o    o o o
```

Figure 2

```
o o o o    o o o
o o o o    o o o
o o o o    o o o
o o o o    o o o

o o o o    o o o
o o o o    o o o
o o o o    o o o
```

Figure 3

```
o o    o o o
o o    o o o
       o o o
```

9. A can of Goober's Mixed Nuts contains almonds, cashews, and peanuts.
   a) If one can contains 9 almonds, x cashews, and 142 peanuts, how many nuts does it contain in all?
   b) If another can contains 160 nuts of which x are almonds and y are cashews, how many peanuts does it contain?

10. Division by zero makes no sense.
    a) Explain why there is no number equal to $\dfrac{2}{0}$.

    b) Is there any number equal to $\dfrac{0}{0}$?

11. Mr. Bunyan is a lumberjack.
    a) If he can cut down 600 trees in an hour, how many trees can he cut down in x hours?
    b) If he can saw up x logs in a day, how many days would it take him to saw up 10,000 logs?

12. Write an expression for each of the following sets of operations.
    a) Multiply x by 5 and then add 1.
    b) Add 3 to x and then square the result.

c) Raise x to the sixth power and then subtract 7.

13. Write each of these products as a sum or difference.
    a) $7(a + 2)$ b) $b(1 - b)$ c) $(c + 9)5$

14. A molecule of propane gas consists of three carbon atoms and eight hydrogen atoms, as shown in the model below.

a) How many of each atom do x molecules of propane contain?
b) Write the total number of atoms in x propane molecules as a sum.
c) How many atoms does one propane molecule contain?
d) Write the total number of atoms in x propane molecules as a product.

Set II

1. Write another problem equivalent to each of the following.
   a) $11 + 11 + 11$ c) $5x$
   b) $2 \cdot 2 \cdot 2 \cdot 2 \cdot 2 \cdot 2 \cdot 2$ d) y^4

2. The pictures below illustrate the steps of a number trick. Tell what is happening in each step.

   Step 1. ☐
   Step 2. ☐ ○
   Step 3. ☐☐☐☐ ○ ○ ○ ○
   Step 4. ☐☐☐☐ ○ ○ ○ ○ ○ ○ ○ ○ ○ ○ ○ ○
   Step 5. ☐ ○ ○ ○
   Step 6. ○ ○ ○

3. Write a number for each of the following:
   a) The difference between a and 5.
   b) The number b cubed.
   c) The sum of 2 and c.
   d) The quotient of 1 and d.

4. If possible, express each of the following as a power of the number given.
   a) 32 as a power of 2.
   b) 3 as a power of 1.
   c) 1,000,000 as a power of 10.

5. During the month of July, there were x shark attacks off the shore of Amity Beach.
   a) If 3 of the attacks were within 50 feet of the shore, how many were farther away?

b) If 5 attacks occurred in August, how many were there in all?

6. Find the missing dimension for each of these rectangles.

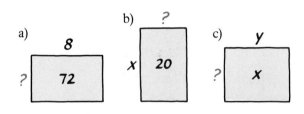

a)
8
? | 72

b) ?
x | 20

c) y
? | x

7. Par on the Shady Acres Golf Course is 72.
   a) If Colonel Bogey's score is x strokes above par, what is his score?
   b) If Miss Birdie's score is y strokes below par, what is her score?
   c) Mr. Bunker's score on the first nine holes is 75 (he has a terrible time with sand traps) and his score on the second nine is x. How many strokes above par is his total score?

8. Find the value of each of these expressions.
   a) $6 \cdot 10^2$
   b) $(6 \cdot 10)^2$
   c) $(2 + 7)(8 - 3)$
   d) $2 + 7 \cdot 8 - 3$

9. The cube numbers are related to the differences of square numbers in an interesting way.
   a) Copy and complete the following table.

$$3^2 - 1^2 = 8 = 2^3$$
$$6^2 - 3^2 = \text{▓▓} = \text{▓▓}^3$$
$$10^2 - 6^2 = \text{▓▓} = \text{▓▓}^3$$
$$15^2 - 10^2 = \text{▓▓} = \text{▓▓}^3$$

   b) Can you guess what the next line of this table is?

10. Write an expression for each of the following sets of operations.
    a) Subtract 6 from x and then multiply by 2.
    b) Divide x by 8 and then add 4.
    c) Cube x and then subtract the result from 150.

11. Write each of these products as a sum or difference.
    a) $8(v + 11)$ c) $x(y + z)$
    b) $(w - 6)3$

12. Show how each of these figures illustrates the distributive rule by writing its area as both a product and a sum.

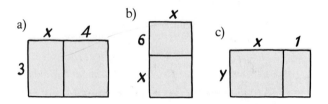

a)
x   4
3

b) x
6
x

c) x   1
y

13. Since going on a diet, Mrs. Uppington has lost 3 kilograms each week.
    a) At this rate, how many kilograms would she lose in x weeks?
    b) If she weighed 200 kilograms before beginning the diet, how much would she weigh after x weeks of it?
    c) If she wants to lose x kilograms, how many weeks will it take her?

14. Find the values of the following expressions for the numbers given.

$x^2 + 5x - 14$ $(x + 7)(x - 2)$
a) if x is 2   d) if x is 2
b) if x is 3   e) if x is 3
c) if x is 10   f) if x is 10

Chapter 2
FUNCTIONS AND GRAPHS

CHICAGO TRIBUNE–NEW YORK NEWS SYNDICATE, INC.

LESSON 1
An Introduction to Functions

How does a person's intelligence as a child relate to his intelligence in later life? From the claim in this cartoon of having the mind of an eight-year-old at the age of four, one might conclude that such a person's mental age will be twice his physical age throughout his life. The table below shows this relationship.

Physical age	4	5	6	7	8	...
Mental age	8	10	12	14	16	...

To each physical age in this table, there corresponds a mental age: 8 corresponds to 4, 10 to 5, and so forth.

A mathematician would say that a person's mental age is a *function* of his physical age.

► A **function** is a pairing of two sets of numbers so that to each number in the first set there corresponds exactly one number in the second set.

One way to represent a function is with a table, as was done above. Another way to represent a function is by writing a *formula*. To show, for example, that a certain person's

mental age is two times his physical age

66

we might let p represent physical age and m represent mental age and simply write

$$m = 2p$$

The two letters in this formula are called *variables:* as p varies in value, so does m. For example, if $p = 15$, then $m = 2(15) = 30$ and if $p = 25$, then $m = 2(25) = 50$.

As the character in the last panel of the cartoon has pointed out, it is also possible to conclude from the claim of having the mind of an eight-year-old at the age of four that the person is merely four years ahead in intelligence. The table below illustrates this possibility.

Physical age	4	5	6	7	8	...
Mental age	8	9	10	11	12	...

A formula stating that someone's mental age is always four years more than his physical age is

$$m = p + 4$$

Although it is unlikely that either of these functions is correct, it is certainly true that a person's intelligence does change in the course of a lifetime. The method by which a person's I.Q. is determined takes this change with age into account.

Exercises

Set I

1. Find the value of each of these expressions:
   a) $12 - 3^2$
   b) $(12 - 3)^2$
   c) $1^3 + 2^3$
   d) $(1 + 2)^3$

2. Write a number for each of the following:
   a) The number that must be added to x to give x.
   b) The number that must be multiplied by x to give x.
   c) The number that must be multiplied by x to give 0.

3. Contrary to popular opinion, ghosts do not last forever. According to the *Gazetteer of*

British Ghosts, they fade away after 400 years.

a) If a ghost has existed for x centuries, how many years has it been around?

b) If a ghost has been haunting a castle for x years, how many years has it to go?

Set II

4. One way to represent a function is with a table. For example, if a function has the formula $y = 3x$, then a partial table for it might look like this:

x	0	1	2	3	4
y	0	3	6	9	12

Copy and complete the tables shown for the functions having these formulas.

a) Formula: $y = x + 5$
   Table:

x	0	1	2	3	4
y	5				

b) Formula: $y = 4x$
   Table:

x	0	2	4	6	8
y		8			

c) Formula: $y = 2x + 3$
   Table:

x	0	1	2	3	4
y	3				

d) Formula: $y = x^2$
   Table:

x	1	2	3	4	5
y					

e) Formula: $y = \dfrac{12}{x}$

   Table:

x	2	4	6	8	10
y	6				

5. To find the value of a complicated expression that includes x, it is helpful to replace each x with parentheses first and then write the number for x in each. For example, suppose a function has the formula

$$y = 2x^2 - x + 3$$

To find the value of y if $x = 5$, we first write $y = 2(\ \ )^2 - (\ \ ) + 3$ and then $y = 2(5)^2 - (5) + 3$. Simplifying, $y = 2 \cdot 25 - 5 + 3 = 48$.

Copy the tables for the following functions and use this method to complete them.

a) Formula: $y = 1 + 3x^2$
   Table:

x	2	3	4	5
y	13			

b) Formula: $y = 2x^3 - x^2$
   Table:

x	2	3	4	5
y	12			

6. Guess a formula for the function represented by each of these tables. Begin each formula with $y =$.

a)
x	1	2	3	4	5
y	6	12	18	24	30

b)
x	0	1	2	3	4
y	7	8	9	10	11

c)
x	5	6	7	8	9
y	1	2	3	4	5

d)

x	2	3	4	5	6
y	4	9	16	25	36

e)

x	2	3	4	5	6
y	5	10	17	26	37

(Hint: Compare table e with table d.)

f)

x	1	2	3	4	5
y	5	11	17	23	29

(Hint: Compare table f with table a.)

g)

x	4	5	6	7	8
y	43	53	63	73	83

h)

x	4	5	6	7	8
y	44	55	66	77	88

i)

x	3	4	5	6	7
y	9	8	7	6	5

j)

x	1	2	4	5	10
y	20	10	5	4	2

7. The perimeter and area of a square are functions of the length of one of its sides.

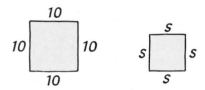

a) What is the perimeter of a square whose sides are 10 units long?

b) Write a formula for the perimeter, p, of a square whose sides are s units long.

c) What is the area of a square whose sides are 10 units long?

d) Write a formula for the area, a, of a square whose sides are s units long.

8. The distance that Miss Winfield travels on her bicycle (she goes at a constant speed) is a function of the time she has been riding. Here is a table for this function.

Number of seconds, s	15	30	45	60
Number of meters traveled, m	165	330	495	660

a) What is Miss Winfield's speed in meters per second?

b) Write a formula for this function.

c) How far would Miss Winfield go in 70 seconds?

9. The population of a city is a function of time. This table shows the population of Grover's Corners every twenty years, starting at 1900.

Year	1900	1920	1940	1960
Population	205	372	620	8145

a) What can you conclude from this table?

b) Do you think there is a formula for this function?

Set III

10. One way to represent a function is with a table. For example, if a function has the formula $y = x + 2$, then a partial table for it might look like this:

x	1	2	3	4	5
y	3	4	5	6	7

Copy and complete the tables shown for the functions having the following formulas.

a) Formula: $y = 5x$
   Table:

x	0	1	2	3	4
y	0				

b) Formula: $y = x - 3$
   Table:

x	3	4	5	6	7
y	0				

c) Formula: $y = 10x + 1$
Table:

x	1	2	3	4	5
y	11	▦	▦	▦	▦

d) Formula: $y = x^3$
Table:

x	1	2	3	4	5
y	▦	▦	▦	▦	▦

e) Formula: $y = x^2 + x$
Table:

x	0	1	2	3	4
y	▦	▦	▦	▦	▦

11. To find the value of a complicated expression that includes x, it is helpful to replace each x with parentheses first and then write the number for x in each. For example, suppose a function has the formula

$$y = x^2 + 4x - 5$$

To find the value of y when $x = 3$, we first write $y = (\quad)^2 + 4(\quad) - 5$ and then $y = (3)^2 + 4(3) - 5$. Simplifying, $y = 9 + 12 - 5 = 16$.

Copy the tables for the following functions and use this method to complete them.

a) Formula: $y = 4x^3 + 2$
Table:

x	0	1	2	3
y	2	▦	▦	▦

b) Formula: $y = 5x^2 - 7x$
Table:

x	2	3	4	5
y	6	▦	▦	▦

12. Guess a formula for the function represented by each of these tables. Begin each formula with $y =$.

a)
x	0	1	2	3	4
y	3	4	5	6	7

b)
x	4	5	6	7	8
y	20	25	30	35	40

c)
x	2	4	6	8	10
y	1	2	3	4	5

d)
x	7	8	9	10	11
y	1	2	3	4	5

e)
x	0	1	2	3	4
y	0	1	8	27	64

f)
x	4	5	6	7	8
y	19	24	29	34	39

(Hint: Compare table f with table b.)

g)
x	0	1	2	3	4
y	2	3	10	29	66

(Hint: Compare table g with table e.)

h)
x	3	4	5	6	7
y	34	44	54	64	74

i)
x	3	4	5	6	7
y	33	44	55	66	77

j)
x	2	3	4	5	6
y	8	7	6	5	4

13. The number of people listening to Senator Claghorn give a speech is a function of the time he has been speaking. This table shows what happened at a speech he gave last week.

Number of minutes the senator had been speaking	0	10	20	30	40
Number of people who were listening	500	384	245	109	7

a) What can you conclude from this table?
b) Do you think there is a formula for this function?

14. The amount of money that Mr. Babbitt makes is a function of the number of hours that he works. Here is a table for this function.

Number of hours worked, h	8	16	24	32	40
Number of dollars earned, d	60	120	180	240	300

a) How much money does Mr. Babbitt make per hour?
b) Write a formula for this function.
c) How much money would Mr. Babbitt make in 50 hours?

15. The number of times that the hour hand of a clock goes around the clock is a function of the time.

a) How many times does the hour hand go around the clock in one day (24 hours)?
b) How many times does it go around the clock in seven days?
c) Write a formula for the number of times, n, that the hour hand goes around the clock in d days.

Set IV

Thousands of meteors enter the earth's atmosphere each year. This shows a piece of one that was found in Saskatchewan, Canada.

As a meteor enters the earth's atmosphere, it rapidly becomes white hot so that it looks like a "falling star." The degree to which it is heated depends on how fast it is traveling. More specifically, the highest temperature reached by the meteor is a function of the speed at which it enters the atmosphere.

The table below shows some approximate values for this function.

Speed of meteor, s, in kilometers per second	5	6	7	8	9	10
Highest temperature reached by meteor, t, in °C	11,250	16,200	22,050	28,800	36,450	

1. There is a pattern in the numbers in the second row of this table. Can you figure out what it is?
2. What would be the highest temperature reached by a meteor if it entered the earth's atmosphere at a speed of 10 kilometers per second?

"I really look forward to your cheery little visits."

LESSON 2
The Coordinate Graph

According to the chart on the wall, this fellow's progress in the hospital doesn't look very encouraging. What the doctor has been drawing each day is called a *coordinate graph*. The patient's health is a function of time, and this is evidently what is being pictured on the graph.

The coordinate graph is one of the simplest, yet most useful, ideas in all of mathematics. Invented in the seventeenth century by a French mathematician and philosopher, René Descartes, it has been used in a wide variety of applications ever since.

To construct a coordinate graph, we begin by drawing two perpendicular lines as shown in the figure at the right. The two lines are labeled with letters, usually x and y as shown, and are called the *axes* of the graph. The horizontal line is ordinarily called the *x-axis* and the vertical line is called the *y-axis*. The point in which they intersect is labeled with a capital O and is called the *origin*. The axes continue indefinitely in each direction from the origin.

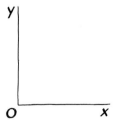

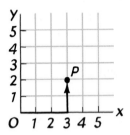

Each axis is numbered like a ruler, beginning with zero at the origin. To locate a point, such as the one named *P* at the left, we move along the *x*-axis until we are directly below the point, counting the units as we go. Then we move directly up to the point itself, again counting the units along the way. The location of the point is given by these two numbers, which are called its *coordinates*. They are written in parentheses and separated by a comma like this: (3, 2). The *x*-coordinate is always given first, because we always move in the direction of the *x*-axis first.

Because a function can be represented by pairs of numbers in a table, these pairs of numbers can be used as coordinates of points to make a picture called the *graph* of the function. For example, to graph the function represented by this table,

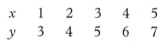

x	1	2	3	4	5
y	3	4	5	6	7

x	y
1	3
2	4
3	5
4	6
5	7

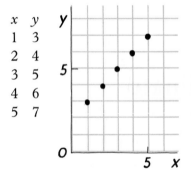

we plot the points (1, 3), (2, 4), (3, 5), (4, 6), and (5, 7). (Writing the table in columns rather than rows, as has been done at the left, makes the paired numbers easier to read.) The numbers in the table follow a simple pattern, which means that the points in the graph also form a simple pattern.

Exercises

Set I

1. Write another expression equivalent to each of the following.

   a) $n + n + n + n + n$
   b) *nnnnn*
   c) 7^4
   d) $7 \cdot 4$

2. Find the missing dimension for each of these rectangles. The expression inside each rectangle represents its area.

a)

b)

c)

3. Write an expression for each of the
   following sets of operations.
   a) Multiply x by 4 and then subtract 9.
   b) Subtract 9 from x and then multiply by 4.

c) Add the squares of a and b.
d) Add a and b and then square the result.

Set II

4. Write the coordinates of each of the seven
   points in this graph (including the origin)
   that is named with a letter.

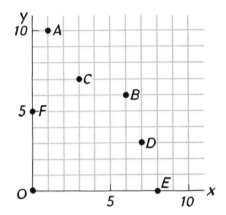

5. On graph paper, draw a separate pair of axes
   extending 8 units in each direction from the
   origin for each part of this exercise. Connect
   the points with straight line segments in the
   order given to form each geometric figure.
   a) Right triangle: (2, 1), (2, 7), (6, 1), and
      (2, 1).
   b) Trapezoid: (3, 6), (7, 6), (8, 3), (1, 3), and
      (3, 6).

6. A certain function is represented by this
   table of numbers.

x	1	2	3	4
y	2	4	6	8

   a) What is a formula for this function?
      Begin your formula with $y =$.

b) How many points are included in the
   table?
c) Graph the function by drawing a pair of
   axes and plotting these points.
d) What do you notice about the points?

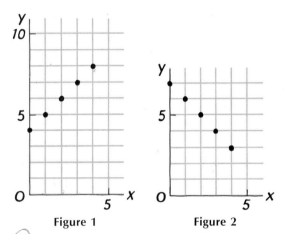

Figure 1   Figure 2

7. A certain function has the graph shown in
   Figure 1, above.
   a) Copy and complete the following table
      for this function.

x	0	1	2	3	4
y	4				

   b) What is a formula for this function?

8. A certain function has the graph shown in
   Figure 2, above.
   a) Copy and complete the following table
      for this function.

x	0	1	2	3	4
y					

b) What is a formula for this function?

9. A function has the formula $y = 8 - x$.
   a) Copy and complete the following table for this function.

x	0	2	4	6	8
y	8				

   b) Graph the function.

10. A function has the formula $y = 3x - 2$.
    a) Copy and complete the following table for this function.

x	1	2	3	4
y				

    b) Graph the function.

Set III

11. The coordinate graph makes it possible to study a geometric figure by means of numbers because each point of the figure can be located by a pair of numbers.

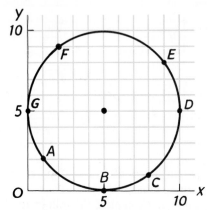

This graph shows a circle whose center is the point $(5, 5)$. What are the coordinates of each point on the circle that is identified by a letter?

12. On graph paper, draw a separate pair of axes extending 8 units in each direction from the origin for each part of this exercise. Connect the points with straight line segments in the order given to form each geometric figure.
    a) Kite: $(3, 5)$, $(5, 7)$, $(7, 5)$, $(5, 1)$, and $(3, 5)$.
    b) Rectangle: $(2, 1)$, $(8, 3)$, $(7, 6)$, $(1, 4)$, and $(2, 1)$.

13. A certain function is represented by this table of numbers.

x	0	1	2	3
y	0	1	4	9

    a) What is a formula for this function? Begin your formula with $y =$.
    b) How many points are included in the table?
    c) Graph the function by drawing a pair of axes and plotting these points.

14. A certain function has the graph shown in the figure below.

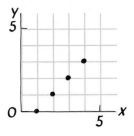

    a) Copy and complete the following table for this function.

x	1	2	3	4
y	0			

    b) What is a formula for this function?

15. A certain function has the graph shown in the figure below.

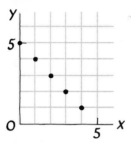

a) Copy and complete the following table for this function.

x	0	1	2	3	4
y					

b) What is a formula for this function?

16. A function has the formula $y = 2x - 3$.
a) Copy and complete the following table for this function.

x	2	3	4	5	6
y	1				

b) Graph the function.

17. A function has the formula $y = \dfrac{x}{2} + 5$.

a) Copy and complete the following table for this function.

x	0	2	4	6	8
y					

b) Graph the function.

Set IV

This exercise requires care and patience but gives an interesting result. Draw a pair of axes extending 15 units in each direction. Connect the points in each set *in order* with straight line segments. After you have connected the points in one set, start all over again with the next. In other words, *do not connect* the last point in each set to the first point in the next one.

Set 1: (9, 8), (9.5, 10), (8, 11), (9, 12), (9, 13), (7, 13), (8.5,14.5), (10, 14), (11, 14.5), (10, 12), (11, 12), (12, 13), (13, 12), and (12, 10).

Set 2: (11, 2.5), (11, 1), (10, 1), (9, 0.5), (9, 0), (12, 0), (12, 2.5), (13, 1), (14, 1), (13, 2), and (13, 2.5).

Set 3: (4.5, 4), (3, 6), (2, 6), (1, 5.5), (0, 4), and (4, 3).

Set 4: (10, 8), (10, 10), (11.5, 11), and (12, 12).

Set 5: (7, 2.5), (7, 1), (5, 1), (4, 0.5), (4, 0), (8, 0), and (8, 2.5).

Set 6: (4, 3), (6, 7), (8, 8), (11, 8), (12, 7), (14, 3), (13, 2.5), (7, 2.5), and (4, 3).

To finish the figure, draw a large dot at (2, 5) and another large dot at (9, 14).

LESSON 3
More on Functions

WIDE WORLD PHOTOS.

Roger Bannister was the first person to run the mile in less than four minutes. He did it in 1954 with a time of 3:59.

The table at the left shows how the record for the mile has changed in the past century. To picture the change, we can plot points corresponding to the pairs of numbers in the table on a graph. A graph in which the years and record times are the respective *x*- and *y*-coordinates of the points is shown at the top of the next page.

In Lesson 1, we defined the word *function* in the following way.

▶ A **function** is a pairing of two sets of numbers so that to each number in the first set there corresponds exactly one number in the second set.

In our example, the first set of numbers is a set of years from 1870 to 1970 and the second set of numbers is the set of record times for those years. The record time for the mile is a function of the year because to each year there corresponds exactly one record time.

If the two columns of numbers in the table at the left were interchanged as shown in the table at the top of the next page, the new pairing of the numbers would *not* be a function because to the record time of 4:16 there corresponds two years, 1900 and 1910.

A convenient way to represent a function is with a formula. Not all functions have formulas, however. The relationship between the year and the record time

Year	Record time
1870	4:29
1880	4:23
1890	4:18
1900	4:16
1910	4:16
1920	4:13
1930	4:10
1940	4:06
1950	4:01
1960	3:55
1970	3:51

Record time	Year
4:29	1870
4:23	1880
4:18	1890
4:16	1900
4:16	1910
4:13	1920
4:10	1930
4:06	1940
4:01	1950
3:55	1960
3:51	1970

for the mile, for example, is too complicated to have a simple formula. And, even if we were able to write a formula for it, there would be no reason to expect it to work in the future.

For a function that does have a formula, it is often useful to be able to graph that function. The most obvious way to do this is to

1. use the formula to make a table and
2. use the table to graph the function.

Here is an example.

EXAMPLE

Graph the function $y = 9 - x^2$.

SOLUTION

First, we make a table of some numbers.

x	0	1	2	3
y	9	8	5	0

Next, we plot the points whose coordinates are the numbers in the table: (0, 9), (1, 8), (2, 5), and (3, 0), as shown in the figure at the right. The points do not lie on a straight line. Whenever this is the case, we will assume that they lie on a smooth curve.* We finish the graph by sketching in the curve.

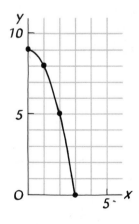

---

*All of the formulas that we study in this course will have graphs that are either straight lines or smooth curves.

Exercises

Set I

1. Tell whether or not the expressions in each of the following pairs are equal.
   a) $(9 + 4) + 3$ and $9 + (4 + 3)$
   b) $(9 - 4) + 3$ and $9 - (4 + 3)$
   c) $(9 + 4) - 3$ and $9 + (4 - 3)$
   d) $(9 - 4) - 3$ and $9 - (4 - 3)$

2. Use the distributive rule to write an expression equivalent to each of the following.
   a) $3(x + y)$ c) $x(2 - x)$
   b) $(x - 7)5$ d) $(y + 1)y^2$

3. One of the world's largest roller coasters, called "The Great American Scream Machine," is in Atlanta, Georgia.
   a) If it has x cars, each of which can hold six people, how many people can ride on it at one time?
   b) If 24 people went for a ride on it and x of them fell off, how many people would be left at the end of the trip?

Set II

4. A certain function has the graph shown at the right.
   a) Make a table for this function.
   b) Write a formula for it. Begin your formula with $y =$.

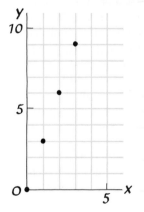

5. The graph shown at the right is *not* the graph of a function because to each value of *x* there does not correspond exactly one value of *y*.

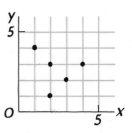

a) For which value of *x* are there two values of *y*?

b) What are they?

6. Which of the following graphs represent functions?

a)
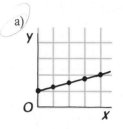

b)
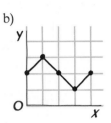

c)
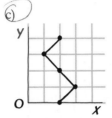

d)
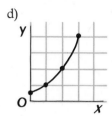

e)
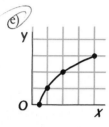

f)
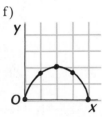

g)
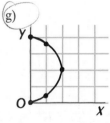

h)
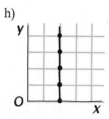

i)
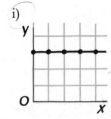

j)
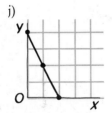

7. Make a table of numbers for each of the following functions and graph each function. In each case, connect the points with a smooth line or curve.

a) $y = x$, letting $x = 0, 1, 2, 3,$ and 4.

b) $y = x + 2$, letting $x = 0, 1, 2, 3,$ and 4.

c) $y = 2x$, letting $x = 0, 1, 2, 3,$ and 4.

d) $y = 2x - 1$, letting $x = 1, 2, 3,$ and 4.

e) $y = x^2$, letting $x = 0, 1, 2,$ and 3.

f) $y = \frac{x}{3}$, letting $x = 0, 3, 6,$ and 9.

g) $y = \frac{x}{3} + 4$, letting $x = 0, 3, 6,$ and 9.

h) $y = \frac{3}{x}$, letting $x = 1, 2, 3, 4,$ and 5.

Set III

8. A certain function has the graph shown in Figure 1.

   a) Make a table for this function.
   b) Write a formula for it.

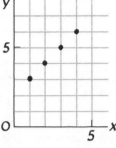

Figure 1

9. The graph shown in Figure 2 is *not* the graph of a function because to each value of x there does not correspond exactly one value of y.

   a) For which value of x are there two values of y?
   b) What are they?

Figure 2

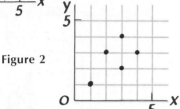

10. Which of the following graphs represent functions?

a)

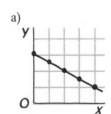

b)

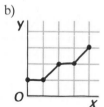

c)

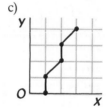

d)

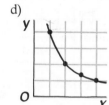

e)

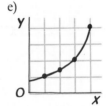

f)

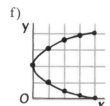

g)

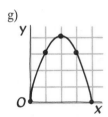

h)

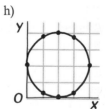

i)

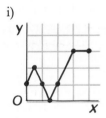

j)

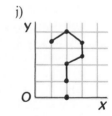

11. Make a table of numbers for each of the following functions and graph each function. In each case, connect the points with a smooth line or curve.

   a) $y = x + 3$, letting $x = 0, 1, 2, 3,$ and 4.
   b) $y = 3x$, letting $x = 0, 1, 2, 3,$ and 4.
   c) $y = 3x - 2$, letting $x = 1, 2, 3,$ and 4.

   d) $y = x^3$, letting $x = 0, 0.5, 1, 1.5,$ and 2.
   e) $y = 2x + 1$, letting $x = 0, 1, 2,$ and 3.
   f) $y = x^2 + 1$, letting $x = 0, 1, 2,$ and 3.

   g) $y = \dfrac{x}{4}$, letting $x = 0, 2, 4, 6,$ and 8.

   h) $y = \dfrac{4}{x}$, letting $x = 1, 2, 3, 4,$ and 5.

Set IV

A musician once drew a picture of part of the New York skyline on a sheet of
graph paper and translated the result into music!

If just the outline of the buildings were considered, as illustrated in the sketch
below, it would not be the graph of a function.

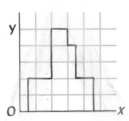

1. Why not?
2. Can you figure out what to remove from the graph so that it does represent a
   function? If you can, do it.

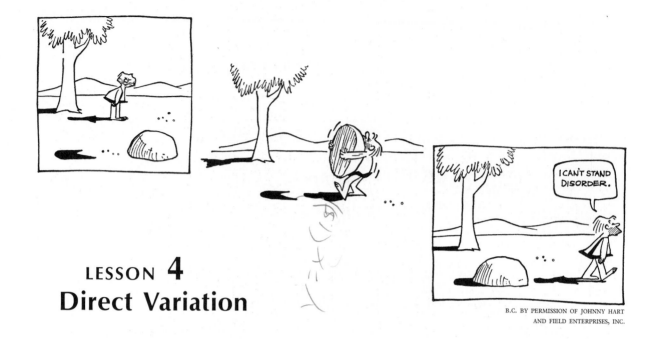

LESSON 4
Direct Variation

At any given time during the day, the length of an object's shadow depends on its height. Here is a table showing the heights of some different objects and the lengths of their shadows at a certain time in the afternoon.

Height of object in feet	1	2	3	4	5	6
Length of shadow in feet	1.5	3	4.5	6	7.5	9

Because to each height there corresponds exactly one shadow length, this is the table of a function. It is easy to see that the two variables in this function, height and shadow length, vary in the same way. If the height is doubled, for example, then the shadow length is also doubled. If the height is tripled, then the shadow length is tripled, and so on.

Because of this, the two variables are said to *vary directly* and the function relating them is called a *direct variation*. The numbers in the table for this "shadow" function have a simple pattern, and so the function can also be represented by a formula. In each case, the length of the shadow is 1.5 times the height of the object. Letting x represent the height and y represent the shadow length, we can write

$$y = 1.5x$$

If the six points in our table are located on a graph (as shown at the left below), they seem to lie along a straight line. It is possible to prove that *every* point whose coordinates fit the formula $y = 1.5x$ lies on this line. Moreover, every point on the line has coordinates that fit the formula. Because of this, we can draw the line through the points and refer to it as the graph of the function.

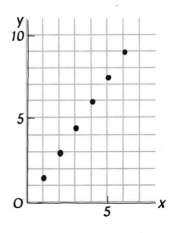
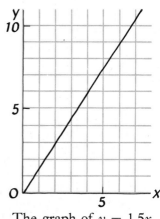

The graph of $y = 1.5x$.

The formula and line give a much more complete picture of the function than do the original table and six points. For example, from the formula we see that, if $x = 0$, then $y = 1.5(0) = 0$. This means that the point $(0, 0)$ is part of the graph; the line intersects the axes at the origin.

▶ In general, a **direct variation** is a function that has an equation of the form

$$y = ax$$

in which a is a fixed number other than zero.

The number a is called the *constant of variation*. In the example of the "shadow" function, the constant of variation is 1.5. The graph of every direct variation is a straight line that intersects the origin.

Exercises

Set I

1. Which figure below illustrates each of the expressions in parts a through d?

   a) $2 \cdot 4^2$
   b) $2^2 \cdot 4$
   c) $(2 + 4)^2$
   d) $2^2 + 4^2$

   Figure 1

   Figure 2

   Figure 3

   Figure 4

2. Guess a formula for the function represented by each of these tables. Begin each formula with $y =$.

   a)

x	2	3	4	5	6
y	4	3	2	1	0

   b)

x	1	2	3	4	5
y	11	21	31	41	51

   c)

x	0	1	2	3	4
y	0	1	8	27	64

3. Miss Brooks, an English teacher at Madison High, has 120 papers to grade.
   a) If it takes her x minutes to read each one, how long will it take her to read the entire set?
   b) If she has finished grading y papers, how many does she still have to read?

Set II

4. The graph of a certain function is shown here.

   a) Copy and complete the following table for this function.

x	0	1	2	3
y				

   b) What happens to y if x is tripled?
   c) What kind of function is this?
   d) Write a formula for the function.
   e) Use your formula to find the value of y when $x = 7$.

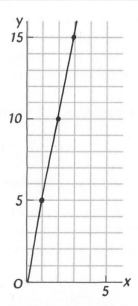

5. The following table represents a direct variation.

x	4	8	12	▓	▓
y	3	▓	▓	15	75

a) Copy and complete the table.
b) By what number do the x-numbers have to be multiplied in order to get the corresponding y-numbers?
c) Write a formula for the function.

6. A direct variation is a function that has an equation of the form $y = ax$.
a) If $x = 0$ in this equation, what can we conclude about y?
b) Why?
c) What are the coordinates of the point for which $x = 0$?
d) What does this indicate about the line that is the graph of a direct variation?

7. Tell whether the function represented by each of the following tables is a direct variation.

a)
x	1	2	3	4	5
y	1	2	3	4	5

b)
x	0	1	2	3	4
y	3	4	5	6	7

c)
x	2	4	6	8	10
y	1	2	3	4	5

d)
x	0	1	2	3	4
y	4	4	4	4	4

e)
x	1	2	3	6
y	6	3	2	1

8. The graph of every direct variation is a straight line. The position of the line is determined by the constant of variation. Make a table for each of the following direct variations. Let $x = 0, 1, 2,$ and 3 in each table.
a) $y = 2x$

b) $y = 3x$
c) $y = 4x$
d) Graph all three functions on one pair of axes. Write each equation along its line.
e) What are the constants of variation for these functions?
f) Which line is the steepest?
g) How are the constants of variation related to the steepness of the lines?

9. The height that a ball bounces varies directly with the height from which it is dropped. A formula for this function for a certain rubber ball is

$$y = 0.5x$$

in which x represents the height from which the ball is dropped in centimeters and y represents the height of the bounce in centimeters.
a) If the rubber ball is dropped from a height of 150 centimeters, how high does it bounce?
b) If the ball bounces 120 centimeters, from what height was it dropped?
c) How is the constant of variation in this formula related to the "bounciness" of the ball?
d) Do you think the constant of variation in the formula for a bouncing ball could be larger than 1? Explain.

10. The number of revolutions that a 45 rpm record makes as it is being played varies directly with the time that it is on the turntable.
a) Write a formula for this function, letting x represent the number of minutes and y represent the number of revolutions.
b) Describe, without drawing it, what the graph of this function would look like.

Set III

11. The graph of a certain function is shown here.

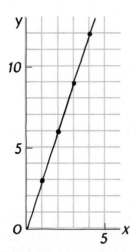

a) Copy and complete the following table for this function.

x	0	1	2	3	4
y					

b) What happens to y if x is doubled?
c) What kind of function is this?
d) Write a formula for the function.
e) Use your formula to find the value of y when x = 10.

12. The following table represents a direct variation.

x	2	4	6		
y	7			42	63

a) Copy and complete the table.
b) By what number do the x-numbers have to be multiplied in order to get the corresponding y-numbers?
c) Write a formula for the function.

13. Tell whether you think the quantities in each of the following relationships vary directly and explain why you believe as you do.

a) The distance of a star and the time it takes its light to reach us.
b) The weight of a cat and its age.
c) The number of cooks and the time needed to make the soup.
d) The number of empty Coke bottles returned to the market and the amount of the refund.

14. Tell whether the function represented by each of the following tables is a direct variation.

a)
x	0	1	2	3	4
y	0	4	8	12	16

b)
x	1	3	5	7	9
y	2	3	4	5	6

c)
x	0	1	2	3	4
y	7	7	7	7	7

d)
x	3	6	9	12	15
y	2	4	6	8	10

e)
x	1	2	3	4	5
y	5	4	3	2	1

15. Each of the lines shown at the top of the next page is the graph of a direct variation. Write a formula for
a) line 1. $y = 2x$
b) line 2. $y = x$
c) line 3. $y = 3x$
d) line 4. $y = 2.5y$
e) What are the constants of variation for these functions?
f) What line is the steepest?
g) How are the constants of variation related to the steepness of the lines?

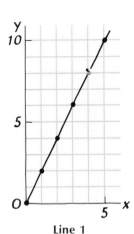

Line 1

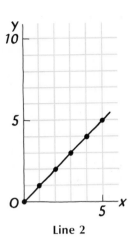

Line 2

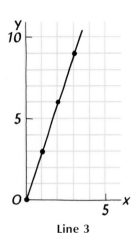

Line 3

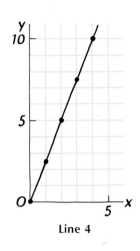

Line 4

16. The volume of a balloon varies directly with the temperature in degrees Kelvin of the air inside it. A formula for this function for a certain balloon is

$$y = 2x$$

in which x represents the temperature in degrees Kelvin and y represents the volume of the balloon in cubic centimeters.
a) If the temperature is 300 degrees Kelvin, what is the volume of the balloon?
b) If the volume of the balloon is 500 cubic centimeters, what is the temperature?
c) How is the constant of variation in this formula related to the size of the balloon?

17. The price of Mrs. See's chocolates varies directly with the weight. One pound costs $3.50.
a) Write a formula for this function, letting x represent the number of pounds bought and y represent the price in dollars.
b) Describe, without drawing it, what the graph of this function would look like.

Set IV

Pacific Stereo sells Maxell UD-90 cassette tape according to the price scale shown in this graph.
1. If it were not for the last three points, this would be the graph of a direct variation. Why do you suppose they are not along the same line as the other points?
2. Explain why it would not be a good idea to buy eleven cassettes.

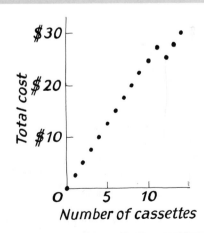

LESSON 5
Linear Functions

The deepest mine in the world, located in Carltonville, South Africa, extends more than two miles below the earth's surface. The walls of the mine are so hot at the bottom that refrigerated air has to be pumped into it to make it possible for the miners to work.

The temperature inside the earth is a function of the depth below the surface. Geophysicists have found that, within the earth's crust, it increases steadily as the depth increases. Near the surface, the temperature is about $20°C$; it increases at the rate of 10 degrees for each kilometer of depth. A table for this function, along with a graph, are shown below.

Depth in kilometers	0	1	2	3	4
Temperature in °C	20	30	40	50	60

Because the graph of the function is a straight line, the function is called *linear.* Although direct variations are linear functions, this one is not a direct variation. As the depth is doubled, the temperature is not doubled: the temperature at a depth of 2 kilometers, 40°C, for example, is not twice the temperature at a depth of 1 kilometer, 30°C. Furthermore, the graph does not intersect the origin because the temperature at a depth of 0 kilometers is not 0°C.

If we let x represent the depth and y represent the temperature, the formula for this function is

$$y = 10x + 20$$

The fact that its graph is a straight line does not depend on the specific numbers in its equation, but rather on the pattern of the equation. The pattern is: "y is equal to the sum of some number times x and some number."

Another function whose equation has the same pattern is

$$y = \frac{1}{2}x + 5$$

Its graph, shown at the right, is also a straight line.

In general, a **linear function** is a function that has an equation of the form

$$y = ax + b$$

in which a and b are constant numbers.

The graph of every linear function is a straight line.

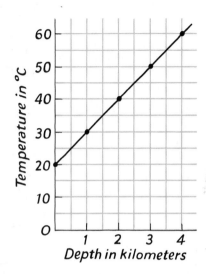

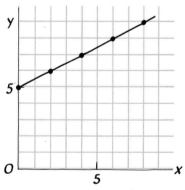

The graph of $y = \frac{1}{2}x + 5$.

Exercises

Set I

1. What is the length marked with a question mark in each of these figures?

a)

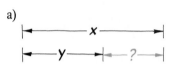

b)

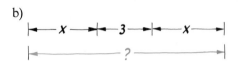

c)

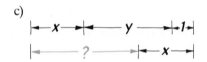

2. There are two pints in a quart and four quarts in a gallon.
   a) How many pints are there in x quarts?
   b) How many quarts are there in x gallons?
   c) How many pints are there in x gallons?

3. Tell, by evaluating both sides of each equation, whether it is true or false.
   a) $3^2 + 4^2 = 5^2$
   b) $3^3 + 4^3 + 5^3 = 6^3$
   c) $3^4 + 4^4 + 5^4 + 6^4 = 7^4$
   $\quad$ 81 $\quad$ 256 625 1296 $\quad$ 2401

Set II

4. If a function is linear, it has an equation of the form $y = ax + b$, in which a and b are constant numbers. These numbers determine the position of the line that is the graph of the function.

   Make a table for each of the following linear functions, letting $x = 0, 1, 2,$ and 3 in each table.

   a) $y = x + 2$
   b) $y = x + 4$
   c) $y = x + 7$

   d) Graph the three functions on one pair of axes. Write each equation along its line.
   e) What do you notice about the lines?
   f) Where does each line meet the y-axis?
   g) Where do you think the graph of the equation $y = x + 10$ would meet the y-axis?

5. By doing exercise 4, you found out how the position of the line that is the graph of

$$y = ax + b$$

changes as b changes. By doing this exercise, you will see the effect of changing a.

   Make a table for each of the following linear functions. Let $x = 0, 1, 2,$ and 3 in each table.
   a) $y = 2x + 1$
   b) $y = 3x + 1$
   c) $y = 4x + 1$
   d) Graph the three functions on one pair of axes. Write each equation along its line.
   e) What do you notice about the lines?
   f) Where does each line meet the y-axis?
   g) Where do you think the graph of the equation $y = 10x + 1$ would meet the y-axis?

6. A function has the formula $y = 0x + 5$.
   a) Make a table for this function, letting $x = 0, 1, 2, 3,$ and 4.
   b) Graph the function.
   c) What do you notice about the graph?
   d) What is a simpler formula for the function?

7. There is a connection between the equation for a linear function, $y = ax + b$, and its table if we let $x = 0, 1, 2, 3,$ and so forth. For example, compare the equation $y = 3x + 5$ and the table

x	0	1	2	3	...
y	5	8	11	14	...

   a) In what way does the number 3 (a in the equation) appear in the table?
   b) Where does the number 5 (b in the equation) appear in the table?

   Use your answers to parts a and b to discover equations for the linear functions having the following tables.

   c)
x	0	1	2	3
y	8	10	12	14

   d)
x	0	1	2	3
y	1	8	15	22

   e)
x	0	1	2	3
y	6	10	14	18

   f)
x	0	1	2	3
y	4	10	16	22

8. The perimeter of a rectangle whose width is 3 is a function of its length.

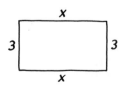

   a) Write a formula for this function, letting y represent the perimeter of the rectangle and x represent its length.

   b) What kind of function is this?
   c) Does the perimeter of a rectangle whose width is 3 vary directly with its length?

9. Spring scales work on the principle that, if a weight is hung from one end of a spring, the

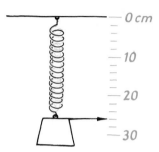

   total length of the spring is a function of the amount of the weight. The equation for a certain spring is

   $$y = 0.5x + 20$$

   in which x is the weight in grams and y is the length of the spring in centimeters.
   a) Copy and complete the following table for this function.

x	0	1	2	3	4
y					

   b) How long is the spring if no weight is attached to it?
   c) How much is the spring stretched if a weight of one gram is hung from it?
   d) How long would the spring be if a weight of 20 grams were hung from it?

Set III

10. If a function is linear, it has an equation of the form $y = ax + b$, in which a and b are constant numbers. These numbers determine the position of the line that is the graph of the function.

Make a table for each of the following linear functions, letting $x = 0, 1, 2$, and 3 in each table.

a) $y = 2x + 1$
b) $y = 2x + 4$
c) $y = 2x + 5$

d) Graph the three functions on one pair of axes. Write each equation along its line.
e) What do you notice about the lines?
f) Where does each line meet the y-axis?
g) Where do you think that the graph of the equation $y = 2x + 8$ would meet the y-axis?

11. By doing exercise 10, you found out how the position of the line that is the graph of

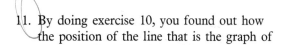

$$y = ax + b$$

changes as b changes. By doing this exercise, you will see the effect of changing a.

Make a table for each of the following linear functions, letting $x = 0, 1, 2$, and 3 in each table.

a) $y = 1x + 3$
b) $y = 2x + 3$
c) $y = 3x + 3$

d) Graph the three functions on one pair of axes. Write each equation along its line.
e) What do you notice about the lines?
f) Where does each line meet the y-axis?
g) Where do you think that the graph of the equation $y = 7x + 3$ would meet the y-axis?

12. A function has the formula $y = 0x + 2$.
a) Make a table for this function, letting $x = 0, 1, 2, 3$, and 4.
b) Graph the function.
c) What do you notice about the graph?
d) What is a simpler formula for the function?

13. There is a connection between the equation for a linear function, $y = ax + b$, and its table if we let $x = 0, 1, 2, 3$, and so forth. For example, compare the equation $y = 4x + 7$ and the table

x	0	1	2	3	...
y	7	11	15	19	...

a) In what way does the number 4, a in the equation, appear in the table?
b) Where does the number 7, b in the equation, appear in the table?

Use your answers to parts a and b to discover equations for the linear functions having the following tables.

c)
x	0	1	2	3
y	2	8	14	20

d)
x	0	1	2	3
y	1	4	7	10

e)
x	0	1	2	3
y	9	14	19	24

f)
x	0	1	2	3
y	5	14	23	32

14. The perimeter of an isosceles triangle whose base is 5 is a function of the length of one of its equal sides.

a) Write a formula for this function, letting *y* represent the perimeter of the triangle and *x* represent the length of one of its equal sides.
b) What kind of function is this?
c) Does the perimeter of an isosceles triangle whose base is 5 vary directly with the length of one of its equal sides?

15. Mr. Scrooge pays his employees a starting salary of $80 per week with weekly raises of 50 cents. As a result, the number of dollars one of his employees earns in a week, *y*, is a function of the number of weeks he has been working, *x*.

a) Copy and complete the following table for this function.

x	1	2	3	4	5
y	80	80.5	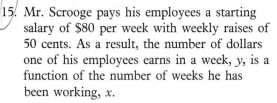		

b) Write a formula for this function.
c) Do an employee's weekly earnings vary directly with the number of weeks he has been working?
d) How much money would someone earn during the 50th week that he works for Mr. Scrooge?

Set IV

The speed at which a certain ant travels is a function of the temperature. This table shows how the two variables are related.

Speed of ant in centimeters per second	2	3	4	5
Temperature in °C	16	22	28	34

1. Can you figure out a formula for this function? (Let *x* represent the speed of the ant and *y* represent the temperature.)
2. How cold must it be for the ant to stay home?

"But I couldn't have been going 70 miles an hour!
I only left home 20 minutes ago!"

LESSON 6
Inverse Variation

The time that it takes someone to drive a certain distance is a function of the speed at which he or she drives. The faster the speed, the shorter the time. Suppose, for example, that the young man in this cartoon wants to travel 120 miles. The times that the trip would take at several different speeds are shown in this table.

Average speed in mph	20	30	40	50	60	70	80
Time the trip takes in hours	6	4	3	2.4	2	1.7	1.5

As one of the variables in this function increases, the other decreases. More specifically, if the average speed is doubled, the time the trip takes is halved. For example, if the average speed changes from 30 to 60 miles per hour, then the time the trip takes changes from 4 hours to 2 hours. If the speed were tripled, the time would be divided by three, and so on.

Two variables that change in this way are said to *vary inversely* and the function relating them is called an *inverse variation*. The time of the trip in this "driving speed" function is found by dividing the distance, 120 miles, by the

96

average speed. If we let x represent the average speed and y represent the time, we can write

$$y = \frac{120}{x}$$

What would a graph of this function look like? The graph shown in Figure 1 (below) includes the seven points represented in the table. It is evident that they do not lie on a straight line. If we use the equation to figure out additional points, such as (25, 4.8), (35, 3.4), (90, 1.3), and (100, 1.2), and join them with the others to make a smooth curve, we get the graph shown in Figure 2. As they do for the other functions that we have studied, the equation and curve give a much more complete picture than do the original table and seven points.

▶ In general, an **inverse variation** is a function that has an equation of the form

$$y = \frac{a}{x}$$

in which a is a constant number other than zero.

The number a is called the *constant of variation*. The graph of every inverse variation is a curve.

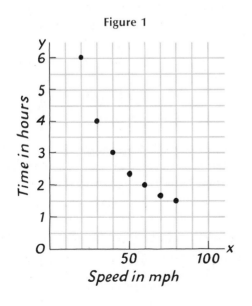

Figure 1

Speed in mph

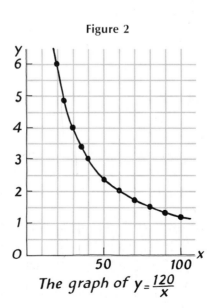

Figure 2

The graph of $y = \frac{120}{x}$

Exercises

Set I

1. Write an expression for each of the following.
   a) The number that is four less than x.
   b) Twice the sum of x and five.
   c) The difference between the cube of x and one.

2. A convenient way to picture a function is with a graph.
   a) What are the two lines labeled x and y called?
   b) What is the point at which these lines intersect called?

   c) What are the pairs of numbers used to locate points on a graph called?

3. Sir Isaac Newton, who was once Master of the Mint in England, assumed that, if the amount of money in circulation is doubled, then prices will also double.
   a) If this is true, how do these two quantities vary with respect to each other?
   b) What would a graph of such a function look like?

Set II

4. The graph of a certain function is shown below.

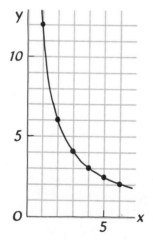

   a) Copy and complete the following table for this function.

x	1	2	3	4	5	6
y	▓	▓	▓	▓	2.4	▓

   b) What happens to y if x is tripled?
   c) What kind of function is this?
   d) Write a formula for the function.
   e) Use your formula to find the value of y when $x = 10$.

5. The following table represents an inverse variation.

x	2	4	6	▓	▓
y	30	▓	▓	5	3

   a) Copy and complete the table.
   b) Into what number do the x-numbers have to be divided in order to get the corresponding y-numbers?
   c) Write a formula for the function.

6. An inverse variation is a function that has an equation of the form $y = \dfrac{a}{x}$.
   a) Write the equation that you would have if

x were replaced by 1 and the result were simplified.

b) What number *cannot* replace x in the equation $y = \dfrac{a}{x}$?

c) Why not?

d) Can the curve that is the graph of an inverse variation touch the y-axis?

7. Tell whether the function represented by each of the following tables is an inverse variation.

a)

x	1	2	3	4	
y	24	12	8	6	

b)

x	0	1	2	3	4
y	10	9	8	7	6

c)

x	1	2	3	4	5
y	32	16	8	4	2

d)

x	2	3	5	6	
y	15	10	6	5	

e)

x	0	1	2	3	4
y	25	16	9	4	1

8. The graph of every inverse variation is a curve. The shape and position of the curve are determined by the constant of variation.

Make a table for each of the following inverse variations, letting $x = 1, 2, 3, 4,$ and 5 in each table.

a) $y = \dfrac{4}{x}$

b) $y = \dfrac{6}{x}$

c) $y = \dfrac{10}{x}$

d) Graph all three functions on one pair of axes. Write each equation along its curve.

e) What are the constants of variation for these functions?

f) How are the constants of variation related to the distances of the curves from the origin?

9. Each of these drawings contains the same number of circles.

○ ○ ○ ○ ○ ○ ○ ○ ○ ○ ○ ○
○ ○ ○ ○ ○ ○ ○ ○ ○ ○ ○ ○

○ ○ ○ ○ ○ ○ ○ ○   ○ ○ ○ ○
○ ○ ○ ○ ○ ○ ○ ○   ○ ○ ○ ○
○ ○ ○ ○ ○ ○ ○ ○   ○ ○ ○ ○
                         ○ ○ ○ ○
○ ○ ○ ○ ○ ○   ○ ○ ○ ○
○ ○ ○ ○ ○ ○   ○ ○ ○ ○
○ ○ ○ ○ ○ ○
○ ○ ○ ○ ○ ○

a) The number of rows is a function of the number of circles in each row. Write a formula for this function, letting x represent the number of circles in each row and y represent the number of rows.

b) What is the constant of variation for this function?

c) How do x and y vary with respect to each other?

10. The time that it takes to run the 100-meter dash varies inversely with the speed of the runner.

a) Copy and complete the following table for this "100-meter dash" function.

Average speed in meters per second	2	4	▦	▦
Time in seconds	▦	▦	20	10

b) Write a formula for this function, letting t represent the time and s represent the average speed.

c) In 1977, the women's world record time for the 100-meter dash, held by Marlies Oelsner of East Germany, was 10.88 seconds. What can you conclude about her average speed?

Set III

11. The graph of a certain function is shown here.

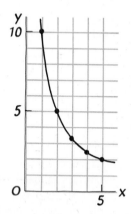

a) Copy and complete the following table for this function.

x	1	2	3	4	5
y	▓	▓	$3\frac{1}{3}$	▓	▓

b) What happens to y if x is doubled?
c) What kind of function is this?
d) Write a formula for the function.
e) Use your formula to find the value of y when $x = 20$.

12. The following table represents an inverse variation.

x	2	4	6	▓	▓
y	24	▓	▓	4	3

a) Copy and complete the table.
b) Into what number do the x-numbers have to be divided in order to get the corresponding y-numbers?
c) Write a formula for the function.

13. Tell whether you think that the quantities in each of the following relationships vary inversely and explain why you believe as you do.
   a) A person's age and the length of his or her attention span.
   b) The time required to swim across a lake and the rate at which you swim.
   c) The number of kilograms of sugar you can buy for $10 and the price of one kilogram.
   d) The number of dogs pulling a sled and the speed at which it goes.

14. Tell whether the function represented by each of the following tables is an inverse variation.

a)
x	1	2	3	4
y	60	30	20	15

b)
x	0	1	2	3	4
y	8	7	6	5	4

c)
x	1	2	3	4	5
y	16	8	4	2	1

d)
x	2	3	6	18
y	9	6	3	1

e)
x	1	2	3	4	5
y	81	27	9	3	1

15. The graph of every inverse variation is a curve. The shape and position of the curve are determined by the constant of variation. Make a table for each of the following inverse variations, letting $x = 1, 2, 3, 4,$ and 5 in each table.

a) $y = \dfrac{5}{x}$

b) $y = \dfrac{8}{x}$

c) $y = \dfrac{12}{x}$

d) Graph all three functions on one pair of axes. Write each equation along its curve.
e) What are the constants of variation for these functions?
f) How are the constants of variation related to the distances of the curves from the origin?

16. Each of these rectangles has the same area.

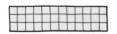

a) The width of each rectangle is a function of its length. Write a formula for this function, letting x represent the length and y represent the width.
b) What is the constant of variation for this function?
c) How do x and y vary with respect to each other?

17. The number of times that a wheel revolves in rolling one kilometer varies inversely with its diameter. If the wheel's diameter is measured in meters, the constant of variation for this function is approximately 318.
a) Write a formula for this function, letting d represent the diameter of the wheel in meters and n represent the number of revolutions it makes in rolling one kilometer.
b) If a wheel has a diameter of 0.5 meter, how many times does it revolve in rolling one kilometer?

$$n = \frac{318}{d}$$

$$n = \frac{318}{.5}$$

Set IV

A lever can be used to move a heavy object with a small force. Suppose that a rock having a mass of 600 kilograms is located one centimeter from the pivot point of a lever. The rock can be

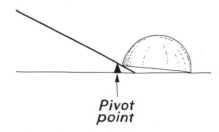

Pivot point

moved by a force of only 6 kilograms if the force is applied 100 centimeters from the other side of the pivot.

1. Given that the force required to move the rock varies *inversely* with the distance it is applied from the other side of the pivot, can you write a formula for this function? Let d represent the distance in centimeters from the pivot and f represent the force in kilograms.

2. If B.C. wants to move the rock described above by pushing at a point 80 centimeters from the pivot, how much force would he have to exert?

Summary and Review

In this chapter, we have become acquainted with the idea of a function and have learned how several types of functions can be pictured with coordinate graphs.

Functions (*Lessons 1 and 3*)   A function is a pairing of two sets of numbers so that to each number in the first set there corresponds exactly one number in the second set. A function can be represented by a table, by a formula, or by a coordinate graph.

The Coordinate Graph (*Lesson 2*)   A coordinate graph contains two perpendicular lines called *axes* that intersect in a point called the *origin*. Points are located on such a graph by means of pairs of numbers called *coordinates*. To graph a function when given its formula, first use the formula to make a table and then use the table to graph it.

Direct Variation (*Lesson 4*)   A direct variation is a function that has an equation of the form $y = ax$, in which a is a constant number other than zero. The number a is called the *constant of variation.* The graph of every direct variation is a straight line that intersects the origin.

Linear Functions (*Lesson 5*)   A linear function is a function that has an equation of the form $y = ax + b$, in which a and b are constant numbers. The graph of every linear function is a straight line.

Inverse Variation (*Lesson 6*)   An inverse variation is a function that has an equation of the form $y = \dfrac{a}{x}$, in which a, the constant of variation, is a number other than zero. The graph of every inverse variation is a curve.

Exercises

Set I

1. Draw a pair of axes extending 8 units in each direction from the origin. Connect the points in the following list with straight line segments in the order given to form a square.

   (2, 1)   (6, 3)   (4, 7)   (0, 5)   (2, 1)

2. Copy and complete the tables for the following functions.

   a) Formula: $y = 7 - x$
      Table:

x	1	2	3	4	5
y					

   b) Formula: $y = x^3 + 1$
      Table:

x	0	1	2	3	4
y					

   c) Formula: $y = x(x - 4)$
      Table:

x	4	5	6	7	8
y					

3. Read the following statements carefully and tell whether each is true or false.

   a) In an inverse variation, if one variable is doubled, then the other is halved.
   b) If one of the coordinates of a point is zero, the point lies on one of the axes.
   c) All linear functions are direct variations.
   d) The graph of every direct variation intersects the origin.
   e) The constant of variation for the function $y = x$ is 0.

4. Guess a formula for the function represented by each of these tables. Begin each formula with $y =$.

   a)
x	6	7	8	9	10
y	1	2	3	4	5

   b)
x	2	3	4	5	6
y	3	5	7	9	11

   c)
x	0	3	6	9	12
y	0	1	2	3	4

   d)
x	2	4	6	8	10
y	5	17	37	65	101

5. The graph of a certain function is shown here.

   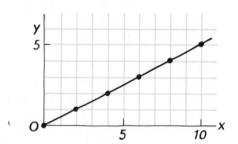

   a) Copy and complete the following table for this function.

x	0	2	4	6	8	10
y						

   b) What happens to y if x is tripled?
   c) What kind of function is this?
   d) Write a formula for the function.
   e) Use your formula to find the value of y when $x = 30$.

6. Make a table of numbers for each of these functions, letting x equal 1, 2, 3, and 4, and graph each one.
   a) $y = x - 1$
   b) $y = 2x + 3$
   c) $y = \dfrac{8}{x}$

S. A. PRENTICE.

in which x represents his age and y represents his height in feet.

a) Copy and complete the following table for this function.

Age in years, x	4	8	12	16
Height in feet, y	▥	▥	▥	▥

b) Graph it.

c) For the ages graphed, this height function seems to be linear. Do you think it would be linear for ages less than 4 and more than 16?

d) Does the fellow's height vary directly with his age for the ages graphed?

7. When lightning strikes, the time interval between the flash that you see and the thunder that you hear depends on the distance that you are from where the lightning struck.

A formula for this function is $y = 3x$, in which x is the distance in kilometers and y is the time interval in seconds.

a) Copy and complete the following table for this function.

x	1	2	3	4
y	▥	▥	▥	▥

b) Graph it.

c) What type of function is it?

d) How long is the time interval between the flash and the thunder if lightning strikes 2.5 kilometers away?

8. A person's height is a function of his or her age. A formula by which to find the height of the fellow in this cartoon from age 4 to age 16 might be

$$y = \frac{1}{4}x + 3$$

"... And what are you going to do when you grow up?"

COPYRIGHT GAHAN WILSON.

9. The amount of heat lost through a windowpane depends on how thick the glass is. A formula for this function for a certain window is

$$y = \frac{12}{x}$$

in which x represents the thickness of the pane in millimeters and y represents the number of units of heat lost.

a) Copy and complete the following table for this function.

Thickness in millimeters, x	3	4	5	6
Units of heat lost, y	▦	▦	▦	▦

b) Graph it.
c) How do the thickness and units of heat lost vary with respect to each other?
d) As one of the two variables in this function becomes very large, what happens to the other?

Set II

1. Draw a pair of axes extending 8 units in each direction from the origin. Connect the points in the following list with straight line segments in the order given to form a parallelogram.

(7, 5)   (5, 2)   (0, 1)   (2, 4)   (7, 5)

2. Copy and complete the tables for the following functions.
a) Formula: $y = 4x - 1$
   Table:

x	1	2	3	4	5
y	▦	▦	▦	▦	▦

b) Formula: $y = x^2 + 3x$
   Table:

x	0	1	2	3	4
y	▦	▦	▦	▦	▦

c) Formula: $y = 2(x + 5)$
   Table:

x	1	2	3	4	5
y	▦	▦	▦	▦	▦

3. Read the following statements carefully and tell whether each is true or false.
a) All direct variations are linear functions.
b) The graph of every linear function intersects the origin.
c) Every point on a coordinate graph is located by a pair of numbers.
d) If one variable in a direct variation is tripled, then so is the other.
e) If the graph of a function is a curved line, it is an inverse variation.

4. Guess a formula for the functions represented by each of these tables. Begin each formula with $y =$.

a)
x	3	4	5	6	7
y	5	4	3	2	1

b)
x	0	2	4	6	8
y	0	4	16	36	64

c)
x	1	2	3	4	5
y	9	19	29	39	49

d)
x	4	6	8	10	12
y	6	9	12	15	18

5. The graph of a certain function is shown below.

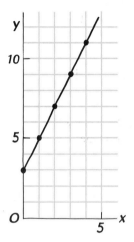

a) Copy and complete the following table for this function.

x	0	1	2	3	4
y	▨	▨	▨	▨	▨

b) Is y doubled if x is doubled?
c) What kind of function is this?
d) Write a formula for the function.
e) Use your formula to find the value of y when $x = 25$.

6. Make a table of numbers for each of these functions, letting x equal 1, 2, 3, and 4, and graph each one.
a) $y = 1.5x$
b) $y = 9 - 2x$
c) $y = x^2 + 1$

7. As a prank, Davy Jones's friends decided to fill his gym locker with water, using a hose. The table at the top of the next column shows the height of water in the locker as a function of time.

Time in seconds	0	10	20	30	40
Height of water in inches	0	5	10	15	20

a) Write a formula for this function, letting x represent the time in seconds and y represent the height of the water in inches.
b) How does the water height vary with respect to the time?
c) What would a graph of this function look like?

8. The relative amount of gold in a ring is measured in karats. A formula for this function is

$$y = \frac{25x}{6}$$

in which x represents the number of karats and y represents the percentage of gold in the ring.

a) Copy and complete the following table for this function.

x	0	6	12	18	24
y	▨	▨	▨	▨	▨

b) How many karats is a ring made of pure gold?
c) Graph this function. (Let each unit on the y-axis represent a 10 percent change in the amount of gold.)
d) What type of function is it?

9. The time that it takes a horse to run a race depends on how fast it runs. A typical formula for this function is

$$y = \frac{180}{x}$$

in which x represents the horse's average speed in meters per second and y represents its time in seconds.

COURTESY OF WALTER MERRICK.

a) Copy and complete the following table for this function.

Speed in meters per second, x	12	14	16	18	20
Time in seconds, y	▓▓▓	12.9	▓▓▓	▓▓▓	▓▓▓

b) Graph it.
c) How do the horse's time and speed vary with respect to each other?
d) Explain, mathematically, why we cannot figure out the horse's time if its speed is 0 meters per second.

Chapter 3

THE INTEGERS

LESSON 1
The Integers

PHOTOGRAPH BY GERHARD OTT.

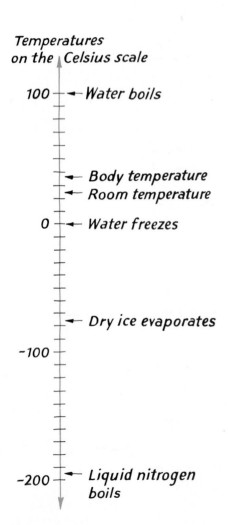

Temperatures on the Celsius scale

100 — ◄— Water boils

◄— Body temperature
◄— Room temperature

0 — ◄— Water freezes

◄— Dry ice evaporates

-100 —

-200 — ◄— Liquid nitrogen boils

The substance boiling in the kettle shown in this photograph is being heated by a block of dry ice! Dry ice is much colder than ordinary ice and usually has a temperature of 78 degrees below zero on the Celsius scale. This is very hot, however, in comparison with the liquid in the kettle: it is liquid nitrogen and its temperature is 196 degrees below zero!

These temperatures are shown on the scale at the left. The Swedish astronomer who invented the Celsius scale assigned the number 0 to the temperature at which water freezes. Temperatures warmer than 0°C are positive and temperatures colder than 0°C are negative. For example, a person's normal body temperature is +37°C and the temperature at which dry ice evaporates is –78°C. Although the symbols for plus and minus are used here, they do not mean to add or subtract. Instead, they indicate whether the temperature is above or below zero on the Celsius scale and are read as *positive* and *negative*.

If we forget about the temperatures and turn such a scale sideways so that the positive numbers are on the right, we get a scale called a *number line* (shown at the top of the next page).

$-4 < -3$

$4 > 3$

Larger

$-10\ -9\ -8\ -7\ -6\ -5\ -4\ -3\ -2\ -1\quad 0\ +1\ +2\ +3\ +4\ +5\ +6\ +7\ +8\ +9\ +10$

Smaller

The numbers shown above the line in this figure are called **integers**. As we read from left to right along the line, the integers get larger. The integer +4, for example, is *larger* than +3 because +4 is to the *right* of +3 on the line. The number –4, on the other hand, is *smaller* than –3 because –4 is to the *left* of –3 on the line. The symbol for "is larger (or more) than" is >, and so "+4 is larger than +3" can be written as "+4 > +3." Turn the symbol around and it means "is smaller (or less) than." To show that –4 is smaller than –3, we write "–4 < –3."

The integers consist of three sets of numbers: the *positive integers,* which are larger than zero; the *negative integers,* which are less than zero; and *zero* itself, which is neither positive nor negative. Because the positive integers are the same as the counting numbers with which everyone becomes familiar as a child, they are usually written without the "positive" symbol: "+3" and "3," for example, mean the same number.

Although we can't count with negative numbers, they are numbers nonetheless. Many measurements with respect to a reference point, such as distance below sea level or time before a certain event, lend themselves to the use of negative numbers. A good understanding of what negative numbers are and how their presence affects calculations will be essential to our work in algebra; so we will study their properties in several lessons.

Exercises

Set I

1. The number $2^5 \cdot 9^2$ is quite unusual.
   a) Find its value.
   b) What do you notice?

2. Find the value of
   $(1 + x)(2 + x)(3 + x)(4 - x)$ if
   a) $x = 1$
   b) $x = 2$
   c) $x = 3$
   d) $x = 4$

3. Mehitabel catches three mice each day.
   a) At this rate, how many days would it take her to catch x mice?
   b) How many mice would she catch in y weeks?
   c) What is the name for the variation of the total number of mice caught with the total number of days?

Set II

4. What do the positive and negative numbers mean in each of the following statements?
   a) The melting point of mercury is –39°C.
   b) The elevation of the Caspian Sea is –29 meters.
   c) The Roman emperor Nero was born in +37.
   d) The class starts in –10 minutes.

5. The fact that +37°C is a higher temperature than +25°C can be expressed in symbols as +37 > +20. Use symbols to express the number relationships in the following statements.

   a) An elevation of +12 meters is higher than an elevation of –15 meters.
   b) A temperature of –196°C is lower than a temperature of –78°C.
   c) The number –3 is to the left of the number +3 on a number line.
   d) The apparent weight of a helium balloon, –22 grams, is more than the apparent weight of a hydrogen balloon, –24 grams.

6. Copy each of the following, replacing each ▓ with either > or <.

   a) 4 ▓ 1   e) 5 ▓ –11
   b) 0 ▓ 9   f) –1 ▓ –6
   c) 7 ▓ –7   g) –12 ▓ 8
   d) –3 ▓ 0   h) –10 ▓ –2

7. The following questions are about these numbers:

   +1   –2   +3   –4   +5   –6   +7   –8

   a) Which number is the largest?
   b) Which is the smallest?
   c) Arrange the numbers in order from smallest to largest.

8. The number of the point midway between 2 and 8 is 5, as the figure below shows.

Use the number line below to find the number of the point midway between each of the pairs of points in parts a through h.

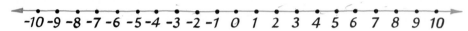

   a) 0 and 6
   b) 0 and –6
   c) 1 and 9
   d) –1 and –9
   e) –2 and 2
   f) 7 and –7
   g) –5 and 3
   h) –3 and 5

9. The distance between –3 and 1 is 4, as the figure below shows.

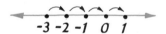

Use the number line in exercise 8 to find the distance between each of the following pairs of points.

   a) 2 and 7   e) –3 and 3
   b) –2 and –7   f) –10 and 10
   c) 0 and 8   g) –6 and 1
   d) 0 and –8   h) –1 and 6

10. Write each of the following statements in symbols, letting x represent the number.
   a) A certain number is less than zero.
   b) The square of a certain number is more than five.
   c) The sum of a certain number and one is less than ten.
   d) The quotient of a certain number and two is more than eight.

Set III

11. What do the positive and negative numbers mean in each of the following statements?
    a) The elevation of Mexico City is +2300 meters.
    b) The Greek scientist Archimedes died in –212.
    c) On September 3, 1929, the Dow Jones industrial average changed –48 points.
    d) Mr. Micawber's net worth is –800 dollars.

12. The fact that an elevation of +48 meters is less than an elevation of +50 meters can be expressed in symbols as $+48 < +50$. Use symbols to express the number relationships in the following statements.
    a) A temperature of +15°C is higher than a temperature of –40°C.
    b) The number –8 is to the left of the number 2 on a number line.
    c) An elevation of 12 meters is more than an elevation of –12 meters.
    d) In counting toward blast-off, the time –60 seconds comes before the time –50 seconds.

13. Copy each of the following, replacing each ▓ with either $>$ or $<$.
    a) 2 ▓ 8   d) –4 ▓ 4   g) –10 ▓ 1
    b) 11 ▓ 5   e) 7 ▓ –9   h) –12 ▓ –8
    c) 0 ▓ –6   f) –1 ▓ –3

14. The following questions are about these numbers:

    0   –1   +2   –3   +4   –5   +6   –7

    a) Which number is the smallest?
    b) Which is the largest?
    c) Arrange the numbers in order from smallest to largest.

15. The distance between –2 and 4 is 6, as the figure below shows.

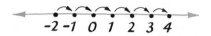

Use the number line below to find the distance between each of the following pairs of points.

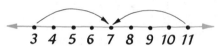

    a) 1 and 5   e) –2 and 2
    b) –1 and –5   f) –6 and 6
    c) 7 and 0   g) –3 and 8
    d) –7 and 0   h) –8 and 3

16. The number of the point midway between 3 and 11 is 7, as the figure below shows.

Use the number line in exercise 15 to find the number of the point midway between each of the following pairs of points.
    a) 0 and 8   e) 5 and –5
    b) 0 and –8   f) –9 and 9
    c) 3 and 7   g) –4 and 6
    d) –3 and –7   h) –6 and 4

17. Write each of the following statements in symbols, letting x represent the number.
    a) A certain number is more than negative two.
    b) The cube of a certain number is less than twenty.
    c) The product of four and a certain number is more than one.
    d) The difference between a certain number and two is less than zero.

$x > -2$

$x^3 < 20$

Set IV

While hidden behind a billboard, Officer Krupke sees three cars pass by in the following order: a Mustang, a Cougar, and a Rabbit. He estimates their respective speeds going east to be 55, –70, and –75 miles per hour.

If the cars pass by Officer Krupke within a minute of each other and they continue traveling on the same street at these speeds,

1. will the Cougar overtake the Mustang?
2. will the Rabbit overtake the Cougar?

   Explain your answers.

"He's lost the scent. Let him smell your keys again, lady."

LESSON 2
More on the Coordinate Graph

To help people find their cars in a vast parking lot, some kind of coordinate system is usually used. We have already become acquainted with a coordinate system for finding points on a graph. This system, which uses the positive numbers and zero, is limited to two directions: to the right and up.

In a parking lot, these directions might correspond to east and north. The location of a car on the lot could be given by saying, for example, that it is four rows east of your starting point and in the sixth space toward the north. The coordinates of its location, then, would be (4, 6).

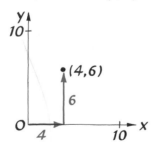

What if the parking lot also extends to the west and south? Suppose that a car is in the fifth row *west* of the starting point and in the third space toward the *south*. What would the coordinates of *its* location be? If we extend the x- and y-axes of a coordinate graph beyond the origin and think of them as number lines, then we get the figure shown at the top of the next page. The system we have been using to locate points still works, but now one or both of the coordinates may be negative. The coordinates of the car five rows west and three spaces toward the south, for example, would be (–5, –3).

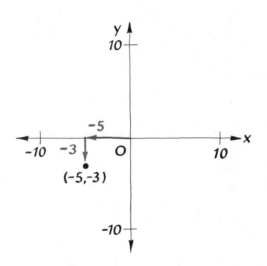

Notice that the origin of this graph is at the center rather than the lower left corner. The graph now has four regions instead of one. The regions are called *quadrants* and are numbered counterclockwise, starting with the upper right, as shown in the figure at the left.

With four directions in which to move from the origin, it is important to remember that the x-coordinate of a point is always given first. This means that, to locate the point, we begin by moving either left or right. The y-coordinate, depending on its sign, then tells us how far to move up or down.

Here is another example of how points are plotted on a graph that contains all four quadrants.

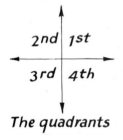

The quadrants

EXAMPLE
Plot the following five points on a coordinate graph: A (6, –2), B (–2, 6), C (–4, –4), D (–3, 0), and E (0, –3).

SOLUTION
The graph is shown at the right. Notice that the x-coordinate of each point tells how many units to the right or left of the y-axis it is and the y-coordinate tells how many units above or below the x-axis it is.

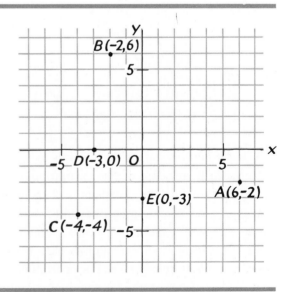

Exercises

Set I

1. The figure shown here is a rectangle.

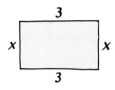

a) How is the perimeter of a rectangle found?
b) What is the perimeter of this rectangle?
c) How is the area of a rectangle found?
d) What is the area of this rectangle?

2. Tell, by evaluating both sides of each equation, whether it is true or false.
a) $1^3 + 2^3 = (1 + 2)^2$
b) $1^3 + 2^3 + 3^3 = (1 + 2 + 3)^2$

c) What do you think is the next equation in this series?
d) Is it true?

3. A snail is crawling along at a steady speed. The distance that it travels is a function of time, as shown in this table.

| Number of minutes, t | 2 | 4 | 6 | 8 |
| Number of meters traveled, d | 1 | 2 | 3 | 4 |

a) How does the distance traveled vary with respect to the time?
b) What is the snail's speed in meters per minute?
c) Write a formula for this function.

Set II

4. This graph shows a curve called an ellipse. Write the coordinates of the points on the ellipse that are named with letters.

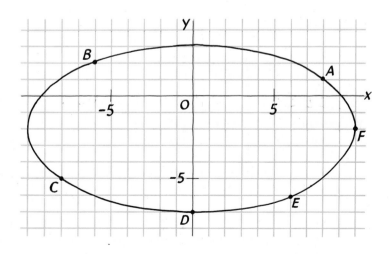

5. On graph paper, draw a pair of axes extending 6 units in each direction from the origin. Then connect the following points in order with straight line segments: (4, 4), (−6, −1), (5, −3), (−3, 5), (−1, −6), and (4, 4).

6. A certain function is represented by this table of numbers.

x	0	1	2	3	4	5
y	0	−1	−2	−3	−4	−5

a) Graph this function by drawing a pair of axes and plotting these points.
b) Write a formula for the function.
c) Draw a line through the points and extend it into the second quadrant.
d) Copy and complete this table by referring to your graph.

x	−5	−4	−3	−2	−1
y					

7. Imagine a point that moves across a coordinate graph so that its y-coordinate is always 3 more than its x-coordinate.
a) Copy and complete this table of some of the point's positions.

x	0	1	2	3	4
y	3				

b) Write a formula for y in terms of x.
c) Plot the points, join them with a line, and extend it into the second and third quadrants.
d) Copy and complete this table of some other positions of the point by referring to your graph.

x	−4	−3	−2	−1
y	−1			

8. A function has the formula $y = 5 - x$.
a) Copy and complete the following table for this function.

x	0	1	2	3	4	5
y						

b) Plot the six points in this table on a graph.
c) Draw a line through the points and extend it into the second and fourth quadrants.
d) Copy and complete these tables by referring to your graph.

x	−4	−3	−2	−1
y				

and

x	6	7	8	9
y				

9. This exercise is about the functions $y = 2x + 1$ and $y = x - 1$.
a) Make a table for each of these functions. In each table, let $x = 1, 2, 3,$ and 4.
b) Graph both functions on the same pair of axes by plotting the points in the tables and joining them with lines.
c) What are the coordinates of the point in which the two lines intersect?

Set III

10. This graph shows a curve called a parabola.

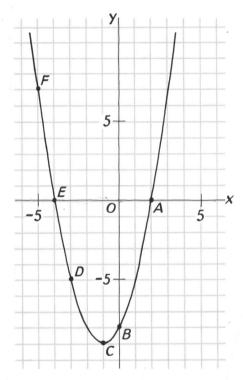

Write the coordinates of the points on the parabola that are named with letters.

11. If both coordinates of a point are positive, it is in the first quadrant. Where is a point if
a) both of its coordinates are negative?
b) its x-coordinate is positive and its y-coordinate is negative?
c) its x-coordinate is negative and its y-coordinate is zero?
d) its x-coordinate is zero and its y-coordinate is negative?

12. A certain function is represented by this table of numbers.

x	0	1	2	3	4	5
y	4	5	6	7	8	9

a) Graph this function by drawing a pair of axes and plotting these points.
b) Write a formula for the function.
c) Draw a line through the points and extend it into the second and third quadrants.
d) Copy and complete this table by referring to your graph.

x	−5	−4	−3	−2	−1
y	−1				

13. Imagine a point that moves across a coordinate graph so that its y-coordinate is always twice its x-coordinate.
a) Copy and complete this table of some of the point's positions.

x	0	1	2	3	4
y		2			

b) Write a formula for y in terms of x.
c) Plot the points, join them with a line, and extend it into the third quadrant.
d) Copy and complete this table of some other positions of the point by referring to your graph.

x	−4	−3	−2	−1
y				

14. A function has the formula $y = x - 4$.
    a) Copy and complete the following table for this function.

x	4	5	6	7	8
y					

    b) Plot the five points in this table on a graph.
    c) Draw a line through the points and extend it into the third and fourth quadrants.
    d) Copy and complete this table by referring to your graph.

x	−4	−3	−2	−1	0	1	2	3
y								

15. This exercise is about functions $y = x + 5$ and $y = 3 - x$.
    a) Make a table for each of these functions. In each table, let $x = 0$, 1, 2, and 3.
    b) Graph both functions on the same pair of axes by plotting the points in the tables and joining them with lines.
    c) What are the coordinates of the point in which the two lines intersect?

Set IV Another Picture Graph

Draw a pair of axes extending from −10 to +10 in each direction. Connect the points in each set with straight line segments in the order given. After you have connected the points in one set, start all over again with the next. In other words, *do not connect* the last point in each set to the first point in the next one.

Set 1. (−1.5, 3) (3, 3) (2, 9) (−1, 9) (−1.5, 3)
   (−3, 2) (2, 2.5) (3, 3) (4, 3) (5, 2) (5, 1)
   (4, 0.5) (3, 0.5) (2, 1) (1, 1) (0, −1)
   (−1.5, 0) (−1, 1) (−3, 1) (−2, 2)

Set 2. (0.5, 0) (3.5, 0) (3.5, −0.5)

Set 3. (−3, 1) (−4, 1) (−2, −1) (0, −1)
   (1, −1.5) (3, −4) (4.5, 0) (4, 0.5)

Set 4. (1, −1.5) (4, −1) (5, −2) (5, −5) (2, −6)
   (−2, −6) (−2.5, −5.5) (−3, −6) (−3.5, −5.5)
   (−4, −6) (−3, −4) (−2, −4) (−2, −6)

Set 5. (−3, −2) (−4, −4) (−1, −4) (−1, −2)

Set 6. (−3, 0) (−4, −1) (−5, −3) (−5, −5)
   (−4, −5.5)

Set 7. (−1, −6) (−1, −8) (−2, −8) (−3, −9)
   (5, −9) (4, −8) (1, −8) (1, −6)

Set 8. (2, −6) (2, −8) (5, −8) (5.5, −9) (4, −9)

Set 9. (5, −3) (7, −2) (8, −3) (7, −3) (8, −4)
   (5, −4)

Set 10. (6.5, −4) (6, −9) (6.5, −9) (7, −4)

LESSON 3
Addition

In 1054, a very bright "star" appeared in the sky. The remnants of this "star" are known as the Crab Nebula, which is one of the strongest sources of radiation in our galaxy. Astronomers have speculated that its energy may result from the destruction of antimatter.

Antimatter has been produced in the laboratory. When an antiparticle collides with an ordinary particle, the two annihilate each other.*

The figure below shows what happens when two protons come in contact with two antiprotons.

$$\text{o o} \quad \bullet \bullet \rightarrow \text{no particles}$$

The result is not four particles , but none. Thinking of the number of protons as positive and the number of antiprotons as negative, the figure shows that

$$+2 + -2 = 0$$

The numbers +2 and -2 are called **opposites** of each other. If we represent a number by the letter x, then its opposite is $-x$ and

$$x + -x = 0$$

If a positive number and a negative number are opposites of each other, their sum is zero. What about the sum of two numbers that are not opposites?

---

* "Anti-Matter" by Geoffrey Burbidge and Fred Hoyle, *Scientific American*, April 1958.

Considering what happens when different numbers of particles and antiparticles come in contact makes it obvious. Look at the examples below.

EXAMPLE 1
What is the sum of –3 and –4?

SOLUTION

Three antiparticles put together with four more make seven antiparticles in all: –3 + –4 = –7.

EXAMPLE 2
What is the sum of –2 and 5?

SOLUTION

Two antiparticles would annihilate two of the five particles, leaving three particles in all: –2 + 5 = 3.

EXAMPLE 3
What is the sum of 1 and –6?

SOLUTION

One particle would annihilate one of the six antiparticles, leaving five antiparticles in all: 1 + –6 = –5.

EXAMPLE 4
What is the sum of 15 and –23?

SOLUTION
Rather than drawing a picture for this problem, we can reason in the following way;

$$15 + -23 = 15 + (-15 + -8)$$
$$= (15 + -15) + -8$$
$$= 0 + -8$$
$$= -8$$

Exercises

Set I

1. Copy each of the following, replacing each ▥ with either $>$ or $<$.
   a) 0 ▥ -10
   b) -25 ▥ -2
   c) 34 ▥ -43

2. Find the value of each of the following expressions.
   a) $5 \cdot 4^2$
   b) $4 \cdot 5^2$
   c) $5^2 - 4^2$
   d) $2^5 - 2^4$

3. The most frequently used pay phone in the United States is in the Greyhound bus terminal in Chicago. It averages 270 calls a day.
   a) The number of calls made on this phone is a function of the time. Write a formula for it, letting x represent the time in weeks and y represent the number of calls.
   b) To what does the constant of variation in your formula refer?

Set II

4. Make some drawings like those on page 122 to illustrate the following addition problems. Use open circles for positive numbers and solid circles for negative numbers.
   a) $2 + 3$
   b) $-7 + -1$
   c) $-4 + 4$
   d) $6 + -2$
   e) $3 + -5$

5. Two numbers are opposites of each other if their sum is zero. What are the opposites of the following numbers?
   a) 8
   b) -3
   c) 1
   d) -15
   e) x
   f) $-y$

6. Find each of the following sums.
   a) $6 + -6$ g) $0 + -10$
   b) $-2 + 2$ h) $-27 + 6$
   c) $-5 + 12$ i) $-4 + -15$
   d) $-11 + -9$ j) $9 + -1$
   e) $3 + -8$ k) $-32 + 0$
   f) $-7 + -7$ l) $-13 + 21$

7. What number should replace ▥ in each of the following equations to make it true?
   a) $5 + $ ▥ $= 0$
   b) $-2 + $ ▥ $= 0$
   c) $-2 + $ ▥ $= -4$
   d) $-7 + 1 = $ ▥
   e) $4 + $ ▥ $= 15$
   f) $-4 + $ ▥ $= -15$
   g) $12 + $ ▥ $= 2$
   h) $-12 + $ ▥ $= -2$
   i) ▥ $+ -3 = 7$
   j) ▥ $+ -3 = -7$
   k) ▥ $+ 5 = -3$
   l) ▥ $+ -5 = -3$

8. Find each of the following sums.
   a) $-3 + -3 + -4$
   b) $-3 + 3 + -4$
   c) $-3 + -3 + 4$
   d) $5 + -5 + -7 + 7$
   e) $5 + -5 + -7 + -7$
   f) $-5 + -5 + -7 + 7$
   g) $9 + -2 + 2 + -8 + -9$
   h) $-1 + -6 + -10 + 6 + -1$
   i) $1 + -6 + -10 + -6 + -1$
   j) $-1 + -3 + -5 + -7 + -9 + -11$
   k) $1 + -3 + 5 + -7 + 9 + -11$
   l) $-1 + 3 + -5 + 7 + -9 + 11$

Set III

9. Make some drawings like those on page 122 to illustrate the following addition problems. Use open circles for positive numbers and solid circles for negative numbers.
   a) $4 + 1$
   b) $-3 + -2$
   c) $5 + -5$
   d) $-4 + 7$
   e) $2 + -8$

10. Two numbers are opposites of each other if their sum is zero. What are the opposites of the following numbers?
   a) 5
   b) -7
   c) 0
   d) -12
   e) a
   f) $-b$

11. Find each of the following sums.
   a) $4 + -4$
   b) $-9 + 9$
   c) $-7 + 17$
   d) $-5 + -8$
   e) $2 + -11$
   f) $-15 + 0$
   g) $-6 + -6$
   h) $-21 + 3$
   i) $-1 + -39$
   j) $0 + -4$
   k) $16 + -7$
   l) $-8 + 25$

12. What number should replace ▥ in each of the following equations to make it true?
   a) $7 + ▥ = 0$
   b) $-3 + ▥ = 0$
   c) $-3 + ▥ = -6$
   d) $-8 + 1 = ▥$
   e) $5 + ▥ = 12$
   f) $-5 + ▥ = -12$
   g) $10 + ▥ = 4$
   h) $-10 + ▥ = -4$
   i) $▥ + -2 = 18$
   j) $▥ + -2 = -18$
   k) $▥ + 7 = -2$
   l) $▥ + -7 = -2$

13. Find each of the following sums.
   a) $-5 + -5 + -2$
   b) $-5 + 5 + -2$
   c) $-5 + -5 + 2$
   d) $8 + -8 + -3 + 3$
   e) $8 + -8 + -3 + -3$
   f) $-8 + -8 + -3 + 3$
   g) $1 + -9 + 9 + -10 + -1$
   h) $-4 + -7 + -6 + 7 + -4$
   i) $4 + -7 + -6 + -7 + -4$
   j) $-1 + -2 + -3 + -4 + -5 + -6$
   k) $1 + -2 + 3 + -4 + 5 + -6$
   l) $-1 + 2 + -3 + 4 + -5 + 6$

Set IV

The sum of the six integers from –2 to 3 inclusive is

$$-2 + -1 + 0 + 1 + 2 + 3 = 3$$

Can you figure out each of the following sums? If you can, explain how you got your answers.

1. The sum of the two hundred and one integers from –100 to 100 inclusive.
2. The sum of the two hundred and one integers from –95 to 105 inclusive.

© 1957 UNITED FEATURE SYNDICATE, INC.

LESSON **4**
Subtraction

In arithmetic, most subtraction problems are limited either to subtracting a smaller number from a larger one or to subtracting a number from itself. In the first case, the answer is positive and, in the second case, it is zero. By having negative numbers to work with, it becomes possible to subtract *any* number from another. Consider, for example, the problem in this cartoon. Thinking in terms of particles, the problem is to take six particles away from four.

○ ○ ○ ○

Although this seems to be impossible, there is a way to do it. We simply add two pairs of particles and antiparticles to the picture:

○ ○ ○ ○ ○ ○
● ●

This doesn't change things because each particle-antiparticle pair adds up to nothing and adding zero to a number does not change it. Now we can take six particles away, leaving the two antiparticles.

● ●

The answer is –2.

▶ In general, whenever a larger number is subtracted from a smaller one, the answer is *negative*.

Here are two more examples.

EXAMPLE 1
Subtract –1 from –7.

● ● ● ● ● ● ✖

SOLUTION
To take one antiparticle from seven antiparticles is easy. We are left with six antiparticles. So $-7 - -1 = -6$.

EXAMPLE 2
Subtract 2 from –5.

● ● ● ● ●

SOLUTION
To take two particles from five antiparticles, we first add two particles and two antiparticles.

● ● ● ● ●   ○ ○
● ●

Taking away the two particles leaves seven antiparticles. So $-5 - 2 = -7$.

Although every subtraction problem can be solved by drawing an appropriate picture and taking away particles or antiparticles, there is another way to deal with subtraction that is generally easier to use. *To every subtraction problem there corresponds an addition problem that has the same answer.* This means that to get the answer to a subtraction problem we don't have to subtract. We can change the problem to an addition problem and add instead.

The method is based on the principle that *subtracting a number gives the same result as adding its opposite.* Here are some examples to illustrate this principle.

Subtraction problem		*Corresponding addition problem*
$4 - 6 = -2$	$\longleftrightarrow$	$4 + -6 = -2$
$8 - -3 = 11$	$\longleftrightarrow$	$8 + 3 = 11$
$-7 - 1 = -8$	$\longleftrightarrow$	$-7 + -1 = -8$

Because of this connection between subtraction and addition, mathematicians prefer to think of subtracting as "adding the opposite" rather than as "taking away."

Exercises

Set I

1. Draw a number line extending from –8 to +8 and use it to find the number of the point midway between each of the following pairs of points.

   a) –8 and 0.
   b) –5 and 5.
   c) –4 and 2.

2. The position of a point is determined by its coordinates. For example, if a point is in the second quadrant, its x-coordinate is negative and its y-coordinate is positive. What can you say about the coordinates of the following points?

   a) A point in the third quadrant.
   b) A point in the fourth quadrant.
   c) A point to the right of the origin on the x-axis.
   d) A point below the origin on the y-axis.

3. The graph of a certain function is shown here.

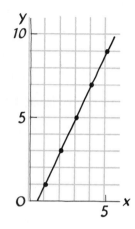

   a) Copy and complete the following table for this function.

x	1	2	3	4	5
y					

   b) What kind of function is it?
   c) Write a formula for it.
   d) Use your formula to find the value of y when x = 25.

Set II

4. Make some drawings like those on page 126 to illustrate the following subtraction problems. Use open circles for positive numbers and solid circles for negative numbers.

   a) 7 − 2
   b) −8 − −5
   c) 3 − 4
   d) −5 − −8
   e) 6 − −1
   f) −2 − 3

5. Because subtracting a number gives the same result as adding its opposite, the subtraction problem 4 − 6 has the same answer as the addition problem 4 + −6. Write the addition problem that corresponds to each of the following subtraction problems and then solve each problem.

   a) 7 − 2
   b) −8 − −5
   c) 3 − 4
   d) −5 − −8
   e) 6 − −1
   f) −2 − 3

6. Find each of the following differences.

   a) 12 − 5
   b) 5 − 12
   c) 12 − −5
   d) −5 − 12
   e) −11 − 9
   f) −11 − −9
   g) 3 − −8
   h) −3 − −8
   i) 0 − 2
   j) 0 − −2
   k) −4 − −4
   l) −4 − 4

7. Find the value of each of the following expressions.

   a) 12 − (5 + 3)
   b) 12 − 5 − 3
   c) 12 − (5 − 3)
   d) 12 − 5 + 3
   e) 7 − (4 + 8)
   f) 7 − 4 − 8
   g) 7 − (4 − 8)
   h) 7 − 4 + 8

i) $2 - (6 + 10)$ k) $2 - (6 - 10)$
j) $2 - 6 - 10$ l) $2 - 6 + 10$

Look again at parts a through l of this exercise and at your answers. What you see there should help you to find an expression without parentheses that is equal to the expression

m) $x - (y + z)$
n) $x - (y - z)$

8. The high and low temperatures in Mudville for four successive months are given in the table below.

	Oct.	Nov.	Dec.	Jan.
High	19	14	5	-2
Low	4	-3	-16	-20

a) The difference between the high and low temperatures in October was 15°. Write the differences between the high and low temperatures for each of the other three months.

b) In which month was the difference between the high and low temperatures the greatest?

9. The amount of money that Freddie the Freeloader bet and the amount that he won at the racetrack on four successive weekends are given in the table below.

Number of dollars bet	20	32	12	18
Number of dollars won	4	15	55	0

a) Freddie's net loss the first week was $16, which can be represented as –16. Write his net winnings or losses for each of the other three weekends as a positive or negative number.

b) Write his net winnings or losses for the four weekends as a sum.

c) How much money did he come out ahead or behind in the four weekends?

Set III

10. Make some drawings like those on page 126 to illustrate the following subtraction problems. Use open circles for positive numbers and solid circles for negative numbers.

a) $8 - 1$ d) $-4 - -6$
b) $-6 - -4$ e) $7 - -2$
c) $2 - 3$ f) $-1 - 5$

11. Because subtracting a number gives the same result as adding its opposite, the subtraction problem $3 - 7$ has the same answer as the addition problem $3 + -7$. Write the addition problem that corresponds to each of the following subtraction problems and then solve each problem.

a) $8 - 1$ d) $-4 - -6$
b) $-6 - -4$ e) $7 - -2$
c) $2 - 3$ f) $-1 - 5$

12. Find each of the following differences.

a) $20 - 9$ g) $13 - -3$
b) $9 - 20$ h) $-13 - -3$
c) $20 - -9$ i) $0 - 8$
d) $-9 - 20$ j) $0 - -8$
e) $-4 - 7$ k) $-5 - -5$
f) $-4 - -7$ l) $-5 - 5$

13. Find the value of each of the following expressions.

a) $15 - (4 + 1)$ g) $3 - (2 - 10)$
b) $15 - 4 - 1$ h) $3 - 2 + 10$
c) $15 - (4 - 1)$ i) $6 - (11 + 5)$
d) $15 - 4 + 1$ j) $6 - 11 - 5$
e) $3 - (2 + 10)$ k) $6 - (11 - 5)$
f) $3 - 2 - 10$ l) $6 - 11 + 5$

Look again at parts a through l of this exercise and at your answers. What you see there should help you to find an expression without parentheses that is equal to the expression

m) $x - (y + z)$

n) $x - (y - z)$

14. The elevations in meters of several very low and very high places on the earth are listed below.

The Dead Sea, Israel-Jordan	–397
Death Valley, California	–86
Mount McKinley, Alaska	6,194
Mount Everest, Nepal-Tibet	8,848

a) How much higher is Mount Everest than Mount McKinley?

b) How much higher is Mount McKinley than Death Valley?

c) How much lower is the Dead Sea than Death Valley?

15. The number of births and deaths in Gopher Prairie for four successive months are given in the table below.

Number of births	7	2	6	5
Number of deaths	3	5	8	1

a) The town's net change in population in the first month is +4. Write the net change in Gopher Prairie's population in each of the other three months as a positive or negative number.

b) Write the net change in the town's population for the four months as a sum.

c) What is the net population change in the four months?

Set IV

There is an old Chinese legend about the Emperor Yu, who lived in about 2200 B.C. He was standing by the Yellow River one day when a turtle appeared on the bank with a pattern of numbers on its back.

This drawing shows how we would write these numbers. The pattern is called a "magic square" because the sum of the three numbers in any row, column, or diagonal is the same: 15.

1. Copy the magic square and then make another pattern by subtracting 10 from each number in it.

2. Is the resulting pattern also a magic square?

3. What do you notice about it?

NATIONAL AERONAUTICS AND SPACE ADMINISTRATION

LESSON 5
Multiplication

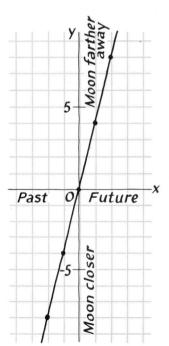

*Graph of the moon
moving away
from the earth.*

This photograph, taken during the flight of Apollo 11, shows the earth from a distance of approximately 240,000 miles. The surface of the moon is in the foreground.

The earth and moon are moving apart at the rate of about 4 feet each year. This means that one year from now the moon will be 4 feet farther away from the earth, two years from now it will be 8 feet farther away, and so on.

The increase in distance between the earth and the moon is a function of time. We can represent this function with a table

Time in years	0	1	2	3	...
Increase in distance in feet	0	4	8	12	...

or by means of a formula. Letting x represent the time and y represent the increase in distance, the formula is

$$y = 4x$$

Because this function is a direct variation, its graph is a straight line that intersects the origin. If we draw the line and extend it into the third quadrant, we get the graph shown at the left.

130

A more complete table for the function is

x	−3	−2	−1	0	1	2	3
y	−12	−8	−4	0	4	8	12

Notice that negative x-numbers represent times in the past and negative y-numbers represent decreases in distance. Three years ago the moon was 12 feet closer to the earth, two years ago it was 8 feet closer, and so on.

Our formula says that each y-number is always 4 times the corresponding x-number, which means that

$$-12 = 4(-3)$$
$$-8 = 4(-2)$$

and so forth. In these examples, the product of a positive number and a negative number is a negative number. This is always true.

The product of a positive number and a negative number is always negative.

What happens when *both* of the numbers being multiplied are negative? To find out, let's imagine that the moon is moving *toward* the earth instead of away from it. Look again at the formula for the moon moving away from the earth:

$$y = 4x$$

The 4 in this formula is the rate at which the moon is moving away: 4 feet per year. If the moon were moving at the same rate but in the *opposite* direction, the 4 would become −4 and the formula would become

$$y = -4x$$

Because the earth and moon would have been farther apart in the past and would be closer in the future, the second row of numbers in our original table would be reversed, as shown here.

x	−3	−2	−1	0	1	2	3
y	12	8	4	0	−4	−8	−12

The graph for this formula and table is shown at the right.

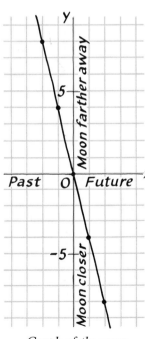

Graph of the moon
moving toward
the earth.

The formula says that each y-number is always -4 times the corresponding x-number, which means that

$$12 = -4(-3)$$
$$8 = -4(-2)$$

and so forth. In these examples, the product of two negative numbers is positive. Although this may seem like a strange result, there are many other patterns that lead to the same conclusion.

The product of two negative numbers is always positive.

Exercises

Set I

1. Find the following sums.
   a) $-2 + -2 + -2 + -2 + -2$
   b) $-333 + -22 + -1 + 1 + 22 + 333$
   c) $-1000 + 100 + -10 + 1$

2. Write an expression for each of the following.
   a) Multiply the integer x by the next larger integer.
   b) Divide the sum of 1 and y by 4.

   c) Divide 1 by the sum of y and 4.
   d) Cube z and subtract the result from 5.
   e) Subtract z from 5 and cube the result.

3. A bottle of Coca-Cola originally cost 5 cents.
   a) At this price, how many dollars would x bottles cost?
   b) How many bottles could be bought for y dollars?

Set II

4. Use repeated addition to show that each of the following equations is true.
   a) $3(4) = 12$
   b) $3(-4) = -12$
   c) $4(-3) = -12$

5. Find each of the following products.
   a) $3(-7)$
   b) $-4(9)$
   c) $-5(-5)$
   d) $7(-8)$
   e) $-10(0)$
   f) $-11(-12)$
   g) $2(-18)$
   h) $-1(-1)$
   i) $-13(3)$
   j) $-15(-20)$

6. The result of multiplying any number by 1 is the same number.
   a) Copy and complete this table showing the result of multiplying numbers by -1.

x	3	2	1	0	-1	-2	-3
$-1 \cdot x$	-3	▓▓	▓▓	▓▓	▓▓	▓▓	▓▓

   b) In general, what is the result of multiplying a number by -1?
   c) What is the result of multiplying x by -1?

d) From your table find a number x for which $-x$ is negative.

e) From your table find a number x for which $-x$ is positive.

f) Does $-x$ always represent a negative number?

7. What number should replace ▨ in each of the following equations to make it true?

a) $-9(12) = $ ▨

b) $9(-12) = $ ▨

c) $4(▨) = 40$

d) $4(▨) = -40$

e) $-4(▨) = 40$

f) $-7(▨) = 0$

g) $-7(▨) = 7$

h) $-32 = 2(▨)$

i) $32 = -2(▨)$

j) $(▨)(-6) = -54$

8. Find each of the following products.

a) $3(5)(7)$

b) $-3(5)(7)$

c) $3(-5)(-7)$

d) $-3(-5)(-7)$

e) $(-2)(-2)(-2)(-2)$

f) $(-2)(-2)(-2)(-2)(-2)$

g) $(-2)(-2)(-2)(-2)(-2)(-2)$

h) $(-1)(-1)(10)(10)$

i) $(-1)(-1)(-1)(10)(10)(10)$

j) $(-5)(-4)(-3)(-2)(-1)$

k) $(-5)(4)(-3)(2)(-1)$

l) $(-5)(4)(-3)(2)(-1)(0)$

9. Which of these symbols, $>$, $=$, or $<$, should replace ▨ in each of the following?

a) $-1(10)$ ▨ $-1(9)$

b) $-2(-3)$ ▨ $4(-5)$

c) $-7(4)(0)$ ▨ $-5(-5)(-3)$

d) $-4(-8)(3)$ ▨ $-3(4)(-8)$

e) $-9(-8)(-7)$ ▨ $-6(-5)(-4)(-3)$

10. Find the values of the following expressions, given that $x = -5$ and $y = -6$.

a) $3x$

b) $-2y$

c) x^2

d) $x + y$

e) $x - y$

f) xy

11. This exercise is about raising a negative integer to different powers.

a) Find the values of $(-3)^2$, $(-3)^3$, $(-3)^4$, $(-3)^5$, and $(-3)^6$.

b) Which powers of -3 are positive?

c) Is the 13th power of -3 positive or negative?

Set III

12. Use repeated addition to show that each of the following is true.

a) $4(5) = 20$

b) $4(-5) = -20$

c) $5(-4) = -20$

13. Find each of the following products.

a) $-2(10)$

b) $7(-6)$

c) $-8(9)$

d) $-6(-1)$

e) $0(-3)$

f) $-12(-7)$

g) $-5(15)$

h) $-4(-4)$

i) $9(-11)$

j) $(-10)(-35)$

14. Simplify each of the following expressions. Assume that x and y represent positive numbers.

a) $0(x)$

b) $0(-y)$

c) $-1(x)$

d) $-1(-y)$

e) $x(-y)$

f) $-x(y)$

g) $-x(-y)$

15. What number should replace ▨ in each of the following equations to make it true?

a) $-11(11) = $ ▨

b) $-11(-11) = $ ▨

c) $3(▨) = 27$

d) $3(\text{▥}) = -27$

e) $-3(\text{▥}) = 27$

f) $-5(\text{▥}) = -35$

g) $-8(\text{▥}) = 0$

h) $-8(\text{▥}) = 8$

i) $-18 = 9(\text{▥})$

j) $18 = -9(\text{▥})$

16. Find each of the following products.

    a) $4(5)(6)$

    b) $-4(5)(6)$

    c) $4(-5)(-6)$

    d) $-4(-5)(-6)$

    e) $(-1)(-1)(-1)(-1)$

    f) $(-1)(-1)(-1)(-1)(-1)$

    g) $(-1)(-1)(-1)(-1)(-1)(-1)$

    h) $(-3)(-3)(10)(10)$

    i) $(-3)(-3)(-3)(10)(10)(10)$

    j) $(-4)(-2)(-2)(-4)$

    k) $(-4)(-2)(2)(4)$

    l) $(-4)(-2)(0)(2)(4)$

17. Which of these symbols, $>$, $=$, or $<$, should replace ▥ in each of the following?

    a) $3(-4)$ ▥ $3(-5)$

    b) $7(-12)$ ▥ $-6(-8)$

    c) $-2(9)(-4)$ ▥ $-9(4)(-2)$

    d) $-4(-7)(-8)$ ▥ $-5(0)(-11)$

    e) $-1(-3)(-5)(-7)$ ▥ $-2(-4)(-6)$

18. Find the values of the following expressions, given that $x = -7$ and $y = -3$.

    a) $2x$ d) $x + y$

    b) $-5y$ e) $x - y$

    c) x^2 f) xy

19. This exercise is about raising negative integers to powers.

    a) Copy and complete this table of squares.

x	-1	-2	-3	-4	-5
x^2	1	▥	▥	▥	▥

    b) Do you think that the square of an integer can ever be negative?

    c) Copy and complete this table of cubes.

x	-1	-2	-3	-4	-5
x^3	▥	▥	▥	▥	▥

    d) What can you conclude about the cube of a negative integer?

Set IV   A Number Puzzle

This number puzzle is similar to one that you may have seen before.

Put a penny in one of the squares in this diagram. Put another penny on a square that is not in the same row or column as the first penny. Put a third penny on a square that is not in the same row or column as either of the first two pennies.

Now look at the numbers under the three pennies and find their product. If you followed the directions correctly, it should be 720, regardless of where you put the pennies. Can you explain why?

12	3	-15
-24	-6	30
8	2	-10

THE WIZARD OF ID BY PERMISSION OF JOHNNY HART AND FIELD ENTERPRISES, INC.

Division

It isn't any trick to divide a number in half, even if the number is negative. The figure below illustrates the number –10 divided in half. (The solid circles represent antiparticles.)

● ● ● ● ● | ● ● ● ● ●

If a group of –10 circles is divided in half, each half contains –5 circles. Dividing a number in half is equivalent to dividing it by 2, and so

$$\frac{-10}{2} = -5$$

Because division has a meaning in terms of multiplication, it is possible to find the answer to the problem $\frac{-10}{2}$ without thinking of circles at all. We simply ask what number must be multiplied by 2 to give –10? The number is –5.

Here is another division problem that includes a negative number. What is 12 divided by –3? We can't answer this by drawing a figure because there is no way to divide 12 circles into –3 groups. Thinking in terms of multiplication, however, the answer is obvious. The answer to the problem $\frac{12}{-3}$ is the number that must be multiplied by –3 to give 12. The number is –4.

Each of the two division problems that we have considered includes one negative number. What happens when *both* numbers are negative? For example, what is –20 divided by –4? It is the number that must be multiplied by –4 to give –20; because 5(–4) = –20, $\frac{-20}{-4} = 5$.

You may have noticed from these examples that, because every division problem has a corresponding multiplication problem, the signs of quotients of positive and negative numbers are *like those of products:*

The quotient of a positive and a negative number is always negative; the quotient of two negative numbers is always positive.

Exercises

Set I

1. Use this number line to find the distance between each of the pairs of points given in parts a through c.

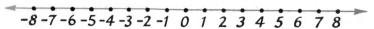

a) 2 and –2
b) –1 and –6
c) –7 and 8

2. Find each of the following differences.
   a) 2 – –2
   b) –1 – –6
   c) 8 – –7

3. On the basis of your answers to exercises 1 and 2, what do you think is the distance between two points numbered x and y on a number line
   a) if x is larger than y?
   b) if x is smaller than y?

Set II

4. Find each of the following quotients.

   a) $\dfrac{-18}{3}$ e) $\dfrac{-49}{7}$

   b) $\dfrac{18}{-3}$ f) $\dfrac{-12}{-12}$

   c) $\dfrac{30}{-2}$ g) $\dfrac{0}{-10}$

   d) $\dfrac{-30}{-2}$ h) $\dfrac{-8}{8}$

5. The result of dividing any number by 1 is the same number.
   a) Copy and complete this table showing the result of dividing numbers by –1.

x	3	2	1	0	–1	–2	–3
$\dfrac{x}{-1}$	–3						

   b) In general, what is the result of dividing a number by –1?
   c) What is the result of dividing x by –1?
   d) Does $-x$ always represent a negative number?

6. The expression $-\dfrac{-6}{3}$ means "the opposite of

$\dfrac{-6}{3}$." Because $\dfrac{-6}{3} = -2$ and the opposite of

-2 is 2, $-\dfrac{-6}{3} = 2$. Use the same reasoning

to simplify each of the following
expressions.

a) $-\dfrac{63}{7}$

e) $-\dfrac{-36}{-6}$

b) $-\dfrac{-12}{4}$

f) $-\dfrac{-44}{11}$

c) $-\dfrac{-16}{-2}$

g) $-\dfrac{9}{9}$

d) $-\dfrac{50}{-5}$

h) $-\dfrac{56}{-8}$

7. Which of these symbols, $>$, $=$, or $<$,
should replace ▧ in each of the following?

a) $\dfrac{-10}{-2}$ ▧ $\dfrac{-10}{2}$

d) $\dfrac{-45}{-3}$ ▧ $\dfrac{45}{3}$

b) $\dfrac{-36}{9}$ ▧ $\dfrac{36}{-9}$

e) $\dfrac{7}{-1}$ ▧ $-\dfrac{7}{-1}$

c) $\dfrac{0}{-8}$ ▧ $\dfrac{0}{8}$

f) $\dfrac{-10}{-10}$ ▧ $\dfrac{9}{9}$

8. What number should replace ▧ in each of
the following equations to make it true?

a) $\dfrac{42}{-3} = $ ▧

g) $\dfrac{20}{▧} = -5$

b) $\dfrac{-42}{-3} = $ ▧

h) $\dfrac{-20}{▧} = 5$

c) $\dfrac{▧}{2} = 9$

i) $\dfrac{-35}{▧} = -7$

d) $\dfrac{▧}{2} = -9$

j) $\dfrac{-35}{▧} = 7$

e) $\dfrac{▧}{-8} = -4$

f) $\dfrac{▧}{-8} = 4$

9. Find the values of the following expressions,
given that $x = -8$ and $y = -4$.

a) $x + 2$

d) $\dfrac{y}{-1}$

b) $\dfrac{x}{2}$

e) xy

c) $y - 1$

f) $\dfrac{x}{y}$

Set III

10. Find each of the following quotients.

a) $\dfrac{15}{-5}$

f) $\dfrac{0}{-12}$

b) $\dfrac{-15}{5}$

g) $\dfrac{-7}{-7}$

c) $\dfrac{-32}{4}$

h) $\dfrac{4}{-4}$

d) $\dfrac{-32}{-4}$

e) $\dfrac{27}{-3}$

11. The expression $-\dfrac{8}{-2}$ means "the opposite of

$\dfrac{8}{-2}$." Because $\dfrac{8}{-2} = -4$ and the opposite of

-4 is 4, $-\dfrac{8}{-2} = 4$. Use the same reasoning

to simplify each of the following
expressions.

a) $-\dfrac{52}{4}$

d) $-\dfrac{-10}{-5}$

g) $-\dfrac{2}{2}$

b) $-\dfrac{-18}{3}$

e) $-\dfrac{-72}{8}$

h) $-\dfrac{42}{-6}$

c) $-\dfrac{36}{-9}$

f) $-\dfrac{-16}{-1}$

12. Simplify each of the following expressions. Assume that neither x nor y is zero.

a) $\dfrac{-x}{-y}$

b) $\dfrac{0}{x}$

c) $\dfrac{0}{-y}$

d) $\dfrac{-x}{1}$

e) $\dfrac{y}{-1}$

f) $\dfrac{-x}{-1}$

g) $\dfrac{y}{-y}$

h) $\dfrac{-x}{-x}$

13. Which of these symbols, $>$, $=$, or $<$, should replace ▦ in each of the following?

a) $\dfrac{18}{-6}$ ▦ $\dfrac{-18}{-6}$

b) $\dfrac{22}{-11}$ ▦ $\dfrac{-22}{11}$

c) $\dfrac{40}{5}$ ▦ $\dfrac{-40}{-5}$

d) $-\dfrac{9}{1}$ ▦ $\dfrac{9}{-1}$

e) $\dfrac{-14}{-2}$ ▦ $-\dfrac{14}{2}$

f) $\dfrac{0}{12}$ ▦ $\dfrac{0}{-12}$

14. What number should replace ▦ in each of the following equations to make it true?

a) $\dfrac{-28}{7} = $ ▦

b) $\dfrac{-28}{-7} = $ ▦

c) $\dfrac{▦}{5} = -5$

d) $\dfrac{▦}{-5} = -5$

e) $\dfrac{▦}{-2} = -11$

f) $\dfrac{▦}{-2} = 11$

g) $\dfrac{-21}{▦} = 3$

h) $\dfrac{-21}{▦} = -3$

15. Find the values of the following expressions, given that $x = -10$ and $y = -2$.

a) $x + 5$

b) $\dfrac{x}{5}$

c) $y - 2$

d) $\dfrac{y}{-2}$

e) xy

f) $\dfrac{x}{y}$

Set IV

You know that a division problem such as $\dfrac{20}{5}$ can be solved by repeated subtraction. The method is shown at the right.

Try using this method to solve these problems.

1. $\dfrac{-20}{-5}$

2. $\dfrac{20}{-5}$

3. Can you explain why the method works in one of the problems and not the other?

Calculation:

$$
\begin{array}{rl}
20 & \\
-\ 5 & 1 \\
\hline
15 & \\
-\ 5 & 2 \\
\hline
10 & \\
-\ 5 & 3 \\
\hline
5 & \\
-\ 5 & 4 \text{ times} \\
\hline
0 & \\
\end{array}
$$

Answer: $\dfrac{20}{5} = 4$

WIDE WORLD PHOTOS.

LESSON 7
Several Operations

The first official world record in the shot put, set in 1876, was slightly less than 31 feet. Since then the record has increased to more than 71 feet.

It is possible, using mathematics, to predict how high and how far the shot will go if the speed and direction in which it is thrown are known. A typical formula, for example, for the upward speed of a shot is

$$v = 25 - 32t$$

in which v represents the speed in feet per second and t represents the time in seconds since the shot was released. By substituting different numbers for t in this formula, we can find the upward speed of the shot at different times.*

When $t = 0$, for example,

$$v = 25 - 32(0)$$
$$= 25 - 0$$
$$= 25$$

At the instant the shot is released, it is traveling upward at a speed of 25 feet per second.

---

* This formula, like others that we have considered in this course, is from a subject other than mathematics. You are not expected to be able to derive these formulas or see immediately why they are true. You should be able, however, to understand how they are being used in the lessons.

Half a second later, $t = 0.5$, and

$$v = 25 - 32(0.5)$$
$$= 25 - 16$$
$$= 9$$

The shot is then moving upward at a speed of 9 feet per second.
One second after the shot is released, $t = 1$, and

$$v = 25 - 32(1)$$
$$= 25 - 32$$
$$= -7$$

According to the formula, the upward speed is now –7 feet per second. What does this mean? That one second after it is released, the shot is moving *down-ward* at a speed of 7 feet per second.

Many practical applications of algebra, like the example we have just con-sidered, require performing several operations that include one or more negative numbers. Because it is important to be able to do such problems accurately, we will practice making calculations with positive and negative numbers in this lesson.

Exercises

Set I

1. Draw a figure to illustrate each of these expressions. Use open circles to represent positive numbers and solid circles to represent negative numbers.
   a) $2 \cdot 3^2 + -7$
   b) $3(4 + -1)$
   c) $5(-2) + 1$

2. Guess a formula for the function represented by each of these tables. Begin each formula with $y =$.

   a)

x	1	2	3	4	5
y	−1	−4	−9	−16	−25

   b)

x	2	3	4	5	6
y	1	0	−1	−2	−3

   c)

x	−1	0	1	2	3
y	6	10	14	18	22

3. When asked to name the smallest number he could think of, Obtuse Ollie said "negative one million."
   a) What is the largest integer smaller than −1,000,000?
   b) What is the smallest integer larger than −1,000,000?

Set II

4. Here are directions for a number trick and part of a table to show what happens if the trick is done with several different numbers.

Think of a number:	7	2	0	-4	-10
Subtract three:	4				
Multiply by two:	8				
Add eight:					
Divide by two:					
Subtract the number that you first thought of:					

a) Copy and complete the table.

b) Show how the trick works by illustrating the steps with boxes and circles. The first two steps are shown below.

Think of a number: ☐

Subtract three: ☐ ● ● ●

(Because subtracting 3 is the same thing as adding –3, 3 solid circles have been drawn to represent adding –3.)

5. Find the value of each of the following expressions. Remember that powers are figured out before multiplications.

a) $5(-3)^2$
b) $-3(5)^2$
c) $2(-4)^3$
d) $-4(-2)^3$
e) $(-1)^3(3)^4$
f) $(-1)^4(3)^3$

6. Find the value of each of these expressions.

a) $2 - 3(4)$
b) $2 + -3(4)$
c) $2 - 3(-4)$
d) $-2 + -3(-4)$
e) $7 - (2 + 12)$
f) $7 - (2 - 12)$
g) $(9 - 1)^2$
h) $(1 - 9)^2$
i) $(1)^2 - (9)^2$
j) $(-1)^2 - (-9)^2$
k) $(-1)^2(-9)^2$
l) $2 + 2(-5)^3$
m) $5 - 2(-2)^3$
n) $(2 - 5)(2)^3$
o) $\dfrac{4^4}{-4}$
p) $\dfrac{(-4)^4}{4}$

7. Find the values of the following expressions, given that $x = -7$ and $y = -3$.

a) $x + 6y$
b) $4x - y$
c) $2(x + y)$
d) $xy - 1$
e) $x^2 - y^2$
f) $(x + y)(x - y)$

8. If an arrow is shot upward at a speed of 50 meters per second, its velocity at any given instant is given by the formula $v = 50 - 10t$, in which t represents the time in seconds and v represents its upward velocity in meters per second. Find the velocity of the arrow after

a) 2 seconds.
b) 5 seconds.
c) 7 seconds.
d) What does your answer to part b mean?
e) What does your answer to part c mean?

9. The formula for converting temperatures in degrees Fahrenheit into degrees Celsius is

$$C = \frac{5(F - 32)}{9}$$

in which F represents the Fahrenheit temperature and C represents the Celsius temperature. Use this formula to find the Celsius temperature corresponding to
a) 32 degrees Fahrenheit.
b) 5 degrees Fahrenheit.
c) –40 degrees Fahrenheit.

Set III

10. Here are directions for a number trick and part of a table to show what happens if the trick is done with several different numbers.

Think of a number:	4	1	0	–5	–12
Multiply by three:	12				
Subtract six:	6				
Divide by three:					
Add seven:					
Subtract the number that you first thought of:					

a) Copy and complete the table.
b) Show how the trick works by illustrating the steps with boxes and circles. (Because subtracting 6 is the same thing as adding –6, 6 solid circles can be drawn in the third step to represent adding –6.)

11. Find the value of each of the following expressions. Remember that powers are figured out before multiplications.
a) $3(-4)^2$
b) $-4(3)^2$
c) $5(-2)^3$
d) $-2(-5)^3$
e) $(-1)^4(6)^2$
f) $(-1)^3(6)^3$

12. Find the value of each of these expressions.
a) $7 - 2(5)$
b) $7 + -2(5)$
c) $7 - 2(-5)$
d) $-7 + -2(-5)$
e) $4 - (3 + 9)$
f) $4 - (3 - 9)$
g) $(5 - 1)^3$
h) $(1 - 5)^3$
i) $(3)^2 - (10)^2$
j) $(-3)^2 - (-10)^2$
k) $(-3)^2(-10)^2$
l) $4 + 4(-2)^3$
m) $(4 + 4)(-2)^3$
n) $4 - 2(-4)^3$
o) $\dfrac{-18}{3} + \dfrac{-30}{2}$
p) $\dfrac{18}{2} - \dfrac{-30}{-3}$

13. Find the values of the following expressions, given that $x = -4$ and $y = -6$.
a) $x + 3y$
b) $7x - y$
c) $-3(x + y)$
d) $xy - y$
e) $x^2 - y^2$
f) $(x - y)(x + y)$

14. If a ball is dropped from the roof of a building 200 meters high, its distance above the roof of a neighboring building 75 meters high is given by the formula $d = 125 - 5t^2$, in which t represents the time in seconds and d represents the distance in meters. Find the distance of the ball above the roof of the second building after

a) 3 seconds.
b) 5 seconds.
c) 6 seconds.
d) What does your answer to part b mean?
e) What does your answer to part c mean?

15. The formula for converting temperatures in degrees Celsius into degrees Fahrenheit is

$$F = \frac{9C}{5} + 32$$

in which C represents the Celsius temperature and F represents the Fahrenheit temperature. Use this formula to find the Fahrenheit temperature corresponding to

a) 100 degrees Celsius.
b) –5 degrees Celsius.
c) –40 degrees Celsius.

Set IV

A man carrying eight identical balloons filled with helium stepped on a scale and found that it read 170 pounds. He let go of the balloons and the scale then read 172 pounds.

1. How could this have happened?
2. On the basis of this information, how much does each balloon seem to have weighed?
3. Do you think that it is possible for something to actually weigh less than nothing?

Gahan Wilson

B.C. BY PERMISSION OF JOHNNY HART AND FIELD ENTERPRISES, INC.

Summary and Review

In this chapter, we have become acquainted with the properties of positive and negative numbers.

The Integers (*Lesson 1*)   The integers can be represented by points evenly spaced along a number line. The integers consist of three sets of numbers: the positive integers (also known as the counting numbers), zero, and the negative integers.

More on the Coordinate Graph (*Lesson 2*)   A complete coordinate graph includes four regions, called *quadrants*, that are numbered counterclockwise, starting with the upper right. Points for which both coordinates are positive are located in the first quadrant. Points having either a positive and a negative coordinate or two negative coordinates lie in one of the other three quadrants. Points for which one coordinate is zero lie on one of the axes.

Addition (*Lesson 3*)   The numbers x and $-x$ are called *opposites* of each other. The sum of a number and its opposite is zero. The addition of positive and negative integers can be pictured in terms of combining groups of particles and antiparticles.

Subtraction (*Lesson 4*)   Although subtraction can be pictured in terms of "taking away," it is more convenient to subtract a number by adding its opposite: $a - b = a + -b$.

Multiplication and Division (*Lessons 5 and 6*) The product or quotient of a positive number and a negative number is always negative. The product or quotient of two negative numbers is always positive.

Several Operations (*Lesson 7*) In performing a series of operations, we work from left to right, first raising to powers, then multiplying and dividing, and finally adding and subtracting. Parentheses usually indicate operations that are to be done first.

Exercises

Set I

1. Read each of the following statements carefully and tell whether it is true or false. If you think a statement is false, give an example to explain why.

   a) The opposite of every number is negative.

   b) If the y-coordinate of a point is positive, the point is above the x-axis.

   c) The product of two negative numbers is always positive.

   d) The sum of two numbers is always more than their difference.

   e) The cubes of some integers are negative.

2. Find each of the following.

   a) The sum of -15 and 3.

   b) The difference between -15 and 3.

   c) The product of -15 and 3.

   d) The quotient of -15 and 3.

   e) The sum of -15 and -3.

   f) The difference between -15 and -3.

   g) The product of -15 and -3.

   h) The quotient of -15 and -3.

3. Write the addition problem that corresponds to each of these subtraction problems.

   a) $3 - -7$

   b) $-8 - 4$

   c) $x - y$

4. Write each of the following statements in symbols, letting x represent the number.

   a) A certain number is more than zero.

   b) The cube of a certain number is equal to the product of the number and nine.

   c) The difference of a certain number and three is less than two.

5. Here are directions for a number trick and part of a table to show what happens if the trick is done with several different numbers.

Think of a number:	2	0	–6	–12
Add six:				
Multiply by two:				
Subtract eighteen:				
Divide by two:				
Add five:				
Subtract the number that you first thought of:				

a) Copy and complete the table.
b) Show how the trick works by illustrating the steps with boxes and circles. (Draw solid circles to represent negative numbers.)

6. What number should replace ▦ in each of the following equations to make it true?
   a) $-9 + ▦ = 2$
   b) $3(▦) = -3$
   c) $▦ + 8 = 0$
   d) $\dfrac{▦}{2} = -6$
   e) $▦ - 7 = -1$
   f) $\dfrac{-10}{▦} = 5$

7. Which of these symbols, $>$, $=$, or $<$, should replace ▦ in each of the following?
   a) $2 + -3 \ ▦ \ 2 - -3$
   b) $-4 + 15 \ ▦ \ -4(15)$
   c) $0(-8) \ ▦ \ 0 - 8$
   d) $-45(39) \ ▦ \ 45(-39)$
   e) $(-708)^3 \ ▦ \ (-78)^2$

8. A point moves across a coordinate graph so that its y-coordinate is always 2 less than its x-coordinate.
   a) Copy and complete this table of some of the point's positions.

x	2	3	4	5
y	0	▦	▦	▦

   b) Write a formula for y in terms of x.
   c) Plot the points, join them with a line, and extend it across your graph.
   d) Copy and complete this table of some other positions of the point by referring to your graph.

x	-2	-1	0	1
y	▦	▦	▦	▦

9. Use a number line to find each of the following.
   a) The number of the point midway between -3 and 7.
   b) The distance between -3 and 7.

c) The number of the point midway between -1 and -9.
d) The distance between 11 and -11.

10. Simplify each of the following expressions.
    a) $-\dfrac{24}{3}$
    b) $-\dfrac{-35}{7}$
    c) $-\dfrac{x}{-1}$
    d) $-\dfrac{-x}{-x}$

11. Find the value of each of these expressions.
    a) $-(1 - 61)$
    b) $-4 + 5(-3)$
    c) $7(2 - 11) + 2(7 - 11)$
    d) $-7(11)(13) - 13(11)(7)$
    e) $(-4)^3 + (-3)^4$

12. Acute Alice and Obtuse Ollie took a test on which the score was found by subtracting the number of wrong answers from the number of correct ones.
    a) Ollie's score was -5. What does that mean?
    b) Alice's score was 3. If she answered 14 questions correctly, how many did she get wrong?
    c) How many points higher was Alice's score than Ollie's?

13. Find the values of the following expressions, given that $x = 2$ and $y = -5$.
    a) $x - 3y$
    b) $11x + y$
    c) $-2(x + y)$
    d) $4 + xy$
    e) $x - y^2$

14. According to the distributive rule for addition, for any three numbers, a, b, and c,

$$a(b + c) = ab + ac$$

Show that this equation is true if
    a) $a = 2$, $b = -4$, and $c = -6$.
    b) $a = -5$, $b = 8$, and $c = -1$.
    c) $a = -7$, $b = -3$, and $c = 9$.

Set II

1. Read each of the following statements carefully and tell whether it is true or false. If you think a statement is false, give an example to explain why.
   a) Zero is an integer.
   b) The sum of two negative numbers is always negative.
   c) If the x-coordinate of a point is negative, the point is below the x-axis.
   d) The squares of some integers are negative.
   e) The sum of a number and its opposite is zero.

2. Use a number line to find each of the following.
   a) The distance between 2 and –10.
   b) The number of the point midway between 2 and –10.
   c) The distance between –7 and –3.
   d) The number of the point midway between –5 and –1.

3. Write each of the following statements in symbols, letting x represent the number.
   a) A certain number is less than negative four.
   b) Twice a certain number is equal to its square.
   c) The quotient of a certain number and two is more than five.

4. The average of a set of numbers is found by adding them and dividing the sum by the number of numbers. Find the average of each of the following sets of numbers.
   a) –7, 16, 21
   b) –1, 3, –5, 7, –9
   c) 12, 34, –56, –78

5. Use repeated addition to show that each of the following equations is true.
   a) $5(-3) = -15$
   b) $4(-x) = -4x$

6. Here are directions for a number trick and part of a table to show what happens if the trick is done with several different numbers.

Think of a number:	3	0	–1	–8
Multiply by four:				
Subtract ten:				
Add the number that you first thought of:				
Divide by five:				
Add seven:				
Subtract the number that you first thought of:				

   a) Copy and complete the table.
   b) Show how the trick works by illustrating the steps with boxes and circles. (Draw solid circles to represent negative numbers.)

7. What number should replace ▊ in each of the following equations to make it true?
   a) $-4(▊) = -12$
   b) $▊ + 6 = -2$
   c) $\dfrac{▊}{-1} = 13$
   d) $-9 + ▊ = 7$
   e) $▊ - 5 = -5$
   f) $\dfrac{8}{▊} = -4$

8. Which of these symbols, $>$, $=$, or $<$, should replace ▊ in each of the following?
   a) $5 - -11 ▊ 5 + 11$
   b) $-3 - 4 ▊ -3(-4)$
   c) $-2(-2)(-2) ▊ -2 + -2 + -2$
   d) $-67(-28) ▊ 67(-28)$
   e) $(495)^2 ▊ (-495)^2$

9. Simplify each of the following expressions.

a) $-\dfrac{28}{7}$

b) $-\dfrac{-54}{-6}$

c) $-\dfrac{-x}{1}$

d) $-\dfrac{x}{-x}$

10. A function has the formula $y = 3 - x$.

a) Copy and complete the following table for this function.

x	0	1	2	3
y				

b) Plot the four points in this table on a graph.

c) Draw a line through the points and extend it into the second and fourth quadrants.

d) Copy and complete these tables by referring to your graph.

x	-3	-2	-1
y			

and

x	4	5	6
y			

11. Find the value of each of these expressions.

a) $5 - (3 - 18)$

b) $-(-25 + 6)$

c) $2(3 - 9) - 3(2 - 9)$

d) $(-5)^2 - (-5)^3$

e) $-3(-7)(37) + 37(7)(-3)$

12. A person's financial worth is found by subtracting his liabilities from his assets.

a) Stan's financial worth is $300. What does that mean?

b) Oliver's financial worth is -$75. If his liabilities are $200, how much are his assets?

c) How much greater is Stan's financial worth than Oliver's?

13. Find the values of the following expressions, given that $x = -1$ and $y = 8$.

a) $7x + y$

b) $x - 3y$

c) $4(x - y)$

d) $x^2 - y$

e) $\dfrac{-y}{x^2}$

14. According to the distributive rule for subtraction, for any three numbers, a, b, and c.

$$a(b - c) = ab - ac$$

Show that this equation is true if

a) $a = 4$, $b = 1$, and $c = -9$.

b) $a = -6$, $b = 2$, and $c = 5$.

c) $a = -7$, $b = -3$, and $c = 8$.

Chapter 4

THE RATIONAL NUMBERS

COURTESY OF THE FRESNO BEE.

The Rational Numbers

This sign was put up along a road near the Fresno airport after it was discovered that many drivers were not slowing down as much as they should. The fact that the number on the sign is not an integer has proved very effective in getting people to pay attention to it. The number $12\frac{1}{2}$ is an example of a *rational number*.

▶ A **rational number** is any number that can be written as the quotient of two integers.

The number $12\frac{1}{2}$ is rational because it can be written as $\frac{25}{2}$. Doing the division indicated by this quotient

$$2\overline{)25.0}^{\,12.5}$$

shows that $12\frac{1}{2}$ can also be written as 12.5. Written this way, it is said to be in *decimal form*.

Rational numbers written as fractions can always be changed to decimal form by carrying out the indicated division. Here is another example.

EXAMPLE 1

Change the rational number $\frac{3}{8}$ to decimal form.

$$\begin{array}{r} 0.375 \\ 8\overline{)3.000} \\ \underline{2\,4} \\ 60 \\ \underline{56} \\ 40 \\ \underline{40} \\ 0 \end{array}$$

SOLUTION

Dividing 8 into 3 as shown at the right, we get 0.375.

Every integer is a rational number because every integer can be written as the quotient of itself and one. For example, 5 is a rational number because $5 = \frac{5}{1}$, and –2 is a rational number because $-2 = \frac{-2}{1}$.

Rational numbers, like integers, can be represented by points on a number line. The number $12\frac{1}{2}$, for example, is located halfway between 12 and 13 on the number line.

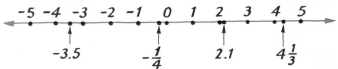

Several other examples of rational numbers on a number line are shown in the figure below.

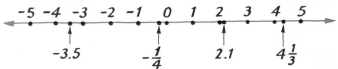

The relative sizes of two rational numbers can be determined by their relative positions on the line.

EXAMPLE 2

Which number is larger: –3.5 or 2.1?

SOLUTION

Because 2.1 is to the right of –3.5 on the line, 2.1 is larger. (We could have answered this question without finding the numbers on a line by reasoning that, because all positive numbers are to the right of zero and all negative numbers are to the left, every positive number is larger than every negative number.)

EXAMPLE 3

Which number is larger: $-\dfrac{1}{4}$ or -4?

SOLUTION

Because $-\dfrac{1}{4}$ is to the right of -4, $-\dfrac{1}{4} > -4$.

Exercises

Set I

1. Find the value of each of these expressions.
   a) $15 - 3 \cdot 4 - 2$
   b) $15 - 3 \cdot (4 - 2)$
   c) $(15 - 3) \cdot 4 - 2$
   d) $(15 - 3) \cdot (4 - 2)$

2. The graph of a certain function is shown here.

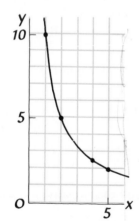

a) Copy and complete the following table for this function.

x	1	2	4	5
y	▓	▓	2.5	▓

b) When x is doubled, what happens to y?
c) What kind of function is this?
d) Write a formula for it, starting with $y =$.

3. Mr. and Mrs. Dinky charge children $1 and adults $2 to skate at their ice-skating rink. One Saturday x children's tickets and y adult tickets were sold.
   a) How many people paid to skate at the Dinky rink that day?
   b) How much money did the Dinkys make from the ticket sales?
   c) If it costs $150 per day to operate the rink, what was the net profit for the day?

Set II

4. Change each of the following rational numbers to decimal form by carrying out the indicated divisions.

a) $\dfrac{7}{5}$ c) $\dfrac{7}{500}$ e) $\dfrac{10}{16}$ g) $\dfrac{21}{3}$ i) $\dfrac{21}{30}$

b) $\dfrac{7}{50}$ d) $\dfrac{1}{16}$ f) $\dfrac{100}{16}$ h) $\dfrac{210}{3}$

5. The number 0.3 is rational because it can be written as the quotient of two integers: $0.3 = \frac{3}{10}$. Show that each of the following numbers is rational by writing it as the quotient of two integers. (Each part has many correct answers.)

a) 4
b) 0
c) 0.7
d) 0.01
e) 6.5
f) –2.9
g) $3\frac{1}{7}$
h) $-8\frac{1}{2}$

6. Use a ruler to draw an accurate number line extending from –8 to +8. A convenient distance to use between each pair of consecutive integers is 1 centimeter.
a) Mark the following points on the line. Write the number of each point beside it.

$3\frac{1}{3}$ $-1\frac{1}{2}$ 0.9   –7.1 $6\frac{1}{5}$ 4.5   –5.8

b) Which of the seven numbers that you have located is the largest?
c) Which number is the smallest?

7. Copy each of the following, replacing each ▥ with either > or <.
a) 4.2 ▥ 4.3
b) –4.2 ▥ –4.3
c) –7.6 ▥ 6.7
d) –7.6 ▥ –6.7
e) 0.05 ▥ 0.5
f) 0.05 ▥ –0.5
g) 2.1 ▥ 2.09
h) –2.1 ▥ –2.09

8. Do the indicated operations.*
a) 12.3 + 0.75
b) 12.3 – 0.75
c) 12.3(0.75)
d) $\frac{12.3}{0.75}$
e) 0.22 + 0.022
f) 0.22 – 0.022
g) 0.22(0.022)
h) $\frac{0.22}{0.022}$

---

*Some decimal calculations are done easily by using an electronic calculator, whereas others are simple enough to do without it. Beginning with this lesson, exercises for which an electronic calculator would be helpful will be marked with an asterisk.

Set III

9. Change each of the following rational numbers to decimal form by carrying out the indicated divisions.

a) $\frac{11}{4}$
b) $\frac{11}{40}$
c) $\frac{11}{400}$
d) $\frac{1}{125}$
e) $\frac{10}{125}$
f) $\frac{100}{125}$
g) $\frac{24}{6}$
h) $\frac{240}{6}$
i) $\frac{24}{60}$

10. The number 1.7 is rational because it can be written as the quotient of two integers: $1.7 = \frac{17}{10}$. Show that each of the following numbers is rational by writing it as the quotient of two integers. (Each part has many correct answers.)

a) 6
b) 1
c) 0.2
d) 0.09
e) 4.5
f) –3.1
g) $7\frac{1}{2}$
h) $-5\frac{1}{5}$

11. Use a ruler to draw an accurate number line extending from –8 to +8. A convenient distance to use between each pair of consecutive integers is 1 centimeter.
   a) Mark the following points on the line. Write the number of each point beside it.

   $$1.6 \quad -2\frac{1}{2} \quad -0.5 \quad 4\frac{1}{3} \quad -5.4 \quad \frac{1}{4} \quad 3.1$$

   b) Which of the seven numbers that you have located is the largest?
   c) Which number is the smallest?

12. Copy each of the following, replacing each ▓ with either > or <.
   a) 2.7 ▓ 2.6
   b) –2.7 ▓ –2.6
   c) 4.5 ▓ –5.4
   d) –4.5 ▓ –5.4
   e) 0.8 ▓ 0.08
   f) –0.8 ▓ 0.08
   g) 3.11 ▓ 3.1
   h) –3.11 ▓ –3.1

13. Do the indicated operations.*
   a) 5.1 + 0.24
   b) 5.1 − 0.24
   c) 5.1(0.24)
   d) $\dfrac{5.1}{0.24}$
   e) 0.99 + 0.099
   f) 0.99 − 0.099
   g) 0.99(0.099)
   h) $\dfrac{0.99}{0.099}$

   * See footnote on page 153.

Set IV

Most rational numbers do not come out "even" if changed to decimal form, but "keep on going," For example, $\dfrac{1}{3} = 0.33333\ldots$, in which the three dots indicate that the threes continue without end. It is obvious from this pattern that the 100th digit after the decimal point will also be a 3.

1. Can you figure out what the 100th digit after the decimal point of $\dfrac{1}{11}$ is?

2. How about the 100th digit after the decimal point of $\dfrac{1}{7}$?

*"But they don't say plus or minus, you'll notice.
You can't get more fiendish than that."*

LESSON **2**
Absolute Value and Addition

According to the sign in this cartoon, the change in the Dow Jones industrial average was 24.04 points. If it had said +24.04, the market would have gone up; –24.04 would indicate that the market had gone down. The sign in the cartoon, however, merely gives the *amount* of the change. This amount is also called the *absolute value* of the change.

A good way to picture the absolute value of a number is with a number line. It is simply the *distance between the number and zero*. The figure below shows that the absolute values of –24.04 and +24.04 are the same: 24.04.

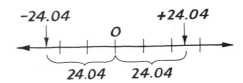

The symbol for absolute value looks like this: | |. The figure above shows that |–24.04| = 24.04 and |+24.04| = 24.04.

Whenever we are dealing with a number in decimal form, we can always write its absolute value by dropping the + or – sign as we have done above. However, we cannot write the absolute value of a variable, such as x, by

155

dropping the + or − sign, because *x* itself has no sign. In this case, we must use the following algebraic definition of **absolute value**.

▶ If *x* is positive or zero, then $|x| = x$. If *x* is negative, then $|x| = -x$ (or the opposite of *x*).

The idea of absolute value is helpful in extending our knowledge of how to compute with the positive and negative numbers that are integers to other numbers as well. In this lesson, we will learn how it applies to the addition of rational numbers.

To add positive and negative integers, we used a particle-antiparticle model. To extend this idea to adding rational numbers, we can think in terms of "lengths" and "antilengths" instead. Here are some examples. (Color is used in the diagrams to represent antilengths.)

EXAMPLE 1
What is the sum of 1.2 and 5.3?

SOLUTION

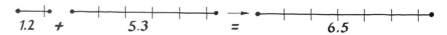

EXAMPLE 2
What is the sum of −3 and −2.4?

SOLUTION

EXAMPLE 3
What is the sum of 4.5 and −6?

SOLUTION

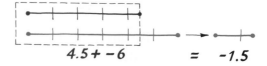

(The dashed line surrounds the lengths that "annihilate" each other.)

EXAMPLE 4

What is the sum of –1.8 and 4.8?

SOLUTION

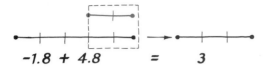

$$-1.8 + 4.8 = 3$$

From examples 1 and 2, we see that the sum of two positive numbers is the positive number given by the sum of their lengths, and the sum of two negative numbers is the negative number given by the sum of their antilengths. In terms of absolute value, this means that *the sum of two numbers having the same sign can be found by adding their absolute values, the answer having the same sign as the numbers.*

Examples 3 and 4 show that, when a positive and a negative number are added, the one that has the shorter length or antilength is "annihilated," leaving as the answer the rest of the number that has the longer length or antilength. This means that *the sum of two numbers having opposite signs can be found by subtracting their absolute values, the answer having the same sign as the number having the larger absolute value.*

Here is another example illustrating this second principle.

EXAMPLE 5

Add 1.5 and –7.6.

SOLUTION

Because one number is positive and the other is negative, we find the difference of their absolute values: $7.6 - 1.5 = 6.1$. Because the absolute value of –7.6 is larger than the absolute value of 1.5, the answer is negative: –6.1.

Exercises

Set I

1. Find the value of $(x - 1)(x - 3)(x - 5)$ if
   a) $x = 1$
   b) $x = 2$
   c) $x = 3$
   d) $x = 4$
   e) $x = 5$

2. Use the distributive rule to write an expression equivalent to each of the following.
   a) $5(x - y)$
   b) $-5(x + y)$
   c) $(x - 1)y$
   d) $-y(1 - x)$

3. On January 22, 1943, a strange thing happened in Spearfish, South Dakota. The temperature rose from $-4°F$ to $45°F$ in only two minutes!
   a) How many degrees was the temperature increase?
   b) Show how this problem can be written as a subtraction problem.
   c) Show how it can be written as an addition problem.

Set II

4. What is the absolute value of each of the following?
   a) $+8$
   b) -5
   c) $+0.6$
   d) -3.4
   e) 0
   f) x if $x > 0$
   g) x if $x < 0$

5. Perform the operations indicated.
   a) $|-1| + |+12|$
   b) $|-1| + |-12|$
   c) $|-9| - |-4|$
   d) $|-4| - |-9|$
   e) $|-3| \cdot |7|$
   f) $|-3| \cdot |-7|$
   g) $|8| \div |-2|$
   h) $|-2| \div |8|$

6. Which of these symbols, $>$, $=$, or $<$, should replace ▓ in each of the following?
   a) $|+4|$ ▓ 4
   b) $|-1.7|$ ▓ -1.7
   c) -2.2 ▓ $+2.2$
   d) $|-2.2|$ ▓ $|+2.2|$
   e) $|-0.5|$ ▓ 0
   f) -10 ▓ -1
   g) $|-10|$ ▓ $|-1|$
   h) $-|3.8|$ ▓ $|-3.8|$

7. Tell whether you think each of the following statements is true for all values of the variables that it contains. If you think that a statement is false for some values of x, give an example of a value of x for which it is false.
   a) $|x| = x$
   b) $|x| + |y| = |x + y|$
   c) $|x| \cdot |y| = |xy|$
   d) $|x^3| = x^3$

8. Find each of the following sums.
   a) $-2 + -8.2$
   b) $-2 + 8.2$
   c) $7 + -4.7$
   d) $-7 + 4.7$
   e) $3.1 + -11.1$
   f) $-5 + 1.05$
   g) $-0.6 + -6.6$
   h) $8.19 + -9$

*9. In adding signed numbers with a calculator, it is convenient to use the following principle: subtracting a number is equivalent to adding its opposite, and adding the opposite of a number is equivalent to subtracting it. In symbols, the fact that

$$x - y = x + -y$$

means that

$$x + -y = x - y$$

For example, to find the answer to the problem

$$1.3 + -3.8$$

on a calculator, we can do the problem

$$1.3 - 3.8$$

instead. Subtracting 3.8 from 1.3, we get −2.5.

Do each of the following problems on a calculator.

a) $3.54 + -0.6$
b) $1.9 + -7.21$
c) $0.875 + -4.6 + -2.38$
d) $5.02 + -0.7 + 3.572 + -11$

*10. (Exercise 9 continued.) If the first number being added is negative, we can begin by subtracting it from zero because

$$0 - x = 0 + -x = -x$$

For example, to find the answer to the problem

$$-6.2 + 0.7$$

on a calculator, we can do the problem

$$0 - 6.2 + 0.7$$

The answer is −5.5.

Do each of the following problems on a calculator.

a) $-7.9 + 4.03$
b) $-0.611 + 2.5$
c) $-8.32 + -10.7 + 0.56$
d) $8.32 + 10.7 + -0.56$
e) $-4.18 + 0.92 + 3 + -7.4$
f) $4.18 + -0.92 + -3 + 7.4$

Set III

11. What is the absolute value of each of the following?

a) +2
b) −9
c) +7.1
d) 0
e) −0.3
f) x if $x > -x$
g) x if $x < -x$

12. Perform the operations indicated.

a) $|+11| + |-4|$
b) $|-11| + |-4|$
c) $|-7| - |-2|$
d) $|-2| - |-7|$
e) $|8| \cdot |-6|$
f) $|-8| \cdot |-6|$
g) $|-15| \div |3|$
h) $|3| \div |-15|$

13. Which of these symbols, $>$, $=$, or $<$, should replace ▒ in each of the following?

a) $|-5|$ ▒ 5
b) $|-8.2|$ ▒ −8.2
c) $+4.3$ ▒ −4.3
d) $|+4.3|$ ▒ $|-4.3|$
e) 0 ▒ $|-6.1|$
f) -7 ▒ −9
g) $|-7|$ ▒ $|-9|$
h) $|-10.6|$ ▒ $-|10.6|$

14. Tell whether you think each of the following statements is true for all values of the variables that it contains. If you think that a statement is false for some values of x, give an example of a value of x for which it is false.

a) $|x| > 0$
b) $|x| - |y| = |x - y|$
c) $|x^2| = x^2$
d) $\dfrac{|x|}{|y|} = \left|\dfrac{x}{y}\right|$

15. Find each of the following sums.

a) $-6.3 + -3$
b) $6.3 + -3$
c) $2.5 + -9.5$
d) $-2.5 + 9.5$
e) $8 + -1.8$
f) $-4.4 + -0.4$
g) $-7 + 1.02$
h) $5.26 + -6$

*16. In adding signed numbers with a calculator, it is convenient to use the following principle: subtracting a number is equivalent to adding its opposite, and adding the opposite of a number is equivalent to subtracting it. In symbols, the fact that

$$x - y = x + -y$$

means that

$$x + -y = x - y$$

For example, to find the answer to the problem

$$4.6 + -5.3$$

on a calculator, we can do the problem

$$4.6 - 5.3$$

instead. Subtracting 5.3 from 4.6, we get –0.7.

Do each of the following problems on a calculator.

a) 2.8 + –5.72
b) 4.11 + –1.3
c) 3.5 + –0.645 + –1.97
d) 0.92 + –8.5 + 12 + –4.023

*17. (Exercise 16 continued.) If the first number being added is negative, we can begin by subtracting it from zero because

$$0 - x = 0 + -x = -x$$

For example, to find the answer to the problem

$$-2.9 + 7.3$$

on a calculator, we can do the problem

$$0 - 2.9 + 7.3$$

The answer is 4.4.

Do each of the following problems on a calculator.

a) –1.37 + 5.8
b) –6.2 + 4.95
c) –9.14 + –0.23 + 7.6
d) 9.14 + 0.23 + –7.6
e) –5.72 + 8 + –0.61 + 9.3
f) 5.72 + –8 + 0.61 + –9.3

Set IV

A spider walks +3 units along a number line and then –4 units more. If we assume that the positive and negative numbers mean that it went first in one direction and then in the opposite direction, it is easy to see that the spider ends up 1 unit from where it started. The point at which it ends depends on the direction in which it headed first.

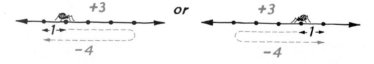

Now suppose that the spider walks x units along a number line and then y units more and that x and y can each be either positive or negative. In terms of $|x|$ and $|y|$,

1. what is the largest distance that the spider can end up from where it started? (Hint: It may help to substitute specific numbers for x and y.)
2. what is the smallest distance that the spider can end up from where it started?

6.27%
PER ANNUM
THE HIGHEST RATE
OF INTEREST
ALLOWED BY LAW

"How high would it be otherwise?"

LESSON 3
More on Operations with Rational Numbers

Suppose that the couple in this cartoon has $1275 to invest in a savings account. How much would the account be worth at the end of a year if they put their money in this bank?

One way to answer this question is to reason like this. An interest rate of 6.27% means that each $100 in the account earns $6.27 in interest per year.

$$\frac{\$1275}{\$100} = 12.75 \qquad 12.75(\$6.27) = \$79.9425$$

$$\begin{array}{r} \$1275 \\ +\$\ \ \ 79.94* \\ \hline \$1354.94 \end{array}$$

The account would be worth $1354.94 at the end of a year.

Finding this answer required making several calculations with rational numbers, all of which are positive. In this lesson, we will learn how to apply the rules that we learned for computing with integers to computing with positive and negative rational numbers.

In learning how to subtract integers, we found that subtracting a number is

_____

* Banks round the interest paid to the nearest cent.

161

equivalent to adding its opposite. By using this principle, we can change subtraction problems into addition problems and then think in terms of addition.

EXAMPLE 1
What is the difference between 7.4 and –1.9?

SOLUTION
Remember that finding the difference between 7.4 and –1.9 means subtracting –1.9 from 7.4.

$$7.4 - -1.9 = 7.4 + 1.9 = 9.3$$

EXAMPLE 2
Subtract –6.5 from –4.1.

SOLUTION

$$-4.1 - -6.5 = -4.1 + 6.5 = 2.4$$

The rules for multiplying and dividing integers also apply to multiplying and dividing rational numbers. Stated in terms of absolute value, they are:

The product or quotient of *two numbers having the same sign* can be found by multiplying or dividing their absolute values, the answer always being *positive*.

The product or quotient of *two numbers having opposite signs* can be found by multiplying or dividing their absolute values, the answer always being *negative*.

EXAMPLE 3
Find the product of –2.5 and –0.64.

SOLUTION

$$(-2.5)(-0.64) = (2.5)(0.64) = 1.6$$

EXAMPLE 4
Divide 9.24 by –2.8.

SOLUTION

$$\frac{9.24}{-2.8} = -\frac{9.24}{2.8} = -3.3$$

Exercises

Set I

1. Show that each of the following numbers is rational by writing it as the quotient of two integers.
   a) −7
   b) 0.3
   c) −1.6
   d) $2\frac{1}{4}$

2. Express each of the following as a power of the indicated number. If a number cannot be expressed as the power indicated, say so.
   a) 16 as a power of −2.
   b) −243 as a power of −3.
   c) −1,000 as a power of −10.
   d) −1,000,000 as a power of −10.

3. On graph paper, draw a pair of axes extending 4 units in each direction from the origin.
   a) Connect the following points with straight line segments in the order given: (2, 3), (−2, 1), (0, −3), (4, −1), (2, 3).
   b) What kind of geometric figure is formed?
   c) The point (3, 1), midway between two consecutive corners of the figure, is called the midpoint of one of its sides. Give the coordinates of the midpoints of the other three sides.

Set II

4. Write the addition problem that corresponds to each of the following subtraction problems and then solve each problem.
   a) 6.1 − −1.9
   b) −2.5 − −2.5
   c) −5 − 7.1
   d) 3.03 − 4

5. Find each of the following differences.
   a) 0.8 − −3
   b) −1.7 − 1.7
   c) −2.64 − −0.4
   d) 4 − 5.8

6. Find each of the following products.
   a) 6(1.5)
   b) 6(−1.5)
   c) −11(9.09)
   d) 1.1(−90.9)
   e) −3.2(−8)
   f) −0.32(−0.8)

7. Find each of the following quotients.
   a) $\dfrac{14.1}{4.7}$
   b) $\dfrac{-14.1}{-4.7}$
   c) $\dfrac{-5.6}{8}$
   d) $\dfrac{-0.56}{0.8}$
   e) $\dfrac{39.2}{-3.5}$
   f) $\dfrac{-3.92}{-35}$

8. What number should replace ▓ in each of the following equations to make it true?
   a) $2.5 + \text{▓} = 0$
   b) $2.5(\text{▓}) = -2.5$
   c) $\text{▓} - 1.6 = 1.6$
   d) $\dfrac{\text{▓}}{1.6} = -1$
   e) $\text{▓} + -3 = 0.3$
   f) $\text{▓}(3) = -0.3$
   g) $7.2 - \text{▓} = 2$
   h) $\dfrac{-7.2}{\text{▓}} = 2$

9. Here are directions for a number trick.

Think of a number:	4.5	1.8	−2.6
Subtract it from 4:	−0.5	▓	▓
Multiply by −3:	▓	▓	▓
Add 6:	▓	▓	▓
Divide by 3:	▓	▓	▓
Subtract the number that you first thought of:	▓	▓	▓

   a) Copy and complete the table to show what happens if the trick is done with some rational numbers.
   b) What number is the result of this trick?

*10. Use a calculator to do each of the following problems. Apply the rules for positive and negative numbers in your calculations.

a) $8.46 - -0.375$

b) $\dfrac{8.46}{-0.375}$

c) $-9.12 + -34.5$

d) $-9.12(-34.5)$

e) $3.9 + -14 - 4.07$

f) $3.9(-14)(-4.07)$

Set III

11. Write the addition problem that corresponds to each of the following subtraction problems and then solve each problem.

a) $4.5 - -5.4$

b) $-7.3 - -7.3$

c) $-2 - 8.9$

d) $1.06 - 6$

12. Find each of the following differences.

a) $5 - -0.2$

b) $-4.1 - 4.1$

c) $-8.73 - -0.3$

d) $2 - 3.9$

13. Find each of the following products.

a) $5(2.8)$

b) $-5(2.8)$

c) $44(-1.01)$

d) $-4.4(10.1)$

e) $-9.6(-3)$

f) $-0.96(-0.3)$

14. Find each of the following quotients.

a) $\dfrac{25.6}{3.2}$

b) $\dfrac{-25.6}{-3.2}$

c) $\dfrac{-6.3}{7}$

d) $\dfrac{-0.63}{0.7}$

e) $\dfrac{-41.4}{-9.2}$

f) $\dfrac{4.14}{-92}$

15. What number should replace ▥ in each of the following equations to make it true?

a) ▥ $+ 1.8 = 0$

b) ▥ $(1.8) = -1.8$

c) ▥ $- 3.5 = 3.5$

d) $\dfrac{▥}{-3.5} = -1$

e) $-4 + ▥ = 0.4$

f) $4(▥) = -0.4$

g) $12.3 - ▥ = 3$

h) $\dfrac{-12.3}{▥} = 3$

16. Here are directions for a number trick.

Think of a number:	1.2	5.1	–3.5
Subtract it from 5:	3.8	▥	▥
Multiply by –2:	▥	▥	▥
Add 4:	▥	▥	▥
Divide by 2:	▥	▥	▥
Subtract the number that you first thought of:	▥	▥	▥

a) Copy and complete the table to show what happens if the trick is done with some rational numbers.

b) What number is the result of this trick?

*17. Use a calculator to do each of the following problems. Apply the rules for positive and negative numbers in your calculations.

a) $2.45 - 6.18$

b) $2.45(-6.18)$

c) $-74.25 + -5.94$

d) $\dfrac{-74.25}{-5.94}$

e) $-21 + 14.3 - 0.37$

f) $-21(14.3)(-0.37)$

Set IV A Calculator Riddle

Acute Alice invited Obtuse Ollie over for dinner one evening and he didn't show up until 11:00 P.M. To find out what got cooked, solve the following problems with a calculator and turn it upside-down to read the answer.

1. $\dfrac{(123)(-778.2)}{-0.18}$

2. $(-7.59)(-1860) - 58(-360.2)$

Ask someone to tell you what time it is and he or she will probably give you an approximate answer. The answer may be to the nearest hour, such as "about noon," to the nearest five or ten minutes, such as "about 12:05," to the nearest minute, such as "12:04," or even to the nearest second. In each case, the number that is the exact time is being approximated by another.

LESSON 4
Approximations

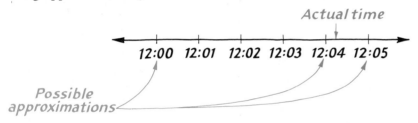

The number that we use as a given approximation depends on the precision in which we are interested. For example, the rational number $\frac{22}{7}$ written in decimal form is

$$3.142857\ldots$$

The first figure at the right shows that this number is closer to 3 than it is to 4, the second figure shows that it is closer to 3.1 than to 3.2, the third figure that it is closer to 3.14 than 3.15, and so forth. Each successive approximation is more precise than the preceding one. We will use the symbol $\approx$ to indicate that an approximation is being used. To the nearest integer, $\frac{22}{7} \approx 3$; to the nearest tenth, $\frac{22}{7} \approx 3.1$; to the nearest hundredth, $\frac{22}{7} \approx 3.14$, and so forth.

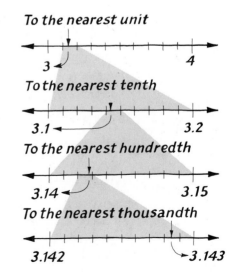

165

In making an approximation of a given number, we generally have to choose between two numbers, one on the left and the other on the right of it on the number line. If the number is exactly midway between the two, one number is as good an approximation of it as the other. In such cases, we will apply the following convention. We will choose the number that is *farther from zero on the line.*

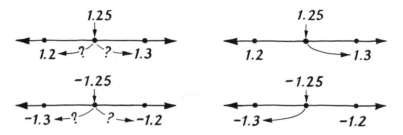

This means that, to the nearest tenth,

$$1.25 \approx 1.3$$

and

$$-1.25 \approx -1.3$$

Ordinarily when we write a number such as "1.3" we mean *exactly* 1.3. In approximate arithmetic, however, "1.3" means "1.3 to the nearest tenth." The number written as "1.30" means "1.30 to the nearest hundredth," "1.300" means "1.300 to the nearest thousandth," and so forth.

Here are some more examples of making approximations of numbers.

EXAMPLE 1
Approximate –11.5 by the nearest integer.

SOLUTION
Because –11.5 is midway between –11 and –12 on a number line, we choose the number farther from 0, which is –12.

EXAMPLE 2

Find an approximation in decimal form of $\frac{5}{12}$ to the nearest hundredth.

$$\begin{array}{r} 0.416\ldots \\ 12\overline{)5.000} \\ \underline{4\,8} \\ 20 \\ \underline{12} \\ 80 \\ \underline{72} \end{array}$$

SOLUTION

Dividing 12 into 5, as shown at the right, and rounding the answer, we get $\frac{5}{12} \approx 0.42$.

Dividing 5 by 12 on a calculator can give an answer rounded to eight places, 0.41666667, but even that is approximate.

Exercises

Set I

1. Which of these symbols, $>$, $=$, or $<$, should replace ▓ in each of the following?
   a) $|-2.5|$ ▓ 2.5
   b) -7.4 ▓ -7.3
   c) $|-7.4|$ ▓ $|-7.3|$
   d) -0.9 ▓ 0
   e) $|-0.9|$ ▓ 0

2. Find the average of each of the following sets of numbers.
   a) $-1, -3, -6, -10$
   b) $-1, -3, -6, 10$
   c) $-1, 3, -6, 10$

3. The Dogpatch Dingos and the Pine Ridge Possums played a football game in which the points were scored by either touchdowns (6 points each) or conversions by kicking (1 point each.)
   The Dingos made 3 touchdowns and x conversions.
   a) What was the Dingos' final score?
   The Possums made y touchdowns and 2 conversions.
   b) What was the Possums' final score?
   The Possums won the game.
   c) Write an inequality expressing this fact.

Set II

4. Approximate each of the following numbers to the nearest integer.
   a) 2.7   c) −5.1   e) 13.49
   b) 2.07   d) −1.5   f) −0.08

5. An important number in mathematics is the number "π": 3.141592 . . . . Round this number to the nearest
   a) integer.   d) thousandth.
   b) tenth.   e) ten-thousandth.
   c) hundredth.   f) hundred-thousandth.

6. Find an approximation in decimal form to the nearest hundredth for each of the following rational numbers.
   a) $\frac{1}{3}$ d) $-\frac{17}{70}$
   b) $\frac{2}{3}$ e) $-\frac{17}{700}$
   c) $-\frac{17}{7}$ f) $-\frac{17}{7000}$

7. Find the answers to the following problems. Express each answer to the nearest tenth.
   a) $-1.23 + -4.56$
   b) $-1.23 - -4.56$
   c) $7.00 - 0.95$
   d) $0.95 - 7.00$
   e) $-8(-1.42)$
   f) $-0.8(-1.42)$
   g) $-0.8(-0.142)$
   h) $\dfrac{-5.5}{12}$
   i) $\dfrac{-5.5}{1.2}$
   j) $\dfrac{1.2}{-5.5}$

8. Find the values of the following expressions. Express each answer to the nearest tenth.
   a) $2(5.4) - 2(6.5)$
   b) $2(5.4 - 6.5)$
   c) $5.4^2 - 6.5^2$
   d) $(5.4 - 6.5)^2$

9. Find the values of the following expressions, given that $x = 7.60$ and $y = -2.31$. Express each answer to the nearest hundredth.
   a) $x + y$
   b) $y - x$
   c) xy
   d) y^2
   e) $\dfrac{x}{y}$

Set III

10. Approximate each of the following numbers to the nearest integer.
    a) 3.6
    b) 3.06
    c) -7.1
    d) -7.5
    e) 0.09
    f) -11.48

11. An important number in mathematics is the number "e": 2.718281 . . . . Round this number to the nearest
    a) integer.
    b) tenth.
    c) hundredth.
    d) thousandth.
    e) ten-thousandth.
    f) hundred-thousandth.

12. Find an approximation in decimal form to the nearest hundredth for each of the following rational numbers.
    a) $\dfrac{4}{9}$
    b) $\dfrac{5}{9}$
    c) $-\dfrac{13}{8}$
    d) $-\dfrac{13}{80}$
    e) $-\dfrac{13}{800}$
    f) $-\dfrac{13}{8000}$

13. Find answers to the following problems. Express each answer to the nearest tenth.
    a) $-1.35 + -6.42$
    b) $-1.35 - -6.42$
    c) $8.00 - 0.85$
    d) $0.85 - 8.00$
    e) $-3(-4.09)$
    f) $-0.3(-4.09)$
    g) $-0.3(-0.409)$
    h) $\dfrac{9.9}{-14}$
    i) $\dfrac{9.9}{-1.4}$
    j) $\dfrac{-1.4}{9.9}$

14. Find the values of the following expressions. Express each answer to the nearest tenth.
    a) $2(1.8) - 2(7.3)$
    b) $2(1.8 - 7.3)$
    c) $1.8^2 - 7.3^2$
    d) $(1.8 - 7.3)^2$

15. Find the values of the following expressions, given that $x = -1.23$ and $y = 8.9$. Express each answer to the nearest hundredth.
    a) $x + y$
    b) $x - y$
    c) xy
    d) x^2
    e) $\dfrac{y}{x}$

Set IV

Several years ago, a painting by the French artist Henri Matisse was accidentally hung upside down in the Museum of Modern Art in New York City. It was reported that 116,000 people had seen the painting before someone noticed that something was wrong.

1. This number is surely not exact, but rather an approximation of the number of people that had seen the painting. How do you suppose it was determined?
2. If the number is accurate, what do you think are the smallest and largest numbers of people that might have seen the painting?

Atlantic Coast Shoreline

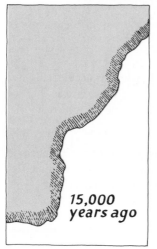

15,000 years ago

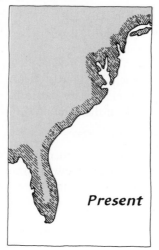

Present

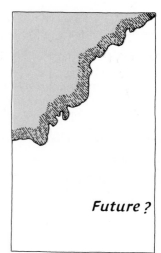

Future ?

LESSON 5
More on Graphing Functions

Geologists think that 15,000 years ago the level of the earth's oceans was more than 100 meters lower than it is today. If this is true, the shorelines of the continents have changed in appearance. The maps above show how the Atlantic coast of the United States is thought to have looked 15,000 years ago, how it looks now, and how it may look in the future.*

The level of the oceans is a function of time. The graph below shows how scientists think that it has changed in the past. The *x*-axis represents time in thousands of years (–40 represents 40,000 years ago, for example) and the *y*-axis represents the changes in level relative to its present level (–50 represents 50 meters below the present level, for example).

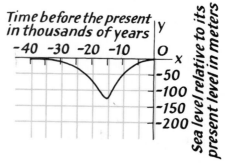

Time before the present in thousands of years

–40 –30 –20 –10

Sea level relative to its present level in meters

–50 –100 –150 –200

---

* "The Continental Shelves" by K. O. Emery, *Scientific American*, September 1969.

As we already know, the graphs of many functions, like this one, include negative numbers. One way in which to draw the graph of a function is by plotting points from a table. If a formula for the function is known, it can be used to make such a table. In this lesson we will make tables and draw the graphs of functions that include both positive and negative numbers.

Here are examples to help you recall how it is done.

EXAMPLE 1
Draw a graph of the function $y = 5 - 2x$. Let $x = -2, -1, 0, 1, 2, 3, 4,$ and 5.

SOLUTION
First we use the formula to find the y-numbers corresponding to these x-numbers. For example, if $x = -2$,

$$y = 5 - 2(-2) = 5 - (-4) = 5 + 4 = 9$$

If $x = -1$,

$$y = 5 - 2(-1) = 5 - (-2) = 5 + 2 = 7$$

and so forth.

After arranging the numbers in a table like this,

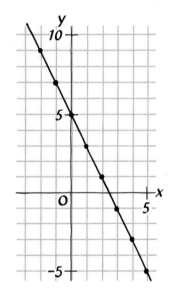

| x | -2 | -1 | 0 | 1 | 2 | 3 | 4 | 5 |
| y | 9 | 7 | 5 | 3 | 1 | -1 | -3 | -5 |

we plot the points having these numbers as their coordinates, $(-2, 9)$, $(-1, 7)$, and so forth, on a graph.

The points seem to lie on a line and so we connect them with one, getting the graph shown at the right.

EXAMPLE 2
Draw a graph of the function $y = x^2$. Let x vary in tenths from 0 to 1.

SOLUTION
Making a table of numbers, we get

| x | 0 | 0.1 | 0.2 | 0.3 | 0.4 | 0.5 | 0.6 | 0.7 | 0.8 | 0.9 | 1.0 |
| y | 0 | 0.01 | 0.04 | 0.09 | 0.16 | 0.25 | 0.36 | 0.49 | 0.64 | 0.81 | 1.00 |

Plotting points having these numbers as coordinates as shown on the next page, we find that they seem to lie on a curve. Connecting the points with a smooth curve gives the result shown.

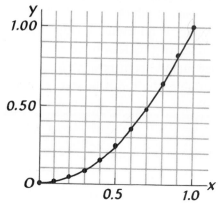

Exercises

Set I

1. Perform the operations indicated.
   a) $0.40 + -1.25$
   b) $0.40 - -1.25$
   c) $0.40 \, (-1.25)$
   d) $\dfrac{0.40}{-1.25}$

2. Arrange the following sets of numbers in order from smallest to largest.
   a) 1.8, 1.08, 1.008
   b) −1.8, −1.08, −1.008
   c) 0.5, 0.55, 0.555
   d) −0.5, −0.55, −0.555

3. Obtuse Ollie counted the change in his pockets and found that he had 7 pennies, x nickels, y dimes, and 2 quarters.
   a) How many coins were there in all?
   b) What was their value in cents?
   c) What was their value in dollars?

Set II

4. A function has the formula $y = 0.5x - 1$.
   a) Copy and complete the following table for this function.

x	0	2	3	5	8
y	−1				

   b) Plot the five points in this table on a graph.
   c) Draw a line through the points and extend it into the third quadrant.
   The point (6, 2) is on the line and its coordinates make the formula $y = 0.5x - 1$

   true because

   $$2 = 0.5(6) - 1$$
   $$= 3 - 1$$
   $$= 2$$

   Parts d and e refer to the following points:

   $$(4, 1), \quad (7, 2.5), \quad (1, 0.5), \quad (-4, -3)$$

   d) Which of these points are on the line in your graph?
   e) Which of the points have coordinates that make the formula $y = 0.5x - 1$ true?

5. A function has the formula $y = x^3$.
   a) Copy and complete the following table for this function.
   Round each y-number to the nearest hundredth.

x	0	0.1	0.2	0.3	0.4	0.5	0.6	0.7	0.8	0.9	1.0
y		0.00	0.01								

(handwritten) 0   0.03 0.06 0.13 0.2   .34   .51   .73   1

   b) Plot the points in this table on a graph. Number the axes in the
   same way as the graph for example 2 on page 172.
   c) Draw a smooth curve through the points.

6. The five lines labeled A, B, C, D, and E in
   the figure at the right are the graphs of five
   functions.
   a) What kind of functions are they?
   The formulas of three of the functions are:

   A: $y = x + 5$
   C: $y = x$
   D: $y = x - 2$

   b) Each formula has the form $y = x + b$, in
   which b is either positive, zero, or
   negative. What does the value of b have
   to do with the position of the line on the
   graph?
   c) What is the formula of function B?
   d) What is the formula of function E?

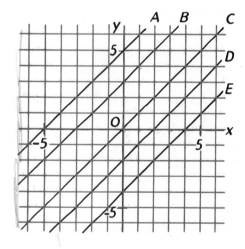

7. Make a table of numbers for each of the
   following functions and draw its graph. In
   each case, connect the points with a line or a
   curve.
   a) $y = -3x$ Let x vary from –2 to 2.
   b) $y = 2x - 5$ Let x vary from –1 to 5.
   c) $y = x^2 - 5$ Let x vary from –3 to 3.
   d) $y = x^2 + x$ Let x vary from –3 to 2.
   e) $y = \dfrac{x}{6}$ Let x vary from –6 to 6.
   f) $y = \dfrac{6}{x}$ Let x vary from –6 to 6.

8. Compare the formulas of the functions for
   exercise 7 with their graphs.
   a) Which functions are linear?
   b) Which are direct variations?
   c) Which functions have graphs that are
   curves?
   d) Which one is an inverse variation?

Set III

9. A function has the formula $y = 4 - 0.5x$.
   a) Copy and complete the following table for this function.

x	0	1	4	5	8
y	4				

   b) Plot the five points in this table on a graph.
   c) Draw a line through the points and extend it into the second quadrant.
   The point (2, 3) is on the line and its coordinates make the formula $y = 4 - 0.5x$

   true because

   $$3 = 4 - 0.5(2)$$
   $$= 4 - 1$$
   $$= 3$$

   Parts d and e refer to the following points:

   (6, 1), (3, 2.5), (−1, 5), (−4, 6)

   d) Which of these points are on the line in your graph?
   e) Which of the points have coordinates that make the formula $y = 4 - 0.5x$ true?

---

10. A function has the formula $y = 1 - x^2$.
    a) Copy and complete the following table for this function.

x	0	0.1	0.2	0.3	0.4	0.5	0.6	0.7	0.8	0.9	1.0
y		0.99	0.96								

    b) Plot the points in this table on a graph. Number the axes in the same way as the graph for example 2 on page 172.
    c) Draw a smooth curve through the points.

---

11. The lines labeled A, B, C, D, and E in the figure at the right are the graphs of five functions.
    a) What kind of functions are they?
    The formulas of three of the functions are:

    A: $y = 2x + 4$
    C: $y = 2x$
    D: $y = 2x - 3$

    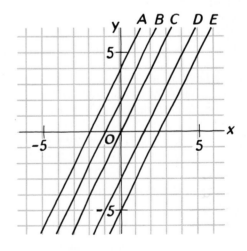

    b) Each formula has the form $y = 2x + b$, in which b is either positive, zero, or negative. What does the value of b have to do with the position of the line on the graph?
    c) What is the formula of function B?
    d) What is the formula of function E?

12. Make a table of numbers for each of the following functions and draw its graph. In each case, connect the points with a line or curve.
    a) $y = -x$ Let x vary from –3 to 3.
    b) $y = 3x - 7$ Let x vary from –1 to 5.
    c) $y = x^3 - 7$ Let x vary from –1 to 2.
    d) $y = 4 - x^2$ Let x vary from –3 to 3.
    e) $y = \dfrac{x}{8}$ Let x vary from –8 to 8.
    f) $y = \dfrac{8}{x}$ Let x vary from –8 to 8.

13. Compare the formulas of the functions for exercise 12 with their graphs.
    a) Which functions are linear?
    b) Which are direct variations?
    c) Which functions have graphs that are curves?
    d) Which one is an inverse variation?

Set IV

This photograph, taken with a strobe light, shows a golf ball falling through the air and then bouncing several times.

A formula for part of its path is

$$y = x(6 - x)$$

1. Make a table for this formula, letting x vary in halves from 0 to 6.
2. Draw a graph.
3. What do you notice?

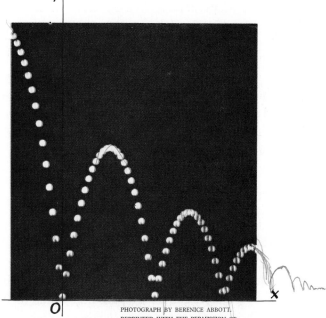

PHOTOGRAPH BY BERENICE ABBOTT, REPRINTED WITH THE PERMISSION OF WILLIAM COLLINS + WORLD PUBLISHING, INC., FROM THE ATTRACTIVE UNIVERSE BY E. G. VALENS (TEXT) AND BERENICE ABBOTT (PHOTOGRAPHS).

". . . Now if you want to spend a little more money. . . ."

Summary and Review

In this chapter, we have extended our knowledge of positive and negative numbers to include the rational numbers.

The Rational Numbers (*Lesson 1*) A rational number is any number that can be written as the quotient of two integers. Because every integer can be written as the quotient of itself and one, every integer is a rational number. A rational number can be changed into decimal form by carrying out the indicated division. A number in decimal form such as 1.2 is rational because it can be written as the quotient of integers, $\dfrac{12}{10}$.

Absolute Value and Addition (*Lesson 2*) The absolute value of a number, represented by the symbol | |, is the distance between the number and zero on a number line. In general, $|x| = x$ if x is positive and $|x| = -x$ (or the opposite of x) if x is negative. The absolute value of 0 is 0.

The sum of two numbers having the same sign can be found by adding their absolute values, the answer having the same sign as the numbers. The sum of two numbers having opposite signs can be found by subtracting their absolute values, the answer having the same sign as the number having the larger absolute value.

More on Operations with Rational Numbers (*Lesson 3*) Subtracting a number is equivalent to adding its opposite.

The product or quotient of two numbers having the same sign can be found by multiplying or dividing their absolute values, the answer always being positive. The product or quotient of two numbers having opposite signs can be found by multiplying or dividing their absolute values, the answer always being negative.

Approximation (*Lesson 4*) The symbol $\approx$ is used to indicate that a number is being used as an approximation of another. If a number is exactly midway between two numbers that might be used as approximations of it, we choose the number that is farther from zero on the line.

More on Graphing Functions (*Lesson 5*) To draw a graph of a function when given its formula, first use the formula to make a table of *x*- and *y*-numbers. Then plot the points having these numbers as their coordinates. Finally connect the points with either a straight line or a curve.

Exercises

Set I

1. Show that each of the following numbers is rational by writing it as the quotient of two integers.
   a) –3
   b) 1
   c) 0.09
   d) $-5\frac{1}{2}$

2. Change each of the following rational numbers to decimal form.
   a) $\frac{9}{8}$
   b) $\frac{-2}{25}$
   c) $\frac{-102}{-12}$

3. Which of these symbols, $>$, $=$, or $<$, should replace ▒ in each of the following?
   a) 3.5 ▒ 3.06
   b) –8.1 ▒ –8
   c) $|{-8.1}|$ ▒ $|{-8}|$
   d) 0.02 ▒ –0.2
   e) 0.02 ▒ $|{-0.2}|$
   f) –7.4 ▒ –7.40

4. Perform the operations indicated.
   a) 8 + –0.3
   b) 0.8 + –3
   c) –4.1 + –5.1
   d) –4.1 – –5.1
   e) –4.1(–5.1)
   f) –2.7 + 0.9
   g) –2.7 – 0.9
   h) –2.7(0.9)
   i) $\frac{-2.7}{0.9}$
   j) 2(–0.16)
   k) (–0.16)^2
   l) $\frac{2}{-0.16}$

5. What number should replace 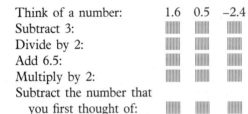 in each of the following equations to make it true?
   a) $1.3 + ▓ = 0$
   b) $1.3(▓) = 0$
   c) $▓ - 4.4 = 4.4$
   d) $\dfrac{-7.5}{▓} = 2.5$
   e) $▓ + -0.1 = 6$
   f) $8(▓) = -0.8$

6. Here are directions for a number trick.

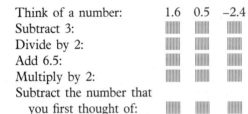

Think of a number:	1.6	0.5	-2.4
Subtract 3:	▓	▓	▓
Divide by 2:	▓	▓	▓
Add 6.5:	▓	▓	▓
Multiply by 2:	▓	▓	▓
Subtract the number that you first thought of:	▓	▓	▓

   a) Copy and complete the table to show what happens if the trick is done with some rational numbers.
   b) What number is the result of this trick?

7. Approximate each of the following numbers as indicated.
   a) 4.09 to the nearest integer.
   b) -1.27 to the nearest tenth.
   c) 0.583 to the nearest hundredth.
   d) -6.5 to the nearest integer.

8. Find an approximation in decimal form to the nearest hundredth for each of the following rational numbers.
   a) $\dfrac{11}{9}$ c) $\dfrac{9}{11}$
   b) $\dfrac{110}{9}$ d) $\dfrac{9}{110}$

9. Make a table of numbers for each of the following functions and draw its graph. In each case, connect the points with either a line or a curve.
   a) $y = -0.5x$ Let x vary from -6 to 6.
   b) $y = x^2 + 1$ Let x vary from -3 to 3.
   c) $y = 4x - 3$ Let x vary from -1 to 3.
   d) $y = \dfrac{-6}{x}$ Let x vary from -6 to 6.

10. Compare the formulas of the functions in exercise 9 with their graphs.
   a) Which functions are linear?
   b) Which one is a direct variation?
   c) Which functions have graphs that are curves?
   d) Which one is an inverse variation?

Set II

1. Show that each of the following numbers is rational by writing it as the quotient of two integers.
   a) -8
   b) 0
   c) 6.1
   d) $-4\dfrac{1}{3}$

2. Change each of the following rational numbers to decimal form.
   a) $\dfrac{-7}{1000}$ b) $\dfrac{-42}{-24}$ c) $\dfrac{13}{-16}$

3. Which of these symbols, $>$, $=$, or $<$, should replace ▓ in each of the following?
   a) $5.08 ▓ 5.7$
   b) $-1.9 ▓ -2$
   c) $|-1.9| ▓ |-2|$
   d) $3.60 ▓ 3.6$
   e) $0.04 ▓ -0.4$
   f) $0.04 ▓ |-0.4|$

4. Perform the operations indicated.
   a) $6 + -0.7$
   b) $0.6 + -7$
   c) $-5.2 + -2.5$
   d) $-5.2 - -2.5$
   e) $-5.2(-2.5)$
   f) $-2.8 + 0.4$
   g) $-2.8 - 0.4$
   h) $-2.8(0.4)$
   i) $\dfrac{-2.8}{0.4}$
   j) $3(-1.2)$
   k) $(-1.2)^3$
   l) $\dfrac{3}{-1.2}$

5. What number should replace ▓ in each of the following equations to make it true?
   a) ▓ $+ 5.7 = 0$
   b) ▓ $(5.7) = 0$
   c) $\dfrac{6.2}{▓} = -3.1$
   d) ▓ $- 2.3 = 2.3$
   e) $-4 + $ ▓ $= 8.4$
   f) $0.5($▓$) = 6$

6. Here are directions for a number trick.

Think of a number:	3.5	–8.1	–0.6
Multiply by –2:	▓	▓	▓
Add 7:	▓	▓	▓
Divide by 2:	▓	▓	▓
Subtract 4.5:	▓	▓	▓
Add the number that you first thought of:	▓	▓	▓

   a) Copy and complete the table to show what happens if the trick is done with some rational numbers.
   b) What number is the result of this trick?

7. Approximate each of the following numbers as indicated.
   a) 5.91 to the nearest integer.
   b) –2.736 to the nearest hundredth.
   c) 8.04 to the nearest tenth.
   d) –4.5 to the nearest integer.

8. Find an approximation in decimal form to the nearest hundredth for each of the following rational numbers.
   a) $\dfrac{12}{7}$
   b) $\dfrac{120}{7}$
   c) $\dfrac{7}{12}$
   d) $\dfrac{7}{120}$

9. Make a table of numbers for each of the following functions and draw its graph. In each case, connect the points with either a line or curve.
   a) $y = x + 2.5$ Let x vary from –6 to 5.
   b) $y = 0.8x$ Let x vary from –5 to 5.
   c) $y = \dfrac{-4}{x}$ Let x vary from –4 to 4.
   d) $y = x^2 - x$ Let x vary from –2 to 3.

10. Compare the formulas of the functions in exercise 9 with their graphs.
    a) Which functions are linear?
    b) Which one is a direct variation?
    c) Which functions have graphs that are curves?
    d) Which one is an inverse variation?

Chapter 5

EQUATIONS IN
ONE VARIABLE

The whetstone of witte,
whiche is the seconde parte of Arithmetike: containyng thextraction of Rootes: The Cossike practise, with the rule of Equation: and the woorkes of Surde Nombers.

Though many stones doe beare greate price,
The whetstone is for exersice
As neadefull, and in woorke as straunge:
Dulle thinges and harde it will so chaunge,
And make them sharpe, to right good vse:
All artesmen knowe, thei can not chuse,
But vse his helpe. yet as men see,
Noe sharpenesse semeth in it to bee.
The grounde of artes did brede this stone:
His vse is greate, and moare then one.
Here if you list your wittes to whette,
Moche sharpenesse therby shall you gette.
Dulle wittes hereby doe greately mende,
Sharpe wittes are fined to their fulle ende.
Now proue, and praise, as you doe finde,
And to your self be not vnkinde.

These Bookes are to bee solde, at the Weste doore of Poules, by Jhon Kyngstone.

The Arte

as their woorkes doe extende) to distincte it onely into twoo partes. Whereof the firste is, when one nomber is equalle vnto one other. And the seconde is, when one nomber is compared as equalle vnto .2. other nombers.

Alwaies willyng you to remēber, that you reduce your nombers, to their leaste denominations, and smalleste formes, before you procede any farther.

And again, if your equation be soche, that the greateste denomination Cossike, be ioined to any parte of a compounde nomber, you shall tourne it so, that the nomber of the greateste signe alone, maie stande as equalle to the reste.

And this is all that neadeth to be taughte, concernyng this woorke.

Howbeit, for easie alteratiō of equations. I will propounde a fewe exāples, bicause the extraction of their rootes, maie the more aptly bee wroughte. And to auoide the tediouse repetition of these woordes: is equalle to: I will sette as I doe often in woorke vse, a paire of paralleles, or Gemowe lines of one lengthe, thus: ═════, bicause noe .2. thynges, can be moare equalle. And now marke these nombers.

1. 14.ze. —+— .15.ℊ ═════ 71.ℊ.
2. 20.ze. ——— .18.ℊ ═════ .102.ℊ.
3. 26.ʒ. —+— 10ze ═════ 9.ʒ ——— 10ze —+— 213.ℊ.
4. 19.ze —+— 192.ℊ ═════ 10ʒ —+— 108ℊ ——— 19ze
5. 18.ze —+— 24.ℊ. ═════ 8.ʒ. —+— 2.ze.
6. 34ʒ ——— 12ze ═════ 40ze —+— 480ℊ ——— 9.ʒ
1. In the firste there appeareth. 2. nombers, that is 14.ze.

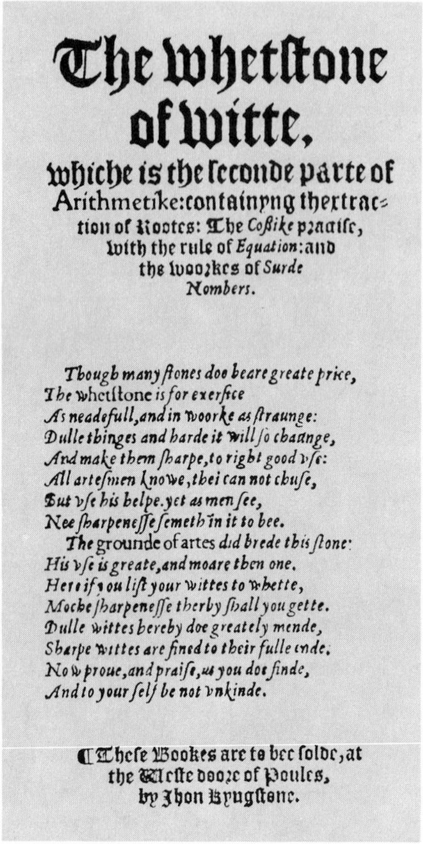

LESSON 1 Equations

The first algebra book in English, called *The Whetstone of Witte*, was written by Robert Recorde and published in 1557. It was in this book that the equal sign was used for the first time in history. The book's title page and the page on which the equal sign first appeared are shown above. Mr. Recorde explained that he chose a pair of parallel lines having the same length to indicate equality because "no two things can be more equal." This symbol has been used ever since to indicate that two expressions represent the same number.

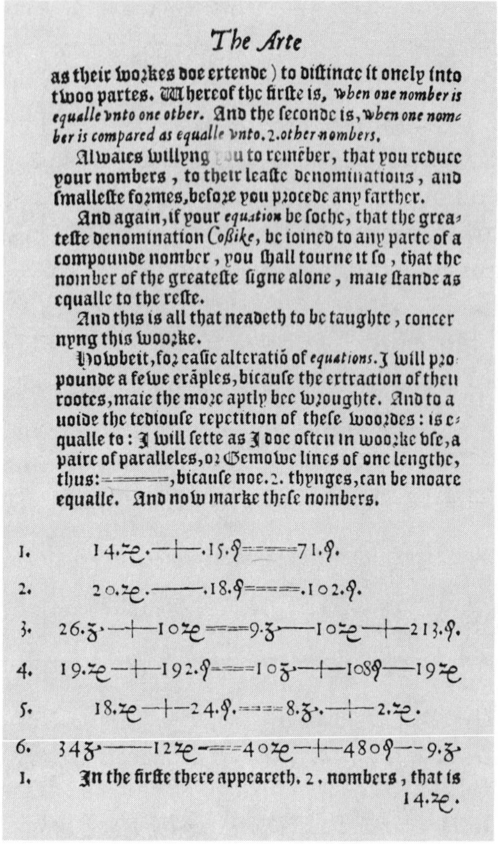

Any mathematical sentence that contains an equal sign is called an **equation.** Equations, like sentences in general, may be true or false. For example, the equation $2 + 3 = 5$ is true, whereas the equation $7 - 1 = 4$ is false.

The equation $x + 6 = 10$, on the other hand, is neither true nor false as it stands. If the letter x, called the **variable** in the equation, is replaced by the number 4, the equation is true. If it is replaced by any other number, however, the equation is false.

A number that can be used to replace a variable in an equation to make it true is called a **solution** of the equation. Some equations, such as $x + 6 = 10$, have only one solution. Others have more than one solution. The equation $x^2 = 9$, for example, has two solutions in that both $3^2 = 9$ and $(-3)^2 = 9$. It is also possible that an equation has no solutions. For example, the equation $x = x + 5$ has no solutions because no number is five more than itself.

To solve an equation means to find all of its solutions. For simple equations it is possible to do this by guessing. To learn how to solve equations whose solutions cannot be easily guessed is a goal of algebra. In this chapter, we will learn some methods for solving certain types of equations containing one variable.

Exercises

Set I

1. An important number in mathematics is the golden mean: 1.618033 .... The *golden mean* is the ratio of the length of one of the diagonals (shown here in green) of a regular

pentagon to the length of one of its sides. Round this number to the nearest
   a) integer.   d) thousandth.
   b) tenth.   e) ten thousandth.
   c) hundredth.

2. Write another expression equivalent to each of the following.
   a) $-8 + -8 + -8 + -8$ c) $3(-x)$
   b) $(-8)(-8)(-8)(-8)$ d) $(-x)^3$

3. The following questions are about the calendar.
   a) How many days are there in an ordinary year and how many days are there in a leap year?
   b) How many days are there in four years if one of them is a leap year?
   c) How many days are there in x years if y of them are leap years?

•

Thirty days hath September, April, June and November. All the rest have thirty-one—except Friday—which has twenty-eight and twenty-nine days each leap year. That old saying merely makes it easier to figure the number of days in each month, but it fails to explain why some months have more days than others.
—*Sarasota (Fla.) Herald-Tribune.*

And that isn't all it fails to explain, either.

THE NEW YORKER.

•

Set II

4. The equation $x - 2 = 8$ can be translated into the sentence, "If 2 is subtracted from a certain number, the result is 8." Translate each of the following equations into sentences.

a) $2x = 8$ c) $\dfrac{x}{6} = 5$ e) $5^x = 25$

b) $x + 3 = 11$ d) $x^3 = 1$

5. Tell whether you think each of the following equations is always true, always false, or true for certain values of the variables and false for others.

a) $2 + 3 \cdot 6 = 20$
b) $5^2 - 4^2 = (5 - 4)^2$
c) $x + 7 = 10$
d) $3(x + 2) = 3x + 6$
e) $x = x - 8$
f) $x^2 + 3 = 12$

6. If possible, find a number or numbers that can replace x in each of the following equations to make it true. If you think that no such number can be found, briefly explain why.

a) $x + 5 = 100$
b) $5x = 100$

c) $12 - x = 3$

d) $\dfrac{12}{x} = 3$

e) $\dfrac{x}{12} = 3$

f) $x + x = 16$
g) $x^2 = 16$
h) $x = 2x$
i) $x = x + 2$
j) $x = x^2$
k) $3 + x = -21$
l) $3x = -21$
m) $10 + x = 0$

n) $10x = 0$
o) $x - 10 = -10$
p) $-10x = -10$
q) $x - x = 0$
r) $1 - x = -1$

s) $\dfrac{1}{x} = -1$

t) $x^2 = -1$
u) $x^3 = -1$
v) $8^x = 64$
w) $4^x = 64$
x) $1^x = 8$

7. The numbers 2 and 5 are both solutions to the equation $x^2 + 10 = 7x$ because $(2)^2 + 10 = 7(2)$ and $(5)^2 + 10 = 7(5)$. Check each of the following numbers to see which are solutions of the equation given.
a) $x^2 + 3 = 4x$ 0, 1, 2, 3
b) $x(x + 2)(x - 5) = 0$ 0, 2, -2, 5, -5
c) $x + 4 = (x + 4)^2$ 0, -1, -2, -3, -4, -5
d) $x(x - 1) = (x + 2)(x - 3) + 6$
                                    1, 3, 8, -2, -9

Set III

8. The equation $x^3 = 27$ can be translated into the sentence, "If a certain number is cubed, the result is 27." Translate each of the following equations into sentences.
a) $3x = 27$
b) $x - 4 = 10$

c) $\dfrac{18}{x} = 6$

d) $x^2 = 25$
e) $4^x = 16$

9. Tell whether you think each of the following equations is always true, always false, or true for certain values of the variables and false for others.
a) $4 \cdot 5 - 9 = 11$
b) $2^3 + 3^3 = (2 + 3)^3$
c) $x - 8 = 6$
d) $2(x + 5) = 2x + 10$
e) $x + 7 = x$
f) $x^2 + x = 30$

10. If possible, find a number or numbers that can replace x in each of the following equations to make it true. If you think that no such number can be found, briefly explain why.

a) $x + 4 = 24$
b) $4x = 24$
c) $60 - x = 5$
d) $\dfrac{60}{x} = 5$
e) $\dfrac{x}{60} = 5$
f) $x + x = 36$
g) $x^2 = 36$
h) $x = 5x$
i) $x = x - 1$
j) $x^3 = x$
k) $-7 + x = 14$
l) $-7x = 14$
m) $x + 9 = 0$
n) $9x = 0$
o) $x - 9 = -9$
p) $-9x = -9$
q) $x + x = 0$
r) $-2 - x = 2$
s) $\dfrac{-2}{x} = 2$
t) $x^2 = -8$
u) $x^3 = -8$
v) $9^x = 81$
w) $3^x = 81$
x) $1^x = 1$

11. The numbers 3 and 8 are both solutions to the equation $x^2 + 24 = 11x$ because $(3)^2 + 24 = 11(3)$ and $(8)^2 + 24 = 11(8)$. Check each of the following numbers to see which are solutions of the given equation.

a) $6x - x^2 = 8$ $1, 2, 3, 4$
b) $x(x - 3)(x + 7) = 0$ $0, 3, -3, 7, -7$
c) $(x + 5)^2 = x + 5$ $0, -1, -2, -3, -4, -5$
d) $x(x - 2) = (x - 1)^2 - 1$ $2, 6, 11, -8, -9$

Set IV

One of the oldest mathematical documents in existence is the Rhind papyrus. It is thought to have been written in about 1650 B.C. A fragment of it is shown at the right.

The papyrus reveals that the ancient Egyptians solved certain algebraic equations by the following method. For example, to find a number for x that would make the equation

$x + \dfrac{x}{3} = 16$ true, they first made a guess.

Suppose that $x = 3$. Then $3 + \dfrac{3}{3} = 4$, which is only one-fourth of 16. They reasoned that the guess must be one-fourth of the correct answer. Because 3 is one-fourth of 12, $x = 12$.

This method works for only certain types of equations. Try it on each of the following by guessing the number indicated.

THE BROOKLYN MUSEUM:
CHARLES EDWIN WILBOUR FOUNDATION.

1. $5x - \dfrac{x}{2} = 36$ Guess 4. Does the method work for this problem?

2. $\dfrac{x}{4} + 4 = 24$ Guess 8. Does the method work for this problem?

CHICAGO TRIBUNE-NEW YORK NEWS SYNDICATE, INC.

LESSON 2
Inverse Operations

In the first panel of this cartoon Irwin is given a dollar and in the second panel he gives the dollar back. Although it wasn't immediately apparent to him, it is easy to see that he is now back where he started. Adding a dollar followed by subtracting a dollar is equivalent to no change. Because subtraction undoes addition, the two operations are called *inverses* of each other.

We used the word "inverse" earlier to refer to a type of variation, but we are now using it in a different sense.

If one operation undoes the effect of another, the operations are inverses of each other. They are called **inverse operations.**

Subtraction is the inverse of addition because, for any number *y* that we *add* to *x* and *then subtract* from the sum, the result is always the same number as that with which we started.

$$x + y - y = x$$

Addition is the inverse of subtraction because, for any number *y* that we *subtract* from *x* and *then add* to the outcome, the result is also always the same number as that with which we started.

$$x - y + y = x$$

To see why this is so, remember that subtracting a number is equivalent to adding its opposite. Because of this, $x - y + y = x + -y + y$. The sum of a number and its opposite is zero, and so $-y + y = 0$, and $x + (-y + y) = x + 0 = x$.

Multiplication and division are also inverse operations. For example, if a number is multiplied by 3 and the result is then divided by 3, the answer is the original number. In general, if a number x is *multiplied* by any number y other than zero and the result *then divided* by y, then

$$\frac{x \cdot y}{y} = x$$

If a number x is *divided* by any number y other than zero and the result *then multiplied* by y, then

$$\frac{x}{y} \cdot y = x$$

Here are examples of problems that require inverse operations.

EXAMPLE 1

What operation on the expression $x - 3$ would give x as the result?

SOLUTION

The expression $x - 3$ means that 3 is being subtracted from x. Because addition is the inverse of subtraction, we can change this expression back to x by adding 3:

$$x - 3 + 3 = x$$

So the answer is, "Add 3."

EXAMPLE 2

What operations on the expression $5x + 1$ would give x as the result?

SOLUTION

The expression $5x + 1$ means that x is being multiplied by 5 and 1 is added to the result.

Multiply by 5: x
  $5x$
Add 1:
  $5x + 1$

Lesson 2: Inverse Operations 187

The easiest way to get back to x is to undo these operations in reverse order as shown here.

Subtract 1: $\begin{array}{c} 5x + 1 \\ 5x \end{array}$

Divide by 5: $\begin{array}{c} 5x \\ x \end{array}$

The answer is, "Subtract 1 and divide the result by 5."

Exercises

Set I

1. What can you conclude about a number if
   a) it is equal to its opposite?
   b) the result of adding it to itself is positive?
   c) the result of multiplying it by itself is positive?
   d) its cube is negative?

2. The graph of a certain function is shown below.

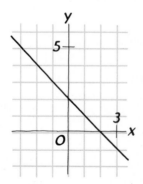

a) Copy and complete the following table for this function.

x	-3	-1	0	2
y	5			

b) What kind of function is it?
c) Write a formula for it.
d) Use your formula to find the value of y if $x = -10$.

3. Five years ago Rip Van Winkle was x years old.
   a) How many years old is he now?
   b) If he has spent twenty years of his life asleep, how many years has he been awake?

Set II

4. What operation is the inverse of each of the following operations?
   a) Addition
   b) Multiplication
   c) Subtraction
   d) Division

5. In each of the following, what operation should be performed on the first expression to give the second?
   a) $5x$ to give x.
   b) $x + 5$ to give x.
   c) $x - 2$ to give x.
   d) $\dfrac{x}{2}$ to give x.
   e) $x + -8$ to give x.
   f) $-8x$ to give x.
   g) $\dfrac{x}{-6}$ to give x.
   h) $x - -6$ to give x.
   i) $9 - x$ to give 9.
   j) $\dfrac{9}{x}$ to give 9. (Assume that x is not zero.)

6. Write a number or expression for each of the following.
   a) The number that must be added to 3 to give 10.
   b) The number that must be added to 3 to give x.
   c) The number from which 5 must be subtracted to give 7.
   d) The number from which 5 must be subtracted to give x.
   e) The number that must be multiplied by 8 to give 32.
   f) The number that must be multiplied by 8 to give x.
   g) The number that must be divided by 2 to give 9.
   h) The number that must be divided by 2 to give x.

7. Write an expression for each of the following sets of operations.
   a) Multiply x by 2 and add 6 to the result.
   b) Add 6 to x and multiply the result by 2.
   c) Divide x by 5 and subtract 1 from the result.
   d) Subtract 1 from x and divide the result by 5.
   e) Add 3 to x and subtract y from the result.
   f) Subtract y from x and add 3 to the result.
   g) Multiply x by 4 and divide the result by y.
   h) Divide x by y and multiply the result by 4.

8. If you have done exercise 7 correctly, your answers should be the same as the expressions below. What operations should be performed on each expression to give back x as the result?
   a) $2x + 6$ e) $(x + 3) - y$
   b) $2(x + 6)$ f) $(x - y) + 3$
   c) $\dfrac{x}{5} - 1$ g) $\dfrac{4x}{y}$
   d) $\dfrac{x - 1}{5}$ h) $4\left(\dfrac{x}{y}\right)$

9. Copy and complete the following tables.

a)
Think of a number:	-3	-2	-1	0
Subtract 5:	-8			
Add 7:	-1			

b)
Think of a number:	-3	-2	-1	0
Add 7:	4			
Subtract 5:	-1			

c)
Think of a number:	-2	-1	0	1
Multiply by 3:				
Subtract 2:				

d)
Think of a number:	-2	-1	0	1
Subtract 2:				
Multiply by 3:				

e) Think of a number: −4   0   4   8
    Divide by 4:
    Add 8:

f) Think of a number: −4   0   4   8
    Add 8:
    Divide by 4:

g) Think of a number: −6   −4   −2   0
    Multiply by 6:
    Divide by 2:

h) Think of a number: −6   −4   −2   0
    Divide by 2:
    Multiply by 6:

i) Which pairs of tables have the same results on the last line?

10. The last lines of tables a and b in exercise 9 are the same because, for any number x,

$$(x - 5) + 7 = (x + 7) - 5$$

Use the results of the rest of the tables to tell whether each of the following is true or false.

a) For any number x,
$(x \cdot 3) - 2 = (x - 2) \cdot 3$. (Look at tables c and d.)

b) For any number x,
$(x \div 4) + 8 = (x + 8) \div 4$. (Look at tables e and f.)

c) For any number x,
$(x \cdot 6) \div 2 = (x \div 2) \cdot 6$. (Look at tables g and h.)

Set III

11. In each of the following, what operation should be performed on the first expression to give the second?

a) $\dfrac{x}{3}$ to give x.

b) $x - 3$ to give x.

c) $7 + x$ to give x.

d) $7x$ to give x.

e) $x + {-5}$ to give x.

f) $-5x$ to give x.

g) $x - {-2}$ to give x.

h) $\dfrac{x}{-2}$ to give x.

i) $8 - x$ to give 8.

j) $\dfrac{8}{x}$ to give 8. (Assume that x is not zero.)

12. Write a number or expression for each of the following.

a) The number that must be added to 6 to give 8.

b) The number that must be added to 6 to give x.

c) The number from which 3 must be subtracted to give 4.

d) The number from which 3 must be subtracted to give x.

e) The number that must be multiplied by 7 to give 70.

f) The number that must be multiplied by 7 to give x.

g) The number that must be divided by 5 to give 11.

h) The number that must be divided by 5 to give x.

13. Write an expression for each of the following sets of operations.

a) Multiply x by 4 and subtract 9 from the result.

b) Subtract 9 from x and multiply the result by 4.

c) Divide x by 7 and add 3 to the result.

d) Add 3 to x and divide the result by 7.

e) Subtract 1 from x and add y to the result.

f) Add y to x and subtract 1 from the result.

g) Multiply x by 5 and divide the result by y.

h) Divide x by y and multiply the result by 5.

14. If you have done exercise 13 correctly, your answers should be the same as the expressions below. What operations should be performed on each expression to give back x as the result?

a) $4x - 9$

b) $4(x - 9)$

c) $\dfrac{x}{7} + 3$

d) $\dfrac{x + 3}{7}$

e) $(x - 1) + y$

f) $(x + y) - 1$

g) $\dfrac{5x}{y}$

h) $5\left(\dfrac{x}{y}\right)$

15. Copy and complete the following tables.

a)
Think of a number:	-2	-1	0	1
Add 3:	1			
Subtract 8:	-7			

b)
Think of a number:	-2	-1	0	1
Subtract 8:				
Add 3:				

c)
Think of a number:	-1	0	1	2
Multiply by 5:				
Add 2:				

d)
Think of a number:	-1	0	1	2
Add 2:				
Multiply by 5:				

e)
Think of a number:	-3	0	3	6
Subtract 10:				
Divide by 3:				

(f)
Think of a number:	-3	0	3	6
Divide by 3:				
Subtract 10:				

g)
Think of a number:	-15	-10	-5	0
Divide by 5:				
Multiply by 4:				

h)
Think of a number:	-15	-10	-5	0
Multiply by 4:				
Divide by 5:				

i) Which pairs of tables have the same results on the last line?

16. The last lines of tables a and b for exercise 15 are the same because, for any number x,

$$(x + 3) - 8 = (x - 8) + 3$$

Use the results of the rest of the tables to tell whether each of the following is true or false.

a) For any number x, $(x \cdot 5) + 2 = (x + 2) \cdot 5$. (Look at tables c and d.)

b) For any number x, $(x - 10) \div 3 = (x \div 3) - 10$. (Look at tables e and f.)

c) For any number x, $(x \div 5) \cdot 4 = (x \cdot 4) \div 5$. (Look at tables g and h.)

Set IV

The following passage is from a nonsense poem by Lewis Carroll called
The Hunting of the Snark.

The Beaver brought paper, portfolio, pens,
    And ink in unfailing supplies:
While strange creepy creatures came out of their dens,
    And watched them with wondering eyes.

So engrossed was the Butcher, he heeded them not,
    As he wrote with a pen in each hand,
And explained all the while in a popular style
    Which the Beaver could well understand.

"Taking Three as the subject to reason about—
    A convenient number to state—
We add Seven, and Ten, and then multiply out
    By One Thousand diminished by Eight.

"The result we proceed to divide, as you see,
    By Nine Hundred and Ninety and Two:
Then subtract Seventeen, and the answer must be
    Exactly and perfectly true.

"The method employed I would gladly explain,
    While I have it so clear in my head,
If I had but the time and you had but the brain—
    But much yet remains to be said."

1. Take 3 as suggested in the poem and carry out the operations described.
2. What do you get for the "answer"?
3. Carry out the same operations, starting out with 4 instead.
4. What is the "answer" this time?
5. What do you think the "answer" will usually turn out to be?
6. Can you explain why?

LESSON 3
Equivalent Equations

If the two bricks and three 1-pound weights on the left-hand pan in the drawing above exactly balance the nine 1-pound weights on the right, how much does each brick weigh?

Because the three 1-pound weights on the left are balanced by three of the 1-pound weights on the right, the two bricks are evidently balanced by the other six 1-pound weights. If the two bricks have equal weights, each one must be balanced by three of the 1-pound weights, and so each one must weigh 3 pounds.

A balanced scale is a good model for understanding algebraic equations because the reasoning that we used to solve this puzzle is identical with the sort of reasoning used to solve equations. If we let x represent the weight in pounds of one of the bricks, then the weight on the pan on the left is $2x + 3$ pounds and the weight on the pan on the right is 9 pounds. Because the weights on the two pans are equal, we can write

$$2x + 3 = 9$$

Compare the method for finding the weight of one of the bricks, illustrated at the left below, with the method used to solve this equation, shown at the right. (The boxes represent the bricks and the circles represent the 1-pound weights.)

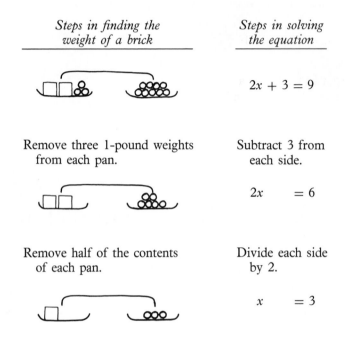

Steps in finding the weight of a brick	Steps in solving the equation
	$2x + 3 = 9$
Remove three 1-pound weights from each pan.	Subtract 3 from each side.
	$2x = 6$
Remove half of the contents of each pan.	Divide each side by 2.
	$x = 3$

All the equations on the right have the same solution because they are **equivalent.** In general, any equation can be transformed into an equivalent one by performing any one of the following operations:

1. Adding the same quantity to each side.
2. Subtracting the same quantity from each side.
3. Multiplying each side by the same quantity other than zero.
4. Dividing each side by the same quantity other than zero.

These operations can be used to solve equations by changing them into simpler equivalent equations. The idea is to "undo" each operation indicated in the equation by performing the inverse operation on both sides. Here are more examples of how it is done.

EXAMPLE 1

Solve the equation $-3x = 12$.

SOLUTION

This equation says that, if a certain number is multiplied by -3, the result is 12. To solve it, we *divide* both sides by -3

$$\frac{-3x}{-3} = \frac{12}{-3}$$

to get the equation

$$x = -4$$

Because this equation is equivalent to the original one, it has the same solution: -4. Checking to be sure that we haven't made a mistake, we substitute -4 for x in the original equation getting

$$-3(-4) = 12$$
$$12 = 12$$

EXAMPLE 2

Solve the equation $x + 7.2 = 0.9$.

SOLUTION

This equation says that, if 7.2 is added to a certain number, the result is 0.9. To solve it, we *subtract* 7.2 from both sides

$$x + 7.2 - 7.2 = 0.9 - 7.2$$

getting the equation

$$x = -6.3$$

Checking -6.3 to see if it makes the original equation true, we get

$$-6.3 + 7.2 = 0.9$$
$$0.9 = 0.9$$

EXAMPLE 3

Solve the equation $4x - 1 = 19$.

SOLUTION

This equation says that the result of multiplying a certain number by 4 and then subtracting 1 is 19. To solve the equation, we undo these operations in reverse order.

Adding 1 to each side of the equation

$$4x - 1 = 19$$

gives the equation

$$4x = 20$$

Dividing both sides of this equation by 4 gives the equation

$$x = 5$$

Because all these equations are equivalent, the solution to the last one, 5, is also the solution to the others. Checking 5 in the original equation, we get

$$4(5) - 1 = 19$$
$$20 - 1 = 19$$
$$19 = 19$$

Exercises

Set I

1. Simplify each of the following expressions.
   a) $4 + -20$
   b) $4 - -20$
   c) $4(-20)$
   d) $\dfrac{4}{-20}$

2. The following questions are about the opposite of a number.
   a) What is the opposite of the number $-x$?
   b) By what number should the number x be multiplied to give its opposite?
   c) From what number should the number x be subtracted to give its opposite?

3. The number of miles that the Cannonball Express can travel in twelve minutes is a function of its speed.

Speed in miles per hour	15	30	45	60
Distance traveled in miles	3	6	9	12

   a) What number can the numbers in the first row of this table be multiplied by to give the numbers in the second row?
   b) Write a formula for this function, letting s represent the speed and d represent the distance traveled.
   c) What kind of function is this?

ich of the following equations for x.
$? = 6$ e) $x + 5 = -5$

4. The diagra
scale: the b   f) $5x = -5$
circles repre^7
g) $\frac{x}{5} = -5$

-7
h) $-x = 9$

1.
11   f the following equations for x.
e) $x - 1.5 = -6$
f) $-1.5x = -6$

2.   l   g) $x + 8 = -0.25$

h) $\frac{x}{8} = -0.25$

3.   ollowing equations by
equivalent equations as
 answers to see if

a) What was done to th equations true.
diagram to obtain wha ld 3; divide by 5.
second?

b) What was done to the sca
second diagram to obtain w
in the third?

c) Write an equation illustrated   of puzzles
diagram.

d) Write an equation illustrated
second diagram.

e) Write an equation illustrated
diagram.   ith a pitcher,
  late, and
f) What could be done to the firs , can you
that you wrote to get the secor nce with a
g) What could be done to the sec
equation that you wrote to get
h) What is the weight of one brick

5. Write the equation that results from
performing the indicated operation or
sides of each of the following equations.
a) $2x + 1 = 6$ Subtract 1.
b) $5x = 0$ Divide by 5.
c) $7x - 4 = 10$ Add 4.

d) $\frac{x}{8} = -9$ Multiply by 8.

e) $3x + 7 = 1$ Subtract 7.
f) $-2x = 12$ Divide by -2.

b) $4x + 9 = 1$ Subtract 9; divide by 4.

c) $\frac{x}{3} - 7 = 2$ Add 7; multiply by 3.

d) $\frac{x}{8} + 1 = -6$ Subtract 1; multiply
by 8.

e) $-2x + 11 = -5$ Subtract 11; divide
by -2.

f) $10 - x = 12$ Subtract 10; divide
by -1.

17. Solve each of the following equations for x.
Check your answers.
a) $4x - 17 = 3$
b) $2x + 9 = -1$ e) $5x - 11 = 4$
c) $1 + 3x = -5$ f) $\frac{x}{5} + 8 = 0$
d) $\frac{x}{7} + 2 = 6$

PUZZLING SCALES

LESSON 4
Equivalent Expressions

One type of question frequently asked on I.Q. tests is illustrated above. The figures are all alike except for one. Can you tell which one is different?*

Now look at the following algebraic expressions:

$$2(1 + x) \qquad 2 + 2x \qquad 2(x + 1) \qquad 2x + 1$$

They are also all "alike" except for one. Can you tell which one is "different"?

Two algebraic expressions are *equivalent* if they are equal for all values of their variables. It is easy to see that the first two expressions are equivalent because of the distributive rule:

$$2(1 + x) = 2 + 2x$$

Furthermore, the first and third expressions are equivalent because the sum of two numbers does not depend upon their order:

$$2(1 + x) = 2(x + 1)$$

It is the last expression that is different.

It is important to be able to recognize algebraic expressions that are equivalent. The following properties of the fundamental operations will be helpful in doing this.

First, you know that the sum of two or more numbers does not depend on either their order or the order in which they are added. These facts are called the *commutative* and *associative* properties of addition.

---

*From *Know Your Own I.Q.* by H. J. Eysenck (Pelican Books, 1962), page 121. Copyright © H. J. Eysenck, 1962. Reprinted by permission of Penguin Books Ltd. The figure in the middle is the one that is different.

▶ **Commutative Property of Addition**
For any two numbers a and b, $a + b = b + a$.

▶ **Associative Property of Addition**
For any three numbers a, b, and c, $(a + b) + c = a + (b + c)$.

Similar properties hold for multiplication.

▶ **Commutative Property of Multiplication**
For any two numbers a and b, $ab = ba$.

▶ **Associative Property of Multiplication**
For any three numbers a, b, and c, $(ab)c = a(bc)$.

Earlier in the course, you learned about a property relating multiplication to the operations of addition and subtraction called the *distributive* rule.

▶ **Distributive Property of Multiplication over Addition and Subtraction**
For any three numbers a, b, and c,

$$a(b + c) = ab + ac$$

and

$$a(b - c) = ab - ac$$

The relations between division, addition, and subtraction are not simple if expressed in terms of the $\div$ sign. However, if division by a is expressed in fractions, we have the simple distributive properties shown below.

▶ **Distributive Property of Division over Addition and Subtraction**
For any three numbers a, b, and c,

$$\frac{b + c}{a} = \frac{b}{a} + \frac{c}{a}$$

and

$$\frac{b - c}{a} = \frac{b}{a} - \frac{c}{a}$$

Here are examples of how these properties are used.

EXAMPLE 1
Simplify the expression $3(4x)$.

SOLUTION
According to the associative property of multiplication,

$$3(4x) = (3 \cdot 4)x = 12x$$

EXAMPLE 2
Add 7 to the expression $2x + 1$.

SOLUTION
According to the associative property of addition,

$$(2x + 1) + 7 = 2x + (1 + 7) = 2x + 8$$

EXAMPLE 3
Divide the expression $10x - 6$ by 2.

SOLUTION
According to the distributive property of division over subtraction,

$$\frac{10x - 6}{2} = \frac{10x}{2} - \frac{6}{2} = 5x - 3$$

In addition to the properties listed in this lesson, our knowledge of the meanings of the operations can be used to recognize equivalent expressions.

EXAMPLE 4
Simplify the expression $9 \cdot x \cdot x$.

SOLUTION
Because $x \cdot x = x^2$, $9 \cdot x \cdot x = 9x^2$.

EXAMPLE 5

Simplify the expression $5x - x$.

SOLUTION

Because $5x = x + x + x + x + x$, $5x - x = x + x + x + x = 4x$.

Another way to do this problem is to use the distributive rule:

$$5x - x = 5x - 1x$$
$$= (5 - 1)x$$
$$= 4x$$

Exercises

Set I

1. Find the value of each of these expressions.
   a) $3 - (9 - 27)$ c) $2 + 6(2 - 6)$
   b) $(1 - 12)(12 - 1)$ d) $(-10)^3 + (-10)^2$

2. Guess a formula for the function represented by each of these tables. Begin each formula with $y =$.

   a)
x	–4	–3	–2	–1	0
y	12	9	6	3	0

   b)
x	0	1	2	3	4
y	–5	–5	–5	–5	–5

   c)
x	–5	–4	–3	–2	–1
y	24	15	8	3	0

3. Baby Snooks said to her father: "I am thinking of a number. If I multiply it by a million and add one, the answer is two."
   a) Write an equation for what she said, letting x represent the number.
   b) Is there such a number, or was Baby Snooks only fooling?

Set II

4. Tell whether or not the following pairs of algebraic expressions are equivalent.
   a) $2x + 5$ and $5 + 2x$
   b) $x \cdot x \cdot x$ and x^3
   c) $x - 9$ and $9 - x$
   d) $x - 9$ and $-9 + x$
   e) $7x^2$ and $7 \cdot x \cdot x$
   f) $2x + 3x$ and $5x$
   g) $6x - x$ and 6
   h) $7 + (1 + x)$ and $8 + x$
   i) $7(1 + x)$ and $7 + x$
   j) $2(6x)$ and $12x$
   k) $3(x - 5)$ and $3x - 15$
   l) $\dfrac{8x + 4}{4}$ and $2x + 1$

5. Use the property named to write another expression equivalent to each of the following.
   a) $8x + 1$ — Commutative property of addition.
   b) $x(x + 3)$ — Commutative property of multiplication.
   c) $x(x + 3)$ — Distributive property of multiplication over addition.
   d) $5 + (2 + y)$ — Associative property of addition.
   e) $5(2y)$ — Associative property of multiplication.
   f) $6(4x - 1)$ — Distributive property of multiplication over subtraction.

g) $\dfrac{3x + 3}{3}$ Distributive property of division over addition.

h) $(7x)x$ Associative property of multiplication.

6. Name the property that is illustrated by each of the following equations.
   a) $4x = x \cdot 4$
   b) $7(x + 3) = 7x + 21$
   c) $(x + 8) + 2 = x + (8 + 2)$
   d) $\dfrac{45 - 5x}{5} = 9 - x$
   e) $3(6x) = (3 \cdot 6)x$
   f) $10 + x = x + 10$

7. Write a simpler expression equivalent to each of the following.
   a) $7x + 3x$
   b) $(x + 7) + 3$
   c) $4(2x)$
   d) $4 \cdot x \cdot x$
   e) $4x + 4x$
   f) $4x - 4x$
   g) $(4x)(4x)$
   h) $9x + x$
   i) $9x - x$
   j) $5x - 4x$
   k) $4x - 5x$

8. Simplify each of the following expressions.
   a) $x + 2x + 4x$ e) $x - 8 + 1$
   b) $x(2x)(4x)$ f) $x - (8 + 1)$
   c) $x + 2x + 4$ g) $5x - 5 - x$
   d) $x(2x)(4)$ h) $5x - (-5x)$

9. Here are directions for a number trick.

   > Think of a number.
   > Subtract 3.
   > Multiply by 2.
   > Add 8.
   > Divide by 2.
   > Subtract the number that
   >   you first thought of.

   a) Show how the trick works by letting x represent the original number and writing an appropriate expression for each of the other steps.
   b) What is the result of this number trick?

Set III

10. Tell whether or not the following algebraic expressions are equivalent.
    a) $5x + 1$ and $1 + 5x$
    b) $5x + x$ and $6x$
    c) x^4 and $x + x + x + x$
    d) $10x^3$ and $10 \cdot x \cdot x \cdot x$
    e) $2x - 1$ and $1 - 2x$
    f) $2x - 1$ and $-1 + 2x$
    g) $3 + (4 + x)$ and $7 + x$
    h) $3(4 + x)$ and $12 + x$
    i) $3(4x)$ and $12x$
    j) $8x - x$ and 8
    k) $4(x - 9)$ and $4x - 36$
    l) $\dfrac{3x + 15}{3}$ and $x + 5$

11. Use the property named to write another expression equivalent to each of the following.
    a) $3(3 + x)$ Distributive property of multiplication over addition.
    b) $2x + 11$ Commutative property of addition.
    c) $(x + 4) + 6$ Associative property of addition.
    d) $x(x - 5)$ Distributive property of multiplication over subtraction.
    e) $x(x - 5)$ Commutative property of multiplication.
    f) $\dfrac{7x + 7}{7}$ Distributive property of division over addition.

g) $2(10x)$ Associative property of multiplication.

h) $(1 + x) + x$ Associative property of addition.

12. Name the property that is illustrated by each of the following equations.
   a) $x + 8 = 8 + x$
   b) $2(7x) = (2 \cdot 7)x$
   c) $5(x + 4) = 5x + 20$
   d) $3x = x \cdot 3$
   e) $\dfrac{9x - 9}{9} = x - 1$
   f) $(x + 6) + 5 = x + (6 + 5)$

13. Write a simpler expression equivalent to each of the following.
   a) $4x + 9x$
   b) $(x + 4) + 9$
   c) $5(3x)$
   d) $5 \cdot x \cdot x \cdot x$
   e) $5x + 5x$
   f) $5x - 5x$
   g) $(5x)(5x)$
   h) $7x + x$
   i) $7x - x$
   j) $3x - 2x$
   k) $2x - 3x$

14. Simplify each of the following expressions.
   a) $5x + 3x + x$
   b) $5x(3x)(x)$
   c) $5 + 3x + x$
   d) $5(3x)(x)$
   e) $x - 7 + 2$
   f) $x - (7 + 2)$
   g) $6x - 4 - 4x$
   h) $6x - (-4x)$

15. Here are directions for a number trick.

   Think of a number.
   Multiply by 3.
   Subtract 6.
   Divide by 3.
   Add 7.
   Subtract the number that
   you first thought of.

   a) Show how the trick works by letting x represent the original number and writing an appropriate expression for each of the other steps.
   b) What is the result of this number trick?

Set IV   An Interesting Number Pattern

1. Copy and complete the following tables.

Table 1

1	$= 1$
$1 + 2$	$= 3$
$1 + 2 + 3$	$= 6$
$1 + 2 + 3 + 4$	$= 10$
$1 + 2 + 3 + 4 + 5$	$= 15$

Table 2

1^3	$= 1$
$1^3 + 2^3$	$= 9$
$1^3 + 2^3 + 3^3$	$= 36$
$1^3 + 2^3 + 3^3 + 4^3$	$= 100$
$1^3 + 2^3 + 3^3 + 4^3 + 5^3$	$= 225$

2. In general, the sum of the cubes of the first x integers can be represented by the expression

$$1^3 + 2^3 + 3^3 + \cdots + x^3$$

By referring to your tables, can you write another expression to which it seems to be equivalent?

3. Add some more lines to each table. Does your answer to exercise 2 still seem to be correct?

DRAWING BY HOFF; © 1958 THE NEW YORKER MAGAZINE, INC.

LESSON 5
More on Solving Equations

"Being a scientist is going to be a lot easier than I thought."

All the equations in Lesson 3 had something in common: the variable appeared only on the left side. Here is an example of an equation in which the variable appears on both sides:

$$5x + 1 = 2x + 7$$

A picture illustrating this equation is shown here.

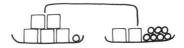

The five boxes and one circle on the left pan of the balance represent the expression $5x + 1$ and the two boxes and seven circles on the right pan represent the expression $2x + 7$.

To solve the equation

$$5x + 1 = 2x + 7$$

for x is equivalent to finding out how many circles will balance one box. One way to do this is shown below. The corresponding steps in solving the equation are shown at the right.

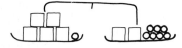

$$5x + 1 = 2x + 7$$

Remove two boxes from each pan.

Subtract $2x$ from each side.

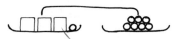

$$3x + 1 = 7$$

Remove a circle from each pan.

Subtract 1 from each side.

$$3x = 6$$

Because three boxes are balanced by six circles, one box would be balanced by two circles.

Divide each side by 3.

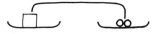

$$x = 2$$

The procedure in solving the equation is to transform it into equivalent equations that become progressively simpler. In the last one, the variable appears by itself on one side of the equal sign: this is always the goal in solving an equation.

To make sure that the equations that we write are equivalent to the original one and to each other, we can do any of the following things:

1. Add the same number to both sides.
2. Subtract the same number from both sides.
3. Multiply both sides by the same number other than zero.
4. Divide both sides by the same number other than zero.
5. Change the form of either side by using the facts that we know about equivalent expressions.

Here are more examples of how these steps can be used to solve equations.

EXAMPLE 1

Solve the equation $3x + 10 = 7x - 2$.

SOLUTION

$$3x + 10 = 7x - 2$$

Because the variable appears on both sides of this equation, we might begin by removing it from one side. One way to do this is to subtract $3x$ from each side:

$$10 = 4x - 2$$

Now we want the variable to be by itself on the right side. Adding 2 to each side, we get

$$12 = 4x$$

Dividing both sides by 4 gives

$$3 = x$$

The solution is 3.

CHECK

Substituting 3 for x in the original equation,

$$3x + 10 = 7x - 2$$

we get

$$3(3) + 10 = 7(3) - 2$$

Simplifying,

$$9 + 10 = 21 - 2$$
$$19 = 19$$

So 3 is correct.

EXAMPLE 2

Solve the equation $4(x - 11) = 6 - x$.

SOLUTION

$$4(x - 11) = 6 - x$$

When an equation contains parentheses, it is usually a good idea to perform the operations that will result in their removal. Using the distributive rule, we get

$$4x - 44 = 6 - x$$

For the variable to be by itself on one side, we can add x to each side:

$$5x - 44 = 6$$

Adding 44 to each side gives

$$5x = 50$$

Dividing each side by 5 gives

$$x = 10$$

The solution is 10.

CHECK
Substituting 10 for x in the original equation,

$$4(x - 11) = 6 - x$$

we get

$$4(10 - 11) = 6 - 10$$

Simplifying,

$$4(-1) = -4$$
$$-4 = -4$$

So 10 is correct.

Exercises

Set I

1. Write each of the following expressions without parentheses.
   a) $2 + (3 - x)$ c) $2 + (-3x)$
   b) $2(3 - x)$ d) $2(-3x)$

2. The "minus" sign has a different meaning in each of the following expressions. Tell what it means in each one.
   a) -5 b) $x - 5$ c) $-x$

3. A function has the equation $y = |x|$.
   a) Make a table for this function, letting $x = -3, -2, -1, 0, 1, 2,$ and 3.
   b) Plot the seven points in your table on a graph. Connect the points.
   c) Is $(0.5, 0.5)$ on your graph?
   d) Is $(-1.5, 1.5)$ on your graph?
   e) Is $(2.5, -2.5)$ on your graph?

Set II

4. The diagrams below are of a balanced scale: the open circles represent positive numbers and the solid circles represent negative numbers.

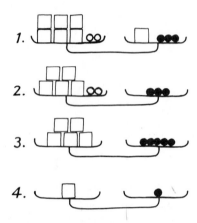

What was done to the scale in
a) the first diagram to get what is shown in the second?
b) the second diagram to get what is shown in the third?
c) the third diagram to get what is shown in the fourth?
   Write an equation illustrated by
d) the first diagram.
e) the second diagram.
f) the third diagram.
g) the fourth diagram.
   What could be done to
h) the first equation that you wrote to get the second?
i) the second equation that you wrote to get the third?
j) the third equation that you wrote to get the fourth?
k) If a box represents x, what does the fourth diagram tell us about x?

5. Make drawings like those in exercise 4 to illustrate how each of the following equations might be solved.
a) $3x + 1 = x + 9$
b) $4x - 3 = 3x - 8$

6. Solve the following equations for x, showing each step on a separate line. Check your answers.

a) $5x - 3x = 8$
b) $5x - 3 = 8$
c) $5x - 3 = 8x$
d) $2x - 11 = 4 - x$
e) $2x - 11 = 4 + x$
f) $2x + 11 = 4 + x$
g) $7x - 6 = 18 + 3x$
h) $7x + 6 = 18 - 3x$

7. Solve the following equations for x. Start by doing the operations required to remove the parentheses. Show each step on a separate line and check your answers.
a) $5(x - 2) = 30$
b) $5(2 - x) = 30$
c) $(x + 3) + x = 17$
d) $(x + 1) + (x + 4) = 0$
e) $3(6 + x) = 2x$
f) $3(6 - x) = 2x$
g) $2(4x - 9) - 6x = 0$
h) $8(x + 5) = 7(x - 7)$

*8. Solve the following equations for x. Express your answers to the nearest tenth.
a) $7.5x - 1.5x = 6.6$
b) $2x + 13.6 = 0.4x$
c) $2x - 13.6 = -0.4x$
d) $4.2(x - 1) = 9.8$
e) $1.7(2x + 5) = 3(x - 8.1)$

Set III

9. The diagrams below are of a balanced scale: the open circles represent positive numbers and the solid circles represent negative numbers.

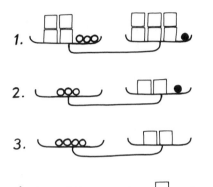

1.
2.
3.
4.

What was done to the scale in
a) the first diagram to get what is shown in the second?
b) the second diagram to get what is shown in the third?
c) the third diagram to get what is shown in the fourth?
   Write an equation illustrated by
d) the first diagram.
e) the second diagram.
f) the third diagram.
g) the fourth diagram.
   What could be done to
h) the first equation that you wrote to get the second?
i) the second equation that you wrote to get the third?
j) the third equation that you wrote to get the fourth?
k) If a box represents x, what does the fourth diagram tell us about x?

10. Make drawings like those in exercise 9 to illustrate how each of these equations might be solved.
a) $5x + 1 = 3x + 7$
b) $2x - 6 = x - 2$

11. Solve the following equations for x, showing each step on a separate line. Check your answers.
a) $7x - 2 = 12$
b) $7x - 2x = 12$
c) $7x - 2 = 12x$
d) $3x + 10 = 4 + x$
e) $3x - 10 = 4 + x$
f) $3x - 10 = 4 - x$
g) $5x + 4 = 16 - 7x$
h) $5x - 4 = 16 + 7x$

12. Solve the following equations for x. Start by doing the operations required to remove the parentheses. Show each step on a separate line and check your answers.
a) $4(x - 3) = 24$
b) $4(3 - x) = 24$
c) $(x + 7) + x = 15$
d) $(x + 2) + (x + 6) = 9$
e) $7(2 - x) = 3x$
f) $7(2 + x) = 3x$
g) $5(6x - 1) + 20x = 0$
h) $2(x + 11) = 9(x - 3)$

*13. Solve the following equations for x. Express your answers to the nearest tenth.
a) $8.2x - 5.2x = 4.5$
b) $4x + 11.7 = 0.3x$
c) $4x - 11.7 = -0.3x$
d) $2.6(x + 1) = -9.4$
e) $1.3(5x - 8) = 6(x + 7.6)$

Set IV

When asked her age, Miss Grundy refused to tell. After being begged for a hint, however, she finally admitted that in 12 years she hoped to be three times as old as she was 72 years ago.

1. If x represents Miss Grundy's present age, how old will she be 12 years from now?

2. How old was she 72 years ago?
3. Write the information given in this problem as an equation.
4. Solve the equation to find out how old Miss Grundy is now.
5. Check your answer by seeing if it agrees with Miss Grundy's hint.

CURRIER & IVES: AMERICAN NATIONAL GAME OF BASEBALL. YALE UNIVERSITY ART GALLERY, WHITNEY COLLECTION OF SPORTING ART.

LESSON 6
Length and Area

A baseball diamond is square in shape, having sides that are 90 feet long. The distance that a player has to run in going around the diamond is called its *perimeter*. This distance, the sum of the lengths of its sides, is

$$90 + 90 + 90 + 90 = 360 \text{ feet}$$

The *area* enclosed by a baseball diamond, on the other hand, is the product of its length and width:

$$90 \cdot 90 = 8,100 \text{ square feet}$$

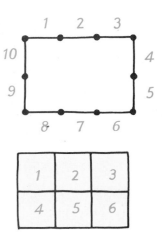

This means that the diamond could be divided into 8,100 squares, each measuring one foot on each side.

The methods for finding the perimeter and area of a baseball diamond can be applied to any rectangle. It is easy to see why if the sides of a rectangle have lengths that are integers. The figures at the right, for example, show why a rectangle 3 centimeters long and 2 centimeters wide has a perimeter of 10 centimeters and an area of 6 square centimeters.

213

▶ The **perimeter** of a rectangle (or any geometric figure bounded by straight line segments) is the sum of the lengths of its sides.

▶ The **area** of a rectangle is the product of its length and width.

Here are examples of how our knowledge of equations can be applied to problems about lengths and areas.

EXAMPLE 1

Find the length and width of this rectangle if its perimeter is 24 units.

SOLUTION

Because the perimeter is the sum of the lengths of its sides (two lengths and two widths), we can write the equation

$$2(x + 3) + 2x = 24$$

Solving this equation for x, we get

$$
\begin{aligned}
2x + 6 + 2x &= 24 \\
4x + 6 &= 24 \\
4x &= 18 \\
x &= 4.5
\end{aligned}
$$

Because $x = 4.5$, $x + 3 = 7.5$. The length is 7.5 centimeters and the width is 4.5 centimeters.

CHECK

$$2(7.5) + 2(4.5) = 15 + 9 = 24 \text{ centimeters}$$

EXAMPLE 2

Find the length of this rectangle if its area is $3x$ square units.

SOLUTION

Because the area is the product of its length and width, we can write the equation

$$5(x - 6) = 3x$$

Solving for x, we get

$$
\begin{aligned}
5x - 30 &= 3x \\
2x - 30 &= 0 \\
2x &= 30 \\
x &= 15
\end{aligned}
$$

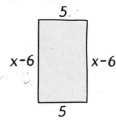

Because $x = 15$, $x - 6 = 9$. The length of the rectangle is 9 centimeters.

CHECK

Substituting 15 for x in the expression $3x$, we get $3(15) = 45$. This agrees with the fact that the length and width of the rectangle are 5 and 9. Their product is 45.

EXAMPLE 3

Find the length of each segment in this diagram if AB and CD are the same length.

SOLUTION

Because AB and CD are equal in length, we can write the equation

$$10 + 2x = 4(x - 7)$$

Solving for x,

$$10 + 2x = 4x - 28$$
$$10 = 2x - 28$$
$$38 = 2x$$
$$19 = x$$

$$EB = 2x = 2(19) = 38$$
$$CD = 4(x - 7) = 4(19 - 7) = 4(12) = 48$$

CHECK

$$AB = AE + EB = 10 + 38 = 48$$

Exercises

Set I

1. Name the property that is illustrated by each of the following equations.
   a) $4(7x) = (4 \cdot 7)x$
   b) $3(2x + 1) = 6x + 3$
   c) $x + -5 = -5 + x$
   d) $\dfrac{8x - 40}{4} = 2x - 10$

2. Solve the following equations for x.
   a) $-2.4 + x = 30$
   b) $-2.4x = 30$
   c) $-2 + 0.4x = 30$
   d) $2 - 0.4x = 30$

3. The number of degrees that the earth turns on its axis is a function of time.

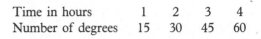

Time in hours	1	2	3	4
Number of degrees	15	30	45	60

a) Write a formula for this function, letting t represent the time in hours and n represent the number of degrees.
b) Through how many degrees does the earth turn in 24 hours?
c) Through how many degrees does the earth turn in $12x$ hours?

Set II

4. Find an expression for the perimeter of each of these figures.

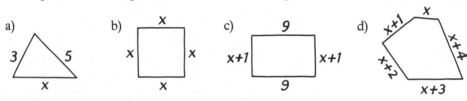

5. Find an expression for the area of each of these rectangles.

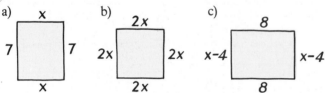

6. The perimeter of the triangle at the right is 19. Use this fact to write an equation and solve it to find the lengths of the triangle's sides. Check your answer.

7. Apply the method that you used in exercise 6 to find the lengths of the sides of the following figures. Check your answers.

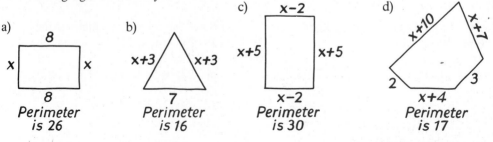

8. Find the lengths of the sides of the following rectangles.
   Check your answers.

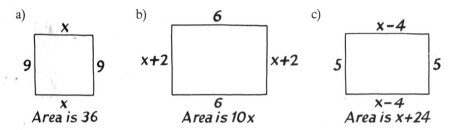

a)

x

9 | | 9

x

Area is 36

b)

6

x+2 | | x+2

6

Area is 10x

c)

x−4

5 | | 5

x−4

Area is x+24

9. Find the length of each segment in each of the following diagrams.
   In each figure, AB and CD are the same length.

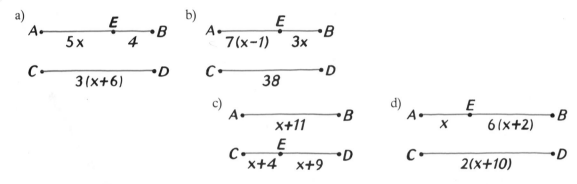

a)

A•————E————•B
  5x   4

C•——————————•D
    3(x+6)

b)

A•————E————•B
  7(x−1)   3x

C•——————————•D
      38

c)

A•——————————•B
     x+11

C•——E————•D
  x+4 x+9

d)

A•————E————•B
   x   6(x+2)

C•——————————•D
    2(x+10)

Set III

10. Find an expression for the perimeter of each of these figures.

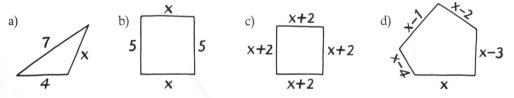

a)

7 / x
4

b)

x
5 | | 5
x

c)

x+2
x+2 | | x+2
x+2

d)

x−1 x−2
x−4   x−3
    x

11. Find an expression for the area of each of these figures.

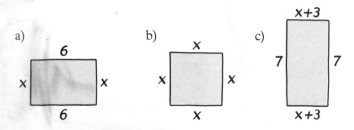

a)

6
x | | x
6

b)

x
x | | x
x

c)

x+3
7 | | 7
x+3

12. The perimeter of the triangle below is 20. Use this fact to write an equation and solve it to find the lengths of the triangle's sides. Check your answer.

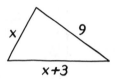

13. Apply the method that you used in exercise 12 to find the lengths of the sides of the following figures. Check your answers.

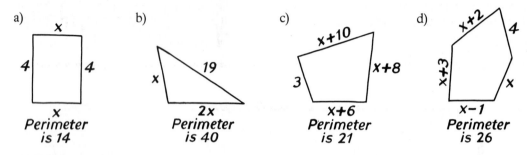

a)

x

4 4

x

Perimeter
is 14

b)

x

19

$2x$

Perimeter
is 40

c)

$x+10$

3 $x+8$

$x+6$

Perimeter
is 21

d)

$x+2$ 4

$x+3$ x

$x-1$

Perimeter
is 26

14. Find the lengths of the sides of the following rectangles. Check your answers.

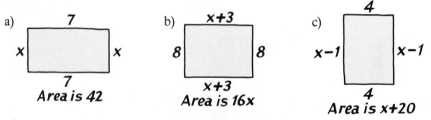

a)

7

x x

7

Area is 42

b)

$x+3$

8 8

$x+3$

Area is 16x

c)

4

$x-1$ $x-1$

4

Area is x+20

15. Find the length of each segment in each of the following diagrams. In each figure, AB and CD are the same length.

a)

A •——E——• B
 $3x$ 11

C •————• D
  $5(x+1)$

b)

A •———E———• B
  $4(x-3)$ $2x$

C •————————• D
      39

c) A •————————• B
        $x+8$

C •——————E————• D
  $x+7$ $x+2$

d) A •————————————• B
        $9(x-2)$

C •————E————• D
  x $3(x+5)$

Set IV

Obtuse Ollie wanted to get a photograph enlarged so that the length and width of the enlargement would be twice those of the original. Because the original cost 40 cents, Ollie assumed that the enlargement would cost 80 cents.

1. Does this seem reasonable? Explain.
2. If the length and width of one rectangle are each three times those of another, how do you think their areas compare?
3. If the length and width of one rectangle are each x times those of another, how do their areas compare?

PHOTOGRAPH BY ROBERT ISHI.

COPYRIGHT 1974 UNIVERSAL PRESS SYNDICATE.

LESSON 7
Distance, Rate, and Time

It isn't possible to solve the homework problem described in this cartoon with the information given. Instead of being told the speed limit and the number of gallons that Mr. Jones's gas tank will hold, we need to know how far it is to Cleveland.

If something moves at a constant speed, there is a simple relationship between the distance it travels, its speed, and the time that it takes to travel that distance. Suppose, for example, that Mr. Jones travels at 50 miles per hour for three hours. A speed of 50 miles per hour means that he travels 50 miles each hour; so in three hours he travels $3 \cdot 50 = 150$ miles.

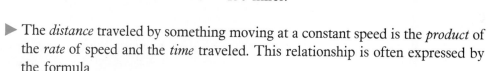

▶ The *distance* traveled by something moving at a constant speed is the *product* of the *rate* of speed and the *time* traveled. This relationship is often expressed by the formula

$$d = rt$$

Suppose that Mr. Jones decides to drive at a speed of 80 miles per hour and that he covers 16 miles at this rate before being stopped by a highway patrol officer. How many minutes did he travel before being stopped?

From the formula $d = rt$, we can write the equation

$$16 = 80t$$

Solving for t by dividing both sides of the equation by 80, we get

$$\frac{16}{80} = t, \quad \text{or} \quad t = 0.2$$

Mr. Jones traveled at this rate for 0.2 hour, or, because there are 60 minutes in an hour, for $0.2(60) = 12$ minutes.

Suppose that, having received a ticket, Mr. Jones slows down and takes 90 minutes to travel 81 miles. If he travels at a steady rate, what is his new rate of speed? From the formula $d = rt$, we can write the equation

$$81 = r(1.5)$$

(If we want to find his speed in miles per *hour*, we must express the time, 90 minutes, in *hours*.) Solving for r by dividing both sides of the equation by 1.5, we get

$$\frac{81}{1.5} = r, \quad \text{or} \quad r = 54$$

Mr. Jones's new rate of speed is 54 miles per hour.

Exercises

Set I

1. Write expressions for the following sets of operations.
   a) Multiply x by 5 and subtract 1 from the result.
   b) Subtract 1 from x and multiply the result by 5.
   c) Divide x by 3 and add 8 to the result.
   d) Add 8 to x and divide the result by 3.

2. Simplify each of the following expressions.
   a) $6x(x)$ d) $6(x + x)$
   b) $6x - x$ e) $x - 6x$
   c) $6(x - x)$ f) $x - 6 + x$

3. Americans consume an average of 15 pounds of oranges per person each year. On the basis of this figure,
   a) how many pounds of oranges would be consumed by x Americans in a year?
   b) how many Americans would consume y pounds of oranges in a year?

Set II

4. Draw a figure like the one on page 220 to illustrate each of the following statements.
   a) If a bus traveled 55 miles per hour for two hours, it would travel 110 miles.
   b) If a truck traveled x miles per hour for four hours, it would travel $4x$ miles.
   c) If an airplane traveled $(x + 10)$ miles per hour for three hours, it would travel $(3x + 30)$ miles.

5. An ostrich can run at a speed of 45 miles per hour.
   a) Copy and complete the following table to show the distances an ostrich can run in various times.

Time in hours	0	1	2	3	4
Distance in miles	0	▓	▓	▓	▓

   b) How does the distance vary with respect to time?
   c) Write a formula for this function, letting d represent the distance and t represent the time.

6. Chicago and London are 3,960 miles apart.
   a) Copy and complete the following table to show how long it would take to fly from one city to the other.

Speed in mph	400	450	500	550
Time in hours	▓	▓	▓	▓

   b) Write a formula for this table, letting r represent the speed and t represent the time.
   c) How does the time vary with respect to the speed?

7. The following table represents six different trips made by Fonzie on his motorcycle.

What number or expression belongs in each of the blanks?

Trip	Distance (miles)	Rate (mph)	Time (hours)
1	a) ▓	60	0.25
2	40	b) ▓	0.8
3	35	70	c) ▓
4	d) ▓	65	x
5	25	e) ▓	y
6	z	75	f) ▓

8. Mr. Fogg plans to travel around the world, a distance of 24,900 miles, in 80 days.
   a) Write a distance-rate-time equation for the trip, letting x represent the rate.
   b) Solve the equation for x to find what Mr. Fogg's average rate of speed will be in miles per day.

*9. The world speed records for free-style swimming are shown in the following table.

Distance	Time
100 meters	49 sec
200 meters	1 min 50 sec
400 meters	3 min 52 sec
800 meters	8 min 2 sec

   a) Find the average speed for each distance in meters per second. Round each answer to the nearest hundredth.
   b) Find the average speed for 100 meters in meters per minute.
   c) Find the average speed for 100 meters in meters per hour.
   d) Find the average speed for 100 meters in kilometers per hour. (One kilometer = 1,000 meters.)

Set III

10. Draw a figure like the one on page 220 to illustrate each of the following statements.
    a) If a train traveled 40 miles per hour for two hours, it would travel 80 miles.
    b) If a ship traveled x miles per hour for three hours, it would travel $3x$ miles.
    c) If a car traveled $2x$ miles per hour for five hours, it would travel $10x$ miles.

11. An eagle can fly at a speed of 90 miles per hour.
    a) Copy and complete the following table to show the distances an eagle can fly in various times.

Time in hours	0	1	2	3	4
Distance in miles	0	▨	▨	▨	▨

    b) How does the distance vary with respect to the time?
    c) Write a formula for this function, letting d represent the distance and t represent the time.

12. Paris and Istanbul are 1,400 miles apart.
    a) Copy and complete the following table to show how long it would take to fly from one city to the other.

Speed in mph	400	500	700	800
Time in hours	▨	▨	▨	▨

    b) Write a formula for this table, letting r represent the speed and t represent the time.
    c) How does the time vary with respect to the speed?

13. The following table represents six different flights made by Mr. Wright in his airplane.

What number or expression belongs in each of the blanks?

Flight	Distance (miles)	Rate (mph)	Time (hours)
1	a) ▨	40	0.5
2	90	b) ▨	2
3	80	60	c) ▨
4	d) ▨	x	1
5	y	e) ▨	3
6	50	z	f) ▨

14. The earth travels 584,000,000 miles each year in its orbit around the sun.
    a) On the assumption that a year contains exactly 365 days, write a distance-rate-time equation for the trip, letting x represent the rate.
    b) Solve the equation for x to find what the earth's average rate of speed in its orbit around the sun is in miles per day.

*15. The world speed records for skating are shown in the following table.

Distance	Time
500 meters	37 sec
1,000 meters	1 min 16 sec
1,500 meters	1 min 55 sec
3,000 meters	4 min 9 sec

    a) Find the average speed for each distance in meters per second. Round each answer to the nearest tenth.
    b) Find the average speed for 500 meters in meters per minute.
    c) Find the average speed for 500 meters in meters per hour.
    d) Find the average speed for 500 meters in kilometers per hour. (One kilometer = 1,000 meters.)

Set IV

Mr. Mercer wants to drive his Raceabout two miles at an average speed of 50 miles per hour.

1. If he drives the first mile at 25 miles per hour, how much time does he have left to drive the second mile?

2. Can he do it? If so, at what speed?

PHOTOGRAPH BY JUDY MACCREADY.

LESSON 8
Rate Problems

Several years ago the British Royal Aeronautical Society offered a prize of $85,000 to the first person who could build a man-powered plane that could fly a round-trip course of one mile. The prize was won by Paul B. MacCready, whose plane, named the Gossamer Condor, successfully completed such a flight in August 1977.*

The plane took 4 minutes and 20 seconds to cover the first part of the flight, flying against a wind of 2 miles per hour. The return trip, in which the plane flew with the same wind, took 3 minutes. When the flight was over, the designer wanted to know the plane's airspeed. We can use the information that we have and some simple algebra to figure it out.

If the Gossamer Condor can fly x miles per hour in still air, its speed against a 2-mile-per-hour wind would be $(x - 2)$ miles per hour and its speed with the

---

* "Science and the Citizen," *Scientific American,* October 1977, p. 74.

same wind would be $(x + 2)$ miles per hour. Its time flying against the wind was 4 minutes and 20 seconds, or 260 seconds. Because there are $60 \cdot 60 = 3600$ seconds in an hour, the plane flew against the wind for $\dfrac{260}{3600}$ hour. Reasoning in the same way, we can conclude that the return trip with the wind took $\dfrac{180}{3600}$ hour.

Assuming that the plane flew at a constant speed during each part of the flight, we can use the formula

$$d = rt$$

to write an expression for the distance covered each way. Flying against the wind, the plane traveled $(x - 2) \cdot \dfrac{260}{3600}$ miles. Returning with the wind, it flew $(x + 2) \cdot \dfrac{180}{3600}$ miles.

Against wind
$(x-2)\dfrac{260}{3600}$

With wind
$(x+2)\dfrac{180}{3600}$

Because the distances for both parts of the flight are the same, we can write the equation,

$$(x - 2) \cdot \frac{260}{3600} = (x + 2) \cdot \frac{180}{3600}$$

Multiplying each side of this equation by 3600, we get

$$260(x - 2) = 180(x + 2)$$

Applying the distributive rule to each side gives

$$260x - 520 = 180x + 360$$

Solving for x gives

$$80x - 520 = 360$$
$$80x = 880$$
$$x = 11$$

The airspeed of the Gossamer Condor was 11 miles per hour.

To solve rate problems such as the one we have just done, it is helpful to draw a diagram to illustrate the distances. The information in the diagram can then be used to write an equation. Here is another example.

Suppose that someone drives for a certain distance at a speed of 60 miles per hour and then slows down to a speed of 50 miles per hour. If it takes two hours to travel 105 miles, how much time was spent traveling at each speed?

If we let x represent the time spent driving at 60 miles per hour, then $60x$ represents the distance in miles traveled at that speed. Because the total time is 2 hours, $(2 - x)$ is the time spent traveling at 50 miles per hour. The distance traveled for that part of the trip would be $50(2 - x)$ miles.

From the figure at the right below illustrating the trip, we see that

$$60x + 50(2 - x) = 105$$

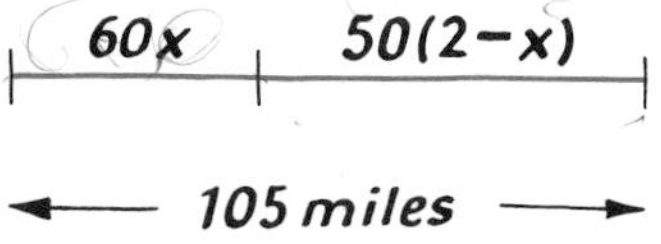

Solving this equation for x, we get

$$60x + 100 - 50x = 105$$
$$10x + 100 = 105$$
$$10x = 5$$
$$x = 0.5$$

If $x = 0.5$, then $2 - x = 2 - 0.5 = 1.5$. The time spent driving at 60 miles per hour is 0.5 hour and the time spent driving 50 miles per hour is 1.5 hours. To find out whether this is correct, we write

$$60(0.5) + 50(1.5) = 105$$
$$30 + 75 = 105$$
$$105 = 105$$

Exercises

Set I

1. Which of these symbols, $>$, $=$, or $<$, should replace ▓ in each of the following?
   a) 0.1 ▓ $(0.1)^2$
   b) $-(11 - 8)$ ▓ $-(8 - 11)$
   c) $\dfrac{-9}{-4}$ ▓ $-\dfrac{9}{4}$
   d) $3(-5)$ ▓ $(-5)^3$

2. Tell whether or not the steps in each of the following will always give the same result.
   a) Subtract 2 from a number and add 5 to the result. Add 5 to a number and subtract 2 from the result.
   b) Multiply a number by 3 and subtract 1 from the result. Subtract 1 from a number and multiply the result by 3.

   c) Add 6 to a number and multiply the result by 2. Multiply a number by 2 and add 12 to the result.

3. A popular quiz show of the 1940s was "Double or Nothing." Each successive question answered correctly doubled the amount of your winnings. An incorrect answer meant that you lost everything. How much money would you have won if, after winning x dollars, you answered
   a) one more question correctly and then quit?
   b) three questions correctly before quitting?
   c) five questions correctly and then missed a question?

Set II

4. Daisy and her boyfriend, Alf Kerazy, decide to have a bicycle race. Because the rates at which Daisy and Alf can ride are 720 and 660 meters per minute, respectively, Alf is given a headstart of two minutes.
   a) If Daisy catches up with Alf in x minutes, how many minutes has Alf been riding? (Give your answer in terms of x.)
   b) Copy and complete the diagram below, which shows the race to the point at which Daisy overtakes Alf.

   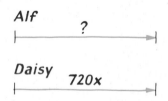

   c) Use the information in this diagram to write an equation.
   d) Solve the equation to find out how long it takes Daisy to catch up.
   e) At this time, how far has each one gone?

5. The largest swimming pool in the world, located in Casablanca, is 480 meters long.

   STERLING PUBLISHING CO., INC.

   Suppose that two swimmers at opposite ends of the pool jump in at the same time and begin swimming toward each other at speeds of 70 meters per minute and 80 meters per minute, respectively.

$$d = r \cdot t \qquad \text{rate} = \text{speed}$$

a) What distance will each one swim in x minutes?

b) Copy and complete the diagram below, which shows the point at which the swimmers will meet. (Let x represent the time that each one has been swimming.)

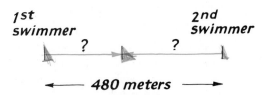

1st swimmer ? ? **2nd swimmer**

← 480 meters →

c) Use the information in this diagram to write an equation.

d) Solve the equation to find how many minutes it will take the swimmers to meet.

e) At the point at which they meet, how far will each swimmer have gone?

6. An airplane flying at a speed of about 100 miles per hour was once overtaken and passed by a flock of sandpipers! Suppose that a flock of these birds can fly at 110 miles per hour with a certain wind and at 90 miles per hour against it. If they fly a certain distance with the wind and the same distance against it, how far can they fly in all if the trip takes two hours?

a) If x represents the time that the flock spends flying with the wind, how much time does it spend flying against it?

b) Draw a diagram showing the distance flown with the wind and the distance returning against it.

c) Use the information in your diagram to write an equation.

d) Solve the equation for x.

e) How far can the flock fly altogether?

Set III

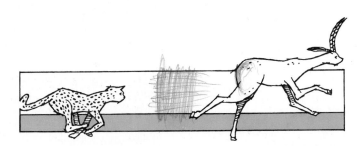

7. A cheetah and an impala spot each other when they are 450 meters apart. The impala begins running directly away at a speed of 20 meters per second. At the same instant, the cheetah takes off after it, running at a speed of 26 meters per second.

a) If x represents the time it takes for the cheetah to overtake the impala, how far will each one have run during this time? (Give your answers in terms of x.)

b) Copy and complete the diagram at the right, which shows the chase to the point where the cheetah catches up.

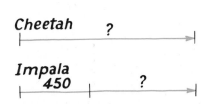

Cheetah ?

Impala 450 ?

c) Use the information in your diagram to write an equation.

d) Solve the equation to find out how long it takes the cheetah to catch up.

e) At this point, how far has each animal run?

8. The Wright brothers took off from an airport at the same time and flew in opposite directions. Their respective speeds were 160 miles per hour and 200 miles per hour.
   a) How far had each one flown after x hours?
   b) Copy and complete the diagram below, which shows the positions of the planes when they were 900 miles apart.

1st plane   Airport   2nd plane
      ?   ?

← 900 miles →

   (Let x represent the time each plane had been flying.)
   c) Use the information in your diagram to write an equation.

d) Solve the equation to find how many hours it took for the planes to be 900 miles apart.
e) At that point, how far had each plane flown?

9. An Explorer Scout paddled his canoe downstream and back in four hours. If his rate down the river was 5 miles per hour and his rate up the river was 3 miles per hour, how far down the river did he go?
   a) If x represents the time that the scout spent going down the river, how much time did he spend going up the river?
   b) Draw a diagram showing the distance traveled down the river and the distance traveled up the river.
   c) Use the information in your diagram to write an equation.
   d) Solve the equation for x.
   e) How far down the river did the scout go?

Set IV

A member of the Cleveland Indians once managed to catch a baseball thrown from a blimp hovering overhead. If the ball was thrown from the blimp at a speed of 10 feet per second, its speed at any given instant afterward is given by the formula

$$r = 10 + 32t$$

in which r is the speed in feet per second and t is the time in seconds.

1. The ball was traveling at a speed of 234 feet per second at the instant it was caught.

How many seconds had passed between the time that the ball was thrown and the time that it was caught?

The distance traveled by a falling object is given by the formula

$$d = 16t^2$$

in which d is the distance in feet and t is the time in seconds that the object has been falling.

2. How far above the ground was the blimp from which the baseball was thrown?

Summary and Review

In this chapter, we have learned how to solve certain types of equations containing one variable and how to use such equations in solving perimeter, area, and rate problems.

Equations (*Lesson 1*) An equation is a mathematical sentence stating that two expressions represent the same number. To solve an equation means to find all of the numbers that can be used to replace the variable in the equation to make it true. Some equations have only one solution, some have more than one, and some have none at all.

Inverse Operations (*Lesson 2*) If one operation undoes the effect of another, the operations are inverses of each other. Addition and subtraction are inverse operations, as are multiplication and division. Inverse operations are useful in solving equations.

Equivalent Equations (*Lesson 3*) An equation can be transformed into an equivalent one (one having the same solutions) by: adding the same quantity to each side; subtracting the same quantity from each side; multiplying each side by the same quantity other than zero; or dividing each side by the same quantity other than zero.

Equivalent Expressions (*Lesson 4*) Equivalent expressions are expressions that are equal for all values of their variables. The following properties can be used to recognize equivalent expressions.

Commutative property of addition.	$a + b = b + a$
Associative property of addition.	$(a + b) + c = a + (b + c)$
Commutative property of multiplication.	$ab = ba$
Associative property of multiplication.	$(ab)c = a(bc)$
Distributive property of multiplication over addition and subtraction.	$a(b + c) = ab + ac$ $a(b - c) = ab - ac$
Distributive property of division over addition and subtraction.	$\dfrac{b + c}{a} = \dfrac{b}{a} + \dfrac{c}{a}$ $\dfrac{b - c}{a} = \dfrac{b}{a} - \dfrac{c}{a}$

More on Solving Equations (*Lesson 5*) To solve an equation, we transform it into equivalent equations that become progressively simpler. In the last one, the variable appears by itself on one side of the equal sign and does not appear at all on the other side.

Length and Area (*Lesson 6*) The perimeter of a rectangle (or any geometric figure bounded by straight line segments) is the sum of the lengths of its sides. The area of a rectangle is the product of its length and width.

Distance, Rate, and Time (*Lesson 7*) The distance traveled by something moving at a constant speed is the product of the rate of speed and the time traveled. This relationship is often expressed by the formula

$$d = rt$$

Rate Problems (*Lesson 8*) To solve rate problems, it is helpful to draw a diagram to illustrate the distances. The information in the diagram can then be used to write an equation.

Exercises

Set I

1. Tell whether each of the following equations is true, false, or neither.
   a) $4^3 - 4^2 = 4$
   b) $\dfrac{6 + 2}{2} = \dfrac{6}{2} + \dfrac{2}{2}$
   c) $x + 7 = 1$
   d) $3(5x) = 15x$

2. If possible, find a number or numbers that can replace x in each of the following equations to make it true. If you think that no such number can be found, briefly explain why.
   a) $x + 6 = -10$
   b) $\dfrac{39}{x} = 3$
   c) $x + x = 16$
   d) $x = x - 2$
   e) $x^2 = 64$
   f) $2^x = 64$

3. Check each of the following numbers to see which are solutions of the equation given.
   a) $x^2 - 3 = 2x$ 1, 2, 3, –1, –2, –3
   b) $x(x + 7) = 0$ 0, 1, 7, –1, –7
   c) $(x + 2)(x - 2) = x^2 - 4$ 2, 3, 4, 5, 6

4. What operations should be performed on each of these expressions to give x as the result?
   a) $4x - 7$
   b) $\dfrac{x + 8}{3}$
   c) $\dfrac{x}{6} - y$
   d) $2(x + y)$

5. Solve each of the following equations for x.
   a) $8 + x = -5$
   b) $-5 + x = 8$
   c) $-5x = 8$
   d) $x - 9 = -7$
   e) $\dfrac{x}{-9} = -7$
   f) $-x = -3$

6. Solve each of the following equations for x. Check your answers.
   a) $3x + 11 = 5$
   b) $\dfrac{x}{2} - 15 = 0$
   c) $-5x + 6 = -9$
   d) $\dfrac{x}{7} + 4 = 1$
   e) $-x + 8 = -10$

7. Use the property named to write another expression equivalent to each of the following.
   a) $(x - 5)2$ Commutative property of multiplication.
   b) $x(x + 8)$ Distributive property of multiplication over addition.
   c) $3 + (7 + x)$ Associative property of addition.
   d) $3(7x)$ Associative property of multiplication.
   e) $\dfrac{12x - 3}{3}$ Distributive property of division over subtraction.

8. Write a simpler expression equivalent to each of the following.
   a) $9x + 2x$
   b) $9(2x)$
   c) $9 + (2 + x)$
   d) $4x - x$
   e) $6 \cdot x \cdot x \cdot x$
   f) $6 + x + x + x$

9. Solve the following equations for x. Check your answers.
   a) $9x \blacksquare x = -24$
   b) $2x \blacksquare 11 = 7x$
   c) $5(x - 3) = 25$
   d) $(x \divideontimes 8) + x = 6$
   e) $4(x - 1) - x = 11$
   f) $10(x + 2) = 2(x + 10)$

*10. Solve the following equations for x.
   a) $3.6x - 2.2x = 4.2$
   b) $8x + 9.3 = 1.8x$
   c) $6.5(x - 2) = 2.6$
   d) $4(0.2x + 3.7) = 0.7x - 6.7$

11. Find expressions for the perimeter *and* area of each of these rectangles.

a)
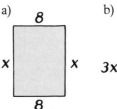

b)
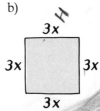

c)
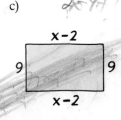

12. Find the lengths of the sides of the following figures.
    Check your answers.

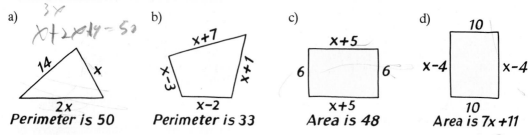

a) $3x$
$x+2x+14=50$

14, x, $2x$
Perimeter is 50

b) $x+7$, $x-3$, $x+1$, $x-2$
Perimeter is 33

c) $x+5$, 6, 6, $x+5$
Area is 48

d) 10, $x-4$, $x-4$, 10
Area is 7x +11

13. Find the length of each segment in each of these diagrams.
    In each figure, AB and CD are the same length.

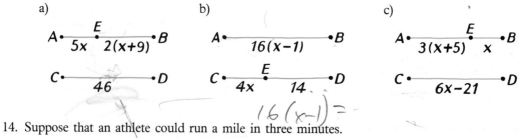

a)
A •——E—— • B
  $5x$ $2(x+9)$
C •————— • D
     46

b)
A •————— • B
   $16(x-1)$
C •——E—— • D
  $4x$ 14
$16(x-1)=$

c)
A •——E—— • B
  $3(x+5)$ x
C •————— • D
     $6x-21$

14. Suppose that an athlete could run a mile in three minutes.

I KNOW IT'S UNORTHODOX, BUT WE GET THREE-MINUTE MILES OUT OF HIM THIS WAY!

CHICAGO TRIBUNE-NEW YORK NEWS SYNDICATE, INC.

a) What would his average speed in miles per hour be?
b) At this rate, how far would he run in x hours?

15. A weasel can run at a speed of nine
    kilometers per hour.
    a) Copy and complete the following table
       to show the distances that a weasel can
       run in various times.

Time in hours	0	1	2	3	4
Distance in km	0	▓	▓	▓	▓

    b) How does the distance vary with respect
       to the time?
    c) Write a formula for this function,
       letting d represent the distance and t
       represent the time.

16. After their boat capsized in the middle of Lake Tippecanoe, Acute Alice and Obtuse Ollie swam in opposite directions toward shore. Alice swam twice as fast as Ollie and, after 3 minutes, they were 1,125 feet apart. Find out how far each one swam by doing each of the following.
   a) If Ollie swam at the rate of x feet per minute, at what rate did Alice swim?
   b) Draw a diagram to represent the problem.
   c) Use the information in your diagram to write an equation.
   d) Solve the equation for x.
   e) How far did each one swim?

Set II

1. Tell whether each of the following equations is true, false, or neither.
   a) $6(4 \cdot 2) = 24(2)$ c) $7(x - 1) = 7x - 7$
   b) $2^2 + 3^2 = 5^2$ d) $x = 4x$

2. If possible, find a number or numbers that can replace x in each of the following equations to make it true. If you think that no such number can be found, briefly explain why.
   a) $2x = -22$
   b) $8 - x = -5$
   c) $x + 3 = x$
   d) $\frac{x}{4} = 12$
   e) $x^2 = 25$
   f) $5^x = 125$

3. Check each of the following numbers to see which are solutions of the equation given.
   a) $x^2 + 2x = 8$ 2, 3, 4, -2, -3, -4
   b) $x(x - 5) = 6$ 0, 1, 6, -1, -6
   c) $(x - 3)(x + 3) = x^2 - 9$ 3, 4, 5, 6, 7

4. What operations should be performed on each of these expressions to give x as the result?
   a) $6(x + 2)$
   b) $\frac{x}{10} - 5$
   c) $3x + y$
   d) $\frac{x - y}{7}$

5. Solve each of the following equations for x.
   a) $3 + x = -4$
   b) $-4 + x = 3$
   c) $-4x = 3$
   d) $x - 8 = -9$
   e) $\frac{x}{-8} = -9$
   f) $-x = 2$

6. Solve each of the following equations for x. Check your answers.
   a) $20 + 4x = 0$
   b) $\frac{x}{3} - 12 = 1$
   c) $-7x + 5 = -9$
   d) $\frac{x}{2} + 8 = 6$
   e) $-x - 10 = 3$

7. Use the property named to write another expression equivalent to each of the following.
   a) $(x + 9) + 3$ Associative property of addition.
   b) $(x + 9)3$ Commutative property of multiplication.
   c) $\frac{8x + 4}{4}$ Distributive property of division over addition.
   d) $(2x)x$ Associative property of multiplication.
   e) $x(x - 10)$ Distributive property of multiplication over subtraction.

8. Write a simpler expression equivalent to each of the following.
   a) $4x + 5x$
   b) $4 + (5 + x)$
   c) $4(5x)$
   d) $7 \cdot x \cdot x$
   e) $7 + x + x$
   f) $10x - x$

9. Solve the following equations for x. Check your answers.
   a) $5x - 3x = 1$
   b) $12 - x = 2x$
   c) $2(x + 8) = 10$
   d) $(x - 7) + x = 9$
   e) $3(x + 4) - 2x = 6$
   f) $12(x - 1) = 2(x + 1)$

*10. Solve the following equations for x.
   a) $5.4x - 4.1x = 6.5$
   b) $7x + 9.5 = 3.2x$
   c) $3.7(x - 1) = 11.1$
   d) $8(0.4x + 2.9) = 1.2x - 5.8$

11. Find expressions for the perimeter *and* area of each of these rectangles.

a)
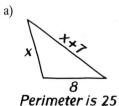

b)
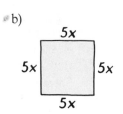

c)
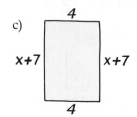

12. Find the lengths of the sides of the following figures. Check your answers

a)
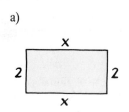
Perimeter is 25

b)

Perimeter is 48

c)
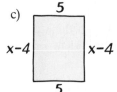
Area is 35

d)
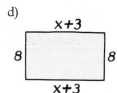
Area is 10x+6

13. Find the length of each segment in each of these diagrams. In each figure, AB and CD are the same length.

a)
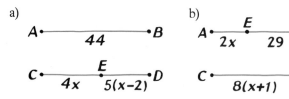

b)
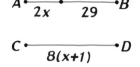

c)
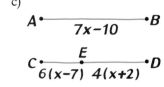

14. Minneapolis and Denver are 700 miles apart.
   a) Copy and complete the following table to show how long it would take to fly from one city to the other.

   b) Write a formula for this table, letting r represent the speed and t represent the time.
   c) How does the time vary with respect to the speed?

Speed in mph	350	400	500
Time in hours			

15. The man in this photograph rowed his boat across the Atlantic Ocean! It took him 71 days to go 1,900 miles.

a) What was his average speed in miles per day? (Round your answer to the nearest integer.)
b) At this rate, how far would he go in x days?

16. In a horse race, Dog Biscuit is 30 feet ahead of the next horse, Beetlebaum. While Dog Biscuit is running at a speed of 46 feet per second, Beetlebaum speeds up to 50 feet per second. Find out how long it will take Beetlebaum to catch up by doing each of the following.
a) If x represents the time it takes, what distance does each horse run during this time? (Give your answers in terms of x.)
b) Draw a diagram to represent the problem.
c) Use the information in your diagram to write an equation.
d) Solve the equation for x.
e) How far did each horse run during this time?

Chapter 6

EQUATIONS IN TWO VARIABLES

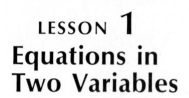

© 1971 UNITED FEATURE SYNDICATE, INC.

LESSON 1
Equations in Two Variables

So far, the equations that you have learned to solve have contained only one variable. However, many practical problems in algebra require solving equations having two variables or even more. Here is an example of such an equation:

$$3x - 2y = 15$$

To solve this equation, we have to find numbers that can replace x and y to make the equation true. The numbers named in the cartoon will work because

$$3(11) - 2(9) =$$
$$33 \ - \ 18 \ = 15$$

A simple way of writing the solution $x = 11$ and $y = 9$ is to write the numbers as if they were the coordinates of a point: $(x, y) = (11, 9)$.

It is easy to find more pairs of numbers that are also solutions of the equation. For example, if $x = 1$, then

$$3(1) - 2y = 15$$
$$3 - 2y = 15$$
$$3 + -2y = 15$$
$$-2y = 12$$
$$y = -6$$

Checking $x = 1$ and $y = -6$ in the equation,

$$3x - 2y = 15$$

we get

$$3(1) - 2(-6) = 3 - (-12)$$
$$= 3 + 12$$
$$= 15$$

So another solution is $(1, -6)$.

It is possible, in fact, to replace x with *any* number and find a number for y that will make the equation true. This means that the equation has an unlimited number of solutions. Some more solutions are: $(5, 0)$, $(6, 1.5)$, and $(-3, -12)$.

Equations that contain more than one variable usually have an unlimited number of solutions. If we restrict the solutions to certain types of numbers, however, their number may be limited. For example, the only pairs of *positive integers* that can replace x and y to make the equation $x + y = 4$ true are $(1, 3)$, $(2, 2)$, and $(3, 1)$.

Notice that the order in which we name a pair of numbers is significant: $(1, 3)$ means that $x = 1$ and $y = 3$, whereas $(3, 1)$ means that $x = 3$ and $y = 1$. Because of this, the solutions to an equation having two variables are called **ordered pairs.**

Exercises

Set I

1. Write the coordinates of each corner of the figure pictured in this graph.

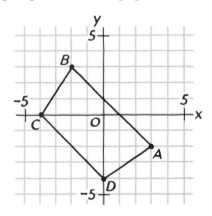

2. The smallest positive integer is 1.
   a) Is there a smallest negative integer? If so, what is it?
   b) Is there a largest negative integer? If so, what is it?
   c) What is the smallest integer that is larger than every negative integer?

3. The rectangles below contain the same number of squares.

a) The number of squares in each row is a function of the number of rows. Copy and complete the following table for this function.

Number of rows, x	1	2	3	4
Number of squares in each row, y				

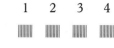

b) How does the number of squares in each row vary with the number of rows?
c) Write a formula for this function beginning with $y =$.

Set II

4. The value of an expression such as $3x + y$ depends on the numbers by which we replace the variables. Replacing x and y with the ordered pair $(4, 5)$, for example, gives

$$3x + y = 3(4) + (5) = 12 + 5 = 17$$

Find the values of the following expressions if x and y are replaced by the ordered pairs given.

$2x + 5y$
a) $(5, 1)$
b) $(1, 5)$
c) $(15, -3)$
d) $(-3, 15)$

$9x - y$
i) $(2, 4)$
j) $(4, 2)$
k) $(5, 11)$
l) $(0, -34)$

$xy - 10$
e) $(3, 4)$
f) $(3, -4)$
g) $(0, 0)$
h) $(-1, -12)$

$x^2 - 4y$
m) $(6, 8)$
n) $(8, 6)$
o) $(7, 0)$
p) $(2, -9)$

5. One of the solutions to the equation $3x + y = 17$ is the ordered pair $(4, 5)$ because

$$3(4) + (5) = 12 + 5 = 17$$

Tell whether or not each of the following ordered pairs is a solution of the given equation.

$2x + 5y = 15$
a) $(5, 1)$
b) $(1, 5)$
c) $(15, -3)$
d) $(-3, 15)$

$9x - y = 34$
i) $(2, 4)$
j) $(4, 2)$
k) $(5, 11)$
l) $(0, -34)$

$xy - 10 = 2$
e) $(3, 4)$
f) $(3, -4)$
g) $(0, 0)$
h) $(-1, -12)$

$x^2 - 4y = 40$
m) $(6, 8)$
n) $(8, 6)$
o) $(7, 0)$
p) $(2, -9)$

6. The following questions are about the rectangle shown here.

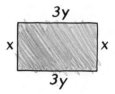

a) Write an expression for its perimeter.
b) Write an equation stating that its perimeter is 30.

If its perimeter is 30, can
c) $x = 3$ and $y = 4$?
d) $x = 6$ and $y = 3$?
e) $x = 1$ and $y = 12$?
f) Write an expression for the area of the rectangle.

g) Write an equation stating that its area is 36.

If its area is 36, can

h) $x = 3$ and $y = 4$?

i) $x = 6$ and $y = 3$?

j) $x = 1$ and $y = 12$?

7. This exercise is about the equation
$y - 4x = 9$.

a) Copy and complete the following table of some of the solutions of this equation.

x	1	0	–1	–2	–3
y	13				

b) What happens to y if x is decreased by 1?

c) Plot the five points having the numbers in this table as their coordinates on a graph.

d) What do you notice about the points?

8. If possible, find every pair of positive integers that can replace x and y in each of the following equations to make it true. If you think that an equation has an unlimited number of such solutions or that it has no solution, say so.

a) $x + y = 6$

b) $2x + y = 7$

c) $y = 4x$

d) $xy = 10$

e) $5y = 11 - x$

f) $x + y = 0$

g) $x = y + 3$

h) $x^2 = y^2 + 3$

Set III

9. The value of an expression such as $x + 7y$ depends on the numbers by which we replace the variables. Replacing x and y with the ordered pair $(5, 2)$, for example, gives

$$x + 7y = (5) + 7(2) = 5 + 14 = 19$$

Find the values of the following expressions if x and y are replaced by the ordered pairs given.

$5x + 6y$

a) $(2, 1)$

b) $(1, 2)$

c) $(8, –4)$

d) $(–4, 8)$

$3x - 10y$

i) $(8, 2)$

j) $(2, 8)$

k) $(–1, 0)$

l) $(0, –1)$

$xy - 4$

e) $(4, 6)$

f) $(4, –6)$

g) $(0, 0)$

h) $(–3, –8)$

$8x + y^2$

m) $(4, 3)$

n) $(3, 4)$

o) $(–8, –8)$

p) $(–13, 12)$

10. One of the solutions to the equation $x + 7y = 19$ is the ordered pair $(5, 2)$ because

$$(5) + 7(2) = 5 + 14 = 19$$

Tell whether or not each of the following ordered pairs is a solution of the given equation.

$5x + 6y = 16$

a) $(2, 1)$

b) $(1, 2)$

c) $(8, –4)$

d) $(–4, 8)$

$3x - 10y = 10$

i) $(8, 2)$

j) $(2, 8)$

k) $(–1, 0)$

l) $(0, –1)$

$xy - 4 = 20$

e) $(4, 6)$

f) $(4, –6)$

g) $(0, 0)$

h) $(–3, –8)$

$8x + y^2 = 40$

m) $(4, 3)$

n) $(3, 4)$

o) $(–8, –8)$

p) $(–13, 12)$

11. This exercise is about the rectangle shown here.

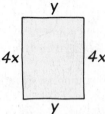

a) Write an expression for its perimeter.
b) Write an equation showing that its perimeter is 36.

If its perimeter is 36, can
c) $x = 4$ and $y = 2$?
d) $x = 8$ and $y = 1$?
e) $x = 2$ and $y = 10$?

f) Write an expression for the area of the rectangle.
g) Write an equation showing that its area is 32.

If its area is 32, can
h) $x = 4$ and $y = 2$?
i) $x = 8$ and $y = 1$?
j) $x = 2$ and $y = 10$?

12. This exercise is about the equation $2x + 3y = 12$.
a) Copy and complete the following table of some of the solutions of this equation.

x	0	3	6	9	12
y					

b) What happens to y if x is increased by 3?
c) Plot the five points having the numbers in this table as their coordinates on a graph.
d) What do you notice about the points?

13. If possible, find every pair of positive integers that can replace x and y to make each of the following equations true. If you think that an equation has an unlimited number of such solutions or that it has no solution, say so.
a) $x + y = 7$ e) $4x = 12 - y$
b) $x + 3y = 8$ f) $x + y = 1$
c) $x = 5y$ g) $x - y = 1$
d) $xy = 9$ h) $x^2 + y^2 = 10$

Set IV

Obtuse Ollie and Acute Alice went to a swap meet and bought some records.

Ollie bought some B. Bumble and the Stingers records, some for 35 cents each and the rest for 45 cents each. He spent $4.25 altogether.

1. Can you figure out how many records Ollie bought? Explain.

Alice bought some Dicky Doo and the Don'ts records, some for 25 cents each and the rest for 50 cents each. She spent $3.75 altogether.

2. Can you figure out how many records Alice bought? Explain.

STERLING PUBLISHING CO., INC.

LESSON 2
Formulas

Perhaps the heaviest human being of all time was Robert Earl Hughes, whose greatest recorded weight was 1,069 pounds. Six feet tall, he weighed 700 pounds when the photograph shown here was taken.

The "normal" weight of a man can be determined by the equation

$$w = 5.5h - 220$$

in which w represents his weight in pounds and h represents his height in inches. According to this equation, a man who is six feet tall (72 inches) should weigh approximately

$$5.5(72) - 220 = 396 - 220 = 176$$

pounds. Given someone's height in inches, it is easy to find his "normal" weight by using the equation as we have done here: replace h with the height and figure out w.

An equation containing two or more variables that is used to solve practical problems is called a **formula.** Most formulas are written so that one variable appears by itself on the left of the equal sign and an expression containing the other variable or variables appears on the right. The variable on the left is called the **subject** of the formula, and the formula tells how to find its value when values for the other variables are known. The subject of the formula

$$w = 5.5h - 220$$

is w; the formula tells how to find w when h is known.

It is frequently useful to change the subject of a formula to one of the other variables. For example, to find the "normal" height for a man whose weight is known, it would be convenient to have a formula that tells how to find h when given w. We can create such a formula by solving the formula

$$w = 5.5h - 220$$

for h in terms of w.

An easy way to get h on the left is to switch the two sides of the equation:

$$5.5h - 220 = w$$

Adding 220 to each side, we get

$$5.5h = w + 220$$

and, dividing each side by 5.5,

$$h = \frac{w + 220}{5.5}$$

This is the formula we were looking for; it tells how to find h when given w. To find the "normal" height for a man who weighs 165 pounds, for example, we can write

$$h = \frac{165 + 220}{5.5} = \frac{385}{5.5} = 70$$

The "normal" height is 70 inches.

Here are some more examples of how to solve a formula for one of its variables.

EXAMPLE 1

The formula for the volume of a box having length ℓ, width w, and height h is $v = \ell w h$. Solve this formula for h.

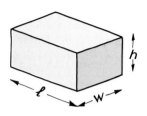

SOLUTION

Switching the sides of this formula, we get

$$\ell w h = v$$

Because h is multiplied by the product ℓw, we have to divide both sides of the equation by this product to get

$$h = \frac{v}{\ell w}$$

EXAMPLE 2

Solve the formula $y = 2x + 1$ for x.

SOLUTION

Switching the sides of this formula, we get

$$2x + 1 = y$$

Subtracting 1 from each side gives

$$2x = y - 1$$

and dividing each side by 2, we get

$$x = \frac{y - 1}{2}$$

Exercises

Set I

1. Find the value of each of these expressions.
   a) $(4 + 5)^2$
   b) $4^2 + 5^2$
   c) $(7 - 3)^2$
   d) $7^2 - 3^2$

2. Solve the following equations for x.
   a) $7x - 5 = 4x + 13$
   b) $3x + (x - 2) = 0$
   c) $5(x + 6) = 8(x - 3)$

3. Charles Lindbergh, the first person to make a solo flight across the Atlantic Ocean, took 33.5 hours to complete the trip. If he had gone 26 miles per hour faster, the flight would have taken 27 hours. Find out how far the flight was by doing the following.
   a) Draw a diagram to represent the problem, letting x represent Captain Lindbergh's actual speed.
   b) Use the information in your diagram to write an equation.
   c) Solve the equation for x.
   d) Find the length of the flight in miles.

Set II

4. Write a formula for each of the following tables. Begin each formula with $y =$.

   a)
x	0	1	2	3	4
y	3	4	5	6	7

   b)
x	8	9	10	11	12
y	0	1	2	3	4

   c)
x	0	1	2	3	4
y	0	4	8	12	16

   d)
x	0	2	4	6	8
y	0	1	2	3	4

   e)
x	0	1	2	3	4
y	10	9	8	7	6

   f)
x	0	1	2	3	4
y	1	6	11	16	21

5. Solve each of the formulas that you have written for exercise 4 for x. Use the tables to check your answers.

6. Solve each of the following equations for the variable indicated.

   $x + y = 3$
   a) For x
   b) For y

   $xy = 6$
   c) For x
   d) For y

   $2x = 5y$
   e) For x
   f) For y

   $4x - y = 8$
   g) For x
   h) For y

7. This exercise is about the rectangle shown at the beginning of the next column.

   a) Write a formula for its area, a, in terms of b and h.
   b) Solve your formula for b.
   c) Solve your formula for h.

8. The average of two numbers, x and y, is the number midway between them on the

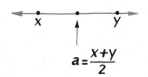

   number line. It is given by the formula

   $$a = \frac{x + y}{2}$$

   a) Solve this formula for x.
   b) Solve this formula for y.
   c) Use the formula above to find the average of 3 and 11.
   d) Check your answer for part c of this exercise in your formula for part a, letting $x = 3$ and $y = 11$.
   e) Use the formula above to find the average of 2 and –8.
   f) Check your answer for part e of this exercise in your formula for part b, letting $x = 2$ and $y = –8$.

9. When a rocket is launched upward, how high it goes depends on its height and speed

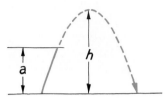

at "burn-out," the moment at which its propellant is used up. A formula for this is

$$h = a + \frac{r^2}{20}$$

in which h is how high the rocket goes in meters, a is its height at burn-out in meters, and r is its speed at burn-out in meters per second.
a) Use this formula to find out how high a rocket goes if it is 40 meters high at burn-out and traveling at a speed of 30 meters per second.
b) Solve the formula for a in terms of h and r.
c) If a rocket is traveling 40 meters per second at burn-out, what would its height at burn-out have to be in order for the rocket to reach a height of 150 meters?

10. In 1978 the formula for the cost of sending a package through the mail at the library rate was

$$c = 4n + 5$$

in which c is the cost in cents and n is the weight of the package in pounds.
a) How much does it cost to send a package weighing 8 pounds by the library rate?
b) Solve this formula for n in terms of c.
c) Check your answer for part a of this exercise in your formula for part b, letting $n = 8$.

11. Psychologists use the formula

$$i = \frac{100\,m}{c}$$

to measure a person's intelligence. Called the intelligence quotient, or I.Q., i is expressed in terms of m, the person's mental age, and c, his or her chronological age, or number of years he or she has lived.
a) Find the I.Q. of someone whose mental age is 18 and whose chronological age is 15.
b) Solve the formula for m in terms of i and c.
c) Check your answer for part a of this exercise in your formula for part b, letting $m = 18$ and $c = 15$.
d) Solve the formula that you wrote for part b of this exercise for c in terms of i and m.

Set III

12. Write a formula for each of the following tables. Begin each formula with $y =$.

a)
x	0	1	2	3	4
y	1	2	3	4	5

b)
x	6	7	8	9	10
y	0	1	2	3	4

c)
x	0	1	2	3	4
y	0	3	6	9	12

d)
x	0	4	8	12	16
y	0	1	2	3	4

e)
x	0	1	2	3	4
y	8	7	6	5	4

f)
x	0	1	2	3	4
y	5	7	9	11	13

13. Solve each of the formulas that you have written for exercise 12 for x. Use the tables to check your answers.

14. Solve each of the following equations for the variable indicated.

$$x + y = 5 \qquad 3x = 4y$$
a) For x e) For x
b) For y f) For y

$$xy = 2 \qquad x - 6y = 12$$
c) For x g) For x
d) For y h) For y

15. The distance that an object moving r meters per second travels in t seconds is given by the formula

$$d = rt$$

a) Solve this formula for r.
b) Solve this formula for t.

16. This exercise is about the rectangle shown here.

a) Write a formula for its perimeter, p, in terms of b and h.
b) Solve your formula for b.
c) Solve your formula for h.

17. The volume of a pyramid having a square base depends on the height of the pyramid and the size of its base. A formula for this is

$$v = \frac{a^2 h}{3}$$

in which v is the volume of the pyramid, a is the length of an edge of its base, and h is its height.
a) Draw a figure of a pyramid, showing a and h.

The pyramids of Giza.

b) Use the formula above to find the volume of a pyramid whose base has edges of length 30 meters if it is 25 meters high.
c) Solve the formula for h in terms of a and v.
d) Find the height of a pyramid whose volume is 1600 cubic meters and whose base has edges of length 20 meters.

18. In 1978, the formula for the cost of sending a letter by first-class mail was

$$c = 11n + 2$$

in which c is the cost in cents and n is the weight of the letter in ounces.
a) How much does it cost to send a letter weighing 3 ounces by first-class mail?
b) Solve this formula for n in terms of c.
c) Check your answer for part a of this exercise in your formula for part b, letting $n = 3$.

19. Anthropologists use the formula

$$c = \frac{100\,w}{\ell}$$

to measure the human head. Called the cephalic index, c is expressed in terms of w, the width of the head from side to side, and ℓ, its length from front to back.

a) Find the cephalic index of someone whose head is 12 centimeters wide and 15 centimeters long.

b) Solve the formula for w in terms of c and ℓ.

c) Check your answer for part a of this exercise in your formula for part b, letting $w = 12$ and $\ell = 15$.

d) Solve the formula that you wrote for part b for ℓ in terms of c and w.

Set IV

Robert Earl Hughes, the man whose photograph appears in this lesson, once weighed 1,069 pounds. According to the formulas given in this lesson, approximately how many feet tall should a man of this weight be?

CHICAGO TRIBUNE–NEW YORK NEWS SYNDICATE, INC.

LESSON 3
Graphing Linear Equations

Broom Hilda's system for winning a game of dice by throwing a seven seems to be based on the principle, "If at first you don't succeed, try and try again." If two dice are thrown, in how many different ways can they land so that the sum of the numbers that turn up is 7?

One way to answer this question is to make a table, letting x represent the number on one die and y represent the number on the other.

x	1	2	3	4	5	6
y	6	5	4	3	2	1

The table shows that there are six ways in all.

The question that we have just answered is equivalent to asking how many different solutions that the equation

$$x + y = 7$$

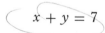

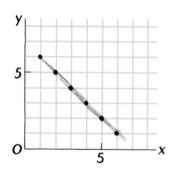

has if the solutions are restricted to pairs of positive integers. The solutions, listed in the table above, are: (1, 6), (2, 5), (3, 4), (4, 3), (5, 2), and (6, 1). If we think of each ordered pair as the coordinates of a point and plot the six points on a graph, we find that they lie along a straight line. The graph is shown at the left.

The six points in this graph show all of the solutions of the equation.

$$x + y = 7$$

that are pairs of positive integers. What would happen if we drew a line through the points and extended it beyond them? The result is at the right. This line shows where *all* of the solutions of the equation would be if we did not restrict them to pairs of positive integers. Point A, for example, has coordinates $(-2, 9)$ and

$$-2 + 9 = 7$$

Points B and C have coordinates $(2.5, 4.5)$ and $(8, -1)$:

$$2.5 + 4.5 = 7 \quad \text{and} \quad 8 + -1 = 7$$

Because the line shows where all of the solutions of the equation are, it is called the graph of the equation.

Other equations having graphs that are straight lines are:

$$x + 4y = -2, \quad 2x - 7y = 0, \quad 5y = 8, \quad \text{and} \quad y - 3x = 5$$

It is possible to prove that the graph of every equation equivalent to an equation of the form

$$ax + by = c$$

is a straight line if a and b are not both zero. For this reason, such equations are called **linear equations in two variables.** A linear equation in two variables written in the form $ax + by = c$ is said to be written in **standard form**.

EXAMPLE 1

Show that each of the equations listed in the preceding paragraph is a linear equation in two variables by writing it in the standard form of such an equation and giving the values of a, b, and c.

SOLUTION

Equation	In standard form	Values of a, b, and c
$x + 4y = -2$	$1x + 4y = -2$	$a = 1, b = 4, c = -2$
$2x - 7y = 0$	$2x + -7y = 0$	$a = 2, b = -7, c = 0$
$5y = 8$	$0x + 5y = 8$	$a = 0, b = 5, c = 8$
$y - 3x = 5$	$-3x + 1y = 5$	$a = -3, b = 1, c = 5$
$(y + -3x = 5)$		

One way to graph a linear equation is to solve the equation for y, use the resulting equation to make a table, plot the points of the table, and draw a line through them.

EXAMPLE 2
Graph the equation $y - 3x = 5$.

SOLUTION
Solving this equation for y by adding $3x$ to each side, we get

$$y = 5 + 3x$$

Choosing some values for x and using this equation to find the corresponding values for y, we get

x	-1	0	1	2
y	2	5	8	11

Plotting the points $(-1, 2)$, $(0, 5)$, $(1, 8)$, and $(2, 11)$ and connecting them with a line, we get the graph at the right.

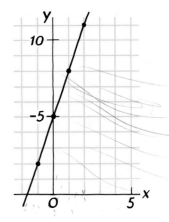

EXAMPLE 3
Graph the equation $x + 4y = 8$.

SOLUTION
Solving this equation for y, we get

$$4y = 8 - x$$

$$y = \frac{8 - x}{4}$$

Choosing some values for x and using this equation to find the corresponding values for y, we get

x	0	1	2	4	8
y	2	1.75	1.5	1	0

Plotting the five points having the pairs of numbers in this table as their coordinates, we get the graph at the right.

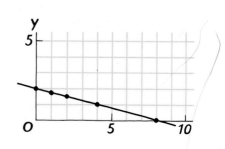

Exercises

Set I

1. Simplify each of the following expressions.
   a) $2 \cdot x \cdot x \cdot x$
   b) $2 + x + x + x$
   c) $2x + 2x + 2x$
   d) $(2x)(2x)(2x)$

2. Guess a formula for the function represented by each of these tables. Begin each formula with $y =$.

   a)
x	-4	-2	0	2	4
y	4	2	0	-2	-4

   b)
x	1	2	3	4	5
y	60	30	20	15	12

   c)
x	0	1	2	3	4
y	0	2	8	18	32

3. Helen of Troy was said to have been so beautiful that her face could launch a thousand ships. A table having this fable as its basis is shown here.

Number of times Helen looked out her window, x	1	2	3	4
Number of ships launched, y	1,000	2,000	3,000	4,000

   a) Write a formula for this function.
   b) Solve the formula for x.
   c) How do x and y vary with respect to each other?

Set II

4. Write each of the following equations in standard form and give the values of a, b, and c.
   a) $6x + y = 2$
   b) $3y + 2x = -7$
   c) $2(x + 4y) = 9$
   d) $4x - 5y = 1$
   e) $3x = 11$
   f) $2y = 8 - x$
   g) $x + 10 = 6y$
   h) $7y - 0.5x = 0$

5. Solve each of the following equations for y in terms of x.
   a) $y - 4x = 3$
   b) $9x = 1 + y$
   c) $x - 10 = 5y$
   d) $8x + 2y = 1$
   e) $3(5x + y) = 0$
   f) $6x - y = 7$

6. This exercise is about the equation $5x + 2y = 15$.
   a) Solve this equation for y in terms of x.
   b) Use your equation to make a table. Choose at least four different values for x.
   c) Plot the points in your table, connect them with a line, and extend the line across the graph.
   d) Does the point $(-1, 10)$ lie on the line?
   e) Is $(-1, 10)$ a solution to the equation $5x + 2y = 15$?
   f) Does the point $(6, -7)$ lie on the line?
   g) Is $(6, -7)$ a solution to the equation $5x + 2y = 15$?

7. What does the graph of the equation $y = 2$ look like? Writing it in standard form, we get $0x + 1y = 2$.
   a) Use the equation in this form to make a table. Choose at least four different values for x.
   b) Plot the points in your table, connect them with a line, and extend the line across the graph.
   c) Does the point $(2, 8)$ lie on the line?
   d) Is $(2, 8)$ a solution to the equation $y = 2$?
   e) Does the point $(8, 2)$ lie on the line?
   f) Is $(8, 2)$ a solution to the equation $y = 2$?

8. Use the methods described in exercises 6
   and 7 to graph the following equations.
   a) $4x + y = 8$
   b) $5y - 1 = 3x$
   c) $x - 2y = 0$
   d) $x = 6$

9. Graph the following equations on one pair
   of axes.
   a) $2x + y = 2$
   b) $2x + y = 8$

c) $2x + y = 0$
d) $2x + y = -4$
e) What do you notice about the graphs?

10. Graph the following equations on one pair
    of axes.
    a) $x - y = 6$
    b) $2x - y = 6$
    c) $3x - y = 6$
    d) $5x - y = 6$
    e) What do you notice about the graphs?

Set III

11. Write each of the following equations in
    standard form and give the values of a, b,
    and c.
    a) $x + 7y = 3$
    b) $9y + 2x = -1$
    c) $4(2x + y) = 5$
    d) $6x - 11y = 0$
    e) $5y = 9$
    f) $3x = 12 - y$
    g) $2y - 13 = -x$
    h) $8y - 1.5x = 2$

12. Solve each of the following equations for y
    in terms of x.
    a) $5x + y = 2$
    b) $4x = y - 8$
    c) $x + 9 = 3y$
    d) $2x + 7y = 10$
    e) $6(x + 2y) = 1$
    f) $9x - y = 5$

13. This exercise is about the equation
    $3x + 2y = 16$.
    a) Solve this equation for y in terms of x.
    b) Use your equation to make a table.
       Choose at least four different values
       for x.
    c) Plot the points in your table, connect
       them with a line, and extend the line
       across the graph.

d) Does the point $(-2, 11)$ lie on the line?
e) Is $(-2, 11)$ a solution to the equation
   $3x + 2y = 16$?
f) Does the point $(7, -3)$ lie on the line?
g) Is $(7, -3)$ a solution to the equation
   $3x + 2y = 16$?

14. What does the graph of the equation $y = 5$
    look like? Writing it in standard form, we
    get $0x + 1y = 5$.
    a) Use the equation in this form to make a
       table. Choose at least four different values
       for x.
    b) Plot the points in your table, connect
       them with a line, and extend the line
       across the graph.
    c) Does the point $(5, -3)$ lie on the line?
    d) Is $(5, -3)$ a solution to the
       equation $y = 5$?
    e) Does the point $(-3, 5)$ lie on the line?
    f) Is $(-3, 5)$ a solution to the equation
       $y = 5$?

15. Use the methods described in exercises 13
    and 14 to graph the following equations.
    a) $2x + y = 9$
    b) $5y - 4 = x$
    c) $x + 3y = 0$
    d) $x = -2$

16. Graph the following equations on one pair of axes.

a) $4x + y = 4$
b) $4x + y = 6$
c) $4x + y = 0$
d) $4x + y = -8$
e) What do you notice about the graphs?

17. Graph the following equations on one pair of axes.

a) $x + y = 8$
b) $x + 2y = 8$
c) $x + 4y = 8$
d) $x + 5y = 8$
e) What do you notice about the graphs?

Set IV

The graph of the equation $|x| + |y| = 4$ is a familiar geometric figure. Can you figure out what it is? (Hint: Four points on the graph are $(4, 0)$, $(0, 4)$, $(0, -4)$, $(-4, 0)$. Check these and then find other points on the graph.)

LESSON 4
Intercepts

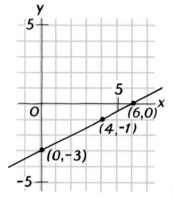

This is a problem from an algebra book written in Portuguese. The first part of the problem is to use the table to "trace a graph." Plotting the three points having the numbers in the table as their coordinates, we find that they lie along a line. The graph at the left shows the points and the line.

Notice that the line crosses the *x*-axis in the point (6, 0) and the *y*-axis in the point (0, –3). The numbers 6 and –3 are called *the* x- *and* y-*intercepts* of the line.

▶ The **intercepts** of a line are the numbers corresponding to the points in which it crosses the coordinate axes.

If we know that the graph of an equation is a line, then we can draw the line by using only its intercepts. We could, for example, have drawn the line in the graph above using only the points (6, 0) and (0, –3). The third point (4, –1) isn't needed to draw the line at all; it is useful, nevertheless, as a check on whether the line that we have drawn is correct.

Suppose that, instead of having been given a table in the problem above, we had been given an equation of the line instead. An equation for the line in standard form is

$$x - 2y = 6$$

To find the intercepts of this line, we can use the fact that, because each crossing point lies on one of the axes, *one of its coordinates is zero*. For example, the x-coordinate is zero where the graph crosses the y-axis. So, if we let $x = 0$ in this equation, we get

$$(0) - 2y = 6$$
$$-2y = 6$$
$$y = -3$$

Thus, the y-intercept is -3. Letting $y = 0$ in the equation, we get

$$x - 2(0) = 6$$
$$x - 0 = 6$$
$$x = 6$$

The x-intercept is 6.

Here are more examples of how to graph a linear equation by finding its intercepts.

EXAMPLE 1

Graph the equation $4x + 3y = 18$.

SOLUTION

First, letting $x = 0$ in this equation, we get

$$4(0) + 3y = 18$$
$$3y = 18$$
$$y = 6$$

This means that the y-intercept is 6. Thus, one crossing point of the line is $(0, 6)$.

Letting $y = 0$ in the equation, we get

$$4x + 3(0) = 18$$
$$4x = 18$$
$$x = 4.5$$

The x-intercept is 4.5; so the other crossing point is $(4.5, 0)$.

Plotting these two points and connecting them with a line, we get the graph at the right. To check our work, we can choose another point on the line to see if its coordinates fit the equation of the line. For example, the point

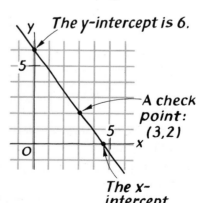

The y-intercept is 6.

A check point: (3,2)

The x-intercept is 4.5.

(3, 2) is on the line:

$$4(3) + 3(2) = 18$$
$$12 + 6 = 18$$
$$18 = 18$$

EXAMPLE 2
Graph the equation $2y = 7$.

SOLUTION
First, we want to let $x = 0$ in this equation. But x does not appear in it. We can make x appear by writing the equation in standard form:

$$0x + 2y = 7$$

Now, letting $x = 0$, we get

$$0(0) + 2y = 7$$
$$2y = 7$$
$$y = 3.5$$

The y-intercept is 3.5 and the line crosses the y-axis in the point $(0, 3.5)$.
   Letting $y = 0$ in either the original equation

$$2(0) = 7$$

or in the equation in standard form

$$0x + 2(0) = 7$$

results in a false equation:

$$0 = 7$$

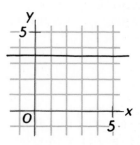

This means that y can never equal 0 in the equation $2y = 7$; so the line does not have an x-intercept. Because the line does not cross the x-axis, it must be parallel to it.
   Drawing a line fitting this description, we get the graph at the right.

It is also possible for a line not to have a y-intercept. In such a case, illustrated at the left at the top of the next page, the line is parallel to the y-axis.

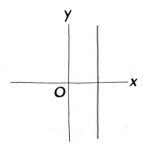

This line does not
have a *y*-intercept.

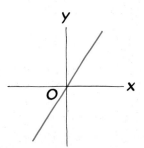

This line crosses both
axes in the same point,
the origin.

Finally, both intercepts of a line can be zero. In this case, illustrated in the right-hand figure, the line goes through the origin. To graph an equation for which both intercepts are zero, we have to choose a second point that is on neither axis.

Exercises

Set I

1. Tell whether each of the following equations is true for all values of x, false for all values of x, or true for some values of x and false for others.
   a) $x + 3x = 4x$
   b) $x + 3x = 4$
   c) $x - 1 = x - 2$
   d) $2(x - 1) = 2x - 2$
   e) $2x - 1 = 2$

2. Find the lengths of the sides of the following figures. Check your answers.

   a)
   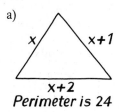
   x $x+1$
   $x+2$
   Perimeter is 24

   b)
   $3x$
   $x-1$ ⬚ $x-1$
   $3x$
   Perimeter is 30

   c)
   $x-5$
   8 ⬚ 8
   $x-5$
   Area is 5x+14

3. When Obtuse Ollie evaluated the following expression on his pocket calculator

   $$5\!\!-\!\!3 \times 7 + 8 \div 2$$

   he got 11 for his answer.
   a) Is this answer correct? If not, what is the correct answer?
   b) Rewrite the problem to show what expression Ollie evaluated.

Set II

4. What are the x- and y-intercepts of each of the lines in the following graphs?

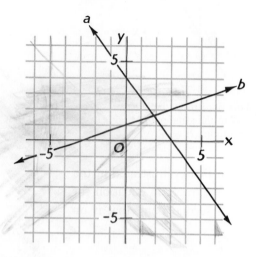

5. Find the x- and y-intercepts of the lines having the following equations.
   a) $2x + 5y = 20$ e) $4x - 9y = -18$
   b) $8x + y = 8$ f) $y = 2x + 6$
   c) $x - 3y = 12$ g) $5x + 0y = 14$
   d) $5x + 7y = 0$ h) $2y - 11 = 0$

6. Find the x- and y-intercepts of the lines having the following equations.
   a) $x - 3y = -6$
   b) $2x - 6y = -12$
   c) $-x + 3y = 6$
   d) $\dfrac{x}{3} - y = -2$

   e) What can you conclude about the graphs of these equations?
   f) Can you explain why?

7. This exercise is about the equation
$$x + \frac{y}{2} = 6$$

   a) Find the x- and y-intercepts of the graph of this equation.

b) Use the intercepts to graph the equation.

c) Does the point $(2, 8)$ lie on the line?

d) Is $(2, 8)$ a solution to the equation
$$x + \frac{y}{2} = 6?$$

e) Does the point $(8, 2)$ lie on the line?

f) Is $(8, 2)$ a solution to the equation
$$x + \frac{y}{2} = 6?$$

8. Graph the following equations by finding their intercepts. If an equation cannot be graphed using the intercepts, graph it by finding other points on the line. Use a third point on each graph to check your answers.
   a) $x + 6y = 6$
   b) $5x + 2y = -10$
   c) $\dfrac{x}{8} + \dfrac{y}{5} = 1$
   d) $\dfrac{x}{3} - \dfrac{y}{4} = 2$
   e) $7x - 7y = 0$
   f) $7x - 7 = 0$

9. Graph the following equations on one pair of axes.
   a) $x + y = 10$
   b) $x + 2y = 10$
   c) $x + 5y = 10$
   d) What do you notice about the graphs?

10. Graph the following equations on one pair of axes.
   a) $3x - 4y = 12$
   b) $3x - 4y = -12$
   c) $3x - 4y = 0$
   d) What do you notice about the graphs?

Set III

11. What are the x- and y-intercepts of each of the lines in the following graphs?

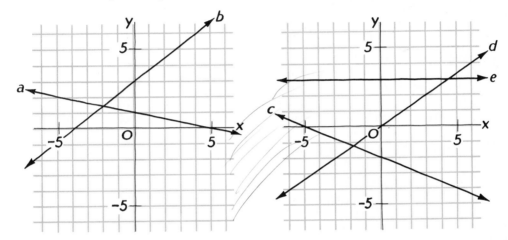

12. Find the x- and y-intercepts of the lines having the following equations.
   a) $3x + 4y = 24$
   b) $x + 7y = 7$
   c) $2x - y = 10$
   d) $6x + 5y = 0$
   e) $8x - 3y = -12$
   f) $y = 4x - 4$
   g) $0x + 2y = 15$
   h) $5x - 9 = 0$

13. Find the x- and y-intercepts of the lines having the following equations.
   a) $2x - y = -8$
   b) $4x - 2y = -16$
   c) $-2x + y = 8$
   d) $x - \dfrac{y}{2} = -4$
   e) What can you conclude about the graphs of these equations?
   f) Can you explain why?

14. This exercise is about the equation
$$x + \frac{y}{3} = 4$$
   a) Find the x- and y-intercepts of the graph of this equation.
   b) Use the intercepts to graph the equation.
   c) Does the point $(1, 9)$ lie on the line?
   d) Is $(1, 9)$ a solution to the equation
$$x + \frac{y}{3} = 4?$$
   e) Does the point $(9, 1)$ lie on the line?
   f) Is $(9, 1)$ a solution to the equation
$$x + \frac{y}{3} = 4?$$

Lesson 4: Intercepts   263

15. Graph the following equations by finding their intercepts. If an equation cannot be graphed using the intercepts, graph it by finding other points on the line. Use a third point on each graph to check your answers.
    a) $5x + y = 5$
    b) $2x + 7y = -14$
    c) $\dfrac{x}{4} + \dfrac{y}{11} = 1$
    d) $\dfrac{x}{2} - \dfrac{y}{3} = 3$
    e) $6x + 6y = 0$
    f) $6x + 6 = 0$

16. Graph the following equations on one pair of axes.
    a) $x - y = 9$
    b) $3x - y = 9$
    c) $9x - y = 9$
    d) What do you notice about the graphs?

17. Graph the following equations on one pair of axes.
    a) $2x - 5y = 10$
    b) $2x - 5y = -10$
    c) $2x - 5y = 0$
    d) What do you notice about the graphs?

Set IV

Here is another exercise from the Portuguese algebra book. Can you figure out what the questions are about and answer each one? (A comma is used in Portuguese where we would place a decimal point.)

> **33.** Os valores que constam do quadro que segue estão relacionados segundo determinada lei:
>
> | x | 0,5 | 1 | 3,5 | 4 | ... |
> |-----|-----|---|-----|---|-----|
> | y | 4 | 5 | 10 | ... | 7 |
>
> *a*) Faça a representação gráfica;
>
> *b*) Escreva a função que relaciona y com x;
>
> *c*) Preencha os espaços que faltam no quadro.

LESSON 5
Slope

There's no doubt about it. Any creatures that can walk up a slope as steep as the one in this cartoon deserve to be called sure-footed. The left-hand diagram at the bottom of this page shows some donkeys walking up hills of differing slopes. The hill represented by line "a" has the gentlest slope and the one represented by line "c" has the steepest one. In all three cases, as we look at the diagram from left to right, the lines representing the hills go *up*. They have *positive* slopes.

The second diagram shows donkeys walking down hills of differing slopes. In each case, as we look at the diagram from left to right, the lines representing the hills go *down*. These lines have *negative* slopes.

These lines have positive slopes.

These lines have negative slopes.

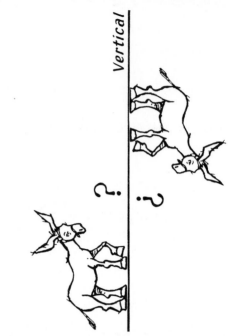

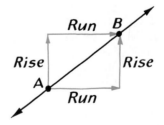

Horizontal
This line has a
slope of zero

If a line is horizontal, it goes neither up nor down. Its slope, being neither positive nor negative, is zero.

If a line is vertical, on the other hand, we often say that it goes up *and* down. Because there is no number that is both positive and negative, the slope of a vertical line is said to be undefined.

The slopes of all lines that are not vertical are given by definite numbers. To find a line's slope, the following method is generally used. First, two points on the line are chosen. As we go from the left point to the right point (the direction in which we read), we move a certain distance vertically, called the *rise,* and a certain distance to the right, called the *run.** The **slope** of the line is found by dividing the rise by the run:

$$\text{slope} = \frac{\text{rise}}{\text{run}}$$

———————

* Or we could move to the right first and then vertically.

EXAMPLE 1

Find the slope of the line through the points $(2, 1)$ and $(6, 4)$.

SOLUTION

Plotting the two points and connecting them with a line, we get the graph shown at the right. The rise in going from $(2, 1)$ to $(6, 4)$ is 3 and the run is 4. Thus the slope is $\frac{3}{4}$, or 0.75.

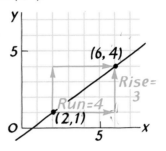

As noted earlier, a line that goes down as we read from left to right has a negative slope. The next example shows how a negative slope can be found.

EXAMPLE 2

Find the slope of the line having the equation $2x + y = 1$.

SOLUTION

First, by graphing this equation, we get the line shown at the right. Any two points on the line can be used to find its slope; it is easiest, however, to choose points whose coordinates are integers. If we choose points A and B, for example, the rise ("drop" might be a better word for this line) is –6 and the run is 3. The slope is

$$\frac{-6}{3} = -2$$

If we had used points C and D instead, the rise would have been –4 and the run 2. The slope, however, is the same as before because

$$\frac{-4}{2} = -2$$

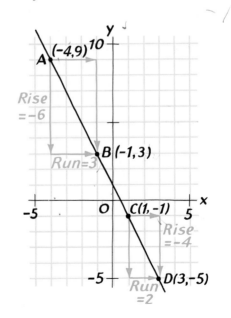

Exercises

Set I

1. Show that each of the following is a rational number by writing it as the quotient of two integers.

   a) -7 b) $1\frac{1}{2}$ c) $-\frac{4}{9}$ d) 6.1

2. Find every pair of *negative* integers that can replace x and y to make these equations true. If you think that an equation has an unlimited number of such solutions, say so. If you think that there aren't any such pairs for an equation, write "none."

   a) $x + y = -3$ c) $xy = 8$
   b) $y = 2x$ d) $y = x^2$

3. Effie Klinker's mother promised to pay her $5 for every A and $2 for every B on her report card. Although her teachers didn't give her any A's or B's, Effie managed to write 752 A's and B's altogether on her card. If x of them were A's,
   a) how many B's did she write?
   b) how much did she expect to get paid for the A's?
   c) how much did she expect to get paid for the B's?
   d) how much did she expect to get paid altogether?

Set II

4. Use the points labeled A and B to find the slopes of the lines in the graphs below.

a)
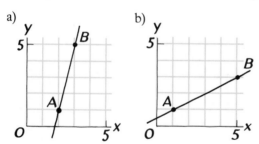

b)

c)

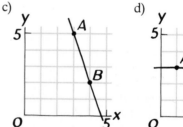

d)
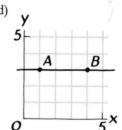

5. Plot the following points, draw a line through each pair, and find its slope.
   a) $(2, 1)$ and $(4, 7)$
   b) $(-3, 5)$ and $(1, 7)$
   c) $(0, 6)$ and $(1, 1)$
   d) $(4, -3)$ and $(4, 5)$

6. Graph the following equations and find the slope of each.
   a) $x - y = 5$
   b) $2x + y = 7$
   c) $3x - 5y = 30$
   d) $2y - 12 = 0$

7. Draw graphs of the following lines. Graph parts a and b on one set of axes and parts c and d on another set.

   a) The line through the origin having a slope of $\frac{1}{6}$.

   b) The line through the origin with a slope of -6.

   c) The line through $(-2, 5)$ with a slope of $-\frac{3}{4}$.

d) The line through $(5, -2)$ with a slope of $\frac{4}{3}$.

e) What is the relation between the lines for parts a and b? for parts c and d?

8. Graph the following linear functions.
   a) $y = 2x + 5$
   b) $y = \frac{1}{3}x + 4$
   c) $y = -1x + 6$
   d) $y = -4x - 3$
   e) $y = 5x$

9. Use your graphs for exercise 8 to give the slope and y-intercept of each function.
   a) $y = 2x + 5$
   b) $y = \frac{1}{3}x + 4$
   c) $y = -1x + 6$
   d) $y = -4x - 3$
   e) $y = 5x$

   Compare your answers for parts a through e with the equations of the functions. Without drawing their graphs, tell what you think are the slopes and y-intercepts of the following functions.
   f) $y = 7x + 2$
   g) $y = -3x - 8$

Set III

10. Use the points labeled A and B to find the slopes of the lines in the graphs below.

a)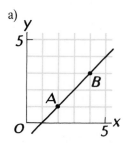

b)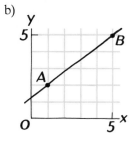

c)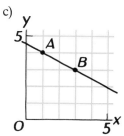

d)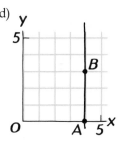

11. Plot the following points, draw a line through each pair, and find its slope.
    a) $(1, 2)$ and $(3, 10)$
    b) $(-4, 1)$ and $(2, 3)$
    c) $(-5, 8)$ and $(0, 6)$
    d) $(-3, 7)$ and $(5, 7)$

12. Graph the following equations and find the slope of each.
    a) $x - y = 2$ c) $2x - 7y = 7$
    b) $3x + y = 9$ d) $4x + 8 = 0$

13. Draw graphs of the following lines. Graph parts a and b on one set of axes and parts c and d on another set.

    a) The line through the origin with a slope of $\frac{1}{5}$.

    b) The line through the origin with a slope of -5.

c) The line through $(4, -1)$ with a slope of $-\dfrac{2}{3}$.

d) The line through $(-4, 1)$ with a slope of $\dfrac{3}{2}$.

e) What is the relation between the lines for parts a and b? for parts c and d?

14. Graph the following linear functions.
 a) $y = 3x + 1$
 b) $y = \dfrac{1}{2}x + 6$
 c) $y = -4x + 7$
 d) $y = -1x - 5$
 e) $y = \dfrac{1}{3}x$

15. Use your graphs for exercise 14 to give the slope and y-intercept of each function.
 a) $y = 3x + 1$
 b) $y = \dfrac{1}{2}x + 6$
 c) $y = -4x + 7$
 d) $y = -1x - 5$
 e) $y = \dfrac{1}{3}x$

Compare your answers for parts a through e with the equations of the functions. Without drawing their graphs, tell what you think are the slopes and y-intercepts of the following functions.
 f) $y = 5x + 3$
 g) $y = -2x - 9$

Set IV

The world record for skiing down the steepest slope is held by Sylvain Saudan. This photograph shows him skiing down the northeast side of Mont Blanc.

1. Find, as accurately as you can, the slope of the mountain in the picture.
2. Approximately how many meters downward does the mountainside go for each 100 meters in a horizontal direction?

STERLING PUBLISHING CO., INC.

LESSON 6
The Slope-Intercept Form

The amaryllis is a remarkable plant. It grows from a bulb to a height of about 18 inches in just a few weeks.

The plant shown in the pictures above grew at a rate of about 3 inches per week. If it was 4 inches high at the beginning, its height, y, while it was growing at this rate would be given by the formula

$$y = 4 + 3x$$

in which x is the number of weeks that had passed. This equation can also be written in the form

$$y = 3x + 4$$

Its graph is shown at the right. Notice that the line has a slope of 3 and that its y-intercept is 4. The equation $y = 3x + 4$, then, includes two pieces of information about the line that is its graph: its slope and its y-intercept.

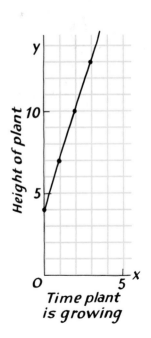

271

We know that any function that has an equation of the form

$$y = ax + b$$

in which a and b are constant numbers is a linear function and that its graph is a straight line. The equation describing the growth of the amaryllis reveals what the numbers a and b mean: a *is the slope of the line and* b *is its* y-*intercept.* Because of this, an equation of the form $y = ax + b$ is said to be in **slope-intercept** form.

An equation in slope-intercept form is very easy to graph. Here are examples of how it is done.

EXAMPLE 1

Graph the equation $y = 2x - 1$.

SOLUTION

Writing this equation in slope-intercept form, we get

$$y = 2x + -1$$

The slope of the line that is the graph of this equation is 2 and its y-intercept is -1. First, we locate the y-intercept. Then, because slope $= \dfrac{\text{rise}}{\text{run}}$, we know that the line rises 2 units for every unit that we move to the right. Plotting additional points and connecting them with a line, we get the graph at the right.

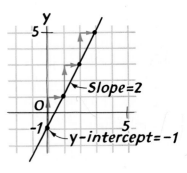

EXAMPLE 2

Graph the equation $y = -\dfrac{3}{4}x$.

SOLUTION

Writing this equation in slope-intercept form, we get

$$y = -\dfrac{3}{4}x + 0$$

The slope of the line is $-\dfrac{3}{4}$ and its y-intercept is 0. This means that it goes through the origin and, because $-\dfrac{3}{4} = \dfrac{-3}{4}$, it falls 3 units for every 4 units that we move to the right. The graph is shown at the right.

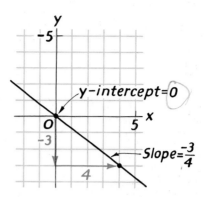

EXAMPLE 3
Graph the equation $y = 2.5$.

SOLUTION
This equation written in slope-intercept form is

$$y = 0x + 2.5$$

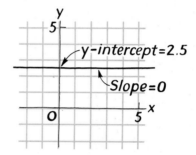

The slope of the line is 0 and its y-intercept is 2.5. This means that the line is horizontal. Its graph is shown at the right.

Exercises

Set I

1. Simplify each of the following expressions.
   a) $2x + (2 + x)$
   b) $2x + (2 - x)$
   c) $2x(2 + x)$
   d) $2x(2 - x)$

2. Solve each of these equations for the variable indicated.

   $x - y = 10 \qquad 3x = y + 2$
   a) For x c) For x
   b) For y d) For y

3. An integer is *even* if it is two times some integer. It is *odd* if it is not.
   a) Which of the following integers are even and which are odd?

$$50 \qquad -5 \qquad 0$$

If x represents an even integer, what type of integer does
b) $x + 1$ represent?
c) $x + 2$ represent?
d) $x - 3$ represent?

If x represents an odd integer, what type of integer does
e) $x + 1$ represent?
f) $x + 2$ represent?
g) $x - 4$ represent?

Set II

4. The following questions are about the equation $y = ax + b$, in which a and b are constant numbers.
   a) What are functions that have equations of this form called?
   b) What kind of graphs do such functions have?
   c) What does the number a represent?
   d) What does the number b represent?

5. What are the slopes and y-intercepts of the lines having the following equations? (If necessary, first write the equation in slope-intercept form.)
   a) $y = 8x + 3$
   b) $y = \frac{1}{2}x + 5$
   c) $y = x - 7$
   d) $y = 6(x - 2)$
   e) $y = x + x$
   f) $y = -x$
   g) $y = 10 - 3x$
   h) $y = 4$

6. Write equations for the following lines.
   a) The line that has a slope of 9 and a y-intercept of 1.
   b) The line that has a slope of $\frac{2}{3}$ and a y-intercept of –6.
   c) The line that has a slope of –4 and contains the point $(0, 7)$.
   d) The line that has a slope of 2.5 and goes through the origin.
   e) The line that has a slope of 0 and intersects the y-axis at –8.
   f) The x-axis.

7. Write equations for the lines labeled a, b, c, and d in the graph below.

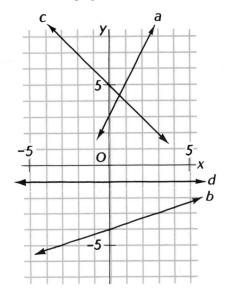

8. Graph the following equations by using their y-intercepts and slopes.
   a) $y = 2x + 5$
   b) $y = 5x + 2$
   c) $y = -3x + 4$
   d) $y = -3x - 4$
   e) $y = \frac{5}{3}x + 1$
   f) $y = \frac{3}{5}x + 1$

9. A point moves so that its y-coordinate is always 5 more than its x-coordinate.
   a) Write an equation for its path.
   b) Graph its path.
   c) Where does it cross the x-axis?
   d) Where does it cross the y-axis?
   e) What is its slope?

10. This exercise is about the rectangle shown below.

a) Write an equation for its perimeter, y, as a function of x.
b) Graph this function.
c) Where does it intersect the x-axis?
d) Where does it intersect the y-axis?
e) What is its slope?

Set III

11. The following questions are about the equation $y = ax$, in which a is a constant number.
   a) What are functions that have equations of this form called?
   b) What kind of graphs do such functions have?
   c) What does the number a represent?
   d) Where do the graphs of such functions intersect the y-axis?

12. What are the slopes and y-intercepts of the lines having the following equations? (If necessary, first write the equation in slope-intercept form.)
   a) $y = 5x + 2$
   b) $y = \frac{1}{3}x + 9$
   c) $y = x - 4$
   d) $y = 2(x - 7)$
   e) $y = x + x + x$
   f) $y = -6x$
   g) $y = 8 - x$
   h) $y = 11$

13. Write equations for the following lines.
   a) The line that has a slope of 4 and a y-intercept of 7.
   b) The line that has a slope of $\frac{3}{5}$ and a y-intercept of –2.
   c) The line that has a slope of –9 and contains the point $(0, 1)$.
   d) The line that has a slope of 0 and intersects the y-axis at –5.
   e) The line that has a slope of 1.6 and goes through the origin.
   f) The line that goes through the point $(0, 3)$ whose slope is undefined.

14. Write equations for the lines labeled a, b, c, and d in the graph below.

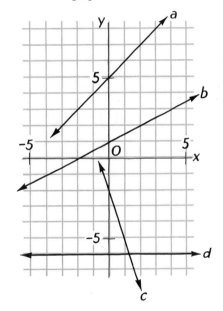

15. Graph the following equations by using their y-intercepts and slopes.
   a) $y = 3x + 4$
   b) $y = 4x + 3$
   c) $y = -2x + 5$
   d) $y = -2x - 5$
   e) $y = \frac{4}{7}x - 1$
   f) $y = \frac{7}{4}x - 1$

16. A point moves so that its y-coordinate is always 2 less than its x-coordinate.

   a) Write an equation for its path.
   b) Graph its path.
   c) Where does it cross the x-axis?
   d) Where does it cross the y-axis?
   e) What is its slope?

17. This exercise is about the rectangle shown below.

   a) Write an equation for its area, y, as a function of x.
   b) Graph this function.
   c) Where does it cross the x-axis?
   d) Where does it cross the y-axis?
   e) What is its slope?

Set IV

The price that the Marx Brothers Bakery charges a grocery store for a carton of its animal crackers depends on how many cartons that the store buys. If the store buys 10 cartons or more, the price is $4 per carton. Otherwise, the price per carton is given by the formula

$$y = \frac{26 - x}{4}$$

in which y is the price in dollars and x is the number of cartons bought.

   1. Draw a graph showing the price charged per carton as a function of the number of cartons bought. Let x vary from 1 to at least 15.
   2. Describe the slope of the graph as you look from left to right.
   3. What can you say about the intercepts of the graph?

Summary and Review

In this chapter, we have learned how to find solutions to equations in two variables and how to graph linear equations in two variables.

Equations in Two Variables (*Lesson 1*) Solutions to equations in two variables are written as ordered pairs. Such equations usually have an unlimited number of solutions. However, if the solutions are restricted to certain types of numbers, their number may be limited.

Formulas (*Lesson 2*) Equations containing two or more variables that are used to solve practical problems are called formulas. It is frequently convenient, before using a formula, to solve it for a particular variable in terms of the others.

Graphing Linear Equations (*Lesson 3*) The graph of any equation having the form $ax + by = c$ is a straight line unless both a and b are zero. Such equations are called linear equations in two variables. One way to graph a linear equation in two variables is to solve it for y, use the resulting equation to make a table, plot the points of the table, and draw a line through them.

Intercepts (*Lesson 4*) The intercepts of a line are the numbers corresponding to the points in which it crosses the coordinate axes. The intercepts can be used to draw the graph of the line.

Slope (*Lesson 5*) The slope of a line is:

> positive if it goes up as we look at it from left to right,
> negative if it goes down,
> zero if it is horizontal, and
> undefined if it is vertical.

It is found by dividing the rise by the run.

The Slope-Intercept Form (*Lesson 6*) A linear equation written in the form $y = ax + b$ is said to be in slope-intercept form: its slope is a and its y-intercept is b. Its graph can be easily determined from these two numbers.

Exercises

Set I

1. Find the values of the following expressions when x and y are replaced by the ordered pairs given.

$$x - 3y \qquad\qquad x^2 + y$$

a) $(15, 1)$ d) $(-7, 0)$
b) $(-2, 7)$ e) $(5, -8)$
c) $(0, -4)$ f) $(-2, 13)$

2. Tell whether or not each of the following ordered pairs is a solution of the equation given.

$$x - 3y = 12 \qquad\qquad x^2 + y = 17$$

a) $(15, 1)$ d) $(-7, 0)$
b) $(-2, 7)$ e) $(5, -8)$
c) $(0, -4)$ f) $(-2, 13)$

3. Find every pair of positive integers that can replace x and y in each of the following equations to make it true. If you think that an equation has an unlimited number of such solutions, say so. If you think that there aren't any such solutions, write "none."

a) $xy = 8$
b) $x = y - 2$
c) $4x + y = 14$
d) $x + y^2 = 1$

4. Solve each of the following equations for the variable indicated.

$$3x = y + 7 \qquad\qquad 6x + y = 1$$

a) For x c) For x
b) For y d) For y

5. The area of a triangle whose base has length b and whose altitude has length h is given by the formula

$$a = \frac{bh}{2}$$

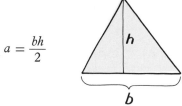

a) Solve this formula for h.
b) Solve this formula for b.
c) Use the formula above to find the area of a triangle whose base is 6 and whose altitude is 4.
d) Check your answer for part c of this exercise in your formula for part a, letting $b = 6$ and $h = 4$.
e) Check your answer for part c of this exercise in your formula for part b.

6. The Silver Shadow Car Rental Company charges for both the number of days that a car is rented and the number of miles that it

COURTESY OF ROLLS ROYCE.

is driven. A formula for the cost in dollars, c, of renting a car from the company is

$$c = 100\,d + 1.5\,m$$

in which d is the number of days and m is the number of miles.

a) How much money does the company charge per day?

b) How much money does it charge per mile?
c) Solve the formula above for d.
d) For how many days would a movie studio have rented a car if it had been driven 200 miles and the rental cost was $700?

7. Write each of the following equations in standard form and give the values of a, b, and c.
a) $8x = y + 10$ c) $x - 7y = 9$
b) $3(x + 2y) = 4$ d) $5x + 1 = 0$

8. Solve each of the following equations for y in terms of x.
a) $6x + y = 12$
b) $x - 8 = 5y$
c) $2x + 7y = 1$
d) $4x - y = 9$

9. Find the x- and y-intercepts of the lines having the following equations.
a) $3x + 5y = 45$
b) $2x - y = 7$
c) $x + 6y = 0$
d) $\dfrac{x}{4} + \dfrac{y}{9} = 1$

10. Plot the following points, draw a line through each pair, and find its slope.
a) $(2, 7)$ and $(6, 10)$
b) $(-1, 6)$ and $(1, -4)$
c) $(5, -3)$ and $(-5, -3)$

11. Draw graphs of the following lines.
a) The line through the origin having a slope of $\dfrac{3}{4}$.
b) The line through the origin having a slope of $\dfrac{4}{3}$.
c) The line through $(1, 8)$ having a slope of -2.
d) The line through $(1, -8)$ having a slope of 2.

$$y = \frac{12 - 6x}{4}$$

12. What are the slopes and y-intercepts of the lines having the following equations?
   a) $y = 6x - 2$

   b) $y = \frac{1}{4}x$

   c) $y = 7 - x$
   d) $y = 3$

13. Write equations for the lines labeled a, b, c, and d in the graph at the right.

14. Graph the following equations.
   a) $6x + 4y = 12$
   b) $2x - y = -10$ e) $y = \frac{2}{5}x - 1$
   c) $x + 5y = 0$
   d) $y = -3x + 7$ f) $x - 6 = 0$

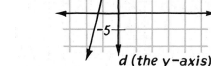

Set II

1. Find the values of the following expressions if x and y are replaced by the ordered pairs given.

$5x + y$	$x^2 - y^2$
a) $(2, 4)$	d) $(6, 2)$
b) $(-3, 7)$	e) $(-3, 0)$
c) $(5, -11)$	f) $(-5, -4)$

2. Tell whether or not each of the following ordered pairs is a solution of the equation given.

$5x + y = 14$	$x^2 - y^2 = 9$
a) $(2, 4)$	d) $(6, 2)$
b) $(-3, 7)$	e) $(-3, 0)$
c) $(5, -11)$	f) $(-5, -4)$

3. Find every pair of positive integers that can replace x and y in each of the following equations to make it true. If you think that an equation has an unlimited number of such solutions, say so. If you think that there aren't any such solutions, write "none."
   a) $x + 7y = 20$ c) $xy = 12$
   b) $y = 3x$ d) $x^2 + y = 0$

4. Solve each of the following equations for the variable indicated.
   $$3x = y + 7 \qquad x + 4y = 9$$
   a) For x c) For x
   b) For y d) For y

5. The density of a substance having mass m and volume v is given by the formula

   $$d = \frac{m}{v}$$

   a) Solve this formula for m.
   b) Solve the formula that you have just written for v.
   c) Use the formula above to find the density of a rock whose mass is 7.5 kilograms and whose volume is 2.5 liters.
   d) Check your answer for part c of this exercise in your formula for part a, letting $m = 7.5$ and $v = 2.5$.
   e) Check your answer for part c of this exercise in your formula for part b.

B.C. BY PERMISSION OF JOHNNY HART AND FIELD ENTERPRISES, INC.

6. Suppose that the person in this cartoon decided to go on her diet when she reached x pounds. A formula for her weight in pounds, w, while on the diet would be

$$w = x - 3y$$

in which x is her weight at the beginning of the diet and y is the number of weeks that she has been on it.

a) How many pounds would she lose per week?
b) If she weighed 156 pounds at the beginning of the diet, how much would she weigh after 10 weeks?
c) Solve the formula above for y.
d) Find the number of weeks she would have to stay on the diet in order to weigh 120 pounds if she weighed 180 pounds at the beginning.

---

7. Write each of the following equations in standard form and give the values of a, b, and c.
   a) $6x = y + 12$
   b) $2(3x + y) = 5$
   c) $4x - y = -1$
   d) $7y - 9 = 0$

8. Solve each of the following equations for y in terms of x.
   a) $y - 5x = 10$
   b) $x + 3 = 4y$
   c) $6x + 2y = 9$
   d) $8x - y = 1$

9. Find the x- and y-intercepts of the lines having the following equations. If a given equation does not have either an x- or a y-intercept, tell which one it lacks.
   a) $2x + 7y = 28$
   b) $5x - y = 9$
   c) $\dfrac{x}{8} + \dfrac{y}{3} = 1$
   d) $4x - 12 = 0$

10. Plot the following points, draw a line through each pair, and find its slope.
    a) $(4, 0)$ and $(9, 2)$
    b) $(-2, 5)$ and $(0, -3)$
    c) $(-6, 1)$ and $(-6, -1)$

11. Draw graphs of the following lines.
    a) The line through the origin having a slope of $\dfrac{2}{5}$.
    b) The line through the origin having a slope of $\dfrac{5}{2}$.
    c) The line through $(3, 6)$ having a slope of -4.
    d) The line through $(-3, 6)$ having a slope of 4.

12. What are the slopes and y-intercepts of the lines having the following equations?
    a) $y = 4x - 5$
    b) $y = \dfrac{1}{2}x$
    c) $y = 11 - x$
    d) $y = 8$

13. Write equations for the lines labeled a, b, c, and d in the graph below.

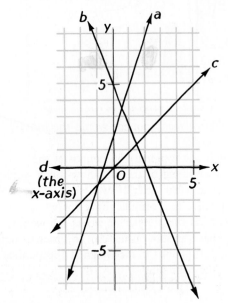

14. Graph the following equations.

a) $4x + 5y = 20$

b) $x - 3y = 9$

c) $2x + 6y = 0$

d) $y = -5x + 8$

e) $y = \dfrac{4}{3}x - 2$

f) $y + 7 = 0$

Chapter 7

SIMULTANEOUS EQUATIONS

LESSON 1
Simultaneous Equations

Don Koehler

David Frost

Mihaly Mesyaros

AMERICAN BROADCASTING COMPANY.

The tallest man in the world is reported to be Don Koehler of Chicago, Illinois. He is shown in this photograph with Mihaly Mesyaros, who works for the Ringling Brothers and Barnum and Bailey Circus and is one of the world's shortest men. Can you figure out how tall each man is from the following clues? The sum of their heights is 131 inches; the difference of their heights is 65 inches.

If we let x and y represent Mr. Koehler's and Mr. Mesyaros's respective heights in inches, these clues can be translated into the equations

$$x + y = 131$$
$$x - y = 65$$

Pictures illustrating these equations are shown below.

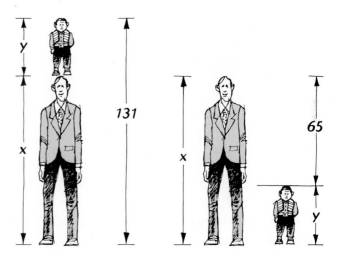

Because we are working with two equations at the same time, we will call them **simultaneous equations**. To solve them, we must find an ordered pair of numbers that can replace the variables in *both* equations to make them true. One way to do this is suggested by the picture below. Although it is merely the two pictures on the facing page drawn as one, it suggests another equation:

$$x + x = 131 + 65$$

This equation contains just one variable and is easy to solve:

$$2x = 196$$
$$x = 98$$

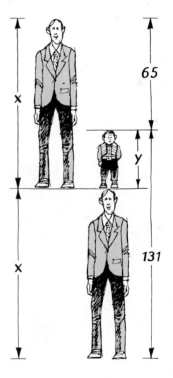

Mr. Koehler is 98 inches, or 8 feet 2 inches, tall.

We can replace x by 98 in either of the original equations to find y:

$$98 + y = 131$$
$$y = 33$$

Mr. Mesyaros is 33 inches, or 2 feet 9 inches, tall.
Checking our answers, we find that

$$98 + 33 = 131$$

and

$$98 - 33 = 65$$

They are correct.

We found the solution to the simultaneous equations

$$x + y = 131$$
$$x - y = 65$$

to be (98, 33) by writing a third equation,

$$x + x = 131 + 65$$

This equation was suggested by a picture illustrating the first two. Is it possible to write this equation without looking at a picture?

The answer is yes. We can get this equation by adding the left sides of the first two equations, adding their right sides, and setting the sums equal.

$$
\begin{aligned}
x + y &= 131 \\
+\quad x - y &= 65 \\
\hline
2x &= 196
\end{aligned}
$$

If the first two equations are true, then this equation must also be true.

We will call this method for solving a pair of simultaneous equations the **addition method**. Here is another example of how it is used.

EXAMPLE

Write a pair of simultaneous equations, the first in terms of addition and the second in terms of subtraction, relating the lengths of the line segments in these diagrams. Then solve the equations for x and y.

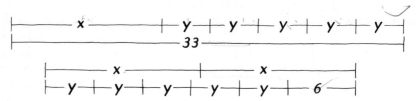

SOLUTION
The equations are:

$$
\begin{aligned}
x + 5y &= 33 \\
2x - 5y &= 6
\end{aligned}
$$

Adding the left and right sides of these equations to make a new equation, we get

$$
\begin{aligned}
x + 5y &= 33 \\
+\quad 2x - 5y &= 6 \\
\hline
3x &= 39
\end{aligned}
$$

If $3x = 39$, then $x = 13$.

To find y, we substitute this value for x in either of the original equations. Substituting it for x in the first equation

$$x + 5y = 33$$

we get

$$
\begin{aligned}
13 + 5y &= 33 \\
5y &= 20 \\
y &= 4
\end{aligned}
$$

Finally, checking these values for x and y in the second equation

$$2x - 5y = 6$$

we get

$$2(13) - 5(4) = 6$$
$$26 - 20 = 6$$
$$6 = 6$$

This shows that our solution, $(13, 4)$, is correct.

Exercises

Set I

1. Simplify each of the following expressions.
   a) $3x + 3x$
   b) $3x - 3x$
   c) $7x + x$
   d) $7x - x$
   e) $5x - 4x$
   f) $4x - 5x$

2. Which of these symbols, $>$, $=$, or $<$, should replace ▓ in each of the following to make it true for all values of x? If none will make the relation true for all values of x, say so.

   a) $x + 4$ ▓ x
   b) $x + 4$ ▓ 2
   c) $x + x$ ▓ $2x$
   d) $x - 3$ ▓ $x - 1$
   e) $x - 3$ ▓ 0
   f) $x - x$ ▓ 0

3. A bumblebee can fly at a speed of 50 meters per minute in still air.
   a) How fast can it fly *with* a wind having a speed of w meters per minute?
   b) How fast can it fly *against* a wind having a speed of w meters per minute?
   c) How far can it fly in x minutes in still air?
   d) How long would it take to fly y meters in still air?

Set II

4. Tell whether or not each of the following ordered pairs is a solution of the simultaneous equations given. (Remember that, to be a solution, the ordered pair must make *both* equations true.)

   $$x + 3y = 10$$
   $$4x - y = 14$$
   a) $(4, 2)$
   b) $(1, 3)$
   c) $(-5, 5)$

   $$x + y = 2$$
   $$x^2 + y^2 = 4$$
   d) $(1, 1)$
   e) $(2, 0)$
   f) $(3, -1)$

   $$x - 2y = 6$$
   $$y = \frac{1}{2}x - 3$$
   g) $(6, 0)$
   h) $(0, -3)$
   i) $(-10, -8)$

5. This exercise is about the simultaneous equations

$$x + 2y = 11$$
$$x - 2y = 3$$

a) Write the equation that results from adding these equations.
b) Solve this equation for x.
c) Substitute your solution for x in the first equation and solve it for y.
d) Write the solution to the simultaneous equations that you have found as an ordered pair.
e) Check your solution by finding out whether it makes the second equation true.

6. The two diagrams below illustrate a pair of simultaneous equations.

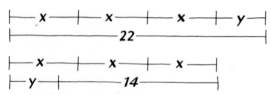

a) Write an equation for the first diagram in terms of addition.
b) Write an equation for the second diagram in terms of subtraction.
c) Solve the equations for x and y.
d) Use the diagrams to check your solution.

Solve the following simultaneous equations. Show your steps and check your answers.

9. $x + y = 21$
   $x - y = 9$

10. $3x + 4y = 2$
    $3x - 4y = 10$

11. $x + y = 3$
    $y - x = 25$

12. $7x - y = 8$
    $3x + y = 42$

13. $8x + 3y = 23$
    $4x - 3y = 7$

14. $5x - 9y = 55$
    $9y + x = 11$

The diagrams below represent pairs of simultaneous equations. Use the method described in exercise 6 to find the lengths of the line segments in each diagram.

7.

8.

Set III

15. Tell whether or not each of the following ordered pairs is a solution of the simultaneous equations given. (Remember that, to be a solution, the ordered pair must make *both* equations true.)

$$5x + y = 16$$
$$x - 2y = 1$$

a) $(3, 1)$
b) $(1, 11)$
c) $(4, -4)$

$$x + y = 3$$
$$x^2 + y^2 = 9$$

d) $(1, 2)$
e) $(0, 3)$
f) $(-2, 5)$

$$2x - y = 10$$
$$x = \frac{1}{2}y + 5$$

g) $(5, 0)$
h) $(0, -10)$
i) $(-2, -14)$

16. This exercise is about the simultaneous equations

$$3x + y = 18$$
$$3x - y = 12$$

a) Write the equation that results from adding these equations.
b) Solve this equation for x.
c) Substitute your solution for x in the first equation and solve it for y.
d) Write the solution to the simultaneous equations that you have found as an ordered pair.
e) Check your solution by finding out whether it makes the second equation true.

17. The two diagrams below illustrate a pair of simultaneous equations.

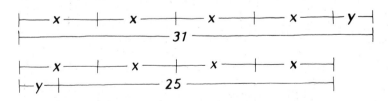

a) Write an equation for the first diagram in terms of addition.
b) Write an equation for the second diagram in terms of subtraction.
c) Solve the equations for x and y.
d) Use the diagrams to check your solution.

The diagrams below represent pairs of simultaneous equations. Use the method described in exercise 17 to find the lengths of the line segments in each diagram.

18. ├── x ──┼── x ──┼── x ──┼─y─┼─y─┼─y─┼─y─┼─y─┤
    ├──────────────── 52 ────────────────┤

    ├── x ──┼── x ──┼── x ──┼── x ──┤
    ├─y─┼─y─┼─y─┼─y─┼─y─┼── 11 ──┤

19. ├── x ──┼── y ──┼── y ──┼── y ──┼── y ──┼── y ──┼── y ──┤
    ├──────────────── 58 ────────────────┤

    ├─x─┼─x─┼─x─┼─x─┼─x─┼─x─┼─x─┼─x─┤
    ├── y ──┼── y ──┼── y ──┼── y ──┼── y ──┼── y ──┼─5─┤

Solve the following simultaneous equations. Show your steps and check your answers.

20. $x + y = 37$
    $x - y = 13$

21. $5x + 3y = 9$
    $5x - 3y = 21$

22. $x - y = 11$
    $y + x = 5$

23. $8x - y = 22$
    $2x + y = 28$

24. $5x + 7y = -14$
    $x - 7y = 14$

25. $4x + y = 1$
    $11y - 4x = 35$

Set IV

The addition method can be used to solve simultaneous equations that contain more than two variables. Can you figure out the lengths of the line segments in the following diagrams by writing and solving three simultaneous equations in three variables? (Hint: Try adding the equations in pairs.)

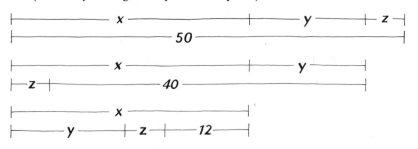

LESSON 2
Solving by Subtraction

Can you solve the following puzzle? Three cats and one kitten weigh 24 pounds, whereas three cats and five kittens weigh 36 pounds. How much does each cat and kitten weigh?

The four extra kittens together evidently weigh 12 pounds. If they have equal weights, each one must weigh 3 pounds. Knowing that each kitten weighs 3 pounds, we can conclude that the three cats together weigh 21 pounds. If they have equal weights, each one weighs 7 pounds.

The method we have used to solve this puzzle illustrates another way to solve a pair of simultaneous equations. If we let x represent the weight of a cat and y represent the weight of a kitten, then the puzzle can be translated into the equations

$$3x + y = 24$$
$$3x + 5y = 36$$

The conclusion that $4y = 12$ can be obtained by *subtracting* the first equation from the second:

$$\begin{array}{r} 3x + 5y = 36 \\ \underline{3x + y = 24} \\ 4y = 12 \end{array}$$

Solving the equation $4y = 12$, we get $y = 3$. Substituting this value for y in the first equation, $3x + y = 24$, gives

$$3x + 3 = 24$$
$$3x = 21$$
$$x = 7$$

This method for solving a pair of simultaneous equations is called the **subtraction method**. It has as its basis the fact that, if two equations are true, the equation that results from subtracting their left sides, subtracting their right sides, and setting the differences equal is also true.

Here is another example of the subtraction method.

EXAMPLE

Use the subtraction method to solve the simultaneous equations

$$6x - 4y = 42$$
$$x - 4y = 17$$

SOLUTION

Before subtracting, it may be helpful to express each equation in terms of addition:

$$6x + -4y = 42$$
$$x + -4y = 17$$

Subtracting the second equation from the first, we get:

$$
\begin{array}{r}
6x + -4y = 42 \\
-\quad \underline{x + -4y = 17} \\
5x \qquad\; = 25 \\
x = \;5
\end{array}
$$

Substituting this value for x in the first equation, $6x + -4y = 42$, we get

$$6(5) + -4y = 42$$
$$30 + -4y = 42$$
$$-4y = 12$$
$$y = -3$$

Finally, checking these values for x and y in the second equation, $x + -4y = 17$, we get

$$5 + -4(-3) = 17$$
$$5 + 12 \quad = 17$$
$$17 \quad\;\; = 17$$

This shows that our solution, $(5, -3)$, is correct.

Exercises

Set I

1. The following questions are about the lines shown in this graph.

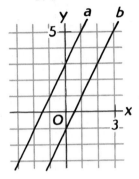

   a) What is the slope of each line?
   b) What is the y-intercept of each line?
   c) What is the equation of line A?
   d) What is the equation of line B?

2. Write each of the following expressions as a sum.
   a) $x - y$ c) $x - y - z$
   b) $y - x$ d) $z - y - x$

3. The Engulf and Devour Corporation pays its employees $10 an hour but fines them 50 cents for each minute that they are late to work.
   a) Write a formula for the weekly pay, p, of someone who works h hours in a given week and is late a total of m minutes.
   b) If Minnie works 40 hours in one week and is late a total of 30 minutes, what is her pay for that week?
   c) Solve your formula of part a of this exercise for h.
   d) If Max is late a total of 80 minutes in one week but wants to earn $450, how many hours must he work?

Set II

4. This exercise is about the diagrams at the right.
   a) By comparing the two diagrams, figure out the length of one of the segments labeled x.
   b) Use your answer to part a to figure out the length of one of the segments labeled y.
   c) Write a pair of simultaneous equations for the diagrams.
   d) Use the subtraction method to solve the equations for x and y.

$$\vdash x \dashv \vdash x \dashv \vdash y \dashv$$
$$\vdash\!\!-\!\!-\!\!-\ 17\ -\!\!-\!\!-\!\!\dashv$$

$$\vdash x \dashv \vdash y \dashv$$
$$\vdash\!\!-\!\!-\ 12\ -\!\!-\!\!\dashv$$

The diagrams below represent pairs of simultaneous equations. Write the equations and solve them to find the lengths of the line segments in each diagram.

5. |— x —|— x —|— x —|— x —|— y —|— y —|— y —|
   |————————————— 57 —————————————|

   |— x —|— x —|— x —|— x —|— y —|
   |————————— 43 —————————|

6. |——— x ———|——— x ———|——— x ———|——— x ———|——— x ———|
   |— y —|— y —|——————————————— 36 ———————————————|

   |——— x ———|
   |— y —|— y —|—2—|

7. This exercise is about the simultaneous equations

$$3x + y = 21$$
$$3x - y = 9$$

Show that each of the following methods for solving these equations results in the same solution.
a) Adding the equations.
b) Subtracting the second equation from the first.
c) Subtracting the first equation from the second.

Solve the following simultaneous equations by subtraction. Show your steps and check your answers.

8. $3x + 8y = 20$
   $3x + y = 13$

9. $4x + 5y = 3$
   $x + 5y = -18$

10. $9x + 7y = 51$
    $9x - 3y = 81$

11. $6x - y = 15$
    $2x - y = 5$

12. $11x - 5y = 23$
    $11x + y = -31$

Solve the following simultaneous equations by either addition or subtraction. Check your answers.

13. $12x + y = 4$
    $8x - y = -64$

14. $5x + y = 21$
    $5x + 9y = 9$

15. $2x + 17y = 50$
    $-2x + 3y = -50$

16. $3x - 4y = 12$
    $10x - 4y = 61$

Set III

17. This exercise is about the diagrams below.

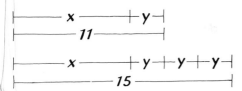

a) By comparing the two diagrams, figure out the length of one of the segments labeled y.

b) Use your answer to part a to figure out the length of one of the segments labeled x.

c) Write a pair of simultaneous equations for the diagram.

d) Use the subtraction method to solve the equations for x and y.

The diagrams below represent pairs of simultaneous equations. Write the equations and solve them to find the lengths of the line segments in each diagram.

18.
```
├─ x ─┼─ x ─┼─ x ─┼─ x ─┼─ x ─┼─ y ─┼─ y ─┤
├──────────────── 62 ────────────────┤
├─ x ─┼─ x ─┼─ y ─┼─ y ─┤
├──────── 38 ────────┤
```

19.
```
├─ x ─┼─ x ─┼─ x ─┼─ x ─┼─ y ─┤
├──────── 23 ────────┤
├─ x ─┼─ x ─┼─ x ─┼─ x ─┤
├─ y ┼ y ┼ y ┼ y ┼ y ┼ y ┼ 2 ┤
```

20. This exercise is about the simultaneous equations

$$x + 4y = 20$$
$$x - 4y = 4$$

Show that each of the following methods for solving these equations results in the same solution.

a) Adding the equations.

b) Subtracting the second equation from the first.

c) Subtracting the first equation from the second.

Solve the following simultaneous equations by subtraction. Show your steps and check your answers.

21. $2x + 5y = 41$
$2x + y = 13$

22. $9x + 4y = 2$
$3x + 4y = -10$

23. $7x + 3y = -32$
$7x - 9y = -44$

24. $8x - y = -20$
$x - y = 8$

25. $3x - 10y = 35$
$3x + 4y = -14$

Solve the following simultaneous equations by either addition or subtraction. Check your answers.

26. $6x - y = 63$
    $11x + y = 73$

27. $4x + 7y = -10$
    $4x + y = -22$

28. $10x + 8y = -57$
    $-10x - 5y = 30$

29. $2x - 9y = 25$
    $13x - 9y = 14$

Set IV

Tweedledum said to Tweedledee: "The sum of your weight and twice mine is 361 pounds." Tweedledee said to Tweedledum: "Contrariwise, the sum of your weight and twice mine is 362 pounds."

How much does each one weigh?

THROUGH THE LOOKING GLASS BY LEWIS CARROLL.

GABRIEL MOULIN STUDIOS, SAN FRANCISCO, 1935.

LESSON 3
More on Solving by
Addition and Subtraction

The Golden Gate Bridge is one of the longest bridges in the world. Extending more than a mile across San Francisco Bay, it is more than twice the length of the longest suspension bridge in existence at the time that it was being planned. This photograph shows the bridge at an early stage of its construction. At this point, catwalks had been hung between the two towers for the workmen, but the cables had not yet been built.

A lot of mathematics must be worked out before work on a structure as immense as this one can begin. To make sure that the bridge would be able to withstand storms from the Pacific, one of the engineers on the project had to solve a system of thirty-three algebraic equations containing from six to thirty variables each!*

* *Golden Gate* by Allen Brown (Doubleday, 1965), p. 35.

From lessons 1 and 2 of this chapter, you learned two ways of solving certain pairs of equations in two variables. The equations are either added or subtracted to produce a third equation that contains just one of the variables.

Here is a pair of simultaneous equations that cannot be immediately solved by either of these methods.

$$4x + y = 28$$
$$2x + 3y = 24$$

Adding these equations results in the equation

$$6x + 4y = 52$$

Subtracting them (the second from the first) results in the equation

$$2x - 2y = 4$$

Although these are both perfectly good equations—they follow from the original equations and give us more clues about x and y—they still contain both variables.

The reason that addition and subtraction do not eliminate x or y is that the original equations contain different numbers of x's (4 and 2) and different numbers of y's (1 and 3). One way to get around this difficulty is to multiply one or both of the equations by numbers such that the resulting equations *do* contain equal numbers of x's or y's. We can do this because we know that an equation can be transformed into an equivalent equation by multiplying each side by the same number other than zero.

Look again at the two equations.

$$4x + y = 28$$
$$2x + 3y = 24$$

If we multiply each side of the second one by 2,

$$2(2x + 3y) = 2(24)$$

we get the equation

$$4x + 6y = 48$$

This equation contains 4 x's, the same number as the first equation. What we have done is illustrated by the diagrams below.

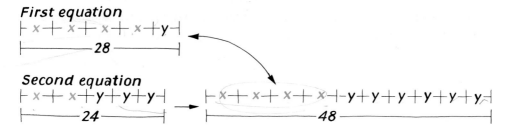

First equation

Second equation

Subtracting the first equation from the new one produces an equation without any x's:

$$\begin{array}{r} 4x + 6y = 48 \\ - \quad 4x + y = 28 \\ \hline 5y = 20 \end{array}$$

Because $5y = 20$, $y = 4$. To find x, we can substitute this value for y in *any one* of the equations that contains both x and y. For example,

$$\begin{array}{rcl} 4x + y &=& 28 \\ 4x + 4 &=& 28 \\ 4x &=& 24 \\ x &=& 6 \end{array}$$

It is easy to confirm that $(6, 4)$ is a solution of both of the original equations.

Here are more examples of how multiplication can be used before addition or subtraction to solve a pair of simultaneous equations.

EXAMPLE 1

Solve the simultaneous equations

$$\begin{array}{rcl} 4x + y &=& 28 \\ 2x + 3y &=& 24 \end{array}$$

by eliminating y first instead of x.

SOLUTION

The first equation contains $1\,y$ and the second contains $3\,y$'s, and so we multiply the first equation by 3:

$$\begin{array}{rcl} 3(4x + y) &=& 3(28) \\ 12x + 3y &=& 84 \end{array}$$

Subtracting the second equation from it,

$$\begin{array}{r} 12x + 3y = 84 \\ - \quad 2x + 3y = 24 \\ \hline 10x \qquad = 60 \end{array}$$

we find that $10x = 60$, and so $x = 6$. Substituting this value for x in one of the original equations gives

$$\begin{aligned} 4x + y &= 28 \\ 4(6) + y &= 28 \\ 24 + y &= 28 \\ y &= \;\; 4 \end{aligned}$$

EXAMPLE 2

Solve the simultaneous equations

$$\begin{aligned} 7x + 4y &= 27 \\ 3x - 5y &= 25 \end{aligned}$$

SOLUTION

One way to solve these equations is to eliminate the y's. Multiplying the first equation by 5 and the second equation by 4 will produce two equations of which each contains 20 y's:

$$\begin{aligned} 5(7x + 4y) = 5(27) &\longrightarrow 35x + 20y = 135 \\ 4(3x - 5y) = 4(25) &\longrightarrow 12x - 20y = 100 \end{aligned}$$

Adding these equations, we get

$$\begin{array}{r} 35x + 20y = 135 \\ + \quad 12x - 20y = 100 \\ \hline 47x \qquad = 235 \end{array}$$

Dividing both sides of this equation by 47, we find that $x = 5$.

Substituting this value for x in the first of the original equations, we get

$$\begin{aligned} 7(5) + 4y &= 27 \\ 35 + 4y &= 27 \\ 4y &= -8 \\ y &= -2 \end{aligned}$$

Substituting these values for x and y in the second of the original equations gives

$$\begin{aligned} 3(5) - 5(-2) &= 25 \\ 15 - (-10) &= 25 \\ 25 \qquad &= 25 \end{aligned}$$

This shows that the solution $(5, -2)$ is correct.

Exercises

Set I

1. Graph the following equations.

   a) $2x + 7y = 7$ c) $y = \frac{1}{3}x + 4$
   b) $3x - y = -6$
      $Y = 3x + 6$ d) $x = 5$

2. A riddle that is thought to be thousands of years old is:

   What we caught we threw away;
   What we couldn't catch, we kept.

The answer is "fleas."

a) If someone has x fleas and manages to get rid of y of them, how many does he or she have left?
b) If someone got rid of z fleas and couldn't catch twice that number, how many fleas did he or she originally have?

Set II

Each of the following diagrams illustrates a pair of equations. Write the equations and tell how the second equation can be obtained from the first.

3. |— x —+— y —+— y —|
   |———— 16 ————|

   |— x —+— x —+— y —+— y —+— y —+— y —|
   |——————— 32 ———————|

4. |— x —+— x —|
   |— y —+—— 5 ——|

   |— x —+— x —+— x —+— x —+— x —+— x —|
   |— y —+— y —+— y —+———— 15 ————|

5. |— x —+— x —+— x —+— x —+— y —+— y —+— y —+— y —|
   |——————— 28 ———————|

   |— x —+— y —|
   |———— 7 ————|

6. Write the equation that results from performing the indicated operation on both sides of each of the following equations.

   a) $x + 4y = 7$ Multiply by 3.
   b) $2x - 5y = 1$ Multiply by 8.
   c) $8x + 2y = 20$ Divide by 2.
   d) $6x - y = -3$ Multiply by -1.
   e) $-x + 3y = 0$ Multiply by -5.
   f) $12x - 8y = 36$ Divide by 4.

7. This exercise is about the simultaneous equations

$$4x + y = 30$$
$$x - 3y = 1$$

Solve these equations by doing the following.

a) Multiply both sides of the first equation by 3.
b) Add your equation to the second one.
c) Solve the resulting equation for x.
d) Substitute the number you got for x in any one of the equations and solve for y.
e) Check your solution to see if it makes both of the original equations true.

Now solve the equations again by doing the following.

f) Multiply both sides of the second equation by 4.
g) Subtract your equation from the first one.
h) Solve the resulting equation for y.
i) Substitute the number you got for y in any one of the equations and solve for x.

Solve the following simultaneous equations. Show your steps and check your answers.

8. $x + y = 7$
   $3x + 2y = 25$

9. $2x + 5y = 29$
   $4x - y = 25$

10. $8x - 3y = 32$
    $7x + 9y = 28$

11. $6x + 6y = 24$
    $10x - y = -15$

12. $5x - 7y = 54$
    $2x - 3y = 22$

13. $1.5x + 2.5y = 16$
    $3x - 1.5y = -33$

Set III

Each of the following diagrams illustrates a pair of equations. Write the equations and tell how the second equation can be obtained from the first.

14. $\vdash$ x $+$ x $+$ y $\dashv$
    $\vdash$ 6 $\dashv$

$\vdash$ x $+$ x $+$ x $+$ x $+$ x $+$ x $+$ y $+$ y $+$ y $\dashv$
$\vdash$ 18 $\dashv$

15. $\vdash$ x $\dashv$
    $\vdash$ y $+$ y $+$ y $+$ y $+$ 11 $\dashv$

$\vdash$ x $\dashv$ x $\dashv$
$\vdash$ y $+$ y $+$ y $+$ y $+$ y $+$ y $+$ y $+$ 22 $\dashv$

16. $\vdash$ x $+$ x $+$ x $+$ x $+$ x $+$ y $+$ y $+$ y $+$ y $+$ y $\dashv$
    $\vdash$ 45 $\dashv$

$\vdash$ x $+$ y $\dashv$
$\vdash$ 9 $\dashv$

17. Write the equation that results from performing the indicated operation on both sides of each of the following equations.
   a) $3x + y = 8$ Multiply by 2.
   b) $5x - 4y = 1$ Multiply by 6.
   c) $6x + 3y = 27$ Divide by 3.
   d) $x - 9y = -10$ Multiply by –1.
   e) $-2x + 7y = 0$ Multiply by –4.
   f) $10x - 15y = 40$ Divide by 5.

18. This exercise is about the simultaneous equations
$$x + 4y = 35$$
$$5x - y = 7$$

Solve these equations by doing the following.
   a) Multiply both sides of the second equation by 4.
   b) Add your equation to the first one.
   c) Solve the resulting equation for x.
   d) Substitute the number you got for x in any one of the equations and solve for y.

e) Check your solution to see if it makes both of the original equations true.

Now solve the equations again by doing the following.
   f) Multiply both sides of the first equation by 5.
   g) Subtract the second equation from your equation.
   h) Solve the resulting equation for y.
   i) Substitute the number you got for y in any one of the equations and solve for x.

Solve the following simultaneous equations. Show your steps and check your answers.

19. $x + 3y = 19$
    $2x + y = 3$

20. $x + 4y = 13$
    $3x - 2y = 25$

21. $7x - 3y = 44$
    $5x + 12y = -11$

22. $x + 9y = 92$
    $8x - 8y = 16$

23. $3x + 7y = 42$
    $4x + 5y = 30$

24. $4.5x - 3y = 48$
    $1.5x + 2.5y = -19$

Set IV

Can you figure out an easy way to solve the pair of simultaneous equations written on the floor in this cartoon?

"I hope you're not a lifer! I can't stand much more of this!"

303

THE WIZARD OF ID BY PERMISSION OF JOHNNY HART AND FIELD ENTERPRISES, INC.

LESSON 4
Graphing Simultaneous Equations

Suppose that a greyhound spots a cat 30 meters away. If the greyhound runs toward the cat at a speed of 15 meters per second and the cat runs away at a speed of 10 meters per second, how long will it take the greyhound to catch up with the cat? (Judging from what happened in this cartoon, we will assume that the cat can take care of itself.)

One way to answer this question is with a graph. If the greyhound runs at the rate of 15 meters per second, the distances it covers in different times are given by the table

x	0	1	2	3	4	. . .
y	0	15	30	45	60	. . .

in which x is the time in seconds and y is the distance in meters. A graph corresponding to this table is shown at the left at the top of the next page. Its equation is

$$y = 15x$$

If the cat runs at the rate of 10 meters per second and starts out 30 meters

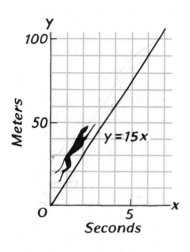 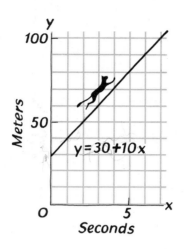 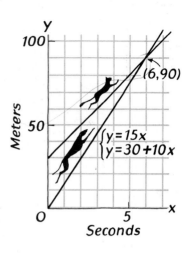

ahead of the greyhound, its distances from the point at which the greyhound begins the chase are given by the table

x	0	1	2	3	4	...
y	30	40	50	60	70	...

The second graph corresponds to this table. Its equation is

$$y = 30 + 10x$$

To find out when and where the greyhound catches up with the cat, we can draw the distance-time graphs of both animals on one pair of axes. This has been done in the third graph, which shows that the lines intersect in the point $(6, 90)$. The coordinates of this point are the solution to the simultaneous equations

$$y = 15x$$
$$y = 30 + 10x$$

The greyhound catches up with the cat in 6 seconds at a spot 90 meters from where the greyhound started.

This problem illustrates a way to picture the solution to a pair of simultaneous equations. The graph of an equation in two variables consists of *those points whose coordinates are solutions to the equation.* This means that, if *two* equations are graphed on *one* pair of axes, their common solution consists of the coordinates of the point or points that are on *both* graphs. In other words, the point or points in which the graphs intersect.

Here are more examples of how the solution to a pair of simultaneous equations can be pictured with a graph.

EXAMPLE 1
Solutions of the equations $x + y = 6$ and $y = 2x$ that are pairs of positive integers are shown on the graphs below. Show how the graphs can be drawn on one pair of

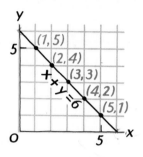 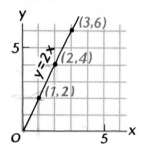

axes to illustrate the solution to the simultaneous equations

$$x + y = 6$$
$$y = 2x$$

SOLUTION
The two graphs drawn on one pair of axes are shown at the right. They intersect in the point (2, 4). The coordinates of this point are the solution to the equations

$$x + y = 6$$
$$y = 2x$$

because

$$2 + 4 = 6$$

and

$$4 = 2(2)$$

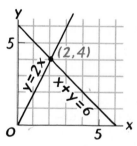

EXAMPLE 2
Find the solution of the simultaneous equations

$$x + 4y = 8$$
$$y = \frac{1}{2}x + 5$$

by graphing them. Check your answer.

SOLUTION

To graph the first equation, we can find its x- and y-intercepts. If $x = 0$, $0 + 4y = 8$; so $y = 2$. If $y = 0$, $x + 4(0) = 8$; so $x = 8$. The line crosses the axes in the points $(0, 2)$ and $(8, 0)$.

Because the second equation is in the form $y = ax + b$, in which $a = \dfrac{1}{2}$ and $b = 5$, we know that it has a slope of $\dfrac{1}{2}$ and that its y-intercept is 5.

Drawing both lines on one pair of axes, we get the graph shown below. The solution of the equations is the pair of coordinates of the point in which the lines intersect: $(-4, 3)$.

Checking this solution in each equation, we get

$$-4 + 4(3) = 8$$
$$-4 + 12 = 8$$
$$8 = 8$$

and

$$3 = \frac{1}{2}(-4) + 5$$
$$3 = -2 + 5$$
$$3 = 3$$

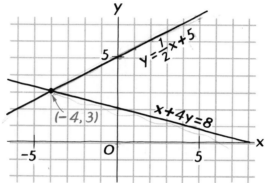

Exercises

Set I

1. Solve each of the following equations for the variable indicated.
   a) For x: $y = ax + b$
   b) For x: $ax + by = c$
   c) For y: $ax + by = c$

2. Find the length of this rectangle's sides labeled x

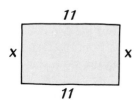

   a) if its perimeter is 36.
   b) if its area is $x + 36$.

   Check your answers.

3. The British eat more candy than do any other people in the world: about eight ounces per person per week.

   About how many ounces of candy
   a) do five people in Britain eat in four weeks?
   b) do x people in Britain eat in y weeks?

   How many weeks would it take
   c) five people to eat 1,000 ounces of candy?
   d) x people to eat z ounces of candy?

Set II

Use the graphs of the following pairs of
simultaneous equations to find their solutions.
Check each answer by showing that it makes
both equations true.

4. $x + 3y = 1$
   $2x - y = 9$

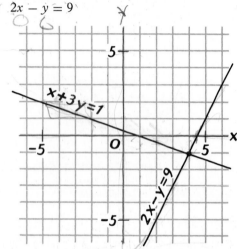

5. $y = 4x + 11$
   $y = 1 - x$

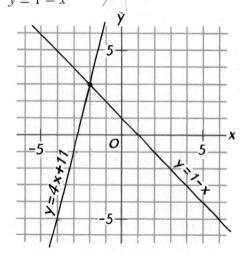

6. $2x - 2y = 3$
   $x = \dfrac{1}{2}y - 1$

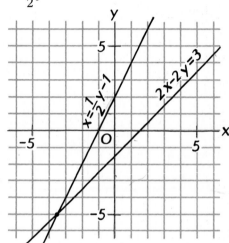

7. This exercise is about the simultaneous
   equations

$$x + 3y = 3$$
$$x + y = 5$$

   a) Graph the two equations on one pair
      of axes.
   b) Write the equation that results from
      adding the two equations.
   c) Graph that equation on the same pair
      of axes.
   d) Write the equation that results from
      subtracting the second of the two
      simultaneous equations from the first.
   e) Graph that equation on the same pair
      of axes.
   f) What do you notice about the four lines?

$y = -1$

8. This exercise is about the simultaneous equations

$$2x + y = 8$$
$$2x - y = 4$$

Follow the instructions given in exercise 7 and answer the question asked in part f.

Find solutions to each of the following pairs of simultaneous equations by graphing them.

Check each solution by seeing if it makes each equation in the pair true.

9. $y = 2x$
   $x + y = 3$

10. $y = x + 4$
    $x + y = -2$

11. $3x - 2y = 6$
    $y = -\frac{1}{2}x - 3$

12. $y = 7 - x$
    $x - y = 2$

Set III

Use the graphs of the following pairs of simultaneous equations to find their solutions. Check each answer by showing that it makes both equations true.

14. $4x + y = 2$
    $x - 2y = 14$

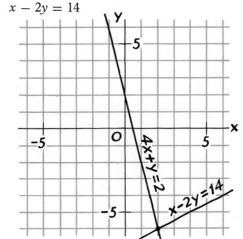

13. $y = x + 8$
    $y = -2x - 1$

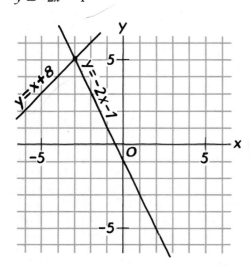

15. $3x = 8y$
    $x = 6y + 5$

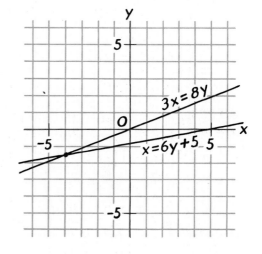

16. This exercise is about the simultaneous equations

$$3x + y = 3$$
$$x + y = 7$$

a) Graph the two equations on one pair of axes.
b) Write the equation that results from adding the two equations.
c) Graph that equation on the same pair of axes.
d) Write the equation that results from subtracting the second of the two simultaneous equations from the first.
e) Graph that equation on the same pair of axes.
f) What do you notice about the four lines?

17. This exercise is about the simultaneous equations

$$x + 2y = 6$$
$$x - 2y = 2$$

Follow the instructions given in exercise 16 and answer the question asked in part f.

Find solutions to each of the following pairs of simultaneous equations by graphing them. Check each solution by seeing if it makes each equation in the pair true.

18. $y = \dfrac{1}{2}x$
   $x + y = 6$

19. $x + 2y = 4$
   $y = x + 5$

20. $2x + 5y = 10$
   $y = 2x - 4$

21. $3x - y = 0$
   $x + y = -4$

Set IV

Suppose that a greyhound spots a cat 40 meters away and begins running toward it at a speed of 10 meters per second. The four graphs below illustrate some of the things that might happen. In each graph, the x-axis represents time in seconds and the y-axis represents distance in meters.

Describe, as specifically as you can, the situation illustrated by each graph.

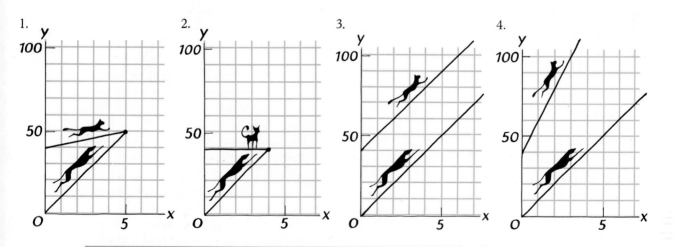

Inconsistent and Equivalent Equations

$$x + y = 5$$
$$y = x + 3$$

$$x + y = 5$$
$$y = -x + 3$$

$$x + y = 5$$
$$y = -x + 5$$

Although the three pairs of simultaneous equations above look quite similar to each other, their solutions are very different. The first pair of equations has exactly one solution, the second pair has no solutions, and the third pair has infinitely many.

To see why, we will graph each pair of equations. The graph of the first pair

$$x + y = 5$$
$$y = x + 3$$

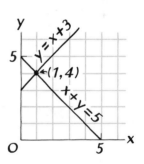

shows that their solution is (1, 4). We know moreover, that this is their *only* solution because two straight lines cannot intersect in more than one point.

The graph of the second pair of equations

$$x + y = 5$$
$$y = -x + 3$$

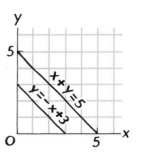

311

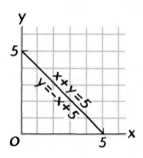

consists of two parallel lines. Because parallel lines do not intersect, this pair of equations has no solutions.

The graph of the third pair of equations

$$x + y = 5$$
$$y = -x + 5$$

consists of just one line because the graphs of the two equations are the same. Because every point on one graph is automatically on the other, this means that these equations have infinitely many solutions.

It is possible to tell that a pair of simultaneous equations has either no solutions or infinitely many without graphing them. Look again at the pair of equations that has no solutions:

$$x + y = 5$$
$$y = -x + 3$$

If we write the second equation in standard form by adding x to each side, we get

$$x + y = 3$$

Now compare this equation to the first one. One equation tells us that the sum of two numbers is 5 and the other equation tells us that their sum is 3. These clues *contradict* each other and the equations are **inconsistent.** If we try to find the solution to these equations by subtraction,

$$\begin{array}{r} x + y = 5 \\ - \ \underline{x + y = 3} \\ 0 \ \ = 2 \end{array}$$

we get a *false* equation. This means that no pair of numbers can satisfy these two equations because, if such a pair existed, then zero and two would be equal.

Look again at the pair of equations that has infinitely many solutions:

$$x + y = 5$$
$$y = -x + 5$$

The first equation can be transformed into the second one by subtracting x from each side. This shows that the two equations are **equivalent.** It follows then, that any solution to one equation is automatically a solution to the other. If we try to find their solution by subtraction,

$$
\begin{array}{r}
x + y = 5 \\
- \quad x + y = 5 \\
\hline
0 \quad = 0
\end{array}
$$

we get an equation that is true, regardless of the values of x and y.

The examples that we have considered illustrate the three things that can happen when we have a pair of linear equations in two variables:

1. The equations can have exactly one solution, in which case their graphs intersect in a single point.

2. The equations can have no solutions, in which case their graphs do not intersect. Such equations are *inconsistent.*

3. The equations can have infinitely many solutions, in which case their graphs are a single line. Such equations are *equivalent.*

Exercises

Set I

1. Draw the following lines on one pair of axes.
   a) The line through $(2, 5)$ having a slope of 1.
   b) The line through $(5, 2)$ having a slope of 1.
   c) The line through $(1, 0)$ having a slope of $\dfrac{2}{5}$.
   d) The line through $(0, 1)$ having a slope of $-\dfrac{5}{2}$.

2. Solve the following pairs of simultaneous equations.
   a) $8x + 3y = 5$
      $x + 3y = 19$
   b) $7x + 2y = 33$
      $4x - 5y = 25$

3. The volume of a balloon depends on the temperature of the gas with which it is filled. The graph of this function for a typical balloon is shown at the right.
   a) What is the approximate volume of the balloon when the temperature is 70°C?
   b) What is the approximate temperature when the volume of the balloon is 110 cubic centimeters?

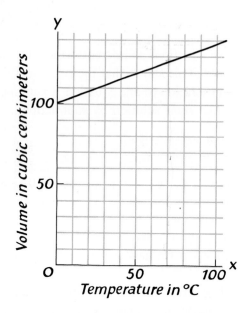

Set II

4. The graphs of three pairs of simultaneous equations are shown below. Tell what you can about the solutions of each.

a)

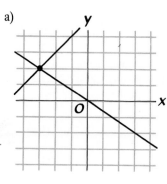

$$y = x + 5$$
$$2x + 3y = 0$$

b)

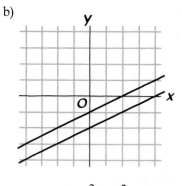

$$x - 2y = 2$$
$$x - 2y = 4$$

c)

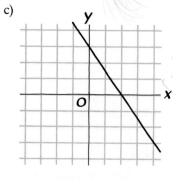

$$3x + 2y = 6$$
$$4y = 12 - 6x$$

The diagrams for exercises 5 through 8 represent pairs of simultaneous equations. For each exercise, do each of the following:
   a) Write the two equations illustrated by the diagram in standard form.
   b) Try to solve them, telling what happens.
   c) If any pair of equations is inconsistent or equivalent, say so.

5.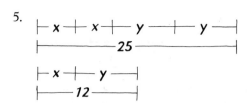

x	x	y	y
25			

x	y
12	

6.

x	y	y
14		

x	x	x	y	y	y	y	y	y
42								

7.

x	x
y + y +	16

x	x	x	x	x
y + y + y + y + y +		40		

8.

x	x	x
y	7	

x	x	x	x	x	x
y	y	10			

9. This exercise is about the simultaneous equations

$$2x + 5y = 10$$
$$2x + 5y = 0$$

   a) Graph the two equations on one pair of axes.
   b) What do you notice?
   c) How many solutions do the equations have?
   d) What are equations like these called?

10. This exercise is about the simultaneous equations

$$3x - 3y = 15$$
$$y = x - 5$$

Follow the instruction given in exercise 9 and answer the questions in parts b through d.

Graph the following pairs of simultaneous equations and tell what you can about the solutions of each.

11. $6x - 2y = 12$ 12. $x - y = 1$ 13. $4x - y = 2$
    $y = 3x - 6$ $x + 3y = -9$ $y = 4x + 1$

Set III

14. The graphs of three pairs of simultaneous equations are shown below. Tell what you can about the solutions of each.

a)

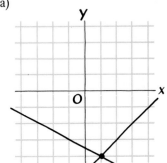

$x + 2y = -7$
$x - y = 5$

b)

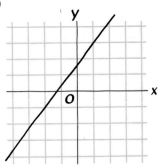

$4x - 3y = -5$
$6y - 10 = 8x$

c)

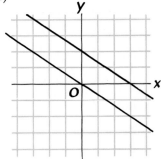

$2x + 3y = 0$
$2x + 3y = 6$

The diagrams for exercises 15 through 18 represent pairs of simultaneous equations. For each exercise, do each of the following:
  a) Write the two equations illustrated by the diagram in standard form.
  b) Try to solve them, telling what happens.
  c) If any pair of equations is inconsistent or equivalent, say so.

15.

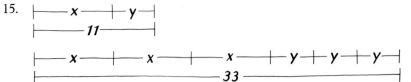

16.

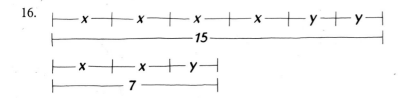

17.

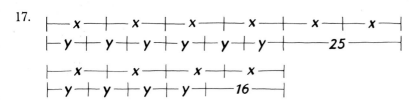

18.

19. This exercise is about the simultaneous equations

$$4x - 2y = 8$$
$$2x - y = 4$$

a) Graph the two equations on one pair of axes.
b) What do you notice?
c) How many solutions do the equations have?
d) What are equations like these called?

20. This exercise is about the simultaneous equations

$$7x - 7y = -14$$
$$y = x - 3$$

Follow the instruction given in exercise 19 and answer the questions asked in parts b through d.

Graph the following pairs of simultaneous equations and tell what you can about the solutions of each.

21. $5x - y = 10$
$y = 5x$

22. $x - 7y = 7$
$2x + 2y = 6$

23. $3x - 4y = -12$
$y = \dfrac{3}{4}x + 3$

Set IV

We have been studying pairs of simultaneous equations whose graphs are lines. The graphs of the simultaneous equations below are curves.

Graph A	*Graph B*
$4x^2 + 4y^2 = 36$	$9x^2 + 4y^2 = 52$
$9x^2 + 9y^2 = 36$	$4x^2 + 9y^2 = 52$

1. Tell what you can about the solutions of each pair of equations.
2. Can you derive a false equation from one pair of equations? If so, what is it?

LESSON 6
Solving by Substitution

For two people to be able to balance each other on a seesaw, it isn't necessary that they have equal weights. The heavier person can simply move closer to the fulcrum, the point at which the seesaw is supported.

It is known from experiments with levers that, for a seesaw to be balanced, the product of the weight on one side and its distance from the fulcrum must be equal to the product of the weight on the other side and its distance from the fulcrum. In the

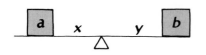

figure above,

$$ax = by$$

in which a and b are the weights and x and y are their distances from the fulcrum.

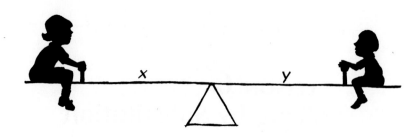

Suppose, for example, that a child weighs twice as much as her little brother and that they want to balance on a seesaw. Say their weights are 40 pounds and 20 pounds. In order for them to balance, their distances from the fulcrum must be such that

$$40x = 20y$$

If the children are seated 12 feet apart (so that $x + y = 12$), how far should each one be from the fulcrum?

The answer to this question is the pair of numbers that can be substituted for x and y to make both of these equations true:

$$40x = 20y$$
$$x + y = 12$$

One way to solve these equations is to graph them: the result is shown at the left.

Although the equations could also be solved by the addition-subtraction method, there is another method that is, for some equations, easier to use. It is called the *substitution method* and it works like this.

Take either equation in the system

$$40x = 20y$$
$$x + y = 12,$$

and solve it for one of the two variables in terms of the other. For example, we might solve the first equation for y by dividing both sides by 20 to get

$$2x = y$$

Now, because this equation says that $2x$ and y are the same number, we can substitute $2x$ for y in the second equation

$$x + y = 12$$

to get

$$x + 2x = 12$$

Solving the resulting equation for x, we get

$$3x = 12$$
$$x = 4$$

To find y, we can substitute this value for x in either of the two equations. For example,

$$2x = y$$
$$2(4) = y$$
$$y = 8$$

The solution is $(4, 8)$.

Here are more examples of how the substitution method can be used to solve a pair of simultaneous equations.

EXAMPLE 1
Write a pair of simultaneous equations illustrated by the diagrams at the right. Solve them by making the indicated substitution to find the lengths of the line segments in each diagram.

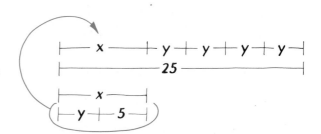

SOLUTION
The equations are

$$x + 4y = 25$$
$$x = y + 5$$

Because the second equation tells us that x and $y + 5$ are the same number, we can substitute $y + 5$ for x in the first equation:

$$x + 4y = 25$$
$$(y + 5) + 4y = 25$$

Solving for y,

$$5y + 5 = 25$$
$$5y = 20$$
$$y = 4$$

Substituting this value for y in the second equation,

$$x = 4 + 5$$
$$x = 9$$

The solution to the simultaneous equations is $(9, 4)$ and the lengths of the segments labeled x and y are 9 and 4, respectively.

EXAMPLE 2
Solve the simultaneous equations

$$2x + 3y = 8$$
$$y - x = 11$$

SOLUTION
In this case, we might begin by solving the second equation for y. Adding x to each side, we get

$$y = 11 + x$$

Substituting $11 + x$ for y in the first equation:

$$2x + 3y = 8$$
$$2x + 3(11 + x) = 8$$
$$2x + 33 + 3x = 8$$
$$5x + 33 = 8$$
$$5x = -25$$
$$x = -5$$

Because we have already shown that $y = 11 + x$, it follows that

$$y = 11 + (-5)$$
$$y = 6$$

The solution is $(-5, 6)$.

Checking these values of x and y in the original equations, we get

$$2x + 3y = 8 \qquad y - x = 11$$
$$2(-5) + 3(6) = 8 \qquad 6 - (-5) = 11$$
$$-10 + 18 = 8 \qquad 6 + 5 = 11$$
$$8 = 8 \qquad 11 = 11$$

Exercises

Set I

1. Use a number line to find each of the following.
   a) The distance between -5 and 6.
   b) The number of the point midway between -9 and 1.
   c) The numbers corresponding to the two points whose distance from 8 is 2.
   d) The numbers corresponding to the two points whose distance from -2 is 7.

2. The following questions are about the axes of a coordinate graph.
   a) What is the equation of the x-axis?
   b) What is its slope?

   c) What is the equation of the y-axis?
   d) What is its slope?

3. In a contest in 1973, a modified Opel station wagon got 376 miles per gallon!
   a) Write a formula for the distance in miles, d, that this car could go on g gallons of gasoline if it continued to get this mileage.
   b) Solve the formula for g in terms of d.
   c) Use your formula for part b to find the amount of gasoline used by the car to travel the 14 miles in the contest. (Round your answer to the nearest hundredth.)

Set II

The diagrams for exercises 4 through 7 represent pairs of simultaneous equations.
Write the equations and solve them by making the indicated substitution to find the lengths of the line segments in each diagram.

4.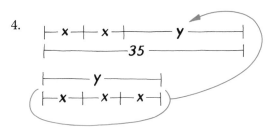

5.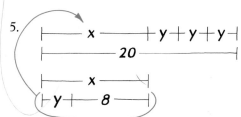

6.

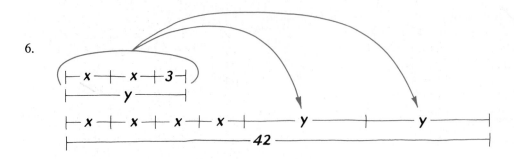

7.

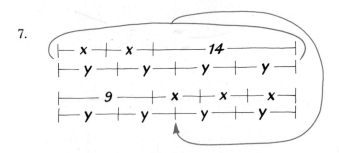

Solve the following simultaneous equations by the substitution method. Show your steps and check your answers.

8. $x = y$
   $2x = y + 8$

9. $y = x - 2$
   $3(x + 1) = 4y$

10. $y = 7x + 10$
    $y = 4 + x$

11. $y = 3x - 1$
    $x + 2y = 33$

12. This exercise is about the simultaneous equations

$$x = 3y + 7$$
$$2x - 6y = 11$$

   a) Try to solve the equations by the substitution method.
   b) What can you conclude from the result?
   c) What would a graph of these equations consist of?

13. This exercise is about the simultaneous equations

$$4y - 8x = 4$$
$$y = 2x + 1$$

Follow the instruction given in exercise 12 and answer the questions asked in parts b and c.

The diagrams for exercises 14 through 16 represent weights balanced on seesaws. For each exercise, do each of the following:

    a) Use the clue and the diagram to write a pair of simultaneous equations.
    b) Solve the equations to find x and y.
    c) Check your solution to see if it makes the seesaw balance.

14. The distance between the weights is 35.

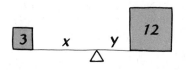

15. The distance between the weights is 20.

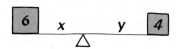

16. The sum of the weights is 24.

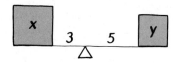

Set III

The diagrams for exercises 17 through 20 represent pairs of simultaneous equations. Write the equations and solve them by making the indicated substitution to find the lengths of the line segments in each diagram.

17.

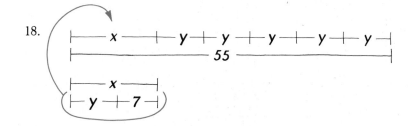

18.

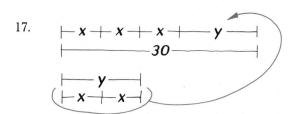

19.

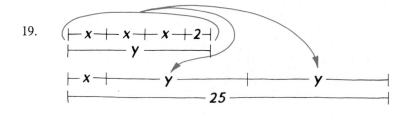

20.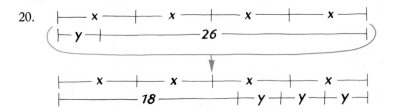

Solve the following simultaneous equations by the substitution method. Show your steps and check your answers.

21. $y = 2x$
$4x - y = 14$

22. $y = x - 7$
$y = 5x - 19$

23. $x = y + 1$
$2x = 3y + 3$

24. $x + 11 = 5y$
$x = 2(3y - 8)$

25. This exercise is about the simultaneous equations

$$x = 2 + 4y$$
$$2x - 8y = 4$$

a) Try to solve the equations by the substitution method.
b) What can you conclude from the result?
c) What would a graph of these equations consist of?

26. This exercise is about the simultaneous equations

$$6x - 3y = 10$$
$$y = 2x$$

Follow the instruction given in exercise 25 and answer the questions asked in parts b and c.

The diagrams for exercises 27 through 29 represent weights balanced on seesaws. For each exercise, do each of the following:
a) Use the clue and the diagram to write a pair of simultaneous equations.
b) Solve the equations to find x and y.
c) Check your solution to see if it makes the seesaw balance.

27. The distance between the weights is 18.

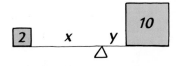

28. The distance between the weights is 21.

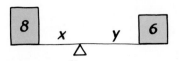

29. The sum of the weights is 27.

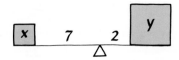

Set IV

A seesaw can balance with more than one weight on each side as long as the sum of the products of the weights on one side and their distances from the fulcrum is equal to the sum of the products of the weights on the other side and their distances from the fulcrum.

Can you figure out how much the duck and rabbit on this seesaw weigh from the following clues?

The sum of the four weights is 40 pounds.
The owl weighs 3 pounds.
The chicken weighs 5 pounds.

$$\left(9 \times (2-1)\right) + \left(7(3)\right) = 9(5) + (13 \times d)$$

$$3 + 5 + a + d = 40$$

$$\left(9d + (7 \times 3)\right) = 45 + (13a)$$

LESSON 7
Mixture Problems

The greatest mathematician of the ancient world was Archimedes, who lived in the Greek city-state of Syracuse in the third century B.C.

The king of Syracuse once came to Archimedes with the following problem. The king had ordered a crown of pure gold but, when the crown arrived, he became suspicious that it might actually be made from a mixture of gold and silver. He asked Archimedes if he could figure out a way to tell whether the crown was pure gold without damaging it.

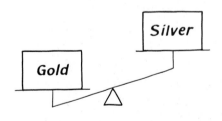

Equal volumes of gold and silver

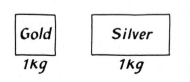

Equal weights of gold and silver

After thinking about the problem for awhile, Archimedes came up with the following ideas. Gold is heavier than silver: a given volume of gold weighs approximately twice as much as the same volume of silver. It follows from this that a given *weight* of silver has approximately twice as much volume as the same *weight* of gold. Specifically, 1 kilogram of gold has a volume of about 50 cubic centimeters, whereas 1 kilogram of silver has a volume of about 100 cubic centimeters.

The volume of the crown could have been found by putting it in a basin of water filled to the brim and measuring the volume of the water that overflows. Suppose that, by doing this, Archimedes found that the crown had a volume of 140 cubic centimeters. By also weighing the crown and finding that it weighed 2 kilograms, Archimedes was able to figure out exactly what it was made of.

First of all, he knew it wasn't pure gold because 2 kilograms of gold would have had a volume of about 100 cubic centimeters and the crown had a volume of 140 cubic centimeters.

If we let x represent the weight of the gold in the crown and y represent the weight of the silver, then

$$x + y = 2$$

because the weight of the crown was 2 kilograms.

Also, because 1 kilogram of gold has a volume of 50 cubic centimeters, x kilograms of gold has a volume of $50x$ cubic centimeters. And, because 1 kilogram of silver has a volume of 100 cubic centimeters, y kilograms of silver has a volume of $100y$ cubic centimeters. The volume of the crown is 140 cubic centimeters, and so

$$50x + 100y = 140$$

To find out what the crown was made of, we have to solve the simultaneous equations

$$x + y = 2$$
$$50x + 100y = 140$$

Multiplying the first equation by 50,

$$50x + 50y = 100$$

and subtracting the result from the second equation, we get

$$50y = 40$$

so that

$$y = \frac{40}{50} = 0.8$$

Substituting this value for y in the first equation, $x + y = 2$, we get

$$x + 0.8 = 2$$

so that

$$x = 1.2$$

The crown contains 1.2 kilograms of gold and 0.8 kilograms of silver.

In solving this problem, we wrote a system of two equations, one dealing with weights and the other with volumes. Any problem about a mixture of two things can be solved in a similar way.

Exercises

Set I

1. This diagram illustrates a pair of simultaneous equations.

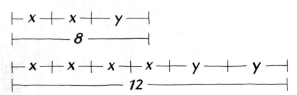

   a) What are the equations?
   b) How many solutions does the pair of equations have?
   c) What single word describes such a pair of equations?
   d) What would the graph of this pair of equations consist of?

2. Write an equation that, together with the equation

$$x + 2y = 5,$$

   makes a pair of simultaneous equations
   a) that are inconsistent.
   b) that are equivalent.
   c) that have the solution (1, 2).

3. The wavelength of a radio wave is a function of its frequency. A formula for this function is

$$w = \frac{300,000}{f}$$

   in which w represents the wavelength in meters and f represents the frequency in kilocycles per second.
   a) What kind of variation is this?
   b) What happens to the wavelength as the frequency of a radio wave increases?
   c) Solve the formula for f.
   d) Find the frequency of a radio wave whose wavelength is 2,000 meters.

Set II

4. A telephone coin box contains 52 coins, of which some are nickels and the rest are dimes. The total value of the coins is $4.50. Find out how many coins of each type it contains by doing each of the following.

   a) Letting x represent the number of nickels and y represent the number of dimes, write an equation relating x, y, and 52.
   b) Express the total value of the nickels in cents in terms of x.
   c) Express the total value of the dimes in cents in terms of y.
   d) Write an equation expressing the fact that the total value of the coins is 450 cents.
   e) Solve the simultaneous equations that you have written for x and y.
   f) How many nickels and how many dimes does the coin box contain?

5. Obtuse Ollie works after school in a health food store. He is supposed to add cranberry juice to apple juice to make 20 liters of cranapple drink. A liter of the apple juice sells for 45 cents and a liter of the cranberry juice sells for 60 cents. A liter of the cranapple drink sells for 48 cents. Find out how many liters of each juice he should use by doing each of the following.
   a) Letting x represent the number of liters of apple juice used and y the number of

liters of cranberry juice, write an equation relating x, y, and 20.

b) In terms of x, what is the value of the apple juice used?

c) In terms of y, what is the value of the cranberry juice used?

d) Write an equation relating the values of the two juices used to the value of the cranapple drink.

e) Solve the simultaneous equations that you have written for x and y.

f) How many liters of each juice should Ollie use?

6. On a trip to the nursery, Ivy got carried away and bought 95 plants, some at $3 each and the rest at $5 each. Before the tax was added, the bill came to $353. Find out how many plants she bought at each price by doing each of the following.

a) Letting x and y represent the numbers of plants costing $3 and $5 respectively, write a pair of simultaneous equations,

one relating the numbers of plants and the other relating their costs.

b) Solve the equations.

c) How many plants at each price did Ivy buy?

7. Christmas tree tinsel is a mixture of lead and tin. Ten cubic centimeters of it weighs 83.4 grams. One cubic centimeter of tin weighs 7.2 grams and one cubic centimeter of lead weighs 11.0 grams. Find out how many grams of each metal are in ten cubic centimeters of tinsel by doing the following.

a) Letting x and y represent the numbers of cubic centimeters of tin and lead respectively, write a pair of simultaneous equations, one relating volumes (numbers of cubic centimeters) and the other relating weights.

b) Solve the equations.

c) How many *grams* of each metal are in ten cubic centimeters of tinsel?

Set III

8. A test contains 42 questions, of which some are worth 2 points and the rest are worth 3 points. A perfect score is 100 points. Find out how many questions of each type are on the test by doing each of the following.

a) Letting x and y represent the numbers of questions worth 2 and 3 points respectively, write an equation relating x, y, and 42.

b) Express the total number of points possible from the 2-point questions in terms of x.

c) Express the total number of points possible from the 3-point questions in terms of y.

d) Write an equation expressing the fact that the total number of points possible on the test is 100.

e) Solve the simultaneous equations that you have written for x and y.

f) How many questions worth 2 points and how many questions worth 3 points are on the test?

9. Acute Alice works Saturdays in a nut shop. She is supposed to add some Spanish peanuts worth 84 cents a pound to 40 pounds of Virginia peanuts worth 71 cents a pound to make a mixture worth 79 cents a pound. Find out how many pounds of Spanish peanuts she should add by doing each of the following.

a) Letting x represent the number of pounds of Spanish peanuts added and y the number of pounds of peanuts in the mixture, write an equation relating x, y, and 40.

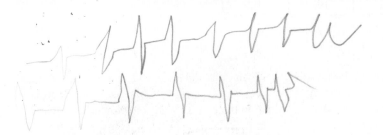

b) In terms of x, how much are the Spanish peanuts worth?

c) In terms of y, how much is the mixture worth?

d) Write an equation relating the worth of the two kinds of peanuts used to the worth of the mixture.

e) Solve the simultaneous equations that you have written for x and y.

f) How many pounds of Spanish peanuts should Alice use?

10. One evening, 1,255 people went to the Orpheum Theatre to see *Gone with the Wind*. The box office receipts totaled $3,680, the price of admission for adults being $3 and that for children being $2. Find out how many tickets of each type the theater sold by doing each of the following.

a) Letting x represent the number of adult tickets sold and y the number of children's tickets, write a pair of simultaneous equations, one relating the numbers of tickets and the other relating their costs.

b) Solve the equations.

c) How many tickets of each type did the theater sell?

11. The metal used in gold coins is an alloy of gold and copper. Ten cubic centimeters of it weighs 172.2 grams. One cubic centimeter of gold weighs 19.3 grams and one cubic centimeter of copper weighs 8.9 grams. Find out how many grams of each metal are in ten cubic centimeters of the alloy by doing each of the following.

a) Letting x and y represent the numbers of cubic centimeters of gold and copper respectively, write a pair of simultaneous equations, one relating volumes (numbers of cubic centimeters) and the other relating weights.

b) Solve the equations.

c) How many *grams* of each metal are in ten cubic centimeters of the alloy?

Set IV

The following problem was invented by a man named Mahavira, who lived in southern India more than a thousand years ago.

The price of 9 citrons and 7 wood apples is 107; the price of 7 citrons and 9 wood apples is 101. Tell me quickly the price of a citron and of a wood apple.

Can you figure out what the two prices are?

$11.2 - .7x$

"Say, I think I see where we went off. Isn't eight times seven fifty-six?"

Summary and Review

In this chapter, we have learned how to solve pairs of simultaneous equations in two variables by addition and subtraction, by graphing, and by substitution. We have also learned how to use simultaneous equations to solve mixture problems.

Simultaneous Equations (*Lesson 1*) To solve a pair of equations in two variables means to find every ordered pair of numbers that can replace the variables in both equations to make them true. We search for these solutions by combining the equations to write an equation that contains only one of the variables. The solution to this equation can then be substituted into one of the original equations to find the other variable.

Solving by Addition and Subtraction (*Lessons 1, 2, and 3*) Some pairs of simultaneous equations can be solved by adding them. The left sides of the two equations are added, the right sides are added, and their sums are set equal.

Other pairs of equations can be solved by subtracting one from the other. The left and right sides of one of the equations are subtracted from the left and right sides of the other, and the differences are set equal.

To solve a pair of simultaneous equations by addition or subtraction, the two equations must contain an equal number of one of the variables. If the numbers are unequal, one or both of the equations can be multiplied on both sides so that the resulting equations have equal numbers of one of the variables.

Graphing Simultaneous Equations (*Lesson 4*) The solution to a pair of simultaneous equations in two variables can be pictured by graphing the equations on the same pair of axes. The pair of coordinates of the point (or points) in which the graphs intersect is the solution to the equations.

Inconsistent and Equivalent Equations (*Lesson 5*) If a contradiction can be derived from a set of equations, the equations are *inconsistent*. Such equations have no solutions. Their graphs consist of two parallel lines.

Equations are *equivalent* if each of them can be derived from the other. Equivalent equations have infinitely many solutions. Their graphs consist of a single line.

Solving by Substitution (*Lesson 6*) To solve a pair of simultaneous equations by substitution, first solve one of the equations for one variable, say x, in terms of the other, say y. The expression that is equal to x can then be substituted for x in the other equation and the resulting equation solved for y. We can also begin by solving for y in one equation and substituting for y in the other equation.

Mixture Problems (*Lesson 7*) Problems about mixtures of two things can be solved by writing a pair of simultaneous equations, each relating a property of the mixture, such as weight, volume, or value.

Exercises

Set I

1. Tell whether or not each of the following ordered pairs is a solution of the simultaneous equations given.

$$7x + 2y = 12$$
$$y = 5x - 11$$
a) $(0, 6)$
b) $(2, -1)$

$$x^2 - 3y = 16$$
$$x + y^2 = 4$$
c) $(4, 0)$
d) $(2, -4)$
e) $(-5, 3)$

2. Write the equation that results from performing the following operations on these equations:

$$2x + y = 6$$
$$2x - 4y = 10$$

a) Adding the two equations.
b) Subtracting the second equation from the first.
c) Multiplying both sides of the first equation by four.
d) Dividing both sides of the second equation by two.

Solve the following simultaneous equations by addition or subtraction.

3. $x + y = 35$
   $x - y = 67$

4. $6x + 11y = 21$
   $6x + y = -9$

5. $3x - 5y = 51$
   $x + 5y = 23$

6. $8x - 7y = 62$
   $4x - 7y = 66$

Solve the following simultaneous equations.

7. $4x + 3y = 31$
   $2x - 9y = 5$

8. $7x - 5y = 40$
   $3x - 2y = 16$

Graph the following pairs of simultaneous equations and tell what you can about the solutions of each.

9. $y = x + 7$
   $y = -\dfrac{1}{2}x + 1$

10. $x - y = 4$
    $y = x - 7$

11. $x + 2y = 10$
    $2x - 3y = 6$

12. $y = 4x$
    $8x - 2y = 0$

Solve the following simultaneous equations by substitution.

13. $x + y = 14$
    $y = 3x$

14. $x = 5y$
    $2x - 7y = 27$

15. $2x + y = -40$
    $8y - 3 = x$

16. $x + 6 = y - 1$
    $3(x + 3) = 2y$

The following diagrams represent weights balanced on seesaws. Write a pair of simultaneous equations for each diagram and solve the equations to find x and y.

17. The distance between the weights is 21.

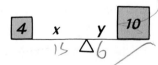

18. The sum of the weights is 20.

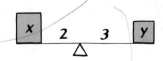

19. Write an equation that, together with the equation

$$x + 4y = 8,$$

makes a pair of simultaneous equations
a) that are inconsistent.
b) that are equivalent.
c) that have the solution $(0, 2)$.

20. Mr. Magoo's wife asked him to buy some 10-cent and 13-cent stamps for her. He couldn't read the numbers of each kind that she had written down but remembered that he was supposed to buy ten dollars worth. If Mrs. Magoo wanted 85 stamps altogether, find out how many of each kind she wanted by doing each of the following.

a) Letting x and y represent the numbers of 10-cent and 13-cent stamps respectively, write a pair of simultaneous equations, one relating the numbers of stamps and the other relating their costs.
b) Solve the equations.
c) How many stamps of each kind did Mrs. Magoo want?

Set II

1. Tell whether or not each of the following ordered pairs is a solution of the simultaneous equations given.

$$x - 6y = 21$$
$$4y = 1 - x$$

a) $(9, -2)$
b) $(21, 0)$

$$2x + y = -5$$
$$x^2 + y^2 = 25$$

c) $(0, -5)$
d) $(-4, 3)$
e) $(-1, -3)$

2. Write the equation that results from performing the following operations on these equations:

$$6x - 3y = 12$$
$$x + 3y = 2$$

a) Adding the two equations.
b) Subtracting the second equation from the first.
c) Dividing both sides of the first equation by three.
d) Multiplying both sides of the second equation by six.

Solve the following simultaneous equations by addition or subtraction.

3. $x - y = 8$
   $x + y = 43$

4. $5x - 7y = 92$
   $5x + y = 4$

5. $9x + 2y = 5$
   $x - 2y = -15$

6. $3x - 11y = 43$
   $12x - 11y = 7$

Solve the following simultaneous equations.

7. $5x + 4y = 53$
   $x - 2y = 5$

8. $4x + 3y = 17$
   $3x + 5y = 10$

Graph the following pairs of simultaneous equations and tell what you can about the solutions of each.

9. $y = x + 3$
   $2x - 3y = -12$

10. $x - 2y = -8$
    $y = \frac{1}{2}x + 4$

11. $x + y = -4$
    $2y - x = 7$

12. $y = -3x + 1$
    $9x + 3y = 9$

Solve the following simultaneous equations by substitution.

13. $x - y = 9$
    $2y = x$

14. $y = x + 3$
    $x + 4y = 2$

15. $3x + 2y = 4$
    $5 - 2x = y$

16. $9x = y + 5$
    $4y - 3x = -20$

The following diagrams represent weights balanced on seesaws. Write a pair of simultaneous equations for each diagram and solve the equations to find x and y.

17. The distance between the weights is 20.

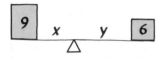

18. The sum of the weights is 26.

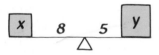

19. Write an equation that, together with the equation

$$3x - y = 5,$$

makes a pair of simultaneous equations
a) that are equivalent.
b) that are inconsistent.
c) that have the solution (2, 1)

20. On a fishing trip, Huckleberry Finn caught 31 fish, some of which were bullheads averaging 1.5 pounds each and the rest of which were catfish averaging 5 pounds each. The entire catch weighed 92 pounds. Find out how many fish of each kind Huckleberry caught by doing each of the following.

    a) Letting x and y represent the numbers of bullheads and catfish respectively, write a pair of equations, one relating the numbers of fish and the other relating their weights.
    b) Solve the equations.
    c) How many fish of each kind did he catch?

Chapter **8**

EXPONENTS

*"Did it ever occur to you that with all the eggs we've laid
there should be more of us?"*

LESSON **1**
Large Numbers

If every chicken egg produced a chicken, there would very quickly be many more chickens in the world than human beings! The number of chicken eggs produced annually throughout the world is about 390,000,000,000. This number is so large that, even though it is easy to name, its size is very difficult to comprehend.* If 390,000,000,000 eggs were packed in cartons and the cartons stacked in a pile 100 feet long and 100 feet wide, the pile would be more than 38 miles high!

The size of a number such as 390,000,000,000 depends on the number of zeros needed to write it. Adding a zero at the end gives a number that is ten times larger. The removal of a zero results in a number that is one-tenth as large. To prevent mistakes that might result from handling so many zeros, large numbers are often written in a form called *scientific notation*.

---

*The name of the number is "three hundred ninety billion." The names of some large numbers that you should know are *million* (1,000,000), *billion* (1,000,000,000) and *trillion* (1,000,000,000,000).

Scientific notation is based on powers of ten. From the table of some of the powers of ten below, it is easy to see that in each case the exponent is the

$$10^2 = 10 \cdot 10 = 100$$
$$10^3 = 10 \cdot 10 \cdot 10 = 1,000$$
$$10^4 = 10 \cdot 10 \cdot 10 \cdot 10 = 10,000$$
$$10^5 = 10 \cdot 10 \cdot 10 \cdot 10 \cdot 10 = 100,000$$
$$10^6 = 10 \cdot 10 \cdot 10 \cdot 10 \cdot 10 \cdot 10 = 1,000,000$$

number of zeros in the number when it is written in decimal form. The number 1,000,000,000,000,000, for example, would be written as 10^{15} because it contains 15 zeros. It also follows that the number 10 can be written as 10^1.

The world's annual production of chicken eggs,

$$390,000,000,000$$

can be written as

$$390 \times 1,000,000,000$$

or

$$390 \times 10^9$$

It can also be written as

$$3.9 \times 100,000,000,000$$

or

$$3.9 \times 10^{11}$$

When it is written in this *last* form, it is written in scientific notation.

▶ A number is in **scientific notation** if it is written in the form

$$a \times 10^b$$

in which a is a number that is at least as large as 1 but less than 10* and b is an integer.

---

* The reason that a is restricted to numbers that are at least 1 but less than 10 is merely a matter of convenience. It is convenient to have exactly one way to write a number in scientific notation. Without the restriction on a, a given number could be written in scientific notation in many ways.

If we compare 390,000,000,000 with its form in scientific notation, 3.9×10^{11}, we see that the decimal point has been moved from the end of the

$$3.\overset{11}{9}\,\overset{10}{0}\,0\,,\overset{9}{0}\,\overset{8}{0}\,\overset{7}{0}\,,\overset{6}{0}\,\overset{5}{0}\,\overset{4}{0}\,,\overset{3}{0}\,\overset{2}{0}\,\overset{1}{0}\,. = 3.9 \times 10^{11}$$

number to just after its first digit. The point has been moved *11* decimal places to the left and the *11* has become the exponent of the 10. The decimal point is placed just after the first digit so that, when the number is written in the form $a \times 10^b$, a will be at least 1 but less than 10. The number of places that the point has been moved to the left becomes the exponent b.

Here are more examples of numbers written in scientific notation.

$$5{,}000 = 5 \times 10^3$$
Decimal form $\qquad 450 = 4.5 \times 10^2 \qquad$ Scientific notation
$$12{,}005 = 1.2005 \times 10^4$$

Exercises

Set I

1. Solve the following formulas for the variables indicated.
   a) For s: $\quad p = 4s$
   b) For p: $\quad i = prt$
   c) For h: $\quad v = \dfrac{a^2 h}{3}$

2. Tell which one of the following properties is illustrated by each equation: commutative property, associative property, distributive property.
   a) $2(7x) = (2 \cdot 7)x$
   b) $2(7 + x) = 14 + 2x$
   c) $7 + x = x + 7$

3. Certain types of bamboo grow at incredible speeds. Measurements taken of one type are presented in this table:

Time in hours, t	0	1	2	3	4	5
Height of plant in inches, h	0	1.5	3	4.5	6	7.5

   a) Write a formula for this function.
   b) What kind of function is this?
   c) How many inches per hour does the bamboo plant represented above grow?
   d) At this rate of growth, how tall would the plant be 24 hours after planting?

Set II

4. Write the name by which each of the following numbers is usually called.
   a) 10^2 d) 10^5
   b) 10^3 e) 10^6
   c) 10^4

5. Write each of the following numbers as a power of ten.
   a) $10 \cdot 10 \cdot 10 \cdot 10 \cdot 10 \cdot 10 \cdot 10 \cdot 10$
   b) 100,000,000,000
   c) One million
   d) Ten thousand

6. Write the number that is ten times as large as each of the following.
   a) 3,400   e) 10^{15}
   b) 5.72   f) 8×10^7
   c) 0.9   g) 0.3×10^{11}
   d) 0.0016

7. Write the number that is one-tenth as large as each of the following.
   a) 750   d) 10^6
   b) 2.8   e) 41×10^9
   c) 0.001

8. Write each of the following numbers in decimal form.
   a) 2×10^3 d) 6.02×10^1
   b) 1×10^9 e) 0.3×10^{12}
   c) 7.5×10^4 f) 0.0084×10^7

9. The number 420×10^5 is not in scientific notation because 420 is larger than 10.
   a) Write 420×10^5 in decimal form.
   b) Write it in scientific notation.

10. Write each of the following numbers in scientific notation.
    a) 30,000   d) 20,100,000
    b) 80   e) 1,984
    c) 720   f) 600,050

11. Write each of the following numbers in scientific notation.
    a) 10^{12} d) 0.3×10^{20}
    b) 50×10^5 e) 0.0076×10^4
    c) 940×10^8

12. The Chinese alphabet contains about forty thousand characters.
    a) Write this number in decimal form.
    b) Write it in scientific notation.

13. Every person shares his or her birthday with approximately 11,000,000 other people.
    a) Write this number in words.
    b) Write it in scientific notation.

14. The mass of the sun is about 3.3×10^5 times the mass of the earth.
    a) Write this number in decimal form.
    b) Write it in words.

15. The thirteen cards in a bridge hand can be arranged in more than 6.227×10^9 different ways.
    a) Write this number in decimal form.
    b) Write it in words.

Set III

16. Write the names of each of the following numbers.
    a) 10^9 e) 10^{13}
    b) 10^{10}
    c) 10^{11}
    d) 10^{12}

17. Write each of the following numbers as a power of ten.
    a) $10 \cdot 10 \cdot 10 \cdot 10$
    b) 1,000,000,000,000,000
    c) One thousand
    d) 10

18. Write the number that is ten times as large as each of the following.
    a) 510
    b) 9.23
    c) 0.4
    d) 0.00077
    e) 10^{12}
    f) 6×10^5
    g) 0.2×10^8

19. Write the number that is one-tenth as large as each of the following.
    a) 6,400
    b) 1.9
    c) 0.05
    d) 10^{10}
    e) 27×10^3

20. Write each of the following numbers in decimal form.
    a) 8×10^2
    b) 1×10^7
    c) 5.2×10^4
    d) 4.04×10^9
    e) 0.6×10^1
    f) 0.0317×10^6

21. The number 0.5×10^4 is not in scientific notation because 0.5 is less than 1.
    a) Write 0.5×10^4 in decimal form.
    b) Write it in scientific notation.

22. Write each of the following numbers in scientific notation.
    a) 5,000,000
    b) 600
    c) 48,000
    d) 1,090
    e) 77
    f) 30,020,000

23. Write each of the following numbers in scientific notation.
    a) 10^9
    b) 80×10^4
    c) 250×10^{10}
    d) 0.4×10^2
    e) 0.00061×10^7

24. Parker Brothers has sold more than seventy million sets of its *Monopoly* game.
    a) Write this number in decimal form.
    b) Write it in scientific notation.

25. One cubic centimeter of smoke from a cigarette contains approximately 5,000,000,000 particles of tar and other pollutants.
    a) Write this number in words.
    b) Write it in scientific notation.

26. Sand dunes cover approximately 9.7×10^5 square miles of the Sahara Desert.
    a) Write this number in decimal form.
    b) Write it in words.

27. Mars is about 1.41×10^8 miles from the sun.
    a) Write this number in decimal form.
    b) Write it in words.

Set IV

To see what one million "looks" like, the children in an elementary school in Pennsylvania collected one million bottle caps. This photograph shows one of the children "buried" in the pile.

Can you show what each of the following numbers "looks" like *in scientific notation?*

1. The number that is the *sum* of a million and a million.
2. The number that is the *product* of a million and a million.
3. The number that is the millionth *power* of a million.

COURTESY OF HALE OBSERVATORIES.

A Fundamental Property of Exponents

The most distant object in the universe that can be seen without a telescope is the Great Galaxy in Andromeda. Shown in this photograph, it is about 2,200,000 light-years from the earth. This means that it takes light about 2,200,000 years to cover the distance between this galaxy and our planet.

Because light travels about 5,900,000,000,000 miles in a year, in 2,200,000 years it travels $(5,900,000,000,000)(2,200,000) = 12,980,000,000,000,000,000$ miles. This means that the Great Galaxy in Andromeda is about 13×10^{18} miles from the earth! The word for 10^{18} is a quintillion. So the Great Galaxy in Andromeda is about 13 quintillion miles from the earth.

The numbers in this calculation are so large that they are hard to work with in decimal form. If the problem is written in scientific notation, it looks like this:

$$(5.9 \times 10^{12})(2.2 \times 10^6)$$

To be able to multiply the numbers in this form, we need to know what to do with the powers of ten. A few simpler examples are worth considering first. Compare the following problems in which powers of ten are multiplied in decimal form and in exponential form.

In decimal form	*In exponential form*
$100 \cdot 1{,}000 = 100{,}000$	$10^2 \cdot 10^3 = 10^5$
$10 \cdot 1{,}000{,}000 = 10{,}000{,}000$	$10^1 \cdot 10^6 = 10^7$
$1{,}000{,}000 \cdot 100{,}000 = 100{,}000{,}000{,}000$	$10^6 \cdot 10^5 = 10^{11}$

To multiply two powers of ten, all we have to do is to add the exponents. In symbols, $10^a \cdot 10^b = 10^{a+b}$, in which a and b are positive integers. Returning to the problem,

$$(5.9 \times 10^{12})(2.2 \times 10^6)$$

we can first rearrange the four numbers being multiplied because the order in which a series of numbers are multiplied does not matter:

$$(5.9)(2.2)(10^{12})(10^6)$$

Multiplying 5.9 and 2.2, we get

$$12.98 \times (10^{12})(10^6)$$

Multiplying 10^{12} and 10^6 by adding the exponents, we get

$$12.98 \times (10^{12+6}) = 12.98 \times 10^{18}$$

Changing this answer to decimal form, we get

$$12{,}980{,}000{,}000{,}000{,}000{,}000$$

In making this calculation, we have applied the fact that $10^a \cdot 10^b = 10^{a+b}$, in which a and b are positive integers. In this equation, a, b, and $a + b$ represent the exponents of the 10's. The 10 is called the *base*.

▶ In the expression x^y, x is the **base** and y is the **exponent.**

The rule of adding exponents also applies to multiplying powers for which the base is a number other than 10. Here are more examples showing why this rule works.

EXAMPLE 1
Show why $2^2 \cdot 2^3 = 2^{2+3}$.

SOLUTION
$2^2 = 2 \cdot 2$ and $2^3 = 2 \cdot 2 \cdot 2$; so
$2^2 \cdot 2^3 = \underbrace{(2 \cdot 2)}_{\text{2 twos}}\underbrace{(2 \cdot 2 \cdot 2)}_{\text{3 twos}} = \underbrace{2 \cdot 2 \cdot 2 \cdot 2 \cdot 2}_{\text{2 + 3 twos}} = 2^{2+3}$

EXAMPLE 2
Show why $x^6 \cdot x^1 = x^{6+1}$.

SOLUTION
$x^6 = x \cdot x \cdot x \cdot x \cdot x \cdot x$ and $x^1 = x$; so
$x^6 \cdot x^1 = \underbrace{(x \cdot x \cdot x \cdot x \cdot x \cdot x)}_{\text{6 x's}}\underbrace{(x)}_{\text{1 x}} = \underbrace{x \cdot x \cdot x \cdot x \cdot x \cdot x \cdot x}_{\text{6 + 1 x's}} = x^{6+1}$.

The rule illustrated by these examples is *the first law of exponents*.

▶ **The First Law of Exponents**
If a and b are positive integers, then $x^a \cdot x^b = x^{a+b}$.

In words, this law says that, to multiply powers of the same base, we add the exponents.
    Here are additional examples illustrating how this law is used.

EXAMPLE 3
Use the table of powers of 6 shown here to find

$$216 \cdot 7{,}776$$

SOLUTION
From the table, we see that $216 = 6^3$ and $7{,}776 = 6^5$.

$$6^3 \cdot 6^5 = 6^{3+5} = 6^8$$

From the table, $6^8 = 1{,}679{,}616$. So $216 \cdot 7{,}776 = 1{,}679{,}616$.

$6^1 = 6$
$6^2 = 36$
$6^3 = 216$
$6^4 = 1{,}296$
$6^5 = 7{,}776$
$6^6 = 46{,}656$
$6^7 = 279{,}936$
$6^8 = 1{,}679{,}616$
$6^9 = 10{,}077{,}696$
$6^{10} = 60{,}466{,}176$

EXAMPLE 4

Find the product of 4×10^4 and 7×10^7, writing the answer in scientific notation.

SOLUTION

$$(4 \times 10^4)(7 \times 10^7) = (4 \cdot 7)(10^4 \cdot 10^7) = 28 \times 10^{11}$$

Because $28 = 2.8 \times 10^1$,

$$28 \times 10^{11} = 2.8 \times 10^1 \times 10^{11} = 2.8 \times 10^{12}$$

Exercises

Set I

1. Guess a formula for the function represented by each of these tables. Begin each formula with $y =$.

   a)
x	1	2	3	4	5
y	–1	–4	–9	–16	–25

   b)
x	2	3	4	5	6
y	10	9	8	7	6

   c)
x	0	1	2	3	4
y	–5	–2	1	4	7

2. Solve the following pairs of simultaneous equations.

   a) $4x - y = 60$
   $3x + y = 59$

   b) $x + 9y = 20$
   $x = y + 5$

3. In one of the first automobile races, held in France in 1887, the winner covered a distance of 20 miles in 74 minutes.

   a) What was his average speed in miles per hour? (Round your answer to the nearest tenth.)

   b) At this rate, how long would it take him to catch up with another car traveling at a speed of 12.2 miles per hour if the other car has a headstart of 2 miles? (Let x represent the time in hours.)

   12 mpm

Set II

4. The product $x \cdot x \cdot x \cdot y \cdot y \cdot y \cdot y$ can be written in exponential form as x^3y^4. Write each of the following products in exponential form.

   a) $x \cdot x \cdot x \cdot x \cdot x \cdot x \cdot x \cdot x$
   b) $y \cdot y \cdot y \cdot y \cdot y$
   c) $x \cdot x \cdot y \cdot y \cdot y \cdot y \cdot y \cdot y \cdot y$
   d) $x \cdot x \cdot x \cdot x \cdot x \cdot x \cdot y$
   e) $x \cdot y \cdot x \cdot y \cdot x \cdot x$

5. Multiply as indicated and express your answers in exponential form.

   a) $x^2 \cdot x^3$
   b) $y \cdot y^9$
   c) $x^4 \cdot x^4$
   d) $y \cdot y^2 \cdot y^3$
   e) $x^4 \cdot y^6 \cdot x^5 \cdot y^3$
   f) $x \cdot y^7 \cdot x \cdot y^7 \cdot x \cdot y^7$
   g) $2x^2 \cdot 8x^8$
   h) $5x^3 \cdot 3x^5$

6. Find x in each of the following equations.
   a) $10^3 \cdot 10^8 = 10^x$
   b) $4^5 \cdot 4^2 = 4^x$
   c) $(-3)^6 \cdot (-3)^4 = (-3)^x$
   d) $y^1 \cdot y^7 = y^x$
   e) $9^x \cdot 9^5 = 9^{12}$
   f) $(-2)^{11} \cdot (-2)^x = (-2)^{14}$
   g) $y^x \cdot y^{10} = y^{30}$
   h) $y^x \cdot y^x = y^{16}$

7. Use the table of powers of 3 at the right to find the following products without doing any multiplying.

   a) $3^3 \cdot 3^4$ d) $6,561 \cdot 6,561$
   b) $3 \cdot 3^{10}$ e) $27 \cdot 81 \cdot 243$
   c) $243 \cdot 19,683$ f) $9 \cdot 9 \cdot 9 \cdot 9 \cdot 9$

   Refer to the table to tell whether each of the following equations is true or false.

   g) $3^2 + 3^5 = 3^7$ i) $3^{11} + 3^4 = 3^{15}$
   h) $3^2 \cdot 3^5 = 3^7$ j) $3^{11} \cdot 3^4 = 3^{15}$

8. Find each of the following products. Express each answer in scientific notation.
   a) $(2 \times 10^3)(4 \times 10^5)$
   b) $(1 \times 10^6)(6 \times 10^1)$
   c) $(5 \times 10^4)(8 \times 10^7)$
   d) $(3 \times 10^2)(7 \times 10^{12})$
   e) $(10)(9 \times 10^4)$
   f) $(0.5)(16 \times 10^{16})$

9. There are approximately 3×10^{18} atoms in a milligram of gold. One kilogram is equal to 10^6 milligrams.
   a) How many atoms are there in a kilogram of gold?
   b) Write this number in decimal form.

$3^1 = 3$
$3^2 = 9$
$3^3 = 27$
$3^4 = 81$
$3^5 = 243$
$3^6 = 729$
$3^7 = 2,187$
$3^8 = 6,561$
$3^9 = 19,683$
$3^{10} = 59,049$
$3^{11} = 177,147$
$3^{12} = 531,441$
$3^{13} = 1,594,323$
$3^{14} = 4,782,969$
$3^{15} = 14,348,907$
$3^{16} = 43,046,721$

10.

b)

Set III

11. The product $x \cdot x \cdot x \cdot x \cdot x \cdot y \cdot y$ can be written in exponential form as $x^5 y^2$. Write each of the following products in exponential form.
    a) $x \cdot x \cdot x \cdot x$
    b) $y \cdot y \cdot y \cdot y \cdot y \cdot y \cdot y \cdot y \cdot y$
    c) $x \cdot x \cdot x \cdot x \cdot x \cdot x \cdot y \cdot y \cdot y$
    d) $x \cdot y \cdot y \cdot y \cdot y \cdot y$
    e) $x \cdot y \cdot y \cdot x \cdot y \cdot y$

12. Multiply as indicated and express your answers in exponential form.
    a) $x^4 \cdot x^2$ e) $x^3 \cdot y^7 \cdot x^9 \cdot y^5$
    b) $y^7 \cdot y$ f) $x^6 \cdot y \cdot x^6 \cdot y \cdot x^6 \cdot y$
    c) $x^5 \cdot x^5$ g) $7x^7 \cdot 3x^3$
    d) $y \cdot y^2 \cdot y^4$ h) $4x^9 \cdot 9x^4$

13. Find x in each of the following equations.
    a) $10^6 \cdot 10^2 = 10^x$
    b) $2^{12} \cdot 2^4 = 2^x$
    c) $(-5)^3 \cdot (-5)^3 = (-5)^x$
    d) $y^8 \cdot y = y^x$
    e) $7^x \cdot 7^3 = 7^{15}$
    f) $(-4)^2 \cdot (-4)^x = (-4)^{10}$
    g) $y^x \cdot y^5 = y^{25}$
    h) $y^x \cdot y^x = y^{36}$

14. Use the table of powers of 4 at the right to find the following products without doing any multiplying.
    a) $4^2 \cdot 4^5$
    b) $4 \cdot 4^8$
    c) $256 \cdot 262{,}144$
    d) $16{,}384 \cdot 16{,}384$
    e) $256 \cdot 1{,}024 \cdot 4{,}096$
    f) $64 \cdot 64 \cdot 64 \cdot 64$

    Refer to the table to tell whether each of the following equations is true or false.

    g) $4^3 + 4^4 = 4^7$
    h) $4^3 \cdot 4^4 = 4^7$
    i) $4^{10} + 4^2 = 4^{12}$
    j) $4^{10} \cdot 4^2 = 4^{12}$

15. Find each of the following products. Express each answer in scientific notation.
    a) $(3 \times 10^4)(2 \times 10^3)$
    b) $(8 \times 10^1)(1 \times 10^8)$
    c) $(6 \times 10^7)(5 \times 10^2)$
    d) $(4 \times 10^5)(9 \times 10^5)$
    e) $(10)(7 \times 10^9)$
    f) $(0.5)(12 \times 10^{12})$

16. There are approximately 3×10^7 cubic kilometers of ice on the earth. One cubic kilometer is equal to 10^9 cubic meters.
    a) How many cubic meters of ice are there on the earth?
    b) Write this number in decimal form.

17. The fastest camera in the world is capable of taking 600 million pictures per second! There are 3,600 seconds in an hour.
    a) Write each of these numbers in scientific notation.
    b) If the camera could operate continuously at this rate, how many pictures could it take in an hour? Express your answer in scientific notation.

$$4^1 = 4$$
$$4^2 = 16$$
$$4^3 = 64$$
$$4^4 = 256$$
$$4^5 = 1{,}024$$
$$4^6 = 4{,}096$$
$$4^7 = 16{,}384$$
$$4^8 = 65{,}536$$
$$4^9 = 262{,}144$$
$$4^{10} = 1{,}048{,}576$$
$$4^{11} = 4{,}194{,}304$$
$$4^{12} = 16{,}777{,}216$$
$$4^{13} = 67{,}108{,}864$$
$$4^{14} = 268{,}435{,}456$$
$$4^{15} = 1{,}073{,}741{,}824$$
$$4^{16} = 4{,}294{,}967{,}296$$

Set IV

How many times does a person breathe in a lifetime? Use the fact that human beings breathe at the rate of about 15 times per minute to find the approximate number of breaths taken by someone who lives to be 80 years old. Round your answer and express it in scientific notation.

$4^{11} = 4,194,304$
$4^{12} = 16,777,216$
$4^{13} = 67,108,864$
$4^{14} = 268,435,456$
$4^{15} = 1,073,741,824$

DRAWING BY CHAS. ADDAMS; © 1973 THE NEW YORKER MAGAZINE, INC.

LESSON **3**

Two More Properties of Exponents

Babies are currently being born at the rate of about 130,000,000 per year! A year contains approximately 530,000 minutes, which means that about $\dfrac{130,000,000}{530,000}$ babies are born each minute. Doing the indicated division and rounding the result, we get 245. About 245 babies are born each minute.

Because the numbers in this calculation are very large, perhaps it would be easier in scientific notation:

$$\frac{1.3 \times 10^8}{5.3 \times 10^5} = \ ?$$

Dividing 1.3 by 5.3 gives 0.245 (approximately). What is 10^8 divided by 10^5?

$$\frac{10^8}{10^5} = \frac{10 \cdot 10 \cdot 10 \cdot 10 \cdot 10 \cdot 10 \cdot 10 \cdot 10}{10 \cdot 10 \cdot 10 \cdot 10 \cdot 10} = 10 \cdot 10 \cdot 10 = 10^3$$

It can be seen that, to divide two powers of ten, we simply subtract the exponents:

$$\frac{10^8}{10^5} = 10^{8-5} = 10^3$$

$$\text{So} \quad \frac{1.3 \times 10^8}{5.3 \times 10^5} = 0.245 \times 10^3 = 245$$

Just as the first law of exponents applies to powers of numbers other than 10, this pattern of subtracting exponents is also true for division problems in other bases. Here are two more examples showing why the pattern works.

EXAMPLE 1

Show why $\dfrac{3^7}{3^2} = 3^{7-2}$.

SOLUTION

$3^7 = 3 \cdot 3 \cdot 3 \cdot 3 \cdot 3 \cdot 3 \cdot 3$ and $3^2 = 3 \cdot 3$; so

$$\frac{3^7}{3^2} = \frac{\overbrace{3 \cdot 3 \cdot 3 \cdot 3 \cdot 3 \cdot 3 \cdot 3}^{7 \text{ threes}}}{\underbrace{3 \cdot 3}_{2 \text{ threes}}} = \underbrace{3 \cdot 3 \cdot 3 \cdot 3 \cdot 3}_{7 - 2 \text{ threes}} = 3^{7-2}$$

EXAMPLE 2

Show why $\dfrac{x^{10}}{x^6} = x^{10-6}$.

SOLUTION

$$\frac{x^{10}}{x^6} = \frac{\overbrace{x \cdot x \cdot x \cdot x \cdot x \cdot x \cdot x \cdot x \cdot x \cdot x}^{10 \; x\text{'s}}}{\underbrace{x \cdot x \cdot x \cdot x \cdot x \cdot x}_{6 \; x\text{'s}}} = \underbrace{x \cdot x \cdot x \cdot x}_{10 - 6 \; x\text{'s}} = x^{10-6}$$

The rule illustrated by these examples is *the second law of exponents.*

▶ **The Second Law of Exponents**
If a, b, and $a - b$ are positive integers and x is any number other than zero, then
$$\frac{x^a}{x^b} = x^{a-b}.$$

In words, this law says that, to divide powers of the same base, we subtract the exponents. In Lesson 4, we will find out about negative exponents and see that this law holds for them as well.

In our study of exponents, we have established two rules: one for *multiplying* two powers of the same base and the other for *dividing* them. Another useful rule helps us find the *power* of a power. Here are examples of how it works.

EXAMPLE 3
Express $(7^2)^3$ as a power of 7.

SOLUTION
By the definition of an exponent, $(7^2)^3 = 7^2 \cdot 7^2 \cdot 7^2$. By the first law of exponents, $7^2 \cdot 7^2 \cdot 7^2 = 7^{2+2+2} = 7^{3 \cdot 2}$, or $7^{2 \cdot 3}$. So $(7^2)^3 = 7^6$.

EXAMPLE 4
Express $(x^6)^4$ as a power of x.

SOLUTION
By the definition of an exponent, $(x^6)^4 = x^6 \cdot x^6 \cdot x^6 \cdot x^6$. By the first law of exponents, $x^6 \cdot x^6 \cdot x^6 \cdot x^6 = x^{6+6+6+6} = x^{4 \cdot 6}$, or $x^{6 \cdot 4}$. So $(x^6)^4 = x^{24}$.

These examples suggest the following rule.

▶ **The Third Law of Exponents**
If a and b are positive integers, then $(x^a)^b = x^{ab}$.

In words, this law says that, to raise a power to a power, we multiply the exponents.

Exercises

Set I

1. Write each of the following numbers in scientific notation.
   a) 186,000
   b) 12
   c) 30×10^{12}
   d) 0.5×10^9

2. This exercise is about the graph at the right.
   a) Write an equation for line a.
   b) Write an equation for line b.
   c) What is the solution to the pair of equations that you have written?

3. A parking meter contains 98 coins, of which some are pennies and the rest are nickels. The total value of the coins in the meter is $1.50.
   a) Use this information to write a pair of simultaneous equations.

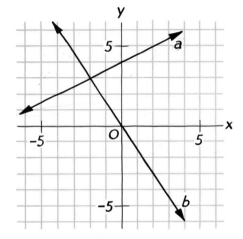

b) Solve the equations to find out how many pennies and how many nickels the meter contains.

Set II

4. Show, as was done in the examples in this lesson, why each of the following equations is true.
   a) $4^5 \cdot 4^2 = 4^7$
   b) $\dfrac{4^5}{4^2} = 4^3$
   c) $(4^5)^2 = 4^{10}$

5. Write each of the following expressions as a single power of x.
   a) $\dfrac{x^{12}}{x^3}$
   b) $\dfrac{x^5}{x^4}$
   c) $\dfrac{x^8}{x}$
   d) $(x^7)^2$
   e) $(x^2)^7$
   f) $(x^6)^6$

6. Find x in each of the following equations.
   a) $\dfrac{10^8}{10^2} = 10^x$
   b) $\dfrac{3^6}{3} = 3^x$
   c) $\dfrac{(-4)^5}{(-4)^4} = (-4)^x$
   d) $\dfrac{9^x}{9^7} = 9^{10}$
   e) $\dfrac{(-2)^9}{(-2)^x} = -2$
   f) $\dfrac{5^{16}}{5^x} = 5^x$
   g) $6^4 \cdot 6^7 = 6^x$
   h) $(6^4)^7 = 6^x$
   i) $11^8 \cdot 11^2 = 11^x$
   j) $(11^8)^2 = 11^x$
   k) $4^x \cdot 4^3 = 4^{12}$
   l) $(4^x)^3 = 4^{12}$
   m) $7^x \cdot 7^x = 7^{36}$
   n) $(7^x)^x = 7^{36}$

7. Use the table of powers of 4 at the right to find the following products or quotients.

a) $\dfrac{4^9}{4^4}$

b) $\dfrac{4^{15}}{4}$

c) $\dfrac{4,194,304}{64}$

d) $\dfrac{268,435,456}{16,384}$

e) $(4^5)^3$

f) $4^5 \cdot 4^3$

g) $(16)^6$

h) $(4,096)^2$

4^1	$= 4$
4^2	$= 16$
4^3	$= 64$
4^4	$= 256$
4^5	$= 1,024$
4^6	$= 4,096$
4^7	$= 16,384$
4^8	$= 65,536$
4^9	$= 262,144$
4^{10}	$= 1,048,576$
4^{11}	$= 4,194,304$
4^{12}	$= 16,777,216$
4^{13}	$= 67,108,864$
4^{14}	$= 268,435,456$
4^{15}	$= 1,073,741,824$
4^{16}	$= 4,294,967,296$

Refer to the table to tell whether each of the following equations is true or false.

i) $4^{10} - 4^8 = 4^2$

j) $\dfrac{4^{10}}{4^8} = 4^2$

k) $(4^7)^2 = 4^{14}$

l) $4^7 \cdot 4^2 = 4^{14}$

8. Find each of the following quotients. Express each answer in scientific notation.

a) $\dfrac{9 \times 10^9}{3 \times 10^3}$

b) $\dfrac{8 \times 10^{15}}{4 \times 10^5}$

c) $\dfrac{4 \times 10^{15}}{8 \times 10^5}$

d) $\dfrac{7 \times 10^8}{10^6}$

e) $\dfrac{6 \times 10^{12}}{2}$

f) $\dfrac{3 \times 10^{10}}{40}$

9. The earth's population two thousand years ago has been estimated to have been one hundred thirty-three million people. It is now more than four billion.

a) Write each of these population numbers in decimal form.

b) Write each in scientific notation.

c) Approximately how many times as many people are living on the earth now compared with two thousand years ago?

10. The earth is about 1.5×10^{13} centimeters from the sun. Light travels at a speed of 3×10^{10} centimeters per second. How long does it take light from the sun to reach the earth?

Set III

11. Show, as was done in the examples in this lesson, why each of the following equations is true.

a) $2^4 \cdot 2^3 = 2^7$

b) $\dfrac{2^4}{2^3} = 2^1$

c) $(2^4)^3 = 2^{12}$

12. Write each of the following expressions as a single power of x.

a) $\dfrac{x^{10}}{x^5}$

b) $\dfrac{x^7}{x^6}$

c) $\dfrac{x^4}{x}$

d) $(x^3)^4$

e) $(x^4)^3$

f) $(x^8)^8$

13. Find x in each of the following equations.

a) $\dfrac{10^{12}}{10^4} = 10^x$ f) $\dfrac{2^{36}}{2^x} = 2^x$

b) $\dfrac{7^5}{7} = 7^x$ g) $9^5 \cdot 9^3 = 9^x$

     h) $(9^5)^3 = 9^x$

c) $\dfrac{(-5)^9}{(-5)^8} = (-5)^x$ i) $4^6 \cdot 4^4 = 4^x$

     j) $(4^6)^4 = 4^x$

d) $\dfrac{3^x}{3^6} = 3^{10}$ k) $11^x \cdot 11^2 = 11^8$

     l) $(11^x)^2 = 11^8$

     m) $6^x \cdot 6^x = 6^{16}$

e) $\dfrac{(-8)^4}{(-8)^x} = -8$ n) $(6^x)^x = 6^{16}$

14. Use the table of powers of 3 at the right to find the following products or quotients.

a) $\dfrac{3^8}{3^3}$

b) $\dfrac{3^{14}}{3}$ e) $(3^2)^6$

     f) $3^2 \cdot 3^6$

c) $\dfrac{531,441}{2,187}$ g) $(27)^5$

     h) $(243)^3$

d) $\dfrac{43,046,721}{6,561}$

Refer to the table to tell whether each of the following equations is true or false.

i) $3^9 - 3^4 = 3^5$

j) $\dfrac{3^9}{3^4} = 3^5$

k) $3^2 \cdot 3^6 = 3^{12}$

l) $(3^2)^6 = 3^{12}$

15. Find each of the following quotients. Express each answer in scientific notation.

a) $\dfrac{8 \times 10^8}{2 \times 10^2}$ d) $\dfrac{5 \times 10^9}{10^5}$

b) $\dfrac{6 \times 10^{11}}{3 \times 10^7}$ e) $\dfrac{4 \times 10^6}{2}$

c) $\dfrac{3 \times 10^{11}}{6 \times 10^7}$ f) $\dfrac{7 \times 10^{12}}{50}$

$3^1 = 3$
$3^2 = 9$
$3^3 = 27$
$3^4 = 81$
$3^5 = 243$
$3^6 = 729$
$3^7 = 2,187$
$3^8 = 6,561$
$3^9 = 19,683$
$3^{10} = 59,049$
$3^{11} = 177,147$
$3^{12} = 531,441$
$3^{13} = 1,594,323$
$3^{14} = 4,782,969$
$3^{15} = 14,348,907$
$3^{16} = 43,046,721$

16. A swarm of locusts once seen flying across the Red Sea was estimated to have consisted of two hundred fifty billion insects and to have covered an area of two thousand square miles.

a) Write each of these numbers in decimal form.

b) Write each in scientific notation.

c) Approximately how many locusts were there in each square mile of the swarm?

17. The sun contains about 1×10^{57} atoms. The volume of the sun is about 8.5×10^{31} cubic inches. How many atoms, on the average, are in each cubic inch of the sun? (Round your answer to the nearest tenth.)

Set IV

$$10{,}000{,}000{,}000{,}000{,}000{,}000{,}000{,}000{,}000{,}000{,}000{,}$$
$$000{,}000{,}000{,}000{,}000{,}000{,}000{,}000{,}000{,}000{,}000{,}$$
$$000{,}000{,}000{,}000{,}000{,}000{,}000{,}000{,}000{,}000{,}000.$$

The number that is written as 1 followed by 100 zeros is called a *googol*. It was named by a young nephew of the American mathematician Edward Kasner when the boy was asked to invent a name for a very large number.

1. Are either or both of the numbers below equal to a googol?

$$10^{100} \qquad 100^{10}$$

2. Write each one in scientific notation.

3. If you think that one number is larger than the other, tell how many times larger it is.

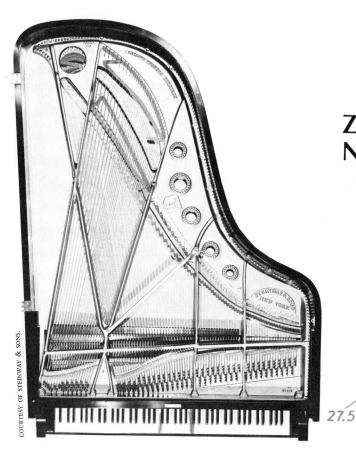

LESSON 4

Zero and Negative Exponents

27.5
55
110
220
440
880
1760
3520

When a piano is tuned, the first note to be tuned is the A above middle C. After is has been tuned to a frequency of 440 cycles per second, the rest of the A's on the keyboard are tuned so that each successive A has twice the frequency of the one before it.* This means that the three A's above "middle A" are tuned as shown here.

$$440 \cdot 2 \qquad = 440 \cdot 2^1 = \quad 880$$
$$440 \cdot 2 \cdot 2 \quad = 440 \cdot 2^2 = 1{,}760$$
$$440 \cdot 2 \cdot 2 \cdot 2 = 440 \cdot 2^3 = 3{,}520$$

*The pitch of a sound (how high or low it sounds) depends on its frequency (the number of vibrations per second). The greater the frequency, the higher the pitch.

COURTESY OF STEINWAY & SONS.

359

From the middle column of this table, it looks as if a formula for the frequencies of these notes is $440 \cdot 2^n$, in which $n = 1, 2$, and 3. If the keyboard continued, it seems reasonable to assume that the formula would continue to work, with $n = 4, 5, 6$, and so on.

What would happen if we tried applying this formula to middle A itself? Because the integer before 1 is 0, this would mean that

$$440 \cdot 2^0 = 440$$

Applying the formula to the A's before middle A would require the use of negative integers, as shown here.

$$440 \cdot 2^{-1} = 220$$
$$440 \cdot 2^{-2} = 110$$
$$440 \cdot 2^{-3} = 55$$
$$440 \cdot 2^{-4} = 27.5$$

Because it doesn't make any sense to think of zero and negative exponents as meaning repeated multiplication, we will have to figure out their meanings from these equations instead.

If $440 \cdot 2^0 = 440$, then it follows that we want 2^0 to equal 1.

If $440 \cdot 2^{-1} = 220$, then it follows that we want 2^{-1} to equal $\frac{1}{2}$.

In the same fashion,

$$2^{-2} = \frac{1}{4} \text{ or } \frac{1}{2^2}$$

$$2^{-3} = \frac{1}{8} \text{ or } \frac{1}{2^3}$$

$$\text{and} \quad 2^{-4} = \frac{1}{16} \text{ or } \frac{1}{2^4}$$

This pattern also extends to bases other than 2. For example, look at this pattern of powers of 5:

$$5^3 = 125$$
$$5^2 = 25$$
$$5^1 = 5$$
$$5^0 = 1$$

$$5^{-1} = \frac{1}{5} = \frac{1}{5^1}$$

$$5^{-2} = \frac{1}{25} = \frac{1}{5^2}, \quad \text{etc.}$$

If the exponent is decreased by 1, the resulting number is the preceding one divided by 5.

This pattern works for all bases except 0; so we can define zero and negative integer exponents in the following way.

▶ **Definition of the Exponent Zero**

$x^0 = 1$, in which x may be any number except 0.

▶ **Definition of Negative Integer Exponents**

$x^{-a} = \dfrac{1}{x^a}$, in which x may be any number except 0.

In addition to being consistent with the patterns we have already observed, these definitions have an even more important consequence. *The laws of exponents that we discovered for positive integer exponents work for all integer exponents.* They are repeated here.

▶ **The First Law of Exponents**

If a and b are integers, then $x^a \cdot x^b = x^{a+b}$.

▶ **The Second Law of Exponents**

If a and b are integers and x is any number other than zero, then $\dfrac{x^a}{x^b} = x^{a-b}$.

▶ **The Third Law of Exponents**

If a and b are integers, then $(x^a)^b = x^{ab}$.

Here are examples in which there are zero or negative exponents.

EXAMPLE 1

Apply the first law of exponents to the expression $x^a \cdot x^0$ to explain why $x^0 = 1$. (Assume that x is not zero.)

SOLUTION

According to the first law of exponents, $x^a \cdot x^0 = x^{a+0} = x^a$. Because the result of multiplying the number x^a by x^0 is the same number, x^a, it follows that x^0 must be 1.

EXAMPLE 2

Apply the second law of exponents to the expression $\dfrac{4^2}{4^3}$ to explain why $4^{-1} = \dfrac{1}{4^1}$.

SOLUTION

According to the second law of exponents, $\dfrac{4^2}{4^3} = 4^{2-3} = 4^{-1}$.

Because $\dfrac{4^2}{4^3} = \dfrac{16}{64} = \dfrac{1}{4}$, this means that 4^{-1} must be equal to $\dfrac{1}{4}$, which is the same as $\dfrac{1}{4^1}$.

EXAMPLE 3

Write $(x^{-4})^{-5}$ as a single power of x.

SOLUTION

According to the third law of exponents, $(x^{-4})^{-5} = x^{(-4)(-5)} = x^{20}$.

Exercises

Set I

1. Find each of the following products. Express each answer in scientific notation.
   a) $(5)(2 \times 10^{10})$
   b) $(1.2 \times 10^{12})(1.2 \times 10^{12})$
   c) $(7 \times 10^7)(9 \times 10^9)$

2. Solve the following pairs of simultaneous equations.
   a) $3x + 4y = 27$ b) $2x - 6y = 11$
   $\ 3x + 5y = 27$ $\ x = 3y + 7$

   $2(3y+7)-6y=11$
   $6y-6y+14=11$

3. There are four consecutive integers whose sum is 2.
   a) If x represents the smallest integer, how would the other three numbers be represented?
   b) Write an equation expressing the fact that the sum of the four numbers is 2.
   c) What are the four numbers?

Set II

4. Write each of the following numbers without using any exponents.

a) 1^{10}
b) 10^1
c) 0^7
d) 7^0
e) $(-1)^6$
f) 6^{-1}
g) $(-2)^5$

h) 5^{-2}
i) 4^3
j) 4^{-3}
k) $(-4)^3$
l) $(-4)^{-3}$
m) $(-8)^{-1}$
n) $(-1)^{-8}$

$5^{10} = 9{,}765{,}625$
$5^9 = 1{,}953{,}125$
$5^8 = 390{,}625$
$5^7 = 78{,}125$
$5^6 = 15{,}625$
$5^5 = 3{,}125$
$5^4 = 625$
$5^3 = 125$
$5^2 = 25$
$5^1 = 5$
$5^0 = 1$
$5^{-1} = 0.2$
$5^{-2} = 0.04$
$5^{-3} = 0.008$
$5^{-4} = 0.0016$
$5^{-5} = 0.00032$
$5^{-6} = 0.000064$
$5^{-7} = 0.0000128$

5. Use the table of powers of 5 at the right to find each of the following.

a) $5^2 \cdot 5^{-3}$
b) $(5^2)^{-3}$
c) $\dfrac{5^2}{5^{-3}}$
d) $\dfrac{5^{-3}}{5^2}$
e) $(0.0016)(15{,}625)$
f) $\dfrac{5}{0.00032}$
g) $(78{,}125)(0.0000128)$
h) $(5^{-1})^{-8}$
i) $(3{,}125)^{-1}$
j) $(9{,}765{,}625)^0$

6. Copy and complete the following table for the function $y = 2^x$ by replacing each ▓ with either an integer or a decimal fraction.

x	4	3	2	1	0	−1	−2	−3
y	16	▓	▓	▓	▓	▓	▓	▓

7. Which of these symbols, $>$, $=$, or $<$, should replace ▓ in each of the following?

a) 3^1 ▓ 4^1
b) 3^0 ▓ 4^0
c) 3^{-1} ▓ 4^{-1}
d) 5^1 ▓ 10^{-1}

e) 6^{-2} ▓ 6^{-3}
f) $(-2)^6$ ▓ $(-3)^6$
g) $(-1)^7$ ▓ 7^{-1}
h) $(-10)^{-1}$ ▓ $(-10)^{-2}$

8. Write each of the following expressions as a single power of x.

a) $x^{-3} \cdot x^5$
b) $(x^{-3})^5$
c) $x^0 \cdot x^{10}$
d) $(x^0)^{10}$
e) $x^4 \cdot x^{-4}$
f) $(x^4)^{-4}$
g) $x^{-6} \cdot x^{-1}$

h) $(x^{-6})^{-1}$
i) $\dfrac{1}{x^8}$
j) $\dfrac{1}{x}$

k) $\dfrac{x^4}{x^9}$
l) $\dfrac{x^3}{x^3}$
m) $\dfrac{x^2}{x^{-5}}$

n) $\dfrac{x^{-8}}{x^4}$
o) $\dfrac{x^{-7}}{x^{-7}}$
p) $\dfrac{x^{-10}}{x}$

q) $\dfrac{x}{x^{-6}}$
r) $\dfrac{x^{-5}}{x^{-15}}$

Set III

9. Write each of the following numbers without using any exponents.

a) 1^8
b) 8^1
c) 0^{12}
d) 12^0
e) $(-1)^9$
f) 9^{-1}
g) $(-2)^6$

h) 6^{-2}
i) 5^3
j) 5^{-3}
k) $(-5)^3$
l) $(-5)^{-3}$
m) $(-4)^{-1}$
n) $(-1)^{-4}$

10. Use the table of powers of 2 at the right to find each of the following.

a) $2^{-3} \cdot 2^2$
b) $(2^{-3})^2$
c) $\dfrac{2^2}{2^{-3}}$
d) $\dfrac{2^{-3}}{2^2}$
e) $(0.0625)(32)$

f) $\dfrac{2}{0.0078125}$
g) $(256)(0.00390625)$
h) $(2^{-6})^{-1}$
i) $(0.5)^{-8}$
j) $(0.03125)^0$

$$2^{10} = 1{,}024$$
$$2^9 = 512$$
$$2^8 = 256$$
$$2^7 = 128$$
$$2^6 = 64$$
$$2^5 = 32$$
$$2^4 = 16$$
$$2^3 = 8$$
$$2^2 = 4$$
$$2^1 = 2$$
$$2^0 = 1$$
$$2^{-1} = 0.5$$
$$2^{-2} = 0.25$$
$$2^{-3} = 0.125$$
$$2^{-4} = 0.0625$$
$$2^{-5} = 0.03125$$
$$2^{-6} = 0.015625$$
$$2^{-7} = 0.0078125$$
$$2^{-8} = 0.00390625$$

11. Copy and complete the following table for the function $y = 10^x$ by replacing each ▓ with either an integer or a decimal fraction.

x	4	3	2	1	0	-1	-2	-3
y	10,000	▓	▓	▓	▓	▓	▓	▓

12. Which of these symbols, $>$, $=$, or $<$, should replace ▓ in each of the following?

a) 6^1 ▓ 5^1
b) 6^0 ▓ 5^0
c) 6^{-1} ▓ 5^{-1}
d) 8^1 ▓ 2^{-1}

e) 4^{-3} ▓ 4^{-2}
f) $(-3)^4$ ▓ $(-2)^4$
g) 9^{-1} ▓ $(-1)^9$
h) $(-7)^{-2}$ ▓ $(-7)^{-1}$

13. Write each of the following expressions as a single power of x.

a) $x^6 \cdot x^{-2}$
b) $(x^6)^{-2}$
c) $x^7 \cdot x^0$
d) $(x^7)^0$
e) $x^{-3} \cdot x^3$
f) $(x^{-3})^3$
g) $x^{-4} \cdot x^{-5}$

h) $(x^{-4})^{-5}$
i) $\dfrac{1}{x^9}$
j) $\dfrac{x^2}{x^2}$

k) $\dfrac{1}{x}$
l) $\dfrac{x^6}{x^{10}}$
m) $\dfrac{x^3}{x^{-8}}$

n) $\dfrac{x^{-7}}{x^7}$
o) $\dfrac{x^{-5}}{x}$
p) $\dfrac{x^{-4}}{x^{-4}}$

q) $\dfrac{x}{x^{-9}}$
r) $\dfrac{x^{-3}}{x^{-11}}$

Set IV

New ideas in mathematics often come from the discovery of patterns. In this lesson, for example, we discovered the meaning of *zero* and the *negative integers* as exponents from a pattern.

The meaning of *fractions* as exponents can also be discovered from a pattern. Look at the pattern shown here.

$$9^1 \cdot 9^1 = 9^{1+1} = 9^2 \quad \text{or} \quad 9 \cdot 9 = 81$$
$$9^{\frac{1}{2}} \cdot 9^{\frac{1}{2}} = 9^{\frac{1}{2}+\frac{1}{2}} = 9^1 \quad \text{or} \quad ? \cdot ? = 9$$
$$9^0 \cdot 9^0 = 9^{0+0} = 9^0 \quad \text{or} \quad 1 \cdot 1 = 1$$

1. What number do you think $9^{\frac{1}{2}}$ means?

2. In general, what number do you think $x^{\frac{1}{2}}$ means?

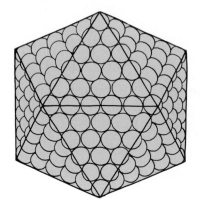

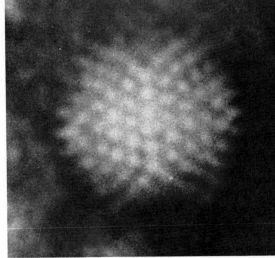

COURTESY OF R. W. HORNE.

LESSON 5
Small Numbers

This remarkable photograph, made with an electron microscope, shows a virus magnified to about a million times its actual size. This virus causes respiratory disease in human beings. The shape of the virus is that of an icosahedron, the geometric solid shown at the left of the photograph.

The virus is so small that its volume is only 0.00000000000000002 cubic centimeters. It is hard to believe that something this small could be photographed or measured. Nevertheless, scientists must work with numbers much smaller than even the volume of a virus. To do so, they usually express such numbers in scientific notation.

We have learned that a number written in scientific notation has the form

$$a \times 10^b$$

in which a is a number that is at least 1 but less than 10 and b is an integer. To see how a small number can be written in scientific notation, consider the

366

numbers 4,000 and 0.004.

$$4,000 = 4 \times 1,000 = 4 \times 10^3$$

$$0.004 = \frac{4}{1,000} = 4 \times \frac{1}{1,000} = 4 \times \frac{1}{10^3} = 4 \times 10^{-3}$$

In the first case, the decimal point has been moved 3 places to the left and the exponent is 3.

$$4,\overset{\frown}{0\,0\,0}, = 4 \times 10^3$$
$$3\ 2\ 1$$

In the second case, the decimal point has been moved 3 places to the right and the exponent is −3.

$$0.\overset{\frown}{0\,0\,4} = 4 \times 10^{-3}$$
$$1\ 2\ 3$$

These examples illustrate the relationship between the number of digits and direction in which the decimal point is moved and the exponent when a number is changed from decimal form to scientific notation. If the decimal point is moved n places to the left, the exponent is n. If it is moved n places to the right, the exponent is $-n$. If the decimal point is not moved at all, the exponent is 0.

Using this relationship to express the volume of the virus in scientific notation, we get

$$0.000\ 000\ 000\ 000\ 000\ 02 = 2 \times 10^{-17}$$
$$1\ 2\ 3\ \ 4\ 5\ 6\ \ 7\ 8\ 9\ \ 10\ 11\ 12\ \ 13\ 14\ 15\ \ 16\ 17$$

Here are more examples of how small numbers are written in scientific notation.

$$3 = 3 \times 10^0$$
$$0.08 = 8 \times 10^{-2}$$
$$0.000027 = 2.7 \times 10^{-5}$$

Exercises

Set I

1. Solve each of the following equations for x.
   a) $10^2 x = 10^5$

   b) $\dfrac{x}{10^2} = 10^5$

   c) $x + 10^2 = 10^5$

2. Graph the following equations.
   a) $y = 3x$
   b) $y = 3$
   c) $x + y = 3$

3. Bees have to travel about one hundred forty thousand miles to make a pound of honey. About three hundred million pounds of honey are produced in the United States and Canada each year.
   a) Write each of these numbers in scientific notation.
   b) About how far do bees have to travel to make the honey produced in the United States and Canada each year?

Set II

4. Write the name that is commonly used to refer to each of the following numbers.
   a) 10^0
   b) 10^{-1}
   c) 10^{-2}
   d) 10^{-3}
   e) 10^{-4}

5. Write the number that is ten times as large as each of the following.
   a) 0.07
   b) 12.345
   c) 10^3
   d) 10^{-3}
   e) 9×10^{-11}
   f) 0.2×10^{-6}

6. Write the number that is one-tenth as large as each of the following.
   a) 0.004
   b) 33.3
   c) 10^8
   d) 10^{-8}
   e) 5×10^{-5}

7. Write each of the following numbers in scientific notation.
   a) $2,000$
   b) 0.0002
   c) 75
   d) 0.75
   e) 0.0314
   f) 0.000000000314
   g) 8
   h) 0.00000100

8. Write each of the following numbers in decimal form.
   a) 6×10^2
   b) 6×10^{-2}
   c) 3×10^{-4}
   d) 3.3×10^{-4}
   e) 1.05×10^8
   f) 1.05×10^{-8}

9. The number 0.06×10^{-3} is not in scientific notation because 0.06 is less than 1.
   a) Write 0.06×10^{-3} in decimal form.
   b) Write it in scientific notation.

10. Write each of the following numbers in scientific notation.
    a) 80×10^3
    b) 80×10^{-3}
    c) 0.4×10^6
    d) 0.4×10^{-6}

11. Which of these symbols, $>$, $=$, or $<$, should replace ▓ in each of the following?
    a) 10^5 ▓ $(10^5)^2$
    b) 10^{-5} ▓ $(10^{-5})^2$
    c) 6×10^{-4} ▓ 7×10^{-4}
    d) 4×10^{-6} ▓ 4×10^{-7}
    e) 8×10^{-10} ▓ 80×10^{-11}
    f) 2×10^{-3} ▓ 3×10^{-2}

12. Find each of the following products and quotients. Express each answer in scientific notation.
    a) $(3 \times 10^{-3})(2 \times 10^{-2})$
    b) $(8 \times 10^6)(5 \times 10^{-6})$
    c) $(7 \times 10^{-9})(9 \times 10^7)$
    d) $\dfrac{8 \times 10^2}{4 \times 10^{-5}}$
    e) $\dfrac{4 \times 10^{-5}}{8 \times 10^2}$
    f) $\dfrac{2 \times 10^{-7}}{5 \times 10^{-7}}$

13. Bacteria with lengths as small as 2×10^{-5} centimeter are the smallest living things that can be seen with an ordinary microscope.
    a) Write this number in decimal notation.
    b) Name it.

14. The "General Sherman," a redwood tree in northern California, is thought to be the heaviest living thing on the earth. Its weight is estimated to be about 4,000,000 pounds. It grew from a seed whose weight was only about 0.00001 pound.
    a) Write each of these numbers in scientific notation.
    b) How many times heavier is the tree now than the seed from which it grew?
    c) Write your answer in words.

PHOTOGRAPH BY J. R. EYERMAN, LIFE MAGAZINE © 1956 TIME INC.

Set III

15. Write each of the following numbers as a power of ten.
    a) 1
    b) $\dfrac{1}{10}$
    c) $\dfrac{1}{100}$
    d) $\dfrac{1}{1,000}$
    e) $\dfrac{1}{10,000}$

16. Write the number that is ten times as large as each of the following.
    a) 0.002
    b) 1.11
    c) 10^5
    d) 10^{-5}
    e) 4×10^{-8}
    f) 0.3×10^{-12}

17. Write the number that is one-tenth as large as each of the following.
    a) 0.9
    b) 12.34
    c) 10^7
    d) 10^{-7}
    e) 8×10^{-10}

18. Write each of the following numbers in scientific notation.
    a) 900
    b) 0.009
    c) 31
    d) 0.31
    e) 0.000222
    f) 0.000000000222
    g) 5
    h) 0.00001000

19. Write each of the following numbers in decimal form.
    a) 4×10^1
    b) 4×10^{-1}
    c) 7×10^{-3}
    d) 7.5×10^{-3}
    e) 2.08×10^6
    f) 2.08×10^{-6}

20. The number 40×10^{-5} is not in scientific notation because 40 is not less than 10.
    a) Write 40×10^{-5} in decimal form.
    b) Write it in scientific notation.

21. Write each of the following numbers in scientific notation.
    a) 50×10^2
    b) 50×10^{-2}
    c) 0.7×10^9
    d) 0.7×10^{-9}

22. Which of these symbols, $>$, $=$, or $<$, should replace ▓ in each of the following?
    a) 10^4 ▓ $(10^4)^3$
    b) 10^{-4} ▓ $(10^{-4})^3$
    c) 9×10^{-5} ▓ 8×10^{-5}
    d) 5×10^{-9} ▓ 5×10^{-8}
    e) 7×10^{-6} ▓ 6×10^{-7}
    f) 2×10^{-1} ▓ 20×10^{-2}

23. Find each of the following products and quotients. Express each answer in scientific notation.
    a) $(2 \times 10^{-4})(4 \times 10^{-2})$
    b) $(5 \times 10^{-8})(6 \times 10^8)$
    c) $(9 \times 10^3)(3 \times 10^{-9})$
    d) $\dfrac{8 \times 10^4}{2 \times 10^{-6}}$
    e) $\dfrac{2 \times 10^{-6}}{8 \times 10^4}$
    f) $\dfrac{3 \times 10^{-5}}{5 \times 10^{-5}}$

24. The carburetor in an automobile engine turns liquid gasoline into a mist of tiny drops, each having a diameter of about 3×10^{-7} centimeter.
    a) Write this number in decimal form.
    b) Write it in words.

25. Radio waves and x rays are two types of electromagnetic radiation. A typical radio wave has a wavelength of 50,000 centimeters. An x-ray, on the other hand, has a wavelength of only about 0.00000001 centimeter.
    a) Write each of these numbers in scientific notation.
    b) How many times as long is the wavelength of a radio wave as that of an x ray? Express your answer in scientific notation.
    c) Write your answer in words.

Set IV

This table shows how a human being compares in mass with several large and small things in the universe. The mass of a human being is roughly "half way" between the masses of two other objects in the table in the sense that one object is as many times as heavy as a human being as the other object is as light.

Can you figure out which objects they are? If you can, show your calculations.

Object	Mass in kg
Sun	2×10^{30}
Earth	6×10^{24}
Moon	7×10^{22}
Whale	1×10^5
Elephant	4×10^3
Human	6×10^1
Flea	3×10^{-4}
Dust particle	1×10^{-6}
Hydrogen atom	2×10^{-27}
Electron	9×10^{-31}

THE BETTMANN ARCHIVE.

LESSON 6
Powers of Products and Quotients

One of the most famous formulas of this century is the result of Albert Einstein's discovery that the amount of energy contained in an object is related to its mass. The formula is

$$E = mc^2$$

in which E represents energy, m represents mass, and c represents the speed of light.

The speed of light is 3×10^8 meters per second. Squaring this number as the formula indicates, we get

$$(3 \times 10^8)^2 = (3 \times 10^8)(3 \times 10^8)$$
$$= (3)(3) \times (10^8)(10^8)$$
$$= 9 \times 10^{16}$$

Substituting this result into the formula gives

$$E = (9 \times 10^{16})m$$

or

$$E = 90{,}000{,}000{,}000{,}000{,}000 m$$

Apparently, the amount of energy contained in even a small amount of matter is tremendous. It can be shown from the result we have obtained, in fact, that the amount of energy equivalent to the mass of a *single glass of water* would be sufficient to run the *entire United States* for several hours!*

In finding the square of the speed of light, we found a power of a product. The results suggests another way to find it:

$$(3 \times 10^8)^2 = (3)^2 \times (10^8)^2$$
$$= 9 \times 10^{16}$$

This pattern is worth remembering. We will call it *the fourth law of exponents.*

▶ **The Fourth Law of Exponents**

If a is an integer, then $(xy)^a = x^a y^a$.

A comparable pattern holds true for a power of a quotient.

▶ **The Fifth Law of Exponents**

If a is an integer and y is not zero, then $\left(\dfrac{x}{y}\right)^a = \dfrac{x^a}{y^a}$.

These properties of exponents, like those studied earlier, work even if the exponent is zero or negative. Here are examples of how they are used.

EXAMPLE 1

Write an expression without parentheses equivalent to $(-4x^2)^3$.

SOLUTION

According to the fourth law of exponents,

$$(-4x^2)^3 = (-4)^3(x^2)^3 = (-64)(x^6) = -64x^6$$

EXAMPLE 2

Find $(3 \times 10^{-5})^4$. Express the answer in scientific notation.

* *Liberal Arts Physics* by John M. Bailey (W. H. Freeman and Company, 1974), pp. 226–227.

SOLUTION

$$(3 \times 10^{-5})^4 = (3)^4 \times (10^{-5})^4$$
$$= 81 \times 10^{-20}$$
$$= 8.1 \times 10^1 \times 10^{-20}$$
$$= 8.1 \times 10^{-19}$$

EXAMPLE 3

Write $\dfrac{2^{-10}}{12^{-10}}$ as a power of 6.

SOLUTION

According to the fifth law of exponents, $\dfrac{2^{-10}}{12^{-10}} = \left(\dfrac{2}{12}\right)^{-10}$.

Simplifying this result and using other properties of exponents that we know, we get

$$\left(\frac{2}{12}\right)^{-10} = \left(\frac{1}{6}\right)^{-10} = (6^{-1})^{-10} = 6^{10}$$

Exercises

Set I

1. Change each of the following numbers to decimal form.

   a) $\dfrac{125}{100}$

   b) $\dfrac{125}{10}$

   c) $\dfrac{100}{125}$

   d) $\dfrac{10}{125}$

2. Write each of the following numbers without using any exponents.
   a) $(-1)^7$
   b) 1^{-7}
   c) $(-7)^1$
   d) 7^{-1}

3. "I am thinking of a certain number. If seven is subtracted from four times the number, the result is the same as the result of multiplying one plus the number by five."
   a) Translate this information into an equation.
   b) Solve the equation to find the number.

Set II

4. Write an expression without parentheses for each of the following.
   a) $(xy)^6$
   b) $(xy)^{-1}$
   c) $(2x)^3$
   d) $(-5y)^4$
   e) $(x^7)^2$
   f) $(xy^6)^5$
   g) $(4x^{-5})^2$
   h) $(-3y^4)^3$
   i) $\left(\dfrac{x}{y}\right)^8$
   j) $\left(\dfrac{y^2}{2}\right)^5$
   k) $\left(\dfrac{1}{x^6}\right)^3$
   l) $\left(\dfrac{7x}{y^7}\right)^2$

5. Which of these symbols, $>$, $=$, or $<$, should replace ▦ in each of the following?
   a) $2^2 + 2^3$ ▦ 2^5
   b) $2^2 \cdot 2^3$ ▦ 2^5
   c) $(2^2)^3$ ▦ 2^5
   d) $2^2 + 3^2$ ▦ 5^2
   e) $2^2 \cdot 3^2$ ▦ 6^2
   f) $2^3 - 2^2$ ▦ 2^1
   g) $\dfrac{2^3}{2^2}$ ▦ 2^1
   h) $6^3 - 2^3$ ▦ 4^3
   i) $6^3 \cdot 2^3$ ▦ 8^3
   j) $\dfrac{6^3}{2^3}$ ▦ 3^3

6. Use the table of powers of 8 at the right to find each of the following.
   a) $2 \cdot 2 \cdot 2 \cdot 4 \cdot 4 \cdot 4$
   b) $4^{-1} \cdot 2^{-1}$
   c) $(8^{-2})^{-2}$
   d) $2 \cdot 8^2 \cdot 4 \cdot 8^2$
   e) $2^2 \cdot 4^2 \cdot 2^4 \cdot 4^4$
   f) $\dfrac{24^2}{3^2}$
   g) $\dfrac{8}{8^3}$
   h) $\dfrac{2^3}{16^3}$

$8^{-3} = 0.001953125$
$8^{-2} = 0.015625$
$8^{-1} = 0.125$
$8^0 = 1$
$8^1 = 8$
$8^2 = 64$
$8^3 = 512$
$8^4 = 4{,}096$
$8^5 = 32{,}768$
$8^6 = 262{,}144$
$8^7 = 2{,}097{,}152$
$8^8 = 16{,}777{,}216$

7. Find each of the following powers. Express each answer in scientific notation.
   a) $(2 \times 10^4)^3$
   b) $(7 \times 10^2)^2$
   c) $(3 \times 10^{-1})^5$
   d) $(5 \times 10^3)^{-1}$
   e) $(1 \times 10^4)^{-4}$
   f) $(2 \times 10^{-6})^{-2}$

8. Find x in each of the following equations.
   a) $3^5 \cdot 12^5 = x^5$
   b) $3^5 \cdot x^5 = 12^5$
   c) $5^3 \cdot 5^x = 5^{12}$
   d) $(5^3)^x = 5^{12}$
   e) $4^6 \cdot 4^6 = 4^x$
   f) $4^6 \cdot 4^6 = 16^x$
   g) $4^x \cdot 4^x = 4^{24}$
   h) $4^x \cdot 6^x = 24^4$
   i) $\dfrac{8^6}{8^2} = 8^x$
   j) $\dfrac{6^8}{2^8} = x^8$
   k) $\dfrac{x^8}{2^8} = 6^8$
   l) $\dfrac{8^x}{8^2} = 8^6$

Set III

9. Write an expression without parentheses for each of the following.
   a) $(xy)^5$
   b) $(xy)^0$
   c) $(7x)^2$
   d) $(-3y)^4$
   e) $(x^3)^9$
   f) $(x^5y)^3$
   g) $(6x^{-1})^2$
   h) $(-2y^2)^5$
   i) $\left(\dfrac{x}{y}\right)^6$
   j) $\left(\dfrac{x^3}{4}\right)^3$
   k) $\left(\dfrac{1}{y^8}\right)^4$
   l) $\left(\dfrac{9x}{y^5}\right)^2$

10. Which of these symbols, $>$, $=$, or $<$, should replace ▦ in each of the following?
   a) $3^2 + 4^2$ ▦ 7^2
   b) $3^2 \cdot 4^2$ ▦ 12^2
   c) $2^3 + 2^4$ ▦ 2^7
   d) $2^3 \cdot 2^4$ ▦ 2^7
   e) $(2^3)^4$ ▦ 2^7
   f) $2^4 - 2^3$ ▦ 2^1
   g) $\dfrac{2^4}{2^3}$ ▦ 2^1
   h) $\dfrac{8^2}{2^2}$ ▦ 4^2
   i) $8^2 - 2^2$ ▦ 6^2
   j) $8^2 \cdot 2^2$ ▦ 10^2

Elliot

11. Use the table of powers of 20 at the right to find each of the following.
a) $4 \cdot 4 \cdot 5 \cdot 5$
b) $5^{-1} \cdot 4^{-1}$
c) $(20^{-2})^{-3}$
d) $2^7 \cdot 4^4 \cdot 5^4 \cdot 10^7$
e) $4^5 \cdot 5^5 \cdot 4^{-2} \cdot 5^{-2}$
f) $\dfrac{40^4}{2^4}$
g) $\dfrac{20}{20^4}$
h) $\dfrac{3^2}{60^2}$

$20^{-4} = 0.00000625$
$20^{-3} = 0.000125$
$20^{-2} = 0.0025$
$20^{-1} = 0.05$
$20^0 = 1$
$20^1 = 20$
$20^2 = 400$
$20^3 = 8,000$
$20^4 = 160,000$
$20^5 = 3,200,000$
$20^6 = 64,000,000$

12. Find each of the following powers. Express each answer in scientific notation.
a) $(3 \times 10^5)^2$
b) $(4 \times 10^4)^3$
c) $(5 \times 10^{-1})^4$
d) $(4 \times 10^5)^{-1}$
e) $(1 \times 10^6)^{-3}$
f) $(2 \times 10^{-7})^{-2}$

13. Find x in each of the following equations.
a) $2^4 \cdot 14^4 = x^4$
b) $2^4 \cdot x^4 = 14^4$
c) $4^2 \cdot 4^x = 4^{14}$
d) $(4^2)^x = 4^{14}$
e) $3^5 \cdot 3^5 = 3^x$
f) $3^5 \cdot 3^5 = 9^x$
g) $3^x \cdot 3^x = 3^{12}$
h) $3^x \cdot 4^x = 12^3$
i) $\dfrac{7^8}{7^2} = 7^x$
j) $\dfrac{8^7}{2^7} = x^7$
k) $\dfrac{x^7}{2^7} = 8^7$
l) $\dfrac{7^x}{7^2} = 7^8$

Set IV

B.C. BY PERMISSION OF JOHNNY HART AND FIELD ENTERPRISES, INC.

Although the question B. C. has asked Peter in this cartoon doesn't seem to make any sense, Peter's answer does.
1. Write Peter's answer as a formula.
2. What kind of variation is this?
3. Can the formula be written in terms of a product rather than a quotient? If so, what would it be?

"His knowledge is doubling every ten years, and it's making him jumpy."

LESSON 7
Exponential Functions

If the knowledge of the fellow in this cartoon is doubling every ten years, then in ten years he will know twice as much as he does now, in twenty years he will know four times as much, in thirty years he will know eight times as much, and so on. Recording the time in decades and representing the man's knowledge now by the number 1, we can construct the following table:

Time in decades, x	0	1	2	3	4	...
Knowledge, y	1	2	4	8	16	...

If the numbers in the second row are written as powers of 2, the table looks like this:

Time in decades, x	0	1	2	3	4	...
Knowledge, y	2^0	2^1	2^2	2^3	2^4	...

With the table written in this form, it is easy to see that, for each value of x,

$$y = 2^x$$

This formula continues to work even if we look backward in time. For example, ten years ago the man knew half as much, twenty years ago, one-fourth as much, thirty years ago, one-eighth as much, and so on.

Time in decades, x	0	-1	-2	-3	$\ldots$
Knowledge, y	$2^0 = 1$	$2^{-1} = \dfrac{1}{2}$	$2^{-2} = \dfrac{1}{4}$	$2^{-3} = \dfrac{1}{8}$	$\ldots$

Because one of the variables in this function is an exponent in its formula, the function is called an *exponential function*.

▶ An **exponential function** is a function that has an equation of the form $y = a \cdot b^x$, in which a and b are positive numbers.

In the "knowledge doubling" function, $a = 1$ and $b = 2$.

What does the graph of an exponential function look like? If the points of the two preceding tables are plotted on one pair of axes and then connected with a smooth curve, we get the graph shown at the left below. As we look at the graph

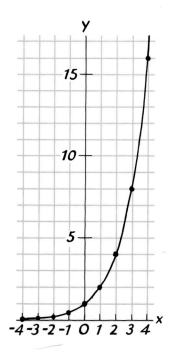

An increasing exponential function:
knowledge doubling

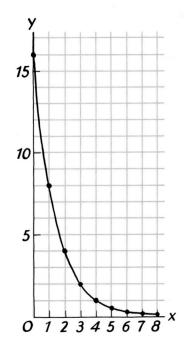

A decreasing exponential function:
radioactive decay

from left to right, the "slope" of the curve becomes steeper and steeper. Many exponential functions have graphs similar to this one.

The "knowledge doubling" function is an example of an *increasing* exponential function. An example of a *decreasing* exponential function is radioactive decay. The atoms of radioactive elements break apart into atoms of other elements at a rate such that the number of atoms remaining is an exponential function of time. For example, it takes ten years for half of the atoms of any given amount of a certain isotope of barium to disintegrate. This means that, if we start with 16 grams of this isotope, after ten years 8 grams of it will remain, after ten more years 4 grams will remain, and so on. A table for the amount of barium remaining as a function of time is shown here. (The graph is shown at the bottom right on the previous page.)

Time in decades, x	0	1	2	3	4	5	6	7	8
Number of grams left, y	16	8	4	2	1	$\frac{1}{2}$	$\frac{1}{4}$	$\frac{1}{8}$	$\frac{1}{16}$

Exponential functions have a wide variety of practical applications. For example, populations can grow exponentially with time. The speed of a chemical reaction is an exponential function of the temperature at which it occurs. Exponential functions apply to compound interest, the healing of wounds, and many other topics.

Exercises

Set I

1. Copy and complete the tables for these functions.

a) Formula: $y = 5 - x$

Table:

x	-4	-2	0	2	4
y					

b) Formula: $y = \dfrac{36}{x^2}$

Table:

x	0	1	2	3	4
y					

c) Formula: $y = (x - 1)(x - 2)$

Table:

x	1	2	3	4	5
y					

2. Tell whether each of the following statements is true for all values of x. If a statement is not true for all values of x, give a value of x for which this is so.

a) $|2x| = 2\,|x|$

b) $|x + 2| = |x| + 2$

c) $|x|^2 = |x^2|$

d) $|2 - x| = 2 - |x|$

e) $\dfrac{|x|}{2} = \left|\dfrac{x}{2}\right|$

3. Of the 2.2×10^8 people living in the United States, approximately 1.1×10^7 are left-handed.

a) Write each of these numbers in decimal form.
b) Write each number in words.
c) What fraction of people living in the United States are left-handed?

Set II

4. Write each of the following in the form $a \cdot b^c$.
   a) $8 \cdot 10 \cdot 10 \cdot 10$
   b) $5 \cdot 3 \cdot 3 \cdot 3 \cdot 3 \cdot 3 \cdot 3 \cdot 3$
   c) $4 \cdot 4 \cdot 11$
   d) $\dfrac{6}{10 \cdot 10 \cdot 10 \cdot 10}$
   e) $\dfrac{2}{7 \cdot 7 \cdot 7 \cdot 7 \cdot 7}$

5. Write in decimal form. $36 \stackrel{?}{=} 9 - 4$
   a) $4 \cdot 5^3$ d) $36 \cdot 3^{-2}$
   b) $0.2 \cdot 10^6$ e) $8 \cdot 1^{-10}$
   c) $9 \cdot 2^0$ f) $4 \cdot 5^{-3}$

6. This exercise is about the exponential function $y = 4^x$.
   a) Copy and complete the following table for this function.

x	0	1	2	3	4
y					

   What happens to y if x is increased by
   b) 1?   d) 3?
   c) 2?

7. This exercise is about the functions

   $$y = 3x, \quad y = x^3, \quad \text{and} \quad y = 3^x$$

   Letting $x = -2, -1, 0, 1,$ and 2, make a table for
   a) $y = 3x$
   b) $y = x^3$
   c) $y = 3^x$

Use your tables to graph each function as indicated below. (Graph each one on a separate pair of axes and connect the points with a smooth line or curve.)
   d) $y = 3x$
   e) $y = x^3$
   f) $y = 3^x$

8. If an automobile worth $6,000 depreciates by one-fourth of its value each year, its value after x years is given by the formula

   $$y = 6,000(0.75^x)$$

   Find the value of the car after
   a) one year.
   b) two years.
   c) three years.
   *d) If you have a calculator, find the value of the car after ten years.

9. A certain type of protozoan is able to divide into two every three hours. Reproducing at this rate, the number of protozoa, y, existing after x hours is an exponential function of time. It is given by the formula

   $$y = 2^{\frac{x}{3}}$$

   a) Make a table for this function, letting $x = 0, 3, 6, 9, 12, 15, 18, 21,$ and 24.
   b) Beginning with one protozoan, how many would there be after 12 hours?
   c) How many after one day?
   *d) If you have a calculator, find out how many there would be after two days.

Set III

10. Write each of the following in the form
    $a \cdot b^c$.
    a) $6 \cdot 10 \cdot 10 \cdot 10 \cdot 10$
    b) $7 \cdot 2 \cdot 2 \cdot 2 \cdot 2 \cdot 2$
    c) $5 \cdot 5 \cdot 5 \cdot 4$
    d) $\dfrac{9}{10 \cdot 10}$
    e) $\dfrac{8}{3 \cdot 3 \cdot 3 \cdot 3 \cdot 3 \cdot 3}$

11. Write in decimal form.
    a) $5 \cdot 2^4$ d) $24 \cdot 2^{-3}$
    b) $11 \cdot 10^5$ e) $4 \cdot 1^{-6}$
    c) $7 \cdot 8^0$ f) $5 \cdot 2^{-4}$

12. This exercise is about the exponential
    function $y = 5^x$.
    a) Copy and complete the following table
       for this function.

x	0	–1	–2	–3	–4
y					

    What happens to y if x is decreased by
    b) 1?
    c) 2?
    d) 3?

13. This exercise is about the functions

    $$y = 2x, \quad y = x^2, \quad \text{and} \quad y = 2^x$$

    Letting $x = -3, -2, -1, 0, 1, 2,$ and 3, make
    a table for
    a) $y = 2x$
    b) $y = x^2$
    c) $y = 2^x$

    Use your tables to graph each function as
    indicated in the next column. (Graph each
    one on a separate pair of axes and connect
    the points with a smooth line or curve.)

    d) $y = 2x$
    e) $y = x^2$
    f) $y = 2^x$

14. If $1,000 is invested at 6% interest and the
    interest is added each year, the amount of
    money after x years is given by the formula

    $$y = 1,000(1.06^x)$$

    What will the amount of money be after
    a) one year?
    b) two years?
    c) three years?
    *d) If you have a calculator, find what the
       amount of money would be after ten
       years.

15. Atoms of a radioactive element slowly break
    apart into atoms of other elements. The
    amount of a radioactive element remaining
    after a given period is an exponential
    function of the time.
      If we start with 80 grams of a certain
    isotope of mercury, the number of grams, y,
    remaining after x days is given by the
    formula

    $$y = 80(0.5^x)$$

    a) Make a table for this function, letting
       $x = 0, 1, 2, 3, 4, 5, 6,$ and 7.
    b) The half-life of a radioactive element is
       the amount of time that it takes half the
       atoms of the element to disintegrate.
       What is the half-life of this isotope of
       mercury?
    c) How much of the mercury would remain
       after one week?
    *d) If you have a calculator, find out how
       much of the mercury would remain after
       two weeks.

Set IV

If hot coffee is poured into a cup and allowed to cool, the difference between the temperature of the coffee and the temperature of its surroundings is an exponential function of time. A table for this function is shown here.*

Time in minutes, x	0	5	10	15	20	25	30	35	40
Temperature of coffee above room temperature in degrees Celsius, y	70	59	49	42	35	29	25	21	17

1. Graph this function.
2. Is the coffee cooling off at a steady rate?
3. Using the information in the table, about how many degrees above room temperature would the coffee be after one hour? (Hint: How long does it take the difference between the temperature of the coffee and that of the room to become half of what it was?)

---

*This table is based on data from "The Amateur Scientist" by Jearl Walker, *Scientific American*, November, 1977.

Summary and Review

In this chapter, we have become acquainted with the properties of exponents, with how to represent large and small numbers in scientific notation, and with the nature of exponential functions.

Large and Small Numbers (*Lessons 1 and 5*)   Very large and very small numbers are frequently written in scientific notation. A number is in scientific notation if it is written in the form $a \times 10^b$, in which a is a number that is at least 1 but less than 10, and b is an integer.

A Fundamental Property of Exponents (*Lesson 2*)   In the expression x^y, x is the base and y is the exponent.

The product of two powers: $x^a \cdot x^b = x^{a+b}$.

Two More Properties of Exponents (*Lesson 3*)

The quotient of two powers: $\dfrac{x^a}{x^b} = x^{a-b}$.

The power of a power: $(x^a)^b = x^{ab}$.

Zero and Negative Exponents (*Lesson 4*)   Patterns with exponents that are positive integers suggest the following definitions of the meanings of zero and the negative integers as exponents.

Definition of the exponent zero: $x^0 = 1$, in which x may be any number except zero.

Definition of negative integer exponents: $x^{-a} = \dfrac{1}{x^a}$, in which x may be any number except zero.

Powers of Products and Quotients (*Lesson 6*)

The power of a product: $(xy)^a = x^a y^a$

The power of a quotient: $\left(\dfrac{x}{y}\right)^a = \dfrac{x^a}{y^a}$

Exponential Functions (*Lesson 7*) An exponential function is a function that has an equation of the form $y = ab^x$, in which a and b are positive numbers. The graph of an exponential function is a smooth curve.

Exercises

Set I

1. Write each of the following numbers in the form indicated.
   a) One hundred thousand as a power of ten.
   b) The number that is ten times as large as 1.35 in decimal form.
   c) The number that is ten times as large as 10^{-8} as a power of ten.
   d) The number that is one-tenth as large as 0.007 in decimal form.
   e) The number that is one-tenth as large as 2×10^6 in scientific notation.

2.

TOLEDO, APR 19 (UPS) PROF. T.B. MURRAY, WORLD FAMOUS PHILOSOPHER-MATHEMATICIAN, WRITING IN THE PRESTIGIOUS LITERARY MAGAZINE, "YOU AND THE IDAHO POTATO," CLAIMS TO HAVE SOLVED THE PROBLEM OF FEEDING 220,000,000 AMERICANS. "THE SOLUTION," WRITES THE PROFESSOR, "IS TO MOVE THE DECIMAL POINT SEVEN DIGITS TO THE LEFT." FEEDING TWENTY-TWO PEOPLE "SHOULD BE A SNAP," PROF. MURRAY SAID.

   a) Write the number 220,000,000 in scientific notation.
   b) Write the number 22 in scientific notation.
   c) What number is a number divided by when its decimal point is moved seven digits to the left?
   d) When you move the decimal point in a number seven digits to the left, how does the exponent of the 10 in the scientific notation form of the number change?

3. Write each of the following numbers in decimal form.
   a) 3×10^4
   c) 6.2×10^{-1}
   b) 3×10^{-4}
   d) 0.085×10^9

4. Write each of the following numbers in scientific notation.
   a) 700,000,000,000
   c) 0.002
   b) 412×10^5
   d) 10.002

5. Write each of the following numbers without using any exponents.
   a) 8^{-1}
   d) $(-6)^2$
   b) 5^{-4}
   e) $(-6)^{-2}$
   c) 12^0
   f) $(-2)^6$

6. Write each of the following expressions as a single power of x.
   a) $x^2 \cdot x^{-8}$
   b) $(x^2)^{-8}$
   e) $\dfrac{x^{-10}}{x}$
   c) $\dfrac{1}{x^5}$
   f) $\dfrac{x^{-7}}{x^{-9}}$
   d) $\dfrac{x^4}{x^{-3}}$

7. Which of these symbols, $>$, $=$, or $<$, should replace ▦ in each of the following?
   a) 10^{-4} ▦ 10^{-5}
   c) 8^{-3} ▦ $(-8)^3$
   b) 16^{-1} ▦ 6^1
   d) $(-9)^0$ ▦ 9^0

8. Write an expression without parentheses for each of the following.
   a) $(4x)^3$
   b) $(7y)^{-1}$
   d) $\left(\dfrac{x}{y^2}\right)^4$
   c) $(-5x^5)^2$

9. Find each of the following products, quotients, or powers. Express each answer in scientific notation.
   a) $(6 \times 10^6)(9 \times 10^9)$
   d) $\dfrac{11 \times 10^{-8}}{2}$
   b) $(2)(7 \times 10^{-3})$
   c) $\dfrac{4 \times 10^1}{5 \times 10^6}$
   e) $(3 \times 10^2)^4$
   f) $(5 \times 10^{-7})^{-1}$

10. Which of these symbols, $>$, $=$, or $<$, should replace ▦ in each of the following?
    a) $2^3 + 3^3$ ▦ 5^3
    b) $2^3 \cdot 3^3$ ▦ 5^3
    f) $\dfrac{3^6}{3^2}$ ▦ 3^4
    c) $3^2 \cdot 3^3$ ▦ 3^5
    d) $(3^2)^3$ ▦ 3^5
    g) $\dfrac{6^3}{2^3}$ ▦ 4^3
    e) $3^6 - 3^2$ ▦ 3^4

11. Find x in each of the following equations.
    a) $3^4 \cdot 3^6 = 3^x$
    e) $\dfrac{7^{12}}{7^3} = 7^x$
    b) $4^3 \cdot 6^3 = x^3$
    c) $5^2 \cdot 5^x = 5^8$
    d) $2^5 \cdot x^5 = 8^5$
    f) $\dfrac{12^7}{3^7} = x^7$

12. Light travels at a speed of about 186,000 miles per second. Sound travels in air at a speed of about 0.2 mile per second.
    a) Write each of these numbers in scientific notation.
    b) How many times the speed of sound is the speed of light?
    c) Write your answer in words.

13. This exercise is about the exponential function $y = 8(0.5^x)$.
    a) Make a table for this function, letting $x = 0, 1, 2, 3,$ and 4.
    b) What happens to y if x is increased by 1?
    c) Graph the function by plotting the points and connecting them with a smooth curve.

14. Suppose that $2,000 is invested at 7% interest per year and the interest is added at the end of each year. The amount of money after x years is given by the formula $y = 2,000(1.07^x)$.
    Find the amount of money after
    a) one year.
    b) two years.
    c) Is the investment growing at a steady rate with respect to time? Explain.

Set II

1. Write each of the following numbers in the form indicated.
   a) Ten trillion as a power of ten.
   b) The number that is ten times as large as 0.036 in decimal form.
   c) The number that is ten times as large as 10^5 as a power of ten.
   d) The number that is one-tenth as large as 82.1 in decimal form.
   e) The number that is one-tenth as large as 7×10^{-4} in scientific notation.

2. Write each of the following numbers in decimal form.
   a) 8×10^3 c) 4.5×10^{-1}
   b) 8×10^{-3} d) 0.0072×10^5

3. Write each of the following numbers in scientific notation.
   a) $50,000,000$ c) 600×10^{-5}
   b) 81×10^9 d) 0.60×10^{-5}

4. Skunks protect themselves by giving off a substance with such a strong odor that as little as 0.000000003 milligram of it can be detected by the human nose.
   a) Write this number in scientific notation.
   b) Write it in words.

5. Write each of the following numbers without using any exponents.
   a) $(-9)^0$ d) $(-3)^{-4}$
   b) 1^{-6} e) 4^{-3}
   c) $(-3)^4$ f) $(-7)^{-1}$

6. Write each of the following expressions as a single power of x.
   a) $x^{-4} \cdot x^9$
   b) $(x^{-4})^9$
   c) $\dfrac{1}{x^3}$ e) $\dfrac{x^{-5}}{x}$
   d) $\dfrac{x^2}{x^{-6}}$ f) $\dfrac{x^{-11}}{x^{-7}}$

7. Which of these symbols, $>$, $=$, or $<$, should replace ▒ in each of the following?
   a) 10^{-7} ▒ 10^{-6}
   b) 3^1 ▒ 13^{-1}
   c) 8^0 ▒ $(-8)^0$
   d) $(-5)^4$ ▒ 5^{-4}

8. Write an expression without parentheses equivalent to each of the following.
   a) $(2x)^5$
   b) $(xy)^{-1}$ d) $\left(\dfrac{x^2}{y}\right)^3$
   c) $(-3y^3)^4$

9. Find each of the following products, quotients, or powers. Express each answer in scientific notation.
   a) $(8 \times 10^8)(4 \times 10^4)$
   b) $(3)(9 \times 10^{-5})$
   c) $\dfrac{2 \times 10^4}{8 \times 10^{-1}}$
   d) $\dfrac{7 \times 10^{-12}}{2}$
   e) $(4 \times 10^3)^3$
   f) $(5 \times 10^{-6})^{-2}$

10. Which of these symbols, $>$, $=$, or $<$, should replace ▒ in each of the following?
    a) $2^2 \cdot 4^2$ ▒ 6^2
    b) $2^2 + 4^2$ ▒ 6^2 f) $\dfrac{2^8}{2^2}$ ▒ 2^6
    c) $(2^2)^4$ ▒ 2^6
    d) $2^2 \cdot 2^4$ ▒ 2^6 g) $2^8 - 2^2$ ▒ 2^6
    e) $\dfrac{8^2}{2^2}$ ▒ 6^2

11. Find x in each of the following equations.
    a) $5^2 \cdot 5^4 = 5^x$
    b) $2^5 \cdot 4^5 = x^5$
    c) $3^x \cdot 3^3 = 3^{12}$
    d) $x^3 \cdot 3^3 = 12^3$ f) $\dfrac{10^6}{2^6} = x^6$
    e) $\dfrac{6^{10}}{6^2} = 6^x$

12. Approximately 33 billion bottles of Coca-Cola are drunk throughout the world in a year.
    a) Write this number in decimal form.
    b) Write it in scientific notation.
    c) Approximately how many bottles of Coke are drunk each *day?* Write your answer in scientific notation, in decimal form, and in words.

13. This exercise is about the exponential function $y = 0.5(2^x)$.
    a) Make a table for this function, letting $x = 0, 1, 2, 3,$ and 4.
    b) What happens to y when x is increased by 1?

    c) Graph the function by plotting the points and connecting them with a smooth curve.

14. If an automobile worth $7,000 depreciates by one-fifth of its value each year, its value after x years is given by the formula $y = 7,000(0.8^x)$.
    Find the value of the car after
    a) one year.
    b) two years.
    c) Is the value of the car decreasing at a steady rate with respect to time? Explain.

MIDTERM REVIEW

Set I

1. Write without parentheses: $8(x - 3)$.

2. True or false: If one variable in a direct variation is tripled, then so is the other.

3. Simplify: $4y - y$.

4. True or false: This pair of simultaneous equations has many different solutions.

$$4x - 4y = 12$$
$$x - y = 3$$

5. Solve for x: $x + 4 = -14$.

6. A marching band contains eighty people arranged in x equal rows. How many people are in each row?

7. True or false: The graph of an inverse variation is a line through the origin.

8. Write as a power of 4: $\dfrac{8^{-1}}{2^{-1}}$.

9. Write without parentheses: $-1(x - 7)$.

10. True or false: These equations are inconsistent.

$$2x - y = 4$$
$$4x - 2y = 6$$

11. Find the value of $(4 - 11)(11 - 4)$.

12. If lemons cost x cents each, how many lemons can you buy for a dollar?

13. Find the product of -3 and 14.

14. True or false: The graph of the equation $x = 1$ is a vertical line.

15. Write as an addition problem: $x - y$.

16. True or false: The sum of two numbers is always larger than either number.

17. Write $\dfrac{1}{25}$ as a power of 5.

18. Solve for x: $x - 18 = -2$.

19. Write as a single power: $\dfrac{x^{-9}}{x}$.

20. What number is the result of subtracting 14 from -9?

21. Solve this formula for p: $i = prt$.

22. True or false: $(12)^0 = (0)^{12}$.

23. Which symbol, $>$, $=$, or $<$, makes the following true?

$$2^{-3} \;\|\|\|\|\; 2^{-2}$$

24. True or false: One is a rational number.

25. If x represents an odd number, what kind of number does $x + 3$ represent?

26. Write an expression for the following: Square x and divide the result by y.

27. How many miles would a donkey walking x miles per hour go in 3 hours?

28. True or false: The square of every integer is positive.

29. What must be done to the equation $x - 7 = -1$ to give the equation $x = 6$?

30. Write 6^{-2} without using any exponents.

31. Where does the line $y = x - 9$ cross the y-axis?

32. Which symbol, $>$, $=$, or $<$, makes the following true?

$$x - 10 \;\|\|\|\|\; x + -10$$

33. What is the slope of the line $y = -x + 17$?

34. Write an equivalent expression without parentheses: $(2x^2)^3$.

35. Write the number that is ten times as large as 10^{-4} as a power of 10.

36. Write an expression for the length marked ? in this figure.

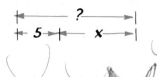

37. Write another expression equivalent to

$$y \cdot y \cdot y \cdot y \cdot y$$

38. True or false: There is no number equal to $\frac{0}{10}$.

39. Which line is steepest?

$$y = x + 7 \qquad y = 4x - 3 \qquad y = 2x - 5$$

40. Write as a single power: $x^5 \cdot x^4$.

Set II

1. Draw a figure with circles to illustrate: $3^2 + 4^2$.

2. Solve for x: $3x + 7 = y$.

3. Simplify: $-3(2x^4)$.

4. Find the value of $x^3 - x + 3$ if $x = -1$.

5. Graph the equation $2x - 5y = 10$.

6. Find the number of the point midway between -4 and 10 on a number line.

7. Write the equation that results from adding these equations:

$$3x + y = -4 \quad \text{and} \quad x - 2y = 5$$

8. Solve for x: $-6x = 21$.

9. Use the formula $f = 1.8c + 32$ to find f if $c = 100$.

10. Solve the following simultaneous equations by graphing:

$$y = -2x$$
$$x - y = 6$$

11. Find the distance between -3.5 and 4.5 on a number line.

12. Graph the function $y = \frac{-8}{x}$.

13. Write the number "fifteen million" in scientific notation.

14. Solve for x: $11 + 2x = 4$.

15. Express 243 as a power of 3.

16. What must be done to the equation $5x - 2 = x + 9$ to give the equation $4x - 2 = 9$?

17. Write in symbols: The cube of a certain number, x, is more than half of the number.

18. Solve for x: $5x - 21 = 8x$.

19. Find every pair of positive integers that make this equation true: $x + 7y = 15$.

20. Find the slope of the line through the origin and the point $(-2, 10)$.

21. Guess a formula for this function:

x	1	4	7	10	13
y	14	11	8	5	2

22. Draw a graph of the function $y = 3^x$.

23. Find the following product:
$(6 \times 10^3)(7 \times 10^5)$. Give your answer in scientific notation.

24. Solve this pair of simultaneous equations:
$$x + 9y = -2$$
$$4x - 9y = 37$$

25. Which symbol, $>$, $=$, or $<$, makes the following true?
$$4^{-5} \ \rule{0.6cm}{0.25cm}\ (-5)^4$$

26. Solve for x: $3 + (x - 1) = 2(5 - 2x)$.

27. Write in decimal form: 3.14×10^5.

28. Is the 8th power of 5 *odd* or *even?*

29. Solve for x: $4(x + 1) = 3(x - 5)$.

30. Find the value of $(-3)^4 + (-2)^3$.

31. Change the rational number $\dfrac{9}{25}$ to decimal form.

32. Write as a power of 18: $2^5 \cdot 9^5$.

33. Solve for y: $3(x + y) = 2y + 1$.

34. Find the value of $x^2 - y$ if $x = 5$ and $y = -5$.

35. Guess a formula for this function:

x	0	1	2	3	4
y	4	7	10	13	16

A bus travels 50 mph from A to B and 40 mph from B to C, covering a distance of 215 miles in all. The whole trip takes 5 hours.

36. Write an expression for the distance marked ? on the diagram above.

37. Use the information in the diagram to write an equation.

38. Find the time, x, that the bus spent in traveling from A to B.

39. Find the value of $x + 2x^3$ if $x = -4$.

40. Solve this pair of simultaneous equations:
$$5x - y = 11$$
$$x = y + 3$$

Set III

1. Write the number "twelve thousand" in scientific notation.

2. Write in symbols: The square of a certain number, x, is less than three times the number.

3. Draw a figure with circles to illustrate: $5 \cdot 2^2$.

4. Write the equation that results from subtracting the second of these equations from the first one:

$$4x + y = 5 \qquad 3x - y = 9$$

5. Find the value of $x^2 + x - 8$ if $x = -2$.

6. Solve for x: $\quad 8x = -18$.

7. Find the number of the point midway between -7 and 3 on a number line.

8. Simplify: $\quad -4(3x^2)$.

9. Use the formula $s = 16t^2 - 24$ to find s if $t = 3$.

10. Graph the equation $3x - 4y = -12$.

11. Solve for x: $\quad 5x + 12 = 1$.

12. Find the distance between -0.5 and 6.5 on a number line.

13. Solve the following simultaneous equations by graphing:

$$y = 3x - 1$$
$$x + y = 7$$

14. Express 64 as a power of 2.

15. Graph the function $y = \dfrac{6}{x}$.

16. Solve for x: $\quad 2x = 9x + 14$.

17. What must be done to the equation $4 - x = 2x + 7$ to give the equation $4 = 3x + 7$?

18. Find every pair of positive integers that make this equation true: $\quad y = 14 - x^2$.

19. Guess a formula for this function:

x	1	2	3	4	5
y	60	30	20	15	12

20. Graph the function $y = 2^x$.

21. Find the slope of the line through the origin and the point $(4, -8)$.

22. Solve this pair of simultaneous equations:

$$3x - 5y = 36$$
$$x + 5y = -28$$

23. Write in decimal form: $\quad 9.5 \times 10^{-4}$.

24. Solve for x: $\quad 7(x - 2) = 6(x + 4)$.

25. Find the following quotient: $\dfrac{2 \times 10^9}{8 \times 10^3}$. Give your answer in scientific notation.

26. Which symbol, $>$, $=$, or $<$, makes the following true?

$$3^{-9} \ \rule[0.3ex]{1.2em}{0.6ex}\ 3^{-10}$$

27. Find the value of $(-2)^5 + (-5)^2$.

28. Solve for x: $\quad 2(x + 4) = (x - 5) - 9$.

29. Change the rational number $\dfrac{3}{40}$ to decimal form.

30. Is the 9th power of 4 *odd* or *even?*

31. Write as a power of 12: $3^6 \cdot 4^6$.

32. Find the value of $x - y^3$ if $x = 4$ and $y = -4$.

33. Solve for x: $2x - 9 = y$.

34. Solve this pair of simultaneous equations:

$$x + 6y = 23$$
$$y = x - 2$$

35. Find the value of $3x^2 + x$ if $x = -2$.

36. Guess a formula for this function:

x	0	1	2	3	4
y	7	9	11	13	15

37. Solve for y: $4(x + y) = 2 + 3y$.

Two bicyclists start from A at the same time and ride in opposite directions. One cyclist arrives at B and the other at C five minutes later. The cyclist going to C rides 40 meters per minute faster than the one going to B. They end up 6,200 meters apart.

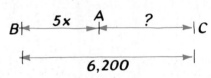

38. Write an expression for the distance marked ? on the diagram above.

39. Use the information in the diagram to write an equation.

40. Find the speed, x, of the bicyclist traveling from A to B.

Chapter 9
POLYNOMIALS

PHOTOGRAPH BY BERENICE ABBOTT REPRODUCED WITH THE PERMISSION OF COLLINS + WORLD PUBLISHING CO., INC., FROM THE ATTRACTIVE UNIVERSE BY E. G. VALENS (TEXT AND DIAGRAMS) AND BERENICE ABBOTT (PHOTOGRAPHS).

LESSON 1
Monomials

A falling object does not travel at a steady rate but moves faster and faster. This photograph, taken with a strobe light, shows several positions of an apple falling through the air. Although the time interval between each flash of light is the same, the distance covered by the apple becomes larger and larger.

A formula relating the distance fallen to the time, discovered by Galileo, is

$$y = 16x^2$$

in which x is the time in seconds and y is the distance in feet. In the first second, then, an object falls $16(1)^2 = 16$ feet; in the first two seconds, it falls $16(2)^2 = 64$ feet; and so forth.

The expression $16x^2$ is called a *monomial* in one variable.

▶ A **monomial** in one variable, say x, is an expression of the form ax^n, in which a is any number and n is a positive integer. The number a is the **coefficient** of the monomial and the integer n is the **degree** of the monomial.

The coefficient of the monomial $16x^2$ is 16 and its degree is 2. Other examples of monomials in one variable are

x^4, whose coefficient is 1 and whose degree is 4,

$-2y^9$, whose coefficient is -2 and whose degree is 9, and

$1.5x$, whose coefficient is 1.5 and whose degree is 1. (Remember that $x = x^1$.)

394

Notice that, according to our definition, zero is a monomial because $0 = 0x^1$, $0 = 0x^2$, $0 = 0x^3$, and so on. Because zero can be written as a monomial of any degree, however, we will not think of it as having a specific degree.

It is useful to define a number by itself, such as 3, or $\frac{2}{5}$, or -8, as a monomial as well. Because a number such as 3 can be written in the form $3x^0$, it seems reasonable to say that its degree is 0. The same reasoning applies to every number. So we will consider the degree of any number except zero to be 0.

If two or more monomials in the same variable have the same degree, their sum or difference can be written as a monomial. For example,

$$3x + 4x = 7x$$
$$6x^2 - x^2 = 5x^2$$
$$2x^3 - 2x^3 = 0$$

Notice that the sum or difference is either a monomial of the same degree or the number zero.

Monomials having different degrees cannot be added or subtracted in this way. For example, the sum of x^2 and x^3 cannot be written as a monomial: there is no way to write $x^2 + x^3$ without using an addition sign.

Although the sum or difference of two monomials cannot always be written as another monomial, their product can. For example, using the first law of exponents, we can write $x^2 \cdot x^3 = x^{2+3} = x^5$. In general, the product of two or more monomials is a monomial whose degree is the sum of the degrees of the monomials being multiplied.

Here are more examples of how monomials are added, subtracted, and multiplied.

EXAMPLE 1
If possible, write the sum $3x^5 + x^5$ as a monomial.

SOLUTION
Because $3x^5$ and x^5 have the same degree, 5, their sum can be written as $4x^5$.
(Note that $3x^5 = x^5 + x^5 + x^5$; $\quad 3x^5 + x^5 = (x^5 + x^5 + x^5) + x^5 = 4x^5$.)

EXAMPLE 2

If possible, write the difference $2x^6 - x^2$ as a monomial.

SOLUTION

Because $2x^6$ and x^2 do not have the same degree (their degrees are 6 and 2), their difference cannot be written as a monomial.

EXAMPLE 3

Write the product $4x^3 \cdot 6x^7$ as a monomial.

SOLUTION

$$4x^3 \cdot 6x^7 = 4 \cdot x^3 \cdot 6 \cdot x^7$$
$$= 4 \cdot 6 \cdot x^3 \cdot x^7$$
$$= 24x^{10}$$

EXAMPLE 4

Write the power $(-2x^4)^3$ as a monomial.

SOLUTION

$$(-2x^4)^3 = (-2)^3(x^4)^3$$
$$= -8x^{12}$$

Exercises

Set I

1. Show how the following number trick works by representing each step by an expression in terms of x.

    Think of a number.
    Add one.
    Multiply by three.
    Subtract nine.
    Divide by three.
    Subtract the number first thought of.
    The result is negative two.

2. Solve the following simultaneous equations.
    a) $y = 2x - 1$
       $3y - 8x = 1$

    b) $7x - 3y = 45$
       $2x + 9y = 3$

3. A mosquito beats its wings about 500 times each second as it flies through the air. At this rate, how long does it take a mosquito to beat its wings once? Express your answer in scientific notation.

Set II

4. What are the coefficients and degrees of the following monomials?
   a) $7x^2$
   b) x^6
   c) $-4x$
   d) 5

5. Tell whether or not each of the following expressions is a monomial.
   a) $\frac{1}{2}x$
   b) x^{-2}
   c) $-2x$
   d) -2
   e) $\frac{2}{x}$
   f) 2^x

6. If possible, write each of the following expressions as a monomial.
   a) $2x + 7x$
   b) $x^2 + 7x$
   c) $x^2 + x^7$
   d) $4x^3 + x^3$
   e) $11x - x$
   f) $11x - 11$
   g) $5x + 5x + 5x$
   h) $x^5 + x^5 + x^5$
   i) $2x^2 + 3x^2 + 4x^2$
   j) $2x^2 + 2x^3 + 2x^4$
   k) $3x^7 - x^7$
   l) $3x^7 - 3$
   m) $4x - 8x$
   n) $x^4 - x^8$
   o) $4x^4 - 8x^4$

7. Multiply as indicated.
   a) $2x \cdot 7x$
   b) $x^2 \cdot 7x$
   c) $x^2 \cdot x^7$
   d) $4x^3 \cdot x^3$
   e) $11x(-x)$
   f) $11x(-11)$
   g) $5x \cdot 5x \cdot 5x$
   h) $x^5 \cdot x^5 \cdot x^5$
   i) $2x^2 \cdot 3x^2 \cdot 4x^2$
   j) $2x^2 \cdot 2x^3 \cdot 2x^4$
   k) $8x^3 \cdot 3x^8$
   l) $-x^5 \cdot 5x$
   m) $(-6x)(x^6)$
   n) $(4x^5)^2$
   o) $(-5x^4)^3$

8. The following questions refer to the monomials $5x^3$ and $20x^3$. What is the coefficient of the monomial that results from
   a) adding them?
   b) multiplying them?

   What is the degree of the monomial that results from
   c) adding them?
   d) multiplying them? adding

9. If possible, write each of the following expressions as a monomial.
   a) $13x^2 + 2x^2$
   b) $13x^2 \cdot 2x^2$
   c) $13x^2 + 2x$
   d) $13x^2 \cdot 2x$
   e) $8x^5 - x^5$
   f) $8x^5(-x^5)$
   g) $8x^5 - x$
   h) $8x^5(-x)$
   i) $4x^4 + 4x^4 + 4x^4$
   j) $4x^4 \cdot 4x^4 \cdot 4x^4$
   k) $5(2x^{10})$
   l) $(2x^{10})^5$

10. Find the perimeter and area of each of these rectangles, simplifying your answers where possible. (Remember that the perimeter of a rectangle is the sum of the lengths of its sides and the area of a rectangle is the product of its length and width.)

    a)
    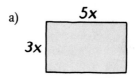
    5x, 3x

    b)
    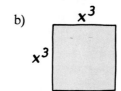
    x^3, x^3

    c)
    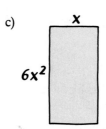
    x, $6x^2$

    d)
    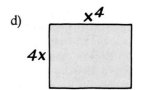
    x^4, 4x

Set III

11. What are the coefficients and degrees of the following monomials?
    a) $4x^3$
    b) x^7
    c) $-6x$
    d) 1

12. Tell whether or not each of the following expressions is a monomial.
    a) $0.5x$
    b) $x + 5$
    c) x^5
    d) $\dfrac{5}{x}$
    e) 5
    f) 5^x

13. If possible, write each of the following expressions as a monomial.
    a) $3x + 5x$
    b) $x^3 + 5x$
    c) $x^3 + x^5$
    d) $2x^7 + x^7$
    e) $9x - x$
    f) $9x - 9$
    g) $4x + 4x + 4x$
    h) $x^4 + x^4 + x^4$
    i) $2x^3 + 3x^3 + 4x^3$
    j) $3x^2 + 3x^3 + 3x^4$
    k) $5x^6 - 5$
    l) $5x^6 - x^6$
    m) $2x - 10x$
    n) $x^2 - x^{10}$
    o) $10x^2 - 2x^{10}$

14. Multiply as indicated.
    a) $3x \cdot 5x$
    b) $x^3 \cdot 5x$
    c) $x^3 \cdot x^5$
    d) $2x^7 \cdot x^7$
    e) $9x(-x)$
    f) $9x(-9)$
    g) $4x \cdot 4x \cdot 4x$
    h) $x^4 \cdot x^4 \cdot x^4$
    i) $2x^3 \cdot 3x^3 \cdot 4x^3$
    j) $3x^2 \cdot 3x^3 \cdot 3x^4$
    k) $5x^5 \cdot 8x^8$
    l) $-x \cdot 6x^6$
    m) $(3x)(-x^3)$
    n) $(7x^4)^2$
    o) $(-4x^7)^3$

15. If possible, write each of the following expressions as a monomial.
    a) $4x^3 + 7x^3$
    b) $4x^3 \cdot 7x^3$
    c) $4x^2 + 7x^3$
    d) $4x^2 \cdot 7x^3$
    e) $10x^6 - x^6$
    f) $10x^6(-x^6)$
    g) $10x^6 - x$
    h) $10x^6(-x)$
    i) $9x^9 + 9x^9 + 9x^9$
    j) $9x^9 \cdot 9x^9 \cdot 9x^9$
    k) $4(5x^8)$
    l) $(5x^8)^4$

16. If possible, express each of the following general sums or products as a single monomial.
    a) $ax^b \cdot cx^d$
    b) $ax^b + cx^d$
    c) $ax^b \cdot cx^b$
    d) $ax^b + cx^b$

17. Find the perimeter and area of each of these rectangles. Simplify your answers where possible.

a)

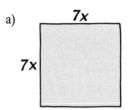

b)

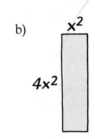

c)

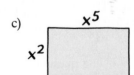

d)

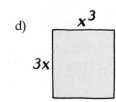

Set IV

When asked to make up an example of a monomial, Acute Alice wrote $10x^2$ and Obtuse Ollie wrote $2x^{10}$. Ollie thought his monomial was larger because it had a larger exponent. Alice thought hers was larger because it had a larger coefficient.

1. Find the value of each monomial if
   a) $x = 0$ b) $x = 1$ c) $x = 2$ d) $x = -1$
   e) $x = -2$

2. Can you draw any conclusion about the relative values of the two monomials for other values of x?

LESSON 2
Polynomials

This photograph shows a litter of 13 kittens. The usual number of kittens in a litter is between 5 and 9. Suppose that a cat had a litter of 5 female kittens and each kitten later produced a litter of 5 kittens. How many cats and kittens would there be altogether?

There would be the original cat (1), her kittens (5), and the kittens' kittens (25). There would be

$$1 + 5 + 25 = 31$$

cats and kittens in all.

Now let's reconsider this problem in terms of a variable. Suppose that a cat had a litter of x female kittens and each kitten later produced a litter of x kittens. How many cats and kittens would there be in all?

There would be the original cat (1), her kittens (x), and the kittens' kittens (x^2). There would be $1 + x + x^2$ cats and kittens in all.

The expression $1 + x + x^2$ is an example of a *polynomial.*

▶A **polynomial** is either a monomial or an expression indicating the addition and/or subtraction of two or more monomials. The monomials are called **terms** of the polynomial.

The expression $1 + x + x^2$ is a polynomial because it is the sum of three monomials: 1, x, and x^2. Although each term of this polynomial is of a different degree (0, 1, and 2 respectively), the degree of the polynomial itself is said to be 2.

▶The **degree** of a polynomial is that of the term having the highest degree.

Other examples of polynomials and their degrees are:

$x^3 + 8$, a polynomial of two terms whose degree is 3,

$2x^5 - x^4 - 1$, a polynomial of three terms whose degree is 5,

$\frac{1}{3}x^7$, a polynomial of one term whose degree is 7, and

-9, a polynomial of one term whose degree is 0.

It is often useful to write polynomials in one variable so that the term having the highest degree is first, the term having the next highest degree is second, and so on. Such a polynomial is said to be written in *descending powers of the variable.* Here are a couple of examples of how this is done.

EXAMPLE 1
Rewrite the polynomial $6x - 3x^2 - 1 + x^3$ in descending powers of x and state its degree.

SOLUTION
The degrees of the terms of the polynomial $6x - 3x^2 - 1 + x^3$ are 1, 2, 0, and 3, respectively. Written in descending powers of x, it becomes $x^3 - 3x^2 + 6x - 1$. The degree of this polynomial is 3.

EXAMPLE 2

Simplify the polynomial $10x + 4x^4 - 2x - x^2 + x^4$ by combining terms of the same degree. Write the result in descending powers of x.

SOLUTION

The degrees of the terms of this polynomial are 1, 4, 1, 2, and 4, respectively. Combining the terms of the same degree,

we get $8x + 5x^4 - x^2$. Written in descending powers of x, this becomes $5x^4 - x^2 + 8x$.

Exercises

Set I

1. Write each of the following numbers in scientific notation.
   a) 0.000004
   b) 9
   c) 125×10^{-7}
   d) 0.32×10^{-5}

2. The following table of numbers represents experimental data.

x	2	5	7	10	11
y	1.0	2.5	3.5	4.0	5.5

   a) Draw a graph for this table.
   b) One value of y seems to be incorrect. Which one do you think it is?
   c) What seems to be its correct value?
   d) Find a formula for the table.

3. This diagram represents two weights balanced on a seesaw.

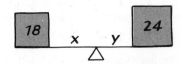

   Write a pair of simultaneous equations for the diagram and solve the equations to find x and y
   a) if the weights are 35 centimeters apart.
   b) if the weights are 70 centimeters apart.
   c) How does your answer to part b compare with your answer to part a?

Set II

4. Tell whether or not each of the following expressions is a polynomial.
   a) $5x^2 + 10x$
   b) $x^4 - x^6$
   c) $x^{-3} + 1$
   d) $-8x$
   e) $\frac{1}{2}x^{10} + \frac{2}{3}x^5 - \frac{3}{4}$
   f) $10^x - 5$

5. Write each of the following polynomials as a sum of monomials.
   a) $x^3 - 4$
   b) $1 + 2x - x^2$
   c) $-5x - 10$

6. Rewrite each of the following polynomials in descending powers of the variable.
   a) $x + x^5$
   b) $3 - 7y + 5y^2$
   c) $3x^3 - x + 2x^4 + 4x^2$
   d) $1 - y^6$
   e) $5x^4 + 20 - x^6 - 2x^8$
   f) $2^5 + y$

7. Use the distributive rule to write each of the following as a <u>sum of monomials</u>.
   a) $2(3x + 8)$
   b) $-5(y^2 - 1)$
   c) $x(x^3 + 7x^2 - 6x)$
   d) $3x(x^2 - 2x + 5)$
   e) $x^3(x^2 - 2x + 5)$
   f) $-4y^2(y^4 - 2)$

8. Like the opposite of a number, the opposite of a polynomial is the polynomial that results from multiplying it by -1. For example, the opposite of $x^3 - 2x$ is

   $$-1(x^3 - 2x) = -x^3 + 2x$$

   Write the opposite of each of the following polynomials.
   a) $x^2 + 5x + 10$ c) $-6x^2 + 11$
   b) $y^5 - 7y$ d) $1 - 2y + 3y^2 - 4y^3$

9. The value of a polynomial in x depends on the value of x. For example, if $x = -3$, then

the value of the polynomial $5x^2 - 9$ is

$$5(-3)^2 - 9 = 5(9) - 9 = 45 - 9 = 36$$

Find the values of each of the following polynomials for the values of the variables given.

$$x^2 + 7x - 1$$
a) $x = 0$
b) $x = 1$
c) $x = 2$
d) $x = -1$
e) $x = -2$

$$15 - 3x^2$$
f) $x = 0$
g) $x = 1$
h) $x = -1$
i) $x = 5$

$$-15 + 3x^2$$
j) $x = 0$
k) $x = 1$
l) $x = -1$
m) $x = 5$

$$x^4 + x^3 + x^2 + x + 1$$
n) $x = 0$
o) $x = 1$
p) $x = -1$
q) $x = 2$
r) $x = -2$
s) $x = 10$
t) $x = -10$

10. Simplify each of the following polynomials by combining terms of the same degree. Write each answer in descending powers of the variable.
    a) $4x + 6x^2 + 2x$
    b) $2x^3 - 2x + 3x^2 - 3x$
    c) $7x^2 - 7x - x^2$
    d) $1 + 8x + x^3 - x - 8$
    e) $x^4 - 4x^2 + 4 - 4x^2$
    f) $x^7 + 7x + x^7 + 7x$

Set III

11. Tell whether or not each of the following expressions is a polynomial.
    a) $2x^4 - x$
    b) $5^x + 5$
    c) $8 - x^8$
    e) $10,000,000$
    f) $3x^{-2} + 2x^{-1} + 1$
    d) $\dfrac{1}{4}x^5 + \dfrac{1}{3}x^4 - \dfrac{1}{2}$

12. Write each of the following polynomials as a sum of monomials.
    a) $3x^2 - 5$
    b) $4 - 4x + x^2$
    c) $-2x - 8$

13. Rewrite each of the following polynomials in descending powers of the variable and state its degree.
    a) $4x + x^4$
    b) $2 - 9y^2 + 3y^5$
    c) $6x^2 - 5x + x^3 - 7$
    d) $10 - 3y^2$
    e) $4^7 + x$
    f) $3y^5 + 24 - 30y - y^6$

14. Use the distributive rule to write each of the following as a sum of monomials.
    a) $4(2x + 5)$
    b) $-3(y^2 - 4)$
    c) $x(x^3 - 8x^2 + 3x)$
    d) $2x(x^2 - x + 6)$
    e) $x^2(x^2 - x + 6)$
    f) $-5y^3(y^5 - 3)$

15. Like the opposite of a number, the opposite of a polynomial is the polynomial that results from multiplying it by -1. For example, the opposite of $3x^2 - x$ is

    $$-1(3x^2 - x) = -3x^2 + x$$

    Write the opposite of each of the following polynomials.

    a) $x^3 + 4x + 6$
    b) $y^4 - 8$
    c) $-5x^2 + 10x$
    d) $4 - 3y + 2y^2 - y^3$

16. The value of a polynomial in x depends on the value of x. For example, if $x = 5$, then the value of the polynomial $2x^3 - 10$ is

$$2(5)^3 - 10 = 2(125) - 10 = 250 - 10 = 240$$

Find the values of each of the following polynomials for the values of the variables given.

$$x^2 - 4x + 6$$
a) $x = 0$
b) $x = 1$
c) $x = 2$
d) $x = -1$
e) $x = -2$

$$20 - 2x^2$$
f) $x = 0$
g) $x = 1$
h) $x = -1$
i) $x = 5$

$$-20 + 2x^2$$
j) $x = 0$
k) $x = 1$
l) $x = -1$
m) $x = 5$

$$x^3 - x^2 + x - 1$$
n) $x = 0$
o) $x = 1$
p) $x = -1$
q) $x = 4$
r) $x = -4$
s) $x = 10$
t) $x = -10$

17. Simplify each of the following polynomials by combining terms of the same degree. Write each answer in descending powers of the variable.
    a) $x^3 + 3x + 3x^3$
    b) $2x^4 + 4x - 2x^3 - 3x$
    c) $8x + x^2 - 10x$
    d) $7x^2 - 6x - 5x^2 - x$
    e) $6x - 6 + x^2 - x + 1$
    f) $x^5 - 5x^4 + 4x^5 + x^4$

Set IV

As an elevator travels from one floor of a building to another, it does not move at a constant speed.* A certain elevator takes four seconds to get from one stop to another. Its speed in feet per second is given by the polynomial

$$-0.5x^4 + 4x^3 - 12x^2 + 16x$$

in which x represents the number of seconds after it starts.

1. Use the polynomial to find the speed of the elevator when $x = 0, 1, 2, 3$, and 4.
2. Do your answers make sense in comparison with the way that you think an elevator moves? Explain.

___

*Adapted from an example in *Calculus* by Gerald Freilich and Frederick P. Greenleaf (W. H. Freeman and Company, 1976), p. 136.

$-8+32-48+32 \qquad -40.5+108-108+48$

$-128+156-102+64$

LESSON 3

Adding and Subtracting Polynomials

In the sixteenth century, King Henry VIII of England decreed that a pound was to be equal to 16 ounces. Since then it has been customary to express many weights in combinations of the two units. For example, if Charlie Brown stepped onto a very precise scale, he might read his weight as "80 pounds, 12 ounces."

If the scale gives Snoopy's weight as "11 pounds, 3 ounces," what would the scale read if both Charlie Brown and Snoopy stepped on it at the same time? We can answer this question by adding the two weights like this:

$$
\begin{array}{r}
80\,\text{lb} + 12\,\text{oz} \\
+\ \underline{11\,\text{lb} + \ \ 3\,\text{oz}} \\
91\,\text{lb} + 15\,\text{oz}
\end{array}
$$

Together, Charlie Brown and Snoopy would weigh "91 pounds, 15 ounces."

Two or more polynomials can be added in the same way. They can be written in a column so that like terms (those of the same degree and variable) are in line. Each set of like terms is then added. For example, suppose that we want to add two polynomials such as $4x^2 + x + 5$ and $x^2 - 7x + 3$. Lining up like terms and adding, we get

$$
\begin{array}{r}
4x^2 + x + 5 \\
+ \quad x^2 + -7x + 3 \\
\hline
5x^2 + -6x + 8
\end{array}
$$

The sum is $5x^2 - 6x + 8$.

The same method can be used to subtract one polynomial from another. Here are more examples of how polynomials are added and subtracted.

EXAMPLE 1
Add $x^3 - 7x^2 - 1$ and $3x^2 + 8x$.

SOLUTION
Before adding, we write each polynomial as a sum and line up like terms.

$$
\begin{array}{r}
x^3 + -7x^2 + -1 \\
+ \quad 3x^2 + 8x \\
\hline
x^3 + -4x^2 + 8x + -1, \quad \text{or} \quad x^3 - 4x^2 + 8x - 1
\end{array}
$$

EXAMPLE 2
Add $2x + 9y$ and $x - 5y$.

SOLUTION
These polynomials contain two variables. We have practiced adding and subtracting polynomials in two variables in our work with simultaneous equations.

$$
\begin{array}{r}
2x + 9y \\
+ \quad x + -5y \\
\hline
3x + 4y
\end{array}
$$

EXAMPLE 3

Add $x^5 + 3x$, $4x^2 - 2$, and $x^2 - 7x$.

SOLUTION

In this example, we will do the problem "horizontally" instead of vertically as we did above.

$$
\begin{aligned}
(x^5 + 3x) + (4x^2 - 2) + (x^2 - 7x) &= \\
x^5 + 3x + 4x^2 - 2 + x^2 - 7x &= \\
x^5 + 4x^2 + x^2 + 3x - 7x - 2 &= \\
x^5 + 5x^2 - 4x - 2 &
\end{aligned}
$$

EXAMPLE 4

Subtract $3x + 10$ from $5x + 4$.

SOLUTION

$$
\begin{array}{r}
5x + 4 \\
-\ \underline{3x + 10} \\
2x + \text{-}6, \quad \text{or} \quad 2x - 6
\end{array}
$$

EXAMPLE 5

Subtract $7x^2 - 9$ from $10x^2 + x + 2$.

SOLUTION

$$
\begin{array}{r}
10x^2 + x + 2 \\
-\ \underline{7x^2 + \text{-}9} \\
3x^2 + x + 11
\end{array}
$$

EXAMPLE 6

Subtract $x^2 - 6$ from $2x^3 - 5x$.

SOLUTION

Doing this problem horizontally and using the fact that subtracting a polynomial is equivalent to adding its opposite, we get

$$
\begin{aligned}
(2x^3 - 5x) - (x^2 - 6) &= \\
(2x^3 - 5x) + (-x^2 + 6) &= \\
2x^3 - 5x - x^2 + 6 &= \\
2x^3 - x^2 - 5x + 6 &
\end{aligned}
$$

Exercises

Set I

1. Add or subtract as indicated.

   a) -13
   $+\ \underline{-9}$

   b) $\ \ 13$
   $+\ \underline{-9}$

   c) -13
   $+\ \underline{\ \ 9}$

   d) -13
   $-\ \underline{-9}$

   e) $\ \ 13$
   $-\ \underline{-9}$

   f) -13
   $-\ \underline{\ \ 9}$

2. What are the slope and y-intercept of each of the following lines?

   a) $y = 2x - 1$ b) $y = x$ c) $x + y = 1$

3. The time that it takes a stopped car to accelerate to a given speed depends on the rate of acceleration. Here is a table of times for a car accelerating to a speed of 60 feet per second.

Rate of acceleration in feet per second per second, x	10	20	30	40	50
Time in seconds, y	6	3	2	1.5	1.2

   a) What kind of variation is this?
   b) Write a formula for y in terms of x.
   c) Find y when $x = 80$.

Set II

4. Add as indicated.

   a) $\ \ \ 5x + 9$
   $+\ \underline{\ \ x - 8}$

   b) $\ \ \ x - 3y$
   $+\ \underline{\ \ x - \ y}$

   c) $\ \ 3x^2 - \ 2$
   $+\ \underline{\ \ x^2 + 10}$

   d) $\ \ \ x^2 + 6x - 1$
   $+\ \underline{2x^2 - 7x + 1}$

   e) $\ \ 5x^3 - 4x^2\qquad\ \ + 11$
   $+\ \underline{\ -x^3\qquad\quad + 4x - \ 7}$

5. Subtract the second polynomial from the first as indicated.

   a) $\ \ 11x + 3$
   $-\ \underline{\ \ 4x - 1}$

   b) $\ \ 2x - 7y$
   $-\ \underline{\ \ x - 7y}$

   c) $\ \ \ x^2 - 3x$
   $-\ \underline{2x^2 + 3x}$

   d) $\ \ 5x^2 + 9x + 6$
   $-\ \underline{3x^2 - \ x + 7}$

   e) $\ \ x^4 - 8x^2 + 4x$
   $-\ \underline{\qquad\ \ x^2 - 5x - 2}$

6. This exercise is about the following polynomials:

   polynomial A: $3x^2 + x - 2$
   polynomial B: $x^2 - 4x + 3$

   a) Find their sum and label it polynomial C.
   b) Find their difference by subtracting polynomial B from polynomial A. Label it polynomial D.
   c) Find the values of the four polynomials if $x = 4$ and if $x = 5$. Use your answers to complete the following table.

x	4	5
polynomial A		
polynomial B		
polynomial C		
polynomial D		

d) What do you observe about the values of the polynomials in your table?

7. Add.
   a) $x + y + 3$ and $x - y - 4$.
   b) $7x - 19$, $2x + 7$, and $11x - 1$.
   c) $x^3 - 3$, $x^2 + 2$, and $x + 1$.
   d) $x - 5y$, $y + 6z$, $z + 7x$, and $x - 8z$.

8. Subtract.
   a) $2x^2 - y$ from $8x^2 + 5y$.

b) $7x + 12$ from $4x - 12$.
c) $6x + 4$ from $6x^2 + 4x$.
d) $x^2 - 1$ from $x^3 - x$.

9. Simplify.
   a) $(7x + 3y) + (8x - y) - (6x + 5y)$
   b) $(4x^2 - 9x + 2) - (x^2 + 3x - 1)$
      $+ (5x^2 + 11x + 7)$
   c) $(1 - x^3) - (x - x^3) - (x^2 - x^3)$

Set III

10. Add as indicated.

    a) $x + 12$
         $+ 4x - 7$

    b) $6x - y$
         $+ 2x - y$

    c) $x^2 - 11x$
         $+ 5x^2 + x$

    d) $3x^2 - 8x + 2$
         $+ x^2 + 8x - 5$

    e) $x^4 + 9x^2 - 3x$
         $+ x^2 + x - 2$

11. Subtract the second polynomial from the first as indicated.

    a) $8x + 5$
         $- x + 17$

    b) $x + 3y$
         $- x - 2y$

    c) $7x^2 - 2$
         $-6x^2 + 2$

    d) $4x^2 + 11x - 8$
         $- 5x^2 + 6x + 7$

    e) $2x^3 - 9x + 1$
         $- x^2 - 9x - 1$

12. This exercise is about the following polynomials:

    polynomial A: $4x^2 - x + 3$
    polynomial B: $x^2 + 3x - 4$

    a) Find their sum and label it polynomial C.
    b) Find their difference by subtracting polynomial B from polynomial A. Label it polynomial D.

c) Find the values of the four polynomials if $x = 3$ and if $x = 5$. Use your answers to complete the following table.

x	3	5
polynomial A		
polynomial B		
polynomial C		
polynomial D		

d) What do you observe about the values of the polynomials in your table?

13. Add.
    a) $x^2 - 2x + 1$ and $x^2 + 2x + 1$.
    b) $5x - 12y$, $x - y$, and $8x + 3y$.
    c) $x + 4$, $x^2 - 3$, and $x^3 - 2$.
    d) $4x^3 + 3x^2$, $2x^2 - x$, $x + 1$, and $3 - x^3$.

14. Subtract.
    a) $4x + y^2$ from $9x - 4y^2$.
    b) $x^2 - 8$ from $x^2 + 18$.
    c) $3x - 1$ from $3x^2 - x$.
    d) $y - 7$ from $x + 3y$.

15. Simplify.
    a) $(5x - 2y) + (x + 9y) - (4x - 3y)$
    b) $(8x^2 + 3x - 4) - (2x^2 - 6x - 3)$
       $+ (x^2 + x + 1)$
    c) $(x^5 - 1) - (x^4 - x) - (x^3 - x^2)$

Set IV A Number Trick

1. Think of a two-digit number in which the units digit is more than the tens digit. Multiply the difference between the digits by nine and add the result to your original number.
2. What do you notice about the result?
3. Can you explain why it happened? (Hint: If x and y represent the tens digit and units digit of the original number, respectively, then the original number can be written as $10x + y$.)

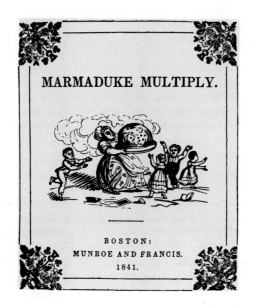

MARMADUKE MULTIPLY.

BOSTON:
MUNROE AND FRANCIS.
1841.

59

8 times 10 are 80.
I think he's pretty weighty.

LESSON 4
Multiplying Polynomials

In the nineteenth century, a book titled *Marmaduke Multiply* was used in many elementary schools to help children learn the multiplication table. The book illustrated each multiplication fact with a picture and a rhyme, beginning with 2 times 1 and ending with 12 times 12.

Even today, most people do not learn the multiplication table beyond this point. To find the product of numbers of which one or both of the numbers is larger than 12, we learn the method illustrated here.

$$\begin{array}{r} 53 \\ \times\ \underline{27} \\ 371 \\ \underline{106} \\ 1431 \end{array}$$

In multiplying 53 by 27, we are, in effect, multiplying the sums $50 + 3$ and $20 + 7$. The steps that we do to find this product are shown in the expanded version of the problem at the top of the next page.

	Short version	Expanded version

53	$50 + 3$	
$\times$ 27	$\times$ 20 + 7	
371	$350 + 21$	371
106	$1000 + 60$	1060
1431		1431

Notice that four multiplications have been done in this example and that the answer is the sum of the four products.

Because the area of a rectangle is found by multiplying its length and width, a rectangle whose length and width are $50 + 3$ and $20 + 7$ can be used to illustrate what is happening. Each of the four multiplications corresponds to finding the area of one of the smaller rectangles in the figure. The answer is the sum of the four areas.

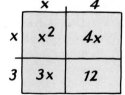

These procedures can be used to multiply polynomials. Examples are shown below.

EXAMPLE 1

Multiply $x + 3$ and $x + 4$.   *all only   like term*

SOLUTION

$$
\begin{array}{r}
x + 4 \\
\times\ x + 3 \\
\hline
3x + 12 \\
x^2 + 4x \\
\hline
x^2 + 7x + 12
\end{array}
$$

(Notice that it is helpful to line up like terms because they must be added later.)

	x	4
x	x^2	$4x$
3	$3x$	12

EXAMPLE 2

Multiply $2x - 7$ and $5x + 1$.

SOLUTION

$$
\begin{array}{r}
5x + 1 \\
\times\ 2x + -7 \\
\hline
-35x + -7 \\
10x^2 + \quad 2x \\
\hline
10x^2 + -33x + -7, \quad \text{or} \quad 10x^2 - 33x - 7
\end{array}
$$

	$5x$	1
$2x$	$10x^2$	$2x$
-7	$-35x$	-7

EXAMPLE 3

Multiply $4x$ and $x^2 + 6x - 5$.

SOLUTION

$$\begin{array}{r} x^2 + 6x - 5 \\ \times \qquad\quad 4x \\ \hline 4x^3 + 24x^2 - 20x \end{array}$$

EXAMPLE 4

Multiply $x + 7$ and $2x^3 + x^2 + 10$.

SOLUTION

$$\begin{array}{r} 2x^3 + x^2 + 0x + 10 \\ \times \qquad\qquad\quad x + 7 \\ \hline 14x^3 + 7x^2 + 0x + 70 \\ 2x^4 + x^3 + 0x^2 + 10x \qquad\quad \\ \hline 2x^4 + 15x^3 + 7x^2 + 10x + 70 \end{array}$$

(Because $2x^3 + x^2 + 10$ is missing a first-degree term, it helps to write one in before multiplying to keep the like terms lined up.)

Although diagrams are helpful in illustrating how two polynomials are multiplied, it is important to be able to carry out the procedure without them.

Exercises

Set I

1. Simplify each expression if possible.
   a) $2 + 5x^3$
   b) $2(5x^3)$
   c) $2(5 + x^3)$
   d) $2(5x)^3$

2. Which symbol, $>$, $=$, or $<$, should replace ▓ in each of the following to make it true for all values of x? If none of them will work for all values of x, explain why not.
   a) x^3 ▓ x^5
   b) x^2 ▓ $(-x)^2$
   c) $2x - 1$ ▓ $2x$
   d) $\dfrac{x}{2}$ ▓ x

3. If a rumor spreads through a large crowd of people, the number of people who have heard it is a function of the time since the rumor began. A typical formula for this function is

$$y = 3^x$$

in which x is the time in minutes since the rumor began and y is the number of people who have heard it.
a) What kind of function is this?
b) According to this formula, how many people will have heard the rumor 5 minutes after it began?
c) How many will have heard it 10 minutes after it began?
d) Does the rumor spread at a constant rate?

Set II

4. When asked to multiply 24 by 15, Obtuse Ollie wrote

$$
\begin{array}{r}
24 \\
\times\ 15 \\
\hline
120 \\
24 \\
\hline
144
\end{array}
$$

a) What is wrong with his method?
b) Draw a diagram like the one on page 413 to illustrate the correct way to do the problem.

5. Do each of the following multiplication problems by arithmetic. Then draw a diagram to illustrate each problem and compare it with your solution by arithmetic.
a)
$$
\begin{array}{r}
37 \\
\times\ 52 \\
\end{array}
$$
b)
$$
\begin{array}{r}
8 \\
\times\ 419 \\
\end{array}
$$
c)
$$
\begin{array}{r}
206 \\
\times\ 93 \\
\end{array}
$$

6. What multiplication problem and what answer are illustrated by each of these diagrams?

7. Make a diagram to illustrate each of these multiplication problems. Then use the diagrams to find the products.
a) $(x + 4)(x + 11)$ c) $a^3(a^2 + 4)$
b) $(5y + 2)(3y - 1)$ d) $(b + 7)(b^2 + 7b - 1)$

8. Multiply as indicated.

a)
$$
\begin{array}{r}
x + 12 \\
\times\ x + 2 \\
\end{array}
$$
g)
$$
\begin{array}{r}
x + 4 \\
\times\ 4 - x \\
\end{array}
$$

b)
$$
\begin{array}{r}
x - 8 \\
\times\ x + 9 \\
\end{array}
$$
h)
$$
\begin{array}{r}
x^3 - 1 \\
\times\ x^3 + 8 \\
\end{array}
$$

c)
$$
\begin{array}{r}
3x + 4 \\
\times\ 2x - 5 \\
\end{array}
$$
i)
$$
\begin{array}{r}
2x^2 - 9x + 1 \\
\times\ x + 5 \\
\end{array}
$$

d)
$$
\begin{array}{r}
6x + 1 \\
\times\ 6x + 1 \\
\end{array}
$$
j)
$$
\begin{array}{r}
x^3 - x^2 + x - 1 \\
\times\ x + 1 \\
\end{array}
$$

e)
$$
\begin{array}{r}
6x + 1 \\
\times\ 6x - 1 \\
\end{array}
$$
k)
$$
\begin{array}{r}
x^3 + 5x - 8 \\
\times\ x - 4 \\
\end{array}
$$

f)
$$
\begin{array}{r}
7x - 3 \\
\times\ 3x - 7 \\
\end{array}
$$
l)
$$
\begin{array}{r}
x^2 + 2x + 3 \\
\times\ x^2 - 2x + 3 \\
\end{array}
$$

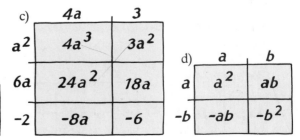

a)

	x	5
$2x$	$2x^2$	$10x$
1	x	5

b)

	$3y$	-7
$6y$	$18y^2$	$-42y$

c)

	$4a$	3
a^2	$4a^3$	$3a^2$
$6a$	$24a^2$	$18a$
-2	$-8a$	-6

d)

	a	b
a	a^2	ab
$-b$	$-ab$	$-b^2$

Set III

9. When asked to multiply 22 by 14, Acute Alice wrote

$$\begin{array}{r} 22 \\ \times\, 14 \\ \hline 88 \\ 22 \\ \hline 902 \end{array}$$

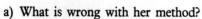

a) What is wrong with her method?
b) Draw a diagram like the one on page 413 to illustrate the correct way to do the problem.

10. Do each of the following multiplication problems by arithmetic. Then draw a diagram to illustrate each problem and compare it with your solution by arithmetic.

a) $\begin{array}{r} 63 \\ \times\, 24 \end{array}$

b) $\begin{array}{r} 5 \\ \times\, 197 \end{array}$

c) $\begin{array}{r} 409 \\ \times\, 38 \end{array}$

11. What multiplication problem and what answer are illustrated by each of these diagrams?

12. Make a diagram to illustrate each of these multiplication problems. Then use the diagrams to find the products.

a) $(x + 6)(x + 5)$
b) $(2y + 3)(7y - 10)$
c) $3a^2(2a^3 - a)$
d) $(b - 5)(b^2 + 5b + 2)$

13. Multiply as indicated.

a) $\begin{array}{r} x + 10 \\ \times\, x + 4 \end{array}$

g) $\begin{array}{r} x + 7 \\ \times\, 7 - x \end{array}$

b) $\begin{array}{r} x - 6 \\ \times\, x + 7 \end{array}$

h) $\begin{array}{r} x^2 - 9 \\ \times\, x^2 + 4 \end{array}$

c) $\begin{array}{r} 4x + 5 \\ \times\, 2x - 3 \end{array}$

i) $\begin{array}{r} x^2 + 6x - 5 \\ \times\, \qquad x + 1 \end{array}$

d) $\begin{array}{r} 8x + 1 \\ \times\, 8x + 1 \end{array}$

j) $\begin{array}{r} x^3 + x^2 + x + 1 \\ \times\, \qquad\qquad x - 1 \end{array}$

e) $\begin{array}{r} 8x + 1 \\ \times\, 8x - 1 \end{array}$

k) $\begin{array}{r} x^3 + 2x^2 - 6 \\ \times\, \qquad\quad x + 3 \end{array}$

f) $\begin{array}{r} 5x - 2 \\ \times\, 2x - 5 \end{array}$

l) $\begin{array}{r} x^2 - 4x + 5 \\ \times\, x^2 + 4x - 5 \end{array}$

a)

	x	8
$3x$	$3x^2$	$24x$
2	$2x$	16

b)

	y^2	$-y$	6
$5y$	$5y^3$	$-5y^2$	$30y$

c)

	a^2	4
a^3	a^5	$4a^3$
$-4a$	$-4a^3$	$-16a$
9	$9a^2$	36

d)

	$2b$	-7
$2b$	$4b^2$	$-14b$
7	$14b$	-49

Set IV

Although he knows he shouldn't, Obtuse Ollie can't resist looking up the answers to his algebra homework before trying to do the problems. Some of the answers to the lesson on multiplying polynomials are shown below. Can you figure out what the problems were? (Hint: Draw some diagrams.)

1. $9x^2 + 12x + 4$
2. $x^2 - 16$
3. $10x^2 + 37x + 7$

LESSON 5

More on Multiplying Polynomials

"Ahem."

When multiplying two numbers in arithmetic, we usually write one below the other. It is also possible, however, to write them in a row.

Consider, for example, the problem 15×48. Writing 15 and 48 as the sums $10 + 5$ and $40 + 8$, and illustrating the problem with a diagram showing area, we get the figure at the left. The figure reminds us that the answer is obtained by multiplying each term of the sum $10 + 5$ by each term of the sum $40 + 8$ and adding the four numbers that result.

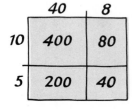

$$(10 + 5)(40 + 8) = 10 \cdot 40 + 10 \cdot 8 + 5 \cdot 40 + 5 \cdot 8$$

Any two sums can be multiplied in the same way. For example, to multiply $x + 5$ and $2x + 6$, we can write

$$
\begin{aligned}
(x + 5)(2x + 6) &= x(2x) + x(6) + 5(2x) + 5(6) \\
&= 2x^2 + 6x + 10x + 30 \\
&= 2x^2 + 16x + 30
\end{aligned}
$$

In these examples, we are multiplying polynomials by exactly the same method that we used in Lesson 4. The only difference is in the way everything is written down. Instead of writing one polynomial below the other and doing the

418

multiplication as we would in arithmetic, we write everything in rows. The principle either way is the same:

▶ To multiply two polynomials, multiply each term of one polynomial by each term of the other and then add the terms produced.

Here are more examples of how polynomials are multiplied by this method.

EXAMPLE 1
Multiply $2x + 7$ and $x^2 - 3x + 2$.

SOLUTION

$$(2x + 7)(x^2 - 3x + 2) = 2x(x^2) - 2x(3x) + 2x(2) + 7(x^2) - 7(3x) + 7(2)$$
$$= 2x^3 - 6x^2 + 4x + 7x^2 - 21x + 14$$
$$= 2x^3 + x^2 - 17x + 14$$

The diagram at the right illustrates the six products that come from multiplying the two terms of $2x + 7$ by the three terms of $x^2 - 3x + 2$.

	x^2	$-3x$	2
$2x$	$2x^3$	$-6x^2$	$4x$
7	$7x^2$	$-21x$	14

EXAMPLE 2
Multiply $x^2 - 7$ and $6x - 1$.

SOLUTION

$$(x^2 - 7)(6x - 1) = x^2(6x) - x^2(1) - 7(6x) - 7(-1)$$
$$= 6x^3 - x^2 - 42x + 7$$

EXAMPLE 3
Multiply $x^2 - 5x + 25$ and $x + 5$.

SOLUTION

$$(x^2 - 5x + 25)(x + 5) = x^2(x) + x^2(5) - 5x(x) - 5x(5) + 25(x) + 25(5)$$
$$= x^3 + 5x^2 - 5x^2 - 25x + 25x + 125$$
$$= x^3 + 125$$

Exercises

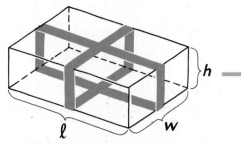

Set I

1. Write each of the following numbers as a power of 12.

   a) 1,728   b) 12   c) 1   d) $\dfrac{1}{144}$

2. Simplify.
   a) $(2x - 5y + 10) + (3x + 4y - 12)$
   b) $(x^2 + 8x - 1) + (6 - 6x)$
   c) $(x + 7y) - (7x - y)$
   d) $(x^4 - 2x^2 + 1) - (x^2 + 4x)$

3. A ribbon is wrapped around a box as shown in this picture. The box is ℓ inches long, w inches wide, and h inches high.
   a) Write a formula for the total length of the ribbon, x, in terms of ℓ, w, and h.
   b) Given that $\ell = 3h$ and $w = 2h$, write a formula for the total length of the ribbon, x, in terms of h only.

Set II

4. Find each of the following products.
   a) $(x + 6)(x + 2)$
   b) $(x + 6)(x - 2)$
   c) $(x - 3)(x - 8)$
   d) $(x - 4)(x + 5)$
   e) $(x + 7)(x + 7)$
   f) $(x + 7)(x - 7)$
   g) $(2x + 8)(3x + 1)$
   h) $(2x - 8)(3x - 1)$
   i) $(6x + 5)(5x - 2)$
   j) $(x - 10)(4x + 9)$
   k) $(x - y)(x - y)$
   l) $(x - y)(x + y)$

5. This exercise is about the following polynomials:

   > polynomial A: $5x - 3$
   > polynomial B: $2x + 1$

   a) Find their product and label it polynomial C.
   b) Find the values of the three polynomials if $x = 1$, if $x = 3$, and if $x = 10$. Use your answers to complete the following table.

	x	1	3	10
polynomial A		▨	▨	▨
polynomial B		▨	▨	▨
polynomial C		▨	▨	▨

   c) What do you observe about the values of the polynomials in your table?

6. Find each of the following products.
   a) $(x^2 + 5)(x^2 - 1)$
   b) $(x - 9)(x^3 + x)$
   c) $(x + 2)(x^2 - 2x + 4)$
   d) $(x^2 + 3x - 1)(x^2 - x + 3)$
   e) $2x(x + 6)(x - 4)$
   f) $(x - 1)(x - 2)(x - 3)$

7. Multiply each of the following polynomials by $x - 1$.
   a) $x + 1$
   b) $x^2 + x + 1$
   c) $x^3 + x^2 + x + 1$
   d) $x^4 + x^3 + x^2 + x + 1$
   e) What do you think the product of $x - 1$ and $x^{10} + x^9 + x^8 + x^7 + x^6 + x^5 + x^4 + x^3 + x^2 + x + 1$ is?

Set III

8. Find each of the following products.
   a) $(x + 5)(x + 3)$
   b) $(x + 5)(x - 3)$
   c) $(x - 9)(x - 4)$
   d) $(x - 6)(x + 7)$
   e) $(x - 8)(x - 8)$
   f) $(x - 8)(x + 8)$
   g) $(4x + 6)(2x + 1)$
   h) $(4x - 6)(2x - 1)$
   i) $(5x + 3)(3x - 4)$
   j) $(8x + 2)(x - 9)$
   k) $(x + y)(x + y)$
   l) $(x + y)(x - y)$

9. This exercise is about the following polynomials:

   polynomial A: $2x + 3$
   polynomial B: $5x - 1$

   a) Find their product and label it polynomial C.
   b) Find the values of the three polynomials if $x = 1$, if $x = 2$, and if $x = 10$. Use your answers to complete the following table.

x	1	2	10
polynomial A	▓	▓	▓
polynomial B	▓	▓	▓
polynomial C	▓	▓	▓

   c) What do you observe about the values of the polynomials in your table?

10. Find each of the following products.
   a) $(x^2 - 2)(x^2 + 6)$
   b) $(x + 4)(x^4 - x)$
   c) $(x - 3)(x^2 + 3x + 9)$
   d) $(x^2 + 2x + 1)(x^2 - x - 2)$
   e) $3x(x - 7)(x + 2)$
   f) $(x + 1)(x + 3)(x + 5)$

11. Multiply each of the following polynomials by $x + 1$.
   a) $x - 1$
   b) $(x^2 - x + 1)(x + 1)$
   c) $x^3 - x^2 + x - 1$
   d) $x^4 - x^3 + x^2 - x + 1$
   e) What do you think the product of $x + 1$ and $x^{10} - x^9 + x^8 - x^7 + x^6 - x^5 + x^4 - x^3 + x^2 - x + 1$ is?

Set IV

Professor Pythagoras is telling his students about an interesting discovery that he made.
   1. To find out what it is, write down three consecutive integers. Square the second number and multiply the first number by the last.

      Do this with several other sets of consecutive integers.
   2. What do you notice?
   3. Can you explain why this will always happen? (Hint: Let x represent the first number.)

AFTER SAM LOYD'S CYCLOPEDIA OF PUZZLES.

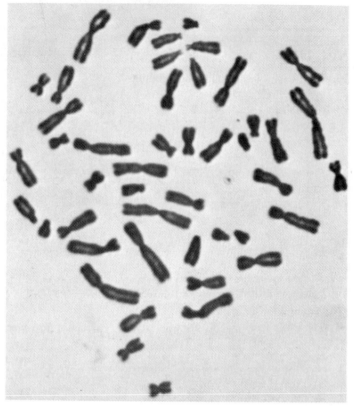

A photograph of human chromosomes. One of these chromosomes contains the gene for eye color.

COURTESY OF MARGERY W. SHAW.

LESSON 6

Squaring Binomials

The color of a person's eyes is determined by a pair of genes inherited from his or her parents. For simplicity, we will assume that these genes are of two types, one corresponding to blue eyes and the other to brown. The gene for brown eyes is dominant, and so a person who inherits one of each type has brown eyes.

If 60 percent of the eye-color genes in a human population are for blue eyes and 40 percent are for brown eyes, what percentage of the population will have eyes of each color? One way to solve this problem is with a diagram. If 60 percent of the genes are for blue eyes, 6 out of every 10 mothers will pass on a gene for blue eyes, as will 6 out of every 10 fathers. The diagram at the top of the next page shows the outcome for 100 births: 36 children (each indicated by ○) will inherit a gene for blue eyes from each parent and will, as a result, have blue eyes. Because the gene for brown eyes is dominant, the 48 children (each indicated by ◑) who inherit a gene for blue eyes from one parent and a gene for brown eyes from the other will have brown eyes. The 16 children (each

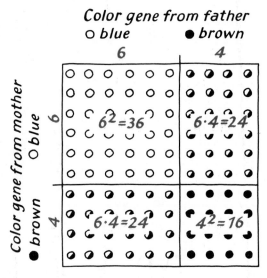

Color gene from father
○ blue   ● brown
6   4

Color gene from mother
○ blue 6
● brown 4

$6^2 = 36$ $6 \cdot 4 = 24$

$6 \cdot 4 = 24$ $4^2 = 16$

(handwritten in margin)

FOIL

$(6 + 4)(6 + 4)$

$6^2 + (6 \cdot 4) + (6 \cdot 4) + y^2$

indicated by ●) who receive a gene for brown eyes from each parent will also have brown eyes. So 36 percent of the population will have blue eyes and 64 percent will have brown eyes.

It is evident from the diagram that

$$(6 + 4)^2 = 6^2 + 2(6 \cdot 4) + 4^2$$

If 6 and 4 are replaced by variables, say x and y, then the diagram looks like the one at the right and the equation looks like this:

$$(x + y)^2 = x^2 + 2xy + y^2$$

A polynomial that contains two terms is called a *binomial* and one that contains three terms is called a *trinomial*. The preceding equation shows that the square of the binomial $x + y$ is a trinomial: $x^2 + 2xy + y^2$. Here are two more examples of this pattern.

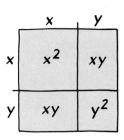

	x	y
x	x^2	xy
y	xy	y^2

EXAMPLE 1
Find the square of 25 by squaring the sum $20 + 5$.

SOLUTION

$$(20 + 5)^2 = (20)^2 + 2(20)(5) + (5)^2$$
$$= 400 + 200 + 25$$
$$= 625$$

(Check this by squaring 25.)

	20	5
20	20^2	$20 \cdot 5$
5	$20 \cdot 5$	5^2

EXAMPLE 2

What is the square of $x + 7$?

SOLUTION

$$(x + 7)^2 = x^2 + 2(7x) + 7^2$$
$$= x^2 + 14x + 49$$

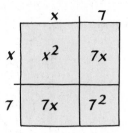

Each of these examples concerns squaring a binomial that is the *sum* of two terms. The diagram below illustrates what happens when a binomial that is the *difference* of two terms is squared.

$$(x - y)^2 = x^2 - 2xy + y^2$$

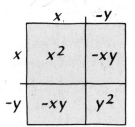

EXAMPLE 3

Find the square of 99 by squaring the difference $100 - 1$.

SOLUTION

According to the pattern above,

$$(100 - 1)^2 = (100)^2 - 2(100)(1) + (1)^2$$
$$= 10{,}000 - 200 + 1$$
$$= 9{,}801$$

(Check this by squaring 99.)

EXAMPLE 4

What is the square of $3x - 5$?

SOLUTION

$$(3x - 5)^2 = (3x)^2 - 2(3x)(5) + (5)^2$$
$$= 9x^2 - 30x + 25$$

The patterns

$$(x + y)^2 = x^2 + 2xy + y^2$$

and

$$(x - y)^2 = x^2 - 2xy + y^2$$

concern multiplying the sum or difference of two terms by itself. The diagram below illustrates what happens if we multiply the sum of two terms by their difference.

$$(x + y)(x - y) = x^2 - y^2$$

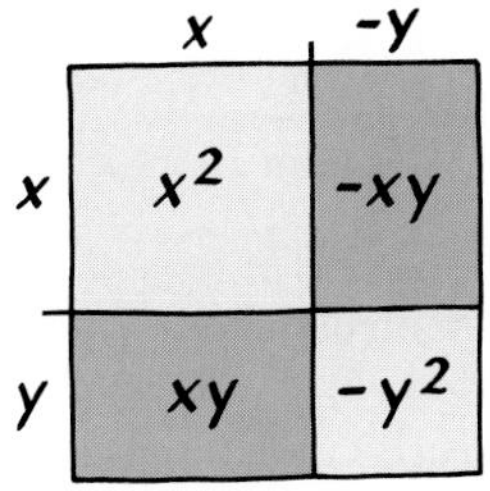

The product of the sum and difference of two terms is equal to the difference of the squares of the terms.

EXAMPLE 5
Find the product of $4x + 9$ and $4x - 9$.

SOLUTION
According to the pattern above,

$$(4x + 9)(4x - 9) = (4x)^2 - (9)^2$$
$$= 16x^2 - 81$$

We have considered the following three patterns in this lesson:

1. The square of the sum of two terms:

$$(x + y)^2 = x^2 + 2xy + y^2$$

2. The square of the difference of two terms:

$$(x - y)^2 = x^2 - 2xy + y^2$$

3. The product of the sum and difference of two terms:

$$(x + y)(x - y) = x^2 - y^2$$

Although these patterns are merely special cases of multiplication, they are frequently found in algebra and should be remembered.

Exercises

Set I

1. Is each of the following statements about monomials true or false?
   a) The sum of two third-degree monomials is a sixth-degree monomial.
   b) The product of two fourth-degree monomials is an eighth-degree monomial.
   c) The difference of a third-degree monomial and a second-degree monomial is a first-degree monomial.
   d) The square of a monomial of zero-degree is another monomial of zero-degree.

2. Solve the following equations for x:
   a) $3(x + 5) = 7 + 4x$
   b) $3(x - 5) = 7 - 4x$
   c) $(8x + 15) - (3x + 8) = 4x - 1$
   d) $8(x + 15) - 3(x + 8) = 4(x - 1)$

3. Match the graphs below with the following equations.
   a) $y = \dfrac{4}{x}$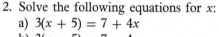

   b) $y = x^2$
   c) $y = 2^x$

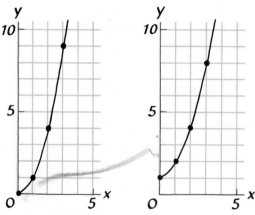

Graph A   Graph B

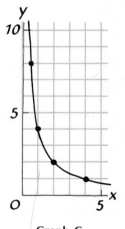

Graph C

Set II

4. Several diagrams are given in this lesson illustrating special multiplication patterns. For example, this diagram illustrates the equation

$$(4 + x)^2 = 16 + 8x + x^2$$

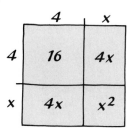

$(4+x)^2 =$

$16 + 8x +$

What equation is illustrated by each of the following diagrams?

a)
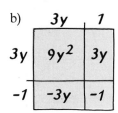

b)
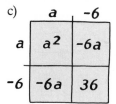

c)
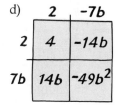

d)

	2	-7b
2	4	-14b
7b	14b	-49b²

5. Make a diagram to illustrate each of the following expressions. Then use the diagram to write each expression as a polynomial.
   a) $(x + 9)^2$ c) $(4a - 3)^2$
   b) $(5x + 2y)^2$ d) $(6a + b)(6a - b)$

6. When a binomial is squared, the result is a trinomial. The first term of the trinomial is found by squaring the first term of the binomial.
   a) How is the last term of the trinomial found?
   b) How is the middle term of the trinomial found?

7. Find the following without making any diagrams.
   a) $(x + 8)^2$
   b) $(x + 8)(x - 8)$
   c) $(12a + 1)^2$
   d) $(3b - 10)^2$
   e) $(7x - y)(7x + y)$
   f) $(2x - 3y)^2$

8. Use the patterns in this lesson to find each of the following. Check your answers by multiplying the long way.
   a) $(40 + 3)^2$ c) $(50 - 7)^2$
   b) $(40 + 3)(40 - 3)$ d) $(50 - 7)(50 + 7)$

9. Write each of the following expressions as a polynomial in descending powers of the variable.
   a) $(3 + x^2)^2$
   b) $(3x^2)^2$
   c) $(4y^5 - 1)^2$
   d) $(4y^5 + 1)(4y^5 - 1)$
   e) $(a^3b^3)^2$
   f) $(a^3 - b^3)^2$

10. Copy each of the following trinomials, replacing ▒ so that the trinomial is the square of a binomial. Then write the binomial of which it is the square.
   a) $x^2 + $ ▒ $ + 16$
   b) $x^2 + 2x + $ ▒
   c) $x^2 - $ ▒ $ + 81$
   d) $x^2 - 10x + $ ▒

Set III

11. What equation is illustrated by each of these diagrams?

a)

	a	8
a	a^2	$8a$
8	$8a$	64

b)

	$2b$	3
$2b$	$4b^2$	$6b$
-3	$-6b$	-9

c)

	1	$-x$
1	1	$-x$
$-x$	$-x$	x^2

d)

	$9y$	-4
$9y$	$81y^2$	$-36y$
4	$36y$	-16

12. Make a diagram to illustrate each of the following expressions. Then use the diagram to write each expression as a polynomial.
    a) $(x + 11)^2$
    b) $(6x + y)^2$
    c) $(10a - 2)^2$
    d) $(a - 5b)(a + 5b)$

13. When a binomial is squared, the result is a trinomial. Copy and complete the following patterns showing how a binomial sum and binomial difference are squared.
    a) $(\square + \triangle)^2 =$
    b) $(\square - \triangle)^2 =$

14. Find the following without making any diagrams.
    a) $(x + 12)^2$
    b) $(x + 12)(x - 12)$
    c) $(5a + 3)^2$
    d) $(2b - 11)^2$
    e) $(x - 9y)(x + 9y)$
    f) $(4x - 5y)^2$

15. Use the patterns in this lesson to find each of the following. Check your answers by multiplying the long way.
    a) $(30 + 9)^2$
    b) $(30 + 9)(30 - 9)$
    c) $(40 - 1)^2$
    d) $(40 - 1)(40 + 1)$

16. Write each of the following as a polynomial.
    a) $(2 + x^3)^2$
    b) $(2x^3)^2$
    c) $(5y^4 - 1)^2$
    d) $(5y^4 - 1)(5y^4 + 1)$
    e) $(a^5b^6)^2$
    f) $(a^5 + b^6)^2$

17. Copy each of the following trinomials, replacing ▓ so that the trinomial is the square of a binomial. Then write the binomial of which it is the square.
    a) $x^2 + $ ▓ $ + 36$
    b) $x^2 + 20x + $ ▓
    c) $x^2 - $ ▓ $ + 49$
    d) $x^2 - 2x + $ ▓

Set IV

Suppose that 50 percent of the genes for eye color in a human population are for blue eyes, 30 percent are for brown eyes, and 20 percent are for green eyes. What percentage of the population would have eyes of each color?

1. To find out, copy and complete the diagram shown here. Assume that the gene for brown eyes is dominant over those for blue and green eyes and that the gene for blue eyes is dominant over that for green eyes.

2. What percentage of the population would have eyes of each color?

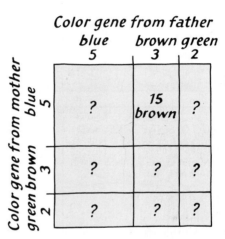

Panel 1: NUMBERS ARE BEAUTIFUL..

Panel 2: I LIKE TWOS THE BEST...THEY'RE SORT OF GENTLE..THREES AND FIVES ARE MEAN, BUT A FOUR IS ALWAYS PLEASANT..I LIKE SEVENS AND EIGHTS, TOO, BUT NINES ALWAYS SCARE ME...TENS ARE GREAT...

Panel 3: HAVE YOU DONE THOSE DIVISION PROBLEMS FOR TOMORROW?

Panel 4: NOTHING SPOILS NUMBERS FASTER THAN A LOT OF ARITHMETIC!

LESSON 7

Dividing Polynomials

Most people learn how to divide one number into another without knowing why the method works. For example, to divide 7 into 2,205, we are taught to write:

$$
\begin{array}{r}
315 \\
7{\overline{\smash{\big)}\,2205}} \\
\underline{21} \\
10 \\
\underline{7} \\
35 \\
\underline{35} \\
0
\end{array}
$$

As you may recall, this method has as its basis repeated subtraction.* Here is the same calculation written out more completely.

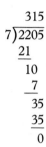

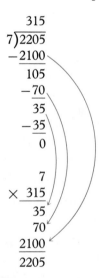

$$
\begin{array}{r}
315 \\
7{\overline{\smash{\big)}\,2205}} \\
-2100 \\
\hline
105 \\
-70 \\
\hline
35 \\
-35 \\
\hline
0
\end{array}
$$

$$
\begin{array}{r}
7 \\
\times\ 315 \\
\hline
35 \\
70 \\
2100 \\
\hline
2205
\end{array}
$$

---

* See page 23.

430

In the first step, we subtract 300 sevens from 2,205 to leave 105. In the next step, we subtract 10 more sevens from 105 to leave 35. In the last step, we subtract 5 sevens from 35 to leave 0. Because we have subtracted (300 + 10 + 5) sevens altogether, the answer is 315.

One way of interpreting what we have done is in terms of a rectangle. Given that one dimension is 7 and that the area is 2,205, we have found that the other dimension is 315. To check this, we can multiply to find out if 315 times 7 is 2,205. Compare this check with the division problem written out in full, and you can see that one is the same as the other in reverse.

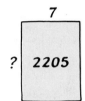

A procedure similar to the one we use to divide one number into another can be used to divide one polynomial into another. To see how it works, we will begin with a multiplication problem. What is $x + 2$ multiplied by $3x + 5$? Finding it the long way and illustrating the procedure with a diagram, we get

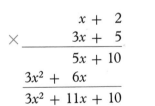

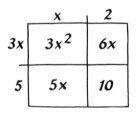

Now we will consider the problem in reverse. What is $3x^2 + 11x + 10$ divided by $x + 2$? Thinking in terms of a rectangle, the problem is to find the other dimension of a rectangle if its area is $3x^2 + 11x + 10$ and one dimension is $x + 2$. Putting part of this information in a diagram, we can see that the other dimension of the rectangle at the upper left must be $3x$.

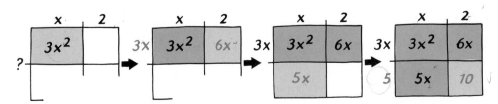

This means that the area of the rectangle at the upper right must be $3x(2) = 6x$. Subtracting the area of these two rectangles from the total area,

$$
\begin{array}{r}
3x^2 + 11x + 10 \\
- \underline{3x^2 + 6x} \\
5x + 10
\end{array}
$$

we see that the area of the rest of the figure is $5x + 10$. Writing $5x$ as the area of the rectangle at the lower left, we see that its other dimension must be 5. So the area of the rectangle at the lower right is 10. All of the area has now been accounted for. So the other dimension of the rectangle is $3x + 5$.

Here is the procedure written as a long division problem. It is broken into steps to make what is being done easier to see.

The result of dividing x into $3x^2$.
$$
3x \\
x + 2 \overline{)3x^2 + 11x + 10}
$$

The result of dividing x into $5x$.
$$
3x + 5 \\
x + 2 \overline{)3x^2 + 11x + 10} \\
\underline{3x^2 + 6x} \\
5x + 10
$$

$3x$ times $x + 2$
$$
3x \\
x + 2 \overline{)3x^2 + 11x + 10} \\
3x^2 + 6x
$$

$$
3x + 5 \\
x + 2 \overline{)3x^2 + 11x + 10} \\
\underline{3x^2 + 6x} \\
5x + 10
$$

The result of subtracting.
$$
3x \\
x + 2 \overline{)3x^2 + 11x + 10} \\
\underline{3x^2 + 6x} \\
5x + 10
$$

5 times $x + 2$
$$
3x + 5 \\
x + 2 \overline{)3x^2 + 11x + 10} \\
\underline{3x^2 + 6x} \\
5x + 10 \\
5x + 10
$$

Quotient

The remainder is 0; so the quotient is $3x + 5$.
$$
3x + 5 \\
x + 2 \overline{)3x^2 + 11x + 10} \\
\underline{3x^2 + 6x} \\
5x + 10 \\
\underline{5x + 10} \\
0
$$

Remainder

Here are two more examples of how polynomials are divided.

EXAMPLE 1
Divide $x^3 + 2x^2 - 8x + 35$ by $x + 5$.

SOLUTION

$$
\begin{array}{r}
x^2 - 3x + 7 \\
x + 5 \overline{\smash{)}\; x^3 + 2x^2 - 8x + 35} \\
\underline{x^3 + 5x^2} \\
-3x^2 - 8x \\
\underline{-3x^2 - 15x} \\
7x + 35 \\
\underline{7x + 35} \\
0
\end{array}
$$

EXAMPLE 2
Divide $x^3 - 64$ by $x - 4$.

SOLUTION
We see neither an x^2 term nor an x term in $x^3 - 64$. Before dividing, it helps to insert these missing terms with zero coefficients so that we can keep like terms lined up.

$$
\begin{array}{r}
x^2 + 4x + 16 \\
x - 4 \overline{\smash{)}\; x^3 + 0x^2 + 0x - 64} \\
\underline{x^3 - 4x^2} \\
4x^2 + 0x \\
\underline{4x^2 - 16x} \\
16x - 64 \\
\underline{16x - 64} \\
0
\end{array}
$$

Exercises

Set I

1. Find each of the following quotients.
   Express each answer in scientific notation.

   a) $\dfrac{6 \times 10^8}{2}$
   b) $\dfrac{7 \times 10^{-1}}{4 \times 10^5}$
   c) $\dfrac{12}{5 \times 10^4}$
   d) $\dfrac{1}{8 \times 10^{-3}}$

2. The graphs of four different pairs of simultaneous equations are shown below.
   Tell what you can about the solutions of each pair of equations.

   a)
   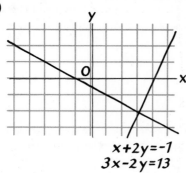
   $x+2y=-1$
   $3x-2y=13$

   b)
   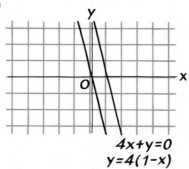
   $4x+y=0$
   $y=4(1-x)$

   c)
   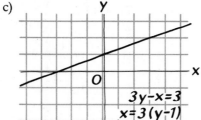
   $3y-x=3$
   $x=3(y-1)$

   d)
   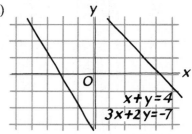
   $x+y=4$
   $3x+2y=-7$

Set II

3. Copy each of these diagrams. Replace each question mark with an appropriate expression.

   a)

	4x	5
?	12x²	?
?	32x	?

   b)

	5x	3
?	5x³	?
?	-10x²	?
?	5x	?

   c)

	3x²	7x	-2
?	18x³	?	?
?	?	?	8

4. Look at your diagrams for exercise 3 to find the answers to the following problems.

a) $\dfrac{12x^2 + 47x + 40}{4x + 5} = ?$

b) $\dfrac{5x^3 - 7x^2 - x + 3}{5x + 3} = ?$

c) $\dfrac{18x^3 + 30x^2 - 40x + 8}{3x^2 + 7x - 2} = ?$

5. Find each of the following quotients by drawing a diagram like those in exercise 3.

a) $\dfrac{14x^2 + 23x + 3}{7x + 1}$

b) $\dfrac{4x^3 - 3x^2 - 23x + 9}{4x + 9}$

6. Find each of the following quotients without using a diagram. Then compare your work with the diagrams you drew for exercise 5.

a) $7x + 1 \overline{)14x^2 + 23x + 3}$

b) $4x + 9 \overline{)4x^3 - 3x^2 - 23x + 9}$

7. Find each of the following quotients.

a) $3x + 2 \overline{)15x^2 + 34x + 16}$

b) $x + 5 \overline{)x^3 + 9x^2 + 14x - 30}$

c) $x^2 - 4x \overline{)3x^3 - 19x^2 + 28x}$

d) $8x - 3 \overline{)8x^3 - 19x^2 + 38x - 12}$

e) $x^3 + 7 \overline{)2x^4 - 3x^3 + 14x - 21}$

f) $2x^2 - 3x + 6 \overline{)2x^4 - 3x^3 + 16x^2 - 15x + 30}$

8. The polynomial $2x^3 + 5$ is missing terms in x^2 and x. Putting these terms in with zero coefficients, we get: $2x^3 + 0x^2 + 0x + 5$.

   Write each of the following polynomials in descending order, putting in any missing terms.

a) $6 + x^2$

b) $8x^3 + 1$

c) $6x^2 + x^4 - 3 + 2x$

d) $10x - x^5$

9. Before an attempt is made to divide one polynomial into another, it is helpful to arrange both of them in descending order with any missing terms inserted. Do this before doing each of the following divisions.

a) Divide $3 + x$ into $x^3 - x + 24$.

b) Divide $x^2 - 1$ into $4x^3 - 5 + 5x^4 - 4x$.

c) Divide $2x - 5$ into $16x^4 - 625$.

Set III

10. Copy each of these diagrams. Replace each question mark with an appropriate expression.

a)

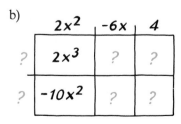

b)

c)

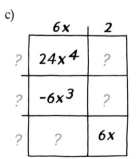

11. Look at your diagrams for exercise 10 to find the answers to the following problems.

a) $\dfrac{14x^2 + 65x + 9}{7x + 1} = ?$

b) $\dfrac{2x^3 - 16x^2 + 34x - 20}{2x^2 - 6x + 4} = ?$

c) $\dfrac{24x^4 + 2x^3 + 16x^2 + 6x}{6x + 2} = ?$

12. Find each of the following quotients by drawing a diagram like those in exercise 10.

a) $\dfrac{12x^2 + 37x + 21}{3x + 7}$

b) $\dfrac{6x^3 + 13x^2 + 3x + 20}{2x + 5}$

13. Find each of the following quotients without using a diagram. Then compare your work with the diagrams you drew for exercise 12.

a) $3x + 7 \overline{)12x^2 + 37x + 21}$

b) $2x + 5 \overline{)6x^3 + 13x^2 + 3x + 20}$

14. Find each of the following quotients.

a) $4x + 7 \overline{)20x^2 + 43x + 14}$

b) $x + 6 \overline{)x^3 + 4x^2 - 9x + 18}$

c) $x^2 - 7x \overline{)x^3 - 12x^2 + 35x}$

d) $5x - 8 \overline{)10x^3 - 21x^2 + 23x - 24}$

e) $x^3 + 4 \overline{)4x^4 - 5x^3 + 16x - 20}$

f) $x^2 + 5x - 2 \overline{)3x^4 + 15x^3 - 2x^2 + 20x - 8}$

15. The polynomial $5x^3 - 7$ is missing terms in x^2 and x. Putting these terms in with zero coefficients, we get: $5x^3 + 0x^2 + 0x - 7$.

Write each of the following polynomials in descending order, putting in any missing terms.

a) $2 + 5x^2$

b) $3x^3 + 4x$

c) $8x - x^3 + 2x^4 + 5$

d) $1 - x^6$

16. Before an attempt is made to divide one polynomial into another, it is helpful to arrange both of them in descending order with any missing terms inserted. Do this before doing each of the following divisions.

a) Divide $4 + x$ into $2x^3 + 9x^2 - 16$.

b) Divide $x^2 - 2$ into $x^3 - 2x + 3x^4 - 12$.

c) Divide $1 + 3x + 9x^2$ into $27x^3 - 1$.

Set IV

Some of the digits in the division problem below have been replaced by asterisks. Can you figure out what each one is?

```
        9**
5*)*****
    **3
    ‾‾‾
    ***
    **5
    ‾‾‾
    ***
    **1
    ‾‾‾
     0
```

Summary and Review

In this chapter, we have learned how to identify polynomial expressions and how to add, subtract, multiply, and divide them.

Monomials (*Lesson 1*)   A monomial in one variable, say x, is an expression of the form ax^n, in which a may be any number and n is a positive integer. The number a is the coefficient of the monomial and the integer n is its degree.

Zero is a monomial but does not have a specific degree. Other numbers are considered to be monomials of degree 0.

Polynomials (*Lesson 2*)   A polynomial is either a monomial or an expression indicating the addition and/or subtraction of two or more monomials. The monomials are called terms of the polynomial.

The degree of a polynomial is that of the term having the highest degree. A polynomial is written in descending powers of the variable if the degrees of its terms get smaller as it is read from left to right.

Adding and Subtracting Polynomials (*Lesson 3*)   Two or more polynomials can be added or subtracted by adding or subtracting like terms (those having the same degree and variable.)

Multiplying Polynomials (*Lessons 4 and 5*)   The multiplication of two polynomials can be illustrated by a rectangle whose dimensions represent the polynomials and whose area represents their product. To multiply two polynomials without a diagram, multiply each term of one polynomial by each term of the other and then add the resulting terms.

Squaring Binomials (*Lesson 6*)   A binomial is a polynomial that contains two terms; a trinomial is one that contains three terms. The square of a binomial is a trinomial. The pattern for the square of a binomial sum is

$$(a + b)^2 = a^2 + 2ab + b^2$$

The pattern for the square of a binomial difference is

$$(a - b)^2 = a^2 - 2ab + b^2$$

The pattern for the product of the sum and difference of two terms is

$$(a + b)(a - b) = a^2 - b^2$$

Dividing Polynomials (*Lesson 7*)   Dividing polynomials is like dividing ordinary numbers. We can consider division in terms of areas and rectangles or simply as an algebraic process. Before the division is carried out, both polynomials should be arranged in descending order with any missing terms inserted.

Exercises

Set I

1. This postage stamp from Israel contains several formulas, one of which we have studied in this chapter. Which one is it and what does it mean?

2. Read each of the following statements carefully and tell whether it is true or false.
   a) The coefficient of the monomial x^4 is 4.
   b) The sum of two monomials can always be written as a monomial.
   c) The expression $2^x + 1$ is a polynomial.
   d) When two binomials are multiplied, the product may be a binomial.
   e) The square of a third-degree polynomial is a sixth-degree polynomial.

3. Find the values of the following polynomials as indicated.

$$x^4 - x^2$$

a) for $x = 4$
b) for $x = -4$
c) for $x = 2$
d) for $x = 3$

$$x^4 - 2x^2 + 1$$

e) for $x = 1$
f) for $x = 10$

$$2x^3 + 3x^2$$

4. If possible, write each of the following as a monomial. If an expression cannot be written as a monomial, say so.
a) $3x^4 - x^4$
b) $3x^4 + 3x$
c) $2(-3x^4)$
d) $(-3x^4)^2$
e) $2(3 - x^4)$
f) $2(-3x)^4$
g) $-3x(x^4)$
h) $3x^4 - 3x^3$

5. Make a diagram to illustrate each of these products. Then use the diagrams to write each one as a sum of monomials.
a) $(3x + 5)^2$
b) $(7x - y)(7x + y)$
c) $(8x + 1)(x - 4)$
d) $(x - 9)(y - 11)$

6. Rewrite each of the following expressions as a polynomial in descending powers of the variable.
a) $5x^3 + 9x + x^4$
b) $6x + x^6 - 6x^2 + 2x^6$
c) $5 - x^5$
d) $3x(4 - x^2) + 5(1 - x)$

7. Write each of the following as a polynomial in descending powers of its variable.
a) $x(x + 4)$
b) $(x + 3)(x + 4)$
c) $8x(x - 2)$
d) $(8x - 1)(x - 2)$
e) $(x + 6)(x - 5)$
f) $(6x - 1)(5x + 1)$
g) $(x + 7)^2$
h) $(7x)^2$
i) $(2x - 3)^2$
j) $(x - 10)(x + 10)$
k) $(x^3 + 3)^2$
l) $(x^3 + 3)(x^3 - 3)$

8. Perform the operations indicated.
a) Add $x - 6y$, $4x + y$, and $y - 5x$.
b) From $3x + x^4$, subtract $x^3 + 3x - 1$.

c) Add $7x^2 - x + 8$ and $x^3 + 9x - 12$.
d) Subtract $5x - 11y$ from $4y + 20x$.
e) Multiply $5x - 2$ by $7x + 10$.
f) Find the product of $2x^3 - x$ and $x^2 - x + 4$.
g) Divide $x^3 + 7x^2 + 11x - 4$ by $x + 4$.
h) Divide $2x - 7$ into $2x^4 - 98x + x^3$.

9. Find expressions for the perimeter and area of each of these rectangles.

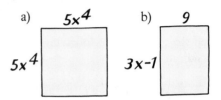

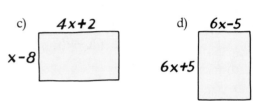

10. Find the missing term in each of the following, given that each is the square of a binomial.
a) $x^2 + \text{▓▓▓} + 144$
b) $x^2 - 30x + \text{▓▓▓}$
c) $x^2 + \text{▓▓▓} + 9y^2$
d) $16x^2 - 40x + \text{▓▓▓}$

11. Perform the operations indicated.
a) $(5x - 3) + (4x + 7)$
b) $(5x - 3)(4x + 7)$
c) $(x^3 + x^2 - 10x - 6) - (x - 3)$
d) $\dfrac{x^3 + x^2 - 10x - 6}{x - 3}$
e) $(x^4 - 1) + (x + 1)$
f) $(x^4 - 1) - (x + 1)$
g) $(x^4 - 1)(x + 1)$
h) $\dfrac{x^4 - 1}{x + 1}$

Set II

1. Read each of the following statements carefully and tell whether it is true or false.
   a) The expression $1 - x^2$ is a polynomial.
   b) The degree of the monomial 1 is 0.
   c) The square of the binomial $x + y$ is $x^2 + y^2$.
   d) The product of two monomials can always be written as a monomial.
   e) A second-degree polynomial in one variable that is in simplest form can have four terms.

2. Find the values of the following polynomials as indicated.

   $x^3 + 10x$
   a) for $x = 3$
   b) for $x = -3$

   $x^3 + 3x^2 + 3x + 1$
   e) for $x = 1$
   f) for $x = 9$

   $2x^4 - 4x^2$
   c) for $x = 2$
   d) for $x = 4$

3. If possible, write each of the following as a monomial.
   a) $x^5 + 5x$ e) $(-4x^5)^2$
   b) $4x^5 - x^5$ f) $2(-4x)^5$
   c) $2(4 - x^5)$ g) $4x^5 + 4x^4$
   d) $2(-4x^5)$ h) $-4x(x^5)$

4. A checkerboard can be used to illustrate how to square a binomial. For example, this picture shows how it might illustrate the fact

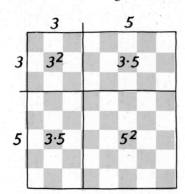

that

$$(3 + 5)^2 = 3^2 + 2(3 \cdot 5) + 5^2$$

   The board can also be used to illustrate the squares of three other binomials. Write an equation like the one above for each. Check each equation to see if it is true.

5. Make a diagram to illustrate each of these products. Then use the diagrams to write each one as a sum of monomials.
   a) $(2x + 7)^2$
   b) $(x + 6y)(x - 6y)$
   c) $(4x - 3)(9x + 7)$
   d) $(3x - 1)(2y + 1)$

6. Rewrite each of the following expressions as a polynomial in descending powers of the variable.
   a) $7x^3 - x^5 + 10$
   b) $8x - x^2 + x^8 - 2x$
   c) $2 - x^4$
   d) $x^2(3x - 1) - (x^2 - x^3)$

7. Write each of the following as a polynomial in descending powers of its variable.
   a) $x(x + 5)$ g) $(x + 8)^2$
   b) $(x + 2)(x + 5)$ h) $(8x)^2$
   c) $6x(x - 3)$ i) $(2x + 9)(2x - 9)$
   d) $(6x - 1)(x - 3)$ j) $(3x - 1)^2$
   e) $(x + 7)(x - 4)$ k) $(x^4 + 4)^2$
   f) $(7x - 1)(4x + 1)$ l) $(x^4 - 4)(x^4 + 4)$

8. Perform the operations indicated.
   a) Add $3x - y$, $y - 5x$, and $2x + 9y$.
   b) Subtract $1 - 3x^2$ from $7x^3 - 3x^2 + x$.
   c) Add $x^2 + 10x - 6$ and $x^4 - x^2 - 4$.
   d) From $6x - 5y + 4$, subtract $3y + 5x + 7$.
   e) Multiply $8x - 15$ by $3x + 8$.
   f) Find the product of $4x^2 - x - 2$ and $x + 5$.

g) Divide $x + 3$ into
   $2x^3 + 11x^2 + 11x - 12$.
h) Divide $16x^4 - 1$ by $2x - 1$.

9. Find expressions for the perimeter and area for each of these rectangles.

a)
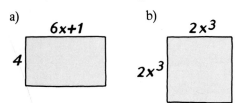

b)
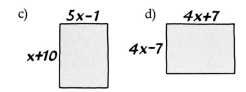

c)

5x-1

x+10

d)

4x+7

4x-7

10. Find the missing term in each of the following, given that each is the square of a binomial.
   a) $x^2 - \text{▊▊▊} + 100$
   b) $x^2 + 50x + \text{▊▊▊}$
   c) $16x^2 - \text{▊▊▊} + y^2$
   d) $9x^2 + 24x + \text{▊▊▊}$

11. Perform the operations indicated.
   a) $(7x - 2) + (3x - 8)$
   b) $(7x - 2)(3x - 8)$
   c) $(x^3 - x^2 - 30x + 50) - (x - 5)$
   d) $\dfrac{x^3 - x^2 - 30x + 50}{x - 5}$
   e) $(4x^2 - 9) + (2x + 3)$
   f) $(4x^2 - 9) - (2x + 3)$
   g) $(4x^2 - 9)(2x + 3)$
   h) $\dfrac{4x^2 - 9}{2x + 3}$

Chapter 10

FACTORING

LESSON 1
Prime and Composite Numbers

In 1978, Laura Nickel and Curt Noll, two 18-year-old students working at California State University in Hayward, discovered that the number $2^{21701} - 1$ is prime. This number, the result of raising 2 to the 21,701st power and subtracting 1, is immense. Shown to be prime with the help of a computer, it contains 6,533 digits in all.

What made this number interesting was the fact that it was the largest prime number known at the time.*

▶ A **prime number** is an integer larger than 1 that cannot be written as the product of smaller positive integers.

The first integer larger than 1 is 2, and 2 is prime. The integers 3, 5, and 7 are also prime. The integers 4, 6, 8, and 9 are not prime because $4 = 2 \cdot 2$, $6 = 2 \cdot 3$, $8 = 2 \cdot 2 \cdot 2$, and $9 = 3 \cdot 3$. Because they can be written as products composed of other integers, they are called *composite*.

▶ A **composite number** is an integer larger than 1 that can be written as the product of smaller positive integers.

When an integer is written as a product of two or more integers, it is said to be *factored* and the numbers in the product are called *factors* of the integer. For example, 15 can be factored as

$$1 \cdot 15 \quad \text{or} \quad 3 \cdot 5 \quad \text{or} \quad -1 \cdot -15 \quad \text{or} \quad -3 \cdot -5$$

Listed in order of increasing size, the factors of 15 are -15, -5, -3, -1, 1, 3, 5, and 15.

From our definition of factor it is apparent that an integer is a factor of a number if and only if it can be divided into the number leaving zero as the remainder. Notice, for example, what happens when 15 is divided by 1, 2, 3, 4, and 5.

$$
\begin{array}{ccccc}
15 & 7 & 5 & 3 & 3 \\
1\overline{)15} & 2\overline{)15} & 3\overline{)15} & 4\overline{)15} & 5\overline{)15} \\
\underline{15} & \underline{14} & \underline{15} & \underline{12} & \underline{15} \\
0 & 1 & 0 & 3 & 0
\end{array}
$$

The remainders show that 1, 3, and 5 are factors of 15 and that 2 and 4 are not.

It follows from our definition of a prime number that a prime number can be

_____

* Curt Noll has continued this work to discover $2^{23209} - 1$, the largest prime number known as of March, 1979.

factored into positive factors in only one way: as the product of itself and 1. Every composite number, on the other hand, can be factored in more than one way. Nevertheless, *there is only one way* (except for order) *in which a composite number can be written as the product of primes.* The method used in the following example to find the prime factors of a number shows how to find the prime factors of a composite number.

EXAMPLE 1

Factor 924 into primes.

SOLUTION

To find the prime factors of 924, we divide by each prime in succession starting with 2 to see how many times that prime is a factor. First, trying factors of 2, we get

$$
\begin{array}{r} 462 \\ 2\overline{)924} \\ 924 \\ \hline 0 \end{array}
\qquad
\begin{array}{r} 231 \\ 2\overline{)462} \\ 462 \\ \hline 0 \end{array}
\qquad
\begin{array}{r} 115 \\ 2\overline{)231} \\ 230 \\ \hline 1 \end{array}
$$

The first factor of 2.　　　The second factor of 2.　　　← Remainder of 1 indicates no third factor of 2.

Next, trying factors of 3, we get

$$
\begin{array}{r} 77 \\ 3\overline{)231} \\ 231 \\ \hline 0 \end{array}
\qquad
\begin{array}{r} 25 \\ 3\overline{)77} \\ 75 \\ \hline 2 \end{array}
$$

The first factor of 3.　　　← No second factor of 3.

Because the prime factors of 77 are 7 and 11, no further division is necessary. Factored into primes, $924 = 2 \cdot 2 \cdot 3 \cdot 7 \cdot 11$, or $2^2 \cdot 3 \cdot 7 \cdot 11$.

The *greatest common factor* of a set of numbers is, as its name indicates, the largest integer that is a factor of all of the numbers. Two numbers whose greatest common factor is 1 are said to be *relatively prime*.

EXAMPLE 2

Find the greatest common factor of 30 and 75.

SOLUTION

Factoring 30 and 75 into primes, we get

$$30 = 2 \cdot 3 \cdot 5$$
$$75 = 3 \cdot 5 \cdot 5$$

Multiplying the common prime factors, 3 and 5, we find that the *greatest* common factor of 30 and 75 is 15.

EXAMPLE 3

Find the greatest common factor of 21 and 110.

SOLUTION

Factoring 21 and 110 into primes, we get

$$21 = 3 \cdot 7$$
$$110 = 2 \cdot 5 \cdot 11$$

These numbers have no common prime factor, but 1 is a factor of every number. So the greatest common factor of 21 and 110 is 1; this means that 21 and 110 are relatively prime.

Exercises

Set I

1. Write each of the following numbers without using any exponents.
   a) 0^5
   b) 5^0
   c) $(-2)^7$
   d) $(-7)^{-2}$

2. Solve each of the following formulas for the variable indicated.
   a) For a: $p = a + b + c$
   b) For f: $w = fd$
   c) For x: $y = ax + b$

3. Neon is an element consisting of two kinds of atoms, some of which weigh 20 units and some of which weigh 22 units. If a sample of 100 atoms of neon weighs 2,018 units, how many atoms are there of each kind in the sample?

Set II

4. Here is a list of the integers from 11 to 20 in which each number has either been labeled "prime" or written as the product of primes.

    Make a similar list for the integers from 21 to 30.

11	prime
12	$2^2 \cdot 3$
13	prime
14	$2 \cdot 7$
15	$3 \cdot 5$
16	2^4
17	prime
18	$2 \cdot 3^2$
19	prime
20	$2^2 \cdot 5$

5. Factor each of the following numbers into primes. List the prime factors in order from smallest to largest, using exponents where possible.
    - a) 100
    - b) 180
    - c) 231
    - d) 2,310
    - e) 816
    - f) 3,264
    - g) 15^2
    - h) 15^6
    - i) $2^3 \cdot 14^3$
    - j) $3^2 \cdot 3^{14}$

6. An integer is a factor of a number if and only if it can be divided into the number leaving zero as the remainder. (Do parts f through h if you have a calculator.)
    - a) Is 5 a factor of 775?
    - b) Is 3 a factor of 2,001?
    - c) Is 8 a factor of 500?
    - d) Is 11 a factor of 12,321?
    - e) Is 46 a factor of 644?
    - *f) Is 19 a factor of 31,350?
    - *g) Is 27 a factor of 277,277?
    - *h) Is 52 a factor of 525,252?
    - i) Is 6 a factor of 6 + 4?
    - j) Is 6 a factor of 6 · 4?
    - k) Is 6 a factor of 6^4?
    - l) Is 6 a factor of 4^6?

7. List, in order from smallest to largest, all of the positive integer factors of each of the following numbers.
    - a) 55
    - b) 32
    - c) 71
    - d) 72
    - e) 5^3
    - f) 3^5

8. Find the greatest common factor of each of the following sets of numbers.
    - a) 15 and 50
    - b) 8 and 96
    - c) 20 and 21
    - d) 36 and 48
    - e) 36 and 90
    - f) 4, 10, and 14
    - g) 7, 10, and 17
    - h) 3^4 and 3^8
    - i) 4^3 and 8^3
    - j) 7^5 and 5^7

Set III

9. Here is a list of the integers from 31 to 40 in which each number has either been labeled "prime" or written as the product of primes.

    Make a similar list for the integers from 41 to 50.

31	prime
32	2^5
33	$3 \cdot 11$
34	$2 \cdot 17$
35	$5 \cdot 7$
36	$2^2 \cdot 3^2$
37	prime
38	$2 \cdot 19$
39	$3 \cdot 13$
40	$2^3 \cdot 5$

10. Factor each of the following numbers into primes. List the prime factors in order from smallest to largest, using exponents where possible.
    - a) 75
    - b) 120
    - c) 182
    - d) 1,820
    - e) 1,782
    - f) 3,564
    - g) 21^2
    - h) 21^5
    - i) $5^4 \cdot 10^4$
    - j) $4^5 \cdot 4^{10}$

11. An integer is a factor of a number if and only if it can be divided into the number leaving zero as the remainder. (Do parts f through h if you have a calculator.)

    a) Is 2 a factor of 12,345?
    b) Is 9 a factor of 909?
    c) Is 4 a factor of 1,776?
    d) Is 12 a factor of 812?
    e) Is 35 a factor of 735?
  *f) Is 17 a factor of 30,702?
  *g) Is 66 a factor of 66,666?
  *h) Is 81 a factor of 818,181?
    i) Is 7 a factor of 7 + 3?
    j) Is 7 a factor of 7 · 3?
    k) Is 7 a factor of 7^3?
    l) Is 7 a factor of 3^7?

12. List, in order from smallest to largest, all of the positive integer factors of each of the following numbers.

    a) 26
    b) 81
    c) 59
    d) 60
    e) 7^2
    f) 2^7

13. Find the greatest common factor of each of the following sets of numbers.

    a) 12 and 48   f) 8, 10, and 18
    b) 12 and 33   g) 9, 10, and 19
    c) 15 and 16   h) 4^3 and 4^6
    d) 28 and 70   i) 3^4 and 6^4
    e) 75 and 175   j) 11^2 and 2^{11}

Set IV

Although the number 11 is prime, the number 111 is not because $111 = 3 \cdot 37$. In fact, every number in the following table is composite.

$$111 = 3 \cdot 37$$
$$1,111 = 11 \cdot 101$$
$$11,111 = 41 \cdot 271$$
$$111,111 = 3 \cdot 7 \cdot 11 \cdot 13 \cdot 37$$
$$1,111,111 = 239 \cdot 4,649$$
$$11,111,111 = 11 \cdot 73 \cdot 101 \cdot 137$$
$$111,111,111 = 3 \cdot 3 \cdot 37 \cdot 333,667$$
$$1,111,111,111 = 11 \cdot 41 \cdot 271 \cdot 9,091$$

1. Judging from this table, what numbers of which all of the digits are 1's have factors of 3?
2. What numbers of this sort have factors of 11?
3. Which one of the numbers do you think would be the most difficult for someone to prove composite without having this table?

LESSON 2
Monomials and Their Factors

Here is a number trick that is somewhat more impressive than the one in this cartoon. Think of a three-digit number and write it down twice to form a six-digit number. Regardless of what number is originally chosen, the resulting six-digit number can be divided evenly by 7, 11, and 13. For example, beginning with the number in the cartoon, 684, we get 684,684. The results of dividing 684,684 by 7, 11, and 13 are shown here.

$$
\begin{array}{r}
97{,}812 \\
7\overline{)684{,}684} \\
63 \\ \hline
54 \\
49 \\ \hline
56 \\
56 \\ \hline
8 \\
7 \\ \hline
14 \\
14 \\ \hline
0
\end{array}
\qquad
\begin{array}{r}
62{,}244 \\
11\overline{)684{,}684} \\
66 \\ \hline
24 \\
22 \\ \hline
26 \\
22 \\ \hline
48 \\
44 \\ \hline
44 \\
44 \\ \hline
0
\end{array}
\qquad
\begin{array}{r}
52{,}668 \\
13\overline{)684{,}684} \\
65 \\ \hline
34 \\
26 \\ \hline
86 \\
78 \\ \hline
88 \\
78 \\ \hline
104 \\
104 \\ \hline
0
\end{array}
$$

Most numbers cannot be divided evenly by even *one* of these numbers, let alone all three. Why is it that, no matter what three-digit number you begin with, the six-digit number formed can be divided evenly by these three numbers?

The reason has to do with the way the six-digit number is produced. Writing down a three-digit number twice is equivalent to multiplying it by 1,001; the number 1,001 is composite, being the product of 7, 11, and 13. If we let x represent the three-digit number first chosen, then $1,001x$ represents the six-digit number formed. Because 1,001 is the product of 7, 11, and 13, $1,001x = 7 \cdot 11 \cdot 13 \cdot x$. No matter what number x represents, $7 \cdot 11 \cdot 13 \cdot x$ can be divided evenly by 7, 11 and 13.

$$
\begin{array}{r}
684 \\
\times\ 1{,}001 \\
\hline
684 \\
000 \\
000 \\
684 \\
\hline
684{,}684
\end{array}
$$

The numbers 7, 11, and 13 are factors of the monomial $1,001x$ because they are factors of its coefficient. Other factors of $1,001x$ are:

> 1 and 1,001,
> 77 (because $7 \cdot 11 = 77$),
> 91 (because $7 \cdot 13 = 91$),
> 143 (because $11 \cdot 13 = 143$),
> x, $7x$, $11x$, $13x$, $77x$, $91x$, $143x$, and $1,001x$.

To be able to recognize factors of monomial expressions is a skill needed in solving many algebra problems. Here is another example of how it is done.

EXAMPLE 1

What are the factors of $6x^2$?

SOLUTION

$6x^2 = 2 \cdot 3 \cdot x \cdot x$, and so three of its factors are 2, 3, and x. Other factors are: 1, 6, x^2, $2x$, $2x^2$, $3x$, $3x^2$, $6x$, and $6x^2$.

In general, the factors of the monomial ax^n include the factors of the coefficient a, the factors of x^n ($x, x^2, \ldots, x^n$), and products of various combinations of these factors.

In Lesson 1, we defined the *greatest common factor of a set of numbers* to be the largest integer that is a factor of all of the numbers and said that two numbers whose greatest common factor is 1 are *relatively prime*. We will now apply these ideas to monomials with integer coefficients.

Consider, for example, the monomials $4x^5$ and $6x^3$. The greatest common factor of their coefficients is 2 and the highest power of x that is a factor of both monomials is x^3 because x^5 is $x^3 \cdot x^2$. It seems reasonable to consider $2x^3$ to be their greatest common factor.

The *greatest common factor of a set of monomials with integer coefficients* is the product of the greatest common factor of their coefficients and the highest power of each variable that is a factor of all of the monomials. Like two numbers, two monomials whose greatest common factor is 1 are called *relatively prime*.

EXAMPLE 2

Find the greatest common factor of $3x^3$ and $7x^7$.

SOLUTION

$$3x^3 = 3 \cdot x \cdot x \cdot x$$
$$7x^7 = 7 \cdot x \cdot x \cdot x \cdot x \cdot x \cdot x \cdot x$$

The greatest common factor is x^3.

EXAMPLE 3

Find the greatest common factor of $6xy$ and $15y^4$.

SOLUTION

$$6xy = 2 \cdot 3 \cdot x \cdot y$$
$$15y^4 = 3 \cdot 5 \cdot y \cdot y \cdot y \cdot y$$

The greatest common factor is $3y$.

EXAMPLE 4

Find the greatest common factor of $9x^2$ and $4y^2$.

SOLUTION

$$9x^2 = 3 \cdot 3 \cdot x \cdot x$$
$$4y^2 = 2 \cdot 2 \cdot y \cdot y$$

The only factor these monomials have in common is 1, and so they are relatively prime.

Exercises

Set I

1. Guess a formula for the function represented by each of these tables. Begin each formula with $y =$.

   a)

x	0	1	2	3	4
y	–7	3	13	23	33

   b)

x	0	1	2	3	4
y	0	2	6	12	20

   c)

x	0	1	2	3	4
y	1	3	9	27	81

2. Write each of the following as a polynomial in descending powers of the variable.
   a) $(5x + 1)^2$
   b) $(5x - 1)^2$
   c) $(5x + 1)(5x - 1)$

3. The largest flag in the world is an American flag displayed each year on the side of the J. L. Hudson store in Detroit. It is 104 feet wide and 235 feet long.
   a) What are its perimeter and area?
   b) If its length and width were each increased by 1 foot, what would its perimeter and area become?
   c) If its length and width were each increased by x feet, what would its perimeter and area become? Express each of your answers as a polynomial.

UNITED PRESS INTERNATIONAL PHOTO.

Set II

4. Write each of the following products or powers as a monomial.
   a) $(4)(14x^2)$
   b) $(7x)(8x)$
   c) $(3x^3)(5x^5)$
   d) $(x^{10})(-x^{10})$
   e) $(6x^4)^2$
   f) $(2x^4)^6$
   g) $(ax^a)(bx^b)$
   h) $(ax^a)^2$

5. Using only positive integer coefficients, factor each of the following monomials into as many different pairs of factors as you can.
   a) 18
   b) $34x$
   c) $25x^2$
   d) $2x^3$

6. Write each of the following monomials as the square of a monomial.
   a) 16
   b) x^{16}
   c) $9x^{10}$
   d) $25x^2y^2$

7. Allowing only positive integers as coefficients, list all of the factors of each of the following.
   a) 24
   b) $15x$
   c) $49x^2$
   d) x^6
   e) $3x^3$

8. Tell what monomial should replace ▍ in each of the following to make it true.
   a) $(4)(4x^4) = $ ▍
   b) $(2x^2)($ ▍$) = 10x^{10}$
   c) (▍$)(6x^3) = 6x^6$

   d) $(-7x)($ ▍$) = 21xy$
   e) $(-x^7)(-y^3) = $ ▍
   f) $(2x^3)(3x^2)($ ▍$) = 12x^{12}$

9. Find the greatest common factor of each of the following sets of monomials.
   a) 15 and 21
   b) $10x^2$ and $45x$
   c) 7 and 17
   d) x^7 and x^{17}
   e) 4 and x
   f) $33x$ and $44y$
   g) $8xy$ and $64xy$
   h) $52x^2y^2$ and $13xy^3$
   i) x^4 and y^4
   j) $6x^2$, $9x$, and 3
   k) $5x^2$, $5x$, and 2
   l) $4x^3$, $12x^2$, and $6x$

Set III

10. Write each of the following products or powers as a monomial.
    a) $(3)(18x^2)$
    b) $(9x)(6x)$
    c) $(4x^4)(2x^2)$
    d) $(x^7)(-x^7)$
    e) $(5x^3)^2$
    f) $(2x^3)^5$
    g) $(ax^b)(cx^d)$
    h) $(ax^b)^3$

11. Using only positive integers as coefficients, factor each of the following monomials into as many different pairs of factors as you can.
    a) 20
    b) $26x$
    c) $9x^2$
    d) $5x^4$

12. Write each of the following monomials as the square of a monomial.
    a) 36
    b) x^{36}
    c) $25x^8$
    d) $9x^2y^2$

13. Using only positive integers as coefficients, list all of the factors of each of the following.
    a) 36
    b) $10x$
    c) $81x^2$
    d) x^8
    e) $2x^4$

14. Tell what monomial should replace ▍ in each of the following to make it true.
    a) $(5)(5x^5) = $ ▍
    b) $(3x^3)($ ▍$) = 9x^9$
    c) (▍$)(8x) = 8x^8$
    d) $(-4x)($ ▍$) = 24xy$
    e) $(-x^4)(-y^6) = $ ▍
    f) $(6x)(x^5)($ ▍$) = 18x^{10}$

15. Find the greatest common factor of each of the following sets of monomials.
    a) 12 and 40
    b) $14x^2$ and $35x$
    c) 9 and 10
    d) x^9 and x^{10}
    e) 2 and x^2
    f) $70x$ and $20y$
    g) $18xy$ and $36xy$
    h) $5x^2y^2$ and $30x^3y$
    i) x^5 and y^5
    j) $2x^2$, $6x$, and 10
    k) $3x^4$, $3x$, and 4
    l) $6x^3$, $15x^2$, and $9x$

Set IV

Here is a number trick similar to the one described in this lesson. Think of a two-digit number and write it down three times to form a six-digit number. Regardless of what number you choose at the beginning, the resulting six-digit number is composite and can be divided evenly by 3, 7, 13, and 37.

   1. Choose a two-digit number and show that this trick works as described.
   2. Explain why the trick works.

STERLING PUBLISHING CO., INC.

Polynomials and Their Factors

Emanuel Zacchini holds the world record for traveling the longest distance through the air after being shot from a cannon. This photograph shows him with his daughter, Florinda, performing their act at the circus.*

The distance of a human cannonball above the ground depends on the time that has passed since he or she was shot upward. For someone who leaves the cannon with an upward speed of 64 feet per second, the distance is given in feet by the polynomial

$$64x - 16x^2$$

in which x represents the number of seconds that have passed.

---

* Mr. Zacchini is now retired, but Florinda has proved to be a human cannonball of the same caliber.

Evaluating this polynomial for different values of x, we get

x	0	1	2	3	4
$64x - 16x^2$	0	48	64	48	0

From this table, we see that after 1 second the distance above the ground is 48 feet, after 2 seconds it is 64 feet, after 3 seconds it is 48 feet, and after 4 seconds the human cannonball is back on the ground. The trip evidently lasts 4 seconds.

There is a way to tell how long the trip would take without bothering to make a table of values as we have done. The method works like this. Clearly, at the beginning and end of the trip, the distance of the human cannonball above the ground is zero. This means that, at these times, the value of

$$64x - 16x^2$$

is zero.

The values of x that will make a polynomial in x equal to zero are not usually obvious. However, in this case, we can perform a few operations that will let us see exactly which values of x will make it equal to zero. Notice that the two terms $64x$ and $16x^2$ have a common factor of $16x$. Factoring $16x$ from each term and applying the distributive rule relating multiplication and subtraction in reverse, we can write

$$64x - 16x^2 =$$
$$16x(4) - 16x(x) =$$
$$16x(4 - x)$$

(Check this by looking at these steps in reverse order.)

We have written $64x - 16x^2$ as the product $16x(4 - x)$. We know that a product is zero if one of its factors is zero. So the product $16x(4 - x)$ is zero only if either x or $(4 - x)$ is zero. Clearly, $(4 - x)$ is zero if x is 4. These numbers, 0 and 4, are the times, then, when the human cannonball is 0 feet above the ground: at the beginning of the trip, 0 seconds, and at the end, 4 seconds.

This is just one example of why it is useful to be able to rewrite a polynomial as a product. In Lessons 1 and 2, we learned how to factor integers and monomials. In this lesson and the following ones in this chapter, we will learn how to factor polynomials in general. Throughout our work, we will restrict the factors to *integers* and *polynomials having coefficients and constant terms that are integers.*

EXAMPLE 1
Factor $5x^2 + 10x$ and illustrate the result by means of the area of a rectangle.

SOLUTION
The greatest common factor of $5x^2$ and $10x$ is $5x$.

$$5x^2 + 10x = 5x(x) + 5x(2)$$
$$= 5x(x + 2)$$

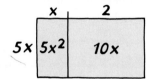

The polynomial $5x^2 + 10x$ represents the area of the rectangle as the sum of the areas of its two parts; $5x(x + 2)$ represents its area as the product of its length and width.

EXAMPLE 2
Factor $x^4 - 2x^3 + 2x^2$.

SOLUTION
The greatest common factor of x^4, $2x^3$, and $2x^2$ is x^2.

$$x^4 - 2x^3 + 2x^2 = x^2(x^2) - x^2(2x) + x^2(2)$$
$$= x^2(x^2 - 2x + 2)$$

Because we are restricting the factors of polynomials to integers and polynomials that have integer coefficients, some polynomials cannot be written as products of polynomials of smaller degree. Positive integers that cannot be written as products of smaller positive integers are called prime. We will define prime polynomials in the same way.

▶A **prime polynomial** is a polynomial that cannot be written as a product of polynomials of lower degree.

A polynomial, like an integer, can be written as a product of prime factors in essentially only one way.

EXAMPLE 3

Factor $6x - 24$ into prime factors.

SOLUTION

The greatest common factor of $6x$ and 24 is 6.

$$6x - 24 = 6(x - 4)$$

Factoring 6 into primes, we get $2 \cdot 3$. Can $x - 4$ be factored? If so, it can be written as a product of polynomials of lower degree. Because $x - 4$ is of first degree, its factors would have to be of zero degree. But polynomials of zero degree are constants and the product of constants cannot be equal to $x - 4$. So $x - 4$ is prime. The prime factors of $6x - 24$ are 2, 3, and $x - 4$:

$$6x - 24 = (2)(3)(x - 4)$$

Exercises

Set I

1. Factor each of the following numbers into primes. Use exponents where possible.
   a) 56
   b) 221
   c) 1,815

2. This exercise is about the polynomials $4x^2 - 2x - 1$ and $2x + 1$.
   a) Find their sum.
   b) Find their difference. (Subtract the second from the first.)
   c) Find their product.

3. The formula for the surface area of a box is

   $$A = 2\ell w + 2wh + 2\ell h$$

   in which A is the area and ℓ, w, and h are the length, width, and height of the box.
   a) Use this formula to find the surface area of the box at the left below.

   b) Use this formula to write an expression for the surface area of the right-hand box. Simplify it as much as you can.
   c) What are the dimensions of the second box if $x = 4$?
   d) Evaluate the expression you wrote for the surface area of the second box when $x = 4$.

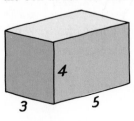
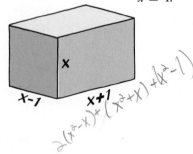
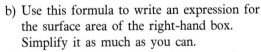

Set II

4. Use the distributive rule to find each of the following products.
   a) $8(2x - 3y)$
   b) $5x(x + 4)$
   c) $x^3(x^2 - 2x)$
   d) $4xy(x - y + 1)$

5. Find the greatest common factor of the terms of the following polynomials.
   a) $14x^2 + 4x$
   b) $x^5 - x^3$
   c) $2x^2 + 3x + 6$
   d) $16x^2 - 40y^2$

6. Factor each of the following polynomials by factoring out the greatest common factor of its terms. Illustrate each problem by means of the area of a rectangle.
   a) $4x + 8$
   b) $x^2 + 5x$
   c) $3x^3 + x^2$
   d) $2x^4 + 12x$

7. Factor each of the following polynomials by factoring out the greatest common factor of its terms.
   a) $3x + 12$
   b) $5x + 5$
   c) $2x + 2y$
   d) $2x + 3y$
   e) $4x - 4$
   f) $10x - 15$
   g) $32x - 8y$
   h) $x^2 + 6x$
   i) $2x^3 + 2$
   j) $3x^2 + 2$
   k) $x^4 - 8x^2$
   l) $x^4 + y^4$

8. Write each of the following polynomials as a monomial by first factoring out the greatest common factor of its terms.
   a) $7x + 3x$
   b) $5x - x$
   c) $x^2 + 8x^2$
   d) $2x^3 - 5x^3$
   e) $4xy + 2xy$
   f) $xy^2 - 9xy^2$

9. When the expression $57 \cdot 64 + 57 \cdot 36$ is written in factored form, $57(64 + 36)$, it is easy to see that it is equal to 5,700, because $64 + 36 = 100$.
   Find the value of each of the following expressions by first factoring it.
   a) $23 \cdot 6 + 23 \cdot 4$
   b) $12 \cdot 31 - 12$
   c) $45 \cdot 36 - 45 \cdot 16$
   d) $99^2 + 99$

10. What polynomial should replace ▓ in each of the following equations?
    a) $(▓)(x^2 + 2x + 4) = 3x^2 + 6x + 12$
    b) $(5x)(▓) = 5x^4 - 30x^2 + 5x$
    c) $8x^6 + 4x^4 - 12x^2 = 4x^2(▓)$
    d) $10xy + 2x^2y - 2xy^2 = 2xy(▓)$

11. Factor each of the following polynomials by factoring out the greatest common factor of its terms.
    a) $12x^3 - 16x^2 + 8x$
    b) $6x^2 + xy + 6y^2$
    c) $10x^{10} - x^5$
    d) $2x^2 + 6x^6 + 10x^{10}$

Set III

12. Use the distributive rule to find each of the following products.
    a) $9(3x + y)$
    b) $4x(2x - 7)$
    c) $x^2(x^4 + 4x)$
    d) $3xy(x + y - 1)$

13. Find the greatest common factor of the terms of the following polynomials.
    a) $6x^2 + 15x$
    b) $x^4 - x^6$
    c) $2x^3 - 6x^2 + 4$
    d) $9x^2 + 16y^2$

14. Factor each of the following polynomials by factoring out the greatest common factor of its terms. Illustrate each problem by means of the area of a rectangle.
    a) $3x + 6$
    b) $x^2 + 4x$
    c) $2x^3 + x^2$
    d) $5x^3 + 15x$

15. Factor each of the following polynomials by factoring out the greatest common factor of its terms.
    a) $5x + 15$
    b) $2x + 2$
    c) $7x + 7y$
    d) $3x - 18$
    e) $3x - 16$
    f) $x + 2y$
    g) $4x - 24y$
    h) $8x + x^2$
    i) $5x^2 + 10$
    j) $2x^5 + 5$
    k) $x^3 - 3x$
    l) $x^2 + y^2$

16. Write each of the following polynomials as a monomial by first factoring out the greatest common factor of its terms.
    a) $2x + 6x$
    b) $4x - x$
    c) $x^3 + 5x^3$
    d) $3x^2 - 8x^2$
    e) $7xy + 5xy$
    f) $x^2y - 10x^2y$

17. When the expression $32 \cdot 26 - 32 \cdot 16$ is written in factored form, $32(26 - 16)$, it is easy to see that it is equal to 320, because $26 - 16 = 10$.

Find the value of each of the following expressions by first factoring it.
    a) $54 \cdot 3 + 54 \cdot 7$
    b) $9 \cdot 49 + 9$
    c) $15 \cdot 28 - 15 \cdot 8$
    d) $101^2 - 101$

18. What polynomial should replace ▨ in each of the following equations?
    a) $(▨)(x^2 + x + 6) = 5x^2 + 5x + 30$
    b) $(4x)(▨) = 4x^5 - 12x^2 + 8x$
    c) $10x^5 + 8x^4 - 2x^3 = 2x^3(▨)$
    d) $21xy - 3x^2y - 3xy^2 = 3xy(▨)$

19. Factor each of the following polynomials by factoring out the greatest common factor of its terms.
    a) $30x^3 + 18x^2 - 12x$
    b) $2x^2y + xy + 2xy^2$
    c) $5x^8 + 8$
    d) $3x^3 + 6x^6 + 9x^9$

Set IV

Here is an interesting pattern. If a number that ends in 5 is squared, the answer ends in 25. Moreover, the rest of the answer can be found in a simple way.

$$1|5^2 = 2|25$$
$$2|5^2 = 6|25$$
$$3|5^2 = 12|25$$
$$4|5^2 = 20|25$$
$$5|5^2 = 30|25$$
$$6|5^2 = 42|25$$
$$7|5^2 = 56|25$$
$$8|5^2 = 72|25$$
$$9|5^2 = 90|25$$

1. Can you explain how, using $65^2 = 4225$ as an example?
2. Show why this is true by representing the original number as $10x + 5$, squaring it, and factoring.

Leonhard Euler

LESSON 4
Factoring Second-Degree Polynomials

No one has ever found a formula for prime numbers. The eighteenth-century Swiss mathematician Leonhard Euler discovered several polynomials that produce prime numbers up to a certain point, however. One of them is

$$x^2 + x + 17$$

If this polynomial is evaluated for each positive integer from 1 through 15,

$$1^2 + 1 + 17 = 19,$$
$$2^2 + 2 + 17 = 23,$$
$$3^2 + 3 + 17 = 29, \quad \text{and so on,}$$

each of the numbers that results is prime.

Unlike Euler's polynomial, the polynomial

$$x^2 + 8x + 7$$

never produces primes when it is evaluated for the positive integers.

$$1^2 + 8 \cdot 1 + 7 = 16,$$
$$2^2 + 8 \cdot 2 + 7 = 27,$$
$$3^2 + 8 \cdot 3 + 7 = 40, \quad \text{and so on.}$$

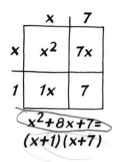

$x^2 + 8x + 7 =$
$(x+1)(x+7)$

The reason is very simple. Look at the diagram at the left. It shows that the polynomial $x^2 + 8x + 7$ and the product $(x + 1)(x + 7)$ are equivalent. This means that the polynomial $x^2 + 8x + 7$ is not prime, even though the greatest common factor of its terms is 1. It can be factored into the product of the polynomials $x + 1$ and $x + 7$, both of lower degree. Evaluating this product for $x = 1$, $x = 2$, $x = 3$, and so on, we get

$$(1 + 1)(1 + 7) = 2 \cdot 8 = 16,$$
$$(2 + 1)(2 + 7) = 3 \cdot 9 = 27,$$
$$(3 + 1)(3 + 7) = 4 \cdot 10 = 40, \quad \text{and so on.}$$

Clearly, for every positive integer x, $(x + 1)(x + 7)$ is the product of two integers both greater than 1. So for every positive integer, $(x + 1)(x + 7)$ and, hence, $x^2 + 8x + 7$ must produce a composite number.

From this example, we see that it may be possible to factor a second-degree polynomial into a product of first-degree polynomials even though its terms have no common factor greater than 1. In this lesson we will learn how to factor such expressions.

The basic idea is simple. Factoring one polynomial into a product of two is the reverse of multiplying two polynomials to get one. Compare the following two problems.

PROBLEM 1
Multiply $x + 1$ and $x + 5$.

SOLUTION
$(x + 1)(x + 5) = x^2 + 6x + 5$.

PROBLEM 2
Factor $x^2 + 6x + 5$.

SOLUTION
From problem 1, we see that $x^2 + 6x + 5 = (x + 1)(x + 5)$.

We knew the answer to problem 2 because of the result of problem 1. Unfortunately, reversing a multiplication problem is not always this easy.

Before we consider another example, look at the diagram shown at the right.

It shows that the product of the first-degree polynomials $x + a$ and $x + b$ and the second-degree polynomial $x^2 + (a + b)x + ab$ are equivalent. Notice where each term of the polynomial $x^2 + (a + b)x + ab$ appears in the figure: the first term, x^2, in the upper left corner, the last term, ab, in the lower right corner, and the middle term, $(a + b)x$, as ax and bx, in the other two corners.

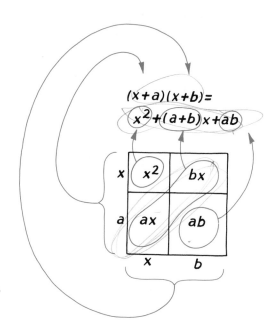

$$(x+a)(x+b)= x^2+(a+b)x+ab$$

Now let's try to apply these observations to factoring the polynomial $x^2 + 9x + 20$ into a product of two first-degree polynomials. Putting the first and last terms into the appropriate corners of a diagram, we get the figure shown at the right. Factoring x^2 as $x \cdot x$, we fill in the dimensions of the "x^2" box as shown in the next figure. Filling in the dimensions of the "20" box is not as easy because 20 can be factored in several different ways: those in which both factors are positive are $1 \cdot 20$, $2 \cdot 10$, and $4 \cdot 5$. If we look at each possibility as shown below, it is clear that the third one is the only one that results in the correct middle term: $9x$.

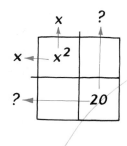

So $x^2 + 9x + 20 = (x + 4)(x + 5)$.

To arrive at this result without drawing any diagrams, we could reason as follows.

Factor the x^2 term:

$$x^2 + 9x + 20 = (x + \quad)(x + \quad)$$

Factor the constant term:

$$x^2 + 9x + 20 = (x + ?)(x + ?)$$

Try combinations of factors:

$20 = 1 \cdot 20$	$20 = 2 \cdot 10$	$20 = 4 \cdot 5$
$(x + 1)(x + 20)?$	$(x + 2)(x + 10)?$	$(x + 4)(x + 5)?$
$1x$	$2x$	$4x$
$20x$	$10x$	$5x$

$x^2 + 21x + 20$	$x^2 + 12x + 20$	$x^2 + 9x + 20$
Wrong	Wrong	Right

So $x^2 + 9x + 20 = (x + 4)(x + 5)$.

Not every second-degree polynomial can be factored like this. Judging from the three results above, for example, we cannot factor $x^2 + 10x + 20$ if the factors of 20 are limited to integers.

Here are more examples of how to factor second-degree polynomials.

EXAMPLE 1
Factor $x^2 + 2x - 3$ as the product of two binomials.

SOLUTION*
Think of $x^2 + 2x - 3$ as $x^2 + 2x + -3$.

$$x^2 + 2x + -3 = (x + \quad)(x + \quad)$$

$$x^2 + 2x + -3 = (x + ?)(x + ?)$$

One of the factors of –3 must be negative.

$$-3 = (1)(-3) \qquad\qquad -3 = (-1)(3)$$
$$(x + 1)(x + -3)? \qquad (x + -1)(x + 3)?$$

$$\underbrace{}_{1x}$$
$$\underbrace{}_{-3x} \qquad\qquad \underbrace{}_{-1x}$$
$$\underbrace{}_{3x}$$

$$x^2 - 2x - 3 \qquad\qquad x^2 + 2x - 3$$

Wrong   Right

So $x^2 + 2x - 3 = (x - 1)(x + 3)$.

EXAMPLE 2

Factor $x^2 - 6x + 8$ as the product of two binomials.

SOLUTION*

Think of $x^2 - 6x + 8$ as $x^2 + -6x + 8$.

$$x^2 + -6x + 8 = (x + \quad)(x + \quad)$$

$$x^2 + -6x + 8 = (x + ?\ )(x + ?\ )$$

In order to get $-6x$ as the middle term, we have to factor 8 into two *negative* factors.

$$8 = (-1)(-8) \qquad\qquad 8 = (-2)(-4)$$
$$(x - 1)(x - 8)? \qquad (x - 2)(x - 4)?$$

$$\underbrace{}_{-1x}$$
$$\underbrace{}_{-8x} \qquad\qquad \underbrace{}_{-2x}$$
$$\underbrace{}_{-4x}$$

$$x^2 - 9x + 8 \qquad\qquad x^2 - 6x + 8$$

Wrong   Right

So $x^2 - 6x + 8 = (x - 2)(x - 4)$.

---

* Most of the work written out in the solutions for Examples 1 and 2 is ordinarily done mentally.

EXAMPLE 3

Factor the following polynomials.

a) $x^2 + 5x + 6$ c) $x^2 + x - 6$
b) $x^2 - 5x + 6$ d) $x^2 - x - 6$

SOLUTION

To factor these polynomials, we think of them as

a) $x^2 + 5x + 6$ c) $x^2 + 1x + -6$
b) $x^2 + -5x + 6$ d) $x^2 + -1x + -6.$

In all four parts, we begin by writing $(x \quad)(x \quad)$.
In parts a and b, 6 can be factored as $(1)(6)$, $(2)(3)$, $(-1)(-6)$, or $(-2)(-3)$.

Because $2 + 3 = 5,$ $x^2 + 5x + 6 = (x + 2)(x + 3).$
Because $-2 + -3 = -5,$ $x^2 - 5x + 6 = (x - 2)(x - 3).$

In parts c and d, -6 can be factored as $(1)(-6)$, $(2)(-3)$, $(3)(-2)$, or $(6)(-1)$.

Because $3 + -2 = 1,$ $x^2 + x - 6 = (x + 3)(x - 2).$
Because $2 + -3 = -1,$ $x^2 - x - 6 = (x + 2)(x - 3).$

Exercises

Set I

1. Find the greatest common factor for each of the following sets of monomials.
   a) 91 and 117
   b) $42x^3$ and $70x$
   c) x^6 and $6x$

2. Find each of the following quotients.
   a) 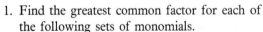 $x^2 + 3\overline{)x^4 - x^3 + 8x^2 - 3x + 15}$
   b) $x - 2\overline{)x^3 - 8}$

3. In the country of Grand Fenwick, 13 simoleons are currently equal in value to one U.S. dollar.
   a) Write a formula for the cost, d, in dollars of something that costs s simoleons.
   b) Write a formula for the cost, s, in simoleons of something that costs d dollars.
   c) How do the costs of things in one monetary system vary with those in the other?

Set II

4. Find each of the following products.
   a) $(x + 1)(x + 15)$ c) $(x + 3)(x + 5)$ e) $(x - 1)(x + 15)$ g) $(x - 3)(x + 5)$
   b) $(x - 1)(x - 15)$ d) $(x - 3)(x - 5)$ f) $(x + 1)(x - 15)$ h) $(x + 3)(x - 5)$

5. Express the area of each of the following figures both as a polynomial in simplest form and as the product of two polynomials.

a)
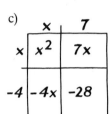

b)
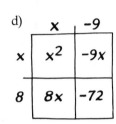

c)

×	x²	7x
×		
-4	-4x	-28

(with × and 7 across the top)

d)

×	x²	-9x
×		
8	8x	-72

(with × and -9 across the top)

6. One of the steps in factoring a second-degree polynomial into a product of two first-degree polynomials is to factor its constant term into the product of two numbers whose sum is a given number.

   Factor each of the following numbers into the product of two numbers having the indicated sum.

   a) 10; sum, 7 f) 42; sum, 17
   b) 10; sum, –7 g) –36; sum, 16
   c) –33; sum, 8 h) –36; sum, 0
   d) –33; sum, –8
   e) 42; sum, 13

7. Tell what polynomial should replace ▓ in each of the following equations to make it true.

   a) $x^2 + 11x + 18 = (x + 2)(▓)$
   b) $x^2 - 9x + 20 = (x - 5)(▓)$
   c) $x^2 + 6x - 7 = (▓)(x - 1)$
   d) $x^2 - 5x - 24 = (▓)(x + 3)$
   e) $x^2 - 18x + 72 = (x - 6)(▓)$
   f) $x^2 - 25 = (▓)(x - 5)$

8. Factor each of the following polynomials as the product of two binomials.
   a) $x^2 + 6x + 5$
   b) $x^2 - 6x + 5$
   c) $x^2 + 12x + 27$
   d) $x^2 + 12x + 32$
   e) $x^2 + 9x - 22$
   f) $x^2 - 9x - 22$
   g) $x^2 + 14x + 49$
   h) $x^2 - 49$
   i) $x^2 - 17x + 60$
   j) $x^2 - 23x + 60$
   k) $x^2 + 11x - 60$
   l) $x^2 - 4x - 60$

9. Use the diagrams below to factor each of the following polynomials.
   a) $x^2 + 3xy + 2y^2$
   b) $x^2 + 3xy - 40y^2$
   c) $xy + 4x + 6y + 24$
   d) $x^4 - 17x^2 + 70$

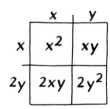

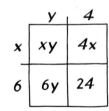

10. Factor each of the following polynomials.
    a) $x^2 + 8xy + 15y^2$
    b) $x^2 + 7xy - 18y^2$
    c) $xy + 4x + 4y + 16$
    d) $x^4 - 9x^2 - 10$

Set III

11. Find each of the following products.
    a) $(x + 1)(x + 14)$
    b) $(x - 1)(x - 14)$
    c) $(x + 2)(x + 7)$
    d) $(x - 2)(x - 7)$
    e) $(x - 1)(x + 14)$
    f) $(x + 1)(x - 14)$
    g) $(x - 2)(x + 7)$
    h) $(x + 2)(x - 7)$

12. Express the area of each of the following figures both as a polynomial in simplest form and as the product of two polynomials.

    a)

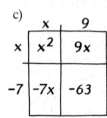

    b)

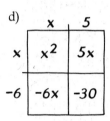

    c)

    d)

13. One of the steps in factoring a second-degree polynomial into a product of two first-degree polynomials is to factor its constant term into the product of two numbers whose sum is a given number.

    Factor each of the following numbers into the product of two numbers having the indicated sum.
    a) 6; sum, 5
    b) 6; sum, –5
    c) –35; sum, 2
    d) –35; sum, –2
    e) 44; sum, 15
    f) 44; sum, 24
    g) –81; sum, 0
    h) –81; sum, 24

14. What polynomial should replace ▓ in each of the following equations to make it true?
    a) $x^2 + 6x + 8 = (x + 4)(▓)$
    b) $x^2 - 12x + 11 = (x - 1)(▓)$
    c) $x^2 + 4x - 45 = (▓)(x - 5)$
    d) $x^2 - 5x - 14 = (▓)(x + 2)$
    e) $x^2 - 15x + 36 = (x - 3)(▓)$
    f) $x^2 - 100 = (▓)(x - 10)$

15. Factor each of the following polynomials as the product of two binomials.
    a) $x^2 + 4x + 3$
    b) $x^2 - 4x + 3$
    c) $x^2 + 16x + 55$
    d) $x^2 + 16x + 60$
    e) $x^2 + 4x - 21$
    f) $x^2 - 4x - 21$
    g) $x^2 - 4x + 4$
    h) $x^2 - 4$
    i) $x^2 - 14x + 48$
    j) $x^2 - 16x + 48$
    k) $x^2 + 13x - 48$
    l) $x^2 - 22x - 48$

16. Use the diagrams below to factor each of the following polynomials.
    a) $x^2 + 6xy + 5y^2$
    b) $x^2 + 3xy - 18y^2$
    c) $xy + 10x + 2y + 20$
    d) $x^4 - 13x^2 + 36$

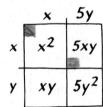

17. Factor each of the following polynomials.
    a) $x^2 + 13xy + 22y^2$
    b) $x^2 + 5xy - 24y^2$
    c) $xy + 5x + 10y + 50$
    d) $x^4 - 3x^2 - 28$

Set IV

Another polynomial that Euler discovered for producing prime numbers was

$$x^2 + x + 41$$

1. Find the value of this polynomial for each of the integers from 1 through 10. If you don't make any mistakes, all of the numbers that you will find are prime.
2. Can you figure out a value of x for which the polynomial $x^2 + x + 41$ does *not* produce a prime number?

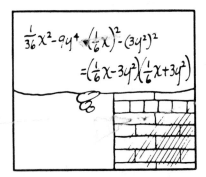

LESSON 5
Factoring the Difference of Two Squares

The solution to the factoring problem in this cartoon is based on a fact that we became acquainted with in learning how to multiply polynomials.* The fact is that

$$(a + b)(a - b) = a^2 - b^2$$

This equation says that the product of the sum and difference of two numbers is equal to the difference of the squares of the numbers. The diagram below shows why.

	a	b
a	a^2	ab
$-b$	$-ab$	$-b^2$

Turned around, the equation can be used to factor polynomials that are the difference of two squares. For example, what are the factors of $x^2 - 25$? Because $25 = 5^2$, this polynomial is the

---

*See page 425.

same as $x^2 - 5^2$. As we have just noted, the difference of the squares of two numbers is equal to the product of the sum and difference of the numbers:

$$a^2 - b^2 = (a + b)(a - b)$$

So $x^2 - 5^2 = (x + 5)(x - 5)$.

Although the difference of two squares can always be factored in this fashion, the sum of two squares cannot. There is no way to complete the diagram at the right without adding another term to the polynomial

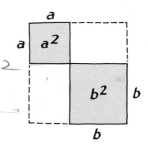

$$a^2 + b^2$$

This means that a polynomial such as $x^2 + 25$ cannot be factored as the product of two first-degree polynomials.

Here are examples to show how other polynomials can be factored.

EXAMPLE 1
Factor $x^4 - 9y^2$.

SOLUTION
Because $x^4 = (x^2)^2$ and $9y^2 = (3y)^2$,

$$x^4 - 9y^2 = (x^2)^2 - (3y)^2$$

This is the difference of two squares and so

$$x^4 - 9y^2 = (x^2 + 3y)(x^2 - 3y)$$

EXAMPLE 2
Factor $x^2 + 36$.

SOLUTION
This polynomial cannot be factored. Although both x^2 and 36 are squares, $x^2 + 36$ is their *sum* and so it cannot be factored as the product of two first-degree polynomials.

only differ one if 2 square

EXAMPLE 3

Factor $2x^2 - 32$.

SOLUTION

Although this polynomial is not the difference of two integer squares, it can be factored because 2 is a factor of both terms.

$$2x^2 - 32 = 2(x^2 - 16)$$

Furthermore, because $x^2 - 16$ *is* the difference of two squares, we can factor it to get $(x + 4)(x - 4)$. So

$$2x^2 - 32 = 2(x + 4)(x - 4)$$

EXAMPLE 4

Factor $3x^2 + 12$.

SOLUTION

Because 3 is a factor of both terms, we can write

$$3x^2 + 12 = 3(x^2 + 4)$$

This cannot be factored further because $x^2 + 4$ is the sum of two squares.

Exercises

Set I

1. Factor each of the following polynomials.
   a) $40x + 5$
   b) $9x - 12y$
   c) $x^2 - 4x$
   d) $2x^3 + 3x^2$

2. Find an approximation in decimal form as indicated for each of the following rational numbers.

   a) $\dfrac{100}{19}$ to the nearest tenth.

   b) $\dfrac{10}{19}$ to the nearest thousandth.

   c) $\dfrac{1}{19}$ to the nearest hundredth.

   d) $\dfrac{1}{190}$ to the nearest hundredth.

3. A volcanic explosion on the island of Krakatoa in 1883 produced a tremendous tidal wave. The wave reached the shore of South Africa in about 12 hours. If its speed had been 90 miles per hour faster, the wave would have reached South Africa in 10 hours instead.

a) Draw a figure and write an equation to find out the speed at which the tidal wave actually traveled in miles per hour. (Represent it by x.)

b) How many miles is the shore of South Africa from Krakatoa?

Set II

4. Find each of the following products.
   a) $(x + 20)(x - 20)$
   b) $(4x - 1)(4x + 1)$
   c) $(5x + y)(5x - y)$
   d) $(x^3 + 3)(x^3 - 3)$
   e) $(x - y^2)(x + y^2)$
   f) $(1 + xy)(1 - xy)$

5. If possible, write each of the following as the square of a monomial.
   a) 64   e) $9x^4$
   b) x^8 f) $4x^9$
   c) $36x$ g) $16x^2y^2$
   d) x^{36} h) $100x^{100}$

6. Express the area of each of the following figures both as a polynomial in simplest form and as the product of two binomials.

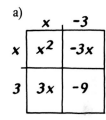

a)

	x	-3
x	x^2	$-3x$
3	$3x$	-9

b)

	$5x$	7
$5x$	$25x^2$	$35x$
-7	$-35x$	-49

c)

	x	$-4y$
x	x^2	$-4xy$
$4y$	$4xy$	$-16y^2$

d)

	x^3	y^2
x^3	x^6	x^3y^2
$-y^2$	$-x^3y^2$	$-y^4$

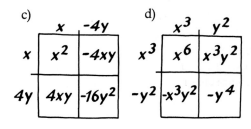

7. Find the values of the following expressions for the values of x indicated.

$x^2 - 9$
a) $x = 3$
b) $x = 7$
c) $x = 10$
d) $x = 0$
e) $x = -2$
f) $x = -5$

$(x + 3)(x - 3)$
g) $x = 3$
h) $x = 7$
i) $x = 10$
j) $x = 0$
k) $x = -2$
l) $x = -5$

8. Factor each of the following polynomials as completely as possible.
   a) $x^2 - 64$ f) $8x^2 - 4$
   b) $x^2 + 25$ g) $5x^2 + 20$
   c) $16x^2 - 9y^2$ h) $x^4 - 49$
   d) $81 - x^2$ i) $36x^{10} - 1$
   e) $8x^2 - 8$ j) $2x^6 - 32$

9. Because the expression $4x^2 - (x + 1)^2$ can be written as the difference of two squares, $(2x)^2 - (x + 1)^2$, it can be factored as the product of a sum and difference: $[2x + (x + 1)][2x - (x + 1)]$. This result can be simplified as shown here:

$$(2x + x + 1)(2x - x - 1) = (3x + 1)(x - 1)$$

Use the same method to factor each of the following expressions.
a) $(x + y)^2 - 9$
b) $(4x + 7)^2 - 36$
c) $25 - (1 - x)^2$
d) $64 - (x - 8)^2$

Set III

10. Find each of the following products.
    a) $(x + 12)(x - 12)$
    b) $(5x - 1)(5x + 1)$
    c) $(x + 3y)(x - 3y)$
    d) $(x^4 + 6)(x^4 - 6)$
    e) $(x^2 - y)(x^2 + y)$
    f) $(xy + 2)(xy - 2)$

11. If possible, write each of the following as the square of a monomial.
    a) 81 e) $25x^4$
    b) x^{10} f) $4x^{25}$
    c) $16x$ g) x^6y^6
    d) x^{16} h) $64x^{64}$

12. Express the area of each of the following figures both as a polynomial in simplest form and as the product of two binomials.

    a)
    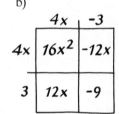

    b)
    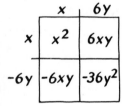

    c)
    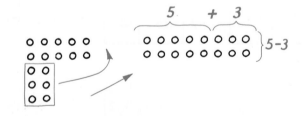

    d)

13. The first two diagrams below illustrate subtracting 3^2 from 5^2. The other two show how the circles that remain can be arranged in a rectangle. The rectangle contains $(5 - 3)$ rows of circles with $(5 + 3)$ circles in each row, or $(5 - 3)(5 + 3)$ circles in all. So

    $$5^2 - 3^2 = (5 - 3)(5 + 3)$$

    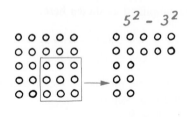

    Draw circles to make figures that illustrate each of the following.
    a) $6^2 - 2^2 = (6 - 2)(6 + 2)$
    b) $4^2 - 1^2 = (4 - 1)(4 + 1)$

14. Factor each of the following polynomials as completely as possible.
    a) $x^2 - 49$
    b) $x^2 + 16$
    c) $9x^2 - 4y^2$
    d) $100 - x^2$
    e) $6x^2 - 6$
    f) $6x^2 - 12$
    g) $4x^2 + 36y^2$
    h) $x^4 - 25$
    i) $64x^{12} - 1$
    j) $3x^6 - 12$

15. Because the expression $9x^2 - (x - 5)^2$ can be written as the difference of two squares, $(3x)^2 - (x - 5)^2$, it can be factored as the product of a sum and difference: $[3x + (x - 5)][3x - (x - 5)]$. This result can be simplified as shown here:

$$(3x + x - 5)(3x - x + 5) =$$
$$(4x - 5)(2x + 5)$$

Use the same method to factor each of the following expressions.
    a) $(x - y)^2 - 16$
    b) $(2x + 8)^2 - 49$
    c) $9 - (3 - x)^2$
    d) $25 - (x - 4)^2$

Set IV

After trying out several examples, Obtuse Ollie has decided that the difference of the squares of two integers is always a composite number. For example,

$$5^2 - 2^2 = 25 - 4 = 21 = 3 \cdot 7,$$
$$10^2 - 6^2 = 100 - 36 = 64 = 4 \cdot 16, \quad \text{and}$$
$$12^2 - 7^2 = 144 - 49 = 95 = 5 \cdot 19.$$

1. Why does his conclusion seem like a reasonable one?
2. After thinking about it for awhile, Acute Alice came up with some exceptions to Ollie's rule. Can you?

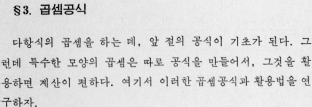

§3. 곱셈공식

다항식의 곱셈을 하는 데, 앞 절의 공식이 기초가 된다. 그런데 특수한 모양의 곱셈은 따로 공식을 만들어서, 그것을 활용하면 계산이 편하다. 여기서 이러한 곱셈공식과 활용법을 연구하자.

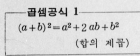

곱셈공식 1
$$(a+b)^2 = a^2 + 2ab + b^2$$
(합의 제곱)

곱셈공식 2
$$(a-b)^2 = a^2 - 2ab + b^2$$
(차의 제곱)

[풀이] [1] $(a+b)^2$
$= (a+b)(a+b)$
$= a^2 + ab + ab + b^2$
$= a^2 + 2ab + b^2$

[2]
$$\begin{array}{r} a + b \\ \times)\ a + b \\ \hline a^2 + ab \\ ab + b^2 \\ \hline a^2 + 2ab + b^2 \end{array}$$

[풀이] [1] $(a-b)^2$
$= (a-b)(a-b)$
$= a^2 - ab - ab + b^2$
$= a^2 - 2ab + b^2$

[2]
$$\begin{array}{r} a - b \\ \times)\ a - b \\ \hline a^2 - ab \\ -ab + b^2 \\ \hline a^2 - 2ab + b^2 \end{array}$$

【주의】 $(a+b)^2$ 을 a^2+b^2, $(a-b)^2$ 을 a^2-b^2 으로 하지 않도록 주의를 해야 한다.

[물음] 1. 오른편 그림을 보고 면적 관계를 이용하여, 합의 제곱공식이 성립함을 증명하여라.

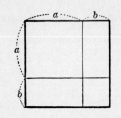

LESSON 6
Factoring Trinomial Squares

This illustration is a page from a Korean algebra book showing how to square a binomial. The column on the left illustrates the result of squaring a binomial *sum:*

$$(a + b)^2 = a^2 + 2ab + b^2$$

The column on the right illustrates the result of squaring a binomial *difference:*

$$(a - b)^2 = a^2 - 2ab + b^2$$

In both cases, the result is a trinomial whose first and last terms are the squares of the terms of the binomial. The middle term is twice the product of the terms of the binomial. Because of this, it is *added* if the binomial is a *sum* and *subtracted* if the binomial is a *difference.*

If these equations are turned around, they can be used to factor polynomials that are trinomial squares.

▶ A trinomial of the form $a^2 + 2ab + b^2$ is the square of the binomial $a + b$:

$$a^2 + 2ab + b^2 = (a + b)^2$$

▶ A trinomial of the form $a^2 - 2ab + b^2$ is the square of the binomial $a - b$:

$$a^2 - 2ab + b^2 = (a - b)^2$$

It is not always easy to see whether or not a trinomial has one of these forms. The following examples show how to find out whether a trinomial is the square of a binomial.

EXAMPLE 1
If possible, write $4x^2 + 12x + 9$ as the square of a binomial.

SOLUTION
The first term is the square of $2x$ and the last term is the square of 3. The middle term is twice the product of $2x$ and 3.

$$4x^2 + 12x + 9 = (2x)^2 + 2(2x \cdot 3) + (3)^2 = (2x + 3)^2$$

EXAMPLE 2
If possible, write $x^2 + 4x + 16$ as the square of a binomial.

SOLUTION
The first and last terms are the squares of x and 4 respectively. However, the middle term is not twice their product. So $x^2 + 4x + 16$ cannot be written as the square of a binomial.

EXAMPLE 3
If possible, write $25x^4 - 10x^2 + 1$ as the square of a binomial.

SOLUTION
The first term is the square of $5x^2$ and the last term is the square of 1. The middle term is twice the product of $5x^2$ and 1.

$$25x^4 - 10x^2 + 1 = (5x^2)^2 - 2(5x^2 \cdot 1) + (1)^2 = (5x^2 - 1)^2$$

Exercises

Set I

1. Multiply as indicated.
   a) $4x(2 - x^3)$ c) $4x(-2x^3)$
   b) $x^4(2 - x^3)$ d) $x^4(-2x^3)$

2. Find the value of each of the following expressions by first factoring it.
   a) $87 \cdot 46 + 87 \cdot 54$
   b) $92 \cdot 29 - 82 \cdot 29$
   c) $123 \cdot 21 - 123$
   d) $235^2 - 35^2$

3. The most expensive metal in the world is californium, which has been sold for $100 per 10^{-7} gram.
   a) At this rate, how much would 1 gram of this metal cost?
   b) One pound is equivalent to about 454 grams. How much money would one pound of californium cost? Write your answer in both scientific notation and in words.

Set II

4. Express the area of each of the following figures both as a polynomial in simplest form and as the square of a binomial.

   a)
	x	3
x	x^2	$3x$
3	$3x$	9

   b)
	x	-10
x	x^2	$-10x$
-10	$-10x$	100

   c)
	$4x$	y
$4x$	$16x^2$	$4xy$
y	$4xy$	y^2

   d)
	x^4	$-y$
x^4	x^8	$-x^4y$
$-y$	$-x^4y$	y^2

5. Find the following squares.
   a) $(x + 7)^2$
   b) $(x - 7)^2$
   c) $(4x - 1)^2$
   d) $(5x + 2y)^2$
   e) $(2x + 5y)^2$
   f) $(x^2 + y^5)^2$

6. If possible, write each of the following as the square of a monomial.
   a) $9x^2$
   b) x^{16}
   c) 400
   d) $121x$
   e) $25x^{36}$
   f) $36x^{25}$
   g) $64x^6y^4$
   h) $49x^4y^9$

7. Find the missing term in each of the following trinomial squares.
   a) $x^2 + \text{\rule{1cm}{0.4cm}} + 64$ c) $9x^2 - \text{\rule{1cm}{0.4cm}} + 16$
   b) $x^2 - 22x + \text{\rule{1cm}{0.4cm}}$ d) $36x^2 + 12x + \text{\rule{1cm}{0.4cm}}$

8. If possible, write each of the following polynomials as the square of a binomial.
   a) $x^2 + 12x + 36$ e) $9x^2 - 60x + 100$
   b) $x^2 - 2xy + y^2$ f) $16x^2 + 40x - 25$
   c) $4x^2 + 4x + 1$ g) $49x^2 + 14xy + y^2$
   d) $x^2 - 9x + 81$ h) $25x^2 - 80xy + 64y^2$

9. Find the values of the following expressions for the values of x indicated.

   $x^2 + 10x + 25$ $(x + 5)^2$
   a) $x = 1$ f) $x = 1$
   b) $x = 3$ g) $x = 3$
   c) $x = 10$ h) $x = 10$
   d) $x = -5$ i) $x = -5$
   e) $x = -9$ j) $x = -9$

10. Factor each of the following polynomials.

a) $x^4 + 22x^2 + 121$

b) $x^6 - 8x^3 + 16$

c) $x^{10} + 2x^5y^5 + y^{10}$

d) $2x^2 - 12x + 18$

e) $5x^4 - 50x^2 + 125$

f) $16x^2 - 8xy^8 + y^{16}$

Set III

11. Express the area of each of the following figures both as a polynomial in simplest form and as the square of a binomial.

a)

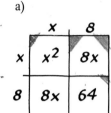

b)

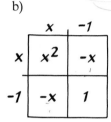

c)

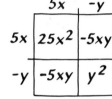

d)

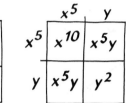

12. Find the following squares.

a) $(x + 5)^2$

b) $(x - 5)^2$

c) $(8x + 1)^2$

d) $(4x + 3y)^2$

e) $(3x + 4y)^2$

f) $(x^3 + y^4)^2$

13. If possible, write each of the following as the square of a monomial.

a) $16x^2$

b) x^{100}

c) 900

d) $81x$

e) $9x^{64}$

f) $64x^9$

g) $25x^2y^5$

h) $36x^6y^6$

14. Find the missing term in each of the following trinomial squares.

a) $x^2 + \text{▨} + 100$ c) $4x^2 - \text{▨} + 1$

b) $x^2 - 18x + \text{▨}$ d) $25x^2 + 20x + \text{▨}$

15. If possible, write each of the following polynomials as the square of a binomial.

a) $x^2 - 8x + 16$ e) $4x^2 + 28x + 49$

b) $x^2 + 2xy + y^2$ f) $25x^2 - 20x + 16$

c) $x^2 - 4x - 4$ g) $36x^2 + 12xy + y^2$

d) $9x^2 + 6x + 1$ h) $16x^2 - 72xy + 81y^2$

16. Find the values of the following expressions for the values of x indicated.

$$x^2 - 12x + 36$$

a) $x = 6$

b) $x = 7$

c) $x = 10$

d) $x = -1$

e) $x = -4$

$$(x - 6)^2$$

f) $x = 6$

g) $x = 7$

h) $x = 10$

i) $x = -1$

j) $x = -4$

17. Factor each of the following polynomials.

a) $x^4 + 14x^2 + 49$

b) $x^8 - 4x^4 + 4$

c) $x^6 - 2x^3y^3 + y^6$

d) $3x^2 - 30x + 75$

e) $72x^2 + 24x + 2$

f) $x^{10} - 10x^5y + 25y^2$

Set IV

Now that you know how to factor a trinomial square, see if you can figure out this problem. The expression

$$x^3 + 15x^2 + 75x + 125$$

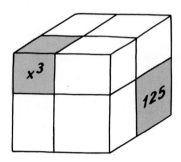

might be called a "polynomial cube" because it is the cube of a binomial.

1. Can you figure out what the binomial is?
2. Show that your answer is correct.

LESSON 7
More on Factoring Second-Degree Polynomials

Leonhard Euler, the man who discovered several polynomials that produce primes up to a certain point, once announced that the number 1,000,009 was prime. Later he discovered that he was wrong and that 1,000,009 can be factored into the product 293 · 3,413. Euler, who was a great mathematician, would never have made this mistake if it were as easy to factor numbers as it is to multiply them.

The same is true of polynomials. Multiplying the two binomials $2x + 3$ and $7x + 10$, for example, is easy:

$$(2x + 3)(7x + 10) = 14x^2 + 20x + 21x + 30$$
$$= 14x^2 + 41x + 30$$

Factoring the polynomial $14x^2 + 41x + 30$ back into the product $(2x + 3)(7x + 10)$, on the other hand, is much more difficult if you don't already know the answer.

In Lesson 4 of this chapter, we learned how to factor polynomials of the form $x^2 + bx + c$. In this lesson, we will apply this method to factoring polynomials of the form $ax^2 + bx + c$, in which a is not 1.

481

To illustrate the steps to be taken, we will factor the polynomial $6x^2 + 17x + 5$. Putting the first and last terms into the appropriate corners of a diagram, we get the figure at the left. Because $6x^2$ can be factored in several ways, we have more possibilities to consider. We begin by factoring $6x^2$ as $6x \cdot x$. Later we may have to try factoring $6x^2$ as $2x \cdot 3x$.

Writing $6x^2$ as $6x \cdot x$ is illustrated by the figure at the left below. There are

now two ways to put the factors of 5 into the picture. The second and third diagrams illustrate these ways. When we fill in the areas of the remaining boxes and add them, we get $31x$ and $11x$ respectively. But the middle term of the polynomial that we are trying to factor is $17x$. So neither of these is what we want.

Factoring $6x^2$ as $2x \cdot 3x$ instead, and filling in the factors of 5 in both ways as before, is illustrated by the next set of figures. This time when we add the areas

of the remaining boxes, we get $13x$ and $17x$. We have found what we were looking for. The last figure shows that

$$6x^2 + 17x + 5 = (2x + 5)(3x + 1)$$

To arrive at this result without drawing any diagrams, we could reason as follows.*

Factor the x^2 term: $6x^2 + 17x + 5 = (6x + \underset{?}{\quad})(x + \underset{?}{\quad})$

_____

* Most of the work written out in this solution is ordinarily done mentally.

Factor the constant term: $\quad 6x^2 + 17x + 5 = (6x + \,^?)(x + \,^?)$

Try combinations of factors:

$$(6x + 1)(x + 5) = 6x^2 + 31x + 5 \quad \text{No.}$$
$$(6x + 5)(x + 1) = 6x^2 + 11x + 5 \quad \text{No.}$$

Factor the x^2 term a different way:
$\quad 6x^2 + 17x + 5 = (2x + \quad)(3x + \quad)$

Factor the constant term: $\quad 6x^2 + 17x + 5 = (2x + \,^?)(3x + \,^?)$

Try combinations of factors:

$$(2x + 1)(3x + 5) = 6x^2 + 13x + 5 \quad \text{No.}$$
$$(2x + 5)(3x + 1) = 6x^2 + 17x + 5 \quad \text{Yes.}$$

So $6x^2 + 17x + 5 = (2x + 5)(3x + 1)$.

Here is another example.

EXAMPLE

Factor $7x^2 + 12x - 4$ as the product of two binomials.

SOLUTION

Think of $7x^2 + 12x - 4$ as $7x^2 + 12x + -4$. Factoring $7x^2$ into $7x \cdot x$, we write

$$7x^2 + 12x + -4 = (7x + \quad)(x + \quad)$$

$$7x^2 + 12x + -4 = (7x + \,^?)(x + \,^?)$$

The number -4 can be factored as $(1)(-4)$, $(-1)(4)$, or $(2)(-2)$. In each case, either factor can be filled in first. For example, if we try 1 and -4, we can write either

$$7x^2 + 12x + -4 = (7x + 1)(x + -4)$$

or

$$7x^2 + 12x + -4 = (7x + -4)(x + 1)$$

All six possibilities are shown below.

$$-4 = (1)(-4) \qquad\qquad -4 = (-1)(4) \qquad\qquad -4 = (2)(-2)$$

$(7x + 1)(x - 4)?$ $\qquad$ $(7x - 1)(x + 4)?$ $\qquad$ $(7x + 2)(x - 2)?$
$7x^2 - 27x - 4$ No. $\qquad$ $7x^2 + 27x - 4$ No. $\qquad$ $7x^2 - 12x - 4$ No.

$(7x - 4)(x + 1)?$ $\qquad$ $(7x + 4)(x - 1)?$ $\qquad$ $(7x - 2)(x + 2)?$
$7x^2 + 3x - 4$ No. $\qquad$ $7x^2 - 3x - 4$ No. $\qquad$ $7x^2 + 12x - 4$ Yes.

So $7x^2 + 12x - 4 = (7x - 2)(x + 2)$.

Exercises

Set I

1. If possible, factor each of the following polynomials.
   a) $2x - 64$ $\qquad$ c) $x^2 + 36$
   b) $x^2 - 64$ $\qquad$ d) $x^2 + 6x$

2. Find x in each of the following equations.
   a) $2^3 \cdot 2^4 = 2^x$ $\qquad$ c) $3^2 + 4^2 = x^2$
   b) $3^2 \cdot 4^2 = x^2$ $\qquad$ d) $(3 + 4)^2 = x^2$

3. Cuthbert J. Twillie makes a lot of money selling snake oil. This table shows how he operates.

	Costs	Sells for
Small bottle	10¢	$1
Large bottle	25¢	$3

One week he sold x small bottles and y large bottles.

a) How much money did he pay for them altogether?

b) How much money did he sell them for altogether?

c) How much money did he make?

Set II

4. Express the area of each of the following figures both as a polynomial in simplest form and as the product of two binomials.

a) $\qquad$ b)

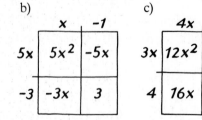

c) $\qquad$ d)

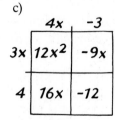

5. Find each of the following products.
   a) $(3x + 1)(x + 12)$
   b) $(3x - 1)(x + 12)$
   c) $(3x + 2)(x + 6)$
   d) $(3x - 2)(x + 6)$
   e) $(3x + 3)(x + 4)$
   f) $(3x - 3)(x + 4)$
   g) $(3x + 6)(x + 2)$
   h) $(3x - 6)(x + 2)$

6. Factor each of the following polynomials as
   the product of two binomials.
   a) $2x^2 + 15x + 7$
   b) $2x^2 + 13x - 7$
   c) $25x^2 - 10x + 1$
   d) $25x^2 - 1$
   e) $3x^2 + 7x - 6$
   f) $3x^2 - 9x + 6$
   g) $8x^2 + 22x + 15$
   h) $8x^2 - 22x + 15$
   i) $6x^2 + 19x - 11$
   j) $6x^2 - 19x - 11$

7. Use the diagrams below to factor each of the
   following polynomials.
   a) $6x^2 - xy - y^2$
   b) $8x^2 - 11xy + 3y^2$
   c) $4xy - 20x + y - 5$
   d) $5x^4 + 16x^2 + 12$

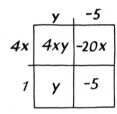

8. Factor each of the following polynomials.
   a) $5x^2 + 16xy + 3y^2$
   b) $8x^2 - 30xy + 7y^2$
   c) $2xy - x + 10y - 5$
   d) $3x^4 + 32x^2 + 20$

Set III

9. Express the area of each of the following figures both as a polynomial in
   simplest form and as the product of two polynomials.

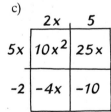

10. Find each of the following products.
    a) $(2x + 1)(x + 18)$ e) $(2x + 3)(x + 6)$
    b) $(2x - 1)(x + 18)$ f) $(2x - 3)(x + 6)$
    c) $(2x + 2)(x + 9)$ g) $(2x + 6)(x + 3)$
    d) $(2x - 2)(x + 9)$ h) $(2x - 6)(x + 3)$

11. Factor each of the following polynomials as the product of two binomials.

   a) $3x^2 + 16x + 5$ e) $2x^2 + 3x - 14$ i) $8x^2 + 26x - 7$
   b) $3x^2 + 14x - 5$ f) $2x^2 + 27x - 14$ j) $8x^2 - 26x - 7$
   c) $16x^2 + 8x + 1$ g) $15x^2 + 19x + 6$
   d) $16x^2 - 1$ h) $15x^2 - 19x + 6$

12. Use the diagrams below to factor each of the following polynomials.

   a) $8x^2 - 2xy - y^2$
   b) $5x^2 - 11xy + 2y^2$
   c) $4xy + 6x - 2y - 3$
   d) $3x^4 + 10x^2 + 8$

	2x	-y
4x	$8x^2$	-4xy
y	2xy	$-y^2$

	5x	-y
x	$5x^2$	-xy
-2y	-10xy	$2y^2$

	2y	3
2x	4xy	6x
-1	-2y	-3

	$3x^2$	4
x^2	$3x^4$	$4x^2$
2	$6x^2$	8

13. Factor each of the following polynomials.

   a) $3x^2 + 22xy + 7y^2$
   b) $10x^2 - 17xy + 3y^2$
   c) $6xy + 6x - 5y - 5$
   d) $2x^4 + 17x^2 + 21$

Set IV

The following problem appeared in *First Lessons in Algebra* by Ebenezer Bailey, published in 1842. According to the title page of the book, the author was principal of the Young Ladies' High School in Boston.

Express the polynomial

$$1 - 4x + 4x^2 + 2y - 4xy + y^2$$

as the square of a trinomial.

Can you solve it? If so, show your method.

Hint: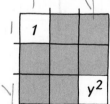

LESSON **8**

Factoring Higher-Degree Polynomials

Because 23 is a prime number, we think of its factors as being limited to 23 and 1. The ant could have had 23 children, on each of whom it spent one dollar. If we do not limit the factors to positive integers, however, then 23 can be factored into the product $(46)(0.5)$. This corresponds to the ant's 46 children, on each of whom it spends half a dollar. In fact, by allowing fractions as factors, every number can be factored in an unlimited number of different ways. For example, it is also true that $23 = (92)(0.25)$; it is possible that the ant could have had 92 children, on each of whom a quarter is spent, and so on.

By the same line of reasoning, every polynomial can be factored in infinitely many ways. For example, $2x + 8$ might be factored into expressions such as $0.5(4x + 16)$ or $8(0.25x + 1)$ as well as $2(x + 4)$. Although this sort of factoring is sometimes useful, we will assume that, when we speak of factoring a polynomial, we are referring to finding factors that are either integers greater than 1 or polynomials having coefficients and constant terms that are integers. This will permit us to continue to say that a polynomial such as $x^2 + 4$ cannot be factored because we are not considering factorizations such as $1(x^2 + 4)$ or $2(0.5x^2 + 2)$.

487

So far in this chapter, we have learned the following ways to factor polynomials.

1. Factor out the greatest factor common to every term of the polynomial.
   Example: $5x^3 + 10x = 5x(x^2 + 2)$.

2. Factor a polynomial of the form $x^2 + bx + c$ by factoring c into two numbers whose sum is b.
   Example: $x^2 + 8x + 15 = (x + 3)(x + 5)$.

3. Factor the difference of two squares of two numbers as the product of the sum and difference of the numbers.
   Example: $4x^2 - 9 = (2x + 3)(2x - 3)$.

4. Factor a trinomial square as the square of a binomial.
   Example: $4x^2 - 12x + 9 = (2x - 3)^2$.

5. Factor a polynomial of the form $ax^2 + bx + c$ by factoring a and c into factors that will produce two products whose sum is b.
   Example: $6x^2 + 13x + 5 = (2x + 1)(3x + 5)$.

We have applied these methods primarily to factoring second-degree polynomials. There are other methods, studied in advanced algebra, for discovering factors of polynomials having higher degree. In this lesson, we will factor some higher-degree polynomials by the methods we already know.

The basic procedure is to begin by factoring out the greatest common factor of the terms of the polynomial.* Then look at the remaining factor to see if it is either a trinomial square or the difference of two squares. If it is, factor it accordingly. If not, look for other combinations of factors that it might have.

The following examples illustrate how this procedure is carried out.

EXAMPLE 1
Factor $x^3 + 4x^2 + 4x$.

SOLUTION
First, the greatest common factor of each term is x, and so we factor it out to get $x(x^2 + 4x + 4)$. The other factor is a trinomial square, and so we can factor it further to get $x(x + 2)^2$ or $x(x + 2)(x + 2)$. Because all three factors, x, $x + 2$, and $x + 2$, are prime, we are finished.

---

* This makes sure that any first-degree factors resulting from further factoring will be prime.

EXAMPLE 2
Factor $3x^3 - 27x$.

SOLUTION
Factoring out the greatest common factor of each term, $3x$, we get $3x(x^2 - 9)$. The factor $x^2 - 9$ is the difference of two squares. Factoring it, we get
$3x(x + 3)(x - 3)$.

EXAMPLE 3
Factor $x^4 + 2x^3 - 15x^2$.

SOLUTION
The greatest common factor of each term here is x^2. Factoring it out, we get
$x^2(x^2 + 2x - 15)$. The factor $x^2 + 2x - 15$ can be factored into $(x - 3)(x + 5)$,
and so
$$x^4 + 2x^3 - 15x^2 = x^2(x - 3)(x + 5)$$

EXAMPLE 4
Factor $2x^5 + 2x^3$.

SOLUTION
The greatest common factor of each term is $2x^3$. Factoring it out gives
$2x^3(x^2 + 1)$. Because $x^2 + 1$ cannot be factored further,
$$2x^5 + 2x^3 = 2x^3(x^2 + 1)$$

Exercises

Set I

1. Which symbol, $>$, $=$, or $<$, should replace ▊ in each of the following?

   a) 1^{100} ▊ 100 d) 100^{-1} ▊ -1
   b) 100^0 ▊ 0 e) 1^{-100} ▊ 1
   c) 0^{100} ▊ 0 f) $(-1)^{100}$ ▊ -1

2. Solve the following pairs of simultaneous equations.

   a) $x = 9 + 5y$ b) $3x + 4y = 41$
      $y = x - 1$ $7x - 2y = 5$

3. It takes the earth about 31,500,000 seconds to complete one orbit around the sun and

our solar system about 2.25×10^8 times as long to complete one orbit around the center of our galaxy.

Find the approximate number of seconds that it takes our solar system to travel once around the center of our galaxy. Express your answer in
   a) scientific notation.
   b) decimal form.

Set II

4. Find each of the following products.
   a) $(5x - 5)(2x + 1)$ c) $5(2x + 1)(x - 1)$
   b) $5(x - 1)(2x + 1)$ d) $(10x + 5)(x - 1)$

5. Find each of the following products.
   a) $(x)(2x)(3x)$
   b) $(x)(2x)(x + 3)$
   c) $(x)(x + 2)(x + 3)$
   d) What property do all three polynomials have in common?

6. Factor each of the following polynomials as completely as possible.
   a) $x^3 + 10x$ f) $2x^5 + 4x^4 - 70x^3$
   b) $4x^3 - 16x$ g) $20x^4 - 45x^2$
   c) $x^4 + 10x^3 + 24x^2$ h) $5x^3 + 10x^2 + x$
   d) $3x^3 - 18x^2 + 15x$ i) $x^4 - 1$
   e) $6x^4 + 6x^3 + 18x^2$ j) $3x^5 + 10x^4 + 3x^3$

7. Find the missing factor in each of the following equations.
   a) $165 = (3)(5)(\text{▓▓})$
   b) $3,337 = (\text{▓▓})(71)$
   c) $4x^3 + x^2 = x^2(\text{▓▓})$
   d) $x^2 - 13x + 36 = (x - 4)(\text{▓▓})$
   e) $15x^2 - x - 2 = (\text{▓▓})(5x - 2)$
   f) $x^3 - 1 = (x - 1)(\text{▓▓})$

8. Factor each of the following polynomials as completely as possible.
   a) $x^3 + 2xy$
   b) $x^3y - xy^3$
   c) $2x^2 - 4xy + 2y^2$
   d) $x^3 + 4x^2y + 3xy^2$
   e) $3xy - 6x + 15y - 30$
   f) $x^4 - 8x^2y^2 - 9y^4$

Set III

9. Find each of the following products.
   a) $(3x - 3)(4x + 1)$ c) $3(4x + 1)(x - 1)$
   b) $3(x - 1)(4x + 1)$ d) $(12x + 3)(x - 1)$
   i) $x^4 - 16$
   j) $7x^5 + 50x^4 + 7x^3$

10. Find each of the following products.
    a) $(x + 1)(x - 1)(x^2 + 1)$
    b) $(x + 2)(x - 2)(x^2 + 4)$
    c) $(x + 3)(x - 3)(x^2 + 9)$
    d) What property do all three polynomials have in common?

11. Factor each of the following polynomials as completely as possible.
    a) $x^3 - 12x$
    b) $5x^3 + 10x^2 + 5x$
    c) $x^4 + 13x^3 + 40x^2$
    d) $2x^3 - 98x$
    e) $4x^3 - 16x^2 + 12x$
    f) $7x^4 + 21x^3 - 35x^2$
    g) $3x^5 + 3x^4 - 18x^3$
    h) $12x^4 - 75x^2$

12. Find the missing factor in each of the following equations.
    a) $455 = (5)(7)(\text{▓▓})$
    b) $1,411 = (\text{▓▓})(17)$
    c) $2x^4 + x^3 = x^3(\text{▓▓})$
    d) $x^2 - 11x + 24 = (x - 3)(\text{▓▓})$
    e) $6x^2 + 5x - 4 = (\text{▓▓})(3x + 4)$
    f) $x^3 + 8 = (x + 2)(\text{▓▓})$

13. Factor each of the following polynomials as completely as possible.
    a) $y^3 - 3xy$
    b) $5x^2 - 5y^2$
    c) $x^3y + 2x^2y^2 + xy^3$
    d) $x^3 - 3x^2y + 2xy^2$
    e) $2xy + 6x - 14y - 42$
    f) $x^4 - 3x^2y^2 - 4y^4$

Set IV

Because it is impossible to factor the sum of two squares according to the rules we have established, it might seem as if it would also be impossible to factor the sum of two cubes. Yet such is not the case.

Given the clue that $x^3 + y^3$ can be factored as the product of a binomial and a trinomial, can you figure out what the factors are? If you can, show your method.

Hint:

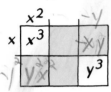

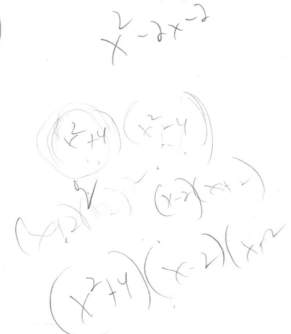

$(x + 2)(x^2 - 2x + 4)$

$x^3 - 2x^2 + 4x$

$2x^2 - 4x + 8$

Summary and Review

In this chapter, we have studied the properties of prime and composite numbers and have learned how to factor integers, monomials, and polynomials.

Prime and Composite Numbers (*Lesson 1*) Every integer larger than 1 is either composite or prime. Integers that can be written as the product of smaller positive integers are composite; those that cannot are prime.

A number is factored when it is written as a product of two or more numbers. The numbers in the product are called factors of the number. Every positive composite number can be written (except for order) as the product of primes in only one way.

The greatest common factor of a set of numbers is the largest integer that is a factor of all of the numbers. The greatest common factor can be found using the prime factors of the numbers. Two numbers whose greatest common factor is 1 are said to be relatively prime.

Monomials and Their Factors (*Lesson 2*) The factors of the monomial ax^n include the factors of the coefficient a, the factors of x^n $(x, x^2, \ldots, x^n)$, and products of various combinations of these factors.

The greatest common factor of a set of monomials is the product of the greatest common factor of their coefficients and the highest power of each variable that is a factor of all of the monomials. Two monomials whose greatest common factor is 1 are called relatively prime.

Polynomials and Their Factors (*Lesson 3*) When we factor a polynomial, we restrict the factors to integers and polynomials having coefficients and constant terms that are integers. The first step in factoring a polynomial is to look for the greatest common factor of its terms and apply the distributive rule in reverse:

$$ab + ac = a(b + c)$$

A prime polynomial is a polynomial that cannot be written as a product of polynomials of lower degree.

Factoring Second-Degree Polynomials (*Lessons 4 and 7*) When two first-degree binomials are multiplied, their product is usually a second-degree trinomial. For example, $(x + a)(x + b) = x^2 + (a + b)x + ab$. To factor a second-degree polynomial, we figure out the factors of its first and last terms and try different combinations of them until we find a combination that will give the middle term of the polynomial.

Factoring the Difference of Two Squares (*Lesson 5*) The difference of the squares of two numbers is equal to the product of the sum and difference of the numbers:

$$a^2 - b^2 = (a + b)(a - b)$$

The sum of two squares cannot be factored algebraically.

Factoring Trinomial Squares (*Lesson 6*) A trinomial of the form $a^2 + 2ab + b^2$ is the square of the binomial $a + b$:

$$a^2 + 2ab + b^2 = (a + b)^2$$

A trinomial of the form $a^2 - 2ab + b^2$ is the square of the binomial $a - b$:

$$a^2 - 2ab + b^2 = (a - b)^2$$

Factoring Higher-Degree Polynomials (*Lesson 8*) Some higher-degree polynomials can be factored by first factoring out the greatest factor common to every term of the polynomial. It may then be possible to factor the remaining factor by the methods we have learned for factoring first- and second-degree polynomials.

Exercises

Set I

1. Can you easily tell whether any of the three numbers in this cartoon are composite? If so, explain how.

2. Factor each of the following numbers into primes. List the prime factors in order from smallest to largest, using exponents where possible.
   a) 112
   b) 185
   c) 1,683
   d) 14^3
   e) $3^4 \cdot 6^4$
   f) $4^3 \cdot 4^6$

3. Find the greatest common factor of each of the following sets of numbers.
   a) 45 and 54
   b) 27 and 127
   c) 501 and 5,001
   d) 30, 45, and 60
   e) 8^3 and 8^4
   f) 3^8 and 4^8

4. List the factors that are positive integers or have positive integral coefficients for each of the following.
   a) 60
   b) $14x$
   c) x^4
   d) $5x^3$

5. Find the greatest common factor for each of the following sets of monomials.
   a) $4x$ and $6x$
   b) $4x$ and x^6
   c) x^4 and x^6
   d) $16x^2$ and $100y^2$
   e) $7xy^2$ and $7x^2y$
   f) $6x^3$, $15x^2$, and $12x$

6. Find the value of each of the following expressions.
   a) $46 \cdot 83 + 54 \cdot 83$
   b) $145^2 - 45^2$
   c) $11 + 79 \cdot 11$
   d) $(120 + 1)(120 - 1)$

7. Find the missing term in each of these trinomial squares.
   a) $16x^2 + 8x + \rule{1cm}{0.4pt}$
   b) $9x^2 - \rule{1cm}{0.4pt} + 49$

8. What should replace ▦ in each of the following equations?
   a) $(4x^2)(▦) = 8x^8$
   b) $(▦)(-5x^3) = 20x^6$
   c) $14x^2 - 21 = (▦)(2x^2 - 3)$
   d) $12x + 6x^4 = (6x)(▦)$
   e) $x^2 + 2x - 35 = (▦)(x + 7)$
   f) $x^2 - 9 = (x - 3)(▦)$
   g) $5x^2 + 23x - 10 = (▦)(x + 5)$
   h) $16x^2 + 8x + 1 = (4x + 1)(▦)$

9. Factor each of the following polynomials as completely as possible.
   a) $6x - 21$
   b) $x^2 + 10x$
   c) $x^2 - 49$
   d) $3x^3 - 2x^2$
   e) $x^2 + 2x + 1$

f) $x^2 + 16x + 39$
g) $x^2 + x - 42$
h) $x^2 + 25$
i) $x^3 - 4x$
j) $16x^2 - 40x + 25$
k) $2x^2 + 15x + 7$
l) $3x^2 + x - 10$
m) $2x^2 + 24x + 72$
n) $x^3 + x^2 - 90x$
o) $5x^3 + 5x$
p) $3x^4 + 9x^3 + 6x^2$

$2(x^2 + 12x + 36)$

10. Factor each of the following polynomials as completely as possible.
   a) $x^2 + 2xy$ e) $4x^2 - 9y^2$
   b) $15x - 5y^3$ f) $x^2 - 2xy^2 + y^4$
   c) $x^2 + 8xy + 7y^2$
   d) $xy - x + 8y - 8$

Set II

1. Factor each of the following numbers into primes. List the prime factors in order from smallest to largest, using exponents where possible.
   a) 143   d) 15^2
   b) 675   e) $2^6 \cdot 10^6$
   c) 1,862   f) $6^2 \cdot 6^{10}$

2. Find the greatest common factor of each of the following sets of numbers.
   a) 14 and 91   e) 7^2 and 7^3
   b) 35 and 135   f) 4^5 and 5^5
   c) 400 and 401
   d) 24, 60, and 84

3. List the factors that are positive integers or have positive integral coefficients for each of the following.
   a) 48   c) x^5
   b) $33x$ d) $9x^2$

4. If a four-digit number is written down twice to form an eight-digit number, the resulting number can be divided evenly by 73 and 137.
   a) Try this with an example.
   b) Explain why it works.

5. Find the greatest common factor for each of the following sets of monomials.
   a) $6x$ and 8   d) $4xy$ and $2x^2y^2$
   b) x^6 and $8x$ e) $5x^3$ and $9y^3$
   c) x^6 and x^8 f) $12x^3$, $18x^2$, and $30x$

6. Find the value of each of the following expressions.
   a) $39 \cdot 56 + 56 \cdot 61$
   b) $(90 + 2)(90 - 2)$
   c) $25 \cdot 81 - 25$
   d) $127^2 - 27^2$

7. Find the missing term in each of these trinomial squares.
   a) $4x^2 - 36x + \text{▨}$
   b) $25x^2 + \text{▨} + 36$

8. What should replace ▨ in each of the following equations?
   a) $(2x^4)(\text{▨}) = 6x^6$
   b) $(\text{▨})(-9x^5) = 18x^{10}$
   c) $24x^2 + 16 = (\text{▨})(3x^2 + 2)$
   d) $40x - 4x^3 = (4x)(\text{▨})$
   e) $x^2 + 5x - 14 = (\text{▨})(x + 7)$
   f) $x^2 + 12x + 36 = (x + 6)(\text{▨})$
   g) $3x^2 + 5x - 28 = (\text{▨})(x + 4)$
   h) $25x^2 - 4 = (5x - 2)(\text{▨})$

9. Factor each of the following polynomials as completely as possible.
   a) $14x + 35$
   b) $x^2 - 8x$
   c) $x^2 - 16$
   d) $4x^4 + 3x^3$

e) $x^2 - 20x + 100$
f) $x^2 + 13x + 22$
g) $x^2 + x - 30$
h) $x^3 - 2x^2 + x$
i) $x^2 + 9$
j) $49x^2 + 28x + 4$
k) $3x^2 + 16x + 5$
l) $2x^2 + x - 28$
m) $3x^2 - 18x + 27$
n) $x^3 + 3x^2 - 40x$
o) $4x^3 + 16x$
p) $2x^4 - 6x^3 + 4x^2$

10. Factor each of the following polynomials as completely as possible.
    a) $3xy + y^3$
    b) $8x - 4y^2$
    c) $x^2 - 10xy + 9y^2$
    d) $xy + 6x - y - 6$
    e) $25x^2 - 4y^2$
    f) $x^4 + 2x^2y + y^2$

Chapter 11
FRACTIONS

LESSON **1**
Fractions

One-half is the simplest example of a fraction. One way to picture $\frac{1}{2}$ is shown in this diagram: if something is divided into two equal parts, each part is called one-half of it.

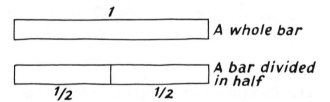

All of the fractions that we first learn about in school can be pictured in the same way. The fraction $\frac{2}{5}$ for example, can be pictured by dividing something into five equal parts and then taking two of them.

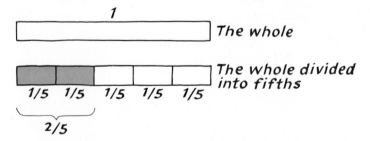

It is for this reason that the top number in a fraction is called its numerator and the bottom number is called its denominator. The denominator tells *what kind*

of part we are taking (fifths in this example) and the numerator tells *how many* of them we are taking (two in the example).

To explain what $\dfrac{a}{b}$ means, in which a and b are positive integers, we simply think of dividing something into b equal parts (each part is called "$\dfrac{1}{b}$th" of the thing) and then taking a of the parts. According to this interpretation, $\dfrac{a}{b}$ means $a\left(\dfrac{1}{b}\right)$.

A more general interpretation of the fraction $\dfrac{a}{b}$ is that it is the indicated *quotient* of the numbers a and b. By using this interpretation, we can represent fractions such as $\dfrac{1}{2}$ or $\dfrac{2}{5}$ in decimal form by carrying out the indicated division:

$$
\begin{array}{r}
0.5 \\
2\overline{)1.0} \\
\underline{1\,0} \\
0
\end{array}
\quad \text{so} \quad \frac{1}{2} = 0.5
\qquad\qquad
\begin{array}{r}
0.4 \\
5\overline{)2.0} \\
\underline{2\,0} \\
0
\end{array}
\quad \text{so} \quad \frac{2}{5} = 0.4
$$

As we have learned from our study of rational numbers, not every fraction has a decimal form as simple as these. The decimal forms of $\dfrac{1}{3}$ and $\dfrac{4}{7}$, for example, never come to an end; so we round them according to the accuracy that we want.

$$
\begin{array}{r}
0.3333\ldots \\
3\overline{)1.0000\ldots} \\
\underline{9} \\
10 \\
\underline{9} \\
10 \\
\underline{9} \\
10 \\
\underline{9} \\
1\ldots
\end{array}
\qquad\qquad
\begin{array}{r}
0.5714\ldots \\
7\overline{)4.0000\ldots} \\
\underline{3\,5} \\
50 \\
\underline{49} \\
10 \\
\underline{7} \\
30 \\
\underline{28} \\
2\ldots
\end{array}
$$

Rounding $\frac{4}{7}$ to the nearest tenth, for example, we get 0.6; to the nearest hundredth, 0.57; to the nearest thousandth, 0.571, and so on.

Because it does not make any sense to divide by 0, it is obvious from looking at a fraction as a quotient that the denominator of a fraction can never be allowed to be 0. This means that, when we write $\frac{a}{b}$, it is understood that $b \neq 0$.

The idea of a fraction as an indicated quotient gives meaning to *all* fractions, not just those consisting of positive integers. A fraction such as $\frac{-2}{-5}$, for example, has meaning even though we cannot divide something into –5 parts and take –2 of them: it means "the number –2 divided by the number –5." Because the quotient of two negative numbers is positive, $\frac{-2}{-5} = \frac{2}{5} = 0.4$. Reasoning in the same way, we can show that, regardless of whether a and b represent positive or negative numbers,

$$\frac{-a}{-b} = \frac{a}{b}, \quad \frac{-a}{b} = -\frac{a}{b}, \quad \text{and} \quad \frac{a}{-b} = -\frac{a}{b}$$

This figure shows that the fractions $\frac{2}{5}$ and $\frac{4}{10}$ represent the same number.

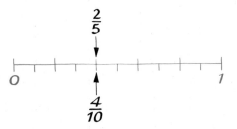

Fractions that represent the same number are called *equivalent*. What fractions other than $\frac{4}{10}$ are equivalent to $\frac{2}{5}$? Some of them are $\frac{6}{15}, \frac{8}{20}, \frac{10}{25},$ and $\frac{12}{30}$. Also, $\frac{-2}{-5}, \frac{-4}{-10}, \frac{-6}{-15},$ and so on. In fact, *any* fraction formed by multiplying the numerator and denominator of a fraction by the same number (other than zero) is equivalent to it.

▶In general, if n is not equal to zero,

$$\frac{na}{nb} = \frac{a}{b}$$

It also follows from this equation that we can *divide* the numerator and denominator of a fraction by the same number other than zero. If the number is a factor of both the numerator and denominator of the fraction, this procedure is called *reducing the fraction to lower terms*. Dividing by the greatest common factor results in a fraction that has been reduced *to lowest terms*. For example, dividing numerator and denominator of $\frac{24}{36}$ by 2 results in an equivalent fraction, $\frac{12}{18}$. Dividing both of them by 12 reduces $\frac{24}{36}$ to lowest terms: $\frac{2}{3}$.

Here are examples of problems about equivalent fractions.

EXAMPLE 1

Write a fraction that is equivalent to $\frac{3}{7}$ and has a denominator of 28.

SOLUTION

Because $28 = 4 \cdot 7$, we multiply both the numerator and the denominator of $\frac{3}{7}$ by 4 to get $\frac{4 \cdot 3}{4 \cdot 7} = \frac{12}{28}$.

EXAMPLE 2

Reduce the fraction $\frac{102}{12}$ to lowest terms.

SOLUTION

Because $102 = 2 \cdot 3 \cdot 17$ and $12 = 2 \cdot 2 \cdot 3$, their greatest common factor is $2 \cdot 3$. $\frac{102}{12} = \frac{6 \cdot 17}{6 \cdot 2} = \frac{17}{2}$. The fraction $\frac{17}{2}$ is in lowest terms because its numerator and denominator have no common factors other than 1.

Exercises

Set I

1. Find each of the following products. Express your answers in scientific notation.
   a) $(2 \times 10^3)(3 \times 10^2)$
   b) $5(2 \times 10^3)$
   c) $(2 \times 10^3)^5$

2. This exercise is about the graph of the equation $3x - y = 12$.
   a) Find the points where it crosses each axis.
   b) Find its slope.
   c) Does the point $(100, 312)$ lie on the graph?

3. If a five-digit number is written down, of which all the digits are the same, the resulting number can be divided evenly by 41 and 271.
   a) Show that this works with an example.
   b) Explain why it works.

Set II

4. Write the coordinate of each lettered point on this number line as a fraction.

5. Use a ruler to draw a number line like the one below. Let 0.5 centimeter represent $\frac{1}{7}$.

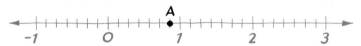

   Mark and label points on the line having the following coordinates.
   (Point A is shown as an example.)

   Point A, $\frac{6}{7}$

   a) Point B, $\frac{10}{7}$

   b) Point C, $\frac{17}{7}$

   c) Point D, $-\frac{4}{7}$

   d) Find the distance between points B and C.
   e) Find the distance between points D and A.

   f) Find the coordinate of the point midway between points A and B.
   g) Find the coordinate of the point midway between points D and A.

6. Express each of the following fractions in decimal form, correct to the nearest hundredth.

   a) $\frac{3}{4}$ d) $\frac{13}{78}$ *g) $\frac{8}{13}$

   b) $\frac{21}{28}$ e) $\frac{2}{11}$ *h) $\frac{17}{7}$ *j) $\frac{123}{246}$

   c) $\frac{1}{6}$ f) $\frac{11}{2}$ *i) $\frac{135}{246}$

7. Between which integers do the following fractions lie?

a) $\dfrac{3}{7}$ c) $-\dfrac{8}{9}$ e) $\dfrac{100}{11}$

b) $\dfrac{7}{3}$ d) $-\dfrac{9}{8}$ f) $\dfrac{1000}{11}$

8. Reduce each of the following fractions to lowest terms.

a) $\dfrac{6}{12}$ c) $\dfrac{15}{51}$ e) $\dfrac{33}{303}$ g) $\dfrac{175}{225}$

b) $\dfrac{8}{12}$ d) $\dfrac{51}{15}$ f) $\dfrac{33}{3003}$ h) $\dfrac{1075}{2025}$

9. Change the following sets of fractions to fractions that have the same denominator.

a) $\dfrac{4}{5}$ and $\dfrac{9}{10}$ d) $\dfrac{5}{6}$ and $\dfrac{5}{8}$

b) $\dfrac{3}{7}$ and $\dfrac{7}{3}$ e) $\dfrac{1}{3}, \dfrac{1}{4},$ and $\dfrac{1}{5}$

c) $\dfrac{1}{24}$ and $\dfrac{1}{36}$ f) $\dfrac{3}{4}, \dfrac{7}{8},$ and $\dfrac{15}{16}$

10. Arrange the following sets of fractions in order of increasing size.

a) $\dfrac{3}{7}, \dfrac{2}{7}, \dfrac{4}{7}$ c) $\dfrac{1}{3}, \dfrac{3}{10}, \dfrac{11}{30}$

b) $\dfrac{7}{3}, \dfrac{7}{2}, \dfrac{7}{4}$ d) $\dfrac{5}{2}, \dfrac{11}{4}, \dfrac{19}{8}$

11. Which fraction in each of the following sets is not equal to the others?

a) $\dfrac{-3}{-7}, \dfrac{3}{7}, -\dfrac{3}{7}$ b) $\dfrac{-8}{5}, -\dfrac{8}{5}, \dfrac{8}{-5}, \dfrac{-8}{-5}$

c) $\dfrac{1}{2}, \dfrac{-1}{-2}, -\dfrac{-1}{-2}$ d) $-\dfrac{-4}{9}, \dfrac{4}{-9}, -\dfrac{4}{-9}$

12. If possible, express each of the following fractions as an integer.

a) $\dfrac{8+4}{8-4}$ c) $\dfrac{3+9}{3-9}$

b) $\dfrac{5-5}{5+5}$ d) $\dfrac{2+2}{2-2}$

13. Find the values of the following expressions for the values of x indicated. Express each answer as an integer.

$$\dfrac{10+x}{2+x} \qquad \dfrac{10x}{2x}$$

a) $x = 2$ f) $x = 2$
b) $x = 6$ g) $x = 6$
c) $x = 0$ h) Why can't a value be
d) $x = -1$ found for $x = 0$?
e) $x = -3$ i) $x = -1$
 j) $x = -3$

14. What number should replace ▓ in each of the following equations to make it true?

a) $\dfrac{2}{3} = \dfrac{8}{▓}$ e) $\dfrac{2+4}{▓+4} = \dfrac{2}{15}$

b) $\dfrac{▓}{25} = \dfrac{7}{5}$ f) $\dfrac{2+2}{▓+2} = \dfrac{2}{15}$

c) $\dfrac{9}{▓} = \dfrac{99}{154}$ g) $\dfrac{6 \cdot 3}{8 \cdot ▓} = \dfrac{6}{8}$

d) $\dfrac{2 \cdot 4}{▓ \cdot 4} = \dfrac{2}{15}$ h) $\dfrac{6-3}{8-▓} = \dfrac{6}{8}$

Set III

15. Write the coordinate of each lettered point on this number line as a fraction.

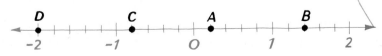

16. Use a ruler to draw a number line like the one below. Let 0.5 centimeter represent $\frac{1}{3}$.

Mark and label points on the line having the following coordinates.
(Point A is shown as an example.)

  Point A, $\frac{4}{3}$

a) Point B, $\frac{8}{3}$

b) Point C, $\frac{13}{3}$

c) Point D, $-\frac{6}{3}$

d) Find the distance between points B and C.
e) Find the distance between points D and A.
f) Find the coordinate of the point midway between points A and B.
g) Find the coordinate of the point midway between points D and A.

17. Express each of the following fractions in decimal form, correct to the nearest hundredth.

a) $\frac{1}{4}$

b) $\frac{9}{36}$

c) $\frac{2}{3}$

d) $\frac{34}{51}$

e) $\frac{5}{11}$

f) $\frac{11}{5}$

*g) $\frac{4}{17}$

*h) $\frac{19}{6}$

*i) $\frac{321}{654}$

*j) $\frac{321}{963}$

18. Between which integers do the following fractions lie?

a) $\frac{2}{9}$

b) $\frac{9}{2}$

c) $-\frac{4}{7}$

d) $-\frac{7}{4}$

e) $\frac{50}{3}$

f) $\frac{500}{3}$

19. Reduce each of the following fractions to lowest terms.

a) $\frac{5}{15}$

b) $\frac{6}{15}$

c) $\frac{27}{72}$

d) $\frac{72}{27}$

e) $\frac{7}{707}$

f) $\frac{77}{7007}$

g) $\frac{15}{245}$

h) $\frac{105}{245}$

20. Change the following sets of fractions to fractions that have the same denominator.

a) $\frac{3}{4}$ and $\frac{5}{8}$

b) $\frac{2}{9}$ and $\frac{9}{2}$

c) $\frac{1}{10}$ and $\frac{1}{15}$

d) $\frac{7}{8}$ and $\frac{7}{6}$

e) $\frac{1}{2}, \frac{1}{3}$, and $\frac{1}{7}$

f) $\frac{2}{3}, \frac{5}{6}$, and $\frac{11}{12}$

21. Arrange the following sets of fractions in order of increasing size.

a) $\frac{8}{11}, \frac{9}{11}, \frac{7}{11}$

b) $\frac{11}{8}, \frac{11}{9}, \frac{11}{7}$

c) $\frac{1}{2}, \frac{11}{20}, \frac{101}{200}$

d) $\frac{2}{3}, \frac{3}{5}, \frac{8}{15}$

22. Which fraction in each of the following sets is not equal to the others?

a) $\dfrac{2}{9}, -\dfrac{2}{9}, \dfrac{-2}{-9}$

c) $\dfrac{5}{8}, \dfrac{-5}{-8}, -\dfrac{-5}{-8}$

b) $\dfrac{-7}{6}, \dfrac{7}{-6}, \dfrac{-7}{-6}, -\dfrac{7}{6}$

d) $-\dfrac{1}{-3}, \dfrac{-1}{3}, -\dfrac{-1}{3}$

23. If possible, express each of the following fractions as an integer.

a) $\dfrac{6+2}{6-2}$

c) $\dfrac{4+8}{4-8}$

b) $\dfrac{3-3}{3+3}$

d) $\dfrac{7+7}{7-7}$

24. Find the values of the following expressions for the values of x indicated. Express each answer as an integer.

$$\dfrac{12+x}{2+x} \qquad \dfrac{12x}{2x}$$

a) $x = 3$ f) $x = 3$
b) $x = 8$ g) $x = 8$
c) $x = 0$ h) Why can't a value be found for $x = 0$?
d) $x = -1$ i) $x = -1$
e) $x = -3$ j) $x = -3$

25. What number should replace 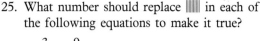 in each of the following equations to make it true?

a) $\dfrac{3}{8} = \dfrac{9}{\text{▓}}$

b) $\dfrac{\text{▓}}{36} = \dfrac{5}{6}$

c) $\dfrac{7}{\text{▓}} = \dfrac{77}{132}$

d) $\dfrac{2 \cdot 6}{\text{▓} \cdot 6} = \dfrac{2}{9}$

e) $\dfrac{2+6}{\text{▓}+6} = \dfrac{2}{9}$

f) $\dfrac{2+2}{\text{▓}+2} = \dfrac{2}{9}$

g) $\dfrac{8 \cdot 4}{6 \cdot \text{▓}} = \dfrac{8}{6}$

h) $\dfrac{8-4}{6-\text{▓}} = \dfrac{8}{6}$

Set IV

The English poet Samuel Taylor Coleridge enjoyed mathematical puzzles. The following one appeared in one of his notebooks.

"Go into an orchard in which there are three gates through all of which you must pass. Take a certain number of apples. Give half of them and half an apple to the man standing by the first gate. Give half of what remain and half an apple to the man by the second gate. Give half of what remain and half an apple to the man by the third gate."

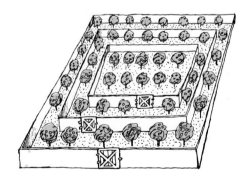

What is the smallest number of apples you could start with and yet never cut one apple? (Hint: The answer is less than 12.)

LESSON 2
Algebraic Fractions

It took a long time for the Hindu-Arabic numerals that we use to write numbers to be accepted throughout the world. The author of a popular arithmetic book published in Germany in the sixteenth century used Roman numerals because he thought his readers would find them easier to understand. This excerpt from the book shows how fractions were represented in it. You may recognize the four fractions at the left: they are $\frac{1}{4}$, $\frac{6}{8}$, $\frac{9}{11}$, and $\frac{20}{31}$.

In Lesson 1, we reviewed two ways to interpret fractions and learned how to transform a given fraction into other fractions that represent the same number. The table below shows, for example, how the fraction $\frac{1}{4}$ can be transformed into other fractions equivalent to it.

If the numerator and denominator of $\frac{1}{4}$ are							
both multiplied by	2	3	10	25	–1	–5	. . .
the fraction that results is	$\frac{2}{8}$	$\frac{3}{12}$	$\frac{10}{40}$	$\frac{25}{100}$	$\frac{-1}{-4}$	$\frac{-5}{-20}$	. . .

All of the fractions that result are equal. Each one has the decimal form 0.25.

The principle that we are applying is so basic to working with fractions that, even though it was stated in Lesson 1, it is repeated in symbols here.

▶ In general, if n is not equal to zero, $\dfrac{na}{nb} = \dfrac{a}{b}$.

We have also already observed that this principle can be used to reduce a fraction by dividing its numerator and denominator by the same number.

In this lesson, we will apply these ideas to the fractions of algebra: fractions that contain one or more variables in their numerator and/or denominator. Examples of algebraic fractions are

$$\frac{x}{2}, \quad \frac{5}{x}, \quad \text{and} \quad \frac{3x}{x+1}$$

The denominator of a fraction cannot be allowed to be zero. Thus, when we write a fraction such as $\dfrac{5}{x}$, the x may be replaced by any number except zero.

For the same reason, in the fraction $\dfrac{3x}{x+1}$, x may be replaced by any number except -1.

The only difference between algebraic fractions and the fractions of arithmetic is that an algebraic fraction contains one or more variables. Their algebraic properties are identical. This means that algebraic fractions, like the fractions of arithmetic, can be transformed into equivalent fractions by multiplying or dividing their numerators and denominators by the same number. For example, the fractions $\dfrac{3x}{x+1}$ and $\dfrac{6x}{2x+2}$ are equivalent because

$$\frac{3x}{x+1} = \frac{2(3x)}{2(x+1)} = \frac{6x}{2x+2}$$

Whenever the numerator and denominator of an algebraic fraction have a common factor other than 1, both numerator and denominator can be divided by that factor to get a simpler fraction. Like the fractions of arithmetic, an algebraic fraction is said to be reduced to lowest terms when its numerator and denominator have no common factor other than 1. The fraction $\dfrac{3x-6}{5x-10}$, for example, can be reduced to the fraction $\frac{3}{5}$ by factoring its numerator and denominator as shown below and dividing both by the common factor:

$$\frac{3x - 6}{5x - 10} = \frac{3(x - 2)}{5(x - 2)} = \frac{3}{5}$$

When we divide numerator and denominator by their common factor, $x - 2$, we are assuming that x is not equal to 2. If it were, $x - 2$ would be equal to 0 and we would be dividing by zero. The possibility that x is equal to 2 is ruled out by the original fraction. Because its denominator, $5x - 10$, cannot be zero, x cannot be equal to 2. When we write $\frac{3x - 6}{5x - 10} = \frac{3}{5}$, then, it is understood that the two fractions are equal for *all values of* x *except 2.*

Here are more examples of problems about algebraic fractions.

EXAMPLE 1

Write a fraction that is equivalent to $\frac{8}{x}$ and has a numerator of $8x$.

SOLUTION

Because $8x = x \cdot 8$, we multiply both the numerator and the denominator of $\frac{8}{x}$ by x

to get $\frac{x \cdot 8}{x \cdot x} = \frac{8x}{x^2}$.

EXAMPLE 2

Reduce the fraction $\frac{x - 1}{x^2 - 1}$ to lowest terms.

SOLUTION

Factoring the numerator and denominator of this fraction and dividing both by the common factor, we get

$$\frac{x - 1}{x^2 - 1} = \frac{1(x - 1)}{(x + 1)(x - 1)} = \frac{1}{x + 1}$$

When we divide numerator and denominator by $x - 1$, we assume that x is not equal to 1. If it were, we would be dividing by zero. The possibility that x

is equal to 1 is ruled out by the original fraction. Because its denominator, $x^2 - 1$, cannot be zero, x cannot be equal to 1.

EXAMPLE 3

If possible, reduce the fraction $\dfrac{2x + 5}{2x}$ to lower terms.

SOLUTION

The numerator of this fraction cannot be factored in any way other than

$1(2x + 5)$: $\dfrac{1(2x + 5)}{2x}$. This means that the numerator and denominator have no

common factor other than 1, and so $\dfrac{2x + 5}{2x}$ cannot be reduced to lower terms.

EXAMPLE 4

Change $\dfrac{3}{x}$ and $\dfrac{x}{6}$ to equivalent fractions that have the same denominator.

SOLUTION

$$\frac{3 \cdot 6}{x \cdot 6} = \frac{18}{6x} \quad \text{and} \quad \frac{x \cdot x}{6 \cdot x} = \frac{x^2}{6x}$$

The fractions are $\dfrac{18}{6x}$ and $\dfrac{x^2}{6x}$.

Exercises

Set I

1. Write each of the following as a power of x. (Assume that x is not zero.)

   a) x b) 1 c) $\dfrac{1}{x}$

2. Perform the operations indicated.
   a) Add $x^2 - x - 2$ and $x - 2$.
   b) Subtract $x - 2$ from $x^2 - x - 2$.
   c) Multiply $x^2 - x - 2$ by $x - 2$.
   d) Divide $x^2 - x - 2$ by $x - 2$.

3. For a family picnic, Mrs. Klutz made 12 sandwiches.
   a) Write a formula for the number of sandwiches, y, that each person gets if there are x people in the family and the sandwiches are shared equally.
   b) What happens to y if x decreases?
   c) How does y vary with respect to x?

"All the more for us."

Set II

4. The denominator of a fraction cannot be allowed to be zero. What values of x, if any, would make the denominator of each of the following fractions equal to zero?

   a) $\dfrac{2}{x}$

   b) $\dfrac{3}{x-1}$

   c) $\dfrac{x}{x+5}$

   d) $\dfrac{2x-1}{7}$

   e) $\dfrac{1}{x^2-4}$

   f) $\dfrac{1}{x^2+4}$

5. Reduce each of the following fractions to lowest terms.

   a) $\dfrac{6x}{16}$

   b) $\dfrac{9x}{9y}$

   c) $\dfrac{x^2}{4x}$

   d) $\dfrac{x}{x^3}$

   e) $\dfrac{xy^2}{x^2y}$

   f) $\dfrac{5(x+1)}{10}$

   g) $\dfrac{x^2}{x(x-2)}$

   h) $\dfrac{3(x+7)}{(x+7)(x+3)}$

6. Change the following sets of fractions to fractions that have the same denominator.

   a) $\dfrac{x}{5}$ and $\dfrac{x}{10}$

   b) $\dfrac{2}{3x}$ and $\dfrac{6}{x}$

   c) $\dfrac{8}{x}$ and $\dfrac{2}{x^2}$

   d) $\dfrac{x}{9}$ and $\dfrac{3}{x}$

   e) $\dfrac{1}{x}$ and $\dfrac{1}{y}$

   f) $\dfrac{2}{x+2}$ and $\dfrac{4}{x+4}$

   g) $\dfrac{2x}{5}$, $\dfrac{5x}{2}$, and $\dfrac{x}{10}$

   h) $\dfrac{3}{x}$, $\dfrac{4}{x^2}$, and $\dfrac{5}{x^3}$

7. If possible, express each of the following fractions as an integer.

   a) $\dfrac{2x+8x}{x}$

   b) $\dfrac{2x+8}{x+4}$

   c) $\dfrac{5x+3x}{5x-3x}$

   d) $\dfrac{5x-3}{5x-3}$

   e) $\dfrac{5x-3}{3-5x}$

   f) $\dfrac{3+5x}{3-5x}$

8. What should replace ▥ in each of the following equations to make it true for all allowable values of x?

a) $\dfrac{7}{x} = \dfrac{21}{▥}$

b) $\dfrac{5x}{x^5} = \dfrac{▥}{x^4}$

c) $\dfrac{3}{x+3} = \dfrac{6}{▥}$

d) $\dfrac{2}{9} = \dfrac{▥}{45x}$

e) $\dfrac{x-1}{x-8} = \dfrac{4x-4}{▥}$

f) $\dfrac{1}{x-5} = \dfrac{▥}{x^2-25}$

9. If possible, reduce each of the following fractions.

a) $\dfrac{x}{3x}$

b) $\dfrac{x}{x+3}$

c) $\dfrac{8x}{2x-4}$

d) $\dfrac{x+12}{x+6}$

e) $\dfrac{4x+8}{5x+10}$

f) $\dfrac{x-4}{3x-12}$

g) $\dfrac{7x}{7x+49}$

h) $\dfrac{2x+2}{x^2-1}$

i) $\dfrac{x^3+x^2}{x+x^2}$

j) $\dfrac{3x+6}{2x+x^2}$

k) $\dfrac{x+4}{x^2-16}$

l) $\dfrac{5x-25}{x^2-10x+25}$

10. Find the values of the following fractions for the values of x indicated. Express each answer in decimal form.

$$\dfrac{x^2+3x-10}{5x-10} \qquad \dfrac{x+5}{5}$$

a) $x = 0$

b) $x = 3$

c) $x = 10$

d) $x = 0$

e) $x = 3$

f) $x = 10$

g) Compare your answers for parts a, b, and c with those for parts d, e, and f. Do you think that you would get the same results for all other values of x? Explain.

Set III

11. The denominator of a fraction cannot be allowed to be zero. What value of x, if any, would make the denominator of each of the following fractions equal to zero?

a) $\dfrac{5}{x}$

b) $\dfrac{2}{x-4}$

c) $\dfrac{x+1}{8}$

d) $\dfrac{x-6}{x+3}$

e) $\dfrac{1}{x^2-1}$

f) $\dfrac{1}{x^2+1}$

12. Reduce each of the following fractions to lowest terms.

a) $\dfrac{8x}{14}$

b) $\dfrac{10y}{5x}$

c) $\dfrac{3x}{x^2}$

d) $\dfrac{x}{x^4}$

e) $\dfrac{x^2y}{xy^2}$

f) $\dfrac{2(x-1)}{12}$

g) $\dfrac{x^2}{x(x+3)}$

h) $\dfrac{6(x-5)}{(x-5)(x+6)}$

13. Change the following sets of fractions to fractions that have the same denominator.

a) $\dfrac{x}{9}$ and $\dfrac{x}{3}$

b) $\dfrac{2}{x}$ and $\dfrac{5}{2x}$

c) $\dfrac{3}{x^2}$ and $\dfrac{7}{x}$

d) $\dfrac{4}{x}$ and $\dfrac{x}{2}$

e) $\dfrac{1}{2x}$ and $\dfrac{1}{3x}$

f) $\dfrac{8}{x+8}$ and $\dfrac{1}{x+1}$

g) $\dfrac{x}{3}, \dfrac{3x}{5},$ and $\dfrac{x}{15}$

h) $\dfrac{2}{x^3}, \dfrac{3}{x^2},$ and $\dfrac{4}{x}$

14. If possible, express each of the following fractions as an integer.

a) $\dfrac{3x+6x}{x}$

b) $\dfrac{3x+6}{x+2}$

c) $\dfrac{7x+5x}{7x-5x}$

d) $\dfrac{7x+5}{7x+5}$

e) $\dfrac{7x+5}{7x-5}$

f) $\dfrac{7x-5}{5-7x}$

15. What should replace ▥ in each of the following equations to make it true for all allowable values of x?

a) $\dfrac{3}{x} = \dfrac{24}{▥}$

b) $\dfrac{4x}{x^4} = \dfrac{▥}{x^3}$

c) $\dfrac{2}{x+2} = \dfrac{6}{▥}$

d) $\dfrac{3}{5} = \dfrac{▥}{35x}$

e) $\dfrac{x+1}{x-4} = \dfrac{4x+4}{▥}$

f) $\dfrac{1}{x+3} = \dfrac{▥}{x^2-9}$

16. If possible, reduce each of the following fractions.

a) $\dfrac{x}{4x}$

b) $\dfrac{x}{x+4}$

c) $\dfrac{2x}{6x-2}$

d) $\dfrac{x-20}{x-5}$

e) $\dfrac{3x+12}{2x+8}$

f) $\dfrac{x-2}{7x-14}$

g) $\dfrac{5x}{5x-55}$

h) $\dfrac{x^2-9}{3x-9}$

i) $\dfrac{x^2-x^3}{x^4}$

j) $\dfrac{x^2+3x}{3+3x}$

k) $\dfrac{x-1}{x^2-1}$

l) $\dfrac{6x+36}{x^2+12x+36}$

17. Find the values of the following fractions for the values of x indicated. Express each answer in decimal form.

$$\dfrac{x^2+5x+6}{10x+20} \qquad \dfrac{x+3}{10}$$

a) $x = 0$

b) $x = 2$

c) $x = 7$

d) $x = 0$

e) $x = 2$

f) $x = 7$

g) Compare your answers for parts a, b, and c with those for parts d, e, and f. Do you think that you would get the same results for all other values of x? Explain.

Set IV

Because the denominator of a fraction cannot be zero, the values of the following fractions cannot be found if $x = 1$.

$$\dfrac{x^2-1}{x-1} \qquad \dfrac{x^3-1}{x-1} \qquad \dfrac{x^4-1}{x-1}$$

If each fraction is reduced, however, the values of the resulting expressions *can* be found if $x = 1$.

1. Do this for each fraction, showing your work. (Hint: Remember that a fraction is an indicated division.)

2. If the fraction $\dfrac{x^{10}-1}{x-1}$ were reduced, what do you think the value of the resulting expression would be if $x = 1$?

"*You're probably all wondering why I called you here today.*"

LESSON 3
Adding and Subtracting Fractions

If you ate two-fifths of a pie for lunch and one-fifth of it at dinner, what fraction of the pie would you have eaten altogether? The answer is clearly three-fifths of the pie, as the diagram below illustrates.

$$\underset{\textbf{\textit{Lunch}}}{\bigcirc} \quad + \quad \underset{\textbf{\textit{Dinner}}}{\bigcirc} \quad = \quad \bigcirc$$

$$\frac{2}{5} \quad + \quad \frac{1}{5} \quad = \quad \frac{3}{5}$$

This example illustrates the principle that, to add two fractions that have the same denominator, we add the numerators and write the sum over the denominator:

$$\frac{a}{b} + \frac{c}{b} = \frac{a + c}{b}$$

513

Now suppose instead that you ate half of the pie for lunch and a third of it at dinner. What fraction of the pie would you have eaten in this case? The next two diagrams illustrate the problem and how it is solved.

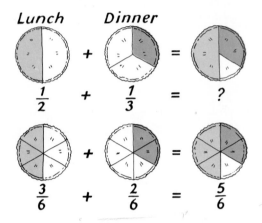

From this example, we see that, to add two fractions that have different denominators, we first replace one or both of the fractions with equivalent ones so that the new fractions have the same denominator. Then we proceed as before:

$$\frac{a}{b} + \frac{c}{d} = \frac{ad}{bd} + \frac{bc}{bd} = \frac{ad + bc}{bd}$$

Here are examples illustrating how these principles are used to add fractions.

EXAMPLE 1

Add $\frac{1}{8}$ and $\frac{3}{8}$.

SOLUTION

Because these fractions have the same denominator,

$$\frac{1}{8} + \frac{3}{8} = \frac{1 + 3}{8} = \frac{4}{8} = \frac{1}{2}$$

EXAMPLE 2

Add $\frac{7x}{12}$ and $\frac{x}{12}$.

SOLUTION

Because these fractions have the same denominator,

$$\frac{7x}{12} + \frac{x}{12} = \frac{7x + x}{12} = \frac{8x}{12} = \frac{2x}{3}$$

EXAMPLE 3

Add $\frac{2}{5}$ and $\frac{2}{9}$.

SOLUTION

Because these fractions do not have the same denominator, we first replace them with equivalent fractions that do.

$$\frac{9 \cdot 2}{9 \cdot 5} = \frac{18}{45} \quad \text{and} \quad \frac{5 \cdot 2}{5 \cdot 9} = \frac{10}{45}$$

and so

$$\frac{2}{5} + \frac{2}{9} = \frac{18}{45} + \frac{10}{45} = \frac{18 + 10}{45} = \frac{28}{45}$$

EXAMPLE 4

Add $\frac{x}{4}$ and $\frac{x}{3}$.

SOLUTION

$$\frac{3 \cdot x}{3 \cdot 4} = \frac{3x}{12} \quad \text{and} \quad \frac{4 \cdot x}{4 \cdot 3} = \frac{4x}{12}$$

and so

$$\frac{x}{4} + \frac{x}{3} = \frac{3x}{12} + \frac{4x}{12} = \frac{3x + 4x}{12} = \frac{7x}{12}$$

Fractions are subtracted by the same method that they are added. These diagrams show, for example, that the problem $\frac{1}{2} - \frac{1}{3}$ can be solved by changing it to the problem $\frac{3}{6} - \frac{2}{6}$.

A couple of examples of how fractions are subtracted are given on the next page.

$$\frac{1}{2} - \frac{1}{3} = \ ?$$

$$\frac{3}{6} - \frac{2}{6} = \frac{1}{6}$$

EXAMPLE 5

Subtract $\frac{2}{7}$ from $\frac{1}{3}$.

SOLUTION

$$\frac{3 \cdot 2}{3 \cdot 7} = \frac{6}{21} \quad \text{and} \quad \frac{7 \cdot 1}{7 \cdot 3} = \frac{7}{21}$$

and so

$$\frac{1}{3} - \frac{2}{7} = \frac{7}{21} - \frac{6}{21} = \frac{1}{21}$$

EXAMPLE 6

From $\frac{x}{2}$, subtract $\frac{x-1}{4}$.

SOLUTION

$$\frac{2 \cdot x}{2 \cdot 2} = \frac{2x}{4}$$

$$\frac{x}{2} - \frac{x-1}{4} = \frac{2x}{4} - \frac{x-1}{4} = \frac{2x - (x-1)}{4} = \frac{2x - x + 1}{4} = \frac{x+1}{4}$$

Although the main purpose of this lesson has been to learn how to add and subtract fractions written as quotients, it is important to realize that arithmetical fractions can also be added and subtracted by changing them to decimal form and adding or subtracting their decimal forms. For example, the problem

$$\frac{2}{5} + \frac{1}{5} = \frac{3}{5}$$

written in decimal form is

$$0.4 + 0.2 = 0.6$$

This is the form in which problems with fractions are done on a calculator.

Exercises

Set I

1. If possible, factor each of the following numbers into primes.
   a) 275   b) 331   c) 5,022

2. Find every pair of positive integers that can replace x and y in each of the following equations to make it true.
   a) $3x + 5y = 27$
   b) $xy + 1 = 40$

3. The world's fastest regularly scheduled train travels between the cities of Osaka and Okayama in Japan, a distance of 112 miles, in just one hour.
   Traveling at this rate, how long would it take it to catch up with another train 8 miles ahead of it if the other train is traveling 80 miles per hour?

UNITED PRESS INTERNATIONAL PHOTO.

Set II

4. Find the following sums and differences, simplifying your answers as much as possible.

   a) $\dfrac{4}{7} + \dfrac{3}{7}$

   b) $\dfrac{8}{15} - \dfrac{2}{15}$

   c) $\dfrac{x}{6} + \dfrac{x}{6}$

   d) $\dfrac{6}{x} + \dfrac{6}{x}$

   e) $\dfrac{8x}{5} + \dfrac{2x}{5}$

   f) $\dfrac{8 + x}{5} + \dfrac{2 + x}{5}$

   g) $\dfrac{5x}{4} - \dfrac{x}{4}$

   h) $\dfrac{5 + x}{4} - \dfrac{x}{4}$

   i) $\dfrac{7x + 1}{8} + \dfrac{3x - 1}{8}$

   j) $\dfrac{7x + 1}{x} + \dfrac{3x - 1}{x}$

   k) $\dfrac{x - 10}{10} + \dfrac{x - 10}{10}$

   l) $\dfrac{10}{x - 10} + \dfrac{10}{x - 10}$

   m) $\dfrac{x}{x - 10} + \dfrac{10}{x - 10}$

   n) $\dfrac{x}{x - 10} - \dfrac{10}{x - 10}$

   o) $\dfrac{3x}{x + 2} - \dfrac{x}{x + 2}$

   p) $\dfrac{x}{x + 2} - \dfrac{x + 3}{x + 2}$

   q) $\dfrac{x^2 + 1}{x^2} - \dfrac{1}{x^2}$

   r) $\dfrac{x^2}{x - 1} - \dfrac{x}{x - 1}$

5. Find the following sums and differences, simplifying your answers as much as possible.

   a) $\dfrac{1}{5} + \dfrac{3}{10}$

   b) $\dfrac{3}{4} - \dfrac{1}{6}$

   c) $\dfrac{2}{x} + \dfrac{1}{4x}$

   d) $\dfrac{x}{7} - \dfrac{x}{11}$

   e) $\dfrac{x^3}{3} - \dfrac{x^2}{2}$

   f) $\dfrac{3}{x^3} - \dfrac{2}{x^2}$

   g) $\dfrac{x}{5} + \dfrac{5}{x}$

   h) $\dfrac{x + 4}{2} + \dfrac{x - 8}{4}$

   i) $\dfrac{5}{2x + 6} - \dfrac{2}{x + 3}$

   j) $\dfrac{x + 3}{2} - \dfrac{2x + 6}{5}$

   k) $\dfrac{x}{x + 1} + \dfrac{x}{x - 1}$

   l) $\dfrac{x + 4}{x - 4} - \dfrac{x - 4}{x + 4}$

6. Write each of the following as an integer or as a single fraction in simplest form.

a) $\dfrac{1}{x} + \dfrac{1}{y}$

b) $\dfrac{x+y}{y} + \dfrac{x-y}{y}$

c) $\dfrac{y}{x+y} + \dfrac{y}{x-y}$

d) $\dfrac{x}{y-x} - \dfrac{y}{y-x}$

e) $\dfrac{y-x}{x} - \dfrac{y-x}{y}$

Set III

7. Find the following sums and differences, simplifying your answers as much as possible.

a) $\dfrac{3}{4} + \dfrac{9}{4}$

b) $\dfrac{7}{10} - \dfrac{3}{10}$

c) $\dfrac{x}{8} + \dfrac{x}{8}$

d) $\dfrac{8}{x} + \dfrac{8}{x}$

e) $\dfrac{2x}{7} + \dfrac{5x}{7}$

f) $\dfrac{2+x}{7} + \dfrac{5+x}{7}$

g) $\dfrac{13x}{6} - \dfrac{x}{6}$

h) $\dfrac{13+x}{6} - \dfrac{x}{6}$

i) $\dfrac{5x+1}{10} + \dfrac{3x-1}{10}$

j) $\dfrac{5x+1}{x} + \dfrac{3x-1}{x}$

k) $\dfrac{x+4}{4} + \dfrac{x+4}{4}$

l) $\dfrac{4}{x+4} + \dfrac{4}{x+4}$

m) $\dfrac{x}{x+4} + \dfrac{4}{x+4}$

n) $\dfrac{x}{x+4} - \dfrac{4}{x+4}$

o) $\dfrac{8x}{x+7} - \dfrac{x}{x+7}$

p) $\dfrac{x}{x+7} - \dfrac{x+8}{x+7}$

q) $\dfrac{x^2-1}{x^2} + \dfrac{1}{x^2}$

r) $\dfrac{x^2}{x+1} + \dfrac{x}{x+1}$

e) $\dfrac{x^2}{2} + \dfrac{x^5}{5}$

f) $\dfrac{2}{x^2} + \dfrac{5}{x^5}$

g) $\dfrac{8}{x} - \dfrac{x}{8}$

h) $\dfrac{x+2}{6} + \dfrac{x-1}{2}$

i) $\dfrac{10}{3x+12} - \dfrac{3}{x+4}$

j) $\dfrac{x+4}{3} - \dfrac{3x+12}{10}$

k) $\dfrac{x}{x-1} - \dfrac{x}{x+2}$

l) $\dfrac{x-5}{x+5} + \dfrac{x+5}{x-5}$

8. Find the following sums and differences, simplifying your answers as much as possible.

a) $\dfrac{1}{4} + \dfrac{1}{12}$

b) $\dfrac{8}{9} - \dfrac{5}{6}$

c) $\dfrac{4}{x} + \dfrac{1}{2x}$

d) $\dfrac{x}{3} - \dfrac{x}{7}$

9. Write each of the following as an integer or as a single fraction in simplest form.

a) $\dfrac{x}{y} + \dfrac{y}{x}$

b) $\dfrac{x-y}{x} + \dfrac{x+y}{x}$

c) $\dfrac{x}{x-y} + \dfrac{x}{x+y}$

d) $\dfrac{y}{x-y} - \dfrac{x}{x-y}$

e) $\dfrac{x-y}{y} - \dfrac{x-y}{x}$

Set IV

Obtuse Ollie forgot how to add fractions. When asked to find the sum

$$\frac{2}{3} + \frac{3}{4}$$

he added the numerators and added the denominators, getting $\frac{5}{7}$.

1. What is the correct sum of these fractions?
2. How does the correct answer compare in size with each of the fractions being added?

3. How does $\frac{5}{7}$ compare in size with each fraction being added?
4. Make up another problem in which the two fractions to be added are equal to each other. Show what Ollie would get by his method and also show the correct answer.
5. How do the correct answer and the answer given by Ollie's method compare in size with each of the fractions in your problem?

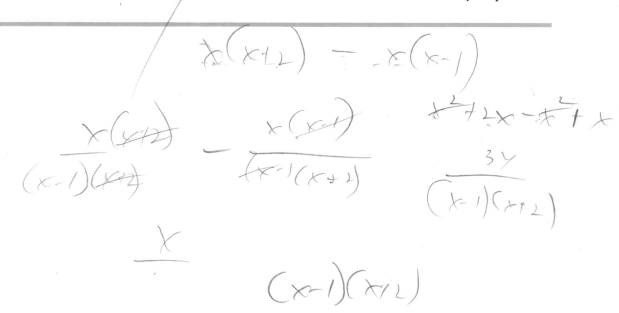

Alfred Tennyson

Charles Babbage

LESSON **4**
More on Addition and Subtraction

The nineteenth-century English poet Alfred Tennyson once wrote

> Every minute dies a man,
> Every minute one is born.

When Charles Babbage, a mathematician of the time, read these words, he wrote to Tennyson saying that, if this were true, the world's population would remain constant. In fact, however, it was increasing. With tongue in cheek, Babbage suggested that the lines be corrected to read:

> Every minute dies a man,
> And one and a sixth is born.

Although this sounds strange and is not very poetic, it does make sense. If the number $1\frac{1}{6}$ is written as a fraction, it is $\frac{7}{6}$. This means that for every six people that die, seven are born. (This is illustrated by the figure below.)

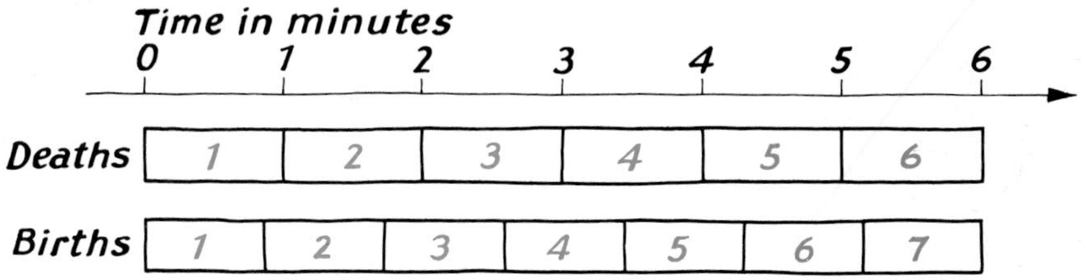

Time in minutes

	0	1	2	3	4	5	6

Deaths

1	2	3	4	5	6

Births

1	2	3	4	5	6	7

When we write a number such as $1\frac{1}{6}$, we do not mean 1 times $\frac{1}{6}$, but rather $1 + \frac{1}{6}$. A number written as the sum of an integer and a fraction is called a *mixed number.* Because $1 = \frac{6}{6}$,

$$1 + \frac{1}{6} = \frac{6}{6} + \frac{1}{6} = \frac{6+1}{6} = \frac{7}{6}$$

This method can be applied to expressing the sum or difference of any integer and fraction as a fraction. First, the integer is written as a fraction having the same denominator as the given fraction. Then the two fractions are added. Here is another example.

EXAMPLE 1

Write the mixed number $4\frac{3}{5}$ as a fraction.

SOLUTION

$$4\frac{3}{5} = 4 + \frac{3}{5} = \frac{4}{1} + \frac{3}{5} = \frac{20}{5} + \frac{3}{5} = \frac{20+3}{5} = \frac{23}{5}$$

By reasoning in the same way that we have in this example, we can express the sum (or difference) of any polynomial and a fraction as a single fraction.

EXAMPLE 2

Write $x - \dfrac{1}{x}$ as a fraction.

SOLUTION

$$x - \frac{1}{x} = \frac{x}{1} - \frac{1}{x} = \frac{x^2}{x} - \frac{1}{x} = \frac{x^2 - 1}{x}$$

EXAMPLE 3

Write $x + 2 + \dfrac{1}{x + 1}$ as a fraction.

SOLUTION

First, we change the polynomial $x + 2$ to a fraction having the denominator $x + 1$.

$$x + 2 = \frac{x + 2}{1} = \frac{(x + 1)(x + 2)}{(x + 1)(1)} = \frac{x^2 + 3x + 2}{x + 1}$$

Now, because $x + 2 = \dfrac{x^2 + 3x + 2}{x + 1}$, it can be added to $\dfrac{1}{x + 1}$.

$$\frac{x^2 + 3x + 2}{x + 1} + \frac{1}{x + 1} = \frac{(x^2 + 3x + 2) + 1}{x + 1} = \frac{x^2 + 3x + 3}{x + 1}$$

The procedure can be reversed, as the next example illustrates.

EXAMPLE 4

Write $\dfrac{8 - x}{4}$ as the difference of an integer and a fraction.

SOLUTION

$$\frac{8 - x}{4} = \frac{8}{4} - \frac{x}{4} = 2 - \frac{x}{4}$$

Exercises

Set I

1. Factor each of the following polynomials as completely as possible.
   a) $39x + 45y$
   b) $2x^2 - 14x + 24$
   c) $x^3y - xy^3$

2. Solve the following equations for x.
   a) $ax + b = 0$
   b) $a(x + b) = 1$
   c) $ax + bx = c$

   [handwritten: $ax + ab = 1$ $a + b = \frac{1}{x} \cdot \frac{c}{1}$]
   [handwritten: $x = \frac{1}{a} - ab$ $\frac{a+b}{c} = \frac{1}{x}$]

3. Cauliflower McPugg won more fights during his career than he lost. If Cauliflower won 60 fights and lost 36 fights,

   a) what fraction of the fights did he win?
   b) what fraction of the fights did he lose?
   If he won x fights and lost y fights,
   c) what fraction of the fights did he win?
   d) what fraction of the fights did he lose?

Set II

4. Write each of the following as a fraction in simplest terms.
   a) $5 + \dfrac{1}{8}$
   b) $8 + \dfrac{1}{5}$
   c) $2\dfrac{3}{7}$
   d) $3\dfrac{2}{7}$
   e) $6 - \dfrac{1}{6}$ *[handwritten: $x = \frac{c}{a+b}$]*
   f) $1 - \dfrac{4}{15}$

5. Write each of the following as a fraction in simplest terms.
   a) $4 + \dfrac{1}{x}$
   b) $x - \dfrac{x}{y}$
   c) $x + \dfrac{x^2}{3}$
   d) $5x - \dfrac{x}{2}$
   e) $1 + \dfrac{x-1}{x}$
   f) $2 - \dfrac{3}{x+6}$
   g) $x - \dfrac{6x}{x+7}$
   h) $x^2 + \dfrac{x^2 - 10}{4}$
   i) $x + 3 + \dfrac{1}{x-3}$
   j) $x + y - \dfrac{2xy}{x+y}$

6. Find the values of the following expressions for the values of x indicated. Express each answer as an integer.

   $$x + 2 + \frac{6+x}{x-3} \qquad \frac{x^2}{x-3}$$

   a) $x = 4$
   b) $x = 6$
   c) $x = 0$
   d) $x = 4$
   e) $x = 6$
   f) $x = 0$

   g) Compare your answers for parts a, b, and c with those for parts d, e, and f. Do you think that you would get the same results for all other values of x? Explain.

7. What should replace ▒ in each of the following equations to make it true for all allowable values of x? Express each answer in simplest form.

   a) $\dfrac{7x + 6}{x} = 7 + ▒$

   b) $\dfrac{2x - 5}{10} = \dfrac{x}{5} - ▒$

   c) $\dfrac{x + 1}{9x} = ▒ + \dfrac{1}{9x}$

   d) $\dfrac{9x}{x + 1} = \dfrac{x}{x + 1} + ▒$

e) $\dfrac{x^3 - 6}{3x^3} = \dfrac{1}{3} - $

f) $\dfrac{3x^3}{x^3 - 6} = \dfrac{x^3}{x^3 - 6} + $

g) $\dfrac{x^2 + x - 7}{x - 7} = \dfrac{x^2 + x}{x - 7} - $

h) $\dfrac{x^2 + x - 7}{x - 7} = \dfrac{x^2}{x - 7} + $

8. Change each of the following to the form described. More than one correct answer is possible.

a) Write $\dfrac{x + y}{5}$ as the sum of two fractions.

b) Write $\dfrac{2x - 1}{6}$ as the difference of two fractions.

c) Write $\dfrac{x + 3}{x - 3}$ as the sum of two fractions.

d) Write $\dfrac{8x + 7}{x}$ as the sum of an integer and a fraction.

e) Write $\dfrac{x - y + z}{x - y}$ as the sum of an integer and a fraction.

f) Write $\dfrac{2x^2 - x - 3}{x + 3}$ as the difference of a fraction and an integer.

9. Simplify each of the following.

a) $\dfrac{5x + 6}{6} + \dfrac{7x}{6}$

b) $\dfrac{4x}{4x + 9} + \dfrac{9}{4x + 9}$

c) $\dfrac{x^2}{x - y} - \dfrac{y^2}{x - y}$

d) $\dfrac{2x^3}{2x - 5} - \dfrac{5x^2}{2x - 5}$

Set III

10. Write each of the following as a fraction in simplest terms.

a) $3 + \dfrac{1}{7}$

b) $7 + \dfrac{1}{3}$

c) $2\dfrac{4}{9}$

d) $4\dfrac{2}{9}$

e) $8 - \dfrac{1}{8}$

f) $1 - \dfrac{5}{12}$

11. Write each of the following as a fraction in simplest terms.

a) $x + \dfrac{1}{6}$

b) $2 - \dfrac{8}{x}$

c) $3x + \dfrac{x}{5}$

d) $x^2 - \dfrac{x}{4}$

e) $9 + \dfrac{x + 1}{x}$

f) $5 - \dfrac{2}{x + 8}$

g) $x + \dfrac{7x}{x - 6}$

h) $x^2 - \dfrac{x^2 + 3}{4}$

i) $x - 5 + \dfrac{1}{x + 5}$

j) $x^2 + x + \dfrac{x}{x - 1}$

12. Find the values of the following expressions for the values of x indicated. Express each answer as an integer.

$$x - 3 + \dfrac{12 - x}{x + 4} \qquad \dfrac{x^2}{x + 4}$$

a) $x = 4$

b) $x = 12$

c) $x = 0$

d) $x = 4$

e) $x = 12$

f) $x = 0$

g) Compare your answers for parts a, b, and c with those for parts d, e, and f. Do you think that you would get the same results for all other values of x? Explain.

13. What should replace in each of the following equations to make it true for all allowable values of x? Express each answer in simplest form.

a) $\dfrac{4x + 9}{x} = 4 + $

b) $\dfrac{6x - 4}{12} = \dfrac{x}{2} - $

c) $\dfrac{x-3}{5x} = $  $- \dfrac{3}{5x}$

d) $\dfrac{5x}{x-3} = \dfrac{3x}{x-3} + $

e) $\dfrac{x^2+8}{4x^2} = \dfrac{1}{4} + $

f) $\dfrac{4x^2}{x^2+8} = \dfrac{x^2}{x^2+8} + $

g) $\dfrac{x^3+x-9}{x-9} = \dfrac{x^3+x}{x-9} - $

h) $\dfrac{x^3+x-9}{x-9} = \dfrac{x^3}{x-9} + $

14. Change each of the following to the form described.

a) Write $\dfrac{x+4}{y}$ as the sum of two fractions.

b) Write $\dfrac{3x-2}{6}$ as the difference of two fractions.

c) Write $\dfrac{x+y}{x-y}$ as the sum of two fractions.

d) Write $\dfrac{5x-9}{x}$ as the difference of an integer and a fraction.

e) Write $\dfrac{x+y+1}{x+y}$ as the sum of an integer and a fraction.

f) Write $\dfrac{3x^2-x-4}{x+4}$ as the difference of a fraction and an integer.

15. Simplify each of the following.

a) $\dfrac{11x+12}{4} - \dfrac{3x}{4}$

b) $\dfrac{x}{x+y} + \dfrac{y}{x+y}$

c) $\dfrac{x^2+10x}{x+5} + \dfrac{25}{x+5}$

d) $\dfrac{3x^3}{3x-1} - \dfrac{x^2}{3x-1}$

Set IV

If a cat and a half eat a fish and a half in a day and a half, how many fish would seven cats eat in a week and a half?

LESSON 5
Multiplying Fractions

The only fish that Mr. Jones caught on his fishing trip wasn't big enough to impress anybody. So he decided to stretch the truth by taking a photograph of it and having the picture enlarged. He began by having a picture the actual size of the fish enlarged $1\frac{1}{2}$ times, but this didn't suit him and so he had the enlargement enlarged $1\frac{1}{2}$ times again. How many times the size of the actual fish was the second enlargement?

The answer can be seen in the figures at the top of the next page. The first enlargement multiplied the original length by $\frac{3}{2}$ and the second enlargement multiplied it by $\frac{3}{2} \cdot \frac{3}{2}$. Measuring the original and final lengths in fourths of the original length, we see that the final length is $\frac{9}{4}$ of the original length.

Actual size of fish

Length x

First enlargement

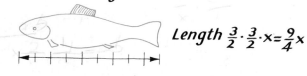

Length $\frac{3}{2} \cdot x$

Second enlargement

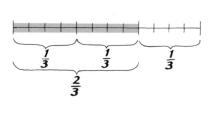

Length $\frac{3}{2} \cdot \frac{3}{2} \cdot x = \frac{9}{4}x$

From this example, we see that $\dfrac{3}{2} \cdot \dfrac{3}{2} = \dfrac{9}{4}$. Notice that we could have gotten this result without the diagram by simply multiplying 3 by 3 to get 9 and 2 by 2 to get 4.

What if the fractions being multiplied have different denominators? The next diagram shows that $\dfrac{2}{3} \cdot \dfrac{4}{5} = \dfrac{2 \cdot 4}{3 \cdot 5} = \dfrac{8}{15}$. Even though the fractions in this

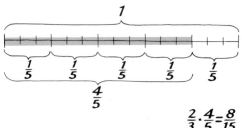

$$\frac{2}{3} \cdot \frac{4}{5} = \frac{8}{15}$$

multiplication problem have different denominators, the same method works. Because of this, it is usually easier to multiply two fractions than it is to add them. Compare the general cases worked out below.

Addition:

$$\frac{a}{b} + \frac{c}{d} = \frac{ad}{bd} + \frac{bc}{bd} = \frac{ad + bc}{bd}$$

Multiplication:

$$\frac{a}{b} \cdot \frac{c}{d} = \frac{ac}{bd}$$

Although the product of two fractions can be expressed as a fraction by multiplying numerators and multiplying denominators, it is usually desirable to express the result in lowest terms. Many times, the easiest way to accomplish this is to merely indicate the multiplications at first and remove any common factors in the numerator and denominator of the product by dividing both numerator and denominator by them.

Here is an example of how this is done.

EXAMPLE 1

Multiply $\dfrac{36}{77}$ and $\dfrac{7}{24}$.

SOLUTION

$$\frac{36}{77} \cdot \frac{7}{24} = \frac{36 \cdot 7}{77 \cdot 24} = \frac{36 \cdot 7}{24 \cdot 77} = \frac{3 \cdot 12 \cdot 7}{2 \cdot 12 \cdot 7 \cdot 11} = \frac{3}{2 \cdot 11} = \frac{3}{22}$$

Here are two more examples of how fractions are multiplied.

EXAMPLE 2

Multiply $\dfrac{x}{5}$ and $\dfrac{10}{x + 1}$.

SOLUTION

$$\frac{x}{5} \cdot \frac{10}{x + 1} = \frac{10x}{5(x + 1)} = \frac{5(2x)}{5(x + 1)} = \frac{2x}{x + 1}$$

EXAMPLE 3

Multiply $\dfrac{x - 2y}{2y}$ and $\dfrac{3x}{3x - 6y}$.

SOLUTION

$$\frac{x - 2y}{2y} \cdot \frac{3x}{3x - 6y} = \frac{(x - 2y)(3x)}{(2y)(3x - 6y)} = \frac{(3)(x)(x - 2y)}{(2)(y)(3)(x - 2y)} = \frac{x}{2y}$$

Although the main purpose of this lesson has been to learn how to multiply fractions written as quotients, it is important to realize that arithmetical fractions can also be multiplied by changing them to decimal form and multiplying their decimal forms. For example, the problem

$$\frac{3}{2} \cdot \frac{3}{2} = \frac{9}{4}$$

written in decimal form is

$$1.5 \cdot 1.5 = 2.25$$

This is the form in which problems with fractions are done on a calculator.

Exercises

Set I

1. What numbers can replace ▓ in the polynomial $x^2 + $ ▓ $x + 18$ so that it can be factored? For each number, write the corresponding factors.

2. Where possible, reduce the following fractions.

   a) $\dfrac{2x - 5}{2x}$

   b) $\dfrac{2x - 8}{2x}$

   c) $\dfrac{2x - 8}{x - 2}$

   d) $\dfrac{2x^2 - 8}{x - 2}$

3. This exercise is about the formula $y = 2^x - 1$.

   a) Use this formula to complete the following table.

x	3	4	5	6	7	8
y	7	▓	▓	▓	▓	▓

   b) Which numbers in the second line of the table are prime?

   c) Do you notice a pattern?

Set II

4. Express the following products as integers or fractions in simplest terms.

   a) $\dfrac{2}{5} \cdot \dfrac{4}{3}$

   b) $\dfrac{1}{2} \cdot \dfrac{1}{6}$

   c) $\dfrac{5}{7} \cdot \dfrac{5}{7}$

   d) $\dfrac{5}{7} \cdot \dfrac{7}{5}$

   e) $\dfrac{9}{4} \cdot \dfrac{8}{3}$

   f) $\dfrac{35}{16} \cdot \dfrac{24}{49}$

5. Write each of the following as an integer or fraction in simplest terms.

   a) $\dfrac{1}{8} + \dfrac{5}{8}$

   b) $\dfrac{1}{8} \cdot \dfrac{5}{8}$

   c) $\dfrac{4}{x} + \dfrac{7}{x}$

   d) $\dfrac{4}{x} \cdot \dfrac{7}{x}$

   e) $\dfrac{2x}{3} + \dfrac{x}{6}$

   f) $\dfrac{2x}{3} \cdot \dfrac{x}{6}$

g) $\dfrac{x}{10} \cdot \dfrac{10}{x}$

h) $\dfrac{x}{10} + \dfrac{10}{x}$

i) $\dfrac{1}{x+1} + \dfrac{1}{x+1}$

j) $\dfrac{1}{x+1} \cdot \dfrac{1}{x+1}$

k) $\dfrac{2}{x^2} + \dfrac{3}{x^3}$

l) $\dfrac{2}{x^2} \cdot \dfrac{3}{x^3}$

m) $\dfrac{6}{x} + \dfrac{6}{x} + \dfrac{6}{x}$

n) $\dfrac{x}{6} + \dfrac{x}{6} + \dfrac{x}{6}$

o) $\dfrac{6}{x} \cdot \dfrac{6}{x} \cdot \dfrac{6}{x}$

p) $\dfrac{x}{6} \cdot \dfrac{x}{6} \cdot \dfrac{x}{6}$

6. Find the following products, expressing your answers in simplest terms.

a) $\dfrac{7}{3} \cdot \dfrac{9}{x}$

b) $\dfrac{1}{5x} \cdot \dfrac{x^2}{6}$

c) $\dfrac{x}{x+2} \cdot \dfrac{x}{x-2}$

d) $\dfrac{2}{x^2} \cdot \dfrac{x^8}{8}$

e) $\dfrac{x+y}{x-y} \cdot \dfrac{x-y}{x+y}$

f) $\dfrac{x+y}{x-y} \cdot \dfrac{x+y}{x-y}$

g) $\dfrac{x}{x+1} \cdot \dfrac{x+1}{x+2} \cdot \dfrac{x+2}{x+3}$

h) $\dfrac{x^2+x}{x^2} \cdot \dfrac{x}{x^3-x}$

7. Find the following products, expressing your answers in simplest terms. (Look for common factors first.)

a) $\dfrac{xy-x}{2x+2y} \cdot \dfrac{x^2+xy}{y-1}$

b) $\dfrac{x^2-10x+25}{5x-25} \cdot \dfrac{5x^2}{x^2-5x}$

c) $\dfrac{3x-12}{x^2-16} \cdot \dfrac{x^2+8x+16}{4x+16}$

d) $\dfrac{x^2+6x+5}{x^2+7x+10} \cdot \dfrac{x^2+7x+12}{x^2+4x+3}$

Set III

8. Express the following products as integers or fractions in simplest terms.

a) $\dfrac{5}{7} \cdot \dfrac{2}{9}$

b) $\dfrac{1}{4} \cdot \dfrac{1}{8}$

c) $\dfrac{3}{10} \cdot \dfrac{3}{10}$

d) $\dfrac{3}{10} \cdot \dfrac{10}{3}$

e) $\dfrac{8}{5} \cdot \dfrac{15}{4}$

f) $\dfrac{22}{39} \cdot \dfrac{26}{99}$

k) $\dfrac{5}{x^5} + \dfrac{4}{x^4}$

l) $\dfrac{5}{x^5} \cdot \dfrac{4}{x^4}$

m) $\dfrac{3}{x} + \dfrac{3}{x} + \dfrac{3}{x}$

n) $\dfrac{x}{3} + \dfrac{x}{3} + \dfrac{x}{3}$

o) $\dfrac{3}{x} \cdot \dfrac{3}{x} \cdot \dfrac{3}{x}$

p) $\dfrac{x}{3} \cdot \dfrac{x}{3} \cdot \dfrac{x}{3}$

9. Write each of the following as an integer or fraction in simplest terms.

a) $\dfrac{5}{9} + \dfrac{1}{9}$

b) $\dfrac{5}{9} \cdot \dfrac{1}{9}$

c) $\dfrac{6}{x} + \dfrac{8}{x}$

d) $\dfrac{6}{x} \cdot \dfrac{8}{x}$

e) $\dfrac{x}{10} + \dfrac{2x}{5}$

f) $\dfrac{x}{10} \cdot \dfrac{2x}{5}$

g) $\dfrac{x}{7} \cdot \dfrac{7}{x}$

h) $\dfrac{x}{7} + \dfrac{7}{x}$

i) $\dfrac{2}{x-2} + \dfrac{2}{x-2}$

j) $\dfrac{2}{x-2} \cdot \dfrac{2}{x-2}$

10. Find the following products, expressing your answers in simplest terms.

a) $\dfrac{1}{7} \cdot \dfrac{14}{x}$

b) $\dfrac{5x}{4} \cdot \dfrac{x^3}{5}$

c) $\dfrac{x+y}{x} \cdot \dfrac{x-y}{x}$

d) $\dfrac{3}{x^3} \cdot \dfrac{x^9}{9}$

e) $\dfrac{x-4}{x+4} \cdot \dfrac{x+4}{x-4}$

f) $\dfrac{x-4}{x+4} \cdot \dfrac{x-4}{x+4}$

g) $\dfrac{x-3}{x-2} \cdot \dfrac{x-2}{x-1} \cdot \dfrac{x-1}{x}$

h) $\dfrac{x}{x-y} \cdot \dfrac{y-x}{x^2}$

11. Find the following products, expressing your answers in simplest terms. (Look for common factors first.)

a) $\dfrac{x^2 + y^2}{(x+y)^2} \cdot \dfrac{x^2 - y^2}{(x-y)^2}$

b) $\dfrac{2x+1}{4x^2 + 4x + 1} \cdot \dfrac{4x^2 + 2x}{2x^2}$

c) $\dfrac{2x-10}{2x+10} \cdot \dfrac{x^2 - 25}{x^2 - 10x + 25}$

d) $\dfrac{x^3 + 3x}{x^2 - 3} \cdot \dfrac{x-3}{x^2 + 3}$

Set IV

This picture is a photograph of a Mercedes Benz 230 SL sports car. Similar in nature to an x-ray, the picture was made with gamma rays from a radioactive element, cobalt-60. The gamma rays were produced as atoms of the cobalt broke apart into atoms of other elements. The cobalt was suspended over the car and five large photographic plates were placed under the car and exposed to the rays for 50 hours.

During that time about 3 out of every 4,000 of the 10^{26} cobalt-60 atoms used broke apart. How many atoms disintegrated while the picture was being taken? Express your answer in scientific notation.

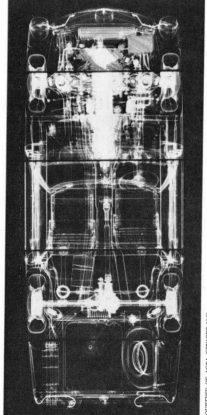

LESSON 6
More on Multiplication

The dollar bill isn't as large as it used to be. In 1929, the dimensions of United States currency notes were reduced from 3 inches by $7\frac{1}{2}$ inches to $2\frac{5}{8}$ inches by $6\frac{1}{8}$ inches. This cartoon was one of many that appeared on the editorial pages of the nation's newspapers at the time.

An obvious advantage of the smaller size is that less paper is needed to produce each bill. Because the dimensions of the larger bill were 3 inches by $7\frac{1}{2}$ inches, it required

$$3 \cdot 7\frac{1}{2} = 3\left(7 + \frac{1}{2}\right) = 21 + \frac{3}{2} = 22\frac{1}{2}$$

square inches of paper. Another way to make this calculation is to write 3 and $7\frac{1}{2}$ as fractions before multiplying.

$$3 \cdot 7\frac{1}{2} = \frac{3}{1} \cdot \frac{15}{2} = \frac{45}{2} = 22\frac{1}{2}$$

The two methods that we have just used can also be applied to finding the amount of paper in the smaller bill. Treating each dimension as a binomial, we have

$$2\frac{5}{8} \cdot 6\frac{1}{8} = \left(2 + \frac{5}{8}\right)\left(6 + \frac{1}{8}\right) = 12 + 2 \cdot \frac{1}{8} + 6 \cdot \frac{5}{8} + \frac{5}{8} \cdot \frac{1}{8}$$

$$= 12 + \frac{2}{8} + \frac{30}{8} + \frac{5}{64}$$

$$= 12 + \frac{32}{8} + \frac{5}{64}$$

$$= 12 + 4 + \frac{5}{64} = 16\frac{5}{64} \text{ square inches}$$

Or, changing each dimension to a fraction before multiplying, we have

$$2\frac{5}{8} \cdot 6\frac{1}{8} = \left(2 + \frac{5}{8}\right)\left(6 + \frac{1}{8}\right) = \frac{21}{8} \cdot \frac{49}{8} = \frac{1029}{64} = 16\frac{5}{64} \text{ square inches}$$

In making calculations that include both integers and fractions, it is helpful to think of the integers as fractions whose denominators are 1. The product of a polynomial and a fraction can be found by the same procedure: the polynomial is first written as a fraction whose denominator is 1.

Here are more examples.

EXAMPLE 1

Find the product of $x + 2$ and $\frac{x}{3}$.

SOLUTION

$$(x + 2) \cdot \frac{x}{3} = \frac{x + 2}{1} \cdot \frac{x}{3} = \frac{x(x + 2)}{3} = \frac{x^2 + 2x}{3}$$

For many purposes, it is more useful to leave an expression such as $\frac{x(x + 2)}{3}$ in factored form rather than to carry out the multiplication.

EXAMPLE 2

Find the product of $4x$ and $\dfrac{4}{x-4}$.

SOLUTION

$$4x \cdot \frac{4}{x-4} = \frac{4x}{1} \cdot \frac{4}{x-4} = \frac{16x}{x-4}$$

The next pair of examples show how a product such as $\left(x + \dfrac{1}{x}\right)\left(x + \dfrac{5}{x}\right)$ can be found in two different ways.

EXAMPLE 3

Find the product of $x + \dfrac{1}{x}$ and $x + \dfrac{5}{x}$ by using the pattern

$(a + b)(c + d) = ac + ad + bc + bd$.

SOLUTION

$$\left(x + \frac{1}{x}\right)\left(x + \frac{5}{x}\right) = x^2 + x \cdot \frac{5}{x} + \frac{1}{x} \cdot x + \frac{1}{x} \cdot \frac{5}{x}$$

$$= x^2 + \frac{5x}{x} + \frac{1x}{x} + \frac{5}{x^2}$$

$$= x^2 + 5 + 1 + \frac{5}{x^2}$$

$$= x^2 + 6 + \frac{5}{x^2}$$

EXAMPLE 4

Find the product of $x + \dfrac{1}{x}$ and $x + \dfrac{5}{x}$ by first changing each to a fraction.

SOLUTION

$$\left(x + \frac{1}{x}\right)\left(x + \frac{5}{x}\right) = \left(\frac{x}{1} + \frac{1}{x}\right)\left(\frac{x}{1} + \frac{5}{x}\right) = \frac{x^2 + 1}{x} \cdot \frac{x^2 + 5}{x}$$

$$= \frac{(x^2 + 1)(x^2 + 5)}{x^2} = \frac{x^4 + 6x^2 + 5}{x^2}$$

Although the answer in Example 4 looks different from that in Example 3, the two are equivalent because

$$\frac{x^4 + 6x^2 + 5}{x^2} = \frac{x^4}{x^2} + \frac{6x^2}{x^2} + \frac{5}{x^2} = x^2 + 6 + \frac{5}{x^2}$$

Exercises

Set I

1. Find each of the following differences, simplifying your answers as much as possible.

    a) $\dfrac{x}{4} - \dfrac{4}{x}$

    c) $3 - \dfrac{x + 6}{x + 2}$

    b) $\dfrac{x}{x - 5} - \dfrac{5}{x - 5}$

2. Find the missing term in each of the following, given that each is the square of a binomial.

    a) $x^2 + \rule{0.8cm}{0.15cm} + \dfrac{1}{9}$

    b) $x^2 - 5x + \rule{0.8cm}{0.15cm}$

3. The mathematics section of the Scholastic Aptitude Test contains 60 questions. Anyone who takes it automatically gets 200 points, with 10 points added for each correct answer and 2.5 points subtracted for each incorrect answer.

    Suppose that someone answers all 60 questions and gets x of them correct.

    a) How many questions did he or she answer incorrectly?

    b) Write a formula for his or her score, y, in terms of x.

    c) Simplify your formula as much as you can.

Set II

4. Express the following products as integers or fractions in simplest terms.

    a) $4 \cdot \dfrac{3}{5}$

    c) $-6 \cdot \dfrac{5}{12}$

    e) $3 \cdot \dfrac{x}{6}$

    g) $\dfrac{1}{10x} \cdot 2x$

    b) $9 \cdot \dfrac{11}{9}$

    d) $\dfrac{10}{7} \cdot 14$

    f) $x \cdot \dfrac{-8}{x}$

    h) $\dfrac{3}{4x^3} \cdot x^2$

5. Write each of the following as an integer or fraction in simplest terms.

a) $\frac{x}{3} \cdot 12$

b) $\frac{x}{3} + 12$

c) $2x \cdot \frac{5}{x}$

d) $2x - \frac{5}{x}$

e) $6 \cdot \frac{x}{x-6}$

f) $6 + \frac{x}{x-6}$

g) $\frac{1}{3x} \cdot x^3$

h) $\frac{1}{3x} - x^3$

i) $(x-1) \cdot \frac{x}{4}$

j) $(x-1) + \frac{x}{4}$

k) $\frac{1}{x+7} \cdot (x+7)$

l) $\frac{1}{x+7} + (x+7)$

6. Find the following products, expressing your answers in simplest terms.

a) $(x-5) \cdot \frac{x}{5}$

b) $8x \cdot \frac{x+1}{x}$

c) $x^3 \cdot \frac{3x}{3+x}$

d) $\frac{4}{x} \cdot (6x^4 - x)$

e) $(x-7) \cdot \frac{x+7}{x^2+49}$

f) $\frac{x+10}{2x} \cdot (2x-1)$

7. Find each of the following products by first multiplying and then simplifying the result.

a) $\left(x + \frac{1}{7}\right)\left(x + \frac{6}{7}\right)$

b) $\left(\frac{2x}{3} + \frac{1}{2}\right)\left(\frac{2x}{3} - \frac{1}{2}\right)$

c) $\left(x - \frac{5}{x}\right)\left(x - \frac{7}{x}\right)$

d) $\left(3x - \frac{2}{5}\right)\left(4x + \frac{1}{5}\right)$

8. Find each of the products in exercise 7 by first changing each expression to a fraction.

Set III

9. Express the following products as integers or fractions in simplest terms.

a) $5 \cdot \frac{2}{7}$

b) $-8 \cdot \frac{3}{8}$

c) $10 \cdot \frac{9}{4}$

d) $\frac{7}{12} \cdot 18$

e) $6 \cdot \frac{x}{2}$

f) $x \cdot \frac{-11}{x}$

g) $\frac{1}{14x} \cdot 7x$

h) $\frac{2}{3x^2} \cdot x^3$

e) $3 \cdot \frac{x}{x+8}$

f) $3 - \frac{x}{x+8}$

g) $\frac{1}{6x} \cdot x^2$

h) $\frac{1}{6x} + x^2$

i) $(x+2) \cdot \frac{x}{2}$

j) $(x+2) + \frac{x}{2}$

k) $\frac{5}{x-10} \cdot (x-10)$

l) $\frac{5}{x-10} + (x-10)$

10. Write each of the following as an integer or fraction in simplest terms.

a) $\frac{2x}{5} \cdot 5$

b) $\frac{2x}{5} + 5$

c) $x \cdot \frac{4}{x^2}$

d) $x - \frac{4}{x^2}$

11. Find the following products, expressing your answers in simplest terms.

a) $(x+3) \cdot \frac{x}{3}$

b) $6x \cdot \frac{1}{x-6}$

c) $x^4 \cdot \frac{4x}{x-4}$

d) $\frac{2}{x} \cdot (5x^3 + x)$

e) $(x+5) \cdot \frac{x-5}{x^2-25}$

f) $\frac{x-1}{7} \cdot (7x+6)$

12. Find each of the following products by first multiplying and then simplifying the result.

a) $\left(x + \frac{2}{3}\right)\left(x + \frac{1}{3}\right)$

b) $\left(\frac{x}{8} + \frac{1}{2}\right)\left(\frac{x}{8} - \frac{1}{2}\right)$

c) $\left(x - \frac{6}{x}\right)\left(x - \frac{4}{x}\right)$

d) $\left(5x - \frac{1}{5}\right)\left(x + \frac{1}{2}\right)$

13. Find each of the products in exercise 12 by first changing each expression to a fraction.

Set IV

Can you simplify each of these expressions?
The three dots indicate that the patterns
continue in the same way, ending as shown.

1. $\left(1 - \frac{1}{1}\right)\left(1 - \frac{1}{2}\right)\left(1 - \frac{1}{3}\right)\left(1 - \frac{1}{4}\right) \cdots \left(1 - \frac{1}{100}\right)$.

2. $\left(1 - \frac{1}{2}\right)\left(1 - \frac{1}{3}\right)\left(1 - \frac{1}{4}\right)\left(1 - \frac{1}{5}\right) \cdots \left(1 - \frac{1}{99}\right)$.

LESSON 7
Dividing Fractions

A pioneer in motion picture photography was Thomas Edison. He produced the first film to be copyrighted in the United States. Several frames of the film, made in 1894 and showing a man sneezing, are reproduced here.

The frames were taken at the rate of 16 per second, and so each frame lasted $\frac{1}{16}$ second and the 12 frames shown ran for only

$$12 \cdot \frac{1}{16} = \frac{12}{16} = \frac{3}{4} \text{ second}$$

How many frames would be required to photograph a sneeze lasting $2\frac{1}{4}$ seconds? One way to figure out the answer to this question is to divide $2\frac{1}{4}$ by $\frac{1}{16}$, because each frame lasts $\frac{1}{16}$ second.

Dividing $2\frac{1}{4}$ by $\frac{1}{16}$ means finding how many times $\frac{1}{16}$ is contained in $2\frac{1}{4}$.

The figure below shows that

$$\frac{2\frac{1}{4}}{\frac{1}{16}} = 36$$

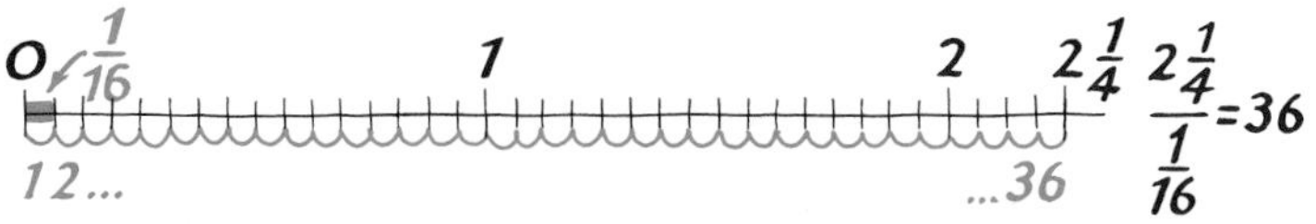

So 36 frames would be required to film a sneeze lasting $2\frac{1}{4}$ seconds.

One way to divide $2\frac{1}{4}$ by $\frac{1}{16}$ without drawing a figure is shown below.

$$\frac{2\frac{1}{4}}{\frac{1}{16}} = \frac{\frac{9}{4}}{\frac{1}{16}} = \frac{\frac{9}{4} \cdot 16}{\frac{1}{16} \cdot 16} = \frac{36}{1} = 36$$

The same procedure can be used to divide any fraction into another. Here it is applied to dividing $\frac{a}{b}$ by $\frac{c}{d}$.

$$\frac{\frac{a}{b}}{\frac{c}{d}} = \frac{\frac{a}{b} \cdot bd}{\frac{c}{d} \cdot bd} = \frac{ad}{bc}$$

Because the result, $\frac{ad}{bc}$, can be written as the product of the fractions $\frac{a}{b}$ and $\frac{d}{c}$, we have shown that

$$\frac{a}{b} \div \frac{c}{d} = \frac{a}{b} \cdot \frac{d}{c}$$

This says that dividing by the fraction $\frac{c}{d}$ is equivalent to multiplying by the fraction $\frac{d}{c}$. The fractions $\frac{c}{d}$ and $\frac{d}{c}$ are called *reciprocals* of each other because their product is 1.

▶ The **reciprocal** of a number is the number that must be multiplied by it to get 1.

The reciprocal of x is $\dfrac{1}{x}$, because $x \cdot \dfrac{1}{x} = 1$. The reciprocal of $\dfrac{x}{y}$ is $\dfrac{y}{x}$, because $\dfrac{x}{y} \cdot \dfrac{y}{x} = 1$.

The idea of the reciprocal of a number is useful because it gives us another way to look at division. *Dividing by a fraction (or any number) is equivalent to multiplying by its reciprocal.*

Here are more examples of division problems that include fractions.

EXAMPLE 1

Divide $\dfrac{2}{3}$ by $\dfrac{4}{9}$.

SOLUTION

$$\frac{2}{3} \div \frac{4}{9} = \frac{2}{3} \cdot \frac{9}{4} = \frac{2 \cdot 3 \cdot 3}{3 \cdot 2 \cdot 2} = \frac{3}{2}$$

EXAMPLE 2

Divide $\dfrac{5}{8}$ by 2.

SOLUTION

$$\frac{5}{8} \div 2 = \frac{5}{8} \cdot \frac{1}{2} = \frac{5}{16}$$

EXAMPLE 3

Divide $\dfrac{x}{6}$ by $\dfrac{x}{x+1}$

SOLUTION

$$\frac{x}{6} \div \frac{x}{x+1} = \frac{x}{6} \cdot \frac{x+1}{x} = \frac{x(x+1)}{6 \cdot x} = \frac{x+1}{6}$$

EXAMPLE 4

Divide $x^2 - y^2$ by $\dfrac{x - y}{9}$.

SOLUTION

$$(x^2 - y^2) \div \frac{x - y}{9} = \frac{(x^2 - y^2)}{1} \cdot \frac{9}{x - y} = \frac{9(x^2 - y^2)}{x - y} = \frac{9(x + y)(x - y)}{x - y}$$
$$= 9x + 9y$$

Exercises

Set I

1. Arrange the fractions in each of these sets in order from smallest to largest. Assume for each set that x represents a positive number. For part c, assume that x is not 1.

   a) $\dfrac{x}{2}, \dfrac{x}{3}, \dfrac{x}{4}$

   b) $\dfrac{2}{x}, \dfrac{3}{x}, \dfrac{4}{x}$

   c) $\dfrac{1}{x}, \dfrac{1}{x + 1}, \dfrac{1}{x - 1}$

   d) $\dfrac{x}{5}, \dfrac{x + 1}{5}, \dfrac{x - 1}{5}$

2. Find the value of each of the following expressions by first writing it as the square of a binomial.

   a) 45^2 (Write 45 as $40 + 5$.)

   b) 54^2

   c) $\left(6\dfrac{1}{3}\right)^2$

   d) $\left(3\dfrac{1}{6}\right)^2$

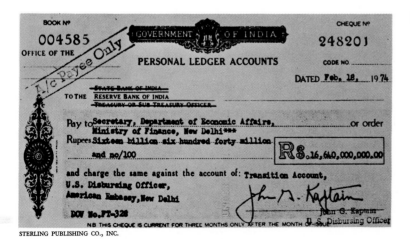

STERLING PUBLISHING CO., INC.

3. The largest check ever written was paid to India by the United States in 1974. It was for 16,640,000,000 Indian rupees.

   a) Write this number in scientific notation.

   b) In 1974, one rupee was equal to approximately 12.3 U.S. cents. Find the approximate value of the check in U.S. dollars at that time.

Set II

4. Write the reciprocals of the following.

   a) 4

   b) $\dfrac{1}{3}$

   c) $\dfrac{5}{7}$

   d) $\dfrac{1}{x}$

   e) $2x + 1$

   f) $\dfrac{x^2}{x - 3}$

5. Express the following quotients as integers or fractions in simplest terms.

   a) $\dfrac{6}{7} \div \dfrac{3}{4}$

   b) $6 \div \dfrac{1}{6}$

   c) $8 \div \dfrac{2}{5}$

   d) $\dfrac{1}{9} \div 10$

   e) $4\dfrac{2}{3} \div \dfrac{7}{12}$

   f) $4\dfrac{5}{8} \div 9\dfrac{1}{4}$

6. Write each of the following as a monomial or fraction in simplest terms.

   a) $\dfrac{x^2}{2} \cdot \dfrac{x}{4}$

   b) $\dfrac{x^2}{2} \div \dfrac{x}{4}$

   c) $\dfrac{x^2}{2} - \dfrac{x}{4}$

   d) $15x \cdot \dfrac{3}{x}$

   e) $15x \div \dfrac{3}{x}$

   f) $\dfrac{6}{x + 1} + \dfrac{4}{x + 1}$

   g) $\dfrac{6}{x + 1} \cdot \dfrac{4}{x + 1}$

   h) $\dfrac{6}{x + 1} \div \dfrac{4}{x + 1}$

   i) $\dfrac{5x - 30}{9} \div \dfrac{5x}{9}$

   j) $\dfrac{5x}{9} \div \dfrac{5x - 30}{9}$

7. Express each of the following quotients as a monomial or fraction in simplest terms.

   a) $\dfrac{1}{x} \div x$

   b) $\dfrac{x}{y^2} \div \dfrac{y}{x^2}$

   c) $\dfrac{3}{x^3} \div \dfrac{6}{x^6}$

   d) $\dfrac{5x + 10}{3} \div \dfrac{5x}{6}$

   e) $\dfrac{x + 1}{8} \div \dfrac{x^2 - 1}{6}$

   f) $\dfrac{x}{x - y} \div \dfrac{x}{y - x}$

8. Express each of the following quotients as a monomial or fraction in simplest terms.

   a) $\dfrac{2x^5}{5x^2} \div \dfrac{(2x)^5}{(5x)^2}$

   b) $\dfrac{12x + 6}{x + 6} \div \dfrac{2x + 1}{2x + 12}$

   c) $\dfrac{x^2 - 9}{x^2 - 6x + 9} \div \dfrac{x - 4}{x + 4}$

   d) $\dfrac{6x - 6y}{(x - y)^3} \div \dfrac{6x + 6y}{(x - y)^2}$

   e) $\dfrac{x^2 - 4x + 3}{x + 1} \div (x - 3)$

   f) $\dfrac{x^4 - 16}{x - 2} \div (x^3 + 2x^2)$

Set III

9. Write the reciprocals of the following.

   a) 6

   b) $\dfrac{1}{8}$

   c) $\dfrac{3}{2}$

   d) x

   e) $\dfrac{1}{x - 1}$

   f) $\dfrac{2x + 5}{2x}$

10. Express the following quotients as integers or fractions in simplest terms.

    a) $\dfrac{2}{5} \div \dfrac{4}{9}$

    b) $5 \div \dfrac{1}{5}$

    c) $6 \div \dfrac{3}{8}$

    d) $\dfrac{7}{4} \div 2$

    e) $2\dfrac{2}{9} \div \dfrac{5}{12}$

    f) $5\dfrac{1}{3} \div \dfrac{8}{9}$

11. Write each of the following as a monomial or fraction in simplest terms.

a) $\dfrac{x}{6} \cdot \dfrac{x^3}{3}$

b) $\dfrac{x}{6} \div \dfrac{x^3}{3}$

c) $\dfrac{x}{6} - \dfrac{x^3}{3}$

d) $2x \cdot \dfrac{10}{x}$

e) $2x \div \dfrac{10}{x}$

f) $\dfrac{x+4}{8} + \dfrac{x-4}{8}$

g) $\dfrac{x+4}{8} \cdot \dfrac{x-4}{8}$

h) $\dfrac{x+4}{8} \div \dfrac{x-4}{8}$

i) $\dfrac{14}{3x+21} \div \dfrac{7}{3x}$

j) $\dfrac{7}{3x} \div \dfrac{14}{3x+21}$

d) $\dfrac{3x-12}{4} \div \dfrac{3x}{2}$

e) $\dfrac{x^2-4}{x} \div \dfrac{x+2}{2x}$

f) $\dfrac{x-y}{x} \div \dfrac{y-x}{y}$

13. Express each of the following quotients as a monomial or fraction in simplest terms.

a) $\dfrac{(3x)^2}{(2x)^3} \div \dfrac{3x^2}{2x^3}$

b) $\dfrac{10x-2}{x-5} \div \dfrac{5x-1}{5x-25}$

c) $\dfrac{x^2+8x+16}{x^2-16} \div \dfrac{x+4}{x-4}$

d) $\dfrac{(x+y)^3}{3x+3y} \div \dfrac{x+y}{3}$

e) $\dfrac{x^2+3x+2}{x} \div (x+2)$

f) $\dfrac{x^4-1}{x+1} \div (x^3-x^2)$

12. Express each of the following quotients as a monomial or fraction in simplest terms.

a) $x \div \dfrac{1}{x}$

b) $\dfrac{x}{y} \div \dfrac{x^2}{y}$

c) $\dfrac{2}{x^2} \div \dfrac{8}{x^8}$

Set IV

This problem once appeared in *Ripley's Believe It Or Not!* The answer given by Ripley, 5, is wrong.

1. What do you think the correct answer to the problem is?

Ripley got the wrong answer because he was confused about the meaning of the word "of."

2. Can you change the problem as restated below by replacing each ▓ with an appropriate symbol of operation so that the answer Ripley gave is correct?

$$\frac{1}{3} \times \frac{1}{2} \ \text{▓} \ \frac{1}{3} \ \text{▓} \ 10$$

What is $\frac{1}{3} \times \frac{1}{2}$ of $\frac{1}{3}$ of 10?

LESSON **8**
Complex Fractions

One of the great scientists of the seventeenth century was a Dutch mathematician named Christiaan Huygens. Among his many projects was a model of the solar system in which the planets were driven by wheels in their orbits around the sun.

Designing the model took a lot of planning because the time that it takes to complete one orbit is different for every planet. For example, the earth goes around the sun almost 30 times in the time that Saturn goes around it once. More exactly, the earth goes around the sun about

$$29 + \cfrac{1}{2 + \cfrac{1}{3}}$$

times in the time that Saturn makes one trip. This complicated number, used by

Huygens in designing the gears in his model, is the sum of an integer, 29, and a *complex fraction,* $\dfrac{1}{2 + \frac{1}{3}}$.

▶ A **complex fraction** is a fraction that has one or more fractions in its numerator or denominator.

Every complex fraction can be written in a simpler form. For example,

$$\frac{1}{2 + \frac{1}{3}} = \frac{1}{\frac{7}{3}} = \frac{3}{7}$$

So the number Huygens used in designing his model, $29 + \dfrac{1}{2 + \frac{1}{3}}$, can be

written as $29\dfrac{3}{7}$ or $\dfrac{206}{7}$. We will refer to a fraction such as $\dfrac{206}{7}$ as a *simple fraction.*

▶ A **simple fraction** is a fraction that does not have any fractions in its numerator or denominator.

Every complex fraction can be transformed into a simple fraction. Although it was helpful to Huygens to write numbers in forms such as $29 + \dfrac{1}{2 + \frac{1}{3}}$ in planning his model of the solar system, simple fractions are usually preferred. Here are examples of how complex fractions can be simplified.

EXAMPLE 1

Simplify $\dfrac{4 + \frac{1}{5}}{2 + \frac{1}{3}}$.

SOLUTION

First we can write the numerator and denominator as single fractions.

$$\frac{4 + \dfrac{1}{5}}{2 + \dfrac{1}{3}} = \frac{\dfrac{20}{5} + \dfrac{1}{5}}{\dfrac{6}{3} + \dfrac{1}{3}} = \frac{\dfrac{21}{5}}{\dfrac{7}{3}}$$

The result, $\dfrac{21}{5} \div \dfrac{7}{3}$, can be written as $\dfrac{21}{5} \cdot \dfrac{3}{7}$.

$$\frac{21}{5} \cdot \frac{3}{7} = \frac{3 \cdot 7 \cdot 3}{5 \cdot 7} = \frac{9}{5}$$

The result is the simple fraction $\dfrac{9}{5}$. For some purposes, it might be useful to write it as $1\dfrac{4}{5}$ or in decimal form, 1.8.

EXAMPLE 2

Simplify $\dfrac{x^2 - 9}{\dfrac{1}{3} - \dfrac{1}{x}}$, given that x is not zero.

SOLUTION

Changing the denominator to a single fraction, we get

$$\frac{x^2 - 9}{\dfrac{1}{3} - \dfrac{1}{x}} = \frac{x^2 - 9}{\dfrac{x}{3x} - \dfrac{3}{3x}} = \frac{x^2 - 9}{\dfrac{x - 3}{3x}}$$

Because the result is $(x^2 - 9) \div \dfrac{x - 3}{3x}$, we have

$$(x^2 - 9) \cdot \frac{3x}{x - 3} = \frac{3x(x^2 - 9)}{x - 3} = \frac{3x(x - 3)(x + 3)}{x - 3} = 3x^2 + 9x$$

The result is a second-degree polynomial.

Exercises

Set I

1. Find the perimeter and area of each of these rectangles.

$$\left(\frac{6x+1}{1}\right)\left(\frac{6x+1}{6}\right)=\frac{36x^2+12x+1}{6}$$

a)

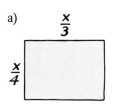

$\frac{x}{4}$ $\frac{x}{3}$

b)

$2-\frac{1}{x}$ $2+\frac{1}{x}$

c)

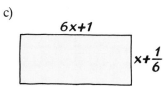

$6x+1$ $x+\frac{1}{6}$

2. Change each of the following to an expression without parentheses. Assume that x is not -1.
   a) $(x + 1)^2$
   b) $(x + 1)^1$
   c) $(x + 1)^0$
   d) $(x + 1)^{-1}$

3. The length of the steps taken by a person on stilts is a function of the height of the stilts. A typical table for this function is shown here.

Height of stilts in feet, x	5	7	9	11
Length of steps in feet, y	4	5	6	7

   a) Graph this function.
   b) What kind of function is it?
   c) Write a formula for it.
   d) According to your formula, how long would each step taken by someone on 21-foot stilts (the highest ever mastered) be?

Set II

4. Write each of the following as a simple fraction in lowest terms.

a) $\dfrac{\frac{1}{3}}{\frac{5}{6}}$

b) $\dfrac{\frac{1}{3}+1}{\frac{5}{6}+1}$

c) $\dfrac{10}{\frac{1}{2}-\frac{1}{7}}$

d) $\dfrac{\frac{2}{5}}{4-\frac{2}{5}}$

e) $4+\dfrac{\frac{2}{5}}{4-\frac{2}{5}}$

f) $\dfrac{\frac{4}{3}-\frac{3}{4}}{\frac{4}{3}+\frac{3}{4}}$

c) $\dfrac{1+\frac{1}{x}}{1-\frac{1}{x^2}}$

f) $\dfrac{\frac{1}{4}+\frac{1}{x}}{x^2-16}$

d) $\dfrac{\frac{x}{2}-\frac{x}{3}}{\frac{x}{4}-\frac{x}{5}}$

g) $\dfrac{x-1-\frac{2}{x}}{1-\frac{2}{x}}$

e) $\dfrac{\frac{7}{x}-1}{1-\frac{x}{7}}$

h) $\dfrac{\frac{1}{x}-\frac{7}{x^2}+\frac{6}{x^3}}{\frac{1}{x}-\frac{4}{x^2}+\frac{3}{x^3}}$

5. Simplify.

a) $\dfrac{\frac{x+2}{3}}{\frac{x+3}{2}}$

b) $\dfrac{x-\frac{3}{8}}{x+\frac{5}{8}}$

6. Write each of the following as a polynomial or simple fraction in lowest terms.

a) $\dfrac{\frac{1}{x}}{y}$

b) $\dfrac{\frac{x}{y}}{\frac{y}{x}}$

c) $\dfrac{\frac{1}{x}-\frac{1}{y}}{\frac{1}{x}\cdot\frac{1}{y}}$

d) $1+\dfrac{1}{x+\frac{1}{x}}$

Set III

7. Write each of the following as a simple fraction in lowest terms.

a) $\dfrac{\frac{3}{2}}{\frac{9}{4}}$

b) $\dfrac{\frac{3}{2}-1}{\frac{9}{4}-1}$

c) $\dfrac{7}{\frac{1}{6}+\frac{1}{8}}$

d) $\dfrac{\frac{4}{7}}{3+\frac{4}{7}}$

e) $3-\dfrac{\frac{4}{7}}{3+\frac{4}{7}}$

f) $\dfrac{\frac{5}{2}+\frac{2}{5}}{\frac{5}{2}-\frac{2}{5}}$ $\dfrac{21}{10}$

8. Simplify. $\dfrac{4x-16}{x-1}$

a) $\dfrac{\frac{4}{x-1}}{\frac{1}{x-4}}$

b) $\dfrac{x+\frac{5}{7}}{x-\frac{2}{7}}$

c) $\dfrac{\frac{x}{5}-1}{\frac{x^2}{25}-1}$

d) $\dfrac{\frac{x}{2}-\frac{x}{4}}{\frac{x}{3}-\frac{x}{6}}$

e) $\dfrac{\frac{x}{8}+1}{1+\frac{8}{x}}$ $\dfrac{x+8}{8}$ $\overline{x+8}$ $\times$

f) $\dfrac{x^2-9}{\frac{1}{3}-\frac{1}{x}}$

$\dfrac{29}{10},\dfrac{10}{21}$

g) $\dfrac{x + 3 - \dfrac{4}{x}}{1 + \dfrac{4}{x}}$

h) $\dfrac{\dfrac{1}{x} + \dfrac{1}{x^2} - \dfrac{12}{x^3}}{\dfrac{1}{x} - \dfrac{1}{x^2} - \dfrac{6}{x^3}}$

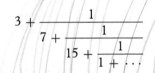

9. Write each of the following as a monomial or simple fraction in lowest terms.

a) $\dfrac{x}{\dfrac{1}{y}}$

b) $\dfrac{\dfrac{x}{y}}{xy}$

c) $\dfrac{x + y}{\dfrac{1}{x} + \dfrac{1}{y}}$

d) $x + \dfrac{1}{x - \dfrac{1}{x}}$

Set IV

In 1770, a Swiss-German mathematician, Johann Lambert, discovered an interesting expression containing a complex fraction that never ends. Part of it is shown below.

$$3 + \cfrac{1}{7 + \cfrac{1}{15 + \cfrac{1}{1 + \cdots}}}$$

1. Change the part of the expression shown here to a simple fraction.

*2. Write that simple fraction in decimal form, correct to the nearest ten millionth.

3. Lambert's expression represents a number so important in mathematics that it has a special name. Do you know what it is?

Summary and Review

In this chapter, we have learned how to simplify fractions and how to add, subtract, multiply, and divide both fractions and polynomials and fractions.

Fractions (*Lessons 1 and 2*) A fraction is the indicated quotient of two numbers or algebraic expressions called its numerator and denominator. The denominator of a fraction cannot be allowed to be zero.

Fractions consisting of integers can be changed to decimal form by dividing the numerator by the denominator.

Fractions that represent the same number are equivalent. Whenever the numerator and denominator of a fraction are multiplied or divided by the same number other than zero, the result is an equivalent fraction: in general, if n is not equal to zero, $\dfrac{a}{b} = \dfrac{na}{nb}$ and $\dfrac{na}{nb} = \dfrac{a}{b}$. To reduce a fraction consisting of integers or polynomials to lowest terms, divide its numerator and denominator by their common factors.

Adding and Subtracting Fractions (*Lessons 3 and 4*) It is often useful to express the sum or difference of two fractions as a single fraction. To do this when the fractions have the same denominator, we add or subtract the numerators and write the result over the denominator.

$$\frac{a}{b} + \frac{c}{b} = \frac{a+c}{b} \qquad \frac{a}{b} - \frac{c}{b} = \frac{a-c}{b}$$

If the fractions have different denominators, we first replace one or both of the fractions with equivalent ones so that the new fractions have the same denominator. Then we proceed as before.

$$\frac{a}{b} + \frac{c}{d} = \frac{ad}{bd} + \frac{bc}{bd} = \frac{ad + bc}{bd} \qquad \frac{a}{b} - \frac{c}{d} = \frac{ad}{bd} - \frac{bc}{bd} = \frac{ad - bc}{bd}$$

To express the sum or difference of a polynomial and a fraction as a fraction, write the polynomial as a fraction having the same denominator as the fraction. Then add or subtract the two fractions.

Multiplying Fractions (*Lessons 5 and 6*) The product of two fractions may be expressed as a single fraction by writing the product of their numerators over the product of their denominators.

$$\frac{a}{b} \cdot \frac{c}{d} = \frac{ac}{bd}$$

To find the product of a polynomial and a fraction, think of the polynomial as a fraction having 1 as its denominator. Then multiply the two fractions.

Dividing Fractions (*Lesson 7*) Dividing by a fraction (or any number) is equivalent to multiplying by its reciprocal:

$$\frac{a}{b} \div \frac{c}{d} = \frac{a}{b} \cdot \frac{d}{c} = \frac{ad}{bc}$$

The reciprocal of a number is the number that must be multiplied by it to get 1. The reciprocal of x is $\frac{1}{x}$; the reciprocal of $\frac{x}{y}$ is $\frac{y}{x}$.

Complex Fractions (*Lesson 8*) A complex fraction is a fraction that has one or more fractions in its numerator or denominator. A simple fraction is a fraction that does not have any fractions in either. Every complex fraction can be transformed into a simple fraction.

Exercises

Set I

1. Although there seem to be six people in the lifeboat in this cartoon, one of them is a dummy.
   a) What fraction of the food and water does the man with the dummy actually deserve?
   b) What fraction of the food and water is he actually getting?
   c) How much more food and water is the man getting than he actually deserves? (Express your answer as a fraction.)

$$\frac{4x^2+2x}{2x+1} + \frac{2x+1}{2x+1} - \frac{4x}{2x+1}$$

$$\frac{(2x+1)(2x-1)}{2x+1}$$

"I just don't understand it, Captain. Equal shares of food and water to all, yet those two thrive while we wither away."

2. Tell whether or not each of the following equations is true for all allowable values of the variables. If an equation is not true for all such values, give an example in which it is not true.

a) $\dfrac{x+4}{y+4} = \dfrac{x}{y}$

b) $\dfrac{4x}{4y} = \dfrac{x}{y}$

c) $\dfrac{x-y}{y-x} = -1$

d) $\dfrac{2x+5}{2x} = 5$

e) $\dfrac{-x}{y} = \dfrac{x}{-y}$

f) $\dfrac{x^2-4}{x-2} = x+2$

3. If the denominator of a fraction is zero, the fraction is meaningless. What value(s) of x would make the denominator of each of these fractions equal to zero?

a) $\dfrac{6}{x+5}$ b) $\dfrac{x}{4}$ c) $\dfrac{1}{x(x-3)}$

4. If possible, reduce each of these fractions.

a) $\dfrac{18}{108}$ d) $\dfrac{3x+6}{x^2+2x}$

b) $\dfrac{x^3}{4x}$ e) $\dfrac{5x+y}{5x-y}$

c) $\dfrac{x+2y}{x}$ f) $\dfrac{x^2+14x+49}{x^2-49}$

5. Which fraction in each pair is larger?

a) $\dfrac{4}{9}$ or $\dfrac{4}{10}$ c) $\dfrac{7}{11}$ or $\dfrac{5}{8}$

b) $\dfrac{9}{4}$ or $\dfrac{10}{4}$ d) $\dfrac{42}{95}$ or $\dfrac{43}{94}$

6. Write each of the following sums or differences as a monomial or fraction in simplest terms.

a) $\dfrac{2}{3x}+\dfrac{10}{3x}$ d) $\dfrac{x}{y}-\dfrac{y}{x}$

b) $\dfrac{x+3}{4}-\dfrac{x+1}{4}$ e) $x^2+\dfrac{x}{5}$

c) $\dfrac{x-1}{x^2}+\dfrac{1}{x}$ f) $2x+1-\dfrac{4x}{2x+1}$

7. Change each of the following to the form described.

a) Write $\dfrac{x+4}{8}$ as the sum of two fractions.

b) Write $\dfrac{3x-7}{3x+7}$ as the difference of two fractions.

c) Write $\dfrac{x^2-x+2}{x^2-x}$ as the sum of an integer and a fraction.

d) Write $\dfrac{10x^2-10}{5x}$ as the difference of a monomial and a fraction.

$$\frac{6x^2-30x}{24x^2-120x}=\frac{x-5}{4x-20}$$

8. Write each of the following products as a monomial or fraction in simplest terms.

a) $\dfrac{x^2}{2}\cdot\dfrac{x^3}{3}$ d) $7x\cdot\dfrac{x-2}{14}$

b) $\dfrac{x+7}{x}\cdot\dfrac{x}{x+7}$ e) $(x-y)\cdot\dfrac{x+y}{x^2}$

c) $\dfrac{x-4}{x}\cdot\dfrac{4}{4-x}$ f) $\dfrac{2x}{4x-20}\cdot\dfrac{3x-15}{6x}$

9. Find each of the following products by first multiplying and then simplifying the result.

a) $\left(x-\dfrac{2}{5}\right)\left(x-\dfrac{3}{5}\right)$ $\dfrac{x^2}{1}-\dfrac{x}{1}+\dfrac{6}{5}$

b) $\left(7+\dfrac{1}{x}\right)\left(9-\dfrac{1}{x}\right)$ $63-\dfrac{7}{x}+\dfrac{9}{x}-\dfrac{1}{x^2}$

10. Find each of the products in exercise 9 by first writing each factor as a single fraction.

11. Write each of the following quotients as a polynomial or fraction in simplest terms.

a) $\dfrac{8x}{y}\div\dfrac{x}{3y}$

b) $\dfrac{x+14}{7}\div\dfrac{x+7}{14}$

c) $\dfrac{x^2-y^2}{6}\div\dfrac{x-y}{2}$

d) $\dfrac{5}{2x^3}\div5x^2$

e) $(3x^3+1)\div\dfrac{1}{x}$

f) $\dfrac{2x^2+5x+2}{x^2-4}\div\dfrac{2x+1}{x-2}$ $\dfrac{(x+1)(x+2)}{x+2}$

12. Simplify.

a) $\dfrac{\dfrac{x}{y}+\dfrac{1}{y}}{y}$

b) $\dfrac{x^2-1}{1-\dfrac{1}{x}}$

c) $\dfrac{\dfrac{x}{2}-\dfrac{x}{6}}{\dfrac{x}{3}+\dfrac{x}{4}}$

d) $x+\dfrac{1}{2+\dfrac{4}{x}}$

Set II

1. The illustration below is from an Italian manuscript written in 1545. The problem in the first circle is

$$\frac{2}{3} \times \frac{3}{4} \quad \frac{8}{9}$$

a) What operation seems to have been performed upon the first two fractions to give the third one?

b) Can you tell what the problems in the other two circles are?

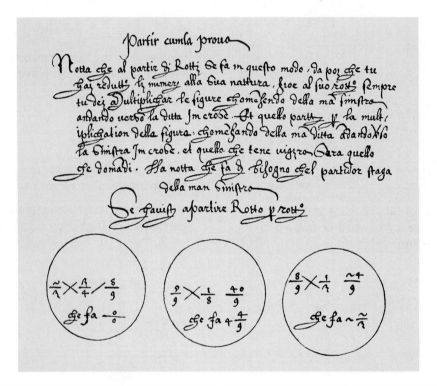

2. Tell whether or not each of the following equations is true for all allowable values of the variables. If an equation is not true for all such values, give an example in which it is not true.

a) $\dfrac{x-2}{y-2} = \dfrac{x}{y}$

b) $\dfrac{x+y}{x-y} = -1$

c) $\dfrac{\dfrac{x}{5}}{\dfrac{y}{5}} = \dfrac{x}{y}$

d) $\dfrac{3x-3y}{3x} = \dfrac{x-y}{x}$

e) $\dfrac{-x}{y} = -\dfrac{x}{y}$

f) $\dfrac{x^2+4}{x+2} = x+2$

3. If the denominator of a fraction is zero, the fraction is meaningless. What value(s) of x would make the denominator of each of these fractions equal to zero?

a) $\dfrac{2}{9-x}$

b) $\dfrac{x-5}{x}$

c) $\dfrac{x+2}{(x+1)(x-1)}$

4. If possible, reduce each of these fractions.

a) $\dfrac{16}{25}$

b) $\dfrac{21}{3x}$

c) $\dfrac{2x - 6}{2x}$

d) $\dfrac{x^2 - x}{5x - 5}$

e) $\dfrac{x - 9y}{x + 9y}$

f) $\dfrac{x^2 + x - 2}{x - 1}$

5. Which fraction in each pair is larger?

a) $\dfrac{3}{14}$ or $\dfrac{3}{15}$

b) $\dfrac{14}{3}$ or $\dfrac{15}{3}$

c) $\dfrac{4}{5}$ or $\dfrac{7}{9}$

d) $\dfrac{15}{60}$ or $\dfrac{16}{61}$

6. Write each of the following sums or differences as a polynomial or fraction in simplest terms.

a) $\dfrac{3x}{4} + \dfrac{5x}{4}$

b) $\dfrac{x^2}{x - 1} - \dfrac{1}{x - 1}$

c) $\dfrac{2}{x} + \dfrac{2}{x + 2}$

d) $\dfrac{6 - x}{x} - \dfrac{6 - y}{y}$

e) $x^2 - \dfrac{1}{x^2}$

f) $x - 3 + \dfrac{6x}{x - 3}$

7. Change each of the following to the form described.

a) Write $\dfrac{5x + 3}{15}$ as the sum of two fractions.

b) Write $\dfrac{x - 2y}{x - 4y}$ as the difference of two fractions.

c) Write $\dfrac{x^3 + x - 6}{x - 6}$ as the sum of an integer and a fraction.

d) Write $\dfrac{3x^2 - 21}{x}$ as the difference of a monomial and a fraction.

8. Write each of the following products as a monomial or fraction in simplest terms.

a) $\dfrac{x^6}{2} \cdot \dfrac{6}{x^2}$

b) $\dfrac{x}{x - 5} \cdot \dfrac{x + 5}{x}$

c) $\dfrac{x - 8}{4 - x} \cdot \dfrac{x - 4}{8 - x}$

d) $x^2 \cdot \dfrac{3 - x}{3x}$

e) $(x + y)^2 \cdot \dfrac{1}{2xy}$

f) $\dfrac{3}{8x - 4} \cdot \dfrac{6x - 3}{4}$

9. Find each of the following products by first multiplying and then simplifying the result.

a) $\left(x + \dfrac{4}{9}\right)\left(x - \dfrac{4}{9}\right)$

b) $\left(3 - \dfrac{2}{x}\right)\left(2 - \dfrac{3}{x}\right)$

10. Find each of the products in exercise 9 by first writing each factor as a single fraction.

11. Write each of the following quotients as a polynomial or fraction in simplest terms.

a) $\dfrac{x^2}{12} \div \dfrac{x}{9}$

b) $\dfrac{x - 10}{10} \div \dfrac{x - 5}{5}$

c) $\dfrac{x^2 - 16}{4} \div \dfrac{x + 4}{16}$

d) $\dfrac{6}{5x} \div 3x^3$

e) $2x^4 \div \dfrac{4}{x^2}$

f) $\dfrac{x^2 + 10x - 11}{x - x^2} \div \dfrac{x + 11}{x}$

12. Simplify.

a) $\dfrac{x}{\dfrac{5}{x} - \dfrac{2}{x}}$

b) $\dfrac{\dfrac{x}{8} + 1}{\dfrac{x}{8} - 1}$

c) $\dfrac{\dfrac{1}{x} - \dfrac{1}{y}}{\dfrac{1}{x^2} - \dfrac{1}{y^2}}$

d) $1 - \dfrac{1}{1 - \dfrac{1}{x}}$

Chapter 12

SQUARE ROOTS

© 1978 UNITED FEATURE SYNDICATE, INC.

LESSON **1**
Squares and Square Roots

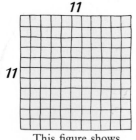

11

11

This figure shows that $11^2 = 121$ and that $11 = \sqrt{121}$.

The idea of a "square root" of a number comes from the idea of numbers as squares. The number 121, for example, is the square of 11 because it is the area of a square each of whose sides is 11 units long. Eleven, on the other hand, is a square root of 121. These ideas are expressed in the following definition.

▶ If $y = x^2$, then y is **the square** of x and x is **a square root** of y.

It follows from this definition that the number 121 has two square roots because both 11^2 and $(-11)^2$ are equal to 121. Every positive number, in fact, has two square roots, each of which is the opposite of the other. The positive root of 121 is represented by the symbol $\sqrt{121}$. The symbol $\sqrt{}$ is called a *radical sign*. The negative root of 121 is represented by the symbol $-\sqrt{121}$. To represent the two square roots of a positive number, a, we write $\sqrt{a}$ and $-\sqrt{a}$. The radical sign gets its name from the Latin word *radix*, which means "root."

Although every positive number has two square roots, many cannot be expressed exactly in decimal form. For example, the positive square root of 50 is a number between 7 and 8 because $7^2 = 49$ and $8^2 = 64$. Furthermore, because 50 is much closer to 49 than it is to 64, it would seem that $\sqrt{50}$ is closer to 7 than it is to 8. The calculations below show that $\sqrt{50}$ is larger than 7.071 but less than 7.072.

$$
\begin{array}{ll}
7^2 = 49 & 8^2 = 64 \\
7.0^2 = 49.00 & 7.1^2 = 50.41 \\
7.07^2 = 49.9849 & 7.08^2 = 50.1264 \\
7.071^2 = 49.999041 & 7.072^2 = 50.013184
\end{array}
$$

No matter how far this table is continued, we will never come to a number in decimal form whose square is exactly 50. For this reason, we have to settle for approximate square roots for most numbers. Approximate square roots can be found in tables, such as the one on page 563, or with a calculator that has a square-root key, or by other methods.

Although positive numbers have two square roots, the number zero has only one, itself. Because there is no positive or negative number whose square is negative, negative numbers do not seem to have any at all. For this reason, mathematicians felt for a long time that negative numbers did not have square roots. Eventually, they devised a new set of numbers, called "imaginary" numbers, that are square roots of negative numbers. Although imaginary numbers have turned out to be much more useful than their name suggests, we will not study them in this course. For our purposes, negative numbers will not have square roots.

Exercises

Set I

1. Solve each of the following equations for x.
   a) $8x + 3 = 3(x + 8)$
   b) $2(3x - 5) - 7x = 11$
   c) $6(x + 4) - 4(x - 6) = 64$

2. Rewrite each of the following polynomials in the form indicated.

   $6x + x^4 - 5$
   a) in descending powers of x.

   $x^2 + xy + y^3 + x^3y^2$
   b) in descending powers of x.
   c) in descending powers of y.

3. The sign in this photograph doesn't make any sense.
   a) According to the sign, how many minutes are you allowed to wait between 9:00 A.M. and 6:00 P.M.?
   b) Write a formula for the number of minutes, y, that you are allowed to wait during a period of x hours.
   c) How does y in your formula vary with respect to x?

Set II

4. List all of the square roots of each of the following numbers.
   a) 49
   b) 4
   c) –4
   d) 900
   e) 0
   f) –25

5. Use the table of squares and square roots on page 563 to express, where possible, each of the following as either an integer or a number in decimal form. Round answers in decimal form to the nearest hundredth.
   a) $(45)^2$
   b) $(-54)^2$
   c) $\sqrt{45}$
   d) $\sqrt{-54}$
   e) $\sqrt{6,084}$
   f) $(6.083)^2$
   g) $(7.2)^2$
   h) $\sqrt{53.29}$

6. Between what two consecutive integers are each of the following square roots?
   a) $\sqrt{10}$
   b) $\sqrt{20}$
   c) $-\sqrt{3}$
   d) $-\sqrt{33}$
   e) $\sqrt{124}$
   f) $\sqrt{1,240}$

7. Find the squares of the following numbers.
   a) 8
   b) 80
   c) 800
   d) 8,000

   Find the square roots of the following

numbers, each correct to the nearest integer.
   e) 36
   f) 360
   g) 3,600
   h) 36,000

*8. The following table lists some approximations of $\sqrt{126}$.

   To the nearest tenth: 11.2
   To the nearest hundredth: 11.22
   To the nearest thousandth: 11.225

   a) Square each of these numbers.
   b) Which approximations are smaller and which are larger than $\sqrt{126}$?

9. Express each of the following as an integer.
   a) $\sqrt{144}$
   b) $-\sqrt{81}$
   c) $(\sqrt{16})^2$
   d) $\sqrt{16^2}$
   e) $(\sqrt{7})^2$
   f) $\sqrt{7^2}$
   g) $\sqrt{(-7)^2}$

10. What symbol, $>$, $=$, or $<$, should replace ▓ in each of the following? (Remember that the square roots listed in the table are approximate.)
   a) $16 \cdot 9$ ▓ 144
   b) $16^2 \cdot 9^2$ ▓ 144^2

c) $\sqrt{16} \cdot \sqrt{9}$ ▨ $\sqrt{144}$
d) $16 + 9$ ▨ 25
e) $16^2 + 9^2$ ▨ 25^2
f) $\sqrt{16} + \sqrt{9}$ ▨ $\sqrt{25}$
g) $16 - 9$ ▨ 7
h) $16^2 - 9^2$ ▨ 7^2
i) $\sqrt{16} - \sqrt{9}$ ▨ $\sqrt{7}$
j) $\sqrt{20} + \sqrt{4}$ ▨ $\sqrt{24}$
k) $\sqrt{20} - \sqrt{4}$ ▨ $\sqrt{16}$
l) $\sqrt{20} \cdot \sqrt{4}$ ▨ $\sqrt{80}$

11. Find an integer equal to each of the following.
   a) $\sqrt{100 - 36}$
   b) $\sqrt{100} - \sqrt{36}$
   c) $\sqrt{4 \cdot 9}$
   d) $\sqrt{4} \cdot \sqrt{9}$
   e) $\sqrt{5^2 + 12^2}$
   f) $\sqrt{5^2} + \sqrt{12^2}$

Set III

12. List all of the square roots of each of the following numbers.
   a) 16
   b) 9
   c) 169
   d) –16
   e) 1
   f) –100

13. Use the table of squares and square roots on page 563 to express, where possible, each of the following as either an integer or a number in decimal form. Round answers in decimal form to the nearest hundredth.
   a) $(38)^2$
   b) $(-83)^2$
   c) $\sqrt{38}$
   d) $\sqrt{-83}$
   e) $\sqrt{5,476}$
   f) $(5.477)^2$
   g) $(1.9)^2$
   h) $\sqrt{3.24}$

14. Between what two consecutive integers are each of the following square roots?
   a) $\sqrt{17}$
   b) $\sqrt{34}$
   c) $-\sqrt{8}$
   d) $-\sqrt{88}$
   e) $\sqrt{300}$
   f) $\sqrt{3,000}$

15. Find the squares of the following numbers.
   a) 11
   b) 110
   c) 1,100
   d) 11,000

   Find the square roots of the following numbers, each correct to the nearest integer.

   e) 4
   f) 40
   g) 400
   h) 4,000

*16. The following table lists some approximations of $\sqrt{172}$.

To the nearest tenth:	13.1
To the nearest hundredth:	13.11
To the nearest thousandth:	13.115

   a) Square each of these numbers.
   b) Which approximations are smaller and which are larger than $\sqrt{172}$?

17. Express each of the following as an integer.
   a) $\sqrt{64}$
   b) $-\sqrt{121}$
   c) $(\sqrt{9})^2$
   d) $\sqrt{9^2}$
   e) $(\sqrt{15})^2$
   f) $\sqrt{15^2}$
   g) $\sqrt{(-15)^2}$

18. What symbol, $>$, $=$, or $<$, should replace ▨ in each of the following? (Remember that the square roots listed in the table are approximate.)
   a) $25 \cdot 4$ ▨ 100
   b) $25^2 \cdot 4^2$ ▨ 100^2
   c) $\sqrt{25} \cdot \sqrt{4}$ ▨ $\sqrt{100}$
   d) $25 + 4$ ▨ 29
   e) $25^2 + 4^2$ ▨ 29^2
   f) $\sqrt{25} + \sqrt{4}$ ▨ $\sqrt{29}$
   g) $25 - 4$ ▨ 21
   h) $25^2 - 4^2$ ▨ 21^2
   i) $\sqrt{25} - \sqrt{4}$ ▨ $\sqrt{21}$

j) $\sqrt{11} + \sqrt{9}$ ▓ $\sqrt{20}$
k) $\sqrt{11} - \sqrt{9}$ ▓ $\sqrt{2}$
l) $\sqrt{11} \cdot \sqrt{9}$ ▓ $\sqrt{99}$

19. Find an integer equal to each of the following.
    a) $\sqrt{9 + 16}$
    b) $\sqrt{9} + \sqrt{16}$
    c) $\sqrt{17^2 - 8^2}$
    d) $\sqrt{17^2} - \sqrt{8^2}$
    e) $\sqrt{4 \cdot 36}$
    f) $\sqrt{4} \cdot \sqrt{36}$

Set IV

Augustus De Morgan, an English mathematician of the nineteenth century, once said:

"I was x years old in the year x^2."

Can you figure out in what year he was born? (Remember that you are living in the *twentieth* century.)

Table of Squares and Approximate Square Roots

Number	Square	Square root	Number	Square	Square root	Number	Square	Square root
1	1	1	41	1681	6.403	81	6561	9
2	4	1.414	42	1764	6.481	82	6724	9.055
3	9	1.732	43	1849	6.557	83	6889	9.110
4	16	2	44	1936	6.633	84	7056	9.165
5	25	2.236	45	2025	6.708	85	7225	9.220
6	36	2.449	46	2116	6.782	86	7396	9.274
7	49	2.646	47	2209	6.856	87	7569	9.327
8	64	2.828	48	2304	6.928	88	7744	9.381
9	81	3	49	2401	7	89	7921	9.434
10	100	3.162	50	2500	7.071	90	8100	9.487
11	121	3.317	51	2601	7.141	91	8281	9.539
12	144	3.464	52	2704	7.211	92	8464	9.592
13	169	3.606	53	2809	7.280	93	8649	9.644
14	196	3.742	54	2916	7.348	94	8836	9.695
15	225	3.873	55	3025	7.416	95	9025	9.747
16	256	4	56	3136	7.483	96	9216	9.798
17	289	4.123	57	3249	7.550	97	9409	9.849
18	324	4.243	58	3364	7.616	98	9604	9.899
19	361	4.359	59	3481	7.681	99	9801	9.950
20	400	4.472	60	3600	7.746	100	10000	10
21	441	4.583	61	3721	7.810			
22	484	4.690	62	3844	7.874			
23	529	4.796	63	3969	7.937			
24	576	4.899	64	4096	8			
25	625	5	65	4225	8.062			
26	676	5.099	66	4356	8.124			
27	729	5.196	67	4489	8.185			
28	784	5.292	68	4624	8.246			
29	841	5.385	69	4761	8.307			
30	900	5.477	70	4900	8.367			
31	961	5.568	71	5041	8.426			
32	1024	5.657	72	5184	8.485			
33	1089	5.745	73	5329	8.544			
34	1156	5.831	74	5476	8.602			
35	1225	5.916	75	5625	8.660			
36	1296	6	76	5776	8.718			
37	1369	6.083	77	5929	8.775			
38	1444	6.164	78	6084	8.832			
39	1521	6.245	79	6241	8.888			
40	1600	6.325	80	6400	8.944			

LESSON 2
Square Roots of Products

$$R_x 9 \quad \ell 9 \quad z/9 \quad \sqrt{9} \quad 9^{1/2}$$

Sir Isaac Newton

Although our symbol for "the positive square root of nine" is $\sqrt{9}$, it has not always been written that way. Some of the symbols that have been used in the past to mean the same thing are shown above. The last of these was used by Sir Isaac Newton, who is famous not only as a scientist, but also as one of the inventors of calculus.

Why did Newton use the exponent $\frac{1}{2}$ to represent square root? To find out, look at the following pattern, which has as its basis the law of exponents.

$$9^3 \cdot 9^3 = 9^{3+3} = 9^6$$
$$9^2 \cdot 9^2 = 9^{2+2} = 9^4$$
$$9^1 \cdot 9^1 = 9^{1+1} = 9^2$$

If $9^{\frac{1}{2}}$ fits this pattern, then

$$9^{\frac{1}{2}} \cdot 9^{\frac{1}{2}} = 9^{\frac{1}{2}+\frac{1}{2}} = 9^1$$

This means that

$$(9^{\frac{1}{2}})^2 = 9$$

564

It follows that $9^{\frac{1}{2}}$ must be a square root of 9. For consistency, $9^{\frac{1}{2}}$ is defined as the positive square root of 9:

$$9^{\frac{1}{2}} = \sqrt{9} = 3$$

This is why Newton used the exponent $\dfrac{1}{2}$ to mean "square root."

► In general, if x is a nonnegative number,

$$x^{\frac{1}{2}} \text{ means } \sqrt{x}$$

When square roots are considered to be powers, the laws of exponents that we already know continue to be true. For example, letting $a = \dfrac{1}{2}$ in the *power of a product* law,

$$(xy)^a = x^a y^a$$

we get

$$(xy)^{\frac{1}{2}} = x^{\frac{1}{2}} y^{\frac{1}{2}}$$

or

$$\sqrt{xy} = \sqrt{x}\sqrt{y}$$

This is a very useful result.

► For any two nonnegative numbers x and y,

$$\sqrt{xy} = \sqrt{x} \cdot \sqrt{y}$$

This law makes it possible to "simplify" the square roots of many numbers by expressing them in terms of the square roots of smaller ones. Consider, for example, $\sqrt{200}$. Although this number is not included in the table on page 563, we can find its approximate value by doing the following:

$$\sqrt{200} = \sqrt{100 \cdot 2} = \sqrt{100} \cdot \sqrt{2} = 10\sqrt{2}$$
$$\text{Because } \sqrt{2} \approx 1.414, \ \sqrt{200} \approx 10(1.414) = 14.14.$$

When $\sqrt{200}$ is written as $10\sqrt{2}$, it is said to be expressed in *simple radical form*.

▶ The square root of an integer is in **simple radical form** if the integer has no factors that are squares of integers other than 1 and –1.

Here is another example of how to express a square root in simple radical form.

EXAMPLE 1
Simplify $\sqrt{48}$.

SOLUTION
The largest factor of 48 that is the square of an integer is 16:

$$\sqrt{48} = \sqrt{16 \cdot 3} = \sqrt{16} \cdot \sqrt{3} = 4\sqrt{3}$$

Because 3 has no factors that are squares of integers other than 1 and –1, $4\sqrt{3}$ is in simple radical form.

The same procedure can be applied to the square roots of monomials that have one or more factors that are squares. In such cases, *we will assume that the variables represent nonnegative numbers.*

▶ The square root of a monomial is in **simple radical form** if the monomial has no factors that are squares of monomials other than 1 and –1.

Here are examples of how square roots of monomials can be simplified.

EXAMPLE 2
Simplify $\sqrt{x^6}$.

SOLUTION
$$\sqrt{x^6} = \sqrt{(x^3)^2} = x^3$$

EXAMPLE 3
Simplify $\sqrt{75x^7}$.

SOLUTION
$$\sqrt{75x^7} = \sqrt{3 \cdot 25 \cdot x \cdot x^6} = \sqrt{25x^6 \cdot 3x} = \sqrt{25x^6} \cdot \sqrt{3x} = 5x^3\sqrt{3x}$$

Exercises

Set I

1. If possible, reduce each of these fractions.

    a) $\dfrac{10xy^2}{12x^2y}$ c) $\dfrac{x^2 + 7x + 6}{x^2 + 3x + 2}$

    b) $\dfrac{4x^2 + 1}{4x + 1}$

2. Find each of these products.
    a) $x \cdot 2x^2 \cdot 3x^3 \cdot 4x^4$
    b) $x(2x^2 + 3x^3 + 4x^4)$
    c) $(x + 2x^2)(3x^3 + 4x^4)$

3. The graph at the right is a rectangle.
    a) Find the slope of each of its sides.
    b) What do you notice about the slopes?

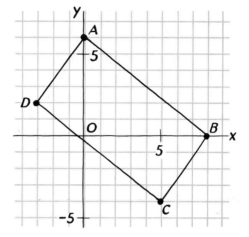

Set II

4. Write each of the following in radical form.
    a) $10^{\frac{1}{2}}$
    b) $(6x)^{\frac{1}{2}}$
    c) $6x^{\frac{1}{2}}$
    d) $(x^3)^{\frac{1}{2}}$
    e) $3x^{\frac{1}{2}}y$
    f) $3xy^{\frac{1}{2}}$
    g) $3(xy)^{\frac{1}{2}}$

    Write each of the following in exponential form.

    h) $\sqrt{8}$
    i) $5\sqrt{x}$
    j) $\sqrt{5x}$
    k) $\sqrt{xy}$
    l) $\sqrt{x}\,y$
    m) $x\sqrt{7y}$
    n) $10x\sqrt{y}$

5. What number should replace ▓ in each of the following equations to make it true?
    a) $\sqrt{9 \cdot 11} = ▓\sqrt{11}$
    b) $\sqrt{7 \cdot 25} = ▓\sqrt{7}$
    c) $\sqrt{16 \cdot 36} = ▓$
    d) $\sqrt{81 \cdot 81} = ▓$
    e) $\sqrt{2 \cdot 2 \cdot 13} = ▓\sqrt{13}$
    f) $\sqrt{17 \cdot 5 \cdot 17} = 17\sqrt{▓}$
    g) $\sqrt{4 \cdot 15 \cdot 100} = ▓\sqrt{15}$
    h) $\sqrt{3 \cdot 3 \cdot 3 \cdot 31} = 3\sqrt{▓}$

6. Write each of the following in simple radical form.
    a) $\sqrt{12}$
    b) $\sqrt{63}$
    c) $\sqrt{32}$
    d) $\sqrt{90}$
    e) $\sqrt{125}$
    f) $\sqrt{396}$
    g) $\sqrt{747}$
    h) $\sqrt{5400}$

7. Find the approximate value of each of the following square roots by first simplifying and then using the table on page 563.
    a) $\sqrt{207}$
    b) $\sqrt{208}$
    c) $\sqrt{700}$
    d) $\sqrt{7000}$

8. What expression should replace ▓ in each of the following equations to make it true? Assume that $x > 0$.
    a) $\sqrt{49x} = ▓\sqrt{x}$
    b) $\sqrt{6x^2} = ▓\sqrt{6}$
    c) $\sqrt{2x^3 \cdot 2x^3} = ▓$
    d) $\sqrt{x^8} = ▓$
    e) $\sqrt{x^5} = ▓\sqrt{x}$
    f) $\sqrt{4x^{36}} = ▓$

9. Where possible, express each of the following as a monomial. Assume that $x > 0$.
   a) $\sqrt{9x^2}$ c) $\sqrt{36x}$
   b) $\sqrt{x^{36}}$ d) $(\sqrt{36x})^2$

10. Write each of the following either as a monomial or in simple radical form. Assume that $x > 0$.
   a) $\sqrt{64x}$ d) $\sqrt{40x^2}$
   b) $\sqrt{x^{64}}$ e) $\sqrt{121x^3}$
   c) $\sqrt{x^7}$ f) $\sqrt{18x^{18}}$

Set III

11. Write each of the following in radical form.
   a) $3^{\frac{1}{2}}$
   b) $(14x)^{\frac{1}{2}}$ e) $7(xy)^{\frac{1}{2}}$
   c) $14x^{\frac{1}{2}}$ f) $7x^{\frac{1}{2}}y$
   d) $(x^5)^{\frac{1}{2}}$ g) $7xy^{\frac{1}{2}}$

   Write each of the following in exponential form.

   h) $\sqrt{6}$
   i) $11\sqrt{x}$
   j) $\sqrt{11x}$
   k) $\sqrt{x^3}$
   l) $4x\sqrt{y}$
   m) $x\sqrt{3y}$
   n) $\sqrt{2xy}$

12. What number should replace ▓ in each of the following equations to make it true?
   a) $\sqrt{4 \cdot 13} = $ ▓ $\sqrt{13}$
   b) $\sqrt{5 \cdot 81} = $ ▓ $\sqrt{5}$
   c) $\sqrt{9 \cdot 49} = $ ▓
   d) $\sqrt{64 \cdot 64} = $ ▓
   e) $\sqrt{3 \cdot 3 \cdot 7} = $ ▓ $\sqrt{7}$
   f) $\sqrt{15 \cdot 11 \cdot 15} = 15\sqrt{▓}$
   g) $\sqrt{16 \cdot 5 \cdot 100} = $ ▓ $\sqrt{5}$
   h) $\sqrt{2 \cdot 2 \cdot 2 \cdot 23} = 2\sqrt{▓}$

13. Write each of the following in simple radical form.
   a) $\sqrt{18}$ e) $\sqrt{600}$
   b) $\sqrt{20}$ f) $\sqrt{284}$
   c) $\sqrt{48}$ g) $\sqrt{343}$
   d) $\sqrt{99}$ h) $\sqrt{825}$

14. Find the approximate value of each of the following square roots by first simplifying and then using the table on page 563.
   a) $\sqrt{147}$ c) $\sqrt{300}$
   b) $\sqrt{148}$ d) $\sqrt{3000}$

15. What expression should replace ▓ in each of the following equations to make it true? Assume that $x > 0$.
   a) $\sqrt{36x} = $ ▓ $\sqrt{x}$
   b) $\sqrt{10x^2} = $ ▓ $\sqrt{10}$
   c) $\sqrt{5x^3 \cdot 5x^3} = $ ▓
   d) $\sqrt{x^6} = $ ▓
   e) $\sqrt{x^{11}} = $ ▓
   f) $\sqrt{9x^{16}} = $ ▓

16. Where possible, express each of the following as a monomial. Assume that $x > 0$.
   a) $\sqrt{81x^2}$
   b) $\sqrt{x^{16}}$
   c) $\sqrt{16x}$
   d) $(\sqrt{16x})^2$

17. Write each of the following either as a monomial or in simple radical form. Assume that $x > 0$.
   a) $\sqrt{100x}$
   b) $\sqrt{x^{100}}$
   c) $\sqrt{x^9}$
   d) $\sqrt{24x^2}$
   e) $\sqrt{81x^3}$
   f) $\sqrt{8x^8}$

Set IV

Orbiting the earth are a large number of communication and weather satellites. In order for a satellite to stay in orbit, it must have a speed given by the formula

$$v = \sqrt{\frac{4 \times 10^{14}}{d}}$$

in which v represents the speed in meters per second and d represents the satellite's distance from the center of the earth in meters.

Can you figure out the approximate speed of these satellites, given that they are 4.2×10^7 meters from the center of the earth? If so, show your method.

Telstar

NATIONAL AERONAUTICS AND SPACE ADMINISTRATION.

LESSON 3
Square Roots of Quotients

If the earth were flat, the distance a person could see in each direction might depend on how good a telescope he or she had. Because the earth is curved, however, it is impossible to see beyond a circle called the horizon. The size of this circle depends on the observer's height above the ground: the greater the height, the larger the circle becomes.

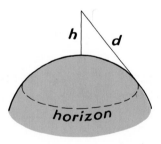

A formula for the distance to the horizon circle, d, is

$$d = \sqrt{\frac{3}{2}h}$$

in which d is measured in miles and h, the observer's height above the ground, is measured in feet. According to this formula, if the bug in this cartoon were one

570

foot above the ground, it could not see any farther than

$$\sqrt{\frac{3}{2}(1)} = \sqrt{\frac{3}{2}} \text{ miles}$$

How far is that? To answer this question, we have to find the square root of a fraction. This is usually not very easy to do. Fortunately, however, the square root of a fraction can be found from the square roots of the numerator and denominator. The method has as its basis the *power of a quotient* law:

$$\left(\frac{x}{y}\right)^a = \frac{x^a}{y^a}$$

Letting $a = \frac{1}{2}$ in this law, we get

$$\left(\frac{x}{y}\right)^{\frac{1}{2}} = \frac{x^{\frac{1}{2}}}{y^{\frac{1}{2}}} \quad \text{or} \quad \sqrt{\frac{x}{y}} = \frac{\sqrt{x}}{\sqrt{y}}$$

This law lets us express the square roots of many fractions as *the quotients of square roots of integers*. For example,

$$\sqrt{\frac{3}{2}} = \frac{\sqrt{3}}{\sqrt{2}}$$

According to the table on page 563, $\sqrt{3} \approx 1.732$ and $\sqrt{2} \approx 1.414$, and so

$$\frac{\sqrt{3}}{\sqrt{2}} \approx \frac{1.732}{1.414} \approx 1.225$$

If you are working with pencil and paper, there is an easier way to get this answer than by dividing 1.414 into 1.732. Before changing the square root of the fraction into a quotient, first rewrite the fraction so that its denominator is a square. Because $\frac{3}{2} = \frac{2 \cdot 3}{2 \cdot 2} = \frac{6}{4}$,

$$\sqrt{\frac{3}{2}} = \sqrt{\frac{6}{4}} = \frac{\sqrt{6}}{\sqrt{4}} = \frac{\sqrt{6}}{2} \approx \frac{2.449}{2} \approx 1.225$$

So a distance of $\sqrt{\frac{3}{2}}$ miles is approximately 1.225 miles.

Lesson 3: Square Roots of Quotients 571

The square root of any fraction can be simplified in the same way. If the denominator of the fraction is not a square, we begin by transforming the fraction into one whose denominator is a square:

$$\sqrt{\frac{a}{b}} = \sqrt{\frac{ab}{b^2}}$$

The square root of the fraction can then be rewritten as the quotient of two square roots:

$$\sqrt{\frac{ab}{b^2}} = \frac{\sqrt{ab}}{\sqrt{b^2}}$$

Because the denominator of this quotient is the square root of a square, it can be rewritten as an integer or monomial:

$$\frac{\sqrt{ab}}{\sqrt{b^2}} = \frac{\sqrt{ab}}{b}$$

Finally, if the numerator contains any factors that are squares of monomials other than 1, it can be changed to simple radical form.

Remember that we do not consider negative numbers to have square roots. Because of this, we will assume that any variable or expression under a radical sign represents a nonnegative number.

Here are more examples of how to simplify square roots of fractions.

EXAMPLE 1

Simplify $\sqrt{\dfrac{7}{9}}$ and find its approximate value.

SOLUTION

Because the denominator of the fraction is already a square, we can write $\sqrt{\dfrac{7}{9}} = \dfrac{\sqrt{7}}{\sqrt{9}} = \dfrac{\sqrt{7}}{3}$. According to the table on page 563, $\sqrt{7} \approx 2.646$, and so $\dfrac{\sqrt{7}}{3} \approx \dfrac{2.646}{3} \approx 0.882$.

EXAMPLE 2

Simplify $\sqrt{\dfrac{5}{12}}$ and find its approximate value.

SOLUTION

Although we could make the denominator of this fraction into a square by multiplying it by 12, $12 \cdot 12 = 144$, multiplying it by 3 is simpler: $12 \cdot 3 = 36$.

$$\sqrt{\frac{5}{12}} = \sqrt{\frac{3 \cdot 5}{3 \cdot 12}} = \sqrt{\frac{15}{36}} = \frac{\sqrt{15}}{\sqrt{36}} = \frac{\sqrt{15}}{6} \approx \frac{3.873}{6} \approx 0.646$$

EXAMPLE 3

Write $\sqrt{\dfrac{3}{x}}$ as a quotient without a square root in the denominator.

SOLUTION

$$\sqrt{\frac{3}{x}} = \sqrt{\frac{x \cdot 3}{x \cdot x}} = \sqrt{\frac{3x}{x^2}} = \frac{\sqrt{3x}}{\sqrt{x^2}} = \frac{\sqrt{3x}}{x}$$

Exercises

Set I

1. Write each of the following sums or products as a single fraction.

   a) $\dfrac{2}{x} + \dfrac{8}{x}$ c) $\dfrac{x}{2} + \dfrac{x}{8}$

   b) $\dfrac{2}{x} \cdot \dfrac{8}{x}$ d) $\dfrac{x}{2} \cdot \dfrac{x}{8}$

2. Solve the following pairs of simultaneous equations.

   a) $x = 7y + 3$ b) $2x + 5y = 23$
   $y = 3x - 7$ $5x - 2y = 14$

3. The amount of heat lost through a windowpane depends on several things, including the thickness of the glass. Here is a table showing the heat loss through several windows of equal size under identical conditions.

Thickness of glass in mm, x	2	4	6	8	10
Heat loss in calories, y	300	150	100	75	60

a) How does the heat lost, y, vary with respect to the thickness of the windowpane, x?

b) Write a formula for y in terms of x.

c) How much heat would be lost through a window 5 millimeters thick under the same conditions?

Set II

4. Write fractions whose denominators are squares, equivalent to the following.

a) $\dfrac{1}{5}$

b) $\dfrac{2}{3}$

c) $\dfrac{11}{18}$

d) $\dfrac{x}{6}$

e) $\dfrac{4}{x^3}$

f) $\dfrac{5}{8x}$

5. What number should replace ▓ in each of the following equations to make it true?

a) $\sqrt{\dfrac{3}{16}} = \dfrac{\sqrt{3}}{▓}$

b) $\sqrt{\dfrac{4}{25}} = ▓$

c) $\sqrt{\dfrac{2}{7}} = \dfrac{\sqrt{14}}{▓}$

d) $\sqrt{\dfrac{1}{18}} = \dfrac{▓}{6}$

6. Write each of the following radicals as a quotient without a square root in the denominator.

a) $\sqrt{\dfrac{5}{9}}$

b) $\sqrt{\dfrac{11}{2}}$

c) $\sqrt{\dfrac{6}{7}}$

d) $\sqrt{\dfrac{1}{8}}$

e) $\sqrt{\dfrac{7}{12}}$

f) $\sqrt{\dfrac{13}{50}}$

7. Find the approximate value of each of the following square roots by first simplifying and then using the table on page 563.

a) $\sqrt{\dfrac{3}{4}}$

b) $\sqrt{\dfrac{1}{10}}$

c) $\sqrt{\dfrac{5}{3}}$

d) $\sqrt{\dfrac{7}{20}}$

8. What expression should replace ▓ in each of the following equations to make it true? Assume that $x > 0$.

a) $\sqrt{\dfrac{x}{64}} = \dfrac{\sqrt{x}}{▓}$

b) $\sqrt{\dfrac{3}{x^2}} = \dfrac{▓}{x}$

c) $\sqrt{\dfrac{14}{x}} = \dfrac{▓}{x}$

d) $\sqrt{\dfrac{x^{16}}{16}} = ▓$

e) $\sqrt{\dfrac{2}{x^5}} = \dfrac{\sqrt{2x}}{▓}$

f) $\sqrt{\dfrac{x+1}{20}} = \dfrac{\sqrt{5x+5}}{▓}$

9. Express each of the following as a quotient without a square root in the denominator. Assume that $x > 0$.

a) $\sqrt{\dfrac{x}{81}}$

b) $\sqrt{\dfrac{3}{x^4}}$

c) $\sqrt{\dfrac{x^6}{25}}$

d) $\sqrt{\dfrac{10}{x}}$

e) $\sqrt{\dfrac{x-2}{3}}$

f) $\sqrt{\dfrac{1}{4x^3}}$

10. Whenever a variable is written under a radical sign, we assume that it represents a nonnegative number. To see why, do the following exercises.

Find the values of $\sqrt{25x^2}$ and $5x$ if

a) $x = 2$

b) $x = 6$

c) $x = 0$

d) $x = -1$

e) $x = -4$

Find the values of $\sqrt{\dfrac{x^2}{25}}$ and $\dfrac{x}{5}$ if

f) $x = 5$

g) $x = 10$

h) $x = -5$

Does $\sqrt{25x^2} = 5x$ if

i) x is a nonnegative number?

j) x is a negative number?

Does $\sqrt{\dfrac{x^2}{25}} = \dfrac{x}{5}$ if

k) x is a nonnegative number?

l) x is a negative number?

Set III

11. Write fractions whose denominators are squares, equivalent to the following.

a) $\dfrac{1}{3}$

b) $\dfrac{4}{7}$

c) $\dfrac{5}{8}$

d) $\dfrac{x}{2}$

e) $\dfrac{9}{x}$

f) $\dfrac{6}{x^3}$

12. What number should replace ▦ in each of the following equations to make it true?

a) $\sqrt{\dfrac{5}{49}} = \dfrac{\sqrt{5}}{▦}$

b) $\sqrt{\dfrac{9}{121}} = ▦$

c) $\sqrt{\dfrac{3}{11}} = \dfrac{\sqrt{33}}{▦}$

d) $\sqrt{\dfrac{1}{50}} = \dfrac{▦}{10}$

13. Write each of the following radicals as a quotient without a square root in the denominator.

a) $\sqrt{\dfrac{7}{16}}$

b) $\sqrt{\dfrac{10}{3}}$

c) $\sqrt{\dfrac{2}{5}}$

d) $\sqrt{\dfrac{1}{32}}$

e) $\sqrt{\dfrac{11}{12}}$

f) $\sqrt{\dfrac{5}{72}}$

14. Find the approximate value of each of the following square roots by first simplifying and then using the table on page 563.

a) $\sqrt{\dfrac{2}{9}}$

b) $\sqrt{\dfrac{7}{10}}$

c) $\sqrt{\dfrac{6}{5}}$

d) $\sqrt{\dfrac{1}{18}}$

15. What expression should replace ▦ in each of the following equations to make it true? Assume that $x > 0$.

a) $\sqrt{\dfrac{x}{144}} = \dfrac{\sqrt{x}}{▦}$

b) $\sqrt{\dfrac{2}{x^6}} = \dfrac{▦}{x^3}$

c) $\sqrt{\dfrac{15}{x}} = \dfrac{▦}{x}$

d) $\sqrt{\dfrac{x^{36}}{36}} = ▦$

e) $\sqrt{\dfrac{5}{x^3}} = \dfrac{\sqrt{5x}}{▦}$

f) $\sqrt{\dfrac{x-1}{32}} = \dfrac{\sqrt{2x-2}}{▦}$

16. Express each of the following as a quotient without a square root in the denominator. Assume that $x > 0$.

a) $\sqrt{\dfrac{x}{25}}$

b) $\sqrt{\dfrac{5}{x^2}}$

c) $\sqrt{\dfrac{x^8}{9}}$

d) $\sqrt{\dfrac{2}{x}}$

e) $\sqrt{\dfrac{x+3}{2}}$

f) $\sqrt{\dfrac{1}{3x^5}}$

17. Whenever a variable is written under a radical sign, we assume that it represents a nonnegative number. To see why, do the following exercises.

Find the values of $\sqrt{16x^2}$ and $4x$ if

a) $x = 3$
b) $x = 10$
c) $x = 0$
d) $x = -2$
e) $x = -5$

Find the values of $\sqrt{\dfrac{x^2}{16}}$ and $\dfrac{x}{4}$ if

f) $x = 4$
g) $x = 8$
h) $x = -4$

Does $\sqrt{16x^2} = 4x$ if

i) x is a nonnegative number?
j) x is a negative number?

Does $\sqrt{\dfrac{x^2}{16}} = \dfrac{x}{4}$ if

k) x is a nonnegative number?
l) x is a negative number?

Set IV

In 1966, Nicholas Piantanida took off in a balloon from Sioux Falls, South Dakota, and reached an altitude of about 123,800 feet before landing in a corn field in Iowa.

Can you use the formula given at the beginning of this lesson and the table on page 563 to estimate the greatest distance in each direction on the earth that he was able to see while taking this trip?

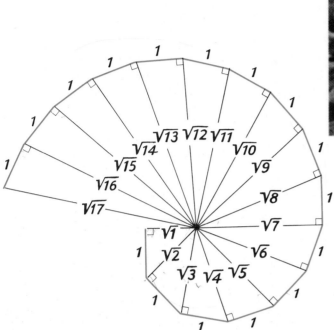

LEONARD LEE RUE III © 1971.

Adding and Subtracting Square Roots

This figure consists of a set of triangles winding around a point. The outer sides of the triangles look very much like a *spiral*, a curve that appears in nature in the horns of a ram. Each triangle was drawn so that it contains a right angle (marked by a small square) and so that its outer side is 1 unit long. It is possible to prove, by means of the Pythagorean Theorem, that the inner sides of the triangles are equal in length to the square roots of the consecutive integers from 1 through 17. As a result, this figure might be called a "spiral of square roots."

If you take a good look at the segments in the figure whose lengths are $\sqrt{2}$, $\sqrt{3}$, and $\sqrt{5}$, you will see that

$$\sqrt{2} + \sqrt{3} > \sqrt{5}$$

In fact, for every pair of positive numbers, x and y,

$$\sqrt{x} + \sqrt{y} > \sqrt{x + y}$$

There is no general equation relating $\sqrt{x}$, $\sqrt{y}$, and $\sqrt{x+y}$, so nothing can be done to "simplify" an expression such as $\sqrt{2} + \sqrt{3}$.

On the other hand, an expression such as $\sqrt{1} + \sqrt{4}$ *can* be simplified because $\sqrt{1} + \sqrt{4} = 1 + 2 = 3$. Furthermore, it is often possible to simplify the sum (or difference) of two square roots by the procedure illustrated by the following examples.

EXAMPLE 1

Simplify $\sqrt{8} + \sqrt{18}$.

SOLUTION

Both 8 and 18 have factors that are squares, and so we can write

$$\begin{aligned}\sqrt{8} + \sqrt{18} &= \sqrt{4 \cdot 2} + \sqrt{9 \cdot 2} \\ &= \sqrt{4}\sqrt{2} + \sqrt{9}\sqrt{2} \\ &= 2\sqrt{2} + 3\sqrt{2} \\ &= 5\sqrt{2}\end{aligned}$$

EXAMPLE 2

Simplify $\sqrt{75} - \sqrt{3}$.

SOLUTION

$$\begin{aligned}\sqrt{75} - \sqrt{3} &= \sqrt{25 \cdot 3} - \sqrt{3} \\ &= \sqrt{25}\sqrt{3} - \sqrt{3} \\ &= 5\sqrt{3} - \sqrt{3} \\ &= 4\sqrt{3}\end{aligned}$$

EXAMPLE 3

Simplify $\sqrt{x} + \sqrt{9x}$.

SOLUTION

$$\begin{aligned}\sqrt{x} + \sqrt{9x} &= \sqrt{x} + \sqrt{9}\sqrt{x} \\ &= \sqrt{x} + 3\sqrt{x} \\ &= 4\sqrt{x}\end{aligned}$$

Exercises

Set I

1. Change each of the following to the form described.

   a) Write $\dfrac{x+6}{2}$ as the sum of an integer and a fraction.

   b) Write $\dfrac{6x}{2}$ as the product of an integer and a fraction.

   c) Write $\dfrac{x^2 - y^2}{xy}$ as the difference of two fractions.

   d) Write $\dfrac{x^2 - y^2}{xy}$ as the product of two fractions.

2. Find x in each of the following equations.

   a) $2^5 \cdot 2^{12} = 2^x$ c) $5^2 + 12^2 = x^2$
   b) $5^2 \cdot 12^2 = x^2$ d) $(5 + 12)^2 = x^2$

3. This figure is called an *isosceles trapezoid*.

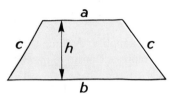

   a) Its perimeter, like that of a rectangle, is the sum of the lengths of its four sides. Write a formula for the perimeter, p, of this figure.

   b) The area of a trapezoid is equal to the product of half its altitude, h, and the sum of its bases, a and b. Write a formula for the area, A, of this figure.

Set II

4. Write a multiplication problem equivalent to each of the following addition problems.

   a) $\sqrt{5} + \sqrt{5} + \sqrt{5}$

   b) $\sqrt{3} + \sqrt{3} + \sqrt{3} + \sqrt{3} + \sqrt{3}$

   c) $\sqrt{x} + \sqrt{x} + \sqrt{x} + \sqrt{x} + \sqrt{x} + \sqrt{x} + \sqrt{x}$

   d) $\sqrt{11x} + \sqrt{11x} + \sqrt{11x} + \sqrt{11x}$

   Write an addition problem equivalent to each of the following multiplication problems.

   e) $2\sqrt{10}$
   f) $5\sqrt{x}$

5. If possible, simplify each of the following sums and differences as shown in the examples of this lesson.

   a) $6\sqrt{2} + 5\sqrt{2}$ g) $\sqrt{12} + 3\sqrt{3}$
   b) $6\sqrt{2} - 5\sqrt{2}$ h) $\sqrt{20} + \sqrt{5}$
   c) $6\sqrt{2} + \sqrt{2}$ i) $8\sqrt{2} - \sqrt{8}$
   d) $6\sqrt{2} - \sqrt{2}$ j) $\sqrt{11} + \sqrt{99}$
   e) $\sqrt{6} + \sqrt{7}$ k) $\sqrt{54} - \sqrt{24}$
   f) $\sqrt{7} - \sqrt{6}$ l) $\sqrt{24} - \sqrt{54}$

6. Use the table on page 563 as necessary to find the value of each of the following expressions. Round any approximate answers to the nearest tenth.

   a) $\sqrt{4} + \sqrt{36}$ c) $\sqrt{11} + \sqrt{5}$
   b) $\sqrt{4 + 36}$ d) $\sqrt{11 + 5}$

e) $\sqrt{10} - \sqrt{9}$
f) $\sqrt{10 - 9}$
g) $2\sqrt{3}$
h) $2\sqrt{3} + 1$
i) $2\sqrt{3 + 1}$
j) $\sqrt{2(3 + 1)}$

7. Simplify.
   a) $(7 + 2\sqrt{5}) + (1 + 3\sqrt{5})$
   b) $(3 + \sqrt{3}) + (6 + \sqrt{6})$
   c) $(18 + \sqrt{18}) + (2 + \sqrt{2})$
   d) $(1 + \sqrt{10}) + (1 - \sqrt{10})$
   e) $(5 + 5\sqrt{2}) - (4 + 4\sqrt{2})$
   f) $(75 + \sqrt{75}) - (27 + \sqrt{27})$

8. If possible, simplify. Assume that the variables represent positive numbers.
   a) $4\sqrt{x} + 16\sqrt{x}$
   b) $x\sqrt{4} + x\sqrt{16}$
   c) $x + \sqrt{x}$

d) $\sqrt{25x} - \sqrt{x}$
e) $64\sqrt{xy} + \sqrt{64xy}$
f) $\sqrt{x} - \sqrt{y}$
g) $\sqrt{28x} + \sqrt{7x}$
h) $\sqrt{72x} - \sqrt{32x}$

9. Solve each of the following equations for x. Leave your answers in radical form but simplify them as much as possible.
   a) $x + \sqrt{8} = \sqrt{18}$
   b) $x + \sqrt{9} = \sqrt{19}$
   c) $x - \sqrt{11} = \sqrt{1100}$
   d) $\sqrt{144} + x = \sqrt{64}$
   e) $3x + \sqrt{32} = \sqrt{200}$
   f) $\sqrt{5x} - \sqrt{45} = \sqrt{125}$

Set III

10. Write a multiplication problem equivalent to each of the following addition problems.
    a) $\sqrt{2} + \sqrt{2} + \sqrt{2} + \sqrt{2} + \sqrt{2} + \sqrt{2}$
    b) $\sqrt{6} + \sqrt{6}$
    c) $\sqrt{x} + \sqrt{x} + \sqrt{x} + \sqrt{x} + \sqrt{x}$
    d) $\sqrt{5x} + \sqrt{5x} + \sqrt{5x} + \sqrt{5x} + \sqrt{5x} + \sqrt{5x}$

    Write an addition problem equivalent to each of the following multiplication problems.
    e) $3\sqrt{7}$
    f) $4\sqrt{x}$

11. If possible, simplify each of the following sums and differences as shown in the examples of this lesson.
    a) $5\sqrt{3} + 4\sqrt{3}$
    b) $5\sqrt{3} - 4\sqrt{3}$
    c) $5\sqrt{3} + \sqrt{3}$
    d) $5\sqrt{3} - \sqrt{3}$
    e) $\sqrt{10} + \sqrt{11}$
    f) $\sqrt{11} - \sqrt{10}$
    g) $\sqrt{18} + 7\sqrt{2}$
    h) $\sqrt{24} + \sqrt{6}$
    i) $9\sqrt{5} - \sqrt{20}$
    j) $\sqrt{13} + \sqrt{52}$
    k) $\sqrt{63} - \sqrt{28}$
    l) $\sqrt{28} - \sqrt{63}$

12. Use the table on page 563 as necessary to find the value of each of the following expressions. Round any approximate answers to the nearest tenth.
    a) $\sqrt{16} + \sqrt{64}$
    b) $\sqrt{16 + 64}$
    c) $\sqrt{40} + \sqrt{9}$
    d) $\sqrt{40 + 9}$
    e) $\sqrt{15} - \sqrt{12}$
    f) $\sqrt{15 - 12}$
    g) $3\sqrt{2}$
    h) $3\sqrt{2} - 1$
    i) $3\sqrt{2 - 1}$
    j) $\sqrt{3(2 - 1)}$

13. Simplify.
    a) $(1 + 4\sqrt{6}) + (4 + 2\sqrt{6})$
    b) $(5 + \sqrt{5}) + (10 + \sqrt{10})$
    c) $(12 + \sqrt{12}) + (3 + \sqrt{3})$
    d) $(8 + \sqrt{2}) + (8 - \sqrt{2})$
    e) $(4 + 4\sqrt{7}) - (3 + 3\sqrt{7})$
    f) $(80 + \sqrt{80}) - (20 + \sqrt{20})$

14. If possible, simplify. Assume that the variables represent positive numbers.
    a) $25\sqrt{x} + 9\sqrt{x}$
    b) $x\sqrt{25} + x\sqrt{9}$
    c) $x - \sqrt{x}$
    d) $\sqrt{49x} + \sqrt{4x}$
    e) $81\sqrt{xy} - \sqrt{81xy}$
    f) $\sqrt{x} + \sqrt{y}$
    g) $\sqrt{44x} + \sqrt{11x}$
    h) $\sqrt{147x} - \sqrt{27x}$

15. Solve each of the following equations for x.
    Leave your answers in radical form but
    simplify them as much as possible.
    a) $x + \sqrt{12} = \sqrt{27}$
    b) $x + \sqrt{10} = \sqrt{25}$
    c) $x - \sqrt{7} = \sqrt{700}$
    d) $\sqrt{121} + x = \sqrt{81}$
    e) $2x + \sqrt{5} = \sqrt{405}$
    f) $\sqrt{3}x - \sqrt{12} = \sqrt{75}$

Set IV

Even though he is repeating algebra for the seventeenth time, Reckless Rex still
thinks that

$$\sqrt{x^2 + y^2} = x + y$$

1. Do you think this is ever true? Explain your answer.
2. Which side of this "equation" is usually larger?

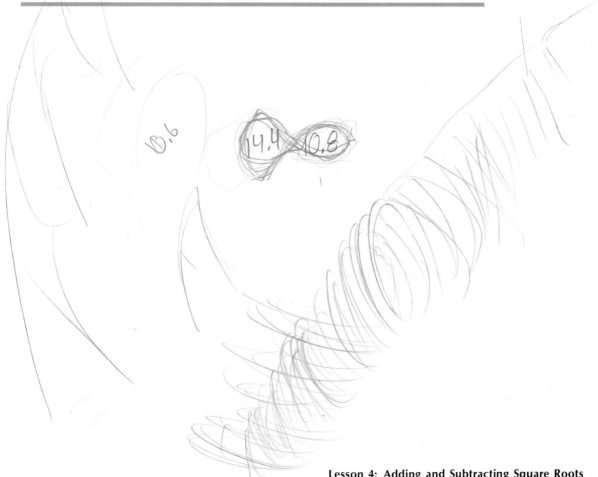

"If only he could think in abstract terms. . . ."

LESSON 5
Multiplying Square Roots

The calculation written on the blackboard by the man in this cartoon contains several operations, including addition, multiplication, and finding a square root. We have already learned how to add and subtract expressions containing square roots. In this lesson we will learn how to multiply them.

Although $\sqrt{2} + \sqrt{3}$ is *not* equal to $\sqrt{5}$, it *is* true that $\sqrt{2} \cdot \sqrt{3} = \sqrt{6}$. This follows from the fact that, if neither x nor y are negative numbers,

$$\sqrt{x}\sqrt{y} = \sqrt{xy}$$

This principle is used in simplifying expressions containing products of square roots. It is often combined with one or more of the following patterns, which you have already used many times.

$$a(b + c) = ab + ac$$
$$(a + b)(c + d) = ac + ad + bc + bd$$
$$(a + b)^2 = a^2 + 2ab + b^2$$
$$(a - b)^2 = a^2 - 2ab + b^2$$
$$(a + b)(a - b) = a^2 - b^2$$

Here are examples of how expressions containing square roots are multiplied.

EXAMPLE 1

Find the product of $\sqrt{3}$ and $\sqrt{48}$.

SOLUTION

$$(\sqrt{3})(\sqrt{48}) = \sqrt{3 \cdot 48} = \sqrt{144} = 12$$

EXAMPLE 2

Find the product of $4\sqrt{x}$ and x^3. Assume that $x > 0$.

SOLUTION

$$(4\sqrt{x})(5\sqrt{x^3}) = 20\sqrt{x^4} = 20x^2$$

EXAMPLE 3

Multiply $\sqrt{2}$ and $\sqrt{8} - \sqrt{6}$.

SOLUTION

Using the distributive rule, we write

$$\begin{aligned}
\sqrt{2}(\sqrt{8} - \sqrt{6}) &= \sqrt{2}\sqrt{8} - \sqrt{2}\sqrt{6} \\
&= \sqrt{16} - \sqrt{12} \\
&= 4 - \sqrt{4 \cdot 3} \\
&= 4 - 2\sqrt{3}
\end{aligned}$$

EXAMPLE 4

Find $(\sqrt{x} + \sqrt{7})^2$. Assume that $x > 0$.

SOLUTION

Treating $(\sqrt{x} + \sqrt{7})^2$ as the square of a binomial, we write

$$\begin{aligned}
(\sqrt{x} + \sqrt{7})^2 &= (\sqrt{x})^2 + 2(\sqrt{x}\sqrt{7}) + (\sqrt{7})^2 \\
&= x + 2\sqrt{7x} + 7
\end{aligned}$$

EXAMPLE 5

Find the product of $\sqrt{10} + \sqrt{3x}$ and $\sqrt{10} - \sqrt{3x}$.

SOLUTION

Because the product of the sum and difference of two numbers is equal to the difference of their squares, we write

$$(\sqrt{10} + \sqrt{3x})(\sqrt{10} - \sqrt{3x}) = (\sqrt{10})^2 - (\sqrt{3x})^2$$
$$= 10 - 3x$$

EXAMPLE 6

Multiply $3 - \sqrt{2}$ and $8 + 5\sqrt{2}$.

SOLUTION

Treating $3 - \sqrt{2}$ and $8 + 5\sqrt{2}$ as binomials, we write

$$(3 - \sqrt{2})(8 + 5\sqrt{2}) = 24 + 3(5\sqrt{2}) - 8\sqrt{2} - \sqrt{2}(5\sqrt{2})$$
$$= 24 + 15\sqrt{2} - 8\sqrt{2} - 10$$
$$= 14 + 7\sqrt{2}.$$

The first line of this solution shows the four terms that are products of the terms of $3 - \sqrt{2}$ and $8 + 5\sqrt{2}$.

Exercises

Set I

1. If possible, factor each of these polynomials.
   a) $4x - 7x^2$
   b) $2x^2 - 15x + 7$
   c) $x^2 + 2x + 17$
   d) $9xy - x^3y^3$

2. Find the square roots of the following numbers, each correct to the nearest integer.

You may want to refer to the table of square roots on page 563.
   a) 10
   b) 100
   c) 1,000
   d) 10,000
   e) 100,000

3. The Great Wall of China, built about 300 B.C., is the longest wall in the world. Spaced along it were 40,000 watchtowers. If these towers had been built 10 meters closer together and the wall had remained the same length, there would have been 48,000 of them instead.

   a) Write an equation for this problem, letting x represent the distance in meters between each pair of watchtowers. (Assume that the towers were evenly spaced along the wall.)

   b) How many meters long is the wall?

© 1978 STERN/BLACK STAR.

Set II

4. Multiply and simplify.
   a) $(\sqrt{3})(\sqrt{27})$
   b) $(\sqrt{20})(\sqrt{2})$
   c) $(\sqrt{11})(\sqrt{99})$
   d) $(\sqrt{5})(12\sqrt{5})$
   e) $(\sqrt{5})(5\sqrt{12})$
   f) $(2\sqrt{10})(18\sqrt{10})$
   g) $(10\sqrt{2})(10\sqrt{18})$
   h) $(\sqrt{3})(\sqrt{14})(\sqrt{42})$
   i) $2(8 + \sqrt{8})$
   j) $\sqrt{2}(\sqrt{8} + 8)$
   k) $\sqrt{7}(\sqrt{7} - 1)$
   l) $\sqrt{3}(\sqrt{27} + \sqrt{3})$

5. Multiply and simplify. Assume that $x > 0$.
   a) $(\sqrt{x})(9\sqrt{x})$
   b) $(\sqrt{x})(\sqrt{x^9})$
   c) $(\sqrt{8x})(\sqrt{18x})$
   d) $(5\sqrt{x^5})(7\sqrt{x^7})$
   e) $4(4 + \sqrt{x})$
   f) $\sqrt{x}(4 + \sqrt{x})$

6. Square each of the following expressions as indicated.
   a) $(\sqrt{3 + 5})^2$
   b) $(\sqrt{3 \cdot 5})^2$
   c) $(3\sqrt{5})^2$
   d) $(5\sqrt{3})^2$
   e) $(3 + \sqrt{5})^2$
   f) $(5 + \sqrt{3})^2$
   g) $(\sqrt{3} + \sqrt{5})^2$
   h) $(\sqrt{50} - \sqrt{18})^2$

7. Multiply and simplify. Assume that the variables represent positive numbers.
   a) $(\sqrt{13} + 2)(\sqrt{13} - 2)$
   b) $(13 + \sqrt{2})(13 - \sqrt{2})$
   c) $(\sqrt{13} + \sqrt{2})(\sqrt{13} - \sqrt{2})$
   d) $(\sqrt{x} - 4)(\sqrt{x} + 4)$
   e) $(x - \sqrt{10})(x + \sqrt{10})$
   f) $(\sqrt{xy} + \sqrt{6})(\sqrt{xy} - \sqrt{6})$

8. Multiply and simplify.
   a) $(5 + \sqrt{2})(1 + 3\sqrt{2})$
   b) $(\sqrt{8} - 4)(\sqrt{8} - 3)$
   c) $(1 + \sqrt{5})(1 + \sqrt{6})$
   d) $(12 - \sqrt{3})(3 + \sqrt{12})$

Set III

9. Multiply and simplify.
    a) $(\sqrt{2})(\sqrt{32})$
    b) $(\sqrt{10})(\sqrt{5})$
    c) $(\sqrt{7})(\sqrt{28})$
    d) $(\sqrt{8})(6\sqrt{8})$
    e) $(\sqrt{8})(8\sqrt{6})$
    f) $(3\sqrt{5})(27\sqrt{5})$
    g) $(5\sqrt{3})(5\sqrt{27})$
    h) $(\sqrt{6})(\sqrt{11})(\sqrt{66})$
    i) $12(3 + \sqrt{3})$
    j) $\sqrt{12}(\sqrt{3} + 3)$
    k) $\sqrt{2}(1 - \sqrt{2})$
    l) $\sqrt{5}(\sqrt{20} - \sqrt{5})$

10. Multiply and simplify. Assume that $x > 0$.
    a) $(25\sqrt{x})(\sqrt{x})$ e) $6(\sqrt{x} + 6)$
    b) $(\sqrt{x^{25}})(\sqrt{x})$ f) $\sqrt{x}(\sqrt{x} + 6)$
    c) $(\sqrt{2x})(\sqrt{8x})$
    d) $(3\sqrt{x^3})(9\sqrt{x^9})$

11. Square each of the following expressions as indicated.
    a) $(\sqrt{2 + 7})^2$ e) $(2 + \sqrt{7})^2$
    b) $(\sqrt{2 \cdot 7})^2$ f) $(7 + \sqrt{2})^2$
    c) $(2\sqrt{7})^2$ g) $(\sqrt{2} + \sqrt{7})^2$
    d) $(7\sqrt{2})^2$ h) $(\sqrt{60} - \sqrt{15})^2$

12. Multiply and simplify. Assume that the variables represent positive numbers.
    a) $(11 + \sqrt{3})(11 - \sqrt{3})$
    b) $(\sqrt{11} + 3)(\sqrt{11} - 3)$
    c) $(\sqrt{11} + \sqrt{3})(\sqrt{11} - \sqrt{3})$
    d) $(2 - \sqrt{x})(2 + \sqrt{x})$
    e) $(\sqrt{5} - x)(\sqrt{5} + x)$
    f) $(\sqrt{x} - \sqrt{y})(\sqrt{x} + \sqrt{y})$

13. Multiply and simplify.
    a) $(6 + \sqrt{5})(2 + 2\sqrt{5})$
    b) $(\sqrt{13} - 3)(\sqrt{13} - 2)$
    c) $(4 - \sqrt{3})(4 + \sqrt{3})$
    d) $(8 - \sqrt{2})(2 + \sqrt{8})$

Set IV

Which of these symbols, $>$, $=$, or $<$, should replace ▓ in each of the following?
1. $\sqrt{1} + \sqrt{2}$ ▓ $\sqrt{1} \cdot \sqrt{2}$
2. $\sqrt{1} + \sqrt{2} + \sqrt{3}$ ▓ $\sqrt{1} \cdot \sqrt{2} \cdot \sqrt{3}$
3. $\sqrt{1} + \sqrt{2} + \sqrt{3} + \sqrt{4}$ ▓ $\sqrt{1} \cdot \sqrt{2} \cdot \sqrt{3} \cdot \sqrt{4}$
4. What general pattern does this suggest?

5. Is that pattern true for
$$\sqrt{1} + \sqrt{2} + \sqrt{3} + \sqrt{4} + \sqrt{5} ▓$$
$$\sqrt{1} \cdot \sqrt{2} \cdot \sqrt{3} \cdot \sqrt{4} \cdot \sqrt{5}?$$
The relation that holds for the sums and products of square roots in part 5 works for all larger numbers.

LESSON 6
Dividing Square Roots

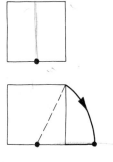

One of the most beautiful temples of ancient Greece was the Parthenon. Built in Athens in the fifth century B.C., its design is thought to be based on a figure called the *golden rectangle*. A golden rectangle can be constructed by first drawing a square and then marking the midpoint of its base as shown in the first figure at the right. With this point as center, part of a circle is drawn that goes through the upper right corner of the square as shown in the second figure. The rectangle is completed as shown in the third figure.

The front of the Parthenon fits almost exactly into a golden rectangle, as the figure below illustrates. It can be shown that the length of a golden rectangle is $\dfrac{2}{\sqrt{5}-1}$ times as long as its width. How many times is that?

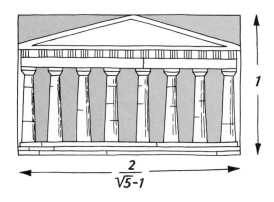

587

Because of the square root in its denominator, this fraction is not easy to evaluate. To make the work easier, we can use the fact that

$$(a + b)(a - b) = a^2 - b^2$$

If we multiply the denominator of the fraction, $\sqrt{5} - 1$, by $\sqrt{5} + 1$, we get

$$(\sqrt{5} + 1)(\sqrt{5} - 1) = (\sqrt{5})^2 - 1^2 = 5 - 1 = 4$$

As you may have guessed from the exercises in Lesson 5, this always happens. If an expression of the form $\sqrt{x} - y$ is multiplied by an expression of the form $\sqrt{x} + y$,

$$(\sqrt{x} - y)(\sqrt{x} + y) = (\sqrt{x})^2 - y^2 = x - y^2$$

the result *does not contain a square root.*

Expressions such as

$$x + \sqrt{y} \quad \text{and} \quad x - \sqrt{y}$$
$$\sqrt{x} + y \quad \text{and} \quad \sqrt{x} - y$$
$$\sqrt{x} + \sqrt{y} \quad \text{and} \quad \sqrt{x} - \sqrt{y}$$

are *conjugates* of each other. The result of multiplying two conjugates is an expression that does not contain any square roots. This means that, if the denominator of a fraction has one of these forms, the fraction can be transformed into one whose denominator does not contain any square roots by multiplying its numerator and denominator by the appropriate conjugate. The procedure of removing square roots from the denominator of a fraction is called *rationalizing the denominator.*

We can rationalize the denominator of the fraction comparing the length and width of a golden rectangle in the following way:

$$\frac{2}{\sqrt{5} - 1} = \frac{2(\sqrt{5} + 1)}{(\sqrt{5} - 1)(\sqrt{5} + 1)}$$
$$= \frac{2(\sqrt{5} + 1)}{5 - 1}$$
$$= \frac{2(\sqrt{5} + 1)}{4}$$
$$= \frac{\sqrt{5} + 1}{2}$$

Because $\sqrt{5} \approx 2.236$, $\sqrt{5} + 1 \approx 3.236$ and $\dfrac{\sqrt{5} + 1}{2} \approx 1.618$. The length of a golden rectangle is approximately 1.618 times as long as its width.

Here are more examples of how to rationalize the denominator of a fraction.

EXAMPLE 1

Rationalize the denominator of $\dfrac{2}{\sqrt{3}}$.

SOLUTION

We can remove the $\sqrt{3}$ from the denominator by multiplying numerator and denominator by $\sqrt{3}$.

$$\frac{2}{\sqrt{3}} = \frac{2\sqrt{3}}{\sqrt{3}\sqrt{3}} = \frac{2\sqrt{3}}{3}$$

EXAMPLE 2

Rationalize the denominator of $\dfrac{1}{3 + \sqrt{6}}$.

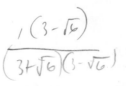

SOLUTION

Multiplying numerator and denominator by the conjugate of $3 + \sqrt{6}$, we get

$$\frac{1}{3 + \sqrt{6}} = \frac{1(3 - \sqrt{6})}{(3 + \sqrt{6})(3 - \sqrt{6})} = \frac{3 - \sqrt{6}}{9 - 6} = \frac{3 - \sqrt{6}}{3}$$

EXAMPLE 3

Rationalize the denominator of $\dfrac{4}{\sqrt{7} - \sqrt{3}}$.

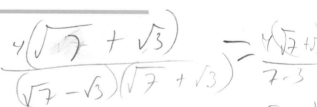

SOLUTION

Multiplying numerator and denominator by the conjugate of $\sqrt{7} - \sqrt{3}$, we get

$$\frac{4}{\sqrt{7} - \sqrt{3}} = \frac{4(\sqrt{7} + \sqrt{3})}{(\sqrt{7} - \sqrt{3})(\sqrt{7} + \sqrt{3})} = \frac{4(\sqrt{7} + \sqrt{3})}{7 - 3}$$

$$= \frac{4(\sqrt{7} + \sqrt{3})}{4} = \sqrt{7} + \sqrt{3}$$

Exercises

Set I

1. Write each of the following differences or quotients as a single fraction in simplest form.

a) $\dfrac{x}{3} - \dfrac{1}{6}$ c) $\dfrac{x+1}{x^2-1} - \dfrac{x-1}{x+1}$

b) $\dfrac{x}{3} \div \dfrac{1}{6}$ d) $\dfrac{1}{x-1} \div \dfrac{1}{x+1}$

2. What symbol, $>$, $=$, or $<$, should replace ▓ in each of the following?

a) $9 + 9$ ▓ 18

b) $9^2 + 9^2$ ▓ 18^2

c) $\sqrt{9} + \sqrt{9}$ ▓ $\sqrt{18}$

d) $9 \cdot 9$ ▓ 81

e) $9^2 \cdot 9^2$ ▓ 81^2

f) $\sqrt{9} \cdot \sqrt{9}$ ▓ $\sqrt{81}$

3. Mr. Orlock has been teaching mathematics at Transylvania High for eighty years. In eight more years, he will be five times as old as he was when he started teaching at the school.

a) If he is x years old now, what expression represents his age when he started teaching?

b) What expression represents his age eight years from now?

c) Use the information given in the problem to write an equation.

d) Solve the equation to find Mr. Orlock's present age.

Set II

4. Divide and simplify.

a) $\dfrac{\sqrt{100}}{\sqrt{4}}$ d) $\dfrac{\sqrt{77}}{\sqrt{7}}$

b) $\dfrac{\sqrt{48}}{\sqrt{3}}$ e) $\dfrac{\sqrt{60}}{\sqrt{5}}$

c) $\dfrac{\sqrt{21}}{\sqrt{84}}$ f) $\dfrac{\sqrt{2}}{\sqrt{10}}$

5. Find the approximate value of each of the following expressions by first rationalizing the denominator and then using the table on page 563. Round each answer to the nearest hundredth.

a) $\dfrac{4}{\sqrt{2}}$ c) $\dfrac{1}{4\sqrt{3}}$

b) $\dfrac{\sqrt{7}}{\sqrt{10}}$ d) $\dfrac{5\sqrt{2}}{2\sqrt{5}}$

6. Divide and simplify.

a) $\dfrac{\sqrt{25x}}{\sqrt{x}}$ c) $\dfrac{\sqrt{3x^3}}{\sqrt{12x}}$

b) $\dfrac{\sqrt{x^{25}}}{\sqrt{x}}$ d) $\dfrac{\sqrt{8x^8}}{\sqrt{2x^2}}$

7. Write the conjugate of each of the following expressions.

a) $8 + \sqrt{2}$ c) $x - \sqrt{y}$

b) $\sqrt{15} - 6$ d) $\sqrt{x} + \sqrt{y}$

8. Rationalize the denominator of each of the following fractions.

a) $\dfrac{1}{3 + \sqrt{5}}$ c) $\dfrac{9}{\sqrt{11} + \sqrt{2}}$

b) $\dfrac{\sqrt{6}}{\sqrt{6} - 2}$ d) $\dfrac{\sqrt{7} + 1}{\sqrt{7} - 1}$

9. Find the approximate value of each of the following expressions by first rationalizing the denominator and then using the table on page 563. Round each answer to the nearest hundredth.

a) $\dfrac{2}{\sqrt{3} - 1}$ b) $\dfrac{10}{\sqrt{6} + \sqrt{5}}$

c) $\dfrac{1}{4 - \sqrt{11}}$ d) $\dfrac{\sqrt{12}}{\sqrt{12} - \sqrt{2}}$

10. Rationalize the denominator of each of the following fractions. Assume that the variables represent positive numbers.

a) $\dfrac{1}{\sqrt{x}}$ c) $\dfrac{x^2 - y}{x - \sqrt{y}}$

b) $\dfrac{1}{\sqrt{x} + 1}$ d) $\dfrac{\sqrt{x} - \sqrt{y}}{\sqrt{x} + \sqrt{y}}$

11. Solve each of the following equations for x. Leave your answers in radical form but simplify them as much as possible.

a) $\sqrt{7x} = \sqrt{35}$ d) $\dfrac{x}{\sqrt{5}} = \sqrt{15}$

b) $2x = \sqrt{44}$

c) $\dfrac{x}{5} = \sqrt{15}$ e) $3x = \sqrt{18}$

                f) $\sqrt{3x} = 18$

Set III

12. Divide and simplify.

a) $\dfrac{\sqrt{144}}{\sqrt{9}}$ d) $\dfrac{\sqrt{26}}{\sqrt{2}}$

b) $\dfrac{\sqrt{45}}{\sqrt{5}}$ e) $\dfrac{\sqrt{108}}{\sqrt{6}}$

c) $\dfrac{\sqrt{10}}{\sqrt{250}}$ f) $\dfrac{\sqrt{21}}{\sqrt{14}}$

13. Find the approximate value of each of the following expressions by first rationalizing the denominator and then using the table on page 563. Round each answer to the nearest hundredth.

a) $\dfrac{9}{\sqrt{3}}$ c) $\dfrac{1}{2\sqrt{5}}$

b) $\dfrac{\sqrt{19}}{\sqrt{2}}$ d) $\dfrac{11\sqrt{6}}{6\sqrt{11}}$

14. Divide and simplify. Assume that $x > 0$.

a) $\dfrac{\sqrt{9x}}{\sqrt{x}}$ c) $\dfrac{\sqrt{2x^3}}{\sqrt{50x}}$

b) $\dfrac{\sqrt{x^9}}{\sqrt{x}}$ d) $\dfrac{\sqrt{x^8}}{\sqrt{4x^2}}$

15. Write the conjugate of each of the following expressions.

a) $6 + \sqrt{5}$ c) $\sqrt{x} + y$

b) $\sqrt{14} - 3$ d) $\sqrt{x} - \sqrt{y}$

16. Rationalize the denominator of each of the following fractions.

a) $\dfrac{2}{3 - \sqrt{2}}$ c) $\dfrac{4}{\sqrt{7} - \sqrt{5}}$

b) $\dfrac{\sqrt{10}}{\sqrt{10} + 3}$ d) $\dfrac{\sqrt{11} - 1}{\sqrt{11} + 1}$

17. Find the approximate value of each of the following expressions by first rationalizing the denominator and then using the table on page 563. Round each answer to the nearest hundredth.

a) $\dfrac{10}{2 + \sqrt{3}}$

b) $\dfrac{3}{\sqrt{5} - \sqrt{2}}$

c) $\dfrac{1}{4 - \sqrt{14}}$

d) $\dfrac{\sqrt{5}}{\sqrt{15} - \sqrt{5}}$ $\dfrac{5 + (\sqrt{3} \cdot \sqrt{5}}{10}$

18. Rationalize the denominator of each of the following fractions. Assume that the variables represent positive numbers.

a) $\dfrac{x}{\sqrt{x}}$

b) $\dfrac{1}{1 - \sqrt{x}}$

c) $\dfrac{x - y^2}{\sqrt{x} + y}$

d) $\dfrac{\sqrt{x} + \sqrt{y}}{\sqrt{x} - \sqrt{y}}$ $(\sqrt{x} + \sqrt{y})$

19. Solve each of the following equations for x. Leave your answers in radical form but simplify them as much as possible.

a) $\sqrt{6x} = \sqrt{12}$

b) $3x = \sqrt{54}$

c) $\dfrac{x}{2} = \sqrt{10}$

d) $\dfrac{x}{\sqrt{2}} = \sqrt{10}$

e) $5x = \sqrt{75}$

f) $\sqrt{5x} = 75$

Set IV

The number relating the dimensions of the golden rectangle, $\dfrac{2}{\sqrt{5} - 1}$, has several remarkable properties. One of them is its relationship to its reciprocal, $\dfrac{\sqrt{5} - 1}{2}$.

1. Find the value of its reciprocal, $\dfrac{\sqrt{5} - 1}{2}$, to three decimal places.

2. Compare your answer to the value derived in this lesson for $\dfrac{2}{\sqrt{5} - 1}$. What seems to be true?

3. Prove your observation by writing an equation and showing that the equation is true.

STERLING PUBLISHING CO., INC.

LESSON *7*
Radical Equations

The first bicycle was built by a Scottish blacksmith named Kirkpatrick Macmillan in 1839. Early bicycles had large front wheels and, being rather unstable, were likely to tip over when going around corners. Even modern bicycles will tip over if they make too sharp a turn at too great a speed.

The greatest speed at which a cyclist can safely take a corner is given by the formula

$$s = 4\sqrt{r}$$

in which s is the speed in miles per hour and r is the radius of the corner in feet. What is the radius of the sharpest corner that a cyclist can safely turn if riding at a speed of 30 miles per hour? Substituting 30 for s in the formula, we get

$$30 = 4\sqrt{r}$$

To solve this equation for r, we can divide both sides by 4, getting

$$\sqrt{r} = \frac{30}{4} = 7.5$$

Squaring both sides of this equation gives

$$r = (7.5)^2 = 56.25$$

The cyclist cannot turn a corner having a radius of less than 56 feet without being in danger of tipping over.

To get this answer, we had to solve an equation containing a square root of the variable. An equation in which the variable appears under a radical sign is a **radical equation.** One of the steps in solving such an equation is to square both sides. Although it is always true that if $a = b$, $a^2 = b^2$, these equations do not necessarily have the same solutions. Consider, for example, the equation, $x = 4$. It obviously has just one solution: 4. The equation that results from squaring both sides, on the other hand, $x^2 = 16$, has *two* solutions: 4 and –4. For this reason, whenever one of the steps in solving an equation consists of squaring both sides, the solutions that result should be checked to see if they are also solutions of the original equation.

Here are examples of how to solve radical equations.

EXAMPLE 1

Solve the equation $\sqrt{x - 1} = 7$.

SOLUTION

Squaring both sides, we get

$$x - 1 = 49$$
$$x = 50$$

Because $\sqrt{50 - 1} = \sqrt{49} = 7$, 50 is a solution to the original equation.

EXAMPLE 2

Solve the equation $\sqrt{x} + 9 = 4$.

SOLUTION

Before squaring both sides of the equation, we must get the square root by itself on one side of the equation. So we subtract 9 from each side:

$$\sqrt{x} = -5,$$
$$x = (-5)^2 = 25$$

But the number 25 is not a solution to the original equation because $\sqrt{25} + 9$ is not equal to 4. So the equation $\sqrt{x} + 9 = 4$ has no solutions.

EXAMPLE 3
Solve the equation $\sqrt{2x - 5} - 3 = 8$.

SOLUTION
Adding 3 to each side, we get

$$\sqrt{2x - 5} = 11$$

Squaring,

$$2x - 5 = 121$$
$$2x = 126$$
$$x = 63$$

Because $\sqrt{2(63) - 5} - 3 = \sqrt{126 - 5} - 3 = \sqrt{121} - 3 = 11 - 3 = 8$, 63 is a solution to the original equation.

EXAMPLE 4
Solve the equation $\sqrt{2}x = 6$.

SOLUTION
This is not a radical equation because the variable does not appear under the radical sign. It can be solved without squaring.

$$x = \frac{6}{\sqrt{2}} = \frac{6\sqrt{2}}{\sqrt{2}\sqrt{2}} = \frac{6\sqrt{2}}{2} = 3\sqrt{2}$$

Exercises

Set I

1. Guess a formula for the function represented by each of these tables. Begin each formula with $y =$.

   a)
x	1	2	3	4	5
y	$\frac{1}{6}$	$\frac{1}{3}$	$\frac{1}{2}$	$\frac{2}{3}$	$\frac{5}{6}$

   b)
x	0	1	2	3	4
y	1	5	25	125	625

   c)
x	0	1	4	9	16
y	0	1	2	3	4

2. Write each of the following as a simple fraction in lowest terms.

   a) $\dfrac{x + \dfrac{1}{2}}{x + \dfrac{1}{4}}$

   b) $\dfrac{\dfrac{x}{y} - \dfrac{y}{x}}{x - y}$

3. In 1908 something from the sky fell on Siberia, causing a tremendous explosion that was heard more than 600 miles away. The object is thought to have weighed about 10^7 kilograms and to have been traveling at a speed of 4×10^4 meters per second.

   a) Use the formula $e = \dfrac{1}{8}mv^2$ to find the energy, e, in calories released when it hit. The letters m and v stand for the mass of the object in kilograms and its velocity in meters per second.

   b) A Mars bar contains about 200 calories. How many Mars bars would contain the energy released in this explosion?

Set II

4. If possible, find the value of each of the following expressions for the value of x indicated.

   a) $\sqrt{12x}$, if $x = 3$

   b) $\sqrt{\dfrac{x}{12}}$, if $x = 3$

   c) $\sqrt{12 + x}$, if $x = 4$
   d) $12 + \sqrt{x}$, if $x = 4$
   e) $\sqrt{3x - 2}$, if $x = 9$
   f) $3\sqrt{x} - 2$, if $x = 9$
   g) $\sqrt{2 - 3x}$, if $x = 9$
   h) $2 - 3\sqrt{x}$, if $x = 9$

5. Find the values of $\sqrt{x^2 + 4x + 4}$ for the values of x indicated.
   a) $x = 1$
   b) $x = 5$
   c) $x = 8$
   d) $x = 10$
   e) How does $\sqrt{x^2 + 4x + 4}$ compare in value with x?
   f) Explain why.
   g) Does the relation that you found work if $x = -10$? Explain why.

6. Check the numbers given in the following exercises to see if they are solutions of the given equations.
   a) Is 32 a solution of $\sqrt{x + 4} = 6$?
   b) Is 1 a solution of $5\sqrt{x} = x + 5$?
   c) Is 4 a solution of $\sqrt{x - 1} = \sqrt{x - 3}$?
   d) Is -9 a solution of $\sqrt{7 - x} = 13 + x$?

7. Solve the following radical equations. Check each solution.
   a) $\sqrt{5x} = 15$
   b) $5\sqrt{x} = 15$
   c) $\sqrt{5 + x} = 15$
   d) $5 + \sqrt{x} = 15$
   e) $2\sqrt{x} + 4 = 0$
   f) $2\sqrt{x + 4} = 0$
   g) $\sqrt{2x} + 4 = 0$
   h) $\sqrt{2x + 4} = 0$
   i) $\sqrt{2x - 7} = \sqrt{x + 9}$
   j) $\sqrt{2x + 7} = \sqrt{x - 9}$
   k) $\sqrt{x + 8} = 3\sqrt{2}$
   l) $\sqrt{8x} = 3\sqrt{2}$
   m) $\sqrt{6 - x} = 11$
   n) $6 - \sqrt{x} = 11$

8. Although the following equations contain square roots, they can be solved without squaring because the variable does not appear under a radical sign. Solve each equation. Leave your answers in radical form but simplify them as much as possible.
   a) $x - \sqrt{10} = 10$
   b) $\dfrac{x}{\sqrt{10}} = 10$
   c) $x + \sqrt{3} = \sqrt{27}$
   d) $\sqrt{3}x = \sqrt{27}$
   e) $2x - \sqrt{8} = 0$
   f) $2(x - \sqrt{8}) = 0$

Set III

9. If possible, find the value of each of the following expressions for the value of x indicated.
   a) $\sqrt{20x}$, if $x = 5$
   b) $\sqrt{\dfrac{20}{x}}$, if $x = 5$
   c) $\sqrt{x - 20}$, if $x = 36$
   d) $\sqrt{x} - 20$, if $x = 36$
   e) $\sqrt{20 - x}$, if $x = 36$
   f) $20 - \sqrt{x}$, if $x = 36$
   g) $\sqrt{10x + 9}$, if $x = 4$
   h) $10\sqrt{x} + 9$, if $x = 4$

10. Find the values of $\sqrt{x^2 + 6x + 9}$ for the values of x indicated.
    a) $x = 1$
    b) $x = 3$
    c) $x = 7$
    d) $x = 10$
    e) How does $\sqrt{x^2 + 6x + 9}$ compare in value with x?
    f) Explain why.
    g) Does the relationship that you found hold for $x = -12$? Explain why.

11. Check the numbers given in the following exercises to see if they are solutions of the given equations.
    a) Is 7 a solution of $\sqrt{x + 9} = 4$?
    b) Is 25 a solution of $2\sqrt{x} - 10 = 0$?
    c) Is 0 a solution of $\sqrt{1 - x} = x - 1$?
    d) Is 14 a solution of $\sqrt{2x + 8} = x - 8$?

12. Solve the following radical equations. Check each solution.
    a) $\sqrt{3x} = 6$
    b) $3\sqrt{x} = 6$
    c) $\sqrt{3 + x} = 6$
    d) $3 + \sqrt{x} = 6$
    e) $4\sqrt{x} - 2 = 0$
    f) $4\sqrt{x - 2} = 0$
    g) $\sqrt{4x} - 2 = 0$
    h) $\sqrt{4x - 2} = 0$
    i) $\sqrt{8x - 1} = \sqrt{5x + 11}$
    j) $\sqrt{8x + 1} = \sqrt{5x - 11}$
    k) $\sqrt{x + 7} = 2\sqrt{7}$
    l) $\sqrt{7x} = 2\sqrt{7}$
    m) $\sqrt{9 - x} = 10$
    n) $9 - \sqrt{x} = 10$

13. Although the following equations contain square roots, they can be solved without squaring because the variable does not appear under a radical sign. Solve each equation. Leave your answers in radical form but simplify them as much as possible.

a) $x + \sqrt{5} = 50$
b) $\sqrt{5}x = 50$
c) $x - \sqrt{2} = \sqrt{32}$
d) $\dfrac{x}{\sqrt{2}} = \sqrt{32}$
e) $3x + \sqrt{18} = 0$
f) $3(x + \sqrt{18}) = 0$

Set IV

The water near the bottom of a river does not move at the same speed as the water at the surface. A formula relating the two speeds is

$$\sqrt{b} = \sqrt{s} - 1$$

in which b is the speed near the bottom and s is the speed at the surface, each measured in miles per hour.

1. Find the speed near the bottom of a river if the speed at the surface is 9 miles per hour.
2. Find the speed at the surface if the speed near the bottom of a river is 7 miles per hour.

DRAWING BY CHON DAY; © 1975 THE NEW YORKER MAGAZINE, INC.

$x = 4\sqrt{2} + \sqrt{8}$

Summary and Review

In this chapter, we have learned what square roots are, how to simplify square roots, how to add, subtract, multiply and divide them, and how to solve equations that contain them.

Square Roots (*Lesson 1*) If $y = x^2$, then y is the square of x and x is a square root of y. Every positive number has two square roots, one of which is the opposite of the other: the symbol $\sqrt{}$ is used to indicate the positive root and the symbol $-\sqrt{}$ to indicate the negative root. Zero has just one square root: itself.

Square Roots of Products (*Lesson 2*) For any two nonnegative numbers, x and y,

$$\sqrt{xy} = \sqrt{x}\sqrt{y}$$

This fact can be used to simplify the square root of any monomial that has one or more factors that are squares. For example,

$$\sqrt{a^2b} = a\sqrt{b}$$

The square root of a monomial is in simple radical form if the monomial has no factors that are squares of monomials other than 1 and –1.

Square Roots of Quotients (*Lesson 3*)   For any two nonnegative numbers, x and y,

$$\sqrt{\frac{x}{y}} = \frac{\sqrt{x}}{\sqrt{y}}$$

This fact can be used to change the square root of a fraction into the quotient of a square root and an integer or polynomial. For example,

$$\sqrt{\frac{a}{b}} = \sqrt{\frac{ab}{b^2}} = \frac{\sqrt{ab}}{b}$$

Adding and Subtracting Square Roots (*Lesson 4*)   There is no general equation relating $\sqrt{x}$, $\sqrt{y}$, and $\sqrt{x + y}$ or $\sqrt{x - y}$. However, square roots can be added or subtracted if they are roots of the same number. For example,

$$a\sqrt{c} + b\sqrt{c} = (a + b)\sqrt{c}$$

Multiplying Square Roots (*Lesson 5*)   Because $\sqrt{x}\sqrt{y} = \sqrt{xy}$, square roots of nonnegative numbers can always be multiplied. For example,

$$(a\sqrt{c})(b\sqrt{d}) = ab\sqrt{cd}$$

Dividing Square Roots (*Lesson 6*)   Because $\dfrac{\sqrt{x}}{\sqrt{y}} = \sqrt{\dfrac{x}{y}}$, square roots of nonnegative numbers can always be divided. Square roots can be removed from the denominator of a fraction by multiplying its numerator and denominator by an appropriate expression. For example,

$$\frac{a}{b\sqrt{c}} = \frac{a\sqrt{c}}{b\sqrt{c}\sqrt{c}} = \frac{a\sqrt{c}}{bc}$$

Removing square roots from the denominator of a fraction is called *rationalizing the denominator*. If the denominator is a sum or difference, it can be rationalized by multiplying the numerator and denominator of the fraction by the conjugate of the denominator. For example,

$$\frac{a}{b + \sqrt{c}} = \frac{a(b - \sqrt{c})}{(b + \sqrt{c})(b - \sqrt{c})} = \frac{ab - a\sqrt{c}}{b^2 - c}$$

Radical Equations (*Lesson 7*) An equation in which the variable appears under a radical sign is a radical equation. One step in solving such an equation is to square both sides. The equation that results does not necessarily have the same solutions as the original equation. For this reason, the solutions obtained should always be checked to see if they are also solutions of the original equation.

Exercises

Set I

1. List all of the square roots of these numbers.
   a) 36
   b) 1
   c) –9
   d) $\frac{1}{4}$

2. Find an integer equal to each of the following.
   a) $(\sqrt{42})^2$
   b) $\sqrt{(-42)^2}$
   c) $\sqrt{12^2 + 16^2}$
   d) $\sqrt{12^2 \cdot 16^2}$
   e) $\sqrt{29^2} - \sqrt{21^2}$
   f) $\sqrt{29^2 - 21^2}$

3. What symbol, $>$, $=$, or $<$, should replace ▓ in each of the following?
   a) $10 + 6$ ▓ 16
   b) $10 \cdot 6$ ▓ 60
   c) $10^2 + 6^2$ ▓ 16^2
   d) $10^2 \cdot 6^2$ ▓ 60^2
   e) $\sqrt{10} + \sqrt{6}$ ▓ $\sqrt{16}$
   f) $\sqrt{10} \cdot \sqrt{6}$ ▓ $\sqrt{60}$

4. Write each of the following expressions in simple radical form.
   a) $\sqrt{50}$
   b) $\sqrt{108}$
   c) $\sqrt{801}$
   d) $\sqrt{1,100}$

5. Write each of the following expressions in simple radical form. Assume that $x > 0$.
   a) $\sqrt{4x}$
   b) $\sqrt{x^4}$
   c) $\sqrt{6x^6}$
   d) $\sqrt{25x^5}$

6. Write each of the following expressions as a quotient without a square root in the denominator.
   a) $\sqrt{\dfrac{5}{16}}$
   b) $\sqrt{\dfrac{3}{50}}$
   c) $\sqrt{\dfrac{x}{11}}$
   d) $\sqrt{\dfrac{1}{x^3}}$

7. Find the value of each of the following expressions. Round any approximate answers to the nearest tenth. (Refer to the table of square roots on page 563.)
   a) $\sqrt{40} - 4$
   b) $\sqrt{40} - \sqrt{4}$
   c) $3\sqrt{7} + 2$
   d) $3\sqrt{7} + 2$

8. If possible, simplify the following sums and differences.
   a) $5\sqrt{3} + 5\sqrt{3}$ d) $\sqrt{33} + \sqrt{17}$
   b) $4\sqrt{7} - \sqrt{7}$ e) $\sqrt{20x} - \sqrt{5x}$
   c) $\sqrt{32} + \sqrt{18}$ f) $\sqrt{10} - \sqrt{5}$

9. Simplify.
   a) $(1 + 2\sqrt{3}) + (2 + 4\sqrt{3})$
   b) $(54 + \sqrt{6}) + (6 + \sqrt{54})$
   c) $(1 + \sqrt{80}) - (1 - \sqrt{20})$

10. Square as indicated.
   a) $(5\sqrt{2})^2$ c) $(\sqrt{5} - 2)^2$
   b) $(5 + \sqrt{2})^2$ d) $(\sqrt{12} + \sqrt{3})^2$

11. Multiply and simplify.
   a) $(\sqrt{2})(\sqrt{18})$ e) $\sqrt{x}(\sqrt{x} + 5)$
   b) $(\sqrt{6})(4\sqrt{6})$ f) $(4 + \sqrt{3})(2 + \sqrt{12})$
   c) $5(7 + \sqrt{7})$ g) $(1 - \sqrt{x})(1 + \sqrt{x})$
   d) $(\sqrt{x})(\sqrt{x^5})$ h) $(\sqrt{x} + 5)(\sqrt{x} - 2)$

12. Rationalize the denominator of each of the following.
   a) $\dfrac{\sqrt{63}}{\sqrt{7}}$ c) $\dfrac{60}{\sqrt{6} - 1}$
   b) $\dfrac{60}{\sqrt{6}}$ d) $\dfrac{\sqrt{2}}{\sqrt{5} + \sqrt{2}}$

13. Find the approximate value of each of the following expressions. Round each answer to the nearest hundredth. (Refer to the table of square roots on page 563.)
   a) $\dfrac{\sqrt{10}}{2}$ c) $\dfrac{2}{\sqrt{11} - \sqrt{10}}$
   b) $\dfrac{2}{\sqrt{10}}$

14. Solve the following equations. Simplify radicals where possible.
   a) $x + \sqrt{2} = 4$
   b) $3(x - \sqrt{5}) = 0$
   c) $\sqrt{7x} = 21$
   d) $\sqrt{7 + x} = 21$
   e) $7 + \sqrt{x} = 21$
   f) $\sqrt{x + 8} = 2\sqrt{x - 1}$
   g) $\sqrt{x - 8} = 2\sqrt{x + 1}$
   h) $\sqrt{x} = \sqrt{9x} + 4$

Set II

1. List all of the square roots of these numbers.
   a) 64
   b) –4   d) $\dfrac{1}{9}$
   c) 0

2. Find an integer equal to each of the following expressions.
   a) $(\sqrt{25})^2$ d) $\sqrt{8^2 \cdot 6^2}$
   b) $\sqrt{(-25)^2}$ e) $\sqrt{34^2} - \sqrt{16^2}$
   c) $\sqrt{8^2 + 6^2}$ f) $\sqrt{34^2 - 16^2}$

3. What symbol, $>$, $=$, or $<$, should replace ▥ in each of the following?
   a) $4 + 5$ ▥ 9
   b) $4 \cdot 5$ ▥ 20
   c) $4^2 + 5^2$ ▥ 9^2 e) $\sqrt{4} + \sqrt{5}$ ▥ $\sqrt{9}$
   d) $4^2 \cdot 5^2$ ▥ 20^2 f) $\sqrt{4} \cdot \sqrt{5}$ ▥ $\sqrt{20}$

4. Write each of the following expressions in simple radical form.
   a) $\sqrt{45}$ c) $\sqrt{104}$
   b) $\sqrt{98}$ d) $\sqrt{60,000}$

5. Write each of the following expressions in simple radical form. Assume that $x > 0$.
   a) $\sqrt{36x}$
   b) $\sqrt{x^{36}}$
   c) $\sqrt{10x^{10}}$
   d) $\sqrt{9x^3}$

6. Write each of the following expressions as a quotient without a square root in the denominator.

   a) $\sqrt{\dfrac{6}{25}}$

   b) $\sqrt{\dfrac{1}{18}}$

   c) $\sqrt{\dfrac{7}{x}}$

   d) $\sqrt{\dfrac{3x}{2}}$

7. Find the value of each of the following expressions. Round any approximate answers to the nearest tenth. (Refer to the table of square roots on page 563.)
   a) $\sqrt{13 + 17}$
   b) $\sqrt{13} + \sqrt{17}$
   c) $2\sqrt{6} - 1$
   d) $2\sqrt{6-1}$

8. If possible, simplify the following sums and differences.
   a) $4\sqrt{5} + 4\sqrt{5}$
   b) $9\sqrt{2} - \sqrt{2}$
   c) $\sqrt{48} + \sqrt{75}$
   d) $\sqrt{55} + \sqrt{11}$
   e) $\sqrt{15} - \sqrt{6}$
   f) $25\sqrt{x} - 16\sqrt{x}$

9. Simplify.
   a) $(4 + 3\sqrt{7}) + (3 + 4\sqrt{7})$
   b) $(32 + \sqrt{2}) + (2 + \sqrt{32})$
   c) $(1 - \sqrt{27}) - (1 + \sqrt{12})$

10. Square as indicated.
    a) $(4\sqrt{3})^2$
    b) $(4 + \sqrt{3})^2$
    c) $(\sqrt{4} - \sqrt{3})^2$
    d) $(\sqrt{8} + \sqrt{2})^2$

11. Multiply and simplify.
    a) $(\sqrt{5})(\sqrt{20})$
    b) $(\sqrt{3})(11\sqrt{3})$
    c) $4(6 - \sqrt{6})$
    d) $(\sqrt{x})(\sqrt{x^7})$
    e) $\sqrt{x}(\sqrt{x} + 7)$
    f) $(2 + \sqrt{2})(3 + \sqrt{8})$
    g) $(\sqrt{x} + 12)(\sqrt{x} - 12)$
    h) $(4 - \sqrt{x})(1 - \sqrt{x})$

12. Rationalize the denominator of each of the following expressions.

    a) $\dfrac{\sqrt{80}}{\sqrt{5}}$

    b) $\dfrac{12}{2\sqrt{3}}$

    c) $\dfrac{12}{2 + \sqrt{3}}$

    d) $\dfrac{\sqrt{6}}{\sqrt{11} - \sqrt{6}}$

13. Find the approximate value of each of the following expressions. Round each answer to the nearest hundredth. (Refer to the table of square roots on page 563.)

    a) $\dfrac{\sqrt{15}}{5}$

    b) $\dfrac{1}{\sqrt{15}}$

    c) $\dfrac{1}{\sqrt{15} - \sqrt{13}}$

14. Solve the following equations. Simplify radicals where possible.
    a) $\sqrt{3x} = 6$
    b) $\sqrt{2}(x - \sqrt{7}) = \sqrt{14}$
    c) $\sqrt{10x} = 40$
    d) $\sqrt{10 + x} = 40$
    e) $10 + \sqrt{x} = 40$
    f) $3\sqrt{x - 1} = \sqrt{x + 7}$
    g) $3\sqrt{x + 1} = \sqrt{x - 7}$
    h) $\sqrt{x} = \sqrt{4x - 8}$

Chapter 13
QUADRATIC EQUATIONS

LESSON 1
Polynomial Equations

A goal in the study of algebra is to learn how to solve equations of various types. Although many equations are simple enough to be solved by intelligent guessing, there are others that can only be solved by algebraic methods. In this chapter we will review some of the methods that we have used in the past and become acquainted with new ones that can be used to solve more difficult equations.

Look at the equations below.

$$2x + 3 = 10 \qquad 5x^2 - 4x + 1 = 0 \qquad x^3 = 6 - x$$

These are examples of *polynomial equations* because both sides of each equation are polynomials. (Remember that a polynomial is either a monomial or an expression built by adding or subtracting monomials.) Furthermore, they are all polynomial equations *in one variable*. Every polynomial equation in one variable can be written so that one side is a polynomial in descending powers of the variable and the other side is zero. For example, our three equations can be written as

$$2x - 7 = 0 \qquad 5x^2 - 4x + 1 = 0 \qquad x^3 + x - 6 = 0$$

A polynomial equation written in this way is said to be in *standard form*.

Polynomial equations written in standard form are classified according to the degree of the polynomial on the left side. The equation

$$5x^2 - 4x + 1 = 0$$

for example, is called a second-degree equation because the degree of its polynomial, $5x^2 - 4x + 1$, is 2.

Polynomial equations are named according to their degree as shown in the following table. (The letters a, b, c, d, e, and f may stand for any number, except that a can never be zero.)

Degree	Standard form of equation	Name
1	$ax + b = 0$	Linear
2	$ax^2 + bx + c = 0$	Quadratic
3	$ax^3 + bx^2 + cx + d = 0$	Cubic
4	$ax^4 + bx^3 + cx^2 + dx + e = 0$	Quartic
5	$ax^5 + bx^4 + cx^3 + dx^2 + ex + f = 0$	Quintic

We have already learned how to solve linear equations. In this chapter, we will develop some methods for finding solutions to quadratic equations. The methods include graphing, factoring, taking square roots, and using the quadratic formula. In the final lesson of the chapter, we will consider examples of higher-degree equations.

Exercises

Set I

1. Change each of the following to the form described.

   a) Write $\dfrac{x+4}{x+9}$ as the sum of two fractions.

   b) Write $\dfrac{4}{9x}$ as the product of two fractions.

   c) Write $\dfrac{x^4}{9}$ as the square of a fraction.

   d) Write $\dfrac{x^4 - 9}{x}$ as the difference of a monomial and a fraction.

2. Find each of the following quotients.
   a) $2x^2\overline{)8x^8 - 2x^2}$
   b) $x + 2\overline{)2x^3 - x^2 - 4x + 12}$
   c) $x^2 - 3x + 9\overline{)x^3 + 27}$

3. The figure at the right represents a race between two porcupines.
   a) How fast is each porcupine running?

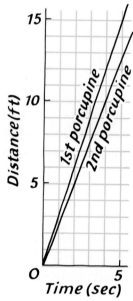

b) If they continue to run at the same rate, when will the faster porcupine be 50 feet ahead of the slower one?

Set II

4. Write each of the following equations in standard form and tell the degree of the equation.
   a) $4x = 11$
   b) $x^3 = 6x^2 - 8$
   c) $(x - 1)(x + 3) = 2$
   d) $3x^5 = x^5 + 1$
   e) $x^2 + 7x - 10 = x^2 + 10$
   f) $x^2(2x^2 + 9) = x$

5. Tell whether each of the following numbers is a solution of the equation given by checking to see if it makes the equation true.

   $x^2 - 4x = 21$

   a) 7
   b) –3

$2x^3 - 128 = 0$
   c) 4
   d) –4

$x^4 = 25$
   e) $\sqrt{5}$
   f) $-\sqrt{5}$

$x^2 - 2x - 2 = 0$
   g) $1 + \sqrt{3}$
   h) $1 - \sqrt{3}$

$x^4 - 5x^2 + 4 = 0$
   i) 1
   j) –1
   k) 2
   l) –2

6. Tell whether or not zero is a solution of each of the following equations.
   a) $5x^2 - x = 0$
   b) $x^4 + 3x^3 - 2x^2 + 1 = 0$
   c) $2x^3 - 7x^2 + 10x = 0$
   d) $x^5 - x^4 + x^3 - x^2 + x - 1 = 0$
   e) How can you tell from looking at a polynomial equation in standard form whether zero is one of its solutions?

7. Solve each of the following equations for x.
   a) $4x + 9 = 3x - 1$
   b) $7(x - 10) = 2x + 25$
   c) $x^2 + 12 = x(x + 3)$
   d) $2(x - 5) = 5(x - 2)$
   e) $x(x^2 - 10) = x^3 + 6$
   f) $x^2 + 5x + 8 = x(x + 4)$
   g) $(x + 9)(x - 3) = x^2 + 15$
   h) $(x + 1)^2 = (x - 7)(x + 7)$

$5 + 4\sqrt{5} + 4$

Set III

8. Write each of the following equations in standard form and tell the degree of the equation.
   a) $5x^2 = 6$
   b) $3(x - 2) = x - 13$
   c) $x^4 - 8x = 2 - x^5$
   d) $9x^3 + x - 5 = 2x^3 + x$
   e) $(x^3 + 1)(x - 4) = 12$
   f) $x^4 + 3x = x^2(x^2 - 3)$

9. Tell whether or not each of the following numbers is a solution of the equation given by checking to see if it makes that equation true.

   $2x^2 - 9x + 4 = 0$
   a) 4
   b) $\dfrac{1}{2}$

   $x^4 - 4 = 0$
   c) $\sqrt{2}$
   d) $-\sqrt{2}$

   $2x^3 = 250$
   e) 5
   f) -5

$x^2 - 4x - 1 = 0$
   g) $\sqrt{5} + 2$
   h) $\sqrt{5} - 2$

$x^4 - 10x^2 + 9 = 0$
   i) 1
   j) -1
   k) 3
   l) -3

10. Tell whether or not the number 1 is a solution of each of the following equations.
   a) $4x^3 - 4 = 0$
   b) $x^2 + 9x - 10 = 0$
   c) $6x^4 - 5x^2 - 2 = 0$
   d) $x^5 - x^4 + x^3 - x^2 + x - 1 = 0$

11. Solve each of the following equations for x.
   a) $6x - 1 = 5x + 2$
   b) $3(x + 7) = x - 11$
   c) $x^2 - 8 = x(x + 4)$
   d) $9(x - 1) = 2(x - 1)$
   e) $x(x^2 + 5) = x^3 + 27$
   f) $x^2 - 4x + 9 = x(x + 6)$
   g) $(x - 2)(x + 4) = x^2 - 18$
   h) $(x - 5)^2 = (x + 6)(x - 6)$

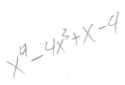

Set IV

Some polynomial equations can be factored in such a way that their solutions are easy to figure out. An example of such an equation is

$$x^4 - 15x^2 - 10x + 24 = 0$$

The left side of this equation can be factored as $(x - 1)(x + 2)(x + 3)(x - 4)$.

1. Can you use this fact to figure out what the solutions of this equation are?
2. Explain your reasoning.

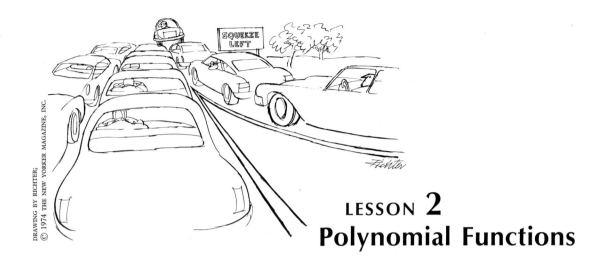

LESSON 2
Polynomial Functions

The average distance between cars on a crowded highway depends on the speed at which the traffic is moving. If the cars are moving very slowly, this distance is relatively small, as anyone who has ever ridden in a car trying to get on a busy freeway knows. The faster the traffic is moving, the greater the average distance between cars becomes.

An approximate formula relating this distance, y, and the speed, x, is

$$y = 0.03x^2 + x + 18$$

in which y is in feet and x is in feet per second. This is the equation of a *second-degree polynomial function*. Its graph is shown at the right.

Earlier in the course you learned how to graph first-degree polynomial functions.* Such functions have equations of the form

$$y = ax + b$$

and, because their graphs are always *straight lines,* they are called *linear.* Second-degree polynomial functions, also called *quadratic* functions, have equations of the form

$$y = ax^2 + bx + c$$

* See pages 90–91.

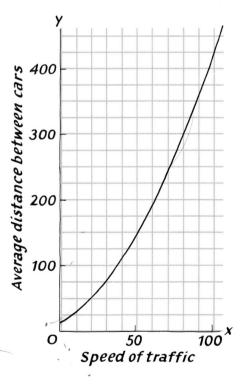

Average distance between cars

Speed of traffic

611

Like the graph for the distance between cars on a highway, the graph of every quadratic function is a curved line. The graphs of higher-degree polynomial functions such as

$$y = ax^3 + bx^2 + cx + d$$

and

$$y = ax^4 + bx^3 + cx^2 + dx + e$$

are also curved.

To graph a polynomial function, we first make a table from its equation. For example, to graph the function

$$y = x^2 - 2x - 3$$

we first make a table such as the one below.

x	−4	−3	−2	−1	0	1	2	3	4
y	21	12	5	0	−3	−4	−3	0	5

To make this table, we first choose some numbers for x. A set of consecutive integers including several negative integers, zero, and several positive integers is usually the most convenient. For our table, we have chosen the integers from −4 to 4 inclusive.

Next, we use the formula for the function to find the corresponding y-numbers. If $x = -4$, for example,

$$y = (-4)^2 - 2(-4) - 3$$
$$= 16 + 8 - 3$$
$$= 21$$

After we have made a table by this method, we draw axes on graph paper before plotting the graph. The graphs at the top of the next page show the function plotted with three choices of scale for the y-axis. For each graph, we plot the points having the ordered pairs of numbers in the table as their coordinates, (−4, 21), (−3, 12), and so on. Then we connect the points with a smooth curve. All three graphs that we have drawn are correct, but only the third has room for all the points in the table. For this reason, the third graph is preferred.

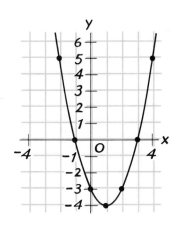

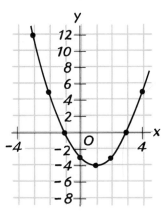

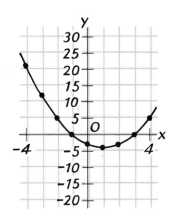

Exercises

Set I

1. Write each of the following expressions in simple radical form.
   a) $\sqrt{124}$
   b) $\sqrt{125}$
   c) $\sqrt{126}$

2. Express each of the following polynomials as the square of a binomial.
   a) $x^2 + 10x + 25$
   b) $4x^2 - 4x + 1$
   c) $x^2 + 16xy + 64y^2$
   d) $9x^2 - 24xy + 16y^2$

3. The temperature of the water in a lake depends on the depth. A typical graph showing the temperatures near the surface of a deep lake is shown at the right.
   a) What kind of function does this seem to be?
   b) What is the y-intercept of the line in this graph?
   c) What does the y-intercept mean in terms of the temperature of the lake?
   d) What is the slope of the line in the graph?

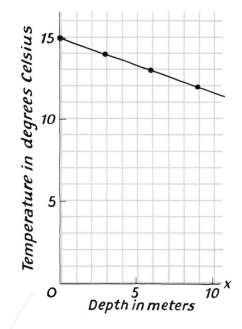

e) What does the slope mean in terms of how the temperature changes as the depth increases?
f) Write an equation for the line.

Set II

4. This exercise is about the function whose equation is

$$y = 2x + 5$$

a) What is its degree?
b) What kind of function is it?
c) What sort of graph does it have?
d) Copy and complete the following table for this function.

x	-3	-2	-1	0	1	2	3
y	-1						

e) Graph the function. Choose a suitable scale for the y-axis so that you have room for all of the points in your table.

5. This exercise is about the function whose equation is

$$y = x^2 - x$$

a) What is its degree?
b) What kind of function is it?
c) What sort of graph does it have?
d) Copy and complete the following table for this function.

x	-4	-3	-2	-1	0	1	2	3	4
y	20								

e) Graph the function. Choose a suitable scale for the y-axis so that you have room for all of the points in your table.

6. Graph the following functions, all on one pair of axes. Let one unit on the y-axis represent 2. (Make a table for each function first.)
a) $y = x^2$
b) $y = x^2 + 6$
c) $y = x^2 - 4$
d) What do you notice about the three graphs?

7. Graph the following functions, each on a separate pair of axes. Let one unit on each y-axis represent 10.
a) $y = x^3$
b) $y = 2x^3$
c) $y = \dfrac{1}{2}x^3$

Graph the following functions, each on a separate pair of axes.
8. $y = x^2 + 4x + 4$
9. $y = 2x^2 - x - 1$
10. $y = x^3 - 10x$
11. $y = x^4 + 2$

Set III

12. This exercise is about the function whose equation is

$$y = 3x - 4$$

a) What is its degree?
b) What kind of function is it?
c) What sort of graph does it have?

d) Copy and complete the following table for this function.

x	-3	-2	-1	0	1	2	3
y	-13						

e) Graph the function. Choose a suitable scale for the y-axis so that you have room for all of the points in your table.

13. This exercise is about the function whose equation is

$$y = x^2 + 2x$$

a) What is its degree?
b) What kind of function is it?
c) What sort of graph does it have?
d) Copy and complete the following table for this function.

x	-4	-3	-2	-1	0	1	2	3	4
y	8								

e) Graph the function. Choose a suitable scale for the y-axis so that you have room for all of the points in your table.

14. Graph the following functions, all on one pair of axes. Let one unit on the y-axis represent 5.

a) $y = 2x^2$
b) $y = 2x^2 + 5$
c) $y = 2x^2 - 10$

15. Graph the following functions, each on a separate pair of axes. Let one unit on each y-axis represent 10.
a) $y = x^3$
b) $y = x^3 + x^2$
c) $y = x^3 + x^2 + x$

Graph the following functions, each on a separate pair of axes.

16. $y = x^2 + 2x + 1$
17. $y = 3x^2 - x + 3$
18. $y = x^3 + 6x$
19. $y = x^4$

Set IV

Suppose that a flea stands in front of the origin of a graph as shown in this picture. It jumps through part of the first quadrant, its path having the equation

$$y = x - \frac{x^2}{12}$$

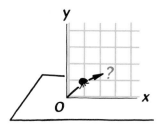

1. Draw its path.
2. Where does the flea take off from the x-axis?
3. Where does it land on the x-axis?
4. How far does the flea jump?
5. How high does it jump?

I'm ve - ry well ac-quain-ted, too, with matters mathe-ma-ti-cal; I

un-der-stand e-qua-tions both the sim-ple and quad-ra-ti-cal, A -

Solving Polynomial Equations by Graphing

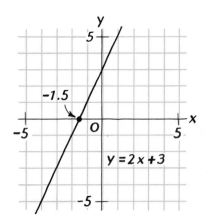

One of the songs in *The Pirates of Penzance* by Gilbert and Sullivan is titled "The Modern Major-General." In it, the Major-General brags of knowing just about everything, including mathematics. The lines of the song shown above boast of his mastery of algebra.

To understand equations "both the simple and quadratical," it is helpful to know what the graphs of their corresponding functions look like. For example, consider the linear equation

$$2x + 3 = 0$$

and the graph of the corresponding linear function

$$y = 2x + 3$$

Solving the equation for x,

$$2x + 3 = 0, \quad 2x = -3, \quad x = -1.5$$

we find that it has one solution: -1.5. This number is the x-coordinate of the point in which the line $y = 2x + 3$ intersects the x-axis. This is a significant result because it is true for all polynomial equations written in standard form.

▶ The solutions of a polynomial equation in standard form are the x-coordinates of the points in which the graph of the corresponding polynomial function intersects the x-axis.

Here is another example illustrating this fact.

EXAMPLE
Find the solutions of
the quadratic equation

$$x^2 - 2x - 3 = 0$$

from the graph of the function

$$y = x^2 - 2x - 3$$

SOLUTION
The graph of $y = x^2 - 2x - 3$
is a curved line. It intersects the
x-axis in two points and so the
equation $x^2 - 2x - 3 = 0$ has
two solutions: -1 and 3.

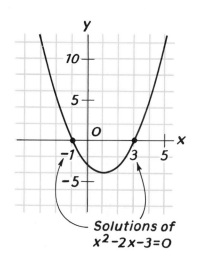

CHECK

$$(-1)^2 - 2(-1) - 3 = 1 + 2 - 3 = 0$$
$$(3)^2 - 2(3) - 3 = 9 - 6 - 3 = 0$$

Exercises

Set I

1. Replace ▓ by a number that will make each of the following the square of a binomial.
   a) $x^2 - 10x + $ ▓
   b) $x^2 + 7x + $ ▓
   c) $4x^2 + 4x + $ ▓
   d) $9x^2 - 24x + $ ▓

2. Express each of the following as either an integer or a radical in simple radical form.
   a) $\sqrt{28}\,\sqrt{7}$ c) $\sqrt{28} - \sqrt{7}$
   b) $\sqrt{28} + \sqrt{7}$ d) $\sqrt{28} \div \sqrt{7}$

3. A method used by scientists to determine the age of very old objects is carbon-14 dating. The basis for this method is the fact that exactly half of any given quantity of carbon-14 remains after 5,700 years.
   a) How old is an object if one-fourth of its carbon-14 atoms remain?

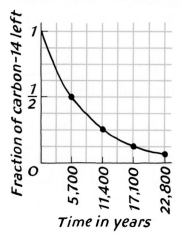

b) How old is an object if one-eighth of its carbon-14 atoms remain?

c) How old is an object if $\dfrac{1}{2^x}$ of its carbon-14 atoms remain?

Set II

4. Use the graphs below to find solutions of the following equations. Check your answers by seeing if they make the equations true.

   a) $3x + 3 = 0$ b) $2x^2 - 9x = 0$ c) $x^3 - 2x^2 + x - 2 = 0$

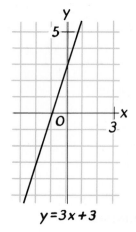

$y = 3x + 3$

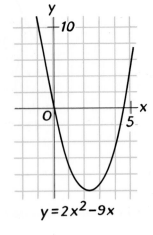

$y = 2x^2 - 9x$

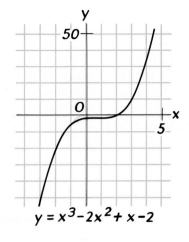

$y = x^3 - 2x^2 + x - 2$

5. Graph the following functions, each on a separate pair of axes.
   a) $y = 2x - 5$
   b) $y = x^2 + x - 12$ d) $y = \frac{1}{2}x^4$
   c) $y = x^3 - 4x$

6. Use the graphs that you drew for exercise 5 to find solutions to the following equations. Check your answers.
   a) $2x - 5 = 0$
   b) $x^2 + x - 12 = 0$ d) $\frac{1}{2}x^4 = 0$
   c) $x^3 - 4x = 0$

7. Graph the following quadratic functions, each on a separate pair of axes.
   a) $y = x^2 - 6x + 7$ c) $y = x^2 + 3x + 6$
   b) $y = x^2 + 4x + 1$ d) $y = 2x^2 + 8x + 7$

8. From the graph of the function

$$y = x^2 - 2x - 2$$

shown in the next column, it is possible to

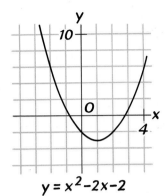

$$y = x^2 - 2x - 2$$

tell that the equation

$$x^2 - 2x - 2 = 0$$

has two solutions: one between –1 and 0 and the other between 2 and 3. Between what integers are the solutions of the following equations? Refer to your graphs for exercise 7 to find out.
   a) $x^2 - 6x + 7 = 0$ c) $x^2 + 3x + 6 = 0$
   b) $x^2 + 4x + 1 = 0$ d) $2x^2 + 8x + 7 = 0$

Set III

9. Use the graphs below to find solutions of the following equations. Check your answers by seeing if they make the equations true.

   a) $\frac{3}{2}x - 3 = 0$ b) $-x^2 + 5x = 0$ c) $x^3 - 19x + 30 = 0$

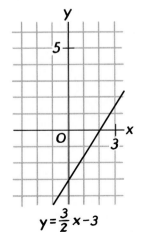

$$y = \frac{3}{2}x - 3$$

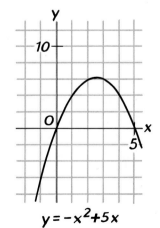

$$y = -x^2 + 5x$$

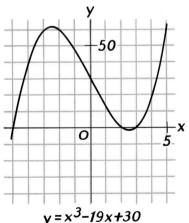

$$y = x^3 - 19x + 30$$

10. Graph the following functions, each on a separate pair of axes.
    a) $y = 3x + 12$ c) $y = x^3 + 8$
    b) $y = 2x^2 - 7x$ d) $y = x^4 + x^2 - 90$

11. Use the graphs that you drew for exercise 10 to find solutions to the following equations. Check your answers.
    a) $3x + 12 = 0$
    b) $2x^2 - 7x = 0$
    c) $x^3 + 8 = 0$
    d) $x^4 + x^2 - 90 = 0$

12. Graph the following quadratic functions, each on a separate pair of axes.
    a) $y = x^2 - 4x - 2$
    b) $y = x^2 + x - 5$
    c) $y = x^2 + 2x + 8$
    d) $y = 2x^2 - 12x + 17$

13. From the graph of the function
$$y = x^2 - 8x + 10$$
    shown in the next column, it is possible to

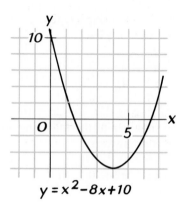

$$y = x^2 - 8x + 10$$

tell that the equation

$$x^2 - 8x + 10 = 0$$

has two solutions: one between 1 and 2 and the other between 6 and 7.

Between what integers are the solutions of the following equations? Refer to your graphs for exercise 12 to find out.
    a) $x^2 - 4x - 2 = 0$ c) $x^2 + 2x + 8 = 0$
    b) $x^2 + x - 5 = 0$ d) $2x^2 - 12x + 17 = 0$

Set IV

From the graph of the function
$$y = x^2 - 2x + 5$$
shown at the right, it seems that the equation
$$x^2 - 2x + 5 = 0$$
has no solutions.

It also seems from this graph that the equation
$$x^2 - 2x + 5 = 4$$
has one solution and that the equation
$$x^2 - 2x + 5 = 8$$
has two solutions.

Can you explain why and tell what the solutions are?

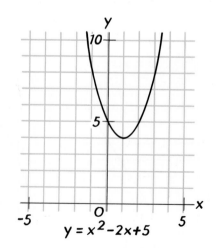

$$y = x^2 - 2x + 5$$

LESSON 4
Solving Quadratic Equations by Factoring

When parentheses are written in algebra as they have been in this cartoon, it is understood that the numbers inside them are to be multiplied. If one of the numbers happens to be zero, we can tell what the product is without knowing any of the other numbers. It must be zero.

$$(\ )(\ )(\ )(0)(\ )(\ )(\ ) = 0$$

The *only* way, in fact, that the product of two or more numbers can be equal to zero is that at least one of the numbers is equal to zero. This is such an important fact that we will refer to it as the *zero-product property*.

▶ **The Zero-Product Property**
If $ab = 0$, then either $a = 0$ or $b = 0$.

This property gives us a way to solve quadratic equations without graphing. To see how it works, we will use the equation

$$x^2 - 2x - 3 = 0$$

621

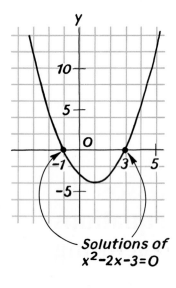

Solutions of
x²-2x-3=0

which was used as an example in the last lesson. We learned there that the solutions to this equation can be found by looking at the points in which the graph of the function

$$y = x^2 - 2x - 3$$

intersects the x-axis. The solutions are -1 and 3.

To find these solutions algebraically, we factor the polynomial $x^2 - 2x - 3$:

$$x^2 - 2x - 3 = 0$$
$$(x + 1)(x - 3) = 0$$

We have changed the left side of the equation into a product and this product is equal to zero. From the zero-product property, we know that at least one of the factors in the product must be equal to zero. If $(x + 1)(x - 3) = 0$, then either

$$x + 1 = 0 \qquad \text{or} \qquad x - 3 = 0$$
$$\text{If } x + 1 = 0, \qquad\qquad \text{If } x - 3 = 0,$$
$$\text{then } x = -1. \qquad\qquad \text{then } x = 3.$$

So *both* the x values, $x = -1$ and $x = 3$, make the product equal to zero. This tells us that the solutions to the equation are -1 *and* 3.

This example illustrates the general method for solving a quadratic equation by factoring. The steps are given below.

To solve a quadratic equation by factoring:
1. Write the equation in standard form: $ax^2 + bx + c = 0$.
2. Factor the polynomial on the left side: ( )( ) = 0.
3. Set each factor equal to zero: ( ) = 0 or ( ) = 0.
4. Solve the resulting equations.

Here is another example.

EXAMPLE
Solve the equation $3x^2 + 20x + 3 = 10$.

SOLUTION

$$3x^2 + 20x + 3 = 10$$
$$3x^2 + 20x - 7 = 0$$
$$(3x - 1)(x + 7) = 0$$

$$3x - 1 = 0 \quad \text{or} \quad x + 7 = 0$$
$$3x = 1 \qquad\qquad x = -7$$
$$x = \frac{1}{3}$$

The solutions are $\frac{1}{3}$ and -7.

Exercises

Set I

1. Write each of the following numbers in decimal form.
   a) 2.1×10^5
   b) 18×10^{-3}
   c) 0.004×10^4

2. Simplify each of the following by writing it either as a quotient without a square root in the denominator or as a product in simple radical form.

   a) $\sqrt{\dfrac{1}{12}}$ c) $\dfrac{10}{\sqrt{5}}$

   b) $\sqrt{\dfrac{9}{x}}$ d) $\dfrac{4x}{\sqrt{x}}$

3. A field goal in basketball counts two points and a successful free throw counts one point.
   a) Write a formula for the score, s, of a team that makes x field goals and y free throws.
   b) Solve the formula for y.
   c) Solve the formula for x.
   d) Can anything be concluded from the fact that a basketball team scores an odd number of points in a game?
   e) Can anything be concluded from the fact that a basketball team scores an even number of points in a game?

Set II

4. What values of x will make each of the following products equal to zero?
   a) $(x - 4)(x - 5)$
   b) $x(x + 8)$
   c) $3(2x - 1)$
   d) $5x(10 - x)$
   e) $(6x + 6)(4x - 12)$
   f) $8(x - 4)(8x + 2)$
   g) $(x + a)(x + b)$
   h) $(ax - 1)(bx - c)$

5. Solve each of the following quadratic equations by the factoring method.
   a) $x^2 - 9 = 0$
   b) $x^2 - 9x = 0$ $($ $)($ $) = 0$
   c) $2x^2 + 2x = 0$
   d) $x^2 + 7x + 12 = 0$

e) $x^2 - 10x + 25 = 0$
f) $x^2 + 5x - 14 = 0$
g) $x^2 + 12x = 0$
h) $x^2 - 64 = 0$
i) $x^2 + 12x - 64 = 0$
j) $9x^2 - 1 = 0$
k) $9x^2 - 8x = 0$
l) $9x^2 - 8x - 1 = 0$

6. Write each of the following equations in standard form and then solve it by factoring.
   a) $x^2 - 16 = 9$
   b) $x(x + 9) = 22$
   c) $(x - 5)^2 = 25$
   d) $(x + 2)(x + 6) = 5$
   e) $(4x - 3)(x + 1) = 1 - 5x$
   f) $3(x^2 - 8) = x(x + 2)$

Set III

7. What values of x will make each of the following products equal to zero?
   a) $x(x - 2)$
   b) $(x + 1)(x + 9)$
   c) $2(5x - 3)$
   d) $7x(4 - x)$
   e) $(2x + 8)(3x - 15)$
   f) $4(x + 3)(4x - 1)$
   g) $(x - a)(x - b)$
   h) $(ax + b)(cx + 1)$

8. Solve each of the following quadratic equations by the factoring method.
   a) $x^2 - 4 = 0$
   b) $x^2 - 4x = 0$
   c) $3x^2 + 3x = 0$
   d) $x^2 + 9x + 14 = 0$

e) $x^2 - 16x + 64 = 0$
f) $x^2 - 2x - 15 = 0$
g) $x^2 + 5x = 0$
h) $x^2 - 36 = 0$
i) $x^2 + 5x - 36 = 0$
j) $4x^2 - 1 = 0$
k) $4x^2 - 3x = 0$
l) $4x^2 - 3x - 1 = 0$

9. Write each of the following equations in standard form and then solve it by factoring.
   a) $x^2 - 36 = 64$
   b) $x(x - 4) = 21$
   c) $(x + 9)^2 = 81$
   d) $(x - 3)(x - 4) = 2$
   e) $(2x + 5)(x - 1) = 1 - x^2$
   f) $3(x^2 - 10) = x(x - 4)$

Set IV

After finishing his algebra assignment, Obtuse Ollie asked Acute Alice to check his answers. Instead, she started with the answers and figured out what the problems were!

The assignment was to solve quadratic equations like those in this lesson. The answers to three of the problems that Ollie solved correctly are listed below. Given that each equation was in standard form, can you figure out what the equations were? If you can, show your method for finding them.

1. 7 and –7
2. 0 and 2
3. 4 and –5

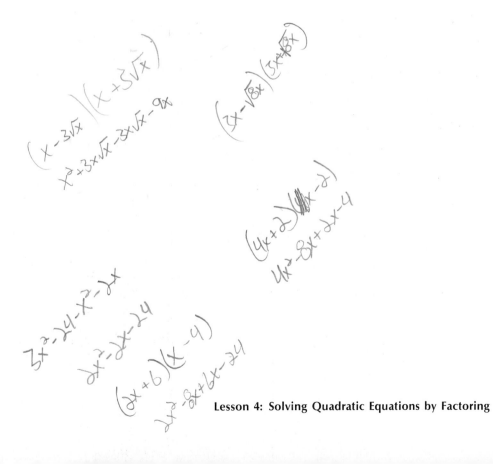

"DID YOU KNOW THAT IF YOUR BELT SNAPPED YOU'D
FALL AT THE RATE OF 32 FEET PER
SECOND PER SECOND?"

LESSON 5
Solving Quadratic Equations
by Taking Square Roots

As an object falls toward the ground, it moves faster and faster.* The distance
that the object has fallen after any given time is given by the formula

$$16x^2 = y$$

in which x is the time in seconds that it has been falling and y is the distance in
feet.

According to this formula, how long would it take an object to fall 400 feet?
Substituting 400 for y, we get

$$16x^2 = 400$$

This is a quadratic equation in one variable. It can be simplified by dividing
both sides by 16:

$$x^2 = \frac{400}{16} = 25$$

---

* See page 394.

626

Evidently, $x = 5$, and so it takes the object 5 seconds to fall 400 feet.

When we recognize that 5 is a solution of this equation, we have, in effect, taken the square root of each side. But every positive number has *two* square roots, each of which is the opposite of the other. The equation

$$x^2 = 25$$

has a second solution: –5. Although this solution does not have meaning for the falling-object problem, it is a solution to the original equation.

▶ In general, if $x^2 = a$, then *either* $x = \sqrt{a}$ *or* $x = -\sqrt{a}$. More briefly, if $x^2 = a$, then $x = \pm\sqrt{a}$. If a is a number greater than 0, the symbol $\pm\sqrt{a}$ represents *two* numbers: the positive square root of a and the negative square root of a.

The most obvious way, then, to solve the equation

$$x^2 = 25$$

is to take the square root of each side,

$$x = \pm\sqrt{25} = \pm 5$$

The square-root method is often more convenient than solving by factoring. Moreover, it can be used to develop a method that will solve *any* quadratic equation.

Here are more examples of how it works.

EXAMPLE 1

Solve the equation $(x + 4)^2 = 9$.

SOLUTION

$$
\begin{aligned}
(x + 4)^2 &= 9 \\
x + 4 &= \pm 3 \\
x &= -4 \pm 3
\end{aligned}
$$

Figuring out these values, we get

$$-4 + 3 = -1 \quad \text{and} \quad -4 - 3 = -7$$

The solutions are –1 and –7.

EXAMPLE 2

Solve the equation $x^2 - 4x + 4 = 5$.

SOLUTION

$$\begin{aligned} x^2 - 4x + 4 &= 5 \\ (x - 2)^2 &= 5 \\ x - 2 &= \pm\sqrt{5} \\ x &= 2 \pm \sqrt{5} \end{aligned}$$

The solutions are $2 + \sqrt{5}$ and $2 - \sqrt{5}$.

Exercises

Set I

1. Multiply and simplify.
   a) $(2\sqrt{x})(5\sqrt{x})$
   b) $(2\sqrt{x})(5 - \sqrt{x})$
   c) $(2 + \sqrt{x})(5 - \sqrt{x})$

2. This exercise is about the graph below.

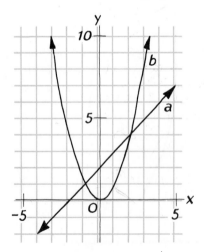

a) Write an equation for line a.
b) Write an equation for curve b.
c) What are the solutions to the pair of simultaneous equations that you have written?

3. The sum of the squares of three consecutive integers is 110.
   a) If x^2 represents the square of the smallest integer, how would the squares of the other two numbers be represented?
   b) Write an equation expressing the fact that the sum of the squares of the three numbers is 110.
   c) Solve the equation by writing it in standard form and factoring.
   d) What are the three integers?

Set II

4. Express the square roots of the following numbers either as integers or in simple radical form.
   a) 144
   b) 30
   c) 12
   d) 150

5. Express the two numbers represented by each of the following expressions in simplest form.
   a) 3 ± 6
   b) -11 ± 7
   c) $\dfrac{8 \pm 2}{5}$
   d) $10 \pm \sqrt{3}$
   e) $\dfrac{3 \pm 12\sqrt{5}}{3}$
   f) $\dfrac{4 \pm \sqrt{8}}{2}$

6. This exercise is about the equation

   $$(x - 3)^2 = 36$$

   a) Solve this equation by writing it in standard form and using the factoring method.
   b) Solve it by the square-root method.
   c) Which method for solving this equation is easier?

7. This exercise is about the equation

   $$(x + 4)^2 = 15$$

   a) Solve it by the square-root method.
   b) Write the equation in standard form and try to solve it by the factoring method.

*8. This exercise is about the equation

   $$(3x - 6)^2 = 45$$

   a) Solve it by the square-root method, simplifying your answers as much as possible.
   b) Express your answers in decimal form, correct to the nearest thousandth. (Refer to the table of square roots on page 563.)
   c) Substitute your answers in decimal form for x in the equation to see if they seem reasonable.

9. Solve the following quadratic equations by the square-root method.
   a) $x^2 = 48$
   b) $(x - 5)^2 = 4$
   c) $(x + 2)^2 = 25$
   d) $(x + 1)^2 = 10$
   e) $(2x - 3)^2 = 81$
   f) $(3x + 4)^2 = 49$

10. Solve the following quadratic equations by expressing the left sides as squares of binomials and using the square-root method.
   a) $x^2 - 6x + 9 = 16$
   b) $x^2 + 16x + 64 = 45$
   c) $4x^2 + 20x + 25 = 25$
   d) $25x^2 - 10x + 1 = 5$

*11. Solve the following quadratic equations. Express each answer in decimal form, correct to the nearest thousandth. (Refer to the table of square roots on page 563.) Check each answer in the original equation to see if it seems reasonable.
   a) $(x - 7)^2 = 5$
   b) $(x + 1)^2 = 11$
   c) $(6x - 5)^2 = 0$
   d) $(8x + 3)^2 = 10$

Set III

12. Express the square roots of the following numbers either as integers or in simple radical form.
    a) 121
    b) 42
    c) 18
    d) 80

13. Express the two numbers represented by each of the following expressions in simplest form.
    a) 2 ± 10
    b) -4 ± 9
    c) $\dfrac{5 \pm 7}{3}$
    d) $6 \pm \sqrt{6}$
    e) $\dfrac{5 \pm 10\sqrt{10}}{5}$
    f) $\dfrac{8 \pm \sqrt{12}}{2}$

14. This exercise is about the equation
    $$(x + 1)^2 = 100$$
    a) Solve this equation by writing it in standard form and using the factoring method.
    b) Solve it by the square-root method.
    c) Which method for solving this equation is easier?

15. This exercise is about the equation
    $$(x - 2)^2 = 3$$
    a) Solve it by the square-root method.
    b) Write the equation in standard form and try to solve it by the factoring method.

*16. This exercise is about the equation
    $$(2x - 8)^2 = 12$$
    a) Solve it by the square-root method, simplifying your answers as much as possible.
    b) Express your answers in decimal form, correct to the nearest thousandth. (Refer to the table of square roots on page 563.)
    c) Substitute your answers in decimal form for x in the equation to see if they seem reasonable.

17. Solve the following quadratic equations by the square-root method.
    a) $x^2 = 50$
    b) $(x - 1)^2 = 16$
    c) $(x + 6)^2 = 0$
    d) $(x - 2)^2 = 15$
    e) $(2x + 1)^2 = 49$
    f) $(5x - 3)^2 = 64$

18. Solve the following quadratic equations by expressing the left sides as squares of binomials and using the square-root method.
    a) $x^2 + 8x + 16 = 25$
    b) $x^2 - 14x + 49 = 10$
    c) $4x^2 + 4x + 1 = 81$
    d) $9x^2 - 12x + 4 = 3$

*19. Solve the following quadratic equations. Express each answer in decimal form, correct to the nearest thousandth. (Refer to the table of square roots on page 563.) Check each answer in the original equation to see if it seems reasonable.
    a) $(x - 5)^2 = 6$ c) $(3x - 8)^2 = 10$
    b) $(x + 4)^2 = 17$ d) $(7x + 1)^2 = 0$

$x^2 - 4x + 4$

Set IV

This picture of George Washington was produced on a typewriter! The picture is square, measuring 4 centimeters on each side and is surrounded by a square frame.

Can you figure out the exact width of the frame, given that it has the same area as the picture? Hint: (length of one side of the frame)^2 = total area of frame and picture.

THE NATIONAL ENQUIRER.

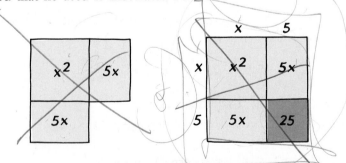

LESSON 6
Completing the Square

Algebra got its name from the title of a book, *Al-jabr wa'l Muqabalah,* by a ninth-century Arab mathematician named al-Khowarizmi. The book explained how to solve various types of linear and quadratic equations. One of the equations solved by al-Khowarizmi was

$$x^2 + 10x = 39$$

The method that he used is illustrated below.

First, picture the left side of the equation, $x^2 + 10x$, as a figure consisting of a square of area x^2 and two rectangles each having an area of $5x$. By adding a square having an area of 25 to the lower right corner, the figure can be made into a large square. This is called *completing the square.*

The area of the large square is $x^2 + 10x + 25$. Completing the square in the equation

$$x^2 + 10x = 39$$

we get

$$x^2 + 10x + 25 = 39 + 25 = 64 \quad \text{or}$$
$$(x + 5)^2 = 64$$

At this point, al-Khowarizmi took the square root of each side of the equation, but ignored the possibility of a negative root. He concluded that

$$x + 5 = 8$$

so that $x = 3$.

It is easy to verify that 3 is a solution to the original equation

$$x^2 + 10x = 39$$

because

$$(3)^2 + 10(3) = 9 + 30 = 39$$

The equation

$$(x + 5)^2 = 64$$

is also true if

$$x + 5 = -8$$

which means that $x = -13$ also is a solution. Checking -13 in the original equation, we get

$$(-13)^2 + 10(-13) = 169 - 130 = 39$$

The equation has two solutions: 3 and -13.

To solve a quadratic equation by completing the square:

1. Write the equation in the form: $ax^2 + bx = -c$.
2. Add a number to each side of the equation that will make the left side the square of a binomial:*

$$ax^2 + bx + (\ \ ) = -c + (\ \ )$$

_____

*Binomial squares are explained on pages 422–425.

3. Take the square root of each side.
4. Solve the resulting equations.

Here are more examples of how the completing-the-square method works.

EXAMPLE 1
Solve the equation $x^2 - 6x - 16 = 0$ by completing the square.

SOLUTION

$$
\begin{aligned}
x^2 - 6x - 16 &= 0 \\
x^2 - 6x &= 16 \\
x^2 - 6x + 9 &= 16 + 9 \\
(x - 3)^2 &= 25 \\
x - 3 &= \pm 5 \\
x &= 3 \pm 5
\end{aligned}
$$

Figuring out these values, we get

$$3 + 5 = 8 \quad \text{and} \quad 3 - 5 = -2$$

The solutions are 8 and –2.

EXAMPLE 2
Solve the equation $x^2 + 14x = -39$.

SOLUTION

$$
\begin{aligned}
x^2 + 14x &= -39 \\
x^2 + 14x + 49 &= -39 + 49 \\
(x + 7)^2 &= 10 \\
x + 7 &= \pm\sqrt{10} \\
x &= -7 \pm \sqrt{10}.
\end{aligned}
$$

The solutions are $-7 + \sqrt{10}$ and $-7 - \sqrt{10}$.

Exercises

Set I

1. The following questions are about the graphs of these functions.

Function A: $y = 3$
Function B: $y = 3x$
Function C: $y = x + 3$

a) Which line has the steepest slope?
b) Which line has a slope of 0?
c) Which lines have a y-intercept of 3?

2. Find the approximate value of each of the following expressions. Let $\sqrt{5} = 2.236$ and round each answer to the nearest hundredth.

a) $\dfrac{1}{\sqrt{5}}$

b) $\dfrac{1}{\sqrt{5} + 2}$

c) $\dfrac{1}{\sqrt{5} - 2}$

"Miss, that person is making a fool of you!"

3. A birdseed manufacturer wants to mix millet worth 28 cents a kilogram and grain worth 18 cents a kilogram to make 500 kilograms of birdseed worth 24 cents a kilogram.

a) Use this information to write a pair of simultaneous equations.
b) Solve the equations to find out how many kilograms of each seed should be used.

Set II

4. Add a term to each of the following binomials to change it into a trinomial square. (Draw diagrams if you need them.)
   a) $x^2 + 8x$
   b) $x^2 - 30x$
   c) $4x^2 - 12x$
   d) $25x^2 + 70x$
   e) $x^2 + 2ax$
   f) $a^2x^2 - 4ax$

5. Here is another example of how to solve a quadratic equation by completing the square. Tell what was done in each lettered step.

$$9x^2 + 12x - 7 = 0$$

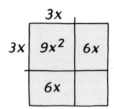

a) $9x^2 + 12x \quad\ = 7$
b) $9x^2 + 12x + 4 = 7 + 4$
   $\quad (3x + 2)^2 \quad\ = 11$
c) $\quad\ 3x + 2 \quad\ = \pm\sqrt{11}$
d) $\quad\quad 3x \quad\quad = -2 \pm \sqrt{11}$
e) $\quad\quad x \quad\quad\ = \dfrac{-2 \pm \sqrt{11}}{3}$

6. This exercise is about the equation

$$x^2 + 6x = 40$$

   a) Solve this equation by writing it in standard form and using the factoring method.
   b) Solve it by completing the square.

*7. This exercise is about the equation

$$x^2 - 10x = 36$$

   a) Solve it by completing the square.
   b) Express your answers in decimal form, correct to the nearest hundredth. (Refer to the table of square roots on page 563.)
   c) Substitute your answers in decimal form for x in the equation to see if they seem reasonable.

8. This exercise is about the equation

$$x^2 - 2x - 4 = 0$$

   a) Graph the function $y = x^2 - 2x - 4$.
   b) Between what integers are the solutions of $x^2 - 2x - 4 = 0$? Use your graph to find out.
   c) Solve the equation by completing the square.
   d) Express your answers in decimal form, correct to the nearest hundredth. (Refer to the table of square roots on page 563.)

9. Solve each of the following quadratic equations by completing the square.
   a) $x^2 + 4x = 77$
   b) $x^2 - 12x + 35 = 0$
   c) $x^2 + 18x = -81$
   d) $x^2 - 2x - 10 = 0$
   e) $x^2 + 16x = 1$
   f) $x^2 - 6x - 9 = 0$
   g) $9x^2 - 6x = 35$
   h) $16x^2 + 40x + 11 = 0$

10. Solve the following equations.
   a) $(x - 2)(x + 6) = 10$
   b) $x(x + 8) = 8(x + 3)$
   c) $x(x - 5) = 3(x + 2)$
   d) $3x(x + 4) = 2(x^2 - 1)$

Set III

11. Add a term to each of the following binomials to change it into a trinomial square. (Draw diagrams if you need them.)
    a) $x^2 - 6x$
    b) $x^2 + 22x$
    c) $9x^2 + 30x$
    d) $4x^2 - 44x$
    e) $x^2 - 2ax$
    f) $a^2x^2 + 8ax$

12. Here is another example of how to solve a quadratic equation by completing the square. Tell what was done in each lettered step.

$$4x^2 + 20x + 19 = 0$$

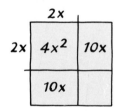

    a) $4x^2 + 20x \qquad = -19$
    b) $4x^2 + 20x + 25 = -19 + 25$
    $\quad\;\; (2x + 5)^2 \qquad = 6$
    c) $\qquad 2x + 5 \quad = \pm\sqrt{6}$
    d) $\qquad\;\; 2x \quad = -5 \pm \sqrt{6}$
    e) $\qquad\quad x \quad = \dfrac{-5 \pm \sqrt{6}}{2}$

13. This exercise is about the equation

$$x^2 + 12x = 45$$

    a) Solve this equation by writing it in standard form and using the factoring method.
    b) Solve it by completing the square.

*14. This exercise is about the equation

$$x^2 - 2x = 29$$

    a) Solve it by completing the square.
    b) Express your answers in decimal form, correct to the nearest hundredth. (Refer to the table of square roots on page 563.)
    c) Substitute your answers in decimal form for x in the equation to see if they seem reasonable.

15. The following questions are about the equation

$$x^2 - 4x + 1 = 0$$

    a) Graph the function $y = x^2 - 4x + 1$.
    b) Between what integers are the solutions of $x^2 - 4x + 1 = 0$? Use your graph to find out.
    c) Solve the equation by completing the square.
    d) Express your answers in decimal form, correct to the nearest hundredth. (Refer to the table of square roots on page 563.)

16. Solve each of the following quadratic equations by completing the square.
    a) $x^2 + 10x = -9$
    b) $x^2 - 2x - 48 = 0$
    c) $x^2 + 12x + 36 = 0$
    d) $x^2 - 8x + 3 = 0$
    e) $x^2 + 18x = 4$
    f) $x^2 - 4x - 16 = 0$
    g) $4x^2 - 4x = 63$
    h) $25x^2 + 30x - 6 = 0$

17. Solve the following equations.
    a) $(x + 9)(x - 1) = 15$
    b) $x(x + 3) = 3(x + 5)$
    c) $x(x - 7) = 5(x - 1)$
    d) $4(x^2 + 2) = 3x(x - 6)$

Set IV

Although this equation looks simple enough, if you try to solve it by completing the square something peculiar happens.

1. What is it and what do you think it means?
2. Does the graph of $y = x^2 + 2x + 4$ show why you might have trouble in solving this equation?

The Quadratic Formula

A famous equation in mathematics is the *quadratic formula.* It can be used to find the solutions to a quadratic equation that has been written in standard form.

The quadratic formula is the result of solving the general quadratic equation

$$ax^2 + bx + c = 0$$

for *x*. We will do it by completing the square. Subtracting *c* from each side, we get

$$ax^2 + bx = -c$$

To make the first term of the equation a simple square and to prevent having to work with fractions, we next multiply each side of the equation by 4*a*:

$$4a(ax^2 + bx) = 4a(-c)$$
$$4a^2x^2 + 4abx = -4ac$$

The first term of this equation is the square of 2*ax*, which is illustrated by the left-hand figure below. Dividing the second term of the equation, 4*abx*, by 2, we get 2*abx*, which is shown as the area of each of the two equal rectangles in the middle figure. To find the other dimension of each rectangle, we divide 2*abx* by 2*ax*, getting *b*. This means that we must add b^2, as shown in the right-hand figure, to complete the square.

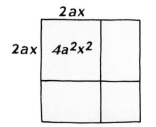

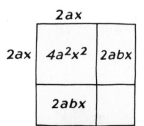

639

Doing this to the equation

$$4a^2x^2 + 4abx = -4ac$$

we get

$$4a^2x^2 + 4abx + b^2 = b^2 - 4ac$$

or

$$(2ax + b)^2 = b^2 - 4ac$$

Taking the square root of each side,

$$2ax + b = \pm\sqrt{b^2 - 4ac}$$

and solving for x,

$$2ax = -b \pm \sqrt{b^2 - 4ac}$$

$$x = \frac{-b \pm \sqrt{b^2 - 4ac}}{2a}$$

This is the quadratic formula. It is so useful that *it should be memorized.*

▶ The Quadratic Formula

If $ax^2 + bx + c = 0$, then $x = \dfrac{-b \pm \sqrt{b^2 - 4ac}}{2a}$.

Here are examples of how this formula can be used to solve quadratic equations.

EXAMPLE 1
Solve the equation $5x^2 + 9x - 2 = 0$ by using the quadratic formula.

SOLUTION
In this equation, $a = 5$, $b = 9$, and $c = -2$. The solutions of the equation are

$$x = \frac{-b \pm \sqrt{b^2 - 4ac}}{2a}$$

Replacing a, b, and c by their values, we get

$$x = \frac{-(9) \pm \sqrt{(9)^2 - 4(5)(-2)}}{2(5)}$$

Simplifying, we get

$$x = \frac{-9 \pm \sqrt{81 + 40}}{10} = \frac{-9 \pm \sqrt{121}}{10} = \frac{-9 \pm 11}{10}$$

$$\frac{-9 + 11}{10} = \frac{2}{10} = \frac{1}{5} \quad \text{and} \quad \frac{-9 - 11}{10} = \frac{-20}{10} = -2$$

The solutions of the equation are $\frac{1}{5}$ and -2.

EXAMPLE 2

Solve the equation $x^2 + 4 = 6x$.

SOLUTION

To find a, b, and c, the equation must be written in standard form:
$x^2 - 6x + 4 = 0$. Because $a = 1$, $b = -6$, and $c = 4$,

$$x = \frac{-(-6) \pm \sqrt{(-6)^2 - 4(1)(4)}}{2(1)}$$

$$x = \frac{6 \pm \sqrt{36 - 16}}{2} = \frac{6 \pm \sqrt{20}}{2} = \frac{6 \pm 2\sqrt{5}}{2} = \frac{2(3 \pm \sqrt{5})}{2} = 3 \pm \sqrt{5}$$

The solutions of the equation are $3 + \sqrt{5}$ and $3 - \sqrt{5}$.

Exercises

Set I

1. Factor each of the following polynomials as completely as you can.
   a) $4x^3 + 36x$
   b) $8x^2 - 80x + 200$
   c) $x^3 + 9x^2 - 10x$
   d) $x^4 - 40x^2 + 144$

2. Solve the following equations.
   a) $\sqrt{2x - 5} = 7$
   b) $x + 1 = \sqrt{x^2 + 4}$
   c) $3\sqrt{14 + x} = \sqrt{6 - x}$

3. The figure at the right consists of 13 squares, each of whose sides is x units long.
a) Write a formula for its perimeter, p.
b) Write a formula for its area, a.

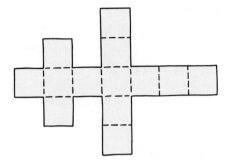

Set II

4. Write each of the following quadratic equations in standard form

$$ax^2 + bx + c = 0$$

and give the values of a, b, and c.
a) $x^2 - 5x + 5 = 2$
b) $4x^2 - 9 = x$
c) $7x^2 = x(x - 3)$
d) $(3x - 1)(x + 5) = x^2 + 6$

5. This exercise is about the equation

$$x^2 - 6x = 12$$

a) Solve this equation by completing the square.
b) Solve it by using the quadratic formula.
c) Express your answers in decimal form, correct to the nearest hundredth. (Refer to the table of square roots on page 563.)
d) Substitute your answers in decimal form for x in the equation to see if they seem reasonable.

6. This exercise is about the equation

$$x^2 + 3x + 1 = 0$$

a) Solve this equation by completing the square.
b) Solve it by using the quadratic formula.
c) Which method for solving this equation is easier?

*7. Solve the following quadratic equations. Express each answer either as an integer or in decimal form, correct to the nearest tenth.
a) $x^2 - 5x - 14 = 0$
b) $x^2 - 5x - 15 = 0$
c) $x^2 - 5x - 16 = 0$
d) $x^2 - 7x + 3 = 0$
e) $x^2 - 8x + 3 = 0$
f) $x^2 - 9x + 3 = 0$

8. Solve the following quadratic equations by using the quadratic formula. Leave answers in radical form but simplify as much as possible.
a) $x^2 - 6x - 16 = 0$
b) $4x^2 + 9x + 2 = 0$
c) $2x^2 - 5x + 1 = 0$
d) $8x^2 - 5 = 0$
e) $3x^2 + x = 0$
f) $x^2 + 10x + 25 = 0$

9. Write each of the following quadratic equations in standard form and solve it by using the quadratic formula. Simplify each solution as much as you can.
a) $x(x + 1) = 15$
b) $4(x^2 - 2) = x$
c) $(x + 5)^2 = x + 17$
d) $(x - 7)(x + 3) = -5$

10. Use the quadratic formula to solve the following equations for x in terms of the other variables.
   a) $x^2 + bx + c = 0$
   b) $dx^2 - x - e = 0$
   c) $ax^2 - 1 = 0$
   d) $fx^2 + gx = 0$

Set III

11. Write each of the following quadratic equations in standard form

$$ax^2 + bx + c = 0$$

and give the values of a, b, and c.
   a) $3x^2 + x + 4 = 6$
   b) $x^2 + 8 = 7x$
   c) $x(2x - 1) = 3x + 1$
   d) $(5x + 3)(x - 1) = x^2 - 3$

12. This exercise is about the equation

$$x^2 - 4x = 15$$
$$x^2 - 4x + 4 = 0 + 4$$

   a) Solve this equation by completing the square.
   b) Solve it by using the quadratic formula.
   c) Express your answers in decimal form, correct to the nearest hundredth. (Refer to the table of square roots on page 563.)
   d) Substitute your answers in decimal form for x in the equation to see if they seem reasonable.

13. This exercise is about the equation

$$x^2 + 7x - 2 = 0$$

   a) Solve this equation by completing the square.
   b) Solve it by using the quadratic formula.
   c) Which method for solving this equation is easier?

*14. Solve the following quadratic equations. Express each answer either as an integer or in decimal form, correct to the nearest tenth.
   a) $x^2 + 2x - 15 = 0$
   b) $x^2 + 2x - 16 = 0$
   c) $x^2 + 2x - 17 = 0$
   d) $x^2 + 8x + 5 = 0$
   e) $x^2 + 9x + 5 = 0$
   f) $x^2 + 10x + 5 = 0$

15. Solve the following quadratic equations by using the quadratic formula. Leave answers in radical form but simplify them as much as possible.
   a) $x^2 + 7x - 18 = 0$
   b) $3x^2 + 5x + 2 = 0$
   c) $x^2 + 3x + 1 = 0$
   d) $6x^2 - 4 = 0$
   e) $2x^2 + 8x = 0$
   f) $x^2 - 12x + 36 = 0$

16. Write each of the following quadratic equations in standard form and solve by using the quadratic formula. Simplify each solution as much as you can.

   a) $x(x - 3) = 12$ c) $(x - 4)^2 = x + 2$
   b) $5(x^2 - 1) = 2x$ d) $(x + 2)(x - 8) = -4$

17. Use the quadratic formula to solve the following equations for x in terms of the other variables.
   a) $ax^2 + x + c = 0$ c) $x^2 - f = 0$
   b) $x^2 - dx - e = 0$ d) $ax^2 - bx = 0$

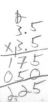

Set IV Solving Quadratic Equations by Stretching a Thread*

The figure below can be used to find the positive solutions to quadratic equations having the form

$$x^2 - bx + c = 0$$

in which b and c are positive. The horizontal scales at the top and bottom represent b and c.

An equation such as

$$x^2 - 7x + 6 = 0$$

can be solved by stretching a thread from 7 on the b-scale to 6 on the c-scale. The numbers of the points in which it crosses the curved scale are solutions to the equation.

1. Do this to find them.
2. Also find them by the quadratic formula.
3. Solve the equation

$$x^2 - 8x + 4 = 0$$

by stretching a thread across the figure.
4. Compare your solutions with those found by the quadratic formula. (Let $\sqrt{3} = 1.73$.)

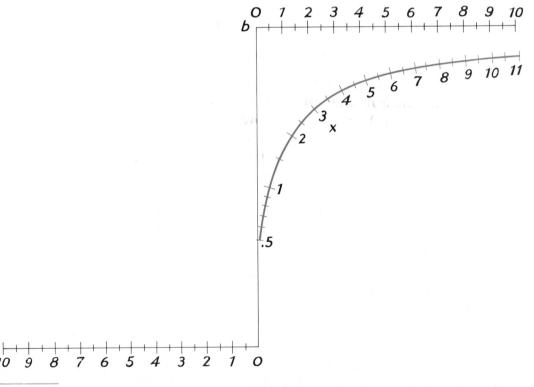

---

*After *More Fun with Mathematics* by Jerome S. Meyer (Fawcett Publications, 1963), pp. 25–26.

COURTESY OF THE FEDERAL RESERVE BANK OF MINNEAPOLIS.

LESSON **8**
The Discriminant

The unusual design of the Federal Reserve Bank building in Minneapolis is the work of the architect Gunnar Birkerts. Two opposite sides of the building have the curve that is shown in the photograph above. This curve is part of a mathematical curve called a *parabola.*

It is easy to write an equation whose graph is a parabola. In fact, *any quadratic function will do.*

▶ The graph of every quadratic function (that is, every function having an equation of the form $y = ax^2 + bx + c$ in which a is not zero) is a **parabola.**

Compare, for example, the graph of the quadratic function $y = 0.2x^2$, shown here, with the curve in the photograph.

In Lesson 3, we learned that the solutions of a polynomial equation in standard form are the *x*-coordinates of the points in which the graph of the corresponding polynomial function intersects the *x*-axis. Because the graph of the function $y = 0.2x^2$ intersects the *x*-axis in just one point, the origin, it follows that the equation $0.2x^2 = 0$ has one solution, zero.

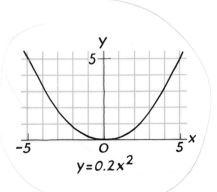

$y = 0.2x^2$

The shape of the graph of a polynomial function depends on its degree. The graph of every *first*-degree function is a *line* (which, as you already know, is why such functions are called *linear*.) Because the line always intersects the *x*-axis in exactly one point, every first-degree equation has exactly one solution. Examples are shown below.

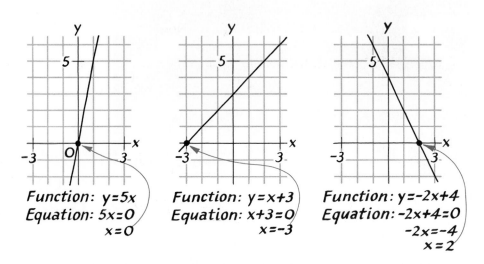

Function: $y=5x$
Equation: $5x=0$
            $x=0$

Function: $y=x+3$
Equation: $x+3=0$
            $x=-3$

Function: $y=-2x+4$
Equation: $-2x+4=0$
            $-2x=-4$
              $x=2$

The graph of every *second*-degree (quadratic) function is a *parabola*. Unlike a line, a parabola can intersect the *x*-axis in either two points, one point, or no points at all. So a quadratic equation can have either two solutions, one solution, or no solutions. Examples are shown below.

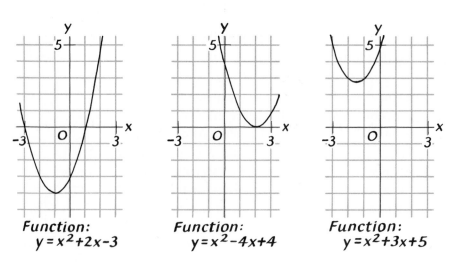

Function:
   $y=x^2+2x-3$

Function:
   $y=x^2-4x+4$

Function:
   $y=x^2+3x+5$

The following table lists the three quadratic equations, together with their solutions, corresponding to the graphs at the bottom of the preceding page.

Equation	Solutions
$x^2 + 2x - 3 = 0$	–3 and 1
$x^2 - 4x + 4 = 0$	2
$x^2 + 3x + 5 = 0$	None

What would happen if we tried solving each of these equations by the quadratic formula?

$$x^2 + 2x - 3 = 0$$
$$a = 1, b = 2, c = -3$$

$$x = \frac{-(2) \pm \sqrt{(2)^2 - 4(1)(-3)}}{2(1)}$$

$$= \frac{-2 \pm \sqrt{4 + 12}}{2}$$

$$x^2 - 4x + 4 = 0$$
$$a = 1, b = -4, c = 4$$

$$x = \frac{-(-4) \pm \sqrt{(-4)^2 - 4(1)(4)}}{2(1)}$$

$$= \frac{4 \pm \sqrt{16 - 16}}{2}$$

$$x^2 + 3x + 5 = 0$$
$$a = 1, b = 3, c = 5$$

$$x = \frac{-(3) \pm \sqrt{(3)^2 - 4(1)(5)}}{2(1)}$$

$$= \frac{-3 \pm \sqrt{9 - 20}}{2}$$

The number of solutions that each equation has becomes clear at this point: *it is determined by the number under the radical sign*. The first equation has *two* solutions because the number under the radical sign is *positive*.

$$x = \frac{-2 \pm \sqrt{16}}{2}$$

$$= \frac{-2 \pm 4}{2}$$

$$\frac{-2 + 4}{2} = \frac{2}{2} = 1 \quad \text{and} \quad \frac{-2 - 4}{2} = \frac{-6}{2} = -3$$

The second equation has *one* solution because the number under the radical sign is *zero*.

$$x = \frac{4 \pm \sqrt{0}}{2}$$

$$= \frac{4 \pm 0}{2} = \frac{4}{2} = 2$$

The third equation has *no* solutions because the number under the radical sign is *negative* and negative numbers do not have square roots in the number system that we are using.

$$x = \frac{-3 \pm \sqrt{-11}}{2}$$

There is no solution because -11 does not have a square root.

Because the nature of the number under the radical sign in the quadratic formula determines the number of solutions that a quadratic equation has, it is called the *discriminant* of the equation.

▶The **discriminant** of the equation

$$ax^2 + bx + c = 0$$

is $b^2 - 4ac$.

$$x = \frac{-b \pm \sqrt{b^2 - 4ac}}{2a}$$

is the *discriminant*
of the equation
$ax^2 + bx + c = 0$

If the discriminant is a *positive* number, the equation has *two* solutions. If it is equal to *zero,* the equation has *one* solution. If it is *negative,* the equation has *no* solutions.

Exercises

Set I

1. Write each of the following as a single power.
   a) $x^3 \cdot x^{-3}$
   b) $(x^3)^{-3}$
   c) $\dfrac{x^3}{x^{-3}}$
   d) $x^3 y^3$
   e) $\dfrac{x^3}{y^3}$

2. Solve the quadratic equation $x^2 + 4x - 45 = 0$ by each of the following methods.
   a) By factoring.
   b) By completing the square.
   c) By the quadratic formula.

3. The following puzzle is from a collection of puzzles published in 1914.*

   "Five years ago sister was four times older than Fido. Now she is only three times as old. How old is Fido?"

   ---
   * *Cyclopedia of Puzzles* by Sam Loyd (Morningside Press, 1914). Reprinted by Pinnacle Books, 1976.

Solve it by doing the following.

   a) Letting x represent Fido's present age and y represent sister's present age, write an equation relating x and y.
   b) Write an equation for the relationship of their ages five years ago.
   c) Solve the pair of simultaneous equations that you have written for x and y.
   d) How old is Fido?

Set II

4. Find the value of the discriminant for each of the following equations and use it to tell how many solutions the equation has.
   a) $x^2 + 4x + 3 = 0$
   b) $x^2 + 4x + 4 = 0$
   c) $x^2 + 4x + 6 = 0$

5. Graph the following functions, each on a separate pair of axes.
   a) $y = x^2 + 4x + 3$
   b) $y = x^2 + 4x + 4$
   c) $y = x^2 + 4x + 6$

   Use your graphs to find solutions to each of the following equations.

   d) $x^2 + 4x + 3 = 0$

   e) $x^2 + 4x + 4 = 0$
   f) $x^2 + 4x + 6 = 0$

6. Find solutions to each of the following quadratic equations by completing the square.
   a) $x^2 + 4x + 3 = 0$
   b) $x^2 + 4x + 4 = 0$
   c) $x^2 + 4x + 6 = 0$

7. Find solutions to each of the following quadratic equations by using the quadratic formula.
   a) $x^2 + 4x + 3 = 0$
   b) $x^2 + 4x + 4 = 0$
   c) $x^2 + 4x + 6 = 0$

8. Find the value of the discriminant for each of the following quadratic equations and use it to tell how many solutions the equation has.
   a) $x^2 - 6x + 7 = 0$ c) $4x^2 + 4x + 1 = 0$
   b) $x^2 + 3x + 6 = 0$ d) $-x^2 + 5 = 0$

9. Use the graphs of the functions given below to estimate the solutions to the following equations.
   a) $x^2 - 6x + 7 = 0$
   b) $x^2 + 3x + 6 = 0$
   c) $4x^2 + 4x + 1 = 0$
   d) $-x^2 + 5 = 0$

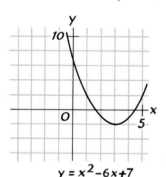

$y = x^2 - 6x + 7$

$y = x^2 + 3x + 6$

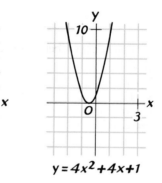

$y = 4x^2 + 4x + 1$

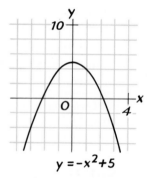

$y = -x^2 + 5$

*10. Use the quadratic formula to find solutions to the following equations. Express each answer in decimal form, correct to the nearest hundredth.
   a) $x^2 - 6x + 7 = 0$ c) $4x^2 + 4x + 1 = 0$
   b) $x^2 + 3x + 6 = 0$ d) $-x^2 + 5 = 0$

Set III

11. Find the value of the discriminant for each of the following equations and use it to tell how many solutions the equation has.
    a) $x^2 - 2x - 3 = 0$
    b) $x^2 - 2x + 1 = 0$
    c) $x^2 - 2x + 2 = 0$

12. Graph the following functions, each on a separate pair of axes.
    a) $y = x^2 - 2x - 3$
    b) $y = x^2 - 2x + 1$
    c) $y = x^2 - 2x + 2$

Use your graphs to find solutions to each of the following equations.

d) $x^2 - 2x - 3 = 0$
e) $x^2 - 2x + 1 = 0$
f) $x^2 - 2x + 2 = 0$

13. Find solutions to each of the following quadratic equations by completing the square.

a) $x^2 - 2x - 3 = 0$
b) $x^2 - 2x + 1 = 0$
c) $x^2 - 2x + 2 = 0$

14. Find solutions to each of the following quadratic equations by using the quadratic formula.

a) $x^2 - 2x - 3 = 0$
b) $x^2 - 2x + 1 = 0$
c) $x^2 - 2x + 2 = 0$

15. Find the value of the discriminant for each of the following quadratic equations and use it to tell how many solutions the equation has.

a) $x^2 - 8x + 10 = 0$ c) $2x^2 + 8x + 2 = 0$
b) $x^2 + 2x + 7 = 0$ d) $4x^2 - 20x + 25 = 0$

---

16. Use the graphs of the functions given below to estimate the solutions to the following equations.

    a) $x^2 - 8x + 10 = 0$
    b) $x^2 + 2x + 7 = 0$
    c) $2x^2 + 8x + 2 = 0$
    d) $4x^2 - 20x + 25 = 0$

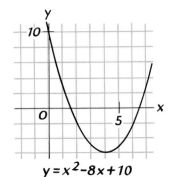

$y = x^2 - 8x + 10$

$y = x^2 + 2x + 7$

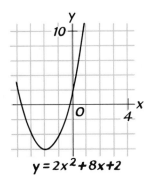

$y = 2x^2 + 8x + 2$

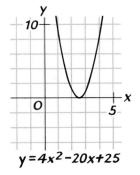

$y = 4x^2 - 20x + 25$

*17. Use the quadratic formula to find solutions to the following equations. Express each answer in decimal form, correct to the nearest hundredth.

a) $x^2 - 8x + 10 = 0$ c) $2x^2 + 8x + 2 = 0$
b) $x^2 + 2x + 7 = 0$ d) $4x^2 - 20x + 25 = 0$

Set IV A Paper-Folding Experiment

Take a sheet of graph paper that is ruled 4 units per inch and cut out a rectangular piece 6 inches long and 4 inches wide. Draw a pair of axes on it as shown in the figure at the left below, labeling the origin and marking scales on the axes as indicated.

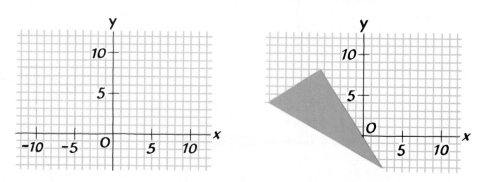

Fold the paper as shown in the second figure so that the lower edge falls along the origin. Make a sharp crease in the paper.

Open the paper flat and fold again at a different angle, being careful that the lower edge again comes to the origin. Repeat this about 20 times, folding the paper at a different angle each time.

1. What do you notice?
2. Trace the curve with your pencil.
3. Copy and complete the following table for the function $y = \dfrac{1}{8}x^2 - 2$.

x	-10	-8	-6	-4	-2	0	2	4	6	8	10
y	10.5										

4. Plot the points having the pairs of numbers in this table as their coordinates on the graph.

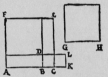
LESSON **9**
Solving Higher-Degree Equations

Jerome Cardan

Although the Babylonians had figured out how to solve certain quadratic equations more than four thousand years ago, general methods for solving cubic equations were not discovered until the sixteenth century. They appeared in print for the first time in a book written by Jerome Cardan, a man who had a remarkable career not only as a mathematician, but also as a doctor and astrologer. One of the pages of Cardan's book is shown here.

Soon after mathematicians learned how to solve cubic equations, they found ways to deal with quartic equations as well. It was discovered that a polynomial equation of degree *n* can have as many as *n* different solutions. This means that a cubic equation, being of third degree, can have as many as three solutions. A quartic equation can have four solutions, a quintic equation can have five, and so on.*

After formulas for solving quadratic, cubic, and quartic equations had been discovered, it was assumed that similar formulas for solving quintic equations and those of even higher degree would also eventually be found. In the nineteenth century, however, a young Norwegian mathematician named Niels Abel proved that it is impossible to solve most polynomial equations of more than the fourth degree by means of ordinary algebraic operations.

Here are examples of how the solutions to equations of degree higher than 2 can be found.

EXAMPLE 1

How many solutions can the equation $x^3 - 3x^2 - 12x + 18 = 0$ have? Use the graph of the function $y = x^3 - 3x^2 - 12x + 18$, shown at the right, to estimate their values.

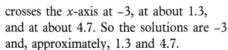

SOLUTION

Because the degree of the equation $x^3 - 3x^2 - 12x + 18 = 0$ is 3, it can have as many as 3 solutions. The graph of the function

$$y = x^3 - 3x^2 - 12x + 18$$

crosses the x-axis at -3, at about 1.3, and at about 4.7. So the solutions are -3 and, approximately, 1.3 and 4.7.

By using a calculator, a solution such as 1.3 can be checked by substituting 1.2, 1.3, and 1.4 for x in the polynomial on the left side of the equation and seeing how close the results come to 0.

$$(1.2)^3 - 3(1.2)^2 - 12(1.2) + 18 = 1.008$$
$$(1.3)^3 - 3(1.3)^2 - 12(1.3) + 18 = -0.473$$
$$(1.4)^3 - 3(1.4)^2 - 12(1.4) + 18 = -1.936$$

From the results shown here, 1.3 seems to be closer than either 1.2 or 1.4 to being correct.

---

*Although an nth-degree equation can have as many as n solutions, it may have fewer. The second-degree equation $x^2 + 1 = 0$, for example, has no solutions.

EXAMPLE 2

Solve the equation $x^4 + 2x^3 - 3x^2 = 0$ by the factoring method.

SOLUTION

Factoring x^2 from each term of the left side, we get

$$x^2(x^2 + 2x - 3) = 0$$

Factoring $x^2 + 2x - 3$, we get

$$x^2(x - 1)(x + 3) = 0$$

So either $x^2 = 0$, $x - 1 = 0$, or $x + 3 = 0$. The solutions to the original equation are 0, 1, and –3. If the function $y = x^4 + 2x^3 - 3x^2$ were graphed, it would cross the x-axis in these three points.

EXAMPLE 3

Solve the equation $3x^3 - 24x^2 + 15x = 0$ by the factoring method.

SOLUTION

Factoring $3x$ from each term of the left side, we get

$$3x(x^2 - 8x + 5) = 0$$

Although there is no obvious way to factor $x^2 - 8x + 5$, we know at this point that either

$$x = 0 \quad \text{or} \quad x^2 - 8x + 5 = 0$$

Solving $x^2 - 8x + 5 = 0$ by the quadratic formula, we can find the other solutions to the original equation. We know that

$$x = \frac{-b \pm \sqrt{b^2 - 4ac}}{2a} \quad \text{and that} \quad a = 1, \quad b = -8, \quad c = 5$$

So

$$x = \frac{-(-8) \pm \sqrt{(-8)^2 - 4(1)(5)}}{2(1)}$$

$$= \frac{8 \pm \sqrt{64 - 20}}{2}$$

$$= \frac{8 \pm \sqrt{44}}{2}$$

$$= \frac{8 \pm 2\sqrt{11}}{2}$$

$$= \frac{2(4 \pm \sqrt{11})}{2}$$

$$= 4 \pm \sqrt{11}$$

The solutions to the original equation are 0, $4 + \sqrt{11}$, and $4 - \sqrt{11}$.

Exercises

Set I

1. Graph the following functions, each on a separate pair of axes. Let one unit on each y-axis represent 10.
   a) $y = 6x$
   b) $y = 3x^2$
   c) $y = 2x^3$

2. This exercise is about the following number trick.

   Think of a number.
   Square it.
   Subtract four.

   Divide the result by the number that is two more than your original number.
   Add nine.
   Subtract the number first thought of.
   The result is seven.

   a) Choose a number and carry out the steps described.
   b) Show how the trick works by letting x represent the number first thought of and carrying out the steps described.
   c) The trick will not work if you begin with -2. Explain why not.

Set II

3. Write a polynomial equation in standard form equivalent to each of the following equations. On the basis of its degree alone, what is the largest number of solutions that it might have?
   a) $x^5 = 8x$
   b) $x(x^2 + 4) = 10$
   c) $x^2 + 3x = x^2 - 1$
   d) $(x^3 + 2)(x - 7) = 0$

4. Tell whether or not each of the following numbers is a solution of the equation

   $$x^4 + x^3 - 11x^2 - 5x + 30 = 0$$

   by checking to see if the number makes it true.
   a) 2
   b) 3
   c) -3
   d) $\sqrt{5}$

5. Use the graphs of the functions given below to estimate the solutions to the following equations.

    a) $3x^3 + 4x^2 - 48x - 64 = 0$
    b) $x^5 - x^4 - 31x^3 + 25x^2 + 150x = 0$

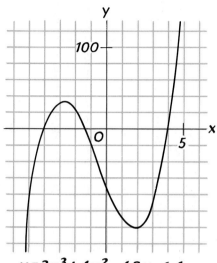

$y = 3x^3 + 4x^2 - 48x - 64$

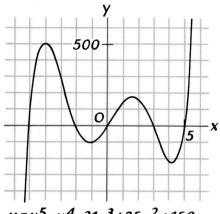

$y = x^5 - x^4 - 31x^3 + 25x^2 + 150x$

6. In exercise 5, you estimated the solutions of polynomial equations by looking at the graphs of the corresponding functions.
    a) Graph the function $y = x^4 - 4x^2 - 5$, from $x = -3$ to $x = 3$.
    b) Use your graph to estimate the solutions of the equation $x^4 - 4x^2 - 5 = 0$.

7. Solve the following equations by factoring, setting each factor equal to zero, and solving the resulting equations.
    a) $x^3 - x^2 - 2x = 0$
    b) $x^3 - x^2 - 3x = 0$
    c) $x^4 - 10x^2 + 9 = 0$
    d) $x^4 - 25 = 0$

8. Solve the following equations.
    a) $5(x^3 + x^2) = 3(x^3 + x)$
    b) $x^2(1 - x^2) = x(6 - x^3) - 4$
    c) $(x^2 + 2)^2 = 4(x^2 + 5)$

Set III

9. Write a polynomial equation in standard form equivalent to each of the following equations. On the basis of its degree alone, what is the largest number of solutions that it might have?
    a) $4x^2 = 11$
    b) $5x + 9 = x - 1$
    c) $x^3(x^4 + 1) = x^3 + x^4$
    d) $(x^3 - 5)^2 = 7$

10. Tell whether or not each of the following numbers is a solution of the equation

$$x^4 + x^3 - 5x^2 - 3x + 6 = 0$$

by checking to see if the number makes it true.
    a) 1   c) –2
    b) –1   d) $\sqrt{3}$

11. Use the graphs of the functions given below to estimate the solutions to the following equations.

     a) $3x^3 - 8x^2 + 3x - 8 = 0$

     b) $x^4 - 11x^2 + 10 = 0$

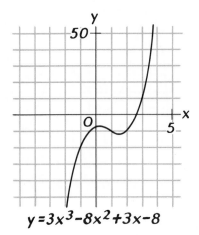

$y = 3x^3 - 8x^2 + 3x - 8$ $y = x^4 - 11x^2 + 10$

12. In exercise 11, you estimated the solutions of polynomial equations by looking at the graphs of the corresponding functions.

    a) Graph the function

       $y = 4x^3 + 8x^2 - x - 2$, from $x = -3$ to $x = 2$.

    b) Use your graph to estimate the solutions of the equation $4x^3 + 8x^2 - x - 2 = 0$.

13. Solve the following equations by factoring, setting each factor equal to zero, and solving the resulting equations.

    a) $x^3 + 2x^2 + x = 0$

    b) $x^3 + 2x^2 - x = 0$

    c) $x^4 - 5x^2 + 4 = 0$

    d) $x^4 - 10x^2 = 0$

14. Solve the following equations.

    a) $2x^2(x - 5) = x(5 - x)$

    b) $x^2(x^2 + 1) + 2 = x^4 + 4x$

    c) $(x^2 + 1)^2 = 2(x^2 + 41)$

Set IV

Although a quartic equation can have as many as four solutions, this quartic equation doesn't have *any*.

$$x^4 + x^2 + 1 = 0$$

Can you explain why not? (Hint: Can the square of a number be negative?)

"Nope; no chess problems until you've finished your quadratic equations!"

Summary and Review

In this chapter, we have studied the properties of polynomial equations in one variable and have learned several different methods for solving quadratic equations.

Polynomial Equations (*Lesson 1*) A polynomial equation is one in which both sides of the equation are polynomials. It is written in standard form if the left side is a polynomial in descending powers of the variable and the right side is zero.

Polynomial equations written in standard form are named according to their degree:

Degree	Name
1	Linear
2	Quadratic
3	Cubic
4	Quartic
5	Quintic

Polynomial Functions (*Lesson 2*)   Functions having equations of the form $y = ax + b$ are called linear because their graphs are straight lines. Quadratic functions have equations of the form $y = ax^2 + bx + c$ and graphs that are parabolas. The graphs of higher-degree polynomial functions are more complicated curves.

To graph a polynomial function, make a table from its equation, plot the points in the table, and connect them with a smooth line or curve.

Solving Polynomial Equations by Graphing (*Lesson 3*)   The solutions of a polynomial equation such as $ax^2 + bx + c = 0$ are the x-coordinates of the points in which the graph of the polynomial function $y = ax^2 + bx + c$ intersects the x-axis.

Solving Quadratic Equations by Factoring (*Lesson 4*)   If the product of two or more numbers is zero, then at least one of the numbers must be equal to zero. This fact, called the zero-product property, can be used to solve certain quadratic equations in the following way.

1. Write the equation in standard form: $ax^2 + bx + c = 0$.
2. Factor the polynomial on the left side: (   )(   ) = 0.
3. Set each factor equal to zero: (   ) = 0   or   (   ) = 0.
4. Solve the resulting equations.

Solving Quadratic Equations by Taking Square Roots (*Lesson 5*)   If $x^2 = a$, then either $x = \sqrt{a}$ or $x = -\sqrt{a}$. This fact can be used to solve quadratic equations written in the form $(x + a)^2 = b$, by taking the square root of each side.

Completing the Square (*Lesson 6*) To solve a quadratic equation by completing the square:

1. Write the equation in the form $ax^2 + bx = -c$.
2. Add a number to each side of the equation that will make the left side the square of a binomial: $ax^2 + bx + (\ \ ) = -c + (\ \ )$.
3. Take the square root of each side.
4. Solve the resulting equations.

The Quadratic Formula (*Lesson 7*) If the general quadratic equation $ax^2 + bx + c = 0$ is solved for x by completing the square, the result is the quadratic formula:

$$x = \frac{-b \pm \sqrt{b^2 - 4ac}}{2a}$$

The solutions to a quadratic equation can be found by substituting its values for a, b, and c in this formula and simplifying the result.

The Discriminant (*Lesson 8*) A quadratic equation has either two solutions, one solution, or no solutions, depending on whether the parabola that is the graph of the corresponding quadratic function intersects the x-axis in two points, one point, or no points.

  The discriminant of the quadratic equation $ax^2 + bx + c = 0$ is the expression $b^2 - 4ac$. If it is a positive number, the equation has two solutions. If it is equal to zero, the equation has one solution. If it is a negative number, the equation has no solutions.

Solving Higher-Degree Equations (*Lesson 9*) A polynomial equation can have as many different solutions as its degree; so cubic and quartic equations can have as many as three and four solutions respectively.

  There is no general algebraic method for solving quintic equations and those of higher degree. However, some polynomial equations of higher degree can be solved by factoring.

Exercises

Set I

1. Write each of the following equations in standard form. What type of polynomial equation is it and what is the largest number of solutions it might have?
   a) $x^3 + 5x = 2$
   b) $(x - 2)(x + 1) = 6$
   c) $x^2(x^3 - 1) = (3 + x)(3 - x)$

2. Tell whether or not each of the following numbers is a solution of the equation given.
   a) Is -3 a solution of $x^3 - 10x = 3$?
   b) Is $\sqrt{7}$ a solution of $2x^4 = 98$?

3. The graph of the function $y = 4x^3 + 8x^2 - 21x$ is shown here.

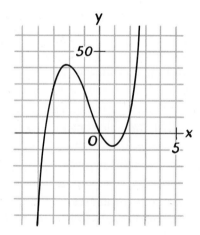

   a) Write an equation whose solutions can be estimated from this graph.
   b) How many solutions does the equation have?
   c) Estimate each of their values to the nearest tenth.

4. Graph the following functions.
   a) $y = x^2 + x - 5$ from $x = -5$ to $x = 5$.
   b) $y = x^4 + 4$ from $x = -3$ to $x = 3$.

5. Use your graphs in exercise 4 to estimate the solutions of the following equations.
   a) $x^2 + x - 5 = 0$
   b) $x^4 + 4 = 0$

6. Solve each of the following quadratic equations by the factoring method. Show all work.
   a) $6x^2 - 5x = 0$
   b) $x^2 - 36 = 0$
   c) $x^2 + 8x = 33$
   d) $(x + 5)(x + 6) = 2$

7. Solve the following quadratic equations by the square-root method. Show all work.
   a) $x^2 = 60$ c) $(2x + 5)^2 = 25$
   b) $(x - 4)^2 = 7$ d) $9x^2 + 6x + 1 = 16$

8. Solve the following quadratic equations by completing the square. Show all work.
   a) $x^2 + 10x = 11$
   b) $x^2 - 6x + 4 = 0$

9. Find the value of the discriminant for each of the following quadratic equations and use it to tell how many solutions the equation has.
   a) $4x^2 + 20x + 25 = 0$
   b) $x^2 - 9x + 21 = 0$
   c) $5x^2 + 2x - 1 = 0$

10. Solve the following quadratic equations by using the quadratic formula. Show all work.
    a) $2x^2 - 11x + 15 = 0$
    b) $x^2 + 4x - 20 = 0$
    c) $6x^2 - 4x = 1$
    d) $3x^2 + 5x = 12$

11. Solve the following cubic and quartic equations by factoring.
    a) $x^3 + x^2 = 12x$
    b) $2x^4 - 10x^2 = 0$

12. Solve the following equations by any method.
   a) $x^2 + 8x + 7 = 0$
   b) $(x + 2)^2 = 36$
   c) $x^2 - 4x = 1$
   d) $x(x + 6) = 27$
   e) $x^2 + x - 5 = 0$
   f) $x^4 - 16x^2 = 0$

Set II

1. Write each of the following equations in standard form. What type of polynomial equation is it and what is the largest number of solutions it might have?
   a) $6x^5 + 3 = 2x$
   b) $(x + 11)(x - 4) = x^2$
   c) $x(x - 9)(x + 9) = 0$

2. Tell whether or not each of the following numbers is a solution of the equation given.
   a) Is 1 a solution of $4x^3 - 64 = 0$?
   b) Is $-\sqrt{3}$ a solution of $x^4 + x^2 = 12$?

3. The graph of the function
   $y = x^4 - 2x^3 - 12$ is shown here.

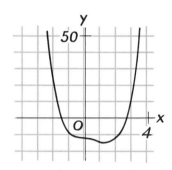

   a) Write an equation whose solutions can be estimated from this graph.
   b) How many solutions does the equation have?
   c) Estimate each of their values to the nearest tenth.

4. Graph the following functions.
   a) $y = x^2 - 3x - 1$ from $x = -3$ to $x = 6$.
   b) $y = x^3 - 4$ from $x = -3$ to $x = 3$.

5. Use your graphs in exercise 4 to estimate the solutions of the following equations.
   a) $x^2 - 3x - 1 = 0$
   b) $x^3 - 4 = 0$

6. Solve each of the following quadratic equations by the factoring method. Show all work.
   a) $x^2 - 64 = 0$
   b) $x^2 + 2x - 35 = 0$
   c) $3x^2 - 1 = 2x$
   d) $(x - 4)(x - 7) = 10$

7. Solve the following quadratic equations by the square-root method. Show all work.
   a) $x^2 = 18$
   b) $(x + 1)^2 = 10$
   c) $(3x - 2)^2 = 49$
   d) $4x^2 - 12x + 9 = 25$

8. Solve the following quadratic equations by completing the square. Show all work.
   a) $x^2 - 2x = 80$
   b) $x^2 + 8x - 6 = 0$

9. Find the value of the discriminant for each of the following quadratic equations and use it to tell how many solutions the equation has.
   a) $6x^2 - x + 3 = 0$
   b) $3x^2 + 8x + 5 = 0$
   c) $x^2 - 12x + 36 = 0$

10. Solve the following quadratic equations by using the quadratic formula. Show all work.
   a) $5x^2 - 4x - 1 = 0$
   b) $x^2 - 8x + 5 = 0$
   c) $9x^2 - 6x + 1 = 0$
   d) $3x^2 + 2x = 4$

11. Solve the following cubic and quartic equations by factoring.
    a) $4x^3 - 24x = 0$
    b) $x^4 - 16 = 0$

12. Solve the following equations by any method.
    a) $x^2 - 7x + 10 = 0$
    b) $(x - 3)^2 = 49$
    c) $x^2 + 2x = 5$
    d) $x(x - 4) = 21$
    e) $x^2 - x - 7 = 0$
    f) $x^3 - 25x = 0$

Chapter 14

THE REAL NUMBERS

CROWN COPYRIGHT. SCIENCE MUSEUM, LONDON.

LESSON 1
Rational Numbers

A popular toy in the nineteenth century gave the illusion of moving pictures by whirling a disc such as the one shown here in front of a mirror. When viewed through the slots around the rim, it shows a woman hitting a man on the head with a stick.

As the disc whirled around, the pictures appeared over and over again in an unchanging sequence. Many rational numbers have decimal forms that behave in the same way.

As we have already learned, a *rational number* is a number that can be written as the quotient of two integers. A rational number written in this form can be changed to decimal form by carrying out the indicated division. Compare the following examples of how this is done.

EXAMPLE 1

Change $\dfrac{37}{16}$ to decimal form.

SOLUTION

```
        2.312500 . . .
16)37.000000
   32
   ──
    5 0
    4 8
    ──
     20
     16
     ──
      40
      32
      ──
       80
       80
      ──
      → 0 0
         0
        ──
        00
         0
        ──
         0 . . .
```

EXAMPLE 2

Change $\dfrac{16}{37}$ to decimal form.

SOLUTION

```
        0.432432 . . .
37)16.000000
    0
    ──
   16 0
   14 8
   ────
    1 20
    1 11
    ────
      90
      74
     ──
     160
     148
     ───
      120
      111
      ───
       90
       74
      ──
      16 . . .
```

In Example 1, we eventually get a remainder of zero. At this point, the decimal form of the number begins repeating zeros. We can indicate this by writing it as $2.3125\overline{0}$; the bar over the zero means that it keeps repeating. The repeating digit, 0, is called the *period* of the number. It is simpler, however, to write the number as 2.3125 and say that the division "ends."

In Example 2, the division will never end because we never get a remainder of zero. After first getting a remainder of 16, followed by remainders of 12 and 9, we get a remainder of 16 again. This, in effect, puts us back to where we started. The result is that the decimal form of $\dfrac{16}{37}$ repeats the three digits 432 over and over again. As in Example 1, the digits that repeat are called the *period* of the number. Rather than writing $\dfrac{16}{37}$ as 0.432432 . . . , we can write it as $0.\overline{432}$, the bar indicating the repetition of the digits 432.

These examples illustrate the fact that, if a rational number is changed to decimal form, it always begins repeating digits. The reverse is also true: every

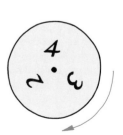

number in decimal form that repeats digits is rational because it can always be expressed as the quotient of two integers. The following examples show how to do this.

EXAMPLE 3
Write 2.08 as the quotient of two integers.

SOLUTION
Because 2.08 means $2 + \dfrac{8}{100}$, we can write

$$2.08 = 2 + \frac{8}{100} = \frac{208}{100}$$

or, reducing to lowest terms, $\dfrac{52}{25}$.

EXAMPLE 4
Write $1.\overline{4}$ as the quotient of two integers.

SOLUTION
To do this, we begin by letting $x = 1.\overline{4}$. If $x = 1.44444\ldots$, then $10x = 14.44444\ldots$. Subtracting the first of these equations from the second, we get

$$
\begin{array}{r}
10x = 14.44444\ldots \\
-\quad x = 1.44444\ldots \\
\hline
9x = 13
\end{array}
$$

or, more briefly,

$$
\begin{array}{r}
10x = 14.\overline{4} \\
-\quad x = 1.\overline{4} \\
\hline
9x = 13
\end{array}
$$

Solving the resulting equation for x, we get $x = \dfrac{13}{9}$. So $1.\overline{4} = \dfrac{13}{9}$.

EXAMPLE 5
Write $0.4\overline{09}$ as the quotient of two integers.

SOLUTION

Let $x = 0.4090909\ldots$. Multiplying by 100 (there are two repeating digits to get rid of this time rather than one), we get $100x = 40.90909\ldots$. Subtracting,

$$100x = 40.90909\ldots$$
$$- \quad x = 0.40909\ldots$$
$$99x = 40.5$$

So $x = \dfrac{40.5}{99} = \dfrac{405}{990}$. Reducing to lowest terms,

$$\frac{405}{990} = \frac{9 \cdot 45}{9 \cdot 110} = \frac{9 \cdot 5 \cdot 9}{9 \cdot 5 \cdot 22} = \frac{9}{22}$$

Examples 4 and 5 illustrate a method that can be used to write any decimal number that repeats digits as the quotient of two integers. The steps are:

1. Write an equation by letting x equal the number.
2. Write a second equation by multiplying both sides of the first equation by 10^n, n being the number of digits in the period of the number.
3. Subtract the first equation from the second.
4. Solve the resulting equation for x.

Exercises

Set I

1. Write expressions for the exact perimeter and area of each of these rectangles.

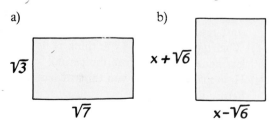

a)

b)

2. The world's longest conveyor belt extends across the Spanish Sahara and is used to carry material from a phosphate mine to a seaport. Material put on the belt at one end arrives at the other end in 6 hours and 15 minutes. If the belt moved 4 kilometers per hour faster, the trip would take 5 hours instead.

STERLING PUBLISHING CO., INC.

a) Draw a figure and write an equation to find out the speed in kilometers per hour at which the belt actually travels. (Represent it by x.)

b) How many kilometers long is the belt?

3. Guess a formula for the function represented by each of these tables. Begin each formula with $y =$.

a)

x	1	2	3	4	5
y	5	4	3	2	1

b)

x	1	2	3	4	5
y	6	3	2	1.5	1.2

c)

x	0	1	2	3	4
y	4	9	16	25	36

d)

x	4	9	16	25	36
y	0	1	2	3	4

Set II

*4. Copy and complete the following table by writing $\frac{1}{x}$ to the first five decimal places in the second column and using a bar as necessary to write $\frac{1}{x}$ in exact decimal form in the third column.

x	$\frac{1}{x}$	$\frac{1}{x}$
1	1.00000	1
2	0.50000	0.5
3	0.33333	$0.\overline{3}$
4		
5		
6		
7		
8		

*5. Change each of the following rational numbers to decimal form.

a) $\frac{1}{5^1}$ b) $\frac{1}{5^2}$ c) $\frac{1}{5^3}$ d) $\frac{1}{5^4}$

Look at your results for parts a through d before answering the following questions about the decimal form of $\frac{1}{5^n}$, in which n is a positive integer.

e) Do you think that it "ends" or does it start repeating digits?

f) What type of digit do you think is its last nonzero digit before the period?

g) How many digits do you think it has before the period?

Change these numbers to decimal form.

h) $\dfrac{1}{5^5}$ i) $\dfrac{1}{5^8}$

j) Do the results agree with your answers to parts e through g?

*6. Change each of the following rational numbers to decimal form.

a) $\dfrac{1}{11}$ b) $\dfrac{1}{111}$ c) $\dfrac{1}{1,111}$

Look at your answers for parts a through c before answering the following questions about the decimal form of $\dfrac{1}{1,111,111,111}$.

d) Do you think that it "ends" or does it start repeating digits?
e) How many digits do you think are in its period?
f) What do you think is the last digit in its period?

7. Write each of the following as the quotient of two integers in lowest terms.
a) 3.3 c) 0.004
b) 2.72 d) 0.4375

8. Use the method explained in this lesson to write each of the following as the quotient of two integers in lowest terms.
a) $0.\overline{6}$ d) $0.08\overline{3}$
b) $2.\overline{5}$ e) $0.5\overline{90}$
c) $0.\overline{45}$ f) $0.2\overline{037}$

9. The following questions refer to the rational number $\dfrac{2}{19}$.

a) How many different numbers, including zero, can there be as remainders if 19 is divided into 2?
b) If zero were one of the remainders, what would you know about the decimal form of $\dfrac{2}{19}$?
c) What is the greatest number of digits that could repeat if zero is not a remainder?

10. If x is replaced by an integer, the expression $\dfrac{x}{2x + 1}$ represents a rational number. For example, if $x = 7$,

$$\frac{x}{2x + 1} = \frac{7}{2(7) + 1} = \frac{7}{15} \quad \text{or} \quad 0.4\overline{6}$$

Find the rational number represented by $\dfrac{x}{2x + 1}$ both as a common fraction and as a decimal number if

a) $x = 1$ c) $x = 5$
b) $x = 2$ d) $x = 50$

e) What happens to the value of $\dfrac{x}{2x + 1}$ as x increases?

Set III

*11. Copy and complete the table at the right by writing $\dfrac{1}{x}$ to the first five decimal places in the second column and using a bar as necessary to write $\dfrac{1}{x}$ in exact decimal form in the third column.

x	$\dfrac{1}{x}$	$\dfrac{1}{x}$
9	0.11111	$0.\overline{1}$
10	0.10000	0.1
11		
12		
13		
14		
15		

*12. Change each of the following rational numbers to decimal form.

a) $\frac{1}{2^1}$ c) $\frac{1}{2^3}$

b) $\frac{1}{2^2}$ d) $\frac{1}{2^4}$

Look at your results for parts a through d before answering the following questions about the decimal form of $\frac{1}{2^n}$, in which n is a positive integer.

e) Do you think that it "ends" or does it start repeating digits?
f) What do you think is its last nonzero digit before the period?
g) How many digits do you think it has before the period?

Change these numbers to decimal form.

h) $\frac{1}{2^5}$ i) $\frac{1}{2^8}$

j) Do the results agree with your answers to parts e through g?

*13. Change each of the following rational numbers to decimal form.

a) $\frac{1}{7}$ d) $\frac{4}{7}$

b) $\frac{2}{7}$ e) $\frac{5}{7}$

c) $\frac{3}{7}$ f) $\frac{6}{7}$

g) What do you notice about the results?

14. Write each of the following as the quotient of two integers in lowest terms.
a) 1.7   c) 0.625
b) 4.16   d) 0.0008

15. Use the method explained in this lesson to write each of the following as the quotient of two integers in lowest terms.
a) $0.\overline{3}$
b) $7.\overline{2}$
c) $0.1\overline{8}$
d) $0.13\overline{8}$
e) $1.4\overline{72}$
f) $0.0\overline{405}$

16. The following questions refer to the rational number $\frac{4}{23}$.

a) How many different numbers, including zero, can there be as remainders if 23 is divided into 4?
b) If zero were one of the remainders, what would you know about the decimal form of $\frac{4}{23}$?
c) What is the greatest number of digits that could repeat if zero is not a remainder?

17. If x is replaced by an integer other than 0, the expression $\frac{x + 3}{x}$ represents a rational number. For example, if $x = 8$,

$$\frac{x + 3}{x} = \frac{8 + 3}{8} = \frac{11}{8} \quad \text{or} \quad 1.375$$

Find the rational number represented by $\frac{x + 3}{x}$ both as a common fraction and as a decimal number if
a) $x = 2$
b) $x = 5$ d) $x = 100$
c) $x = 10$ e) $x = 10,000$

f) What happens to the value of $\frac{x + 3}{x}$ as x increases?

Set IV

The decimal form of $\frac{1}{17}$, written to 100 places, is

0.0588235294117647058823529411764705882352941176470705
88235294117647058823529411764705882352941176470588

1. How many digits are in the period of this number?
2. Write the digits of the period down in order and draw a line at the halfway point to separate them into two equal groups.
3. The digits in the second group are related in a simple way to those in the first group. What is it?

$\sqrt{2}$ = 1.4142135623 7309504880 1688724209 6980785696 7187537694 8073176679 7379907324 7846210703 8850387534 3276415727
3501384623 0912297024 9248360558 5073721264 4121497099 9358314132 2266592750 5592755799 9505011527 8206057147
0109559971 6059702745 3459686201 4728517418 6408891986 0955232923 0484308714 3214508397 6260362799 5251407989
6872533965 4633180882 9640620615 2583523950 5474575028 7759961729 8355752203 3753185701 1354374603 4084988471
6038689997 0699004815 0305440277 9031645424 7823068492 9369186215 8057846311 1596668713 0130156185 6898723723
5288509264 8612494977 1542183342 0428568606 0146824720 7714358548 7415565706 9677653720 2264854470 1585880162
0758474922 6572260020 8558446652 1458398893 9443709265 9180031138 8246468157 0826301005 9485870400 3186480342
1948972782 9064104507 2636881313 7398552561 1732204024 5091227700 2269411275 7362728049 5738108967 5040183698
6836845072 5799364729 0607629969 4138047565 4823728997 1803268024 7442062926 9124859052 1810044598 4215059112
0249441341 7285314781 0580360337 1077309182 8693147101 7111168391 6581726889 4197587165 8215212822 9518488472
0896946338 6289156288 2765952635 1405422676 5323969461 7511291602 4087155101 3515045538 1287560052 6314680171
2740265396 9470240300 5174953188 6292563138 5188163478 0015693691 7688185237 8684052287 8376293892 1430065586
9568685964 5951555016 4472450983 6896036887 3231143894 1557665104 0883914292 3381132060 5243362948 5317049915
7717562285 4974143899 9188021762 4309652065 6421182731 6726257539 5947172559 3463723863 2261482742 6222086711
5583959992 6521176252 6989175409 8815934864 0083457085 1814722318 1420407042 6509056532 3333984364 5786579679
6519267292 3998753666 1721598257 8860263363 6178274959 9421940377 7753681426 2177387991 9455139723 1274066898
3299898953 8672882285 6378697749 6625199665 8352577619 8939322845 3447356947 9496295216 8891485492 5389047558
2883452609 6524096542 8893945386 4662574492 7556381964 4103169798 3306185201 9379384940 0571563337 2054806854
0575867999 6701213722 3947582142 6306585132 2174088323 8294728761 7393647467 8374319600 0159218880 7347857617
2522118674 9042497736 6929207311 0963697216 0893370866 1156734585 3348332952 5467585164 4710757848 6024636008
3444911481 8587655554 2864551233 1421992631 1332517970 6084365597 0435285641 0087918500 7603610091 5946567067
6883605571 7400767569 0509613671 9401324935 6052401859 9910506210 8163597726 4313806054 6701029356 9971042425
1057817495 3105725593 4984451126 9227803449 1350663756 8747760283 1628296055 3242242695 7534529028 8387684464
2917328271 0888318087 0253398523 3812274999 0812371892 5407264753 6785030482 1591801886 1671089728 6922920119
7599880703 8185433325 3646021108 2299279293 0728717807 9988809917 6741774108 9830608003 2631181642 7988231171
5436386966 1702999934 1616148786 8601804550 5553986913 1151860103 8637532500 4558186044 ....

LESSON 2

Irrational Numbers

We know that, when rational numbers are changed to decimal form, they eventually begin a repeating pattern of digits. The first thousand digits of $\sqrt{2}$ shown above have no such pattern. In fact, when 1,000,082 digits of $\sqrt{2}$ were calculated by a computer at Columbia University in 1971, no simple pattern of *any* sort was found.

If there is a pattern to the decimal form of $\sqrt{2}$, it cannot be one of repeating digits because $\sqrt{2}$ is not a rational number. The Greek mathematician Hippasus is credited with having proved that the $\sqrt{2}$ is not a rational number in the fifth century B.C. by showing that it cannot be written as the quotient of two integers.

▶ An **irrational number** is a number that cannot be written as the quotient of two integers.

The number $\sqrt{2}$, then, is an irrational number. Other square roots that are irrational are $\sqrt{3}$, $\sqrt{5}$, $\sqrt{6}$, $\sqrt{7}$, $\sqrt{8}$, and $\sqrt{10}$. The square roots $\sqrt{1}$, $\sqrt{4}$, and $\sqrt{9}$, on the other hand, are rational because $\sqrt{1} = 1$, $\sqrt{4} = 2$, and $\sqrt{9} = 3$. In general, if an integer is not the square of an integer, its square roots are irrational.

The same is true for the square roots of rational numbers that are not integers. Compare the numbers in the following lists.

Irrational square roots: $\sqrt{\dfrac{1}{2}}, \sqrt{\dfrac{1}{3}}, \sqrt{\dfrac{2}{3}}, \sqrt{\dfrac{3}{4}}, \sqrt{\dfrac{4}{7}}, \ldots$

Rational square roots: $\sqrt{\dfrac{1}{4}}, \sqrt{\dfrac{1}{9}}, \sqrt{\dfrac{4}{9}}, \sqrt{\dfrac{1}{16}}, \sqrt{\dfrac{9}{16}}, \ldots$

If a rational number is written as a fraction in lowest terms, the numerator and denominator of the fraction must be squares of integers in order for its square roots to be rational.

To find the decimal form of an irrational number such as $\sqrt{2}$ is not easy. There are, however, numbers whose decimal forms are easily described that are irrational. For example, the number

$$0.123456789101112131415\ldots$$

whose digits are formed by writing the counting numbers in succession must be irrational. This follows from the fact that it is not a repeating decimal.

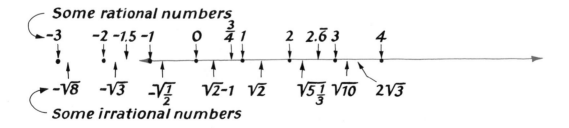

Irrational numbers, like rational numbers, can be assigned to points on a number line. A number line showing some examples of both types of numbers is shown above. Numbers such as $2\sqrt{3}$ and $\sqrt{2} - 1$ are irrational because, if they could be expressed as the quotient of two integers, then so could $\sqrt{3}$ and $\sqrt{2}$.

Exercises

Set I

1. This exercise is about the number line shown here.

a) Write the coordinate of point A as a fraction in lowest terms.
b) Write the coordinate of point B as a fraction in lowest terms.
c) Find the distance between points A and B.
d) Find the coordinate of the point midway between points A and B.

2. Tell whether or not each of the following equations is true for all values of x and y. If an equation is not always true, give an example for which it is false.
a) $2(x^2 + y^2) = 2x^2 + 2y^2$
b) $(x^2 + y^2)^2 = x^4 + y^4$
c) $\sqrt{x^2 + y^2} = x + y$

3. The graph of one of the following polynomial functions is shown here.
A. $y = x^3 + 5x^2 - 20x$
B. $y = x^3 + 5x^2 - 4x - 20$
C. $y = x^4 + 5x^3 - 4x - 20$
D. $y = x^4 + 5x^3 - 4x^2 - 20x$
a) Which function is it?
b) Explain how you decided which function is shown.

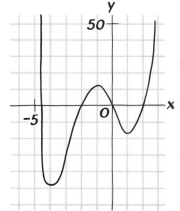

Set II

4. Draw an accurate number line and mark the points corresponding to the following numbers on it.
a) -4
b) 2.5
c) -1.6
d) $\sqrt{9}$
e) $\dfrac{5}{6}$
f) $\sqrt{3}$
g) $-\dfrac{7}{3}$
h) $0.\overline{3}$
i) $-\sqrt{\dfrac{9}{4}}$
j) $-\sqrt{10}$

k) Which of the numbers in parts a through j are irrational?

5. Is each of the following square roots rational or irrational?
a) $\sqrt{9}$
b) $\sqrt{90}$
c) $\sqrt{900}$
d) $\sqrt{9,000}$
e) $\sqrt{250}$
f) $\sqrt{25}$
g) $\sqrt{2.5}$
h) $\sqrt{0.25}$

6. If possible, find the answers to each of the following problems in exact decimal form. Tell whether each answer is rational or irrational.

a) $\sqrt{4} + \sqrt{5}$

b) $\sqrt{2} \cdot \sqrt{8}$

c) $\dfrac{\sqrt{70}}{\sqrt{7}}$

d) $\dfrac{\sqrt{49}}{2}$

e) $3 \cdot \sqrt{12}$

f) $\sqrt{3 \cdot 12}$

g) $\sqrt{12} - \sqrt{3}$

h) $\dfrac{1}{\sqrt{25} - \sqrt{9}}$

i) $\sqrt{\dfrac{1}{9} - \dfrac{1}{25}}$

j) $\sqrt{\dfrac{1}{9}} - \sqrt{\dfrac{1}{25}}$

7. Two rational numbers close to $\sqrt{2}$ are $\dfrac{7}{5}$ and $\dfrac{17}{12}$.

a) Change $\dfrac{7}{5}$ to decimal form.

b) Is $\dfrac{7}{5}$ smaller or larger than $\sqrt{2}$? (Refer to the decimal form of $\sqrt{2}$ given at the beginning of this lesson.)

c) Express $\left(\dfrac{7}{5}\right)^2$ as a rational number.

d) Is $\left(\dfrac{7}{5}\right)^2$ smaller or larger than 2?

e) Change $\dfrac{17}{12}$ to decimal form.

f) Is $\dfrac{17}{12}$ smaller or larger than $\sqrt{2}$?

g) Express $\left(\dfrac{17}{12}\right)^2$ as a rational number.

h) Is $\left(\dfrac{17}{12}\right)^2$ smaller or larger than 2?

*8. Each of the following decimal numbers might be used as an approximation of $\sqrt{35}$. Find the square of each number, each correct to the nearest thousandth.

a) 6

b) 5.9

c) 5.92

d) 5.916

e) 5.9161

f) Each number in parts a through e is rational. Explain why.

g) Use your answers to parts a through e to tell which of the numbers are smaller than $\sqrt{35}$ and which are larger.

h) Is it possible that a decimal number carried out to a larger number of decimal places could be exactly equal to $\sqrt{35}$? Explain.

9. Solve each of the following equations by using the quadratic formula. Simplify your answers as much as possible but do not round them.

a) $4x^2 - 5x + 1 = 0$

b) $4x^2 - 6x + 1 = 0$

c) $4x^2 - 4x + 1 = 0$

d) $4x^2 - 8x + 1 = 0$

e) What must be true about the discriminant, $b^2 - 4ac$, of a quadratic equation in order for the solutions to the equation to be rational? (Assume that a, b, and c are integers.)

Set III

10. Draw an accurate number line and mark the points corresponding to the following numbers on it.

a) 3

b) −1.5

c) 2.8

d) $\sqrt{16}$

e) $-\dfrac{2}{5}$

f) $\sqrt{5}$

g) $\dfrac{9}{2}$

h) $\sqrt{\dfrac{1}{9}}$

i) $-0.\overline{6}$

j) $-\sqrt{15}$

k) Which of the numbers in parts a through j are irrational?

11. Is each of the following square roots rational or irrational?

a) $\sqrt{10}$
b) $\sqrt{100}$
c) $\sqrt{1,000}$
d) $\sqrt{10,000}$
e) $\sqrt{64}$
f) $\sqrt{6.4}$
g) $\sqrt{0.64}$
h) $\sqrt{0.064}$

12. If possible, find the answers to each of the following problems in exact decimal form. Tell whether each answer is rational or irrational.

a) $\sqrt{3} \cdot \sqrt{3}$
b) $\sqrt{5} - \sqrt{1}$

c) $\dfrac{\sqrt{24}}{\sqrt{6}}$

d) $\dfrac{\sqrt{75}}{3}$

e) $\sqrt{2 \cdot 8}$
f) $2 \cdot \sqrt{8}$
g) $\sqrt{8} - \sqrt{4}$

h) $\dfrac{1}{\sqrt{25} - \sqrt{16}}$

i) $\sqrt{\dfrac{1}{16}} - \sqrt{\dfrac{1}{25}}$

j) $\sqrt{\dfrac{1}{16} - \dfrac{1}{25}}$

13. Two rational numbers close to $\sqrt{3}$ are $\dfrac{7}{4}$ and $\dfrac{19}{11}$.

a) Change $\dfrac{7}{4}$ to decimal form.

b) Is $\dfrac{7}{4}$ smaller or larger than $\sqrt{3}$?
($\sqrt{3} = 1.73205080\ldots$)

c) Express $\left(\dfrac{7}{4}\right)^2$ as a rational number.

d) Is $\left(\dfrac{7}{4}\right)^2$ smaller or larger than 3?

e) Change $\dfrac{19}{11}$ to decimal form.

f) Is $\dfrac{19}{11}$ smaller or larger than $\sqrt{3}$?

g) Express $\left(\dfrac{19}{11}\right)^2$ as a rational number.

h) Is $\left(\dfrac{19}{11}\right)^2$ smaller or larger than 3?

*14. Each of the following decimal numbers might be used as an approximation of $\sqrt{42}$. Find the square of each number, each correct to the nearest thousandth.

a) 6
b) 6.5
c) 6.48
d) 6.481
e) 6.4807
f) Each number in parts a through e is rational. Explain why.
g) Use your answers to parts a through e to tell which of the numbers are smaller than $\sqrt{42}$ and which are larger.
h) Is it possible that a decimal number carried out to a larger number of decimal places could be exactly equal to $\sqrt{42}$? Explain.

15. Solve each of the following equations by using the quadratic formula. Simplify your answers as much as possible but do not round them off.

a) $9x^2 - 10x + 1 = 0$
b) $9x^2 - 8x + 1 = 0$
c) $9x^2 - 6x + 1 = 0$
d) $9x^2 - 9x + 1 = 0$
e) What must be true about the discriminant, $b^2 - 4ac$, of a quadratic equation in order for the solutions to the equation to be rational? (Assume that a, b, and c are integers.)

Set IV

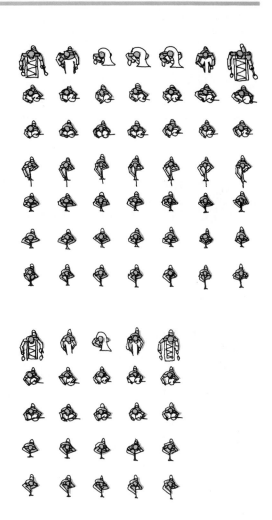

The figures at the right represent two marching bands as seen from overhead. Each band is arranged in a square formation.

Notice that the first band has *almost* twice as many people in it as the second. If the first band had *exactly* twice as many people, then it would follow that $\sqrt{2}$ is a rational number.

Can you explain why?

(Hint: Suppose that there are x people in the first row of the larger band and y people in the first row of the smaller band. Then x^2 represents the number of people in the larger band and y^2 represents the number of people in the smaller band. If

$$\frac{x^2}{y^2} = 2$$

then to what number would $\frac{x}{y}$ be equal?)

LESSON 3
More Irrational Numbers

There are many other types of irrational numbers besides square roots. One of them can be used to solve the problem illustrated in this cartoon.

Although it doesn't seem possible that someone could figure out the distance of the sun from the earth from simply knowing that there are 365 days in a year, the two ideas are connected. The great seventeenth-century astronomer Johannes Kepler discovered that the distance of a planet from the sun is a function of the length of its year. An approximate formula for this function is

$$y = \sqrt[3]{6x^2}$$

in which y represents the distance of the planet from the sun in millions of miles and x represents the number of earth-days that it takes the planet to travel once around the sun.

This function contains the *cube root* of a number. In the same way that the idea of "square root" comes from the idea of numbers as squares, the idea of "cube root" comes from the idea of numbers as cubes.

▶ If $y = x^3$, then y is called the *cube* of x and x is called a **cube root** of y.

The number 2, for example, is a cube root of 8 because $2^3 = 8$. The number –5 is a cube root of –125 because $(-5)^3 = -125$.

You know that positive numbers have two square roots, zero has one square root, and negative numbers do not have square roots. *Every* number, on the other hand, has exactly *one* cube root, represented by the symbol

$$\sqrt[3]{}$$

To use Kepler's formula to find the distance from the earth to the sun,

$$y = \sqrt[3]{6x^2}$$

we substitute 365 for *x*, getting

$$y = \sqrt[3]{6(365)^2} = \sqrt[3]{6(133,225)} = \sqrt[3]{799,350}$$

The approximate cube root of 799,350 can be found by either using a calculator or looking through a table of cubes similar to the table of squares on page 563.* From part of such a table, shown at the right, we see that, because 799,350 is closest to 804,357, $\sqrt[3]{799,350}$ is approximately 93. So *y*, the distance of the earth from the sun, is about 93 million miles.

Number	Cube
91	753,571
92	778,688
93	804,357
94	830,584

The idea of a root of a number can be extended beyond square roots and cube roots to *fourth roots, fifth roots,* and so on.

▶ In general, if $y = x^n$, then *y* is called the *nth power* of *x* and *x* is called an **nth root** of *y*.

For example, the number 3 is a fourth root of 81 because $3^4 = 81$. The number –2 is a fifth root of –32 because $(-2)^5 = -32$.

The symbol $\sqrt[n]{x}$ is used to represent the *largest nth root* of *x*. Although, for example, both 3 and –3 are fourth roots of 81,

$$\sqrt[4]{81} = 3$$

because 3 is larger than –3.

---

*If an integer is not the *n*th power of an integer, its *n*th roots are irrational. The number 799,350 is not the cube of an integer; so the cube root of 799,350 is irrational.

We noted in Lesson 2 that, if a rational number is written as a fraction in lowest terms, the numerator and denominator of the fraction must be squares of integers in order for its square roots to be rational. This is true of *n*th roots in general:

▶ If a rational number is written as a fraction in lowest terms, the numerator and denominator of the fraction must be *n*th powers of integers in order for its *n*th roots to be rational.

Exercises

Set I

1. If possible, reduce each of these fractions.

   a) $\dfrac{x - 2y}{3x - 6y}$

   b) $\dfrac{3x^5}{5x^3}$

   c) $\dfrac{x^2 + y^2}{x + y}$

2. Solve the following pairs of simultaneous equations.

   a) $5x + 6y = 2$
      $3x + 6y = 2$

   b) $x^2 + y = 0$
      $y = x - 2$

3. If a cannonball is shot into the air, its distance above the ground at any point along its path is a function of time. A typical formula for this function is

$$y = -5x^2 + 30x$$

in which y is the distance above the ground in meters and x is the time in seconds that the ball has been in the air.

a) How far above the ground is the ball 3 seconds after it was shot?

b) When is the ball 25 feet above the ground?

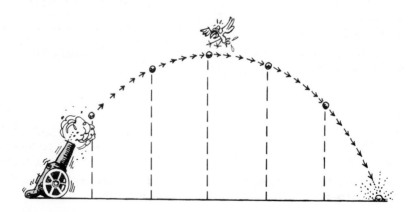

Set II

4. Copy and complete the following table of powers.

x	x^2	x^3	x^4	x^5
1				
2				
3				
4				
5				

5. Use your table to tell whether each of the following numbers is rational or irrational. Express each of the rational numbers in simplest form.

a) $\sqrt{8}$

b) $\sqrt[3]{8}$

c) $\sqrt{25}$

d) $\sqrt[3]{25}$

e) $\sqrt{81}$

f) $\sqrt[4]{81}$

g) $\sqrt[3]{\dfrac{1}{27}}$

h) $\sqrt{\dfrac{8}{18}}$

i) $\sqrt[5]{\dfrac{1}{100}}$

j) $\sqrt[4]{\dfrac{16}{625}}$

6. A table listing some of the roots of 5 and 0.5 is shown below. Each root is rounded to the nearest thousandth.

n	$\sqrt[n]{5}$	$\sqrt[n]{0.5}$
2	2.236	0.707
3	1.710	0.794
4	1.495	0.841
5	1.380	0.871
10	1.175	0.933
100	1.016	0.993
1,000	1.002	0.999

a) What happens to $\sqrt[n]{5}$ as n gets larger?

b) What number does $\sqrt[n]{5}$ get very close to when n is very large?

c) What happens to $\sqrt[n]{0.5}$ as n gets larger?

d) What number does $\sqrt[n]{0.5}$ get very close to when n is very large?

7. The graph of the function

$$y = x^2$$

can be used to estimate square roots of numbers. For example, from the graph shown here, we see that the square roots of

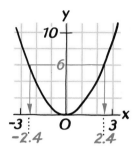

6 are approximately –2.4 and 2.4.

a) Some of the points used to graph this function include $(1, 1)$, $(3, 9)$, and $(-2, 4)$. What is the relationship of the y-coordinate of each point to the x-coordinate of the point?

b) What is the relationship of the x-coordinate of each point to the y-coordinate of the point?

c) Use the scales shown in the figure above. Plot points extending from $x = -5$ to $x = 5$ to enlarge the graph of $y = x^2$.

Use your graph to estimate the square roots of the following numbers, each to the nearest tenth.

d) 10

e) 18

f) 22

g) Judging from your graph, how many square roots does a positive number have?

h) Judging from your graph, how many square roots does a negative number have?

8. The graph of the function

$$y = x^5$$

can be used to estimate fifth roots of numbers.
   a) Graph this function, letting 1 unit on the y-axis represent 20 and plotting points extending from $x = -3$ to $x = 3$.

   Use your graph to estimate the fifth roots of the following numbers, each to the nearest tenth.

   b) 20
   c) 80
   d) –10
   e) Judging from your graph, how many fifth roots does a positive number have?
   f) Judging from your graph, how many fifth roots does a negative number have?

9. List as many of the roots indicated for each of the following numbers as you can. If you think that a number has no such roots, write "none."
   a) Cube roots of 1,000.
   b) Cube roots of –8.
   c) Fourth roots of 625.
   d) Fourth roots of –16.
   e) Fifth roots of 1.
   f) Fifth roots of –243.

10. Which of these symbols, $>$, $=$, or $<$, should replace ▦ in each of the following?
    a) $\sqrt[3]{10}$ ▦ $\sqrt[4]{10}$
    b) $\sqrt[5]{3}$ ▦ $\sqrt[5]{2}$
    c) $\sqrt[4]{0}$ ▦ $\sqrt[4]{0}$
    d) $\sqrt[4]{8}$ ▦ 2
    e) $\sqrt[3]{-8}$ ▦ -1
    f) $\sqrt[3]{\dfrac{1}{27}}$ ▦ $\sqrt[3]{-27}$
    g) $\sqrt{\dfrac{4}{9}}$ ▦ $\dfrac{4}{9}$
    h) $\sqrt[4]{\dfrac{1}{16}}$ ▦ $\sqrt{\dfrac{1}{16}}$

Set III

11. Copy and complete the following table of powers.

x	x^2	x^3	x^4	x^5
1	▦	▦	▦	▦
2	▦	▦	▦	▦
3	▦	▦	▦	▦
4	▦	▦	▦	▦
5	▦	▦	▦	▦

12. Use your table to tell whether each of the following numbers is rational or irrational. Express each of the rational numbers in simplest form.
    a) $\sqrt{9}$
    b) $\sqrt[3]{9}$
    c) $\sqrt{27}$
    d) $\sqrt[3]{27}$
    e) $\sqrt{16}$
    f) $\sqrt[4]{16}$
    g) $\sqrt[5]{\dfrac{1}{25}}$
    h) $\sqrt{\dfrac{4}{8}}$
    i) $\sqrt[4]{\dfrac{3}{243}}$
    j) $\sqrt[3]{\dfrac{64}{125}}$

13. A table listing some of the roots of 2 and 0.2 is shown below. Each root is rounded to the nearest thousandth.

n	$\sqrt[n]{2}$	$\sqrt[n]{0.2}$
2	1.414	0.447
3	1.260	0.585
4	1.189	0.669
5	1.149	0.725
10	1.072	0.851
100	1.007	0.984
1,000	1.001	0.998

   a) What happens to $\sqrt[n]{2}$ as n gets larger?
   b) What number does $\sqrt[n]{2}$ get very close to when n is very large?
   c) What happens to $\sqrt[n]{0.2}$ as n gets larger?
   d) What number does $\sqrt[n]{0.2}$ get very close to when n is very large?

14. The graph of the function

$$y = x^3$$

can be used to estimate cube roots of numbers. For example, from the graph

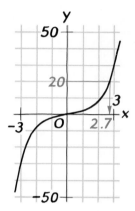

shown here, we see that the cube root of 20 is approximately 2.7.

a) Some of the points used to graph this function include $(1, 1)$, $(3, 27)$, and $(-2, -8)$. What is the relationship of the y-coordinate of each point to the x-coordinate of the point?

b) What is the relationship of the x-coordinate of each point to the y-coordinate of the point?

c) Use the scales shown in the figure above. Plot points extending from $x = -5$ to $x = 5$ to enlarge the graph of $y = x^3$.

Use your graph to estimate the square roots of the following numbers, each to the nearest tenth.

d) 40

e) –80

f) 110

g) Judging from your graph, how many cube roots does a positive number have?

h) Judging from your graph, how many cube roots does a negative number have?

15. The graph of the function

$$y = x^4$$

can be used to estimate fourth roots of numbers.

a) Graph this function, letting 1 unit on the y-axis represent 10 and plotting points extending from $x = -3$ to $x = 3$.

Use your graph to estimate the fourth roots of the following numbers, each to the nearest tenth.

b) 10

c) 40

d) 50

e) Judging from your graph, how many fourth roots does a positive number have?

f) Judging from your graph, how many fourth roots does a negative number have?

16. List as many of the roots indicated for each of the following numbers as you can. If you think that there are no such roots, write "none."

a) Cube roots of –27.

b) Cube roots of 0.

c) Fourth roots of 10,000.

d) Fourth roots of –1.

e) Fifth roots of –1.

f) Fifth roots of 1,024.

17. Which of these symbols, $>$, $=$, or $<$, should replace ▓ in each of the following?

a) $\sqrt{2}$ ▓ $\sqrt[3]{2}$

b) $\sqrt[4]{10}$ ▓ $\sqrt[4]{12}$

c) $\sqrt[3]{-1}$ ▓ $\sqrt[5]{-1}$

d) $\sqrt[3]{6}$ ▓ 2

e) $\sqrt[3]{-6}$ ▓ -2

f) $\sqrt{\dfrac{1}{64}}$ ▓ $\sqrt[3]{\dfrac{1}{64}}$

g) $\sqrt{\dfrac{9}{25}}$ ▓ $\dfrac{9}{25}$ h) $\sqrt[5]{\dfrac{1}{32}}$ ▓ $\sqrt[5]{-32}$

Set IV

Mercury, the planet closest to the sun, travels once around it in 88 earth-days. Use the formula given on page 680 to find out how many million miles from the sun Mercury is.

Number	Cube
31	29,791
32	32,768
33	35,937
34	39,304
35	42,875
36	46,656
37	50,653
38	54,872
39	59,319
40	64,000

NATIONAL AERONAUTICS AND SPACE ADMINISTRATION.

PHOTOGRAPH BY JOHN TIMBERS.

LESSON 4
Pi

This photograph shows eighty people seated in a circle in an unusual way. Instead of sitting on chairs or on the ground, each person is sitting on the knees of the person behind him!

Although a circle is a very simple geometric figure, the relationship of the circumference of a circle (the distance around it) to its diameter (the distance across it) is very complex. The figure at the right shows that the circumference is a little more than 3 times as long as the diameter. More accurately, it is about 3.14 times as long but even this number is not quite correct. The *exact* number of times that the diameter of a circle will go into its circumference is usually represented by the symbol π, a letter of the Greek alphabet. Called "pi," this number has been proved to be irrational, which means that it cannot be written out completely in decimal form. Nevertheless, it has been calculated, like $\sqrt{2}$, to more than one million decimal places.

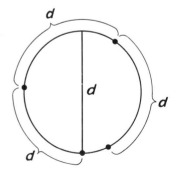

The fact that the circumference of a circle is π times as long as its diameter can be written as the formula

$$c = \pi d$$

in which c represents the circumference and d the diameter. As the figures below illustrate, the diameter of a circle is 2 times as long

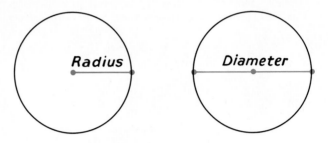

as its radius. Because $d = 2r$, we can also write the circumference formula as

$$c = \pi(2r) = 2\pi r$$

The formula for the area of a circle also contains the number π. The formula is

$$a = \pi r^2$$

in which a represents the area and r the radius.

For practical use in finding the circumference and area of a circle, π is often approximated by 3.14.* Here are examples of how the circumference and area formulas are used.

EXAMPLE 1
Find the exact circumference of a circle whose radius is 14 centimeters.

SOLUTION
Using the circumference formula, $c = 2\pi r$, we get

$$c = 2\pi(14) = 28\pi$$

The exact circumference is 28π centimeters.

---

* The decimal form of π to ten places is 3.1415926535.

EXAMPLE 2

Find the approximate area of a circle whose radius is 30 meters.

SOLUTION

Using the area formula, $a = \pi r^2$, we get

$$a \approx (3.14)(30)^2 = (3.14)(900) = 2{,}826 \text{ square meters}$$

EXAMPLE 3

Find the exact radius of a circle whose area is 24π.

SOLUTION

Substituting 24π for a in the area formula, $a = \pi r^2$, we get

$$24\pi = \pi r^2$$

Dividing both sides of this equation by π,

$$24 = r^2$$

and taking square roots, we get

$$r = \pm\sqrt{24} = \pm 2\sqrt{6}$$

Because the radius of a circle is a positive number, the answer is $2\sqrt{6}$.

Exercises

Set I

1. Write each of the following sums or products as a single fraction.

   a) $3 + \dfrac{1}{x}$

   b) $3 \cdot \dfrac{1}{x}$

   c) $\dfrac{2}{x^2} + \dfrac{4}{x^4}$

   d) $\dfrac{2}{x^2} \cdot \dfrac{4}{x^4}$

2. Write each of the following as the quotient of two integers in lowest terms.

   a) 0.2

   b) $0.\overline{2}$

   c) 0.27

   d) $0.\overline{27}$

3. The following problem is from a collection written in the eighth century called *Problems for the Quickening of the Mind.*

   "A dog chasing a rabbit jumps 9 feet every time the rabbit jumps 7. If the rabbit has a headstart of 150 feet, in how many leaps does the dog overtake the rabbit?"

a) Write an equation for this problem, letting x represent the number of leaps made by each animal.

b) Solve the equation to find the number of leaps in which the dog catches up with the rabbit.

Set II

4. The ancient Egyptians used the number $\left(\frac{4}{3}\right)^4$ as an approximation of π.

   a) Write $\left(\frac{4}{3}\right)^4$ as the quotient of two integers.

   b) Change the result to decimal form, correct to 4 decimal places.

   c) Is it *smaller* or *larger* than π?

   d) How is it possible to know that π is not equal to $\left(\frac{4}{3}\right)^4$ without doing any dividing?

5. The following questions concern the relationship of the circumference of a circle to its radius.

   a) Copy and complete the following table.

Radius	1	2	3	4	5
Circumference	2π				

   b) Is the circumference of a circle doubled if its radius is doubled?

   c) How do the circumference and radius of a circle vary with respect to each other?

6. Find the exact circumference *and* area of a circle whose radius is
   a) 0.5 cm.
   b) $\sqrt{2}$ cm.   d) $\frac{3}{\pi}$ cm.
   c) π cm.

7. Find the exact radius of a circle whose
   a) circumference is 9π.
   b) circumference is 14.
   c) area is 9π.
   d) area is 20π.

8. The diameter of each circle in this figure is equal to half the side of the square.

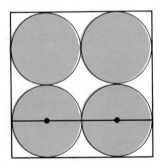

   Find the area of the square if the radius of each circle is
   a) 10.
   b) x.

   Find the green area (in terms of π) if the radius of each circle is

   c) 10.
   d) x.

9. The diameter of the larger circle in this figure is twice the diameter of the smaller circle.

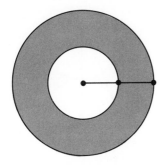

Find the green area (in terms of π) if the radius of the smaller circle is
a) 3.
b) x.
c) What fraction of the area of the larger circle is in green?

*10. In each of the following problems, let $\pi = 3.14$ and round your answer to the nearest integer.

a) Find the approximate circumference of a Hula Hoop if its radius is 18 inches.
b) Find the approximate radius of a person's neck if its circumference is 14 inches.
c) Find the approximate area of a circular sand dollar whose radius is 4.1 centimeters.
d) Find the approximate diameter of a circular radar screen if its area is 1,260 square centimeters.

Set III

11. Claudius Ptolemy, an astronomer who lived in the second century, used the number $3\dfrac{17}{120}$ as an approximation of π.

a) Write $3\dfrac{17}{120}$ as the quotient of two integers.
b) Change the result to decimal form.
c) Is it *smaller* or *larger* than π?
d) How is it possible to know that π is not equal to $3\dfrac{17}{120}$ without doing any dividing?

12. The following questions concern the relationship of the area of a circle to its radius.

a) Copy and complete the following table.

Radius	1	2	3	4	5
Area	π				

b) Is the area of a circle doubled if its radius is doubled?
c) Does the area of a circle vary directly with its radius?

13. Find the exact circumference *and* area of a circle whose radius is
a) 0.1 cm.
b) $\dfrac{\pi}{2}$ cm.
c) $\sqrt{3}$ cm.
d) $\dfrac{6}{\pi}$ cm.

14. Find the exact radius of a circle whose
    a) circumference is 25π.
    b) circumference is 8.
    c) area is 25π.
    d) area is 12π.

15. The diameter of the circle in this figure is equal to the side of the square.

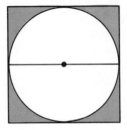

Find the green area (in terms of π) if the radius of the circle is
a) 5.
b) x.

Find the difference between the perimeter of the square and the circumference of the circle (in terms of π) if the radius of the circle is
c) 5.
d) x.

16. The diameter of the larger circle in this figure is twice the diameter of each of the smaller circles.

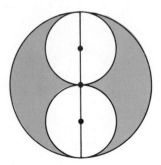

Find the green area (in terms of π) if the radius of one of the smaller circles is
a) 4.
b) x.
c) What fraction of the area of the large circle is in green?

*17. In each of the following problems, let $\pi = 3.14$ and round your answer to the nearest integer.
    a) Find the approximate circumference of a bicycle wheel if its radius is 13.5 inches.
    b) Find the approximate diameter of a rope whose circumference is 20 centimeters.
    c) Find the approximate area of a pizza whose radius is 18 centimeters.
    d) Find the approximate diameter of a circular swimming pool if its area is 320 square feet.

Set IV

Can you figure out what the following sentences have to do with the number π?
   "May I have a large container of coffee?"
   "How I want a drink, alcoholic of course."

"Nine and three-quarters...nine and seven-eighths..."

LESSON 5
The Real Numbers

The first set of numbers that everyone becomes familiar with consists of the *counting numbers:*

$$1, 2, 3, 4, 5, \ldots$$

The counting numbers are part of a larger set of numbers called the *integers:*

$$\ldots, -5, -4, -3, -2, -1, 0, 1, 2, 3, 4, 5, \ldots$$

and the integers are part of yet a larger set of numbers called the *rational numbers.* The referee in this cartoon is "counting" with rational numbers.

The rational numbers, together with the irrational numbers, make up an even larger set of numbers called the *real numbers.* As you have already learned, the rational numbers can be distinguished from the irrational numbers by their decimal forms. The decimal forms of rational numbers repeat endlessly. The

The real numbers

Rational numbers	Irrational numbers
Integers	
Counting numbers	

decimal forms of irrational numbers do not repeat.

An important property of the real numbers is *closure*.

▶ A set of numbers is said to be **closed** with respect to an operation if, whenever the operation is performed on numbers in the set, the result is also a number in the set.

Whenever two real numbers are added, subtracted, multiplied, or divided (except for division by zero, which is meaningless), the result is a real number. For this reason, the real numbers are *closed* with respect to these operations.

Another important property of the real numbers is *order*.

▶ A set of numbers is said to be **ordered** if for any two numbers a and b in the set either $a < b$, $a = b$, or $a > b$.

We have used this property of numbers throughout the course. To picture the order of numbers, we identify them with the points on a line. It is possible to set up a correspondence between the real numbers and the points on a line so that to every number there corresponds a point and to every point there corresponds a number. An example of a correspondence between some real numbers and some points on a line is shown by the figure below.

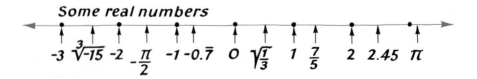

Some real numbers

One way to tell which of two real numbers is larger is to compare their decimal forms. Here is an example.

EXAMPLE

Arrange the following rational numbers in order from smallest to largest: 3.14, 3.1$\overline{4}$, and 3.$\overline{14}$.

SOLUTION

Writing each number to four decimal places, we get 3.1400, 3.1444, and 3.1414. Clearly, 3.1400 is the smallest and 3.1444 is the largest. So the numbers arranged in increasing order are: 3.14, 3.$\overline{14}$, and 3.1$\overline{4}$.

Exercises

Set I

1. Change each of the following to the form described. (Each part has many possible answers.)

   a) Write $\dfrac{x + 6}{x + 2}$ as the sum of two fractions.

   b) Write $\dfrac{8 - x}{4}$ as the difference of an integer and a fraction.

   c) Write $\dfrac{5}{xy}$ as a product of two fractions.

2. Solve the following equations.
   a) $5x = 2(x + 1)$
   b) $5x = 2(x^2 + 1)$
   c) $5x^2 = 2(x + 1)$

3. While traveling in Africa, Captain Spaulding caught 37 animals, of which some were monkeys and the rest were weasels. If he sold each monkey for $20 and each weasel for $5, making $425 altogether, how many animals of each type did he catch?

Set II

4. The following questions are about these numbers:

   $8, 0.8, 0.\overline{8}, 0^8, \sqrt{8}, \sqrt[3]{8}, -8, \dfrac{1}{8}, 8\pi, 8^2$

   a) Which ones are counting numbers?
   b) Which ones are integers but not counting numbers?
   c) Which ones are rational numbers but not integers?
   d) Which ones are real numbers but not rational?

5. The counting numbers are closed with respect to addition because the sum of any two counting numbers is also a counting number.

   Tell whether or not the counting numbers are closed with respect to each of the following operations. If an answer is no, give an example to show why.
   a) Subtraction. (Is the difference of two counting numbers always a counting number?)
   b) Multiplication.
   c) Division.

d) Cubing.

e) Taking square roots.

6. Tell whether or not the even integers are closed with respect to each of the following operations. If an answer is no, give an example to show why.
   a) Addition.
   b) Subtraction.
   c) Multiplication.
   d) Division (excluding zero).
   e) Cubing.

7. Tell whether or not the rational numbers are closed with respect to each of the following operations. If an answer is no, give an example to show why.
   a) Addition.
   b) Subtraction.
   c) Multiplication.
   d) Division (excluding zero).
   e) Squaring.
   f) Taking square roots.

8. Arrange each of the following sets of numbers in order from smallest to largest.
   a) 0.6, 0.009, 0.07, 0.0008
   b) $0.18\overline{3}$, $0.1\overline{83}$, $0.\overline{183}$, 0.183
   c) $\sqrt{10}$, $\sqrt[3]{10}$, $\sqrt[4]{10}$
   d) $\sqrt[3]{1}$, $\sqrt[3]{-1}$, $\sqrt[3]{2}$, $\sqrt[3]{-2}$

   e) π, $\dfrac{\pi}{2}$, $-\pi$, $-\dfrac{\pi}{2}$

9. Use the table below to find the rational number, to the nearest hundredth, closest to the irrational numbers in parts a through e.

x	$\sqrt{x}$	$\sqrt[3]{x}$	$\sqrt[4]{x}$	$\sqrt[5]{x}$
1	1	1	1	1
2	1.414	1.260	1.189	1.149
3	1.732	1.442	1.316	1.246
4	2	1.587	1.414	1.320
5	2.236	1.710	1.495	1.380
6	2.449	1.817	1.565	1.431
7	2.646	1.913	1.627	1.476
8	2.828	2	1.682	1.516
9	3	2.080	1.732	1.552
10	3.162	2.154	1.778	1.585

a) $\sqrt[3]{6}$

b) $\dfrac{\sqrt{3}}{4}$

c) $\sqrt[5]{7} - \sqrt[5]{8}$

d) $\dfrac{1}{\sqrt{10}}$

e) $\dfrac{-3 - \sqrt{2}}{2}$

10. Solve the following equations for x and tell what kind of numbers the solutions are.
    a) $7x - 22 = 0$
    b) $\sqrt{3}x = \sqrt{12} - \sqrt{3}$
    c) $x + \sqrt{2} = \sqrt{8}$
    d) $2x^2 - 10 = 0$
    e) $x(x - 4) = 5$
    f) $\sqrt{x} + 9 = 16$

Set III

11. The following questions are about these numbers:

    25, $\dfrac{1}{25}$, -25, 2.5, $\sqrt{25}$,

    $2.\overline{5}$, $\sqrt[3]{25}$, 25π, $-\dfrac{2}{5}$, 2^5

    a) Which ones are counting numbers?
    b) Which ones are integers but not counting numbers?

    c) Which ones are rational numbers but not integers?
    d) Which ones are real numbers but not rational?

12. The integers are closed with respect to addition because the sum of any two integers is also an integer.

    Tell whether or not the integers are closed with respect to each of the following

operations. If an answer is no, give an example to show why.
a) Subtraction. (Is the difference of two integers always an integer?)
b) Multiplication.
c) Division (excluding zero).
d) Cubing.
e) Taking square roots.

13. Tell whether or not the odd integers are closed with respect to each of the following operations. If an answer is no, give an example to show why.
a) Addition.
b) Subtraction.
c) Multiplication.
d) Division.
e) Squaring.

14. Tell whether or not the irrational numbers are closed with respect to each of the following operations. If an answer is no, give an example to show why.
a) Addition.
b) Subtraction.
c) Multiplication.
d) Division (excluding zero).
e) Squaring.
f) Taking square roots.

15. Arrange each of the following sets of numbers in order from smallest to largest.
a) 0.002, 0.3, 0.0005, 0.04
b) 0.719, 0.71$\overline{9}$, 0.7$\overline{19}$, 0.$\overline{719}$
c) $\dfrac{2}{3}, \dfrac{6}{7}, \dfrac{4}{5}, \dfrac{8}{9}$
d) $\sqrt{5}, -\sqrt{5}, \sqrt[4]{5}, -\sqrt[4]{5}$
e) $\pi, \sqrt{\pi}, \sqrt[3]{\pi}$

16. Use the table in exercise 9 on page 696 to find the rational number, to the nearest hundredth, closest to each of the following irrational numbers.
a) $\sqrt[4]{3}$
b) $\dfrac{\sqrt{10}}{5}$
c) $\sqrt[3]{6} - \sqrt[3]{5}$ e) $\dfrac{3 - \sqrt{7}}{2}$
d) $\dfrac{1}{\sqrt{2}}$

17. Solve the following equations for x and tell what kind of numbers the solutions are.
a) $3x + 14 = 0$
b) $x + \sqrt{5} = \sqrt{20}$
c) $\sqrt{2}x = \sqrt{32} - \sqrt{2}$
d) $4x^2 - 15 = 0$
e) $x(x + 4) = 21$
f) $\sqrt{x} - 1 = 8$

Set IV

Can you arrange, without using a calculator, the numbers in each of the following sets in order from smallest to largest? Explain your reasoning. (The expression a^{b^c} is evaluated by first finding b^c and then raising a to that power.)

1. $1^{2^3}, \quad 1^{3^2}, \quad 2^{1^3}, \quad 2^{3^1}, \quad 3^{1^2}, \quad 3^{2^1}$
2. $2^{3^4}, \quad 2^{4^3}, \quad 3^{2^4}, \quad 3^{4^2}, \quad 4^{2^3}, \quad 4^{3^2}$

Summary and Review

In this chapter, we have become acquainted with the properties of the real numbers, both rational and irrational.

Rational Numbers (*Lesson 1*)   A rational number is a number that can be written as the quotient of two integers. When a rational number is changed to decimal form by carrying out the indicated division, it always falls into a repeating pattern of digits. Moreover, every number in decimal form that has a repeating decimal pattern is rational because it can always be expressed as the quotient of two integers.

Irrational Numbers (*Lessons 2 and 3*)   An irrational number is a number that cannot be written as the quotient of two integers. The decimal form of an irrational number does not have a repeating pattern of digits.

If $y = x^n$, then x is an nth root of y. The largest nth root of y is represented by the symbol $\sqrt[n]{y}$. If an integer is not the nth power of an integer, its nth roots are irrational. If a rational number is written as a fraction in lowest terms, the numerator and denominator of the fraction must be nth powers of integers in order for its nth roots to be rational.

Pi (*Lesson 4*)   If the circumference of a circle is divided by its diameter, the result is the irrational number π.

The circumference of a circle is given by the formula $c = 2\pi r$, in which r represents the radius of the circle. The area of a circle is given by the formula $a = \pi r^2$. A useful approximation for π is 3.14.

The Real Numbers (*Lesson 5*)   The rational numbers together with the irrational numbers make up the set of numbers called the real numbers. The real numbers are closed with respect to addition, subtraction, multiplication, and division. The real numbers have a definite order and can be put in a one-to-one correspondence with the points on a line.

Exercises

Set I

1. Tell whether each of the following statements is true or false.
   a) Every fraction is a rational number.
   b) The square roots of every positive integer are irrational.
   c) To every point on a number line, there corresponds a real number.
   d) Every number has exactly one cube root.
   e) Pi is an irrational number.

2. Change the following rational numbers to decimal form.
   a) $\dfrac{16}{5}$ c) $\dfrac{8}{11}$
   b) $\dfrac{9}{40}$ d) $\dfrac{5}{24}$

3. Write each of the following as the quotient of two integers in lowest terms.
   a) 10.3
   b) 10.$\overline{3}$
   c) 0.85
   d) 0.$\overline{85}$

4. The following questions are about the number in the cartoon.
   a) What kind of number is it?
   b) What do you think the "next number" after it is? Explain your answer.

5. If possible, find the rational number represented by

$$\frac{5x}{x-2}$$

   if
   a) $x = 1$ c) $x = 3$
   b) $x = 2$ d) $x = 4$

6. Arrange the following numbers in order from smallest to largest.

$$0.345, \ 0.34\overline{5}, \ 0.3\overline{45}, \ 0.\overline{345}$$

7. List as many of the roots indicated for each of the following numbers as you can. If you think that there are no such roots, write "none."
   a) Cube roots of 125.
   b) Fourth roots of –16.
   c) Fifth roots of –100,000.
   d) Tenth roots of 1.

8. Which of these symbols, $>$, $=$, or $<$, should replace ▨ in each of the following?
   a) $\sqrt{5}$ ▨ $\sqrt[4]{5}$
   b) $\sqrt[3]{-9}$ ▨ $\sqrt[3]{-7}$
   c) $\sqrt[5]{100}$ ▨ 3
   d) $\sqrt[4]{\dfrac{1}{16}}$ ▨ $\sqrt[4]{\dfrac{1}{81}}$

9. A table listing some of the roots of 0.9 and 1.1 is shown below. Each root is rounded to the nearest thousandth.

n	$\sqrt[n]{0.9}$	$\sqrt[n]{1.1}$
2	0.949	1.049
3	0.965	1.032
4	0.974	1.024
5	0.979	1.019
10	0.990	1.010
100	0.999	1.001

   a) What happens to $\sqrt[n]{0.9}$ as n gets larger?
   b) What number does $\sqrt[n]{0.9}$ get very close to when n is very large?
   c) What happens to $\sqrt[n]{1.1}$ as n gets larger?
   d) What number does $\sqrt[n]{1.1}$ get very close to when n is very large?

10. The third-century Chinese mathematician Liu Hui used the number $3\dfrac{7}{50}$ as an approximation of π.
    a) Write this number as the quotient of two integers.
    b) Change the result to decimal form.
    c) Is it smaller or larger than π?
       ($\pi = 3.14159265\ldots$)

11. Find exact answers to each of the following.
    a) The circumference of a circle whose radius is 0.8.
    b) The area of a circle whose radius is $2\sqrt{3}$.
    c) The radius of a circle whose circumference is 30.
    d) The radius of a circle whose area is 17π.

12. The length of the rectangle in this figure is twice its width and twice the diameter of each circle.

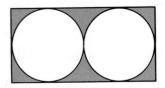

   Find the area of the rectangle if the radius of each circle is
   a) 2.
   b) x.
      Find the green area (in terms of π) if the radius of each circle is
   c) 2.
   d) x.

13. In each of the following problems, let $\pi = 3.14$ and round your answer to the nearest integer.
    a) Find the approximate circumference of the earth if its radius is 4,000 miles.
    b) Find the approximate area of a circular putting green whose diameter is 6 meters.

14. The following questions are about these numbers:

    $$2.7,\ 2.\overline{7},\ \sqrt{27},\ \sqrt[3]{27},\ -27,\ \frac{2}{7},\ 2^7,\ 27\pi$$

    a) Which ones are counting numbers?
    b) Which ones are integers but not counting numbers?
    c) Which ones are rational numbers but not integers?
    d) Which ones are real numbers but not rational?

15. Tell whether or not you think the negative integers are closed with respect to each of the following operations. If an answer is no, give an example to show why.
    a) Addition.
    b) Subtraction.

c) Multiplication.
d) Squaring.
e) Cubing.

16. Solve the following equations for x and tell what kind of numbers the solutions are.
    a) $4x + 19 = 0$
    b) $x - \sqrt{3} = \sqrt{48}$
    c) $36x^2 - 7 = 0$
    d) $\sqrt{x} + 6 = 2$

Set II

1. Tell whether each of the following statements is true or false.
   a) Every integer is a rational number.
   b) The square root of a number can be a repeating decimal.
   c) Every positive number has two fourth roots.
   d) To every point on a number line, there corresponds a rational number.
   e) The circumference of a circle is more than six times as long as its radius.

2. Change the following rational numbers to decimal form.
   a) $\dfrac{10}{4}$
   b) $\dfrac{7}{25}$
   c) $\dfrac{44}{9}$
   d) $\dfrac{2}{13}$

3. Write each of the following as the quotient of two integers in lowest terms.
   a) 0.06
   b) $0.0\overline{6}$
   c) 1.125
   d) $1.\overline{125}$

4. The following questions refer to the rational number $\dfrac{6}{17}$.
   a) How many different numbers, including zero, can there be as remainders if 17 is divided into 6?

   b) If zero were one of the remainders, what would you know about the decimal form of $\dfrac{6}{17}$?

   c) What is the greatest number of digits that could make up the period of $\dfrac{6}{17}$?

5. If possible, find the rational number represented by
   $$\frac{x^2 + 1}{x^2 - 1}$$
   if
   a) $x = 0$
   b) $x = 1$
   c) $x = 2$
   d) $x = 3$

6. Arrange the following numbers in order from smallest to largest.
   $$0.027,\ 0.02\overline{7},\ 0.0\overline{27},\ 0.\overline{027}$$

7. List as many of the roots indicated for each of the following numbers as you can. If you think that there are no such roots, write "none."
   a) Cube roots of –64.
   b) Fourth roots of 81.
   c) Fifth roots of 0.
   d) Seventh roots of 128.

8. Which of these symbols, $>$, $=$, or $<$, should replace ▦ in each of the following?

   a) $\sqrt[4]{12}$ ▦ $\sqrt[3]{12}$
   b) $\sqrt[5]{1}$ ▦ $\sqrt{1}$
   c) $\sqrt[4]{80}$ ▦ 3
   d) $\sqrt[3]{\dfrac{1}{1000}}$ ▦ $\sqrt[3]{-1000}$

9. Solve each of the following equations by using the quadratic formula. Simplify your answers as much as possible but do not round them.

   a) $x^2 - 6x + 5 = 0$
   b) $x^2 - 6x + 6 = 0$
   c) $x^2 - 6x + 7 = 0$
   d) $x^2 - 6x + 9 = 0$
   e) What must be true about the discriminant, $b^2 - 4ac$, of a quadratic equation in order for the solutions to the equation to be rational? (Assume that a, b, and c are integers.)

10. The Italian mathematician Leonardo Fibonacci used $\dfrac{864}{275}$ as an approximation of π.

   a) Change this fraction to decimal form.
   b) Is it smaller or larger than π? ($\pi = 3.14159265\ldots$)
   c) How is it possible to know that π is not equal to this fraction without doing any dividing?

11. Find exact answers to each of the following.
   a) The circumference of a circle whose radius is $\sqrt{5}$.
   b) The area of a circle whose radius is 0.3.
   c) The radius of a circle whose circumference is 14.
   d) The radius of a circle whose area is 200π.

12. The diameter of the larger circle in this figure is twice the diameter of the smaller circle.

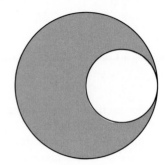

   Find the green area (in terms of π) if the radius of the smaller circle is
   a) 5.
   b) x.
   c) What fraction of the area of the larger circle is shaded?

13. In each of the following problems, let $\pi = 3.14$ and round your answer to the nearest integer.
   a) Find the approximate circumference of a phonograph record if its radius is 6 inches.
   b) Find the approximate area of a circular card table whose radius is 50 centimeters.

14. The following questions are about these numbers:

   $$9,\ 0.9,\ 0^9,\ \sqrt{9},\ \sqrt[3]{9},\ \frac{1}{9},\ -9,\ 9\pi$$

   a) Which ones are counting numbers?
   b) Which ones are integers but not counting numbers?
   c) Which ones are rational numbers but not integers?
   d) Which ones are real numbers but not rational?

15. Tell whether or not you think the odd counting numbers are closed with respect to each of the following operations. If an answer is no, give an example to show why.
    a) Addition.
    b) Subtraction.
    c) Multiplication.
    d) Division.
    e) Squaring.

16. Solve the following equations for x and tell what kind of numbers the solutions are.
    a) $5x - 12 = 0$
    b) $x + \sqrt{6} = \sqrt{96}$
    c) $3x^2 - 18 = 0$
    d) $\sqrt{x} + 2 = 13$

Chapter 15

FRACTIONAL EQUATIONS

RKO GENERAL PICTURES.

LESSON 1
Ratio and Proportion

Although King Kong seemed to be an ape of tremendous size on the movie screen, some of the models of him used in the film were less than two feet tall. To disguise this fact, these models were filmed on sets that had been built at a very small scale.

The dimensions of a model and building for a set built at a scale of $\frac{1}{25}$ are given in the following table:

	Height of model	Actual height
King Kong	2 feet	50 feet
Building	8 feet	200 feet

The scale number, $\frac{1}{25}$, is the *ratio* of the numbers 1 to 25.

▶ The **ratio** of the numbers a to b is the number $\frac{a}{b}$.

For such a set to be convincing, the ratios of the dimensions of the objects on it to the corresponding dimensions of the objects they represent must be the same. In other words, the corresponding dimensions must be *proportional*.

▶ A **proportion** is an equation stating that two ratios are equal.

For the King Kong set, we can write the proportion $\frac{2}{50} = \frac{8}{200}$. In general, a proportion is an equation of the form $\frac{a}{b} = \frac{c}{d}$, in which a, b, c, and d are the first, second, third, and fourth terms of the proportion, respectively. The second and third terms, b and c, are the *means,* and the first and fourth terms, a and d, are the *extremes* of the proportion.

In the proportion for the King Kong set, the means are 50 and 8 and the extremes are 2 and 200. Notice that $50 \cdot 8 = 2 \cdot 200$. It is easy to show that this is true, not just for this proportion, but for *any* proportion:

▶ In a proportion, the product of the means is equal to the product of the extremes.

This gives us a convenient way to solve equations that are proportions. Here are examples.

EXAMPLE 1

Solve for x: $\frac{7}{3} = \frac{x}{6}$.

SOLUTION

Multiplying means and extremes, we get

$$3x = 42$$

Dividing both sides by 3,

$$x = 14$$

Checking our answer by substituting it into the original equation,

$$\frac{7}{3} = \frac{14}{6}$$

we see that it is correct because the fractions are equivalent.

EXAMPLE 2

Solve for x: $\dfrac{1}{2x} = \dfrac{x}{10}$.

SOLUTION

Multiplying means and extremes, we get the quadratic equation,

$$2x^2 = 10$$

Dividing both sides by 2,

$$x^2 = 5$$

and taking square roots,

$$x = \pm\sqrt{5}$$

The equation has two solutions: $\sqrt{5}$ and $-\sqrt{5}$.
    Checking the first solution, $\sqrt{5}$, we get

$$\frac{1}{2\sqrt{5}} = \frac{1 \cdot \sqrt{5}}{2\sqrt{5} \cdot \sqrt{5}} = \frac{\sqrt{5}}{10}$$

The second solution can be checked in the same way.

EXAMPLE 3

Solve for x: $\dfrac{4x + 1}{2x - 3} = \dfrac{3}{5}$.

SOLUTION

Multiplying means and extremes,

$$3(2x - 3) = 5(4x + 1)$$
$$6x - 9 = 20x + 5$$
$$-14x - 9 = 5$$
$$-14x = 14$$
$$x = -1$$

Checking, we get

$$\frac{4(-1) + 1}{2(-1) - 3} = \frac{-4 + 1}{-2 - 3} = \frac{-3}{-5} = \frac{3}{5}$$

Exercises

Set I

1. Arrange the following numbers in order from smallest to largest. ($\pi = 3.14159\ldots$)

   a) π

   b) $3\dfrac{1}{7}$

   c) 3.14

   d) 3.1416

2. Find the following quotients.

   a) $\dfrac{3x^3 + 2x^2 - 7x + 2}{x + 2}$

   b) $\dfrac{x^4 + 64x}{x^2 - 4x + 16}$

3. This is an aerial photograph of a large cornfield in Colorado. Water from a well in the center of the field is pumped into a pipe 440 yards long. The pipe slowly rotates about the center, giving the field its circular shape.

   a) Find the exact area of the field in square yards.

   b) Find the approximate area of the field, giving your answer to the nearest thousand square yards.

PHOTOGRAPH BY HAROLD DUKE AND DALE F. HEERMANN.

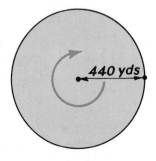

440 yds

Set II

4. An American roulette wheel contains 38 compartments, of which 18 are red (white in this illustration), 18 are black, and 2 are green. Give each of the following ratios in simplest terms.
   a) The ratio of red to the total number of compartments.
   b) The ratio of green to the total number of compartments.
   c) The ratio of red to green.
   d) The ratio of black to red.

5. If possible, simplify the following ratios.

   a) $\dfrac{4x}{6x}$

   b) $\dfrac{4x + 1}{6x + 1}$

   c) $\dfrac{4x + 2}{6x + 2}$

   d) $\dfrac{4x + 4}{6x + 6}$

   e) $\dfrac{x + 4}{x + 6}$

   f) $\dfrac{x^4}{x^6}$

   g) $\dfrac{4x}{x^4}$

   h) $\dfrac{6 + x}{6x}$

6. Solve the following equations. Check your answers.

   a) $\dfrac{x}{8} = \dfrac{3}{4}$

   b) $\dfrac{6}{x} = \dfrac{5}{11}$

   c) $\dfrac{2}{x - 1} = \dfrac{5}{x}$

   d) $\dfrac{x + 5}{2} = \dfrac{4x - 3}{7}$

   e) $\dfrac{2}{x + 1} = \dfrac{6}{3x + 3}$

   f) $\dfrac{4}{x} = \dfrac{x}{25}$

   g) $\dfrac{1}{x - 3} = \dfrac{1}{4x - 12}$

   h) $\dfrac{2x - 1}{2x + 1} = \dfrac{9}{4}$

   i) $\dfrac{x + 7}{3} = \dfrac{5}{x - 7}$

   j) $\dfrac{x + 2}{x - 4} = \dfrac{x - 5}{x + 3}$

7. Ethel is exactly six years older than Lucy.
   a) Copy and complete the following table of their ages.

Lucy's age	1	2	3	6	10	20	30
Ethel's age	7						

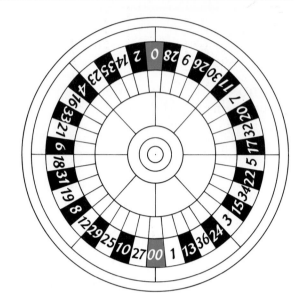

   b) Find the ratio in decimal form of Ethel's age to Lucy's age for each pair of ages listed in the table.
   c) Write a formula for the ratio of Ethel's age to Lucy's age when Lucy is n years old.
   d) As Ethel and Lucy grow older, what happens to the ratio of their ages?
   e) Will the ratio of their ages ever become equal to 1? Explain.

8. The amount of gas used by an idling car is proportional to the time.
   a) Copy and complete the following table.

Time in minutes	1	5		
Amount of gas in ounces		8	10	20

   b) What is the ratio of the amount in ounces of gas used to the time in minutes?
   c) How does the amount of gas used vary with respect to the time?
   d) Write a formula for the amount in ounces of gas used, g, in terms of the time in minutes, t.

Set III

9. This exercise refers to the cards at the right. Give each of the following ratios in simplest terms.
   a) The ratio of spades to the total number of cards.
   b) The ratio of hearts to the total number of cards.
   c) The ratio of diamonds to clubs.
   d) The ratio of nines to fours.

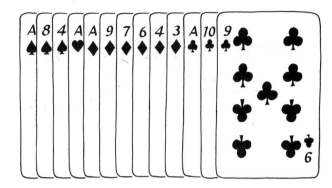

10. If possible, simplify the following ratios.

   a) $\dfrac{6x}{10x}$ e) $\dfrac{x-6}{x-10}$

   b) $\dfrac{6x-1}{10x-1}$ f) $\dfrac{x^6}{x^{10}}$

   c) $\dfrac{6x-2}{10x-2}$ g) $\dfrac{6x}{x^6}$

   d) $\dfrac{6x-6}{10x-10}$ h) $\dfrac{10+x}{10x}$

11. Solve the following equations. Check your answers.

   a) $\dfrac{x}{6}=\dfrac{4}{3}$ f) $\dfrac{x}{4}=\dfrac{9}{x}$

   b) $\dfrac{7}{x}=\dfrac{9}{2}$ g) $\dfrac{3x+1}{3x-1}=\dfrac{1}{8}$

   c) $\dfrac{5}{3x}=\dfrac{1}{x+2}$ h) $\dfrac{1}{2x-4}=\dfrac{1}{x-2}$

   d) $\dfrac{5}{x-5}=\dfrac{10}{2x-10}$ i) $\dfrac{x-3}{10}=\dfrac{4}{x+3}$

   e) $\dfrac{3x-8}{5}=\dfrac{x+9}{2}$ j) $\dfrac{x-1}{x+5}=\dfrac{x}{x+8}$

12. The table below lists the frequencies in cycles per second of some of the notes in a musical scale.

Note	C	D	E	F	G	A	B
Frequency	264	297	330	▥	396	▥	495

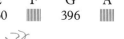

Find the ratio of the frequencies of the following notes in simplest terms.
a) E to C
b) G to E
c) G to C

Find the frequency of

d) F, if its ratio to the frequency of C is $\dfrac{4}{3}$.

e) A, if its ratio to the frequency of B is $\dfrac{8}{9}$.

13. The speed of a train is proportional to the number of clicks per minute made by its wheels on the rails.
   a) Copy and complete the following table.

Number of clicks per minute	90	120	▥	225
Speed of train in miles per hour	30	▥	60	▥

   b) What is the ratio of the speed of the train in miles per hour to the number of clicks per minute?
   c) How does the speed of the train vary with respect to the number of clicks per minute?
   d) Write a formula for the speed of the train in miles per hour, s, in terms of the number of clicks per minute, n.

Set IV

The *natural scale* of a map is the ratio of 1 unit of length on the map to the number of identical units of length that it represents. A natural scale of $\frac{1}{1,000}$, for example, would mean that 1 inch on a map represents an actual distance of 1,000 inches.

Can you figure out the natural scale of this map, given that the distance between Oxnard Street and Burbank Boulevard is 0.5 mile and that 1 mile = 5,280 feet?

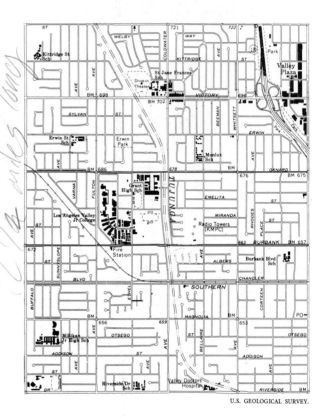

U.S. GEOLOGICAL SURVEY.

LESSON 2
Equations Containing Fractions

One of the first people to use symbols to write and solve equations was a Greek mathematician named Diophantus. Although historians do not know when Diophantus was born or when he died, we do know exactly how long he lived!

The reason for this is that one of his admirers made up a riddle about his life.

According to the riddle, one-sixth of Diophantus' life was spent in childhood. He grew a beard after one-twelfth more and married after one-seventh more. Five years later a son was born who lived to be half as old as his father. Diophantus died four years after his son. How many years did Diophantus live?

Letting x represent the answer to this equation, we can write the equation,

$$\frac{x}{6} + \frac{x}{12} + \frac{x}{7} + 5 + \frac{x}{2} + 4 = x$$

A convenient way to solve this equation, or any equation that contains fractions, is to begin by multiplying both sides by a number or expression that will clear it of all of the fractions. The simplest number that can be used for this purpose is the least common denominator of the fractions. In this case, it is $12 \cdot 7 = 84$.

Multiplying both sides of the equation by 84, we get

$$84\left(\frac{x}{6} + \frac{x}{12} + \frac{x}{7} + 5 + \frac{x}{2} + 4\right) = 84x$$

$$\frac{84x}{6} + \frac{84x}{12} + \frac{84x}{7} + 84(5) + \frac{84x}{2} + 84(4) = 84x$$

713

Simplifying the result, we get

$$14x + 7x + 12x + 420 + 42x + 336 = 84x$$
$$75x + 756 = 84x$$
$$756 = 9x$$
$$x = \frac{756}{9} = 84$$

This method can be applied to solving any equation containing fractions, as the following examples illustrate.

EXAMPLE 1

Solve for x: $\quad \dfrac{x}{4} + \dfrac{x-1}{3} = 2$

SOLUTION
The least common denominator of the fractions is 12. Multiplying both sides of the equation by 12, we get

$$12\left(\frac{x}{4}\right) + 12\left(\frac{x-1}{3}\right) = 12(2)$$

$$\frac{12x}{4} + \frac{12(x-1)}{3} = 24$$

$$3x + 4(x-1) = 24$$
$$3x + 4x - 4 = 24$$
$$7x = 28$$
$$x = 4$$

Checking this result in the original equation, we get

$$\frac{4}{4} + \frac{4-1}{3} = 1 + 1 = 2$$

EXAMPLE 2

Solve for x: $\quad \dfrac{1}{6} = \dfrac{10}{x} - \dfrac{1}{2}$.

SOLUTION

The least common denominator of the fractions is $6x$. Multiplying both sides by $6x$, we get

$$6x\left(\frac{1}{6}\right) = 6x\left(\frac{10}{x}\right) - 6x\left(\frac{1}{2}\right)$$

$$\frac{6x}{6} = \frac{60x}{x} - \frac{6x}{2}$$

$$x = 60 - 3x$$
$$4x = 60$$
$$x = 15$$

Show that this result is correct by checking it in the original equation.

Exercises

Set I

1. Find the following sums and products, giving each answer in scientific notation.
   a) $(3 \times 10^4) + (5 \times 10^4)$
   b) $(3 \times 10^4)(5 \times 10^4)$
   c) $(4 \times 10^3) + (5 \times 10^4)$
   d) $(4 \times 10^3)(5 \times 10^4)$

2. Write each of the following in decimal form.
   a) $\dfrac{27}{1000}$
   c) $\dfrac{1}{37}$
   b) $\dfrac{1}{36}$
   d) $\dfrac{3}{110}$
   e) $\dfrac{1}{44}$

© KING FEATURES SYNDICATE INC. 1967.

3. Suppose that Mr. Dithers is now 42 years older than Dagwood and that nine years ago he was three times as old as Dagwood was at the time.

   a) Use this information to write a pair of simultaneous equations, letting x represent Mr. Dithers's present age and y represent Dagwood's present age.
   b) Solve the equations to find out each one's present age.

Set II

4. Express each of the following products as a polynomial in simplest form.

a) $4\left(5 + \dfrac{x}{4}\right)$

b) $6\left(\dfrac{x}{2} - \dfrac{1}{3}\right)$

c) $x\left(\dfrac{8}{x} + 2\right)$

d) $5x\left(\dfrac{x}{5} - \dfrac{3}{x}\right)$

e) $9\left(\dfrac{x-1}{9} + \dfrac{x+9}{3}\right)$

f) $2x^2\left(\dfrac{3}{2} + \dfrac{2}{x} - \dfrac{1}{x^2}\right)$

5. Write the equation that results from multiplying both sides of the equation

$$\frac{x}{6} - 2 = \frac{x}{8}$$

a) by 6.
b) by 8.
c) by 24.
d) Which one of the three equations that you have written would be the easiest to solve?
e) Why?

6. Solve the following equations. Check your answers.

a) $\dfrac{2}{x} = 10$

b) $\dfrac{6}{x-1} = 4$

c) $\dfrac{11}{x} + 2 = \dfrac{5}{x}$

d) $\dfrac{x}{3} = \dfrac{x}{2} - 7$

e) $\dfrac{1}{3} - \dfrac{4}{x} = \dfrac{1}{x}$

f) $\dfrac{8}{x} + \dfrac{1}{8x} = 0$

g) $\dfrac{5}{x+3} - \dfrac{2}{3} = 1$

h) $\dfrac{7x-2}{x} + \dfrac{x+2}{x} = x$

i) $\dfrac{x-2}{5} + \dfrac{x+5}{2} = \dfrac{x}{10}$

j) $\dfrac{3x+1}{4} - \dfrac{x}{5} = x - 2$

7. Solve the following pairs of simultaneous equations. Check your answers.

a) $\dfrac{x}{3} + \dfrac{y}{2} = 9$

$\dfrac{x}{5} - \dfrac{y}{4} = 1$

b) $\dfrac{x+1}{7} - \dfrac{y+5}{3} = 3$

$\dfrac{x-3}{2} + \dfrac{y-1}{8} = 0$

Set III

8. Express each of the following products as a polynomial in simplest form.

a) $3\left(\dfrac{x}{3} + 7\right)$

b) $10\left(\dfrac{1}{5} - \dfrac{x}{2}\right)$

c) $x\left(3 + \dfrac{4}{x}\right)$

d) $6x\left(\dfrac{x}{6} - \dfrac{2}{x}\right)$

e) $8\left(\dfrac{x+5}{2} + \dfrac{x-1}{8}\right)$

f) $4x^2\left(\dfrac{1}{4} - \dfrac{2}{x} - \dfrac{3}{x^2}\right)$

9. Write the equation that results from multiplying both sides of the equation

$$3x - \frac{1}{4} = \frac{x}{6}$$

a) by 4.
b) by 6.
c) by 12.
d) Which one of the three equations that you have written would be the easiest to solve?
e) Why?

10. Solve the following equations. Check your answers.

a) $\dfrac{3}{x} = 12$

b) $\dfrac{x}{x + 1} = 7$

c) $\dfrac{9}{x} - 4 = \dfrac{1}{x}$

d) $\dfrac{1}{x} + \dfrac{5}{2} = \dfrac{2}{x}$

e) $\dfrac{x}{4} = \dfrac{x}{5} + 1$

f) $\dfrac{10}{x - 3} - \dfrac{6}{x - 3} = 8$

g) $\dfrac{2}{5x} - \dfrac{5}{x} = 0$

h) $\dfrac{15}{x + 4} + \dfrac{1}{3} = 2$

i) $\dfrac{x - 1}{x} + \dfrac{6x + 1}{x} = x$

j) $\dfrac{x + 8}{2} - \dfrac{x - 4}{8} = \dfrac{x}{4}$

11. Solve the following pairs of simultaneous equations. Check your answers.

a) $\dfrac{x}{2} + \dfrac{y}{5} = 20$

$\dfrac{x}{6} - \dfrac{y}{2} = 1$

b) $\dfrac{x - 6}{4} + \dfrac{y + 1}{3} = 2$

$\dfrac{x + 8}{3} + \dfrac{y - 2}{9} = 3$

Set IV

If a watermelon weighs $\dfrac{9}{10}$ of its weight and $\dfrac{9}{10}$ of a pound, how many pounds does it weigh?

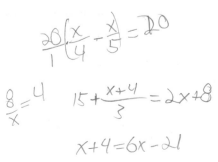

"But I digress."

LESSON 3
More on Fractional Equations

The topic of discussion in this class seems somehow to have changed from mathematics to football. Before the professor began diagramming the play, he had written a number of equations containing fractions on the board. One of them, the quadratic formula, is often needed in the solution of such equations. Here is an example illustrating how it is used.

EXAMPLE 1

Solve for x: $\dfrac{3}{x} + 1 = \dfrac{x}{2}$.

SOLUTION

Multiplying both sides of the equation by $2x$ to clear it of fractions, we get

$$2x\left(\frac{3}{x}\right) + 2x(1) = 2x\left(\frac{x}{2}\right)$$

$$6 + 2x = x^2$$

This is a quadratic equation. Writing it in standard form, we get

$$x^2 - 2x - 6 = 0$$

Because the polynomial on the left side of this equation has no simple factors, we use the quadratic formula:

$$x = \frac{-b \pm \sqrt{b^2 - 4ac}}{2a} \qquad a = 1, \quad b = -2, \quad c = -6$$

$$x = \frac{-(-2) \pm \sqrt{(-2)^2 - 4(1)(-6)}}{2(1)}$$

$$= \frac{2 \pm \sqrt{4 + 24}}{2}$$

$$= \frac{2 \pm \sqrt{28}}{2} = \frac{2 \pm 2\sqrt{7}}{2} = 1 \pm \sqrt{7}$$

The solutions of the equation are $1 + \sqrt{7}$ and $1 - \sqrt{7}$. Because $\sqrt{7}$ is irrational ($\sqrt{7} = 2.645\ldots$), the solutions are irrational and, for practical purposes, the best that we can do is approximate them. To the nearest hundredth,

$$1 + \sqrt{7} = 1 + 2.645\ldots \approx 3.65$$
$$1 - \sqrt{7} = 1 - 2.645\ldots \approx -1.65$$

One way to check approximate solutions such as these is to substitute them into the original equation and evaluate each side with a calculator to see if the two sides are approximately equal. Checking 3.65 in this way, we get

$$\frac{3}{3.65} + 1 \approx 1.82 \quad \text{and} \quad \frac{3.65}{2} \approx 1.83$$

Although a fractional equation may lead to a polynomial equation having more than one solution, the solutions may not be solutions of the original equation. The next example shows how this can happen.

EXAMPLE 2

Solve for x: $x - \dfrac{12}{x+2} = \dfrac{6x}{x+2}$.

SOLUTION

Multiplying both sides by $x + 2$, we get

$$(x+2)x - 12 = 6x$$
$$x^2 + 2x - 12 = 6x$$
$$x^2 - 4x - 12 = 0$$

Factoring the left side,

$$(x-6)(x+2) = 0$$

Either $x - 6 = 0$ or $x + 2 = 0$ so either $x = 6$ or $x = -2$. But -2 cannot be a solution of the original equation because it would make the denominators of the fractions equal to zero. The number 6, on the other hand, is a solution because

$$6 - \frac{12}{6+2} = \frac{6(6)}{6+2}$$

$$6 - \frac{12}{8} = \frac{36}{8}$$

$$6 - 1.5 = 4.5$$

Because the steps in solving a fractional equation may lead to numbers that are not solutions of the equation, it is always important to check each one.

Exercises

Set I

1. Write each of the following numbers without using any exponents.
   a) $(-1)^{100}$
   b) -1^{100}
   c) 100^0
   d) 100^{-1}

2. Solve the following equations.
   a) $4x + 5 = 3$
   b) $\sqrt{4x + 5} = 3$
   c) $(4x + 5)^2 = 3$

3. The amount of punch in a bowl is a function of its depth. A typical formula is

$$v = \frac{\pi d^2 (100 - d)}{3,000}$$

in which d is the depth of the punch in centimeters and v is the volume of the punch in liters.
a) Find the exact volume of the punch in the bowl if its depth is 10 centimeters.
b) Find the approximate volume to the nearest tenth of a liter.

Set II

4. Solve the following equations. Check your answers.

a) $\dfrac{x + 2}{3} = \dfrac{7}{x - 2}$

b) $\dfrac{3}{2x - 1} + 1 = x$

c) $\dfrac{x}{10} - \dfrac{4}{x} = \dfrac{3}{10}$

d) $x + \dfrac{x}{x - 6} = \dfrac{6}{x - 6}$

5. Solve the following equations. Express each answer in decimal form correct to the nearest hundredth. Use a calculator or refer to the table on page 563.

a) $\dfrac{4}{x - 4} = \dfrac{x}{x + 1}$

b) $\dfrac{3}{x} + \dfrac{1}{x + 2} = 1$

c) $\dfrac{x}{3} + \dfrac{1}{9} = \dfrac{x + 1}{7}$

d) $\dfrac{3x}{x - 1} - \dfrac{1}{x^2 - 1} = 2$

6. It takes Obtuse Ollie, working at a steady rate, 80 minutes to wax his car. Acute Alice can wax the car in 60 minutes. Find how long it would take them to do the job working together by answering each of the following.

a) What fraction of the job can Ollie do in one minute? What fraction of the job can Alice do in one minute?

Suppose that x represents the number of minutes that it takes them to wax the car if they work together.

b) Write expressions representing the fraction of the job that Ollie does in the x minutes and the fraction of the job that Alice does in the x minutes.

c) Write an equation expressing the fact that the sum of the two fractions of the job done by Ollie and Alice during the x minutes is 1.

d) Solve the equation to find how long it would take them working together to wax the car.

7. Miss Marple rode her bicycle from her house to Inspector Craddock's at an average speed of 15 kilometers per hour and back home again at an average speed of 10 kilometers per hour. Find her average speed for the roundtrip by doing each of the following.

a) Letting x represent the distance one way in kilometers, write expressions representing the time that Miss Marple spent going each way. Use the fact that time = $\dfrac{\text{distance}}{\text{rate}}$.

b) Write an expression for the total time spent on the roundtrip.

c) Find her average speed by using the fact that average rate = $\dfrac{\text{total distance}}{\text{total time}}$.

Set III

8. Solve the following equations. Check your answers.

a) $\dfrac{x-1}{5} = \dfrac{7}{x+1}$

b) $\dfrac{1}{3x+2} + x = \dfrac{2}{3}$

c) $\dfrac{1}{x} - \dfrac{x}{8} = \dfrac{1}{4}$

d) $x - \dfrac{4x}{x+3} = \dfrac{12}{x+3}$

9. Solve the following equations. Express each answer in decimal form correct to the nearest hundredth. Use a calculator or refer to the table on page 563.

a) $\dfrac{x}{x+3} = \dfrac{3}{x-1}$

b) $\dfrac{2}{x-4} + \dfrac{1}{x} = 2$

c) $\dfrac{2x}{x+5} - \dfrac{3}{x-2} = 1$

d) $\dfrac{6}{x^2-4} + \dfrac{x}{x+2} = 3$

10. It takes Mr. Gildersleeve, working at a steady rate, 20 minutes to mow his front lawn. His nephew Leroy can mow the lawn in 15 minutes. Find how long it would take them to do the job working together by answering each of the following.
a) What fraction of the lawn can Mr. Gildersleeve mow in one minute? What fraction of the lawn can Leroy mow in one minute?

Suppose that x represents the number of minutes that it takes them to mow the lawn if they work together.

b) Write expressions representing the fraction of the lawn that Mr. Gildersleeve mows in the x minutes and the fraction of the lawn that Leroy mows in the x minutes.
c) Write an equation based upon the fact that the sum of the two fractions of the lawn mowed by Mr. Gildersleeve and Leroy during the x minutes is 1.
d) Solve the equation to find how long it would take them working together to mow the lawn.

11. It took Obtuse Ollie 1 hour to paddle a canoe 8 kilometers up a river and back again. If Ollie can go 18 kilometers per hour in still water, find the speed of the current by answering each of the following.

a) Letting x represent the speed of the current, write expressions representing the rate of the canoe with the current and against it.
b) Using the fact that time = $\dfrac{\text{distance}}{\text{rate}}$, write expressions representing the time that Ollie spent going with the current and against it.
c) Use the fact that the total time is 1 hour to write an equation.
d) Solve the equation to find the speed of the current.

Set IV

The following problem appears in the Rhind papyrus, thought to have been written about 1650 B.C.

> "If a certain number, two-thirds of it, half of it, and a seventh of it are added together, the result is 97. What is the number?"

Can you write an equation for this problem and solve it to find the number?

B.C. BY PERMISSION OF JOHNNY HART AND FIELD ENTERPRISES, INC.

LESSON 4
Solving Formulas

The warmer it is, the faster crickets chirp. The Fahrenheit temperature can be found fairly accurately, in fact, by the method described by B.C. in the cartoon above. Written as a formula, it is

$$F = \frac{n}{4} + 40$$

in which F represents the temperature in degrees Fahrenheit and n represents the number of chirps per minute.

Now that the United States is changing over to the metric system, temperatures will be measured on the Celsius scale instead. How can the temperature in degrees Celsius be found by counting cricket chirps? A formula relating the two temperature scales is

$$C = \frac{5}{9}(F - 32)$$

in which C represents the temperature in degrees Celsius and F represents the temperature in degrees Fahrenheit.

To find a formula for the Celsius temperature in terms of cricket chirps, we can use the fact that $F = \dfrac{n}{4} + 40$ to substitute $\dfrac{n}{4} + 40$ for F in the Celsius-Fahrenheit formula:

$$C = \frac{5}{9}(F - 32)$$

$$C = \frac{5}{9}\left(\frac{n}{4} + 40 - 32\right)$$

$$C = \frac{5}{9}\left(\frac{n}{4} + 8\right)$$

$$C = \frac{5}{36}n + \frac{40}{9}$$

Unfortunately, this formula is much more complicated than the one for the Fahrenheit temperature. We can simplify it by observing that $\dfrac{5}{36} \approx \dfrac{1}{7}$ and $\dfrac{40}{9} \approx 4$:

$$C \approx \frac{n}{7} + 4$$

The Fahrenheit temperature formula is not exact to begin with; so the Celsius formula is about as accurate.

Although there is no such thing as "metric crickets," we can use the Celsius formula to find the Celsius temperature from the number of chirps per minute. For example, if a cricket chirps 140 times per minute, we have

$$C = \frac{140}{7} + 4$$
$$= 20 + 4$$
$$= 24$$

The temperature is about 24°C.

Many formulas, like the ones in this lesson, contain fractions. More examples are given on the next page.

EXAMPLE 1

Solve for x in terms of a and b: $\dfrac{a}{x} = \dfrac{x}{b}$. Assume that a and b are positive numbers.

SOLUTION

Because this is a proportion, we can multiply means and extremes, getting

$$x^2 = ab$$

If a and b are positive numbers, then so is ab. So we can take square roots, getting

$$x = \pm\sqrt{ab}$$

EXAMPLE 2

Solve for x in terms of a and b: $\dfrac{a}{x} + \dfrac{a}{b} = \dfrac{b}{x}$.

SOLUTION

Multiplying both sides by bx to get rid of the fractions, we get

$$\frac{bxa}{x} + \frac{bxa}{b} = \frac{bxb}{x}$$

$$ba + xa = b^2$$

Subtracting ba (which is the same as ab) from both sides,

$$xa = b^2 - ab$$

and dividing by a, we find that

$$x = \frac{b^2 - ab}{a}$$

Exercises

Set I

1. Write each of the following as a single fraction.

    a) $\dfrac{1}{2x} - \dfrac{1}{x^2}$ c) $\dfrac{x^2}{7-x} + x$

    b) $\dfrac{1}{2x} \cdot \dfrac{1}{x^2}$ d) $\dfrac{x^2}{7-x} \div x$

2. Solve the following equations.

    a) $\dfrac{1}{2x} = \dfrac{1}{x^2}$

    b) $\dfrac{1}{2x} \cdot \dfrac{1}{x^2} = 0$

    c) $\dfrac{x^2}{7-x} = x$

    d) $\dfrac{x^2}{7-x} \div x = 0$

HALE OBSERVATORIES.

3. The planet Saturn is surrounded by three rings, the largest of which has a diameter of 2.7×10^5 kilometers. Find the approximate circumference of this ring, expressing your answer in scientific notation.

Set II

4. The batting average of a baseball player is given by the formula

    $$a = \dfrac{h}{b}$$

    in which a represents the average, h represents the number of hits, and b represents the number of times at bat.

    a) Solve this formula for h in terms of the other variables.
    b) Ty Cobb went to bat 11,400 times, making a batting average of 0.367. How many hits did he make? (Because the batting average is approximate, the number of hits given by the formula is also approximate. Round this number to the nearest integer.)

5. The figure below shows that the graph of the function $y = 2x + 6$ crosses the x-axis at -3.

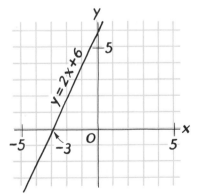

    a) Find a formula for the coordinate of the point in which the function $y = ax + b$

crosses the *x*-axis by solving the equation $ax + b = 0$ for x in terms of a and b.

b) Check your formula by seeing if it gives the correct number for the line in the graph above.

c) Use your formula to find the coordinate of the point in which the function $y = 5x - 3$ crosses the *x*-axis.

6. Solve the following equations for *x*.

a) $\dfrac{x}{2} + 5 = 0$ d) $\dfrac{1}{x} - \dfrac{1}{4} = 0$

b) $\dfrac{x}{7} + 3 = 0$ e) $\dfrac{1}{x} - \dfrac{1}{9} = 0$

c) $\dfrac{x}{a} + b = 0$ f) $\dfrac{1}{x} - \dfrac{1}{a} = 0$

7. Solve the following equations for *x* in terms of the other variables. Simplify your answers as much as possible.

a) $\dfrac{a}{x} = \dfrac{a}{b}$ c) $\dfrac{x - a}{a} = \dfrac{a}{b}$

b) $ax - b = 1$ d) $\dfrac{1}{x + a} = \dfrac{1}{a}$

e) $\dfrac{a}{x} + \dfrac{b}{x} = c$

f) $\dfrac{1}{x + a} = \dfrac{1}{2x - b}$

g) $\dfrac{x}{a - 1} = \dfrac{a + 1}{x}$

h) $a^2 + b^2 = c^2 x$

8. The formula for the Fahrenheit temperature, F, as a function of the number of chirps, n, made by a cricket in a minute is

$$F = \frac{n}{4} + 40$$

a) What is the Fahrenheit temperature when a cricket makes 60 chirps per minute?

b) Solve the formula for n in terms of F.

c) How many chirps per minute does a cricket make when the temperature is 72°F?

d) The formula does not work for temperatures below a certain number. What is it?

Set III

9. A person's I.Q. is given by the formula

$$i = \frac{100m}{p}$$

in which i represents the I.Q., m represents the mental age, and p represents the physical age.

a) Solve this formula for m in terms of the other variables.

b) Mozart is thought to have had an I.Q. of 165. What was his mental age when he was 12 years old?

10. The figure at the right shows that the graph of the function $y = 3x - 6$ crosses the *x*-axis at 2.

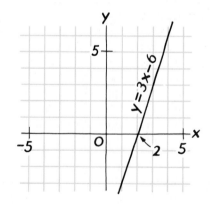

a) Find a formula for the coordinate of the point in which the function $y = ax - b$ crosses the *x*-axis by solving the equation $ax - b = 0$ for x in terms of a and b.

b) Check your formula by seeing if it gives the correct number for the line in the graph above.

c) Use your formula to find the coordinate of the point in which the function $y = 2x - 9$ crosses the x-axis.

11. Solve the following equations for x.

a) $\dfrac{x}{3} - 4 = 1$ d) $\dfrac{1}{x} + \dfrac{1}{2} = 0$

b) $\dfrac{x}{7} - 10 = 1$ e) $\dfrac{1}{x} + \dfrac{1}{8} = 0$

c) $\dfrac{x}{a} - b = 1$ f) $\dfrac{1}{x} + \dfrac{1}{a} = 0$

12. Solve the following equations for x in terms of the other variables. Simplify your answers as much as possible.

a) $\dfrac{x}{a} = \dfrac{a}{b}$ c) $\dfrac{x + a}{b} = \dfrac{b}{a}$

b) $ax + b = 0$ d) $\dfrac{1}{x - a} = \dfrac{1}{b}$

e) $\dfrac{a}{x} - \dfrac{b}{x} = a$ g) $\dfrac{x}{a} = \dfrac{a}{x}$

f) $x + \dfrac{1}{b} = \dfrac{a}{b}$ h) $a^2 - b^2 = abx$

13. The formula for the Celsius temperature, C, as a function of the number of chirps, n, made by a cricket in a minute is

$$C = \dfrac{n}{7} + 4$$

a) What is the Celsius temperature when a cricket makes 84 chirps per minute?

b) Solve the formula for n in terms of C.

c) How many chirps per minute does a cricket make when the temperature is 20°C?

d) The formula does not work for temperatures below a certain number. What is it?

Set IV

In giving the temperature forecast for Little America, the radio announcer forgot to tell whether it was in degrees Fahrenheit or degrees Celsius. Strange as it may seem, it didn't matter.

Can you figure out why not and what the temperature was as well?

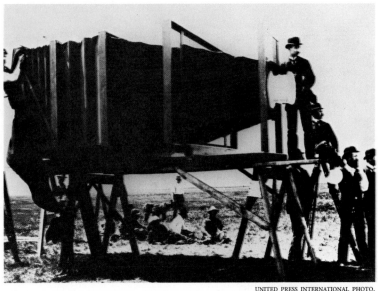

LESSON 5
More on
Solving Formulas

The largest movable camera ever built was nine feet high and twenty feet long. Pictured in this photograph, it was made in Chicago in 1900 and required fifteen people to operate. It had a lens with a focal length of ten feet, which is sixty times the focal length of an ordinary camera lens.

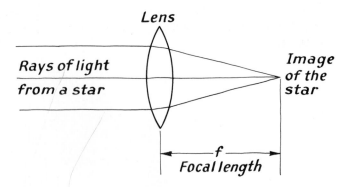

The focal length of a lens can be determined by focusing the light of a distant object, such as a star. The distance between the lens and the image of the star, illustrated in the diagram above, is the focal length, f, of the lens.

In taking a picture of a closer object, such as a person, the lens bends the light rays to form an image, as shown in the diagram below. A simple equation relates the distance from the lens of the object being photographed, a, the distance from the lens of the image, b, and the focal length, f:

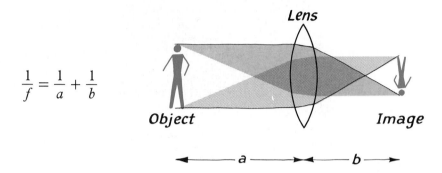

$$\frac{1}{f} = \frac{1}{a} + \frac{1}{b}$$

This equation is used in designing cameras and optical equipment and in determining the magnification of lenses.

Before using this equation, it is convenient to solve for one of the three variables, f, a, and b, in terms of the other two. For example, to solve for f, we might first clear the equation of fractions by multiplying both sides by fab:

$$\frac{fab}{f} = \frac{fab}{a} + \frac{fab}{b}$$
$$ab = fb + fa$$

Notice that f is now in *two* terms of the equation. At this point, we can factor it from both of them to get

$$ab = f(b + a)$$

Dividing both sides of the equation by $b + a$, we get

$$\frac{ab}{b + a} = f \quad \text{or} \quad f = \frac{ab}{a + b}$$

To solve an equation for one of its variables, it is frequently necessary at some point in the solution to factor the variable from the terms containing it. Here are more examples.

EXAMPLE 1

Solve for x in terms of a, b, and c: $\quad ax = bx + c$.

SOLUTION

Subtracting bx from both sides in order to get the terms containing x on the left side of the equation, we get

$$ax - bx = c$$

Factoring x from the left side, we get

$$x(a - b) = c$$

Dividing both sides by $a - b$,

$$x = \frac{c}{a - b}$$

Checking this solution in the original equation, we get

$$a\left(\frac{c}{a - b}\right) = b\left(\frac{c}{a - b}\right) + c$$

$$\frac{ac}{a - b} = \frac{bc}{a - b} + c$$

$$ac = bc + c(a - b)$$
$$ac = bc + ac - bc$$
$$ac = ac$$

EXAMPLE 2

Solve for x in terms of a and b: $\quad \dfrac{a + x}{b - x} = \dfrac{b}{a}$.

SOLUTION

Because this is a proportion, we can multiply means and extremes to get

$$a(a + x) = b(b - x)$$
$$a^2 + ax = b^2 - bx$$

To get the terms containing x on the left side, we subtract a^2

$$ax = b^2 - bx - a^2$$

and add bx

$$ax + bx = b^2 - a^2$$

Factoring x from the left side,

$$x(a + b) = b^2 - a^2$$

and dividing by $a + b$,

$$x = \frac{b^2 - a^2}{a + b}$$

This result can be simplified by factoring the numerator of the fraction

$$x = \frac{(b - a)(b + a)}{a + b}$$

and reducing it to get

$$x = b - a$$

Show that this solution is correct by checking it in the original equation.

Exercises

Set I

1. This exercise is about the function
   $y = x^3 - 2x$.
   a) What kind of function is it?
   b) Graph it from $x = -3$ to $x = 3$.

2. This exercise is about the equation
   $x^3 - 2x = 0$.

   a) What is the largest number of solutions
   that it might have?
   b) Use your graph for exercise 1 to
   estimate its solutions.
   c) Use algebraic methods to find its exact
   solutions.

3. Some world records in the hurdles are shown in the following table.

Distance in meters	110	200	400
Time in seconds	13.1	21.9	47.8

a) Draw a pair of axes, letting the x-axis represent distance and the y-axis represent time. Let 1 inch on the x-axis represent 100 meters and 1 inch on the y-axis represent 10 seconds. Plot the three points corresponding to these records as accurately as you can. Two of the points that you have plotted should seem to be in line with the origin. Draw that line.

b) What kind of function does this graph suggest?

c) The hurdles for the three distances are of different heights. On the basis of your graph, which distance do you think has the lowest hurdles?

*d) Find the average record speed for each distance, each to the nearest tenth of a meter per second.

"AT LEAST KICK THEM TO ONE SIDE OR THE OTHER, FOSTER!"

Set II

4. Solve the following equations for x.

a) $\dfrac{x-2}{x-5} = \dfrac{5}{2}$

b) $\dfrac{x-10}{x-3} = \dfrac{3}{10}$

c) $\dfrac{x-a}{x-b} = \dfrac{b}{a}$

d) $4x - 1 = \dfrac{x+1}{4}$

e) $8x - 1 = \dfrac{x+1}{8}$

f) $ax - 1 = \dfrac{x+1}{a}$

5. Solve the following equations for x in terms of the other variables. Simplify your answers as much as possible.

a) $ax = 1 - x$

b) $x - a = bx$

c) $\dfrac{x+1}{a} = \dfrac{x-1}{b}$

d) $\dfrac{x}{a} - \dfrac{x}{b} = c$

e) $\dfrac{1}{x} = \dfrac{1}{a} + b$

f) $x^2 - a^2 = b^2$

6. The area of a trapezoid can be found by multiplying the sum of the lengths of its

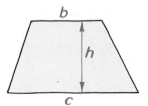

bases, b and c, by its altitude, h, and dividing the result by 2.

a) Find the area of a trapezoid whose bases are 5 and 8 units and whose altitude is 6 units.

b) Write a formula for the area of a trapezoid, a, in terms of b, c, and h.

c) Solve your formula for h in terms of a, b, and c.

d) Find the altitude of a trapezoid whose area is 24 square units and whose bases are 3 and 7 units.

*7. A woman's blood pressure depends on her age. The normal blood pressure in millimeters of mercury of a woman who is A years old is given by the formula

$$p = \frac{A(A + 5)}{100} + 107$$

a) Copy and complete the following table. Round each number to the nearest integer.

Age, A	20	35	50	65
Normal blood pressure, p	112			

b) What happens to a woman's blood pressure as she grows older?

c) Does it change at a steady rate?

8. The formula relating the focal length of a camera lens, f, to the distances of the object and image from the lens, a and b, is

$$f = \frac{1}{\dfrac{1}{a} + \dfrac{1}{b}}$$

a) Find the focal length of a lens for which $b = 8$ centimeters if $a = 12$ centimeters.

b) Solve the formula for a in terms of b and f.

c) Find a for a lens for which $b = 10$ centimeters and $f = 6$ centimeters.

Set III

9. Solve the following equations for x.

a) $\dfrac{4 - x}{1 + x} = \dfrac{1}{4}$

b) $\dfrac{2 - x}{7 + x} = \dfrac{7}{2}$

c) $\dfrac{a - x}{b + x} = \dfrac{b}{a}$

d) $3x - 1 = \dfrac{x - 1}{3}$

e) $9x - 1 = \dfrac{x - 1}{9}$

f) $ax - 1 = \dfrac{x - 1}{a}$

10. Solve the following equations for x in terms of the other variables. Simplify your answers as much as possible.

a) $ax - b = x$

b) $\dfrac{x - a}{b} = \dfrac{x + b}{a}$

c) $ax + a = bx + b$

d) $\dfrac{x}{a} + \dfrac{x}{b} = 1$

e) $x^2 + a^2 = b^2$

f) $\dfrac{1}{x} = a - \dfrac{1}{b}$

11. The radius of the circle in this figure can

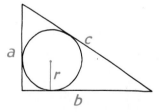

be found by subtracting the longest side of the triangle, c, from the sum of the other two sides, a and b, and dividing the result by 2.

a) Find the radius of the circle if the sides of the triangle are 6, 8, and 10 units.

b) Write a formula for the radius of the circle, r, in terms of a, b, and c.

c) Solve your formula for c in terms of a, b, and r.

d) Find the longest side of the triangle if its other sides are 20 and 21 and the radius of the circle is 6.

*12. A man's blood pressure depends on his age. The normal blood pressure in millimeters of mercury of a man who is A years old is given by the formula

$$p = \frac{A(3A - 10)}{500} + 120$$

a) Copy and complete the following table. Round each number to the nearest integer.

Age, A	20	35	50	65
Normal blood pressure, p	122	▓	▓	▓

b) What happens to a man's blood pressure as he grows older?

c) Does it change at a steady rate?

13. The formula relating the focal length of a camera lens, f, to the distances of the object and image from the lens a and b is

$$f = \frac{1}{\dfrac{1}{a} + \dfrac{1}{b}}$$

a) Find the focal length of a lens for which $b = 9$ centimeters if $a = 21$ centimeters.

b) Solve the formula for b in terms of a and f.

c) Find b for a lens for which $a = 20$ centimeters and $f = 4$ centimeters.

Set IV The Golden Rectangle

The golden rectangle has influenced artists as well as architects.* In the drawing by Leonardo da Vinci shown here, the proportions of part of

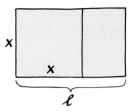

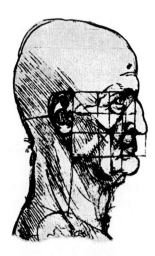

the face seem to be based on a system of golden rectangles.

When a golden rectangle is divided into a square and small rectangle as shown above, the dimensions of the small rectangle have the same ratio as the corresponding dimensions of the golden rectangle.

1. Write a proportion expressing this fact.

2. Solve the proportion for x in terms of ℓ.

* 3. Find the value of x if $\ell = 10$. Express your answer in decimal form, correct to the nearest hundredth.

———————

*An example of the golden rectangle in architecture is on page 587.

Summary and Review

In this chapter, we have learned how to solve equations that contain fractions and how to solve formulas containing two or more variables for one of the variables.

Ratio and Proportion (*Lesson 1*) The ratio of the numbers a to b is the number, $\frac{a}{b}$. A proportion is an equation stating that two ratios are equal. An easy way to solve equations that are proportions is to use the fact that the product of the means of a proportion is equal to the product of the extremes.

Equations Containing Fractions (*Lessons 2 and 3*) An equation that contains fractions can be solved by first multiplying both sides by a number that will clear the equation of all of the fractions. The simplest way to do this is to multiply by the least common denominator of the fractions.

A fractional equation may have more than one solution, depending on the degree of the equation that results when all fractions are cleared. The solutions of the resulting equation, however, may not necessarily be solutions of the original equation and should therefore be checked to see if they make it true.

Solving Formulas (*Lessons 4 and 5*) The methods used to solve fractional equations can also be used to solve formulas containing two or more variables for one of the variables. If the variable being solved for is in more than one term of the equation, it should ordinarily be factored out of the terms containing it.

Exercises

Set I

"IT'S AN UNUSUALLY SMALL CROWD FOR
A SATURDAY NIGHT."

1. There are two "large" people in this cartoon and fifteen people altogether.
   a) Write the ratio of the number of large people to the number of small people.
   b) Suppose that later in the evening there are four "large" people in the bar and twenty-six small ones. Write a proportion to show that the ratio of the number of large people to the number of small people has not changed.
   c) What are the means of your proportion?
   d) What are the extremes of your proportion?
   e) How does the product of the means compare with the product of the extremes?

2. If possible, simplify the following ratios.

   a) $\dfrac{5x}{15x}$

   b) $\dfrac{2x - 2y}{3x - 3y}$

   c) $\dfrac{x + 12}{x + 2}$

   d) $\dfrac{4x^4}{5x^5}$

3. Express each of the following products as a polynomial in simplest form.

   a) $8\left(\dfrac{x}{8} - 2\right)$

   b) $12\left(\dfrac{1}{3} + \dfrac{x}{4}\right)$

   c) $x\left(7 - \dfrac{3}{x}\right)$

   d) $2x\left(\dfrac{x + 5}{2} + \dfrac{10}{x}\right)$

4. Solve the following equations.

a) $\dfrac{5}{6} = \dfrac{x}{3}$

b) $\dfrac{1}{x-7} = 2$

c) $\dfrac{3x}{x-1} = \dfrac{6x}{2x-2}$

d) $\dfrac{x}{x+4} = \dfrac{x-1}{x+2}$

e) $\dfrac{x}{3} - 5 = \dfrac{x}{6}$

f) $\dfrac{7}{x} + \dfrac{1}{8} = \dfrac{5}{x}$

g) $\dfrac{1}{3x+2} + \dfrac{1}{x+2} = 0$

h) $\dfrac{x-3}{15} = \dfrac{1}{x+3}$

i) $\dfrac{1}{x} - \dfrac{1}{x-1} = 5$

j) $2 - \dfrac{4}{3x+1} = x$

5. The volume of a cone can be found by the formula

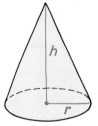

$$v = \dfrac{\pi r^2 h}{3}$$

in which r represents the radius of its base and h represents its altitude.

a) Find the exact volume of a cone for which $r = 6$ centimeters and $h = 15$ centimeters.

b) Solve the formula for h in terms of the other variables.

c) Find the altitude of a cone for which $v = 256\pi$ cubic centimeters and $r = 8$ centimeters.

6. Solve the following equations for x in terms of the other variables. Simplify your answers as much as possible.

a) $\dfrac{a+1}{x} = \dfrac{x}{a-1}$

b) $a^2 x + a^2 = a$

c) $2\pi x = a\pi$

d) $\dfrac{a}{x} = \dfrac{b}{x-1}$

e) $\dfrac{1}{a} - \dfrac{b}{x} = b$

7. A temperature in degrees Fahrenheit can be changed into degrees Celsius by using the formula

$$C = \dfrac{5}{9}(F - 32)$$

a) The normal body temperature of a human being is 98.6°F. Use the formula above to change this temperature into degrees Celsius.

b) Solve the formula for F in terms of C.

c) The melting point of sugar is about 160°C. Use your formula to change this temperature into degrees Fahrenheit.

8. The number of different connections that can be made through a telephone switchboard can be found by the formula

$$n = \dfrac{x(x-1)}{2}$$

in which x represents the number of telephones.

a) How many different connections are possible for a switchboard containing 20 telephones?

b) Find a formula for the number of telephones that a switchboard contains if n different connections are possible by solving the formula above for x in terms of n. (Remember that x cannot be negative.)

*c) How many telephones does a switchboard contain if 3,160 different connections are possible?

Set II

1. This is a picture of an old African game called Wari.

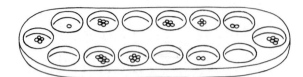

The board consists of a set of bowls that contain varying numbers of stones as the game is played.

Give each of the following ratios in simplest terms.
   a) The ratio of the number of bowls containing stones to the total number of bowls.
   b) The ratio of the number of bowls containing stones to the number of empty bowls.
   c) The ratio of the total number of bowls to the number of empty bowls.

2. If possible, simplify the following ratios.

   a) $\dfrac{4}{12x - 4}$

   b) $\dfrac{2x + 10}{5x + 10}$ d) $\dfrac{x^2 + 1}{x^2 - 1}$

   c) $\dfrac{x - 6}{3x - 18}$

3. Express each of the following products as a polynomial in simplest form.

   a) $5\left(9 + \dfrac{x}{5}\right)$

   b) $14\left(\dfrac{x}{2} - \dfrac{1}{7}\right)$

   c) $x\left(\dfrac{4}{x} + 6\right)$

   d) $3x\left(\dfrac{x - 1}{3} - \dfrac{8}{x}\right)$

4. Solve the following equations.

   a) $\dfrac{10}{7} = \dfrac{7}{x}$

   b) $\dfrac{1}{x - 1} = 4$

   c) $\dfrac{4}{x + 5} = \dfrac{8}{2x + 10}$

   d) $\dfrac{x}{x - 3} = \dfrac{x + 5}{x - 1}$

   e) $\dfrac{x}{2} + 9 = \dfrac{x}{4}$

   f) $\dfrac{1}{2x} - \dfrac{1}{6} = \dfrac{3}{x}$

   g) $\dfrac{1}{4x - 3} + \dfrac{1}{x - 3} = 0$

   h) $\dfrac{2x + 1}{13} = \dfrac{3}{2x - 1}$

   i) $\dfrac{1}{x + 2} - \dfrac{1}{x} = 10$

   j) $\dfrac{5}{x} - \dfrac{x - 1}{2} = \dfrac{5}{4}$

5. The volume of an ostrich egg can be found by the formula

   $$v = \frac{4\pi a^2 b}{3}$$

   in which a and b represent the lengths shown in the figure.
   a) Find the exact volume of an ostrich egg for which $a = 2$ centimeters and $b = 3$ centimeters.
   b) Solve the formula for b in terms of the other variables.
   c) Find the length of an ostrich egg ($2b$) for which $v = 48\pi$ cubic centimeters and $a = 3$ centimeters.

6. Solve the following equations for x in terms of the other variables. Simplify your answers as much as possible.

a) $\dfrac{x}{a - b} = \dfrac{a + b}{b^2}$

b) $ax - a = a^2$

c) $\pi x^2 = a\pi$ (Assume that $a > 0$.)

d) $\dfrac{a}{x + 1} = \dfrac{b}{x}$

e) $\dfrac{1}{x} - \dfrac{1}{a} = b$

7. A temperature in degrees Fahrenheit can be changed into degrees Kelvin by using the formula

$$K = \frac{5}{9}(F + 460)$$

a) The freezing point of water is $32°F$. Use the formula above to change this temperature into degrees Kelvin.

b) Solve the formula for F in terms of K.

c) The temperature of a campfire is approximately 1,075 degrees Kelvin. Use your formula to change this temperature into degrees Fahrenheit.

8. If three people work together on a job, the amount of time that it takes them is less than the amount of time that it would take any one of them to do the job alone. If they can do the job alone in a, b, and c hours respectively, and it takes them x hours working together, then the times are related by the equation

$$\frac{x}{a} + \frac{x}{b} + \frac{x}{c} = 1$$

a) How long would it take three people working together to do a job if they can do it alone in 3, 4, and 6 hours respectively?

b) Solve the equation for x in terms of a, b, and c.

c) Check your equation to see if it gives the correct answer for part a.

Chapter 16
INEQUALITIES

© 1975 BY NEA, INC.

LESSON 1
Inequalities

". . . And you haven't heard the best part. Wait till I tell you about gas mileage . . ."

The gas mileage of a car depends on where the car is driven. A car that gets as little as 18 miles per gallon in the city, for instance, might get as much as 34 miles per gallon on the open road. Letting x represent the car's gas mileage, this variation can be written in symbols as

$$18 \leq x \leq 34$$

It can also be represented on a number line by shading the points from 18 to 34 inclusive.

15   20   25   30   35

The symbol $\leq$ means *is less than or equal to* and is a combination of the symbol for equality, $=$, and one of the symbols for inequality, $<$. In like manner, the symbol $\geq$ means *is greater than or equal to*. When we write $18 \leq x \leq 34$, we are indicating that "18 is less than or equal to x" *and* "x is less than or equal to 34," or, as the number line indicates, that "x lies between 18 and 34 inclusive."

We now have five symbols for comparing numbers. Their meanings are illustrated in the following table.

Symbol	Meaning	For points on a line	Picture
$a = b$	a is equal to b	a and b are identical	
$a < b$	a is less than b	a is to the left of b	
$a > b$	a is greater than b	a is to the right of b	
$a \leq b$	a is less than *or* equal to b	a is to the left of *or* identical to b	
$a \geq b$	a is greater than *or* equal to b	a is to the right of *or* identical to b	

A mathematical sentence that contains any of the symbols $<$, $>$, $\leq$, or $\geq$ is called an **inequality.**

Whether an inequality containing one or more variables is true or false generally depends on what numbers we replace the variables with. For example, the numbers that can replace x in the inequality $x \geq 5$ to make it true are 5 and all numbers greater than 5. These numbers can be pictured by drawing a solid circle at 5 on a number line and shading the points to the right of 5.

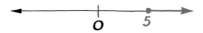

The numbers that can replace x in the inequality $x < 2$ to make it true are all numbers less than 2. These numbers can be pictured by drawing an open circle at 2 on a number line (to show that 2 is not included) and shading the points to the left of 2.

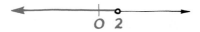

Sometimes we may combine two equations or inequalities to form a third one. For example, if $x < y$ and $y < z$, then it follows that $x < z$. It is easy to see with a number line why this is true. The inequalities $x < y$ and $y < z$ mean that x is to the left of y and y is to the left of z.

If x is to the left of y and y is to the left of z, then x must be to the left of z.

You have had experience writing simple inequalities throughout your study of algebra. In this chapter, we will explore the properties of inequalities further. Here are more examples.

EXAMPLE 1
Write two inequalities illustrated
by this figure.

SOLUTION
Because -5 is to the left of x, we can write $-5 < x$.
Because x is to the right of -5, we can write $x > -5$.
(Notice that it is also true that $-5 \leq x$ and $x \geq -5$.)

EXAMPLE 2
Show by means of a number line the numbers that make
the inequality $1 \leq x < 6$ true.

SOLUTION
This inequality allows x to be any value between
1 and 6, and so we begin by marking 1 and 6
with open circles and shading the points between
them. The inequality also allows x to be equal to
1, and so we fill in the circle at 1 to show that 1
is included.

EXAMPLE 3
If $x > y$ and $y = z$, what can be concluded about x and z?

SOLUTION
Illustrating these relationships on a number line,
we can conclude that $x > z$ (or that $z < x$).

Exercises

Set I

1. Factor each of the following polynomials.
   a) $x^3 - 4x^2$
   b) $x^3 - 4x$
   c) $4x^3 - x^2$
   d) $4x^3 - x$

2. Guess a formula for the function represented by each of these tables. Begin each formula with $y =$.

   a)
x	1	2	3	4	5
y	2	6	12	20	30

   b)
x	0	1	2	3	4
y	1	2	9	28	65

   c)
x	1	2	3	4	5
y	1	$\frac{1}{4}$	$\frac{1}{9}$	$\frac{1}{16}$	$\frac{1}{25}$

   d)
x	4	5	6	7	8
y	$\frac{1}{4}$	$\frac{2}{5}$	$\frac{1}{2}$	$\frac{4}{7}$	$\frac{5}{8}$

3. The longest ski lift in the world is in New South Wales, Australia. Suppose that it carries a skier up the mountain at a speed of 6 miles per hour and that he or she skis down the mountain at a speed of 54 miles per hour.
   a) If the round trip takes 40 minutes not counting the time spent at the top and the skier spends x minutes riding up in the lift, what represents the time spent skiing down the mountain?

   b) Draw a diagram to represent the problem.
   c) Use the information in your diagram to write an equation and solve it for x.
   d) How many miles long is the run?

Set II

4. Which of these symbols, $=$, $>$, or $<$, should replace ▓ in each of the following?

   a) 9 ▓ 8
   b) -9 ▓ -8
   c) $\frac{1}{9}$ ▓ $\frac{1}{8}$
   d) $-\frac{1}{9}$ ▓ $-\frac{1}{8}$
   e) $\frac{1}{9}$ ▓ $-\frac{1}{8}$
   f) -2 ▓ 4
   g) $(-2)^2$ ▓ 4
   h) $(-2)^3$ ▓ 4
   i) $(-2)^4$ ▓ 4
   j) 0.5 ▓ 0.3
   k) $(0.5)^2$ ▓ 0.3
   l) $(-0.5)^2$ ▓ -0.3
   m) $\sqrt[3]{64}$ ▓ $\sqrt{64}$
   n) $\sqrt{1}$ ▓ $\sqrt[4]{1}$
   o) $\sqrt{0.01}$ ▓ $(0.01)^2$

5. Write two inequalities illustrated by each of the following figures.

a)   b)   c)   d)

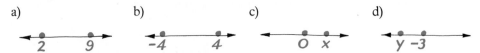

6. Tell which figure below matches each equation or inequality in parts a through f. The open circles indicate that the numbers corresponding to them are not included.

a) $x \geq 3$
b) $3 \leq x \leq 7$
c) $x < 7$
d) $3 < x < 7$
e) $x = 3$ or $x = 7$
f) $x < 3$ or $x > 7$

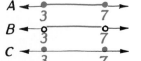

7. Which of these symbols, $>$, $=$, or $<$, should replace in each of the following to make it true?

$2x$ ▦ $5x$

a) if $x = 10$
b) if $x = 1$
c) if $x = 0$
d) if $x = -1$
e) if $x = -10$
f) if $x > 0$
g) if $x < 0$

$2 + x$ ▦ $5 + x$

h) if $x = 10$
i) if $x = 1$
j) if $x = 0$
k) if $x = -1$
l) if $x = -10$
m) if $x > 0$
n) if $x < 0$

8. Which of these symbols, $>$, $=$, or $<$, should replace ▦ in each of the following to make it true?

$2 - x$ ▦ $5 - x$

a) if $x = 10$
b) if $x = 1$
c) if $x = 0$
d) if $x = -1$
e) if $x = -10$
f) if $x > 0$
g) if $x < 0$

$\dfrac{2}{x}$ ▦ $\dfrac{5}{x}$

h) if $x = 10$
i) if $x = 1$
j) if $x = 0$
k) if $x = -1$
l) if $x = -10$
m) if $x > 0$
n) if $x < 0$

9. Which of these symbols, $=$, $>$, $<$, $\geq$, or $\leq$, should replace ▦ in each of the following to make it true for all values of x? If none will, state this.

a) x ▦ $x + 1$
b) $3x$ ▦ x
c) x^2 ▦ 0
d) x^2 ▦ x
e) $\sqrt{x}$ ▦ x
f) $6 - x$ ▦ $4 - x$

10. Draw figures to decide which of these symbols, $=$, $>$, or $<$, should replace ▦ in each of the following to make it true for all values of the variables. If none will, state this.

a) If $a > b$ and $b > c$, then a ▦ c.
b) If $d < e$ and $e = f$, then d ▦ f.
c) If $g < h$ and $h > i$, then g ▦ i.
d) If $j = k$ and $j > l$, then k ▦ l.
e) If $m > n$ and $o < n$, then m ▦ o.
f) If $p > q$ and $r > q$, then p ▦ r.

Set III

11. Which of these symbols, $=$, $>$, or $<$, should replace ▒ in each of the following?
    a) 4 ▒ 5
    b) −4 ▒ −5
    c) $\dfrac{1}{4}$ ▒ $\dfrac{1}{5}$
    d) $-\dfrac{1}{4}$ ▒ $-\dfrac{1}{5}$
    e) $\dfrac{1}{4}$ ▒ $-\dfrac{1}{5}$
    f) −3 ▒ −9
    g) $(-3)^2$ ▒ −9
    h) $(-3)^2$ ▒ 9
    i) $(-3)^3$ ▒ 9
    j) 0.2 ▒ 0.4
    k) $(0.2)^2$ ▒ 0.4
    l) $(-0.2)^2$ ▒ −0.4
    m) $\sqrt[3]{1{,}000}$ ▒ $\sqrt{100}$
    n) $\sqrt{16}$ ▒ $\sqrt[4]{16}$
    o) $\sqrt{0.16}$ ▒ 0.16

12. Write two inequalities illustrated by each of the following figures.
    a)
    c)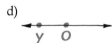
    b)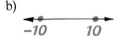
    d)

13. Tell which figure below matches each equation or inequality in parts a through f. The open circles indicate that the numbers corresponding to them are not included.

    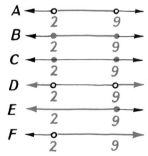

    a) $x > 2$
    b) $2 < x < 9$
    c) $x \le 9$
    d) $2 \le x \le 9$
    e) $x = 2$ or $x = 9$
    f) $x < 2$ or $x > 9$

14. Which of these symbols, $>$, $=$, or $<$, should replace ▒ in each of the following to make it true?

    $4x$ ▒ $3x$ $\dfrac{4}{x}$ ▒ $\dfrac{3}{x}$
    a) if $x = 12$
    b) if $x = 1$
    c) if $x = 0$
    d) if $x = -1$
    e) if $x = -12$
    f) if $x > 0$
    g) if $x < 0$
    h) if $x = 12$
    i) if $x = 1$
    j) if $x = 0$
    k) if $x = -1$
    l) if $x = -12$
    m) if $x > 0$
    n) if $x < 0$

15. Which of these symbols, $>$, $=$, or $<$, should replace ▒ in each of the following to make it true?

    $4 + x$ ▒ $3 + x$ $4 - x$ ▒ $3 - x$
    a) if $x = 12$
    b) if $x = 1$
    c) if $x = 0$
    d) if $x = -1$
    e) if $x = -12$
    f) if $x > 0$
    g) if $x < 0$
    h) if $x = 12$
    i) if $x = 1$
    j) if $x = 0$
    k) if $x = -1$
    l) if $x = -12$
    m) if $x > 0$
    n) if $x < 0$

16. Which of these symbols, $=$, $>$, $<$, $\ge$, or $\le$, should replace ▒ in each of the following to make it true for all values of x? If none will, state this.
    a) x ▒ $x - 1$
    b) x ▒ $2x$
    c) $\sqrt{x}$ ▒ 0
    d) x^3 ▒ x
    e) x^2 ▒ $(-x)^2$
    f) $5 - x$ ▒ $7 - x$

17. Draw figures to decide which of these symbols, $=$, $>$, or $<$, should replace ▒ in each of the following to make it true for all values of the variables. If none will, state this.
    a) If $a = b$ and $b > c$, then a ▒ c.
    b) If $d < e$ and $e < f$, then d ▒ f.
    c) If $g > h$ and $h < i$, then g ▒ i.
    d) If $j < k$ and $j = l$, then k ▒ l.
    e) If $m < n$ and $o > n$, then m ▒ o.
    f) If $p < q$ and $r < q$, then p ▒ r.

Set IV

All animals are equal
But some animals are more equal than others.
GEORGE ORWELL, *Animal Farm*

One way in which animals are *not* equal is in the length of their life-spans. The typical life-span of a horse is more than twice that of a dog. Dogs, on the other hand, live longer than either cows or pigs. If cows outlive pigs by at least eight years and the typical life-span of a pig is ten years, can you draw any conclusion about the life-span of a horse?

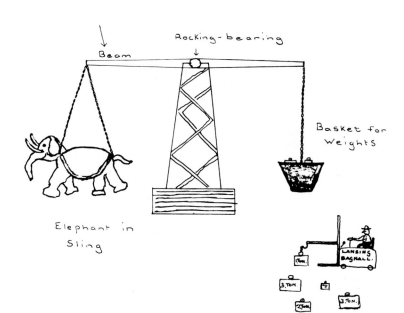

Beam

Rocking-bearing

Basket for Weights

Elephant in Sling

LESSON 2
Solving Linear Inequalities

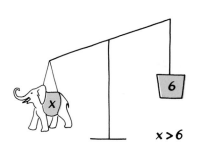

x > 6

If you were a zoo keeper and wanted to find out how heavy an elephant was, how would you do it? When some children were asked this question, one of them made the drawing shown above.* His idea was to attach the elephant to one end of a giant scale and have a fork-lift operator load weights on the other end until the two ends balance.

The weights shown in the drawing vary from 1 to 3 tons. Suppose that when 6 tons have been put into the weight basket, the scale looks like the first figure at the right and when 1 more ton has been added, the scale looks like the second figure. The weight of the elephant, x, is more than 6 tons but less than 7: in symbols, $6 < x < 7$.

Just as the idea of a balanced scale is useful in understanding equations, the idea of an unbalanced scale is useful in understanding inequalities. Suppose, for example, that a 3-pound weight is put on one side of a scale and a 2-pound weight on the other side to illustrate the inequality

$$3 > 2$$

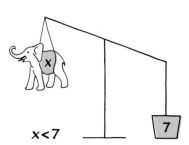

x < 7

* "Elephant-17" from *Children Solve Problems* by Edward de Bono. Copyright © 1972 by The Cognitive Trust. Reprinted by permission of Harper & Row, Publishers, Inc., and A. P. Watt & Son.

751

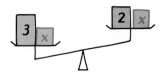

If we *add* the same amount of weight to each side, say x pounds, then clearly the scale will remain unbalanced in the same direction:

$$3 + x > 2 + x$$

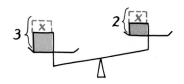

Similarly, if we *take away* the same amount of weight from each side, the scale will remain unbalanced in the same direction:

$$3 - x > 2 - x$$

These observations illustrate the following fact.

▶ Adding the same number to or subtracting the same number from both sides of an inequality results in an inequality having the same direction.

It might seem reasonable to think that multiplying or dividing an inequality by the same number would also produce an inequality having the same direction but this isn't necessarily the case. Look at the pattern at the left. Although $3 > 2$, $3x > 2x$ *only if* x *is a positive number*. If $x = 0$, then $3x = 2x$. And if x is a negative number, $3x < 2x$. The diagrams of the scales at the lower left show that 3 pounds weighs more than 2 pounds, but a balloon that can lift 3 pounds weighs less than a balloon that can lift only 2 pounds.

$$3 > 2$$
$$3 \cdot 3 > 3 \cdot 2$$
$$2 \cdot 3 > 2 \cdot 2$$
$$1 \cdot 3 > 1 \cdot 2$$
$$0 \cdot 3 = 0 \cdot 2$$
$$-1 \cdot 3 < -1 \cdot 2$$
$$-2 \cdot 3 < -2 \cdot 2$$
$$-3 \cdot 3 < -3 \cdot 2$$

▶ Multiplying both sides of an inequality by the same *positive* number results in an inequality having the *same* direction. Multiplying both sides of an inequality by the same *negative* number results in an inequality having the *opposite* direction.

The same rules hold for division.

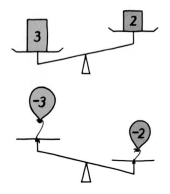

▶ Dividing both sides of an inequality by the same *positive* number results in an inequality having the *same* direction. Dividing both sides of an inequality by the same *negative* number results in an inequality having the *opposite* direction.

These rules can be used to solve inequalities in the same way that we solve equations. Examples are given on the next page.

EXAMPLE 1

Solve the inequality $x - 4 > 9$.

SOLUTION

Adding 4 to each side, we get

$$x > 13$$

This tells us that all numbers greater than 13 are solutions of $x - 4 > 9$.

EXAMPLE 2

Solve the inequality $-5x + 3 < 38$.

SOLUTION

Subtracting 3 from each side, we get

$$-5x < 35$$

Dividing each side by -5, we get

$$x > -7$$

(Remember that dividing by a negative number reverses the direction of the inequality.) All numbers greater than -7 are solutions of $-5x + 3 < 38$.

EXAMPLE 3

Solve the inequality $0 \leq \frac{x}{6} - 1 \leq 2$.

SOLUTION

This inequality tells us that $\frac{x}{6} - 1$ is greater than or equal to 0 and less than or equal to 2. Adding 1, we get

$$1 \leq \frac{x}{6} \leq 3$$

Multiplying by 6 gives

$$6 \leq x \leq 18$$

This tells us that all numbers between 6 and 18 inclusive are solutions of $0 \leq \frac{x}{6} - 1 \leq 2$.

Exercises

Set I

1. If possible, write each of the following as a single power.
   a) $x^2 \cdot x^4$
   b) $x^2 + x^4$
   c) $(x^2)^4$
   d) $x^4 - x^2$
   e) $x^4(-x^2)$
   f) $\dfrac{x^2}{x^4}$

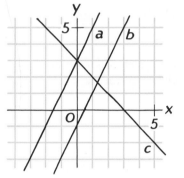

2. This exercise is about the graph at the right.
   a) Write an equation for line a.
   b) Write an equation for line b.
   c) Write an equation for line c.
   d) What do lines a and b have in common?
   e) What do lines a and c have in common?

THE WIZARD OF ID BY PERMISSION OF JOHNNY HART AND FIELD ENTERPRISES, INC.

3. Suppose that the fellow in this cartoon has $1200 invested, some in a long-term savings account at 7 percent interest and the rest in a regular savings account at 5 percent interest. If he earns $77 in interest at the end of a year, how much does he have invested in each account? (Hint: In one year, x dollars invested at 7 percent interest earns $0.07x$ dollars.)

Set II

4. The figure below shows that $-3 < 2$ and that $-3 + 1 < 2 + 1$.

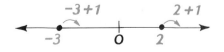

   Draw a similar figure showing that
   a) $-4 < -1$ and $-4 - 2 < -1 - 2$.
   b) $5 > 3$ and $2(5) > 2(3)$.
   c) $5 > 3$ and $-1(5) < -1(3)$.
   d) $-2 < 1$ and $-3(-2) > -3(1)$.

5. What happens to the direction of an inequality if
   a) the same number is added to both sides?
   b) both sides are multiplied by the same negative number?
   c) both sides are divided by the same positive number?

6. Tell whether or not each of the following numbers is a solution of the inequality given.

   $x + 9 > 6$
   a) 1
   b) −1
   c) −3

   $x - 2 \leq 5$
   d) 6
   e) −6
   f) 7
   g) −7
   h) 10
   i) −10

   $4x + 1 < 13$ $11 - 2x \geq 21$
   j) 2   m) −5
   k) 3   n) −4
   l) −4   o) 4

7. What was done to both sides of the first inequality to give the second in each of the following?
   a) $x - 7 < 2$ to give $x < 9$
   b) $3x \geq 15$ to give $x \geq 5$
   c) $x + 10 > 0$ to give $x > -10$
   d) $-6x < 12$ to give $x > -2$
   e) $-x \leq 8$ to give $x \geq -8$
   f) $\dfrac{x}{-4} > 1$ to give $x < -4$

8. Solve each of the following inequalities for x.
   a) $x + 8 < 1$
   b) $8x < 1$ f) $\dfrac{x}{-2} \leq 9$
   c) $x - 5 > -15$
   d) $-5x > -15$ g) $0 < x - 7 < 7$
                                h) $-5 \leq x + 3 \leq 5$
   e) $\dfrac{x}{2} \leq -9$

9. Solve each of the following inequalities for x.
   a) $4x + 7 > -5$ e) $1 < 2x - 9 < 11$
   b) $-x + 6 \leq 8$ f) $17 - 3x > 2$
   c) $\dfrac{x}{5} - 3 \geq 0$ g) $\dfrac{9x + 1}{10} < \dfrac{9x}{10} + 1$
   d) $\dfrac{x}{3} > \dfrac{x + 1}{7}$ h) $-5 \leq \dfrac{x}{-4} + 5 \leq 5$

Set III

10. The figure below shows that $-4 < 1$ and that $-4 - 2 < 1 - 2$.

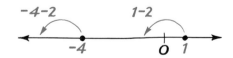

Draw a similar figure showing that
   a) $-5 < -2$ and $-5 + 1 < -2 + 1$.
   b) $4 > 1$ and $2(4) > 2(1)$.
   c) $4 > 1$ and $-2(4) < -2(1)$.
   d) $-3 < 0$ and $-1(-3) > -1(0)$.

11. What happens to the direction of an inequality if
   a) the same number is subtracted from both sides?
   b) both sides are multiplied by the same positive number?
   c) both sides are divided by the same negative number?

12. Tell whether or not each of the following numbers is a solution of the inequality given.

$$x - 4 < 10$$

a) 13
b) 14
c) –14

$$x + 7 \geq 2$$

d) 0
e) 2
f) –7
g) 7
h) –5
i) –9

$$3x - 2 > 25$$

j) 10
k) 9
l) –9

$$8 - 5x \leq 12$$

m) –1
n) 0
o) 2

13. What was done to both sides of the first inequality to give the second in each of the following?

a) $x + 3 > 0$ to give $x > -3$
b) $7x \leq 28$ to give $x \leq 4$
c) $x - 9 < -3$ to give $x < 6$

d) $-2x > 22$ to give $x < -11$
e) $-x \geq 5$ to give $x \leq -5$
f) $\frac{x}{-5} < 0$ to give $x > 0$

14. Solve each of the following inequalities for x.

a) $x + 10 > 7$
b) $10x > 7$
c) $x - 3 < -12$
d) $-3x < -12$
e) $\frac{x}{4} \geq -5$
f) $\frac{x}{-4} \geq 5$
g) $8 < x + 8 < 11$
h) $-1 \leq x - 6 \leq 4$

15. Solve each of the following inequalities for x.

a) $5x - 2 < -17$
b) $-x + 8 \geq 11$
c) $\frac{x}{6} + 1 \leq 0$
d) $\frac{x - 5}{2} > \frac{x}{4}$
e) $4 < 3x - 11 < 10$
f) $13 - 2x < 1$
g) $-3 \leq \frac{x}{-5} + 3 \leq 3$
h) $\frac{7x + 1}{8} - 1 > \frac{7x}{8}$

Set IV

Games played with dice have been popular for thousands of years.

When a die is thrown, any number from one through six can turn up. This fact can be represented by the inequality

$$1 \leq x \leq 6$$

in which x represents the number that appears.

Suppose that three dice are thrown at one time, as in the game of Chuck-a-luck.

1. Letting s represent the sum of the numbers that turn up, write an inequality to show its possible values.
2. How many integer solutions does your inequality have?
3. How many different sums can turn up when three dice are thrown?

Suppose that n dice are thrown at one time.

4. Letting s represent the sum of the numbers that turn up, write an inequality in terms of n to show its possible values.
5. Write an expression in terms of n for the number of different sums that can turn up when n dice are thrown.

LESSON 3
More on Solving Inequalities

A good poker player knows a great deal about the mathematical probabilities of the game: the chances of being dealt different hands and the odds against improving them as the game is played. For example, a given play is favorable if the following inequality is true:*

$$\frac{wp}{c} > 1$$

In this inequality, w represents the potential worth of the pot, p represents the probability of winning it, and c represents the potential cost of playing the hand.

If, for example, the pot is potentially worth \$100, the probability of winning it is 0.5, and it may cost \$40 to play the hand, then

$$\frac{100(0.5)}{40} = 1.25$$

---

* *Total Poker* by David Spanier (Simon and Schuster, 1977), p. 40.

757

Because $1.25 > 1$, the hand is worth playing. On the other hand, if the pot is potentially worth \$300, the probability of winning it is 0.1, and it may cost \$50 to play the hand, then

$$\frac{300(0.1)}{50} = 0.6$$

Because $0.6 < 1$, the player should fold.

The inequality for the favorability of a poker hand contains three variables. Like a formula, an inequality in two or more variables may be solved for any one of them in terms of the others. For example, to solve the poker-hand inequality

$$\frac{wp}{c} > 1$$

for p, we can first clear it of fractions by multiplying both sides by c:

$$wp > c$$

Dividing both sides by w gives

$$p > \frac{c}{w}$$

(The direction of the inequality is not changed by doing either of these things because c and w, the cost of playing the hand and the potential worth of the pot, are both positive numbers.)

Here are more examples of how to solve inequalities.

EXAMPLE 1

Solve the inequality $\dfrac{10}{x} - 5 \le 0$, given that $x > 0$. Draw a figure to illustrate its solutions.

SOLUTION
Adding 5 to each side gives

$$\frac{10}{x} \le 5$$

Because $x > 0$, multiplying each side by x gives

$$10 \leq 5x$$

Finally, dividing by 5 we get

$$2 \leq x, \quad \text{or} \quad x \geq 2$$

Because all positive numbers greater than or equal to 2 are solutions of the inequality, we draw a solid circle at 2 and shade the line to the right of 2.

EXAMPLE 2

The length of each side of a triangle is less than the sum of the lengths of the other two sides. Use this information to write three inequalities for the triangle shown here and solve them for x.

SOLUTION
The three inequalities are:

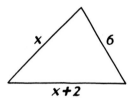

1. $x < 6 + (x + 2)$
2. $6 < x + (x + 2)$
3. $x + 2 < x + 6$

Solving each one, we get

1. $x < 6 + (x + 2)$
   $x < 8 + x$
   $0 < 8$

2. $6 < x + (x + 2)$
   $6 < 2x + 2$
   $4 < 2x$
   $2 < x$

3. $x + 2 < x + 6$
   $\quad 2 < 6$

The only inequality that tells us anything about x is the second one: x must be greater than 2.

Exercises

Set I

1. If possible, simplify each of the following.
   a) $\sqrt{20x} + \sqrt{5x}$
   b) $\sqrt{20x} - \sqrt{5x}$
   c) $\sqrt{20x}\sqrt{5x}$
   d) $\sqrt{20x} \div \sqrt{5x}$

2. Solve the following pairs of simultaneous equations.

   a) $\dfrac{x}{2} + 3y = 12$ b) $\dfrac{x}{4} - \dfrac{y}{5} = 0$

      $\dfrac{x}{2} - 3y = 3$ $x = y + 1$

3. The cost of flying an airplane depends on the speed at which it is flown. A typical formula is

$$y = 0.004x^2 + 800$$

   in which y is the cost in dollars per hour and x is the speed of the plane in kilometers per hour.

   a) What kind of function is this?
   b) Find the cost of flying the plane at a speed of 500 kilometers per hour for one hour.
   c) Find the cost of flying the plane at a speed of 1,000 kilometers per hour for one hour.

Set II

4. Solve each of the following inequalities and tell which figure at the right below illustrates its solutions.

   a) $10 - x < 8$
   b) $2(x - 6) > 3 - x$

   c) $\dfrac{x + 7}{3} \le 4$

   d) $1 + \dfrac{4}{x} > \dfrac{6}{x}$, if $x > 0$

   e) $x < 2(x - 1) < x + 3$
   f) $(x + 2)^2 \ge x^2 + 5x - 1$

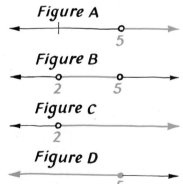

Figure A

Figure B

Figure C

Figure D

5. Solve each of the following inequalities and draw a figure to illustrate its solutions.

   a) $\dfrac{x}{15} < \dfrac{1}{5}$ b) $x > \dfrac{x}{2} - 4$ c) $\dfrac{2}{x} \ge 1$, if $x > 0$ d) $\dfrac{x}{3} + \dfrac{x}{6} \le 0$

e) $\dfrac{1}{x} - \dfrac{9}{x} < 8$, if $x < 0$

f) $\dfrac{x + 12}{x} > 5$, if $x < 0$

g) $\dfrac{x - 1}{5} - 1 \geq \dfrac{x}{4}$

h) $(x - 4)(x + 1) > x(x - 4)$

6. Solve the following inequalities for x in terms of the other variables.

a) $ax - b > 0$, if $a > 0$

b) $\dfrac{x}{a} < \dfrac{1}{b}$, if $a < 0$

c) $a \leq x + a \leq b$

d) $\dfrac{1}{x} + a \geq b$, if $x > 0$ and $a > b$

7. The length of each side of a triangle is less than the sum of the lengths of the other two sides. Use this information to write three inequalities for each of the following triangles and solve them for x.

a)
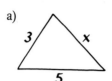

c)
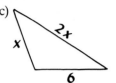

b)
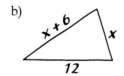

d)
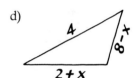

Set III

8. Solve each of the following inequalities and tell which figure at the right below illustrates its solutions.

a) $6 \geq 5 - x$

b) $5 > \dfrac{x + 12}{3}$

c) $3(x + 3) < 2(x + 6)$

d) $\dfrac{2}{x} < 1 - \dfrac{1}{x}$, if $x > 0$

e) $(x - 1)^2 \geq x(x - 3)$

f) $x \leq \dfrac{5x + 1}{4} \leq x + 1$

Figure A

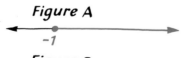

-1

Figure B

-1   3

Figure C

3

Figure D

3

9. Solve each of the following inequalities and draw a figure to illustrate its solutions.

a) $\dfrac{x}{6} > \dfrac{2}{3}$

c) $\dfrac{10}{x} \leq 2$, if $x > 0$

e) $\dfrac{6}{x} - \dfrac{4}{x} < 1$, if $x < 0$

g) $\dfrac{x + 5}{2} - 5 > \dfrac{x}{3}$

b) $x < \dfrac{x}{4} + 9$

d) $\dfrac{x}{5} - \dfrac{x}{2} \geq 3$

f) $\dfrac{8 - x}{x} < 7$, if $x > 0$

h) $(x + 2)(x - 3) \leq x(x - 2)$

10. Solve the following inequalities for x in terms of the other variables.

a) $ax + b < 1$, if $a > 0$

b) $\dfrac{x}{b} > \dfrac{1}{a}$, if $b < 0$

c) $a \geq x + b \geq b$

d) $a - \dfrac{1}{x} \leq b$, if $x > 0$ and $a < b$

11. The length of each side of a triangle is less than the sum of the lengths of the other two sides. Use this information to write three inequalities for each of the following triangles and solve them for x.

a)

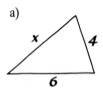

b)

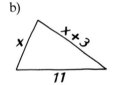

c)

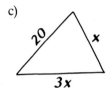

d)

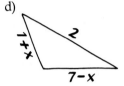

Set IV

Suppose that in a poker game you get a hand with which you are just as likely to lose as win. The probability of winning such a hand is 0.5.

Using the inequality given in this lesson, can you figure out how the potential worth of the pot, w, should compare with the potential cost of playing the hand, c, in order for it to be favorable to play?

BEER DRINKERS SHORTED BY DRAM A CAN, SUIT SAYS

Attorney Jay S. Bulmash, saying that he is acting on behalf of all California beer drinkers, charged Monday in a Superior Court suit that Anheuser-Busch Co. short-weights its beer one dram a can.

A dram is one-sixteenth of an ounce.

Bulmash contended in the civil suit that the company commits fraud and breach of warranty by falsely promising its 12-and 16-ounce cans of Busch Bavarian beer each contain a full 12 or 16 ounces.

Beer drinkers are being overcharged at least $259,000 a year because of the shortage, Bulmash alleged.

Within the last four years, Bulmash said, he has purchased at least 24 cans of beer and found each about a dram short.

The suit seeks monetary damages for drinkers and attorney's fees for Bulmash.

LESSON **4**
Absolute Value and Inequalities

Although the label on a can of beer may indicate that the can contains 12 ounces, it is unreasonable to expect that every can will contain exactly that amount. Whenever something is produced in large quantities, there are always variations in the results.

Because these variations are unavoidable, a certain *tolerance* is always allowed in producing the product. Suppose that for 12-ounce cans of beer this tolerance is 0.1 ounce. Then cans filled with as little as 11.9 ounces or as much as 12.1 ounces would be acceptable. This variation can be represented on a number line by shading the part of the line from 11.9 to 12.1 inclusive.

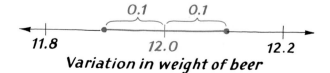

Variation in weight of beer

If we represent the weight of beer actually contained in a 12-ounce can as x ounces, then the can is acceptable if

$$11.9 \leq x \leq 12.1$$

We can indicate this variation in another way by using absolute value. If a 12-ounce can actually contains x ounces, then the amount of the error is

$$|x - 12|$$

ounces and, because this amount is limited to 0.1 ounce or less,

$$|x - 12| \leq 0.1$$

If a 12-ounce can of beer actually contained 11.8 ounces, then the error would be

$$|11.8 - 12| = |-0.2| = 0.2$$

ounces and, because $0.2 > 0.1$, it would not be considered acceptable.

From this example, we see that the inequalities

$$11.9 \leq x \leq 12.1 \quad \text{and} \quad |x - 12| \leq 0.1$$

are equivalent. The figure and statement below put this idea in more abstract form.

▶ If $|x - a| \leq b$, then $a - b \leq x \leq a + b$.

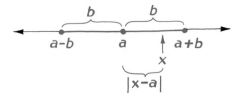

In our analysis of the variations that occur in manufacturing a product, we wrote an inequality that contains an absolute value. Here are more examples of equations and inequalities containing absolute values.

EXAMPLE 1
Solve for x: $|x| < 2$.

SOLUTION
This inequality says that the absolute value of x is less than 2. On a number line, this means that the distance between x and 0 is less than 2. As the figure below shows, the numbers whose distance from 0 is less than 2 are between –2 and 2.

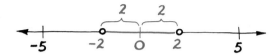

The open circles indicate that –2 and 2 are not included. The solutions of the inequality $|x| < 2$, then, are all numbers x such that $-2 < x < 2$.

EXAMPLE 2
Solve for x: $|x - 1| = 5$.

SOLUTION
This equation says that the distance between x and 1 on a number line is equal to 5. From the figure shown below, we see that there are two such numbers: –4 and 6.

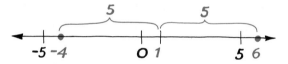

It is possible to figure this out without thinking of a number line. The absolute value of $x - 1$ is either $x - 1$ or $-(x - 1)$. In the first case,

$$|x - 1| = 5$$
$$x - 1 = 5$$
$$x = 6$$

In the second case,

$$|x - 1| = 5$$
$$-(x - 1) = 5$$
$$x - 1 = -5$$
$$x = -4$$

Checking these solutions, we write

$$|6 - 1| = 5, \quad |5| = 5, \quad 5 = 5,$$

and

$$|-4 - 1| = 5, \quad |-5| = 5, \quad 5 = 5$$

EXAMPLE 3

Solve for x: $|x - 3| > 4$.

SOLUTION

This inequality says that the distance between x and 3 on a number line is more than 4. From the figure shown below, we see that this is true of all numbers x such that either $x < -1$ or $x > 7$.

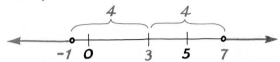

To solve this without thinking of a number line, we can reason as follows. The absolute value of $x - 3$ is equal to either $x - 3$ or $-(x - 3)$.

In the first case,

$$|x - 3| > 4$$
$$x - 3 > 4$$
$$x > 7$$

In the second case,

$$|x - 3| > 4$$
$$-(x - 3) > 4$$
$$x - 3 < -4$$
$$x < -1$$

Exercises

Set I

1. Solve the following quadratic equations.
   a) $x^2 + 5x = 0$
   b) $2x^2 - 24 = 0$
   c) $(x - 1)^2 = 9$

2. Simplify.

   a) $\dfrac{\dfrac{1}{x} + \dfrac{1}{y}}{\dfrac{1}{x} \cdot \dfrac{1}{y}}$

   b) $\dfrac{\dfrac{x}{y} - 1}{\dfrac{x}{y} + 1}$

3. Each year on Annie's birthday, Daddy Warbucks gives her $7 plus $3 for each year of her age.
   a) How much money does Annie get when she is ten?
   b) Write a formula for m, the amount of money Annie receives in dollars, in terms of a, her age in years.
   c) Solve your formula for a in terms of m.
   d) How old is Annie when Daddy Warbucks gives her $100 for her birthday?

Set II

4. What are the coordinates of the two points on a number line that are
   a) 15 units from 0?
   b) 6 units from 11?
   c) 11 units from 6?
   d) 8 units from –2?
   e) 3 units from x?
   f) a units from b? (Assume that a is positive.)

5. Tell which figure below matches each equation or inequality in parts a through d.

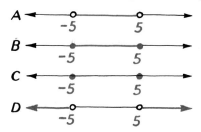

   a) $|x| = 5$
   b) $|x| < 5$
   c) $|x| > 5$
   d) $|x| \le 5$

6. Tell whether or not each of the following numbers is a solution of the equation or inequality given.

   $|x + 6| = 10$
   a) 4
   b) –4
   c) –16

   $|x - 15| \le 1$
   d) 13
   e) 14
   f) 15
   g) 16
   h) 17

   $|x + 9| > 0$
   i) 9
   j) –10
   k) –9
   l) –100

7. Solve for x.
   a) $|x| = 8$
   b) $|x| = 0$
   c) $|x| = -1$
   d) $|x| < 3$
   e) $2|x| \le 10$
   f) $|x| \ge 4$
   g) $|x| > -7$
   h) $\dfrac{|x|}{5} > 6$

8. Draw a figure to illustrate each of the following equations and inequalities.
   a) $|x| = 3$
   b) $|x - 5| = 3$
   c) $|x + 2| = 3$ (Hint: $|x + 2| = |x - (-2)|$.)
   d) $|x| < 4$
   e) $|x - 1| < 4$
   f) $|x + 4| < 4$ (See hint for part c.)
   g) $|x| \ge 2$
   h) $|x - 3| \ge 2$
   i) $|x + 3| \ge 2$

9. Solve for x.
   a) $|x| - 7 = 2$
   b) $|x - 7| = 2$
   c) $|x| - 4 < 9$
   d) $|x - 4| < 9$
   e) $|x| - 1 \ge 11$
   f) $|x - 1| \ge 11$
   g) $|x + 5| \le 8$
   h) $|x| + 8 > 5$

10. When Acute Alice measured Obtuse Ollie's height, she got 172 centimeters but may have made an error of as much as 1 centimeter. Suppose that x represents Ollie's actual height in centimeters.

a) Write an inequality indicating the numbers that x lies between.
b) Write an equivalent inequality using absolute value.

Set III

11. What are the coordinates of the two points on a number line that are
    a) 9 units from 0?
    b) 4 units from 10?
    c) 10 units from 4?
    d) 2 units from –5?
    e) 7 units from x?
    f) y units from z? (Assume that y is positive.)

12. Tell which figure below matches each equation or inequality in parts a through d.

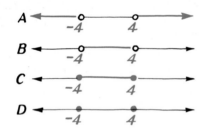

a) $|x| < 4$ c) $|x| > 4$
b) $|x| = 4$ d) $|x| \leq 4$

13. Tell whether or not each of the following numbers is a solution of the equation or inequality given.

    $|x + 3| = 12$
    a) 9
    b) –15
    c) –9

    $|x - 8| < 2$
    d) 8
    e) 9
    f) 10
    g) 6
    h) 7

    $|x - 1| \geq 15$
    i) 14
    j) –14
    k) 100
    l) 0

14. Solve for x.
    a) $|x| = 6$
    b) $|x| = -5$
    c) $|x| < 10$
    d) $3|x| \leq 12$
    e) $|x| \geq 1$
    f) $|x| > -8$
    g) $|x| > 0$
    h) $\dfrac{|x|}{4} \geq 7$

15. Draw a figure to illustrate each of the following equations and inequalities.
    a) $|x| = 2$
    b) $|x - 3| = 2$
    c) $|x + 1| = 2$ (Hint: $|x + 1| = |x - (-1)|$.)
    d) $|x| \leq 5$
    e) $|x - 2| \leq 5$
    f) $|x + 5| \leq 5$ (See hint for part c.)
    g) $|x| > 3$
    h) $|x - 4| > 3$
    i) $|x + 4| > 3$

16. Solve for x.
    a) $|x| - 5 = 4$
    b) $|x - 5| = 4$
    c) $|x| - 10 \leq 3$
    d) $|x - 10| \leq 3$
    e) $|x| - 6 > 2$
    f) $|x - 6| > 2$
    g) $|x| + 7 < 1$
    h) $|x + 1| \geq 7$

17. When Acute Alice stepped on the drugstore scale, it read 105 pounds. The scale, however, might have been as much as 2 pounds off. Suppose that x represents Alice's actual weight in pounds.
    a) Write an inequality indicating the numbers that x lies between.
    b) Write an equivalent inequality using absolute value.

Set IV

Parts for a machine that are mass-produced usually differ in size to a small degree: some are slightly larger than called for and some slightly smaller. Suppose that, in an assembly for a wristwatch, the allowable errors for the parts are as shown below:

Part	Allowable error in tenths of a millimeter
A	$\lvert a \rvert \leq 4$
B	$\lvert b \rvert \leq 1$
C	$\lvert c \rvert \leq 6$
D	$\lvert d \rvert \leq 3$
E	$\lvert e \rvert \leq 2$

The watch will run properly if, when the five parts are assembled,

$$\lvert a + b + c + d + e \rvert \leq 8$$

Otherwise, the assembly must be thrown away.
Suppose that watches are put together only from parts within the allowable errors.

1. If the errors in the parts for a certain watch assembly are $a = 3$, $b = -1$, $c = 4$, $d = -3$, and $e = 2$, will the assembly work properly?
2. What if the errors are $a = 2$, $b = 1$, $c = 5$, $d = 1$, and $e = 0$?
3. Suppose that the errors in four parts of an assembly are $a = -1$, $b = 0$, $c = 5$, and $d = -3$. Can any acceptable part E be used to complete the assembly so that it will work properly?
4. Answer the question asked in part 3 if the errors are $a = 4$, $b = 1$, $c = 6$, and $d = -2$.

Summary
and Review

THE LESSER OF TWO WEEVILS

DOUG'

In this chapter, we have learned how to solve inequalities and have extended our knowledge of absolute value.

Inequalities (*Lesson 1*) An inequality is a mathematical sentence that contains any of the symbols $<$ (is less than), $>$ (is greater than), $\leq$ (is less than or equal to), or $\geq$ (is greater than or equal to).

Solving Linear Inequalities (*Lessons 2 and 3*) An inequality can be changed into an equivalent one by adding the same number to or subtracting the same number from each side. Multiplying or dividing both sides of an inequality by the same positive number results in an inequality having the same direction. Multiplying or dividing both sides of an inequality by the same negative number reverses the direction of the inequality.

By applying these rules, inequalities containing variables can be solved in the same way that equations are solved.

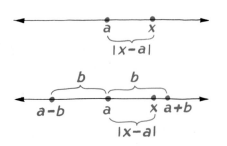

Absolute Value and Inequalities (*Lesson 4*) The absolute value of the expression $x - a$, written $|x - a|$, may be interpreted as the distance between x and a on a number line. If $|x - a| \leq b$, then $a - b \leq x \leq a + b$.

To solve an equation or inequality containing an absolute value, we can use the fact that $|x - a|$ is either $x - a$ or $-(x - a)$.

Exercises

Set I

1. Which of these symbols, $=$, $>$, or $<$, should replace ▓ in each of the following?
   a) -2 ▓ -12
   b) $-\dfrac{1}{2}$ ▓ $-\dfrac{1}{12}$
   c) 0.18 ▓ 0.4
   d) 0.18 ▓ $(0.4)^2$
   e) $|-7|$ ▓ 7
   f) $(-1)^4$ ▓ $(-1)^5$
   g) $\sqrt[3]{-27}$ ▓ -3
   h) $\sqrt{100}$ ▓ $\sqrt[3]{125}$

2. Write an inequality in terms of x to represent each of the following figures.

   a)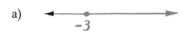
   -3

   b)
   2 7

   c)
   9

   d)
   -4 1

3. Which of these symbols, $=$, $>$, $<$, $\geq$, or $\leq$, should replace ▓ in each of the following to make it true for all values of x? If none will, say so.
   a) $x + 2$ ▓ x
   b) x ▓ $10x$
   c) $|x|$ ▓ x
   d) $\dfrac{x}{4}$ ▓ $\dfrac{x}{3}$

4. Which of these symbols, $=$, $>$, or $<$, should replace ▓ in each of the following to make it true for all values of the variables? If none will, say so.
   a) If $a < b$ and $b < c$, then a ▓ c.
   b) If $d = e$ and $e > f$, then d ▓ f.
   c) If $g > h$ and $i > h$, then g ▓ i.
   d) If $j > k$ and $l = k$, then j ▓ l.

5. Tell whether or not each of the following numbers is a solution of the inequality given.

   $x - 1 < 7$
   a) 8
   b) 0
   c) -6

   $2x \geq -10$
   d) -5
   e) -10
   f) 4

   $5x + 9 \leq 0$
   g) -2
   h) -3
   i) 1

6. Solve each of the following inequalities for x.
   a) $x + 4 > 3$
   b) $4x > 3$
   c) $-2x < 16$
   d) $9 - x \leq 1$
   e) $-2 < x - 4 < 2$
   f) $5x + 3 > -7$
   g) $\dfrac{x}{3} - 8 \leq 0$
   h) $\dfrac{x}{12} > \dfrac{1}{4}$
   i) $x < \dfrac{x}{5} - 4$
   j) $9 > 8 - \dfrac{3}{x}$, if $x > 0$
   k) $9 > 8 - \dfrac{3}{x}$, if $x < 0$
   l) $\dfrac{x - 5}{3} - \dfrac{x}{6} \leq 2$

7. Solve the following inequalities for x in terms of the other variables.
   a) $ax - b < 1$, if $a > 0$
   b) $\dfrac{x}{a} > \dfrac{b}{c}$, if $a < 0$
   c) $ax - bx \leq c$, if $a > b$
   d) $0 < x - a < a + b$

8. What are the coordinates of the two points on a number line that are
   a) 5 units from 0?
   b) 9 units from 2?
   c) 10 units from -6?
   d) 4 units from x?

9. Draw a figure to illustrate each of the following equations and inequalities.
   a) $|x| = 4$
   b) $|x| < 3$
   c) $|x| \geq 6$
   d) $|x - 5| = 2$
   e) $|x - 7| \leq 1$
   f) $|x + 1| > 7$

10. Solve for x.
    a) $4|x| = 20$
    b) $\dfrac{|x|}{3} < 5$
    c) $|x| + 7 = 10$
    d) $|x + 7| = 10$
    e) $|x| - 2 \geq 6$
    f) $|x - 2| \geq 6$
    g) $|x + 1| < 0$
    h) $|x + 1| > 0$

Set II

1. Which of these symbols, $=$, $>$, or $<$, should replace ▓ in each of the following?
   a) $-15 \ \text{▓} \ -6$
   b) $-\dfrac{1}{15} \ \text{▓} \ -\dfrac{1}{6}$
   c) $0.7 \ \text{▓} \ 0.50$
   d) $(0.7)^2 \ \text{▓} \ 0.50$
   e) $|-9| \ \text{▓} \ -9$
   f) $1^5 \ \text{▓} \ 1^7$
   g) $\sqrt[3]{10} \ \text{▓} \ 2$
   h) $\sqrt[4]{81} \ \text{▓} \ \sqrt{9}$

2. Write an inequality in terms of x to represent each of the following figures.

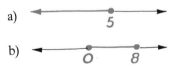

   a)
   5
   b)
   O   8
   c)
   -2
   d)
   -6   4

3. Which of these symbols, $=$, $>$, $<$, $\geq$, or $\leq$, should replace ▓ in each of the following to make it true for all values of x? If none will, say so.
   a) $x \ \text{▓} \ x - 5$
   b) $4x \ \text{▓} \ x$
   c) $|x| \ \text{▓} \ 0$
   d) $x^2 \ \text{▓} \ x^3$

4. Which of these symbols, $=$, $>$, or $<$, should replace ▓ in each of the following to make it true for all values of the variables? If none will, say so.
   a) If $a > b$ and $b = c$, then $a \ \text{▓} \ c$.
   b) If $d < e$ and $e > f$, then $d \ \text{▓} \ f$.
   c) If $g = h$ and $h < i$, then $g \ \text{▓} \ i$.
   d) If $j > k$ and $l < k$, then $j \ \text{▓} \ l$.

5. Tell whether or not each of the following numbers is a solution of the inequality given.

   $x + 5 > 8$
   a) 5
   b) 3
   c) -13

   $4x - 7 \geq 1$
   g) 0
   h) 2
   i) 7

   $3x \leq -12$
   d) 4
   e) -4
   f) -5

6. Solve each of the following inequalities for x.
   a) $x + 5 < 2$
   b) $5x < 2$
   c) $-3x > 30$
   d) $7 - x \geq 4$
   e) $0 < x - 6 < 1$
   f) $2x + 15 > 3$
   g) $\dfrac{x}{10} - 2 \leq 0$

h) $\frac{x}{3} < \frac{1}{6}$

i) $x > \frac{x}{3} + 8$

j) $5 > 4 - \frac{7}{x}$, if $x > 0$

k) $5 > 4 - \frac{7}{x}$, if $x < 0$

l) $\frac{x+1}{2} - \frac{x}{8} \leq 5$

7. Solve the following inequalities for x in terms of the other variables.

a) $ax + b > 0$, if $a > 0$

b) $\frac{x}{b} < \frac{a}{c}$, if $b < 0$

c) $ax - b \geq cx$, if $a > c$

d) $0 > x + a > a - b$

8. What are the coordinates of the two points on a number line that are
a) 8 units from 0?
b) 15 units from 3?
c) 7 units from –5?
d) x units from y?

9. Draw a figure to illustrate each of the following equations and inequalities.

a) $|x| = 1$ d) $|x - 6| = 4$
b) $|x| \leq 5$ e) $|x - 8| < 3$
c) $|x| > 2$ f) $|x + 3| \geq 8$

10. Solve for x.

a) $3|x| = 24$ e) $|x| + 8 > 3$
  f) $|x + 8| > 3$
b) $\frac{|x|}{7} < 2$ g) $|x + 2| \leq 0$
  h) $|x + 2| \geq 0$
c) $|x| - 9 = 1$
d) $|x - 9| = 1$

Chapter **17**

NUMBER SEQUENCES

COURTESY OF THE GQ-SECURITY PARACHUTE INC. ALAN LEVINSON, PRESIDENT.

LESSON 1
Number Sequences

Despite its dangers, sky diving can be an exhilarating experience, as the expression on the free-faller's face in the above photograph shows. In 1960, Captain Joseph W. Kittinger jumped from a balloon at an altitude of almost 20 miles and fell more than 16 miles before pulling the rip cord of his parachute.

The distances in feet that he traveled during the first few seconds of his fall were

$$16 \quad 48 \quad 80 \quad 112 \quad 144 \quad \ldots$$

in which the dots indicate that the numbers continue. These numbers are an example of a *number sequence*.

▶ A **number sequence** is an ordered set of numbers with one number for each counting number.

Rewriting the numbers in the "falling" sequence to show this pairing, we get

$$\begin{array}{ccccc} 16 & 48 & 80 & 112 & 144 \quad \ldots \\ \text{1st} & \text{2nd} & \text{3rd} & \text{4th} & \text{5th} \end{array}$$

(The counting numbers in this case represent the seconds that have passed since the moment at which the man jumped from the balloon.)

The numbers in a sequence are called its **terms.** To represent the terms of a number sequence, we will use the notation

$$t_1 \quad t_2 \quad t_3 \quad t_4 \quad t_5 \quad \ldots \quad t_n$$

in which t_n represents the nth term. For the "falling" sequence, $t_1 = 16$, $t_2 = 48$, and so forth. Notice that each successive term of this sequence can be obtained by adding 32 to the preceding term.

16   48   80   112   144 $\ldots$
   $+32$ $+32$ $+32$ $+32$

▶ A number sequence in which each successive term may be found by *adding the same number* is called an **arithmetic sequence.**

The differences between successive terms in an arithmetic sequence are the same. For example, the sequence above, this difference is 32.

Number sequences can be formed in many different ways. Consider, for example, the sequence

$$4 \quad 20 \quad 100 \quad 500 \quad 2{,}500 \quad \ldots$$

Each successive term of this sequence is obtained by multiplying the preceding term by 5.

4   20   100   500   2,500 $\ldots$
   $\times 5$ $\times 5$ $\times 5$ $\times 5$

▶ A number sequence in which each successive term may be found by *multiplying by the same number* is called a **geometric sequence.**

The ratios of successive terms in a geometric sequence are the same. In the sequence above, this ratio is $\frac{20}{4} = 5$.

Arithmetic and geometric sequences have many interesting applications, some of which we will study in this chapter. In this lesson, we will consider some of the many ways in which number sequences in general can be formed.

EXAMPLE 1
What number should replace ▓ in the following sequence? What kind of sequence is it?

$$11 \quad 8 \quad 5 \quad 2 \quad ▓$$

SOLUTION
Each successive term in this sequence is 3 less than the term before it. Subtracting 3 from 2, we get –1 for the missing term. Because subtracting 3 from each term to get the next one is equivalent to adding –3, the sequence is arithmetic.

EXAMPLE 2
A formula for the nth term of a certain sequence is $t_n = 3 \cdot 7^n$. Find the first three terms of this sequence. What kind of a sequence is it?

SOLUTION
To find the first term, we let $n = 1$: $t_1 = 3 \cdot 7^1 = 3 \cdot 7 = 21$. Letting $n = 2$ and $n = 3$ to find the next two terms, we get

$$t_2 = 3 \cdot 7^2 = 3 \cdot 49 = 147 \quad \text{and}$$
$$t_3 = 3 \cdot 7^3 = 3 \cdot 343 = 1{,}029$$

The first three terms of the sequence are 21, 147, and 1,029. Because each successive term is 7 times the preceding term, the sequence is geometric.

EXAMPLE 3
Write a formula for the nth term of this sequence:

$$\frac{1}{2} \quad \frac{2}{3} \quad \frac{3}{4} \quad \frac{4}{5} \quad \frac{5}{6} \quad \cdots$$

SOLUTION

Each term of this sequence is a fraction whose denominator is 1 more than its numerator. Because the numerator is the number of the term, the formula is $t_n = \dfrac{n}{n+1}$. This is not an arithmetic sequence because the differences between successive terms are not the same; nor is it a geometric sequence because the ratios of successive terms are not the same. There are many sequences that are neither arithmetic nor geometric but that still follow a simple pattern.

Exercises

Set I

1. Write an inequality in terms of x to represent each of the following figures.

   a)

   b)

   c)

2. Solve the following equations.

   a) $\dfrac{x+1}{5} = \dfrac{2x-1}{9}$

   b) $\dfrac{1}{x} + \dfrac{1}{4} = \dfrac{1}{4x}$

   c) $\dfrac{x}{8} - \dfrac{2}{x} = 1$

3. One way to "lose weight" is to travel in a space ship away from the earth. The weight in pounds of someone who is d thousand miles above the earth can be found from the formula

   $$y = \dfrac{16x}{(d+4)^2}$$

   in which x is his or her weight in pounds on the earth.

   a) Show that this formula gives the correct result when $d = 0$.

   Obtuse Ollie's weight on the earth's surface is 200 pounds. How much would he weigh if

   b) he were 4 thousand miles above the earth?

   c) he were 16 thousand miles above the earth?

   d) Solve the formula above for x in terms of y and d.

   e) If Acute Alice were 2 thousand miles above the earth, she would weigh 48 pounds. How much does she weigh on the earth's surface?

Set II

4. Tell whether each of the following sequences is arithmetic, geometric, both, or neither. If a sequence is arithmetic, what is the common difference of the terms? If a sequence is geometric, what is the common ratio?
   a) 4 11 18 25 . . .
   b) 3 6 12 24 . . .
   c) 1 3 6 10 . . .
   d) 15 13 11 9 . . .
   e) 10 100 1,000 10,000 . . .
   f) 2 6 24 120 . . .
   g) 8 8 8 8 . . .
   h) 16 24 36 54 . . .
   i) 7 10.5 14 17.5 . . .
   j) 4 –6 8 –10 . . .
   k) 1 –3 9 –27 . . .
   l) $\dfrac{1}{2}$ $\dfrac{1}{3}$ $\dfrac{1}{4}$ $\dfrac{1}{5}$ . . .
   m) $\dfrac{1}{2}$ $\dfrac{1}{4}$ $\dfrac{1}{8}$ $\dfrac{1}{16}$ . . .
   n) $\dfrac{1}{2}$ $\dfrac{2}{3}$ $\dfrac{5}{6}$ 1 . . .
   o) $\dfrac{1}{2}$ $\dfrac{1}{3}$ $\dfrac{2}{9}$ $\dfrac{4}{27}$ . . .

5. What number do you think should replace ▦ in each of the following sequences?
   a) 6 10 14 18 ▦
   b) 4 12 36 108 ▦
   c) 3 5 8 12 ▦
   d) 9 1 –7 –15 ▦
   e) 96 48 24 12 ▦
   f) 1 8 27 64 ▦
   g) $\dfrac{2}{1}$ $\dfrac{3}{2}$ $\dfrac{4}{3}$ $\dfrac{5}{4}$ ▦
   h) $\dfrac{1}{4}$ $\dfrac{1}{2}$ $\dfrac{3}{4}$ 1 ▦
   i) 250 25 2.5 0.25 ▦
   j) 3 –15 75 –375 ▦
   k) $\sqrt{7}$ $2\sqrt{2}$ 3 $\sqrt{10}$ ▦
   l) 4.1 3 1.9 0.8 ▦

6. A formula for the nth term of a certain sequence is

$$t_n = 5 + 2n$$

   a) Find the first five terms of this sequence.
   b) From the second term on, each term of this sequence can be found from the preceding term. How are they found?
   c) What type of sequence is it?

7. A formula for the nth term of a certain sequence is

$$t_n = 5 \cdot 2^n$$

   a) Find the first five terms of this sequence.
   b) From the second term on, each term of this sequence can be found from the preceding term. How are they found?
   c) What type of sequence is it?

8. What number should replace each ▦ in each of the following sequences

$$\text{▦} \quad 10 \quad 50 \quad \text{▦}$$

   a) if it is arithmetic?
   b) if it is geometric?

$$3 \quad \text{▦} \quad \text{▦} \quad 24$$

   c) if it is arithmetic?
   d) if it is geometric?

9. Write a formula for the nth term of each of the following sequences.
   a) 4 5 6 7 8 . . .
   b) 1 4 9 16 25 . . .
   c) 5 10 15 20 25 . . .
   d) –8 –8 –8 –8 –8 . . .
   e) 4 16 64 256 1,024 . . .
   f) –5 –4 –3 –2 –1 . . .
   g) 7 $\dfrac{7}{2}$ $\dfrac{7}{3}$ $\dfrac{7}{4}$ $\dfrac{7}{5}$ . . .
   h) 1 $\sqrt{2}$ $\sqrt{3}$ 2 $\sqrt{5}$ . . .

Set III

10. Tell whether each of the following sequences is arithmetic, geometric, both, or neither. If a sequence is arithmetic, what is the common difference of the terms? If it is geometric, what is the common ratio?
    a) 2   6   18   54   . . .
    b) 7   16   25   34   . . .
    c) 2   5   9   14   . . .
    d) 20   16   12   8   . . .
    e) 1   0.1   0.01   0.001   . . .
    f) 4   16   48   96   . . .
    g) 6   8.5   11   13.5   . . .
    h) 8   12   18   27   . . .
    i) 2   2   2   2   . . .
    j) 1   –4   16   –64   . . .
    k) 3   –6   9   –12   . . .

    l) $\dfrac{1}{6}$ $\dfrac{1}{7}$ $\dfrac{1}{8}$ $\dfrac{1}{9}$ . . .

    m) $\dfrac{1}{6}$ $\dfrac{1}{3}$ $\dfrac{1}{2}$ $\dfrac{2}{3}$ . . .

    n) $\dfrac{1}{5}$ $\dfrac{1}{10}$ $\dfrac{1}{20}$ $\dfrac{1}{40}$ . . .

    o) $\dfrac{1}{3}$ $\dfrac{1}{4}$ $\dfrac{3}{16}$ $\dfrac{9}{64}$ . . .

11. What number do you think should replace ▦ in each of the following sequences?
    a) 7   14   28   56   ▦
    b) 3   9   15   21   ▦
    c) 1   4   9   16   ▦
    d) 162   54   18   6   ▦
    e) 58   47   36   25   ▦
    f) 5   10   16   23   ▦

    g) $\dfrac{3}{4}$ $\dfrac{5}{8}$ $\dfrac{1}{2}$ $\dfrac{3}{8}$ ▦

    h) $\dfrac{1}{3}$ $\dfrac{1}{4}$ $\dfrac{1}{5}$ $\dfrac{1}{6}$ ▦

    i) –2   8   –32   128   ▦
    j) 5 $\sqrt{26}$ $3\sqrt{3}$ $2\sqrt{7}$ ▦
    k) 420   42   4.2   0.42   ▦
    l) 16   40   100   250   ▦

12. A formula for the nth term of a certain sequence is

    $$t_n = 4 + 3n$$

    a) Find the first five terms of this sequence.
    b) From the second term on, each term of this sequence may be found from the preceding term. How are they found?
    c) What type of sequence is it?

13. A formula for the nth term of a certain sequence is

    $$t_n = 4 \cdot 3^n$$

    a) Find the first five terms of this sequence.
    b) From the second term on, each term of this sequence may be found from the preceding term. How are they found?
    c) What type of sequence is it?

14. What number should replace each ▦ in each of the following sequences?

    ▦   6   18   ▦

    a) if it is arithmetic?
    b) if it is geometric?

    2   ▦   ▦   128

    c) if it is arithmetic?
    d) if it is geometric?

15. Write a formula for the nth term of each of the following sequences.
    a) 2   4   6   8   10   . . .
    b) 0   1   2   3   4   . . .
    c) 1   8   27   64   125   . . .
    d) 9   10   11   12   13   . . .
    e) 5   25   125   625   3,125   . . .
    f) 6   6   6   6   6   . . .
    g) 17   27   37   47   57   . . .

    h) $\dfrac{1}{4}$ $\dfrac{1}{2}$ $\dfrac{3}{4}$ 1 $\dfrac{5}{4}$ . . .

Set IV

Number sequences are frequently included in intelligence tests.

1   8   11   69   88   96   101 |||| |||| ||||

1. Can you figure out what the three missing terms of this sequence should be?
2. How is the sequence formed?

(Hint: What does it look like this way?)

The Toronto Star

FRIDAY, MAY 30, 1975

LESSON 2
Arithmetic Sequences

Several years ago, the students at a school in the province of Ontario wanted to build the world's largest human pyramid. To do so, they all lay down on the ground and formed the pyramid shown above.

The pyramid had 16 rows of people, with 1 person in the first row, 2 in the second row, 3 in the third, and so on. How many people did it contain in all?

The numbers of people in the rows form the arithmetic sequence

1   2   3   4   5   6   7   8   9   10   11   12   13   14   15   16

To find out how many there are altogether, we have to find the sum of these 16 numbers. One way to do this would be to add the numbers in order from left to right. Another way is suggested by the figures at the top of the next page. To find the number of people in a pyramid of any number of rows, imagine the pyramid copied upside down beside itself.

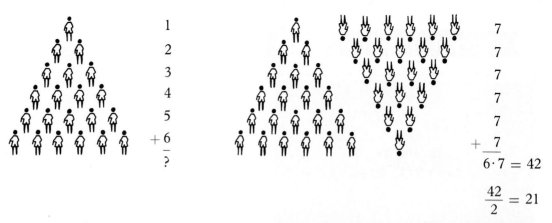

$$6 \cdot 7 = 42$$

$$\frac{42}{2} = 21$$

The figures illustrate this for a pyramid of 6 rows. Instead of having to add 6 different numbers, placing the pyramid upside down beside itself changes the problem into one of adding 6 numbers all of which are the same. Because this is equivalent to multiplying this number by 6, all that we have to do here is multiply 6 by 7 and then divide the result by 2 to make up for counting everything twice.

Applying this method to finding the number of people in a pyramid of 16 rows, we reason as follows. If the pyramid were turned upside down beside itself, each row would have 17 people (1 + 16, 2 + 15, 3 + 14, etc.). Multiplying the number of rows, 16, by the number of people in each row, 17, we get $16 \times 17 = 272$. Dividing by 2 to make up for counting everyone twice, we get

$$\frac{272}{2} = 136$$

There are 136 people in a pyramid of 16 rows.

This method for finding the sum of the terms of a sequence can be applied to any arithmetic sequence. The figure below and the one on the next page illustrate why. If the first term of an arithmetic sequence is t_1 and d is the common difference between successive terms, then the second term is $t_1 + d$, the third term is $t_1 + d + d$, and so forth. The figure below represents an arithmetic sequence of seven terms in which each row represents one term. Each successive row is longer than the preceding row by the same amount: d.

t_1							t_1
t_1	d						t_2
t_1	d	d					t_3
t_1	d	d	d				t_4
t_1	d	d	d	d			t_5
t_1	d	d	d	d	d		t_6
t_1	d	d	d	d	d	d	$+t_7$
							$?$

To find a formula for the sum of the terms of this arithmetic sequence, $t_1, t_2, t_3, \ldots, t_7$, we copy the figure upside down beside itself as shown in the next figure.

t_1	d d d d d d	t_1	$t_1 + t_7$
t_1	d d d d d d	t_1	$t_2 + t_6$
t_1	d d d d d d	t_1	$t_3 + t_5$
t_1	d d d d d d	t_1	$t_4 + t_4$
t_1	d d d d d d	t_1	$t_5 + t_3$
t_1	d d d d d d	t_1	$t_6 + t_2$
t_1	d d d d d d	t_1	$+\ \dfrac{t_7 + t_1}{(t_1 + t_7)7}$

$$S_n = \frac{(t_1 + t_7)7}{2}$$

Each row of the resulting figure has the same length: $t_1 + t_7$. Multiplying this sum by the number of rows, 7, gives a result that is twice the sum of the terms in the sequence. Dividing the result by 2 gives us the answer.

This procedure will work for any arithmetic sequence, no matter how many terms it contains. All we have to do is to think of the terms written down twice as shown and then:

1. find the sum of the first and last terms,
2. multiply this sum by the number of terms, and
3. divide by 2.

Stating this result as a formula in which n represents the number of terms, t_1 and t_n represent the first and last terms, and S_n represents the sum of the n terms, we have

$$S_n = \frac{(t_1 + t_n)n}{2}$$

With this shortcut, we can find the sum of the terms of an arithmetic sequence without even having to write them down.

EXAMPLE 1

Find the sum of the even numbers from 2 through 100.

SOLUTION

We want to find the sum of the sequence

$$2 \quad 4 \quad 6 \quad 8 \quad 10 \quad \ldots \quad 100$$

Because this sequence is arithmetic (each successive even number is 2 more than the preceding one), we can use the formula

$$S_n = \frac{(t_1 + t_n)n}{2}$$

The first term, t_1, is 2 and the last term, t_n, is 100. Furthermore, $n = 50$ because 100 is the fiftieth even number in the sequence of even numbers starting with 2.

$$S_n = \frac{(2 + 100)50}{2} = \frac{102 \cdot 50}{2} = \frac{102}{2} \cdot 50 = 51 \cdot 50 = 2,550$$

The sum of the even numbers from 2 through 100 is 2,550.

To use our method, we must know the last term of the sequence that we are adding. To find the last term, it would be useful to have a formula for the nth term of an arithmetic sequence. Look again at the figure representing an arithmetic sequence of seven terms.

$$
\begin{array}{l}
t_1 \\
t_2 = t_1 + d \\
t_3 = t_1 + 2d \\
t_4 = t_1 + 3d \\
t_5 = t_1 + 4d \\
t_6 = t_1 + 5d \\
t_7 = t_1 + 6d
\end{array}
$$

From this figure, we can see that the nth term of an arithmetic sequence is given by the formula

$$t_n = t_1 + (n - 1)d$$

in which t_1 represents the first term and d represents the common difference between successive terms. Using this formula, we can write down any term of an arithmetic sequence if we know the first term and the common difference.

EXAMPLE 2
Find the seventeenth term of the sequence

$$123 \quad 119 \quad 115 \quad 111 \quad \ldots$$

SOLUTION
Because the first term, t_1, is 123 and the common difference, d, is –4, we can write

$$t_{17} = t_1 + (17 - 1)d$$
$$= 123 + 16(-4) = 123 - 64 = 59$$

The seventeenth term of the sequence is 59.

Exercises

Set I

1. Solve each of the following inequalities for x.
   a) $x + 7 < 1$
   b) $-3x > 15$
   c) $2 < x - 4 < 8$
   d) $5x - 6 \geq 4$

2. Tell whether each of the following statements is true or false.
   a) Every positive integer is a counting number.
   b) Zero is an integer.
   c) Every integer is a rational number.
   d) Every square root is an irrational number.

3. The span from a pianist's thumb to his little finger was measured five times. The measurements in centimeters were

$$22.60 \quad 22.81 \quad 22.75 \quad 22.87 \quad 23.27$$

   a) Find the average of these measurements by adding them and dividing their sum by 5.
   b) Subtract the average from each of the five numbers. (Your first answer should be –0.26.)
   c) Add the five differences.
   d) Is the answer to part c due to chance alone? Make up another set of five measurements and make the same calculations with them. What do you conclude from this about the sum of the differences between the average and the numbers averaged?

Set II

4. This bar graph represents the first five terms of an arithmetic sequence.

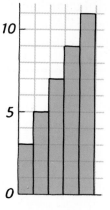

a) What are the five terms represented?
b) What is their common difference? Draw bar graphs of the following arithmetic sequences. (Represent negative terms by drawing bars that extend downward from the x-axis.)
c) 5   5.5   6   6.5   7
d) 7   7   7   7   7
e) 7   4   1   –2   –5

5. Graph the following functions, letting x vary from 1 to 5.
a) $y = 2x + 1$.
b) $y = 0.5x + 4.5$
c) $y = 7$
d) $y = -3x + 10$
e) To what kind of functions do arithmetic sequences correspond?

6. A formula for the nth term of a certain arithmetic sequence is

$$t_n = 8 + (n - 1)3$$

a) Find the first three terms of this sequence.
b) What is their common difference?
c) Find the tenth term of the sequence.
d) Find the hundredth term.
e) What is the sum of the first ten terms?
f) What is the sum of the first one hundred terms?

7. A formula for the nth term of a certain arithmetic sequence is

$$t_n = 6 - (n - 1)10$$

a) Find the first three terms of this sequence.
b) What is their common difference?
c) Find the tenth term of the sequence.
d) Find the hundredth term.
e) What is the sum of the first ten terms?
f) What is the sum of the first one hundred terms?

8. What number should replace each ▦ in each of the following arithmetic sequences?
a) 5   11   17   ▦
b) 12   ▦   28   36
c) 3   ▦   ▦   45
d) ▦   14   9   ▦
e) 6   ▦   ▦   7
f) 15   ▦   0   ▦
g) –2   ▦   ▦   ▦   10
h) $\frac{1}{2}$ ▦ $\frac{2}{3}$ $\frac{3}{4}$

9. Write a formula for the nth term of each of the following arithmetic sequences and use it to find the indicated term.
a) 2   7   12   . . . ;   6th term
b) 16   25   34   . . . ;   10th term
c) 92   89   86   . . . ;   21st term
d) 20   9   –2   . . . ;   12th term
e) –6   –6   –6   . . . ;   43rd term
f) –70   –58   –46   . . . ;   101st term

10. Use the formula for the sum of the terms of an arithmetic sequence to find each of the following sums. (The three dots represent all of the terms between those given.)
a) $1 + 9 + 17 + 25 + 33 + 41$
b) $22 + 25 + 28 + 31 + 34 + 37 + 40 + 43 + 46 + 49$
c) $101 + 97 + 93 + 89 + 85$
d) $1 + 2 + 3 + 4 + \cdots + 50$
e) $6 + 12 + 18 + 24 + \cdots + 144$
f) $-98 + -96 + -94 + \cdots + -2$

11. Find each of the following sums by figuring out the first and last terms and then using the formula for the sum of the terms of an arithmetic sequence.
   a) The sum of the first 8 terms of 21   25   29   ...
   b) The sum of the first 15 terms of 3   10   17   ...
   c) The sum of the first 11 terms of 7   1   –5   ...
   d) The sum of the first 50 terms of 40   39.5   39   ...

12. If 5 points are marked at equal intervals on a circle and each point is connected with straight lines to the rest of the points, 10 lines are required in all. The figures below show one way in which they can be drawn.

   +   +   +   =

4   +   3   +   2   +   1   =   10

   a) How many lines would be required to connect 6 points in the same way?
   b) How many lines would be required to connect 12 points in the same way?

13. Obtuse Ollie decided to go jogging every day of his summer vacation. He decided to run 0.5 mile the first day, 0.75 mile the second day, 1 mile the third day, and so on, increasing his distance by 0.25 mile each day. If his vacation lasted 85 days and Ollie kept this up every day,
   a) how far would he jog on the eighty-fifth day?
   b) how far would he have jogged altogether?

Set III

14. The bar graph below represents the first five terms of an arithmetic sequence.

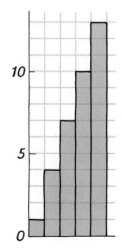

   a) What are the five terms represented?
   b) What is their common difference?

   Draw bar graphs of the following arithmetic sequences. (Represent negative terms by drawing bars that extend downward from the x-axis.)

   c) 3   3.5   4   4.5   5
   d) 4   4   4   4   4
   e) 5   3   1   –1   –3

15. Graph the following functions, letting x vary from 1 to 5.
   a) $y = 3x - 2$
   b) $y = 0.5x + 2.5$
   c) $y = 4$
   d) $y = -2x + 7$
   e) To what kind of functions do arithmetic sequences correspond?

16. A formula for the nth term of a certain arithmetic sequence is

$$t_n = 7 + (n - 1)5$$

   a) Find the first three terms of this sequence.
   b) What is their common difference?
   c) Find the tenth term of the sequence.
   d) Find the hundredth term.
   e) What is the sum of the first ten terms?
   f) What is the sum of the first one hundred terms?

17. A formula for the nth term of a certain arithmetic sequence is

$$t_n = 9 - (n - 1)4$$

   a) Find the first three terms of this sequence.
   b) What is their common difference?
   c) Find the tenth term of the sequence.
   d) Find the hundredth term.
   e) What is the sum of the first ten terms?
   f) What is the sum of the first one hundred terms?

18. What number should replace each ▌ in each of the following arithmetic sequences?
   a) 8   15   22   ▌
   b) 11   ▌   17   20
   c) 6   ▌   ▌   39
   d) ▌   15   11   ▌
   e) 3   ▌   0   ▌
   f) 4   ▌   ▌   5

g) –1   ▌   ▌   ▌   19
h) $\dfrac{3}{10}$ ▌ $\dfrac{1}{2}$ $\dfrac{3}{5}$

19. Write a formula for the nth term of each of the following arithmetic sequences and use it to find the indicated term.
   a) 1   5   9   ... ;   8th term
   b) 23   30   37   ... ;   11th term
   c) 80   78   76   ... ;   25th term
   d) 15   6   –3   ... ;   16th term
   e) –100   –87   –74   ... ;   52nd term
   f) –11   –11   –11   ... ;   90th term

20. Use the formula for the sum of the terms of an arithmetic sequence to find each of the following sums. (The three dots represent all of the terms between those given.)
   a) $3 + 12 + 21 + 30 + 39 + 48$
   b) $17 + 19 + 21 + 23 + 25 + 27 + 29 + 31 + 33$
   c) $84 + 78 + 72 + 66 + 60$
   d) $1 + 2 + 3 + 4 + \cdots + 32$
   e) $7 + 14 + 21 + 28 + \cdots + 175$
   f) $-99 + -96 + -93 + \cdots + -3$

21. Find each of the following sums by figuring out the first and last terms and then using the formula for the sum of the terms of an arithmetic sequence.
   a) The sum of the first 10 terms of 9   15   21   ...
   b) The sum of the first 13 terms of 31   33   35   ...
   c) The sum of the first 21 terms of 14   6   –2   ...
   d) The sum of the first 100 terms of 60   59.5   59   ...

22. If a football league has 6 teams, for each team to play all of the other teams, 15 games must be played. The diagram on the next page shows one way in which they can be counted. (A-B represents the game played between teams A and B, and so forth.)

```
A-B
      B-C
A-C         C-D
      B-D         D-E
A-D         C-E         E-F
      B-E         D-F
A-E         C-F
      B-F
A-F

  5  +  4  +  3  +  2  +  1  =  15
```

How many games would be required in order for each team in a football league to play all of the other teams if
a) there are 7 teams in the league?
b) there are 10 teams in the league?

23. Acute Alice had a choice of two jobs for her summer vacation. The first job paid $15 per day. The second job paid $1 the first day, $1.50 the second day, $2 the third day, and so on, with a raise of $0.50 on each successive day. If Alice could work 60 days on either job,
a) how much would she get paid for the first job?
b) how much would she get paid for the second job?

Set IV

The following problem is from *Olney's Complete Algebra*, published in 1870.

"If 100 oranges are placed in a line, exactly 2 yards from each other, and the first 2 yards from a basket, what distance must a boy travel, starting from the basket, to gather them up singly, and return with each to the basket?"

Can you solve it? If you can, show your work.

LESSON 3
Geometric Sequences

According to legend, the game of chess was invented for a king of Persia by one of his servants. The king was so pleased that he asked the servant what he would like as a reward. The man's request seemed very reasonable. He asked that one grain of wheat be placed on the first square of the chessboard, two grains on the second square, four grains on the third, and so on, each square having twice as many grains as the square before. The king was surprised, thinking that the servant had asked for very little. He was even more surprised when he found out how much wheat the man actually wanted.

The numbers of grains of wheat on the squares form a geometric sequence:

$$1 \quad 2 \quad 4 \quad 8 \quad 16 \quad 32 \quad 64 \quad 128 \quad \ldots$$

Because there are 64 squares on a chessboard, the inventor's request was for as many grains of wheat as the sum of the first 64 terms of this sequence. To begin, let's figure out what the last term of the sequence would be.

The first four terms can be written as

$$t_1 = 1 = 1 \cdot 2^0$$
$$t_2 = 2 = 1 \cdot 2^1$$
$$t_3 = 4 = 1 \cdot 2^2$$
$$t_4 = 8 = 1 \cdot 2^3$$

From this pattern, we see that the nth term of the sequence is

$$t_n = 1 \cdot 2^{n-1}$$

so that the sixty-fourth term is

$$t_{64} = 1 \cdot 2^{63}$$

The terms of every geometric sequence follow the same sort of pattern. If t_1 represents the first term and r represents the common ratio of successive terms, we have

$$t_n = t_1 \cdot r^{n-1}$$

To find a formula for the sum of the terms of an arithmetic sequence, we wrote the sequence twice, added, and divided by 2. Although this procedure won't work for a geometric sequence, a slightly different method will. To see how it works, we will apply it to finding the sum of the first five terms of the chessboard sequence:

$$1 + 2 + 4 + 8 + 16$$

We will represent this sum as S_5. The figure below represents S_5.

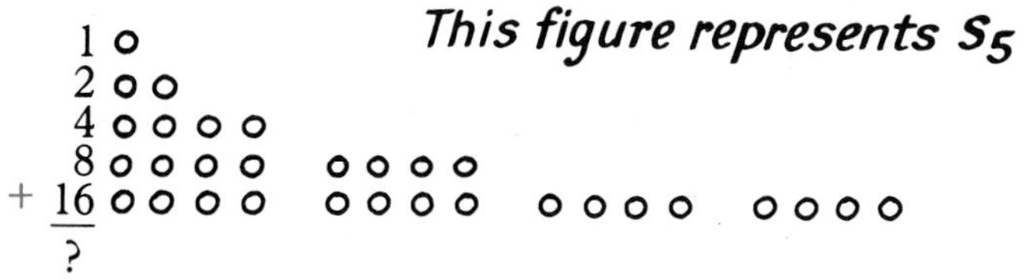

This figure represents S_5

We multiply each term in this sequence by the common ratio of terms, 2, to get a new sequence the sum of whose terms is $2S_5$.

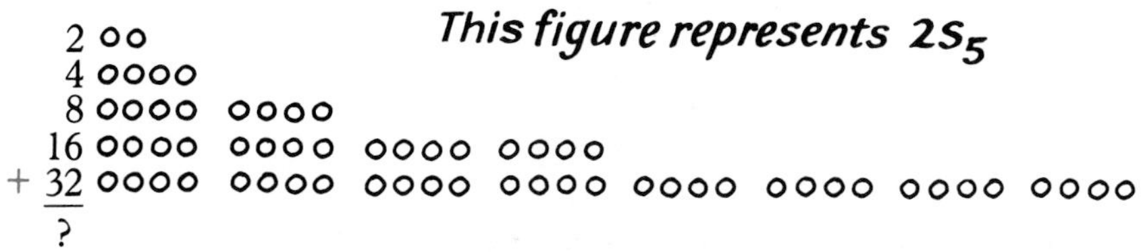

This figure represents $2S_5$

The first four terms of the new sequence are identical to the last four terms of the original sequence. This makes it easy to subtract the original sequence from the new one as shown below.

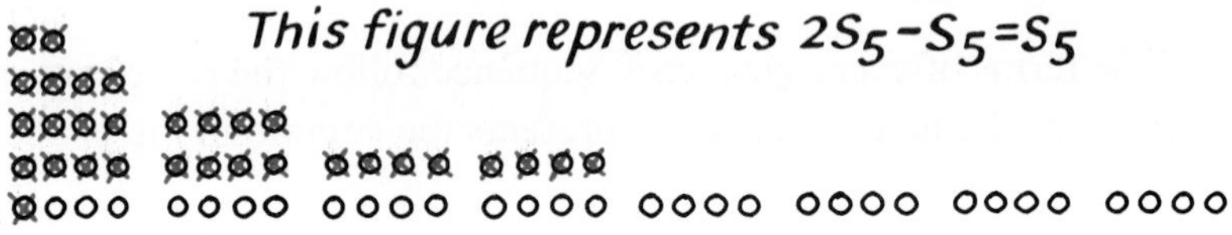

This figure represents $2S_5 - S_5 = S_5$

$32 - 1 = 31$

The result is 1 less than the last term of the new sequence; so $S_5 = 32 - 1 = 31$.

In a similar way, we can find a formula for the sum of the terms of any geometric sequence. Letting n represent the number of terms, t_1 represent the first term, and r represent the ratio of successive terms, we can write

$$S_n = t_1 + t_1 r + t_1 r^2 + t_1 r^3 + \cdots + t_1 r^{n-2} + t_1 r^{n-1}$$

Multiplying each term by r, we get

$$r \cdot S_n = t_1 r + t_1 r^2 + t_1 r^3 + t_1 r^4 + \cdots + t_1 r^{n-1} + t_1 r^n$$

Subtracting the original sum from the new sum, we get

$$
\begin{array}{rl}
r \cdot S_n = & t_1 r + t_1 r^2 + t_1 r^3 + \cdots + t_1 r^{n-1} + t_1 r^n \\
- \quad S_n = & t_1 + t_1 r + t_1 r^2 + t_1 r^3 + \cdots + t_1 r^{n-1} \\
\hline
r S_n - S_n = -t_1 & \hspace{4cm} + t_1 r^n
\end{array}
$$

Factoring each side of the resulting equation,

$$r S_n - S_n = t_1 r^n - t_1$$

we get

$$S_n(r - 1) = t_1(r^n - 1)$$

Dividing both sides by $r - 1$ gives a formula for the sum of the terms of the sequence:

$$S_n = \frac{t_1(r^n - 1)}{r - 1}$$

To find the number of grains of wheat that would be required to cover all 64 squares of the chessboard, we can use the following values

$$t_1 = 1, \quad r = 2, \quad \text{and} \quad n = 64$$

in this formula to get

$$S_{64} = \frac{1(2^{64} - 1)}{2 - 1} = 2^{64} - 1$$

The number of grains of wheat, then, is $2^{64} - 1$. If we raise 2 to the 64th power and subtract 1, this number turns out to be

$$18,446,744,073,709,551,615$$

Far from having made a modest request, the servant had asked for more wheat than is produced on the earth in two thousand years!

We have developed two formulas in this lesson, one for the nth term of a geometric sequence

$$t_n = t_1 \cdot r^{n-1}$$

and the other for the sum of the first n terms of a geometric sequence

$$S_n = \frac{t_1(r^n - 1)}{r - 1}$$

Here are more examples of how to use these formulas.

EXAMPLE 1

Find an expression for the twelfth term of the sequence

$$3 \quad 21 \quad 147 \quad 1,029 \quad \ldots$$

SOLUTION

Because the first term, t_1, is 3 and the ratio is 7, we write

$$t_{12} = 3 \cdot 7^{11}$$

Like the number of grains of wheat, this number is very large. We can either leave it in this form or use a calculator to write it as

$$5,931,980,229$$

EXAMPLE 2

Find an expression for the sum of the first ten terms of the sequence

$$2 \quad 6 \quad 18 \quad 54 \quad \ldots$$

SOLUTION

Because the first term, t_1, is 2, the ratio is 3, and $n = 10$, we write

$$S_{10} = \frac{2(3^{10} - 1)}{3 - 1} = 3^{10} - 1$$

We can either leave the answer in this form or use a calculator to convert it into

$$59,048$$

Exercises

Set I

1. If possible, tell which of these symbols, $>$, $=$, or $<$, should replace ▓ in each of the following to make it true.
   a) If $a = b$ and $b > c$, then a ▓ c.

   b) If $d < e$ and $f < d$, then e ▓ f.
   c) If g ▓ h and $h = i$, then $g < i$.
   d) If $j > k$ and k ▓ l, then $j = l$.

2. Write a formula for the *n*th term of each of the following sequences.
   a) –3 –6 –9 –12 –15 . . .
   b) 8 9 10 11 12 . . .
   c) 2 4 8 16 32 . . .
   d) $\dfrac{0}{2}$ $\dfrac{1}{3}$ $\dfrac{2}{4}$ $\dfrac{3}{5}$ $\dfrac{4}{6}$ . . .

3. For a dollar, Moby Dick's Fish Store, will sell you either five minnows and five guppies or eleven minnows and one guppy.
   a) Use this information to write a pair of simultaneous equations, letting *x* and *y* represent the respective prices in cents of a minnow and a guppy.
   b) Solve the equations for *x* and *y*.
   c) How much would three minnows and seven guppies cost?

Set II

4. The bar graph below represents the first four terms of a geometric sequence.

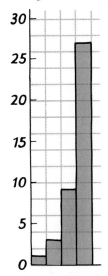

   a) What are the four terms represented?
   b) What is their common ratio?
   Draw bar graphs of the following geometric sequences.
   c) 0.5 2.5 12.5
   d) 3 3 3 3
   e) 16 8 4 2 1

5. Graph the following functions, letting *x* vary as indicated.

   a) $y = \dfrac{1}{10}(5)^x$ (1 to 3)

   b) $y = 3(1)^x$ (1 to 4)

   c) $y = 32\left(\dfrac{1}{2}\right)^x$ (1 to 5)

   d) From the results obtained in exercises 4 and 5, what kind of functions correspond to geometric sequences?

6. A formula for the *n*th term of a certain geometric sequence is

   $$t_n = \dfrac{1}{2} \cdot 4^{n-1}$$

   a) Find the first three terms of this sequence.
   b) What is the ratio of two successive terms?
   c) Write an expression for the tenth term of the sequence.
   d) Write an expression for the hundredth term.

7. A formula for the nth term of a certain geometric sequence is

$$t_n = 15 \cdot \left(\frac{1}{3}\right)^{n-1}$$

a) Find the first three terms of this sequence.
b) What is the ratio of two successive terms?
c) Write an expression for the tenth term of the sequence.
d) Write an expression for the hundredth term.

8. What number should replace each ▦ in each of the following sequences?
a) 2   20   200   ▦
b) 8   ▦   18   27
c) 1   ▦   ▦   64
d) ▦   –15   45   ▦
e) 7^{-1} ▦   ▦ 7^2
f) $\frac{4}{5}$ $\frac{2}{5}$ $\frac{1}{5}$ ▦

9. Write a formula for the nth term of each of the following geometric sequences.
a) 6   12   24   . . .
b) 64   8   1   . . .
c) 5   –5   5   . . .
d) 4   14   49   . . .
e) 2 $2\sqrt{6}$ 12   . . .
f) 300   –30   3   . . .

10. Use the formula for the sum of the terms of a geometric sequence and the table on page 801 to find each of the following sums.
a) The sum of the first six terms of

2   6   18   . . .

b) The sum of the first eight terms of

4   20   100   . . .

c) The sum of the first ten terms of

1   –2   4   . . .

d) The sum of the first seven terms of

12   48   192   . . .

11. The number of ancestors that a person can have in the ten generations preceding his own is the sum of a geometric sequence containing ten terms.
a) Write the ten terms of this sequence, beginning with 2, the number of a person's parents.
b) Use the formula for the sum of the terms of a geometric sequence and the table on page 801 to find the number of ancestors in ten generations.
c) Write, in simplest form, a formula for the number of ancestors in the nth generation back.
d) Write a formula for the total number of ancestors in n generations.

12. Suppose that the inventor of chess had asked for 1 grain of wheat on the first square of the board, 3 grains of wheat on the second square, 9 grains on the third square, and so on.

a) Write a sequence of the numbers of grains that would be required to cover the first eight squares of the board.
b) Use the formula for the sum of the terms of a geometric sequence and the table of powers on page 801 to find the number of grains that there would be altogether on the first eight squares.
c) Write an expression for the number of grains of wheat that would be required for the sixty-fourth square.
d) Write an expression for the total number of grains that would be required to cover the board.

Set III

13. The bar graph below represents the first four terms of a geometric sequence.

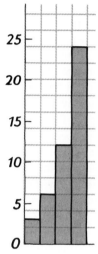

a) What are the four terms represented?
b) What is their common ratio?

Draw bar graphs of the following geometric sequences.

c) 0.5   2   8   32
d) 10   10   10   10

e) 27   9   3   1

14. Graph the following functions, letting x vary from 1 to 4.

a) $y = \frac{1}{8}(4)^x$

b) $y = 10(1)^x$

c) $y = 81\left(\frac{1}{3}\right)^x$

d) From the results obtained in exercises 4 and 5, what kind of functions correspond to geometric sequences?

15. A formula for the nth term of a certain geometric sequence is

$$t_n = \frac{1}{3} \cdot 6^{n-1}$$

a) Find the first three terms of this sequence.
b) What is the ratio of two successive terms?
c) Write an expression for the tenth term of this sequence.
d) Write an expression for the hundredth term.

16. A formula for the nth term of a certain geometric sequence is

$$t_n = 10 \cdot \left(\frac{1}{2}\right)^{n-1}$$

a) Find the first three terms of this sequence.
b) What is the ratio of two successive terms?
c) Write an expression for the tenth term of this sequence.
d) Write an expression for the hundredth term.

17. What number should replace each ▓ in each of the following geometric sequences?
a) 500   50   5   ▓
b) 16   ▓   100   250
c) ▓   –8   32   ▓
d) 7   ▓   ▓   56

e) $\frac{9}{10}$ $\frac{3}{10}$ $\frac{1}{10}$ ▓

f) 6^{-2} ▓   ▓ 6^1

18. Write a formula for the nth term of each of the following geometric sequences.
a) 5   15   45   . . .
b) 1   –7   49   . . .
c) 32   8   2   . . .
d) 4   18   81   . . .
e) 9   –6   4   . . .
f) 3 $3\sqrt{5}$ 15   . . .

19. Use the formula for the sum of the terms of a geometric sequence and the table on page 801 to find each of the following sums.
a) The sum of the first ten terms of

    1   2   4   . . .

b) The sum of the first seven terms of

    5   30   180   . . .

c) The sum of the first six terms of

$$4 \quad -12 \quad 36 \quad \ldots$$

d) The sum of the first eight terms of

$$20 \quad 100 \quad 500 \quad \ldots$$

20. Although the sequence

$$12 \quad 12 \quad 12 \quad 12 \quad 12 \quad 12 \quad 12 \quad 12 \quad 12$$

is geometric, its terms cannot be added by using the formula for the sum of the terms of a geometric sequence.
   a) To find out why, try to use the formula. What happens?
   b) What is the sum of the terms of this sequence?

c) Write a formula for the sum of this sequence if it has n terms.

21. Acute Alice got a chain letter in the mail. Directions in the letter said to make four copies of it and send them to four of her friends with instructions to do the same.
   a) Write a sequence starting with 1 and containing six terms to show how many letters there would be in each step of the chain if everyone cooperated.
   b) Use the formula for the sum of the terms of a geometric sequence and the table of powers on page 801 to find the number of letters that there would be altogether.
   c) Write a formula for the number of letters in the nth step of the chain.
   d) Write a formula for the number of letters in a chain of n steps.

Set IV

In a gambling system called the "martingale," the rule is to double your bet each time you lose.

1. If someone following this system started with a bet of $1 and lost seven times in a row, how much money would he lose altogether?
2. According to the system, how much money should he bet the eighth time?
3. If he won the eighth bet, how much money ahead or behind would he be on the eight bets altogether?

4. Suppose that someone following this system started with a bet of $3 and lost x times in a row. How much money would he lose altogether?
5. According to the system, how much money should he bet the $(x + 1)$th time?
6. If he won the $(x + 1)$th bet, how much money ahead or behind would he be on the $x + 1$ bets altogether?
7. These exercises seem to indicate that someone who uses the martingale system will always come out ahead. Why isn't this necessarily the case?

Table of Powers

x	2	3	4	5	6
x^2	4	9	16	25	36
x^3	8	27	64	125	216
x^4	16	81	256	625	1,296
x^5	32	243	1,024	3,125	7,776
x^6	64	729	4,096	15,625	46,656
x^7	128	2,187	16,384	78,125	279,936
x^8	256	6,561	65,536	390,625	1,679,616
x^9	512	19,683	262,144	1,953,125	10,077,696
x^{10}	1,024	59,049	1,048,576	9,765,625	60,466,176

REPRINTED WITH THE PERMISSION OF RALPH MORSE.

LESSON 4
Infinite Geometric Sequences

Although a typical racehorse is about 64 inches tall, a breeder in Georgia has succeeded in producing an adult horse that is only 19 inches tall! The three horses in this photograph are, from back to front, 60 inches, 37 inches, and 25 inches tall respectively.

Each year the breeder mates his smallest horses, hoping to get horses that are even smaller. He hopes to eventually get a horse that is only 12 inches tall.

Suppose that it is discovered that the heights of the horses in several successive generations form the following geometric sequence:

$$50 \quad 45 \quad 40.5 \quad 36.45 \quad \ldots$$

If this pattern were to continue indefinitely, would it be possible to get a horse less than 12 inches tall?

Because the sequence is geometric, we can use the formula

$$t_n = t_1 \cdot r^{n-1}$$

to find the height of the horses in any given generation. For example, because $t_1 = 50$ and $r = 0.9$, the horse in the tenth generation would be

$$t_{10} = 50(0.9)^9 \approx 17.4$$

inches tall. Those in the twentieth generation, if it were possible to continue that far, would be

$$t_{20} = 50(0.9)^{19} \approx 6.8$$

inches tall! The diagram below shows the heights of the horses in 20 successive generations.

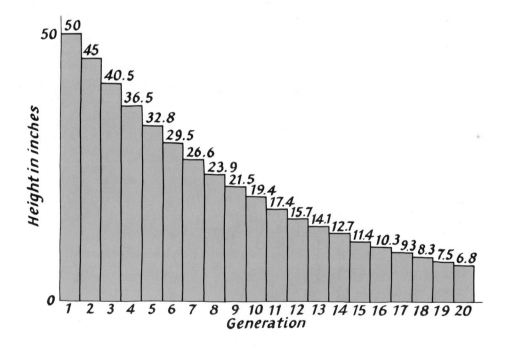

By continuing this sequence, we can make the terms smaller and smaller. For example,

$$t_{50} = 50(0.9)^{49} \approx 0.3,$$
$$t_{100} = 50(0.9)^{99} \approx 0.001, \quad \text{and}$$
$$t_{200} = 50(0.9)^{199} \approx 0.00000004.$$

The terms get closer and closer to zero because their ratio is less than 1. *For every geometric sequence having ratio* r *such that* $|r| < 1$, t_n *gets closer and closer to zero as* n *gets larger and larger.* Here is another example illustrating this fact.

EXAMPLE 1
Draw a bar graph illustrating the first eight terms of a geometric sequence whose first term is 20 and whose ratio is –0.6.

SOLUTION
The first eight terms of this sequence can be found using the table of powers on page 801 and the formula

$$20(-0.6)^{n-1} = 20(-1)^{n-1} \cdot \frac{6^{n-1}}{10^{n-1}}$$

More conveniently, they can be found by a calculator and rounded to the nearest tenth to give

$$20 \quad -12 \quad 7.2 \quad -4.3 \quad 2.6 \quad -1.6 \quad 0.9 \quad -0.6$$

Drawing bars that extend downward from the x-axis to represent the negative terms, we get the graph shown here.

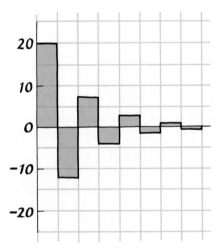

Another interesting property of every geometric sequence having ratio r such that $|r| < 1$ concerns the sum of its terms, S_n. Consider, for example, the sequence

$$8 \quad 4 \quad 2 \quad 1 \quad 0.5 \quad 0.25 \quad 0.125 \quad \ldots$$

As we go farther and farther along this sequence, not only do its terms get closer and closer to zero, but their sum gets closer and closer to a fixed number. Using the formula for the sum of the terms of a geometric sequence,

$$S_n = \frac{t_1(r^n - 1)}{r - 1}$$

with $t_1 = 8$ and $r = 0.5$, we find that

$$S_{10} = \frac{8(0.5^9 - 1)}{0.5 - 1} \approx 15.96875$$

$$S_{20} = \frac{8(0.5^{19} - 1)}{0.5 - 1} \approx 15.99997$$

$$S_{50} = \frac{8(0.5^{49} - 1)}{0.5 - 1} \approx 16.00000$$

As we add more and more terms of this sequence, their sum gets closer and closer to 16.

It is possible to discover this without making a lot of calculations. Using the distributive rule on the numerator of our formula for S_n, we get

$$S_n = \frac{t_1 r^n - t_1}{r - 1}$$

As n gets larger and larger, r^n gets closer and closer to zero. As a result, $t_1 r^n$ gets closer and closer to zero, which means that S_n gets closer and closer to

$$\frac{-t_1}{r - 1}, \quad \text{or} \quad \frac{t_1}{1 - r}$$

This tells us that the "sum" of the terms of an infinite geometric sequence having ratio r such that $|r| < 1$ is given by the formula

$$S = \frac{t_1}{1 - r}$$

Applying this formula to the infinite sequence

$$8 \quad 4 \quad 2 \quad 1 \quad 0.5 \quad 0.25 \quad 0.125 \quad \ldots$$

in which $t_1 = 8$ and $r = 0.5$, we get

$$S = \frac{8}{1 - 0.5} = \frac{8}{0.5} = 16$$

Here is an example showing how this formula can be used to express a repeating decimal as the quotient of two integers.

EXAMPLE 2
Express $0.\overline{18}$ as the quotient of two integers.

SOLUTION
The number 0.181818. . . can be written as

$$0.18 + 0.0018 + 0.000018 + \cdots, \text{ or as}$$

$$\frac{18}{100} + \frac{18}{10,000} + \frac{18}{1,000,000} + \cdots$$

Because the terms being added form an infinite geometric sequence whose first term is $\frac{18}{100}$ and whose ratio is $\frac{1}{100}$, we can use the formula

$$S = \frac{t_1}{1 - r}$$

getting

$$S = \frac{\dfrac{18}{100}}{1 - \dfrac{1}{100}} = \frac{\dfrac{18}{100}}{\dfrac{99}{100}} = \frac{18}{99} = \frac{2}{11}$$

Checking this answer by dividing 2 by 11, we get

$$\frac{2}{11} = 0.181818. . .$$

This same method can be used to express any repeating decimal as a quotient of integers.

Exercises

Set I

1. Solve each of the following for x.
   a) $|x| < 1$
   b) $|x| + 2 = 5$
   c) $|x - 4| = 6$
   d) $3|x| \leq 12$

2. This exercise is about the arithmetic sequence

$$7 \quad 11 \quad 15 \quad 19 \quad \ldots$$

   a) Write a formula for the nth term of this sequence.
   b) Use your formula to find the 21st term of the sequence.
   c) Find the sum of the first 21 terms of the sequence.

3. The average weight of the monsters in Loch Ness is thought to depend on the number of monsters that the lake contains.*

   _____
   *R. W. Shelton and S. R. Kerr, "Population Density of Monsters in Loch Ness," *Limnology and Oceanography,* September 1972.

Here is a table showing how the average weight and the number of monsters may be related:

Number of monsters, n	1	2	3	4	5
Average weight in kilograms, w	3,000	1,500	1,000	750	600

   a) How does the average weight, w, vary with respect to the number of monsters, n?
   b) Write a formula for w in terms of n.
   c) What would be the average weight of the monsters if there were eight of them in the lake?

Set II

4. Find the ratio for each of the following geometric sequences either as an integer or in decimal form.
   a) 5 6 7.2 8.64 . . .
   b) 10 9 8.1 7.29 . . .
   c) 8.5 8.5 8.5 8.5 . . .
   d) 12 –6 3 –1.5 . . .
   e) 7 –7 7 –7 . . .
   f) 15 –21 29.4 –41.16 . . .
   g) –4 –1.2 –0.36 –0.108 . . .
   h) –2 –2.02 –2.0402 –2.060602 . . .

5. The following questions refer to the sequences in exercise 4.
   a) In which sequences are the terms getting farther and farther from zero? What do the ratios for these sequences have in common?
   b) In which sequences are the terms getting closer and closer to zero? What do the ratios for these sequences have in common?

*6. Draw a bar graph to illustrate the first six terms of each of the following geometric sequences. Round each term to the nearest tenth.
a) $t_1 = 4$; $r = 1.2$
b) $t_1 = 4$; $r = 0.8$
c) $t_1 = 5$; $r = -1.1$
d) $t_1 = 5$; $r = -0.9$

7. Use the formula for the sum of the terms of an infinite geometric sequence to find each of the following sums.
a) $54 + 18 + 6 + 2 + \cdots$
b) $\dfrac{1}{2} + \dfrac{1}{6} + \dfrac{1}{18} + \dfrac{1}{54} + \cdots$
c) $48 + 36 + 27 + 20.25 + \cdots$
d) $\dfrac{1}{3} - \dfrac{1}{15} + \dfrac{1}{75} - \dfrac{1}{375} + \cdots$

8. Express each of the following repeating decimals as the quotient of two integers by treating it as the sum of the terms of an infinite geometric sequence. Reduce your answers to lowest terms.
a) 0.333333. . .
b) 0.454545. . .
c) 0.216216. . .
d) 0.077777. . .

9. The figure below represents an infinite sequence of squares formed by joining the midpoints of the sides of each square to form the next smaller square. Each side of

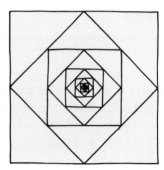

the largest square is 4 centimeters long and each successive square has half the area of the one before it.
a) Find the areas of the five largest squares.
b) Find the sum of their areas.
c) Find the sum of the areas of all of the squares in the figure, assuming that they go on infinitely.

10. A rubber ball is thrown upward, reaching a height of 25 meters. If it rebounds on each bounce to a height that is $\dfrac{3}{5}$ of its previous height, how far will the ball travel before it stops?

Set III

11. Find the ratio for each of the following geometric sequences either as an integer or in decimal form.
a) 10   8   6.4   5.12   . . .
b) 2   2.2   2.42   2.662   . . .
c) 5   –3   1.8   –1.08   . . .
d) 4.2   4.2   4.2   4.2   . . .
e) 9   4.5   2.25   1.125   . . .
f) –3.5   3.5   –3.5   3.5   . . .
g) –20   –28   –39.2   –54.88   . . .
h) –1   –0.99   –0.9801   –0.970299   . . .

12. The following questions refer to the sequences in exercise 11.
a) In which sequences are the terms getting farther and farther from zero? What do the ratios for these sequences have in common?
b) In which sequences are the terms getting closer and closer to zero? What do the ratios for these sequences have in common?

*13. Draw a bar graph to illustrate the first six terms of each of the following geometric sequences. Round each term to the nearest tenth.
a) $t_1 = 6$; $r = 1.1$
b) $t_1 = 6$; $r = 0.9$
c) $t_1 = 3$; $r = -1.2$
d) $t_1 = 3$; $r = -0.8$

14. Use the formula for the sum of the terms of an infinite geometric sequence to find each of the following sums.
a) $32 + 16 + 8 + 4 + \cdots$
b) $\dfrac{1}{4} + \dfrac{1}{8} + \dfrac{1}{16} + \dfrac{1}{32} + \cdots$
c) $75 + 30 + 12 + 4.8 + \cdots$
d) $\dfrac{1}{5} - \dfrac{1}{20} + \dfrac{1}{80} - \dfrac{1}{320} + \cdots$

15. Express each of the following repeating decimals as the quotient of two integers by treating it as the sum of the terms of an infinite geometric sequence. Reduce your answers to lowest terms.
a) 0.666666...
b) 0.727272...
c) 0.148148...
d) 0.022222...

16. The figure below represents an infinite sequence of triangles formed by joining the midpoints of the sides of each triangle to form the next smaller triangle. The area of

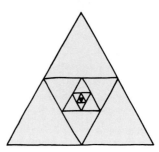

the largest triangle is 12 square centimeters and each successive triangle has one fourth the area of the one before it.
a) Find the areas of the five largest triangles.
b) Find the sum of their areas.
c) Find the sum of the areas of all of the triangles in the figure, assuming that they go on infinitely.

17. A rubber ball is thrown upward, reaching a height of 35 meters. If it rebounds on each bounce to a height that is $\dfrac{2}{3}$ of its previous height, how far will the ball travel before it stops?

Set IV

Obtuse Ollie gave Acute Alice a twenty-pound box of chocolates for her birthday. Alice was so overwhelmed by Ollie's generosity that she asked Ollie to help himself to as many chocolates as he wanted.

Ollie, accepting the offer, got out a chessboard and asked Alice to put 64 chocolates on the first square, 32 chocolates on the second square, 16 on the third square, and so on, each square having half as many chocolates as the square before.

Having heard the story about the inventor of chess, Alice didn't think there would be enough chocolates in the box to fulfill Ollie's request.

How many chocolates did Ollie ask for? Round your answer to the nearest integer.

Summary and Review

In this chapter, we have learned about sequences. We have learned how to identify arithmetic and geometric sequences, how to find their terms, and how to find their sums.

Number Sequences (*Lesson 1*) A number sequence is an ordered set of numbers with one number for each counting number. The numbers in a sequence are called its terms.

Number sequences in which each successive term is found by adding the same number are called arithmetic. The differences between successive terms in an arithmetic sequence are always the same.

Number sequences in which each successive term is found by multiplying by the same number are called geometric. The ratios of successive terms in a geometric sequence are always the same.

Arithmetic Sequences (*Lesson 2*) The nth term of an arithmetic sequence is given by the formula

$$t_n = t_1 + (n - 1)d$$

in which t_1 represents the first term and d represents the difference between successive terms.

The sum of the first n terms of an arithmetic sequence is given by the formula

$$S_n = \frac{(t_1 + t_n)n}{2}$$

in which t_1 and t_n represent the first and last terms.

Geometric Sequences (*Lesson 3*)   The nth term of a geometric sequence is given by the formula

$$t_n = t_1 \cdot r^{n-1}$$

in which t_1 represents the first term and r represents the ratio of each successive term to the preceding one.

The sum of the first n terms of a geometric sequence is given by the formula

$$S_n = \frac{t_1(r^n - 1)}{r - 1}$$

in which t_1 represents the first term and r represents the ratio.

Infinite Geometric Sequences (*Lesson 4*)   For every geometric sequence having ratio r such that $|r| < 1$, the terms get closer and closer to zero as n gets larger and larger.

The "sum" of the terms of an infinite geometric sequence having ratio r such that $|r| < 1$ is given by the formula

$$S = \frac{t_1}{1 - r}$$

in which t_1 represents the first term. This formula can be used to express a repeating decimal as the quotient of two integers.

Exercises

Set I

1. Tell whether each of the following sequences is arithmetic, geometric, both, or neither. If a sequence is arithmetic, tell the common difference. If it is geometric, tell the common ratio.
   a) 1.5   6   24   96 . . .
   b) 7   9.5   12   14.5 . . .
   c) $\sqrt{10}$ $\sqrt{11}$ $\sqrt{12}$ $\sqrt{13}$ . . .
   d) $\dfrac{5}{6}$ $\dfrac{3}{4}$ $\dfrac{2}{3}$ $\dfrac{7}{12}$ . . .
   e) 10^{-4} 10^{-3} 10^{-2} 10^{-1} . . .
   f) $\dfrac{1}{5}$ $\dfrac{1}{10}$ $\dfrac{1}{15}$ $\dfrac{1}{20}$ . . .

2. What number do you think should replace ▦ in each of the following sequences?
   a) 50   39   28   17   ▦
   b) $\dfrac{1}{4}$ $\dfrac{2}{9}$ $\dfrac{3}{16}$ $\dfrac{4}{25}$ ▦
   c) 8   12   18   27   ▦
   d) 3   5   8   13   ▦
   e) $\dfrac{1}{3}$ $\dfrac{1}{2}$ $\dfrac{2}{3}$ $\dfrac{5}{6}$ ▦
   f) 27   –9   3   –1   ▦

3. What number should replace ▦ in each of the following sequences?

   8   28   ▦

   a) if it is arithmetic?
   b) if it is geometric?

   9   ▦   1

   c) if it is arithmetic?
   d) if it is geometric?

4. Write the first four terms of the sequences having the following formulas for their nth terms.
   a) $t_n = 3n - 1$
   b) $t_n = n^4$
   c) $t_n = 4^n$
   d) $t_n = 5(n + 2)$
   e) $t_n = 5 \cdot 2^n$
   f) $t_n = n + \dfrac{1}{n}$
   g) Which sequences in parts a through f are arithmetic?
   h) Which sequences in parts a through f are geometric?

5. Write a formula for the nth term of each of the following sequences and use it to find the indicated term.
   a) 5   6   7   8   9   . . . ;   100th term
   b) –2   –4   –6   –8   –10   . . . ;   15th term
   c) 5   25   125   625   3,125   . . . ;   7th term
   d) $\dfrac{1}{6}$ $\dfrac{1}{3}$ $\dfrac{1}{2}$ $\dfrac{2}{3}$ $\dfrac{5}{6}$ . . . ;   24th term
   e) 1   8   27   64   125   . . . ;   10th term
   f) 4   28   196   1,372   9,604   . . . ;   6th term
   g) 1 $\dfrac{1}{4}$ $\dfrac{1}{9}$ $\dfrac{1}{16}$ $\dfrac{1}{25}$ . . . ;   14th term
   h) 0   1 $\sqrt{2}$ $\sqrt{3}$ 2   . . . ;   50th term

6. Use the formula for the sum of the terms of an arithmetic sequence to find each of the following sums.
   a) $5 + 11 + 17 + 23 + 29 + 35 + 41 + 47$
   b) $30 + 17 + 4 + -9 + -22 + -35 + -48$
   c) $3 + 6 + 9 + 12 + \cdots + 123$
   d) $46 + 47 + 48 + 49 + \cdots + 74$

7. Use the formula for the sum of the terms of a geometric sequence and the table on page 801 to find each of the following sums.

a) The sum of the first eight terms of

$$3 \quad 12 \quad 48 \quad \ldots$$

b) The sum of the first seven terms of

$$8 \quad 40 \quad 200 \quad \ldots$$

c) The sum of the first six terms of

$$12 \quad -36 \quad 108 \quad \ldots$$

8. Find the ratio for each of the following geometric sequences in decimal form.

a) 25   0.25   0.0025   . . .
b) 10   –7   4.9   . . .
c) 3.6   5.4   8.1   . . .

9. Use the formula for the sum of the terms of an infinite geometric sequence to find each of the following sums.

a) $15 + \dfrac{5}{2} + \dfrac{5}{12} + \dfrac{5}{72} + \cdots$

b) $250 + 150 + 90 + 54 + \cdots$

c) $\dfrac{1}{3} - \dfrac{1}{6} + \dfrac{1}{12} - \dfrac{1}{24} + \cdots$

10. Express each of the following repeating decimals as the quotient of two integers by treating it as the sum of the terms of an infinite geometric sequence. Reduce your answers to lowest terms.

a) 0.888888. . .
b) 0.363636. . .
c) 0.003003. . .

11. Find each of the following sums.

a) $1 + 3$
b) $1 + 3 + 5$
c) $1 + 3 + 5 + 7$
d) $1 + 3 + 5 + 7 + 9$
e) What type of numbers are all of the sums?
f) Write a formula for the nth term of the sequence

$$1 \quad 3 \quad 5 \quad 7 \quad 9 \quad \ldots$$

Simplify the formula as much as you can.

g) Write a formula for the sum of the first n terms of this sequence. Simplify the formula as much as you can.

*12. Suppose that you borrowed $1,000 at 18 percent interest compounded annually and that you did not pay back any money on the loan until ten years later. The amounts of money that you would owe at the end of each year of the loan would form a geometric sequence in which the first term was $1,180 and the ratio was 1.18.

a) Find the second through tenth terms of this sequence, rounding each number to the nearest dollar.

b) Draw a bar graph with a bar representing the amount borrowed, $1,000, and ten bars representing the amount owed at the end of each of the ten years.

c) Write a formula representing the amount of money owed after n years.

13. A case of splitting hares.

Obtuse Ollie built a cage for his pet rabbit. One day he noticed the rabbit look out of the window at one end of the cage, run across the cage to look out of the window at the opposite end, and then back again.

Suppose that the rabbit kept doing this, taking 4 seconds to run across the cage the first time, 2 seconds the second time, 1 second the third, and so on, so that each trip took half the time of the preceding one. When would the rabbit be looking out of both windows at the same time?

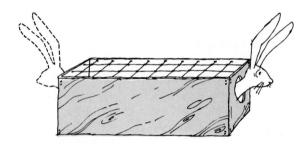

Set II

1. Tell whether each of the following sequences is arithmetic, geometric, both, or neither. If a sequence is arithmetic, tell the common difference. If it is geometric, tell the common ratio.
   a) $-7 \quad -1 \quad 5 \quad 11 \quad \ldots$
   b) $1^3 \quad 2^3 \quad 3^3 \quad 4^3 \quad \ldots$
   c) $8 \quad 20 \quad 50 \quad 125 \quad \ldots$
   d) $\dfrac{1}{6} \quad \dfrac{1}{5} \quad \dfrac{1}{4} \quad \dfrac{1}{3} \quad \ldots$
   e) $-3.5 \quad -3.5 \quad -3.5 \quad -3.5 \quad \ldots$
   f) $9^{-1} \quad 9^{-2} \quad 9^{-3} \quad 9^{-4} \quad \ldots$

2. What number do you think should replace ▦ in each of the following sequences?
   a) $5 \quad -10 \quad 20 \quad -40 \quad$ ▦
   b) $3 \quad 10.5 \quad 18 \quad 25.5 \quad$ ▦
   c) $4 \quad \sqrt{17} \quad 3\sqrt{2} \quad \sqrt{19} \quad$ ▦
   d) $\dfrac{7}{10} \quad \dfrac{3}{5} \quad \dfrac{1}{2} \quad \dfrac{2}{5} \quad$ ▦
   e) $10 \quad 3 \quad 0.9 \quad 0.27 \quad$ ▦
   f) $2 \quad 8 \quad 18 \quad 32 \quad$ ▦

3. What number should replace ▦ in each of the following sequences

$$12 \quad 18 \quad ▦$$

   a) if it is arithmetic?
   b) if it is geometric?

$$40 \quad ▦ \quad 10$$

   c) if it is arithmetic?
   d) if it is geometric?

4. Write the first four terms of the sequences having the following formulas for their nth terms.
   a) $t_n = n^3$
   b) $t_n = 3^n$
   c) $t_n = 5n + 6$
   d) $t_n = 8 - n$
   e) $t_n = \dfrac{n}{n + 1}$
   f) $t_n = 4 \cdot 10^n$
   g) Which sequences in parts a through f are arithmetic?
   h) Which sequences in parts a through f are geometric?

5. Write a formula for the nth term of each of the following sequences and use it to find the indicated term.
   a) 1 4 9 16 25 . . . ; 11th term
   b) −6 −5 −4 −3 −2 . . . ; 50th term
   c) 1 $\frac{1}{2}$ $\frac{1}{3}$ $\frac{1}{4}$ $\frac{1}{5}$ . . . ; 35th term
   d) −4 −8 −12 −16 −20 . . . ; 20th term
   e) 3 9 27 81 243 . . . ; 7th term
   f) $\sqrt{3}$ 2 $\sqrt{5}$ $\sqrt{6}$ $\sqrt{7}$ . . . ; 98th term
   g) $\frac{1}{8}$ $\frac{1}{4}$ $\frac{3}{8}$ $\frac{1}{2}$ $\frac{5}{8}$ . . . ; 24th term
   h) 3 18 108 648 3,888 . . . ; 6th term

6. Use the formula for the sum of the terms of an arithmetic sequence to find each of the following sums.
   a) $2 + 15 + 28 + 41 + 54 + 67 + 80$
   b) $25 + 17 + 9 + 1 + −7 + −15 + −23 + −31$
   c) $2 + 4 + 6 + 8 + \cdots + 150$
   d) $59 + 58 + 57 + 56 + \cdots + 21$

7. Use the formula for the sum of the terms of a geometric sequence and the table on page 801 to find each of the following sums.
   a) The sum of the first seven terms of

   $$2 \quad 6 \quad 18 \quad . . .$$

   b) The sum of the first five terms of

   $$15 \quad 90 \quad 540 \quad . . .$$

   c) The sum of the first ten terms of

   $$6 \quad −12 \quad 24 \quad . . .$$

8. Find the ratio for each of the following geometric sequences in decimal form.
   a) 32 3.2 0.32 . . .
   b) 4 4.8 5.76 . . .
   c) 10 −4.5 2.025 . . .

9. Use the formula for the sum of the terms of an infinite geometric sequence to find each of the following sums.
   a) $40 + 8 + \frac{8}{5} + \frac{8}{25} + \cdots$
   b) $81 + 54 + 36 + 24 + \cdots$
   c) $\frac{1}{2} - \frac{1}{12} + \frac{1}{72} - \frac{1}{432} + \cdots$

10. Express each of the following repeating decimals as the quotient of two integers by treating it as the sum of the terms of an infinite geometric sequence. Reduce your answers to lowest terms.
    a) 0.111111. . .
    b) 0.545454. . .
    c) 0.123123. . .

11. The first eight terms of the arithmetic sequence

    $$199 \quad 409 \quad 619 \quad . . .$$

    are prime numbers.
    a) Find the five terms following 619.
    b) Write a formula for the nth term of the sequence.
    c) Find the eleventh term of the sequence.
    d) Is the eleventh term prime or composite? Justify your answer.

12. In a national backgammon tournament, 512 players make it to the first playoff for the championship. The number of games in each playoff to determine the final winner form a geometric sequence in which the first term is 256 and each successive term is half the preceding one.
    a) List all of the terms of this sequence.
    b) How many playoffs are required to determine the final winner?
    c) How many games are played altogether?

13. Acute Alice bought a new house plant that was 30 centimeters tall when she brought it home from the store. In each successive month, the plant grew by an amount equal to $\frac{2}{5}$ of its previous height.

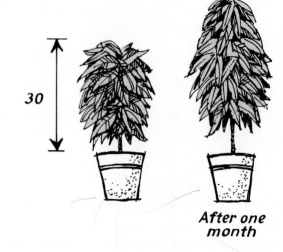

After one month

a) How *much taller* was it at the end of the first month?
b) How *tall* was it at the end of the first month?
c) At the rate of growth described, how long would it take the plant to become 60 centimeters tall? Justify your answer.

Final Review

Test I

1. Arrange in order from smallest to largest:

$$0.01, \quad 0.2, \quad 0.003$$

2. Express in terms of addition: $x - y$.
3. Simplify: $7x - x$.
4. What symbol should replace ▥ in x^2 ▥ 0?
5. Multiply: $2x^2 \cdot 3x^3$.
6. True or false: The expression $x^5 + x - 1$ is a polynomial.
7. Simplify: $\sqrt{50}$.
8. If $a > b$ and $b > c$, what can you conclude about a and c?
9. Add $x - y + 9$ and $x + y - 1$.
10. Solve for x: $(x - 8)^2 = 25$.

11. Write $4\frac{1}{5}$ as a fraction.

12. Write an expression for the perimeter of this rectangle.

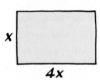

4x

13. Write an expression for its area.
14. Factor $4x^2 - 1$.
15. Find the greatest common factor of $15x^3$ and $21x^2$.
16. True or false: The number 87 is prime.
17. Write $x^2 - 6xy + 9y^2$ as the square of a binomial.

18. Which fraction is larger: $\frac{2}{3}$ or $\frac{7}{11}$?

19. Square as indicated: $(2x - 5)^2$

20. Reduce $\frac{8x}{20}$ to lowest terms.

21. True or false: If x and y are positive numbers, $\sqrt{\dfrac{x}{y}} = \dfrac{\sqrt{x}}{\sqrt{y}}$.

22. Write the coordinate of point P as a fraction.

23. Simplify: $\sqrt{18} + \sqrt{8}$.

24. Simplify: $\dfrac{\dfrac{x}{2} + 1}{\dfrac{x}{2} - 1}$.

25. Solve for x: $\sqrt{x} - 7 = 4$.
26. Graph the equation $3x - 2y = 6$.

27. Multiply: $\dfrac{x}{2} \cdot \dfrac{x}{4}$.

28. Write the conjugate of $x - \sqrt{y}$.
29. Find the value of $x^4 - x^2 + 1$ if $x = 3$.
30. True or false: A cubic equation can have three solutions.

31. Solve for x: $\dfrac{7}{x - 3} = 5$.

32. Write as the quotient of a square root and an integer: $\sqrt{\dfrac{x}{8}}$.

33. Write in descending powers of x:
$$8 + x - 4x^2.$$

34. Solve for x: $-7x > 35$.

35. True or false: The equation $\dfrac{x}{2} = \dfrac{x}{3} + 1$ is a proportion.

36. Factor $x^3y - xy^3$ as completely as you can.

37. Divide: $\dfrac{x^3 + x^2 - 10x - 6}{x - 3}$.

38. Use the formula $v = \dfrac{1}{3}\pi r^2 h$ to find v if $r = 4$ and $h = 6$. Let $\pi = 3.14$.

39. Express the repeating decimal $0.272727\ldots$ as the quotient of two integers.

40. Write an inequality in terms of x to represent this figure.

41. Reduce: $\dfrac{x-3}{5x-15}$.

42. Factor $x^2 + x - 42$.

43. Simplify: $\dfrac{x}{x+2} + \dfrac{2}{x+2}$.

44. Solve for x: $3x - \sqrt{45} = 0$.

45. Write $\dfrac{x-y}{x+y}$ as the difference of two fractions.

46. Write $\dfrac{5}{11}$ in decimal form.

47. Guess a formula for this function:

x	-2	-1	0	1	2
y	2	1	0	1	2

48. Write in scientific notation: 0.0051.

49. Solve for x: $x^2 - 3x = 0$.

50. Find the following quotient: $\dfrac{7 \times 10^6}{5 \times 10^2}$.
    Give your answer in scientific notation.

51. Multiply: $x(x+1)(x+2)$.

52. What must replace ▦ to make $x^2 - ▦ + 49$ the square of a binomial?

53. Multiply: $(x-7)(x+2)$.

54. Solve for x: $\dfrac{2}{x} = \dfrac{5}{12}$.

55. Simplify: $|-9| + |-5|$.

56. Write the first three terms of the sequence whose nth term is given by the formula

$$t_n = \dfrac{1}{n^2}$$

57. Solve the following pair of simultaneous equations:

$$2x - y = 10$$
$$5x + y = 39$$

58. Is the number $\sqrt{5}$ *rational* or *irrational?*

59. Solve for x in terms of the other variables: $\dfrac{a}{x} = \dfrac{1}{b}$.

60. Express as a single fraction: $\dfrac{1}{x} + \dfrac{1}{3x}$.

61. Graph the function $y = x^2 - 4$.
    Let $x = -3, -2, -1, 0, 1, 2,$ and 3.

62. Divide: $\dfrac{x^2}{8} \div \dfrac{x}{4}$.

63. What happens to the direction of an inequality if the same negative number is added to both sides?

64. Solve for x: $|x| + 4 = 13$.

65. Factor $4x^3 - 6x^2$ into prime factors.

66. Find the sum of the first 40 counting numbers: $1 + 2 + 3 + 4 + \cdots + 40$.

67. True or false: If $x < 0$, $|x| = -x$.

68. Solve for x: $x^2 + 4x - 1 = 0$.

69. Solve for x in terms of the other variables: $ax + b < 0$, if $a > 0$.

The number of chocolate mints that you can buy for a dollar depends on the cost of each one.

70. Write a formula for this function, letting x represent the cost of each mint in cents and y represent the number that can be bought for a dollar.

71. How does the number that you can buy for a dollar vary with respect to the cost?

72. Sketch a graph of this type of function.

Acute Alice swam across a swimming pool at a rate of 1.5 meters per second and back at a rate of 1.2 meters per second. The return trip took 20 seconds longer.

73. Draw a diagram to represent this information. Let x represent the time corresponding to the faster rate.
74. Use the information in your diagram to write an equation.
75. How long is the swimming pool?

Packages sent by air parcel post cannot be more than 100 inches in length and girth combined. The girth is the distance around the package, shown as the green rectangle in this figure.

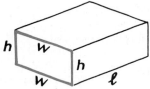

76. Write the parcel post rule as an inequality.
77. Write an inequality to show the possible lengths of a package that has a width of 20 inches and a height of 15 inches.

The number of bacteria in a colony is an exponential function of the time. A typical formula is

$$y = 10 \cdot 2^x$$

in which y is the number of bacteria after x hours.

78. How many bacteria are there at the beginning (when $x = 0$)?
79. How many bacteria are there after five hours?
80. How long does it take the number of bacteria in the colony to double?

Test II

1. True or false: Every number has two square roots.

2. Write the reciprocal of $\frac{x}{9}$. (Assume that x does not equal zero.)
3. What are the slope and y-intercept of the line $y = x + 7$?

4. Reduce to lowest terms: $\frac{7}{35x}$.

5. Simplify: $\sqrt{48} - \sqrt{3}$.
6. Solve for x: $(x + 7)^2 = 64$.

7. Write $\frac{9}{8}$ in decimal form.

8. By what name are second-degree equations called?

9. What number do you think should replace ▨ in the following sequence?

   17   12   7   2   ▨

10. Solve for x: $|x| \le 8$.
11. Simplify: $\sqrt{x^3}$.
12. Arrange in order from smallest to largest:

    $0.\overline{35}, \quad 0.3\overline{5}, \quad 0.35$

13. What symbol should replace ▨ in x ▨ $x + 3$?
14. Solve the following pair of simultaneous equations:

    $x + 4y = 13$
    $y = x - 3$

15. Find the following product:
$(9 \times 10^3)(4 \times 10^5)$. Give your answer in scientific notation.

16. Write as a single fraction: $\dfrac{x}{2} - \dfrac{x}{5}$.

17. Write 2^{-3} without using an exponent.

18. Solve for x: $\dfrac{x}{16} = \dfrac{4}{x}$.

19. Guess a formula for this function:

x	0	1	2	3	4
y	1	2	4	8	16

20. Solve for x:
$(2x - 1)(x + 5) = 2(x^2 + 4) - 1$.
21. Find the product of ax^b and cx^d.
22. Write 370,000 in scientific notation.
23. Simplify: $|-6| \cdot |7|$.
24. Solve for x: $x^2 + 5 = 21$.
25. Write a formula for the nth term of this sequence:

$$3 \quad 9 \quad 27 \quad 81 \quad \cdots$$

26. Express the ratio of white squares to green squares in this pattern in simplest form.

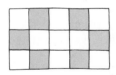

27. Solve for x: $|x - 3| < 7$.
28. Factor 350 into primes.
29. Simplify: $(6x - y) + (2x + 9y) - (7x - 3y)$.
30. Is the number $0.\overline{17}$ *rational* or *irrational*?

31. Write $\dfrac{2x + 5}{x}$ as the sum of an integer and a fraction.

32. Find the distance between points A and B.

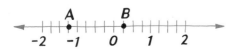

33. Find the coordinate of the point midway between A and B.

34. Solve for x: $\dfrac{4}{x} - \dfrac{2}{9} = \dfrac{1}{x}$.

35. Multiply: $(x - 1)(x^2 + x + 1)$.

36. Reduce $\dfrac{4x}{4x + 8}$.

37. Solve for x: $x^2 - 6x + 2 = 0$.

38. Divide: $\dfrac{x}{3y} \div \dfrac{2x}{y}$.

39. Write $1 - \dfrac{3}{x}$ as a fraction.

40. True or false: For all values of x, $|x| \geq 0$.
41. Solve for x in terms of the other variables:
$$\dfrac{x}{a} - b = \dfrac{1}{a}.$$

42. True or false: The graph of the equation $y = x^2$ is a curved line.

43. Divide: $\dfrac{x^3 + 5x^2 + x - 12}{x + 4}$.

44. What must replace ▓ to make $x^2 + 20x + ▓$ the square of a binomial?
45. Graph the function $y = x^3 + 2$.
Let $x = -2, -1, 0, 1,$ and 2.
46. Simplify and write in descending powers of x: $5x^2 - 5x - x^2 + x^5$.
47. Solve for x: $x^3 + 3x^2 - 10x = 0$.

48. Divide and simplify: $\dfrac{\sqrt{72}}{\sqrt{6}}$.

49. If $a < b$ and $b = c$, what can you conclude about a and c?
50. Factor $15xy + 9y$ as completely as you can.
51. Solve for x in terms of the other variables: $a + bx > 1$, if $b < 0$.

52. Simplify: $\dfrac{9x}{3x - 1} - \dfrac{3}{3x - 1}$.

53. What happens to the direction of an inequality if both sides are multiplied by the same negative number?
54. Multiply: $(3x + 1)(2x + 1)$.

55. Square: $(6\sqrt{5})^2$.
56. Factor $x^6 - 4$.
57. Write $9x^{16}$ as the square of a monomial.
58. Simplify: $\dfrac{\dfrac{1}{x} \cdot \dfrac{1}{y}}{\dfrac{1}{x} + \dfrac{1}{y}}$.
59. Write an inequality in terms of x to represent this figure.

60. Solve for x: $2x - 1 < 17$.
61. Factor $x^2 - 10x + 21$.
62. Solve for x: $x - \sqrt{2} = \sqrt{8}$.
63. Multiply: $\dfrac{x + 10}{5} \cdot 15$.
64. Between what two consecutive integers is $\sqrt{15}$?
65. Use the graph below to find solutions to the equation $x^3 + x^2 - 6x = 0$.

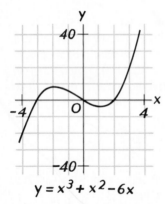

$$y = x^3 + x^2 - 6x$$

66. Factor $x^3 + 2x^2y + xy^2$ as completely as you can.
67. Solve for x: $\sqrt{x + 9} = 2$.
68. Subtract $3x - 2$ from $x^2 + 2x$.
69. What values of x will make the product $(x - 5)(x + 2)$ equal to zero?
70. Square as indicated: $(x^3 + 1)^2$.
71. What happens to the value of $\dfrac{x + 2}{x}$ as x gets larger and larger?

72. Multiply: $\dfrac{1}{x} \cdot \dfrac{x^2}{5}$.
73. List the fourth roots of 16.
74. This figure is an isosceles triangle. Write a formula for its perimeter.

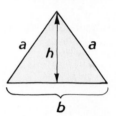

75. The area of a triangle is equal to the product of half its altitude, h, and its base, b. Write a formula for the area, A, of the isosceles triangle.

Mr. Baggins bought 15 packets of seeds for a vegetable garden. Some of the packets cost 20 cents each and the rest cost 30 cents each. Altogether they cost $3.40.

76. Use this information to write a pair of simultaneous equations.
77. How many packets at each price did Mr. Baggins buy?

A person's weight on the moon is less than what it is on the earth. This table shows some typical values.

Weight on earth in pounds	112	126	140	154	168
Weight on moon in pounds	16	18	20	22	24

78. Write a formula for this function, letting x represent the weight on earth and y represent the weight on the moon.
79. How does a person's weight on the moon vary with respect to his or her weight on the earth?
80. If this function were graphed, what would the graph look like?

ANSWERS
to the Set II Exercises

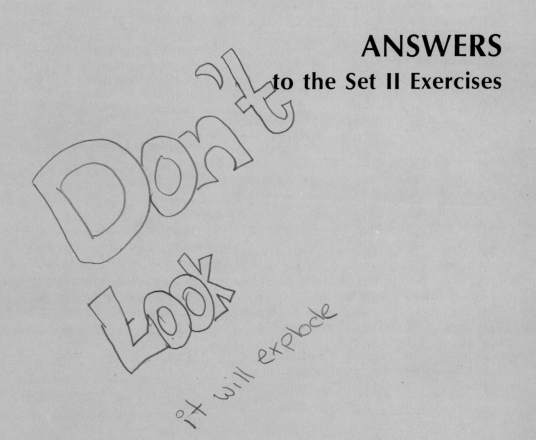

Don't
Look
it will explode

ANSWERS
to the Set II Exercises

Chapter 1, Lesson 1

11. a) $10 + 7$ or 17. b) $x + 7$. c) $10 + y$.
d) $x + y$. e) $4 + 8$ or 12. f) $4 + z$. g) $2 + 5 + 1$
or 8. h) $x + 5 + 1$ or $x + 6$. i) $2 + y + 1$ or
$y + 3$. j) $x + y + 1$. **12.** a) $9 + 4$. b) 13.
c) $x + 5$. d) 7. e) 9. **13.** a) $11 + 4 + 5$ or 20.
b) $x + 6$. c) $x + y$. d) $5 + 3 + x$ or $8 + x$.
e) $x + 1 + y + 1$ or $x + y + 2$. f) $x + y + z$.

14. a) $\square\circ\circ\circ\circ\circ\circ$ and $\circ\circ\circ\circ\circ\circ\square$

b) $\circ\circ\square\circ\circ\circ\circ\circ$ and $\square\circ\circ\circ\circ\circ\circ\circ$

c) $\square\circ\circ\circ\circ\square$ and $\circ\circ\circ\circ\square\square$

15. a) $8 + y + 2$ or $y + 10$.
b) $9 + y + 2$ or $y + 11$. c) $x + 3 + 2$ or $x + 5$.
d) $x + 0 + 2$ or $x + 2$. e) $6 + 2 + 2$ or 10.
16. a) 44. b) $39 + x$. c) $39 + x + 6$ or $x + 45$.
d) $x + 5$. e) $x + y$. f) $x + y + z$.

Chapter 1, Lesson 2

11. a) $10 - 7$ or 3. b) $6 - x$. c) $x - 6$.
d) $11 - 3$ or 8. e) $x - 1$. f) $x - y$. g) $4 - x$.
h) $x - 4$. **12.** a) $12 - 7$ or 5. b) $14 - x$.
c) $x - 3$. d) $y - x$. e) $9 - 2 - 3$ or 4.
f) $x - y - 1$. **13.** a) 2. b) 3. c) 10. d) The
value of $x - 4$ gets larger. e) 12. f) 11. g) 5.
h) The value of $15 - x$ gets smaller. **14.** a) 8.
b) 10. c) 8. d) 10. e) Each expression is
$x + y - 3$. **15.** a) 4. b) $7 - x$. c) 6.
d) $14 - y$. **16.** a) $7,000 - x$ pounds.
b) $7,000 + y$ pounds. **17.** a) 24 cents. b) $y - x$
cents. c) $x + 30$ cents. d) $95 - y$ cents.

Chapter 1, Lesson 3

11. a)
```
○ ○ ○
○ ○ ○            ○ ○ ○ ○
○ ○ ○  for 4·3 and  ○ ○ ○ ○  for 3·4.
○ ○ ○            ○ ○ ○ ○
```
b) $\square\square\square\square\square$ c)

Chapter 1, Lesson 4

12. a) $5 \cdot 6$ or 30. b) $5 + 6$ or 11. c) $5x$.
d) $5 + x$. e) xy. f) $x + y$. g) xx. h) $8x$.
i) $x - 8$. j) $2 + 7 + x$ or $9 + x$. k) $2 \cdot 7 \cdot x$ or
$14x$. l) $10 + y + 3$ or $y + 13$. m) $10 \cdot y \cdot 3$ or
$30y$. n) $4 + x + y$. o) $4xy$. **13.** a) $6 \cdot 2$.
b) $2 \cdot 6$. c) $5x$. d) $11 \cdot 7$. e) $x \cdot 7$ or $7x$. f) xy.
g) $17 + 17 + 17$. h) $x + x + x + x$.
i) $2 + 2 + \cdots + 2$ (y of them).
j) $z + z + \cdots + z$ (y of them). **14.** a) $7 \cdot 8$ or
56. b) $10x$. c) xy. d) xx. **15.** a) 140.
b) $354x$. **16.** a) $7x$. b) $24x$. c) $1,440$.
d) $1,440x$. e) $10,080x$. f) $100x$. g) $1,200x$.
17. a) 165. b) $11x$.

Chapter 1, Lesson 4

9. a) $\dfrac{12}{3}$ or 4. b) $12 - 3$ or 9. c) $\dfrac{7}{x}$. d) $\dfrac{x}{7}$. e) $\dfrac{x}{2}$.

f) $x \cdot 2$ or $2x$. g) $\dfrac{10}{x}$. h) $10 - x$. i) $\dfrac{x}{y}$. j) xy.

10. a) $\dfrac{12}{3} = 4$ and $\dfrac{12}{4} = 3$. b) $\dfrac{16}{4} = 4$. c) $\dfrac{18}{3} = 6$

and $\dfrac{18}{6} = 3$. **11.** a) $5 \cdot 3 = 15$. b) $23 \cdot 4 = 92$.

c) $12 \cdot 0 = 0$. d) $(7.5)(1) = 7.5$. e) $10 \cdot 7 = x$.
f) $x \cdot 12 = 36$. g) $4x = 20$. h) $y \cdot 2 = x$.
12. a) 45. b) 63. c) 108. d) The value of $9x$ gets
larger. e) 1. f) 5. g) 25. h) The value of $\dfrac{x}{4}$ gets

larger. i) 15. j) 6. k) 0.5. l) The value of $\dfrac{30}{x}$

gets smaller. **13.** a) 20. b) $\dfrac{300}{x}$. **14.** a) $170x$

dollars. b) 600. c) $\dfrac{x}{170}$. **15.** a) $12x$ inches or x

feet. b) 50. c) $\dfrac{x}{12}$. **16.** a) 10.6. b) $\dfrac{159}{x}$.

Chapter 1, Lesson 5

11. a) "x squared" and "x to the second power."
b) An exponent. **12.** a) 3^2. b) 5^2. c) x^2.
d) 4^3. e) x^3. **13.** a) 7^2. b) 2^6. c) x^3. d) x^8.
e) 3^x. f) x^y. **14.** a) 7^4. b) 4^7. c) x^6. d) 2^{12}.
e) 2^x. f) x^y. g) $8 \cdot 8 \cdot 8 \cdot 8 \cdot 8$. h) $x \cdot x \cdot x$.
i) $3 \cdot 3 \cdot \ldots \cdot 3$ (x of them). j) $y \cdot y \cdot \ldots \cdot y$ (x of
them). **15.** a) 2,401. b) 49^2. **16.** a) 3^6.
b) 2^6. c) 4^3. d) 8^2. e) 10^4. f) 10^9. g) Because
all powers of 1 are equal to 1. **17.** a) 512.
b) 14,641. c) 2,187. d) 390,625. e) x. f) x^2.

Chapter 1, Lesson 6

4. a) The sum of a number and zero is the number.
b) The difference between a number and zero is the
number. c) The product of a number and zero is
zero. d) The product of a number and one is the
number. e) The quotient of zero and a nonzero
number is zero. f) The quotient of a number and
zero is not defined. g) The quotient of a number
and one is the number. **5.** a) 45. b) 0. c) 11.
d) 1. **6.** a) $1^2 = 1$ because 1 times 1 is 1. b) 1.
c) 1. **7.** a) 0. b) 10 has a "higher value" than 1.
8. a) x. b) 0. c) x. d) $x + 1$ cannot be
simplified. e) x. f) 0. g) $\frac{x}{0}$ is not defined.
h) x. **9.** a) $x + y$. b) x. c) y. d) 0.
e) $x + y$. f) 0. g) x. h) y. **10.** a) 1. b) 0.
c) Even.

Chapter 1, Lesson 7

4. a) Figure 4. b) Figure 5. c) Figure 2.
d) Figure 6. e) Figure 3. f) Figure 4.
g) Figure 1. **5.** a) 50. b) 32. c) 48. d) 19.
e) 9. f) 400. g) 57. h) 57. i) 27. j) 27. k) 18.
l) 12. m) 18. n) 19. o) 52. p) 39. q) 531.
6. a) $x^2 + y^2$. b) $10 - 5x$. c) $\frac{x}{5} - 10$. d) $8x^3$.
e) $y^4 - y$. f) $\frac{12}{x} + 2$. g) $x + xy$. **7.** a) 2.
b) 26. c) 128. d) 458. **8.** a) 19. b) 9.
c) 101. d) 900. e) 84. f) 82. **9.** a) 845 cents or
$8.45. b) $80x + 95y$ cents.

Chapter 1, Lesson 8

4. a) Yes. b) No. c) Yes. d) No. e) Yes.

f) Yes. g) No. h) Yes. **5.** a) 63. b) 441.
c) 15. d) 23. e) 20. f) 48. g) 1. h) 24. i) 5.
j) 9. k) 6. l) 3. m) 9. n) 9. o) 5. p) 80.
q) 225. **6.** a) Figure 2. b) Figure 3.
c) Figure 1. d) Figure 5. e) Figure 4.
f) Figure 6. g) Figure 1. h) Figure 6.
7. a) $(x - 5)^3$. b) $x \cdot 6 + y$ or $6x + y$.
c) $(y + 6)x$. d) $\frac{10}{x} - y$. e) $\frac{10 - y}{x}$.
f) $(x + 2)(x + 7)$. g) $\frac{x - y}{2x}$. h) $11 - (3x)^2$.
i) $(11 - 3x)^2$. j) $(x^3 + y^3)8$ or $8(x^3 + y^3)$.
8. a) 0. b) 9. c) 105. d) 2,585. e) 0. f) 9.
g) 105. h) 2,585.

Chapter 1, Lesson 9

4. a) $3(6 + 2) = 3(6) + 3(2)$.
b) $4(7 - 3) = 4(7) - 4(3)$.
c) $5(1 + 8) = 5(1) + 5(8)$.
d) $6(5 - 1) = 6(5) - 6(1)$. **5.** a) $4x^3$. b) $7(2x)$.
c) $3(x + 1)$. d) $9(x + y)$. e) $x^4 + x^4$.
f) $3x + 3x + 3x + 3x + 3x$.
g) $(x + 7) + (x + 7) + (x + 7) + (x + 7)$.
6. a) $3(x + 5) = (x + 5) + (x + 5) + (x + 5)$
$\qquad = x + x + x + 5 + 5 + 5$
$\qquad = 3x + 15$
b) $2(x + y) = (x + y) + (x + y)$
$\qquad = x + x + y + y$
$\qquad = 2x + 2y$
c) $4(x^2 + 1)$
$\qquad = (x^2 + 1) + (x^2 + 1) + (x^2 + 1) + (x^2 + 1)$
$\qquad = x^2 + x^2 + x^2 + x^2 + 1 + 1 + 1 + 1$
$\qquad = 4x^2 + 4$
7. a) $8x + 24$. b) $5y - 10$. c) $x^2 + x$.
d) $xy - y^2$. e) $2x + 18$. f) $4y + xy$. g) $7y - 7x$.
h) $x^2 - 6x$. i) $10x^2 + 40$. j) $x^4 - x$.
8. a) $\begin{array}{r} 72 \\ \times 43 \\ \hline 216 \\ 2880 \\ \hline 3096 \end{array}$
b) $43 \cdot 72 = (40 + 3)72$
$\qquad = 40 \cdot 72 + 3 \cdot 72$
$\qquad = 2880 + 216$
$\qquad = 3096$
c) $\begin{array}{r} 43 \\ \times 72 \\ \hline 86 \\ 3010 \\ \hline 3096 \end{array}$
d) $72 \cdot 43 = (70 + 2)43$
$\qquad = 70 \cdot 43 + 2 \cdot 43$
$\qquad = 3010 + 86$
$\qquad = 3096$

9. a) $4(x + 5)$ and $4x + 20$. b) $x(10 + x)$ and
$10x + x^2$. c) $3(x + y + 3)$ and $3x + 3y + 9$.
d) $x(x^2 + x + 1)$ and $x^3 + x^2 + x$. **10.** a) $x + y$.
b) $2(x + y)$. c) $2x$. d) $2y$. e) $2x + 2y$.

Chapter 1, Review

1. a) $3 \cdot 11$. b) 2^7. c) $x + x + x + x + x$.
d) $y \cdot y \cdot y \cdot y$.
2. Step 1. Think of a number.
Step 2. Add 1.
Step 3. Multiply by 4.
Step 4. Add 8.
Step 5. Divide by 4.
Step 6. Subtract the number that you first
thought of.

3. a) $a - 5$. b) b^3. c) $2 + c$. d) $\dfrac{1}{d}$. **4.** a) 2^5.

b) It is impossible to express 3 as a power of 1.
c) 10^6. **5.** a) $x - 3$. b) $x + 5$. **6.** a) 9.

b) $\dfrac{20}{x}$. c) $\dfrac{x}{y}$. **7.** a) $72 + x$. b) $72 - y$.

c) $3 + x$. **8.** a) 600. b) 3,600. c) 45. d) 55.
9. a) $\quad 3^2 - 1^2 = 8 = 2^3$
$\qquad 6^2 - 3^2 = 27 = 3^3$
$\qquad 10^2 - 6^2 = 64 = 4^3$
$\qquad 15^2 - 10^2 = 125 = 5^3$
b) $21^2 - 15^2 = 216 = 6^3$

10. a) $(x - 6)2$ or $2(x - 6)$. b) $\dfrac{x}{8} + 4$.

c) $150 - x^3$. **11.** a) $8v + 88$. b) $3w - 18$.
c) $xy + xz$. **12.** a) $3(x + 4) = 3x + 12$.
b) $(6 + x)x = 6x + x^2$. c) $y(x + 1) = yx + y$.

13. a) $3x$. b) $200 - 3x$ kilograms. c) $\dfrac{x}{3}$.

14. a) 0. b) 10. c) 136. d) 0. e) 10. f) 136.

Chapter 2, Lesson 1

4. a)

x	0	1	2	3	4
y	5	6	7	8	9

b)

x	0	2	4	6	8
y	0	8	16	24	32

c)

x	0	1	2	3	4
y	3	5	7	9	11

d)

x	1	2	3	4	5
y	1	4	9	16	25

e)

x	2	4	6	8	10
y	6	3	2	1.5	1.2

5. a)

x	2	3	4	5
y	13	28	49	76

b)

x	2	3	4	5
y	12	45	112	225

6. a) $y = 6x$. b) $y = x + 7$. c) $y = x - 4$.
d) $y = x^2$. e) $y = x^2 + 1$. f) $y = 6x - 1$.
g) $y = 10x + 3$. h) $y = 11x$. i) $y = 12 - x$.

j) $y = \dfrac{20}{x}$. **7.** a) 40. b) $p = 4s$. c) 100.

d) $a = s^2$. **8.** a) 11 meters per second.
b) $m = 11s$. c) 770 meters. **9.** a) The population
is increasing. b) No.

Chapter 2, Lesson 2

4. O $(0, 0)$; A $(1, 10)$; B $(6, 6)$; C $(3, 7)$; D $(7, 3)$;
E $(8, 0)$; F $(0, 5)$.
5.
a) 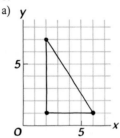 b)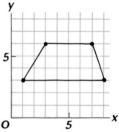

6. a) $y = 2x$. b) 4. c)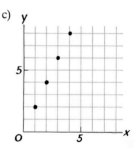

d) They lie on a straight line.
7. a)

x	0	1	2	3	4
y	4	5	6	7	8

b) $y = x + 4$.
8. a)

x	0	1	2	3	4
y	7	6	5	4	3

b) $y = 7 - x$.
9. a)

x	0	2	4	6	8
y	8	6	4	2	0

b)

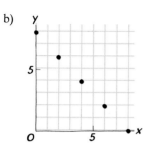

c)

x	0	1	2	3	4
y	0	2	4	6	8

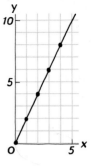

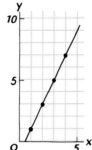

10. a)

x	1	2	3	4
y	1	4	7	10

d)

x	1	2	3	4
y	1	3	5	7

b)

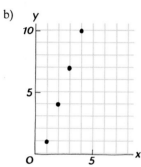

e)

x	0	1	2	3
y	0	1	4	9

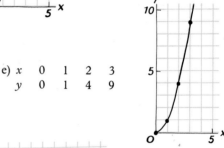

Chapter 2, Lesson 3

4. a)

x	0	1	2	3
y	0	3	6	9

b) $y = 3x$.　**5.** a) $x = 2$.　b) $y = 1$ and
$y = 3$.　**6.** a) Yes.　b) Yes.　c) No.　d) Yes.
e) Yes.　f) Yes.　g) No.　h) No.　i) Yes.　j) Yes.
7.

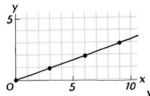

f)

x	0	3	6	9
y	0	1	2	3

a)

x	0	1	2	3	4
y	0	1	2	3	4

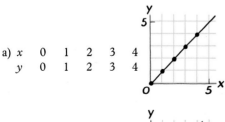

g)

x	0	3	6	9
y	4	5	6	7

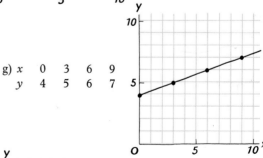

b)

x	0	1	2	3	4
y	2	3	4	5	6

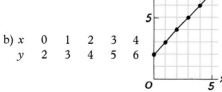

h)

x	1	2	3	4	5
y	3	1.5	1	0.75	0.6

Chapter 2, Lesson 4

4. a)

x	0	1	2	3
y	0	5	10	15

b) y is tripled.   c) A direct variation.   d) $y = 5x$.
e) 35.

5. a)

x	4	8	12	20	100
y	3	6	9	15	75

b) $\dfrac{3}{4}$.   c) $y = \dfrac{3}{4}x$.   **6.** a) $y = 0$.

b) $y = a \cdot 0 = 0$.   c) $(0, 0)$.   d) The line goes through the origin.   **7.** a) Yes.   b) No.   c) Yes.
d) No.   e) No.

8. a)

x	0	1	2	3
y	0	2	4	6

b)

x	0	1	2	3
y	0	3	6	9

c)

x	0	1	2	3
y	0	4	8	12

d)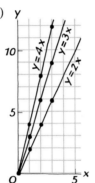

e) 2 for $y = 2x$; 3 for $y = 3x$; 4 for $y = 4x$.
f) $y = 4x$.   g) The steepness increases as the constant of variation increases.   **9.** a) 75 centimeters.   b) 240 centimeters.   c) The bounciness increases as the constant of variation increases.
d) No; in that case the ball would bounce higher and higher.   **10.** a) $y = 45x$.   b) A steep line going through the origin.

Chapter 2, Lesson 5

4. a)

x	0	1	2	3
y	2	3	4	5

b)

x	0	1	2	3
y	4	5	6	7

c)

x	0	1	2	3
y	7	8	9	10

d)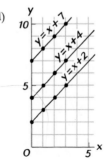

e) They are parallel.   f) $(0, 2)$ for $y = x + 2$; $(0, 4)$ for $y = x + 4$; $(0, 7)$ for $y = x + 7$.   g) At $(0, 10)$.

5. a)

x	0	1	2	3
y	1	3	5	7

b)

x	0	1	2	3
y	1	4	7	10

c)

x	0	1	2	3
y	1	5	9	13

d)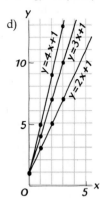

e) They all go through $(0, 1)$.   f) $(0, 1)$.   g) $(0, 1)$.
6. a)

x	0	1	2	3	4
y	5	5	5	5	5

b)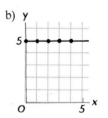

c) It is a horizontal straight line.   d) $y = 5$.
7. a) The difference in successive values of y is 3.
b) Below $x = 0$.   c) $y = 2x + 8$.   d) $y = 7x + 1$.
e) $y = 4x + 6$.   f) $y = 6x + 4$.
8. a) $y = 2x + 6$.   b) Linear.   c) No.
9. a)

x	0	1	2	3	4
y	20	20.5	21	21.5	22

b) 20 centimeters.   c) 0.5 centimeter.
d) 30 centimeters.

Chapter 2, Lesson 6

4. a)

x	1	2	3	4	5	6
y	12	6	4	3	2.4	2

b) y is divided by 3.   c) An inverse variation.
d) $y = \dfrac{12}{x}$.   e) 1.2.

5. a)

x	2	4	6	12	20
y	30	15	10	5	3

b) 60.   c) $y = \dfrac{60}{x}$.   **6.** a) $y = a$.   b) 0.   c) We cannot divide by 0.   d) No.   **7.** a) Yes.   b) No.
c) No.   d) Yes.   e) No.

8. a)

x	1	2	3	4	5
y	4	2	1.3	1	0.8

b)

x	1	2	3	4	5
y	6	3	2	1.5	1.2

c)

x	1	2	3	4	5
y	10	5	3.3	2.5	2

d)

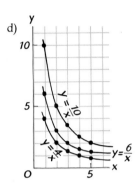

e) 4 for $y = \dfrac{4}{x}$; 6 for $y = \dfrac{6}{x}$; 10 for $y = \dfrac{10}{x}$.

f) The larger the constant of variation, the farther the curve is from the origin.   **9.** a) $y = \dfrac{24}{x}$.

b) 24. c) Inversely.

10. a)

2	4	5	10
50	25	20	10

b) $s = \dfrac{100}{t}$. c) It is approximately 9.19 meters per second.

Chapter 2, Review

1.

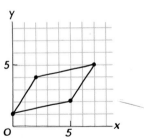

2. a)

x	1	2	3	4	5
y	3	7	11	15	19

b)

x	0	1	2	3	4
y	0	4	10	18	28

c)

x	1	2	3	4	5
y	12	14	16	18	20

3. a) True. b) False. c) True. d) True.
e) False.   **4.** a) $y = 8 - x$. b) $y = x^2$.

c) $y = 10x - 1$. d) $y = \dfrac{3}{2}x$ or $y = 1.5x$.

5. a)

x	0	1	2	3	4
y	3	5	7	9	11

b) No. c) A linear function. d) $y = 2x + 3$.
e) 53.

6. a)

x	1	2	3	4
y	1.5	3	4.5	6

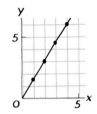

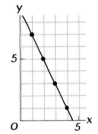

b)

x	1	2	3	4
y	7	5	3	1

c)

x	1	2	3	4
y	2	5	10	17

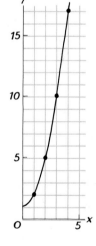

7. a) $y = \dfrac{x}{2}$. b) The water height varies directly with respect to the time. c) A straight line going through the origin.

8. a)

x	0	6	12	18	24
y	0	25	50	75	100

b) 24 karats.

c)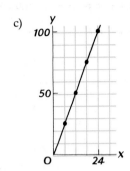

d) A direct variation.

9. a)

x	12	14	16	18	20
y	15	12.9	11.25	10	9

b)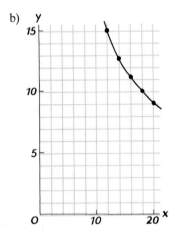

c) Inversely.   d) It is impossible to divide by 0.

Chapter 3, Lesson 1

4. a) 39 degrees below zero on the Celsius scale.
b) 29 meters below sea level.   c) 37 years after the birth of Christ.   d) 10 minutes ago.
5. a) $+12 > -15$.   b) $-196° < -78°$.   c) $-3 < +3$.
d) $-22 > -24$.   **6.** a) $4 > 1$.   b) $0 < 9$.
c) $7 > -7$.   d) $-3 < 0$.   e) $5 > -11$.   f) $-1 > -6$.
g) $-12 < 8$.   h) $-10 < -2$.   **7.** a) $+7$.   b) -8.
c) $-8, -6, -4, -2, +1, +3, +5, +7$.   **8.** a) 3.   b) -3.
c) 5.   d) -5.   e) 0.   f) 0.   g) -1.   h) 1.   **9.** a) 5.
b) 5.   c) 8.   d) 8.   e) 6.   f) 20.   g) 7.   h) 7.   .
10. a) $x < 0$.   b) $x^2 > 5$.   c) $x + 1 < 10$.

d) $\frac{x}{2} > 8$.

Chapter 3, Lesson 2

4. A $(8, 1)$; B $(-6, 2)$; C $(-8, -5)$; D $(0, -7)$; E $(6, -6)$;
F $(10, -2)$.
5.

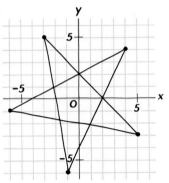

6. a)

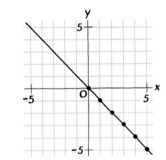

b) $y = -x$.   c) See graph in part a.

d)

x	-5	-4	-3	-2	-1
y	5	4	3	2	1

7. a)

x	0	1	2	3	4
y	3	4	5	6	7

b) $y = x + 3$

c)

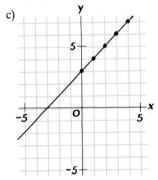

d)

x	-4	-3	-2	-1
y	-1	0	1	2

8. a)

x	0	1	2	3	4	5
y	5	4	3	2	1	0

b and c)

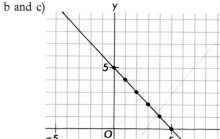

d)

x	-4	-3	-2	-1
y	9	8	7	6

and

x	6	7	8	9
y	-1	-2	-3	-4

9. a)

x	1	2	3	4
y	3	5	7	9

and

x	1	2	3	4
y	0	1	2	3

b)

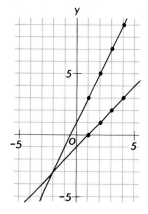

c) Intersect at $(-2, -3)$.

Chapter 3, Lesson 3

4. a) o o   o o o $\longrightarrow$ o o o o o

b) ● ● ● ● ● ● ● ●   ● $\longrightarrow$ ● ● ● ● ● ● ● ● ●

c) ● ● ● ● o o o o $\longrightarrow$

d) o o o o o o ● ● $\longrightarrow$ o o o o

e) o o o ● ● ● ● ● $\longrightarrow$ ● ●

5. a) -8. b) 3. c) -1. d) 15. e) $-x$. f) y.
6. a) 0. b) 0. c) 7. d) -20. e) -5. f) -14.
g) -10. h) -21. i) -19. j) 8. k) -32. l) 8.
7. a) -5. b) 2. c) -2. d) -6. e) 11. f) -11.
g) -10. h) 10. i) 10. j) -4. k) -8. l) 2.
8. a) -10. b) -4. c) -2. d) 0. e) -14. f) -10.
g) -8. h) -12. i) -22. j) -36. k) -6. l) 6.

Chapter 3, Lesson 4

4. a)   o o o o o o ⊠ ⊠

b)   ● ● ● ⊠ ⊠ ⊠ ⊠ ⊠

c)   ⊠ ⊠ ⊠ ⊠
　　　　●

d)   ⊠ ⊠ ⊠ ⊠ ⊠ ⊠ ⊠ ⊠
　　　　　　o o o

e)   o o o o o o o
　　　　　　　⊠

f)   ● ● ● ● ●
　　　⊠ ⊠ ⊠

5. a) $7 + -2 = 5$. b) $-8 + 5 = -3$.
c) $3 + -4 = -1$. d) $-5 + 8 = 3$. e) $6 + 1 = 7$.
f) $-2 + -3 = -5$. **6.** a) 7. b) -7. c) 17.
d) -17. e) -20. f) -2. g) 11. h) 5. i) -2. j) 2.
k) 0. l) -8. **7.** a) 4. b) 4. c) 10. d) 10.
e) -5. f) -5. g) 11. h) 11. i) -14. j) -14.
k) 6. l) 6. m) $x - y - z$. n) $x - y + z$.
8. a) Nov., 17; Dec., 21; Jan., 18. b) Dec.
9. a) $15 - 32 = -17$; $55 - 12 = 43$; $0 - 18 = -18$.
b) $-16 + -17 + 43 + -18$. c) He came out $8
behind $(-16 + -17 + 43 + -18 = -8)$.

Chapter 3, Lesson 5

4. a) $3(4) = 4 + 4 + 4 = 12$.
b) $3(-4) = -4 + -4 + -4 = -12$.
c) $4(-3) = -3 + -3 + -3 + -3 = -12$. **5.** a) -21.
b) -36. c) 25. d) -56. e) 0. f) 132. g) -36.
h) 1. i) -39. j) 300.
6. a)

x	3	2	1	0	-1	-2	-3
$-1 \cdot x$	-3	-2	-1	0	1	2	3

b) The opposite of the number. c) $-x$. d) 3 or 2 or 1. e) -1 or -2 or -3. f) No.
7. a) -108. b) -108. c) 10. d) -10. e) -10.
f) 0. g) -1. h) -16. i) -16. j) 9.   **8.** a) 105.
b) -105. c) 105. d) -105. e) 16. f) -32. g) 64.
h) 100. i) $-1,000$. j) -120. k) -120. l) 0.
9. a) $<$. b) $>$. c) $>$. d) $=$. e) $<$.
10. a) -15. b) 12. c) 25. d) -11. e) 1. f) 30.
11. a) $(-3)^2 = 9$; $(-3)^3 = -27$; $(-3)^4 = 81$;
$(-3)^5 = -243$; $(-3)^6 = 729$. b) The even powers.
c) Negative.

Chapter 3, Lesson 6

4. a) -6. b) -6. c) -15. d) 15. e) -7. f) 1.
g) 0. h) -1.
5. a)

x	3	2	1	0	-1	-2	-3
$\dfrac{x}{-1}$	-3	-2	-1	0	1	2	3

b) The opposite of the number. c) $-x$. d) Not
necessarily.   **6.** a) -9. b) 3. c) -8. d) 10.
e) -6. f) 4. g) -1. h) 7.   **7.** a) $>$. b) $=$.
c) $=$. d) $=$. e) $<$. f) $=$.   **8.** a) -14. b) 14.
c) 18. d) -18. e) 32. f) -32. g) -4. h) -4.
i) 5. j) -5.   **9.** a) -6. b) -4. c) -5. d) 4.
e) 32. f) 2.

Chapter 3, Lesson 7

4. a)

7	2	0	-4	-10
4	-1	-3	-7	-13
8	-2	-6	-14	-26
16	6	2	-6	-18
8	3	1	-3	-9
1	1	1	1	1

b) Think of a number: □
Subtract three: □ ● ● ●
Multiply by two: □□ ● ● ● ● ● ●
Add eight: □□ ○ ○
Divide by two: □ ○
Subtract the number that
  you first thought of: ○

5. a) 45. b) -75. c) -128. d) 32. e) -81.
f) 27.   **6.** a) -10. b) -10. c) 14. d) 10. e) -7.
f) 17. g) 64. h) 64. i) -80. j) -80. k) 81.
l) -248. m) 21. n) -24. o) -64. p) 64.

7. a) -25. b) -25. c) -20. d) 20. e) 40.
f) 40.   **8.** a) 30 meters per second. b) 0 meters
per second. c) -20 meters per second. d) The
arrow is at its highest point. e) The arrow is coming
down.   **9.** a) 0 degrees. b) -15 degrees.
c) -40 degrees.

Chapter 3, Review

1. a) True. b) True. c) False. d) False.
e) True.   **2.** a) 12. b) -4. c) 4. d) -3.
3. a) $x < -4$. b) $2x = x^2$. c) $\dfrac{x}{2} > 5$.
4. a) 10. b) -1. c) -22.
5. a) $5(-3) = -3 + -3 + -3 + -3 + -3 = -15$.
b) $4(-x) = -x + -x + -x + -x = -4x$.
6. a)

3	0	-1	-8
12	0	-4	-32
2	-10	-14	-42
5	-10	-15	-50
1	-2	-3	-10
8	5	4	-3
5	5	5	5

b) Think of a number: □
Multiply by four: □□□□
Subtract ten: □□□□ ●●●●● / ●●●●●
Add the number that you
  first thought of: □□□□□ ●●●●● / ●●●●●
Divide by five: □ ● ●
Add seven: □ ○ ○ ○ ○ ○
Subtract the number that
  you first thought of: ○ ○ ○ ○ ○

7. a) 3. b) -8. c) -13. d) 16. e) 0. f) -2.
8. a) $=$. b) $<$. c) $<$. d) $>$. e) $=$.
9. a) -4. b) -9. c) x. d) 1.
10. a)

x	0	1	2	3
y	3	2	1	0

b and c)

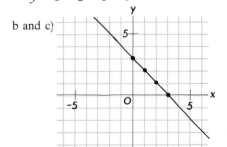

d)
x	-3	-2	-1
y	6	5	4

and

x	4	5	6
y	-1	-2	-3

11. a) 20. b) 19. c) 9. d) 150. e) 0.

12. a) His assets are \$300 more than his liabilities.
b) \$125. c) \$375. **13.** a) 1. b) -25. c) -36.
d) -7. e) -8.

14. a) $4(1 - -9) = 4(10) = 40$
$\quad\quad 4(1) - 4(-9) = 4 + 36 = 40$
b) $-6(2 - 5) = -6(-3) = 18$
$\quad\quad -6(2) - (-6)(5) = -12 + 30 = 18$
c) $-7(-3 - 8) = -7(-11) = 77$
$\quad\quad -7(-3) - (-7)(8) = 21 + 56 = 77$

Chapter 4, Lesson 1

4. a) 1.4. b) 0.14. c) 0.014. d) 0.0625.

e) 0.625. f) 6.25. g) 7. h) 70. i) 0.7. **5.** a) $\dfrac{4}{1}$.

b) $\dfrac{0}{1}$. c) $\dfrac{7}{10}$. d) $\dfrac{1}{100}$. e) $\dfrac{65}{10}$. f) $-\dfrac{29}{10}$.

g) $\dfrac{22}{7}$. h) $-\dfrac{17}{2}$.

6. a)

b) $6\dfrac{1}{5}$. c) -7.1. **7.** a) $4.2 < 4.3$.

b) $-4.2 > -4.3$. c) $-7.6 < 6.7$. d) $-7.6 < -6.7$.
e) $0.05 < 0.5$. f) $0.05 > -0.5$. g) $2.1 > 2.09$.
h) $-2.1 < -2.09$. **8.** a) 13.05. b) 11.55.
c) 9.225. d) 16.4. e) 0.242. f) 0.198.
g) 0.00484. h) 10.

Chapter 4, Lesson 2

4. a) 8. b) $-(-5)$ or 5. c) 0.6. d) $-(-3.4)$ or 3.4.
e) 0. f) x. g) $-x$. **5.** a) 13. b) 13. c) 5.
d) -5. e) 21. f) 21. g) 4. h) 0.25. **6.** a) $=$.
b) $>$. c) $<$. d) $=$. e) $>$. f) $<$. g) $>$.
h) $<$. **7.** a) False; example, $|-1| \neq -1$. b) False;
example, $|-1| + |2| \neq |-1 + 2|$. c) True. d) False;
example, $|(-1)^3| \neq (-1)^3$. **8.** a) -10.2. b) 6.2.
c) 2.3. d) -2.3. e) -8. f) -3.95. g) -7.2.
h) -0.81. **9.** a) 2.94. b) -5.31. c) -6.105.
d) -3.108. **10.** a) -3.87. b) 1.889. c) -18.46.
d) 18.46. e) -7.66. f) 7.66.

Chapter 4, Lesson 3

4. a) $6.1 + 1.9 = 8$. b) $-2.5 + 2.5 = 0$.
c) $-5 + -7.1 = -12.1$. d) $3.03 + -4 = -0.97$.
5. a) 3.8. b) -3.4. c) -2.24. d) -1.8. **6.** a) 9.
b) -9. c) -99.99. d) -99.99. e) 25.6. f) 0.256.
7. a) 3. b) 3. c) -0.7. d) -7. e) -11.2.
f) 0.112. **8.** a) -2.5. b) -1. c) 3.2. d) -1.6.
e) 3.3. f) -0.1. g) 5.2. h) -3.6.
9. a)
4.5	1.8	-2.6
-0.5	2.2	6.6
1.5	-6.6	-19.8
7.5	-0.6	-13.8
2.5	-0.2	-4.6
-2	-2	-2

b) -2. **10.** a) 8.835. b) -22.56. c) -43.62.
d) 314.64. e) -14.17. f) 222.222.

Chapter 4, Lesson 4

4. a) 3. b) 2. c) -5. d) -2. e) 13. f) 0.
5. a) 3. b) 3.1. c) 3.14. d) 3.142. e) 3.1416.
f) 3.14159. **6.** a) 0.33. b) 0.67. c) -2.43.
d) -0.24. e) -0.02. f) 0. **7.** a) -5.8. b) 3.3.
c) 6.1. d) -6.1. e) 11.4. f) 1.1. g) 0.1. h) -0.5.
i) -4.6. j) -0.2. **8.** a) -2.2. b) -2.2. c) -13.1.
d) 1.2. **9.** a) 5.29. b) -9.91. c) -17.56.
d) 5.34. e) -3.29.

Chapter 4, Lesson 5

4. a)
x	0	2	3	5	8
y	-1	0	0.5	1.5	3

b and c)

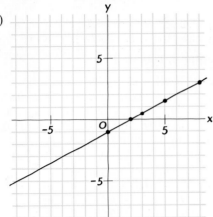

b)

x	−1	0	1	2	3	4	5
y	−7	−5	−3	−1	1	3	5

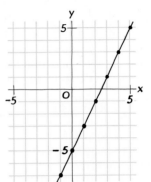

d) (4, 1), (7, 2.5), and (−4, −3).   e) (4, 1), (7, 2.5), and (−4, −3).

5. a)

x	0	0.1	0.2	0.3	0.4	0.5	0.6	0.7	0.8	0.9	1.0
y	0	0.00	0.01	0.03	0.06	0.13	0.22	0.34	0.51	0.73	1.00

b and c)

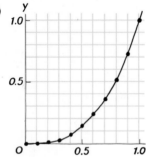

c)

x	−3	−2	−1	0	1	2	3
y	4	−1	−4	−5	−4	−1	4

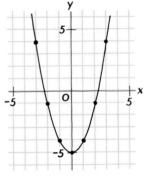

6. a) Linear functions.   b) (0, b) is the point in which the graph crosses the y-axis.   c) $y = x + 3$.
d) $y = x − 4$.

7. a)

x	−2	−1	0	1	2
y	6	3	0	−3	−6

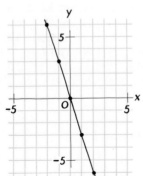

d)

x	−3	−2	−1	0	1	2
y	6	2	0	0	2	6

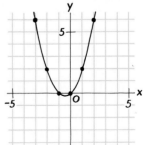

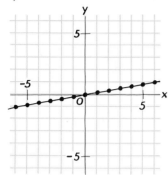

e)

x	-6	-5	-4	-3	-2	-1	
y	-1	-0.8	-0.7	-0.5	-0.3	-0.2	
(x con'd)	0	1	2	3	4	5	6
(y con'd)	0	0.2	0.3	0.5	0.7	0.8	1

f)

x	-6	-5	-4	-3	-2	-1	
y	-1	-1.2	-1.5	-2	-3	-6	
(x con'd)	0	1	2	3	4	5	6
(y con'd)	—	6	3	2	1.5	1.2	1

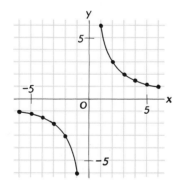

8. a) $y = -3x$, $y = 2x - 5$, and $y = \dfrac{x}{6}$. b) $y = -3x$ and $y = \dfrac{x}{6}$. c) $y = x^2 - 5$, $y = x^2 + x$, and $y = \dfrac{6}{x}$. d) $y = \dfrac{6}{x}$.

Chapter 4, Review

1. a) $\dfrac{-8}{1}$. b) $\dfrac{0}{1}$. c) $\dfrac{61}{10}$. d) $-\dfrac{13}{3}$.

2. a) -0.007. b) 1.75. c) -0.8125.   **3.** a) <.
b) >. c) <. d) =. e) >. f) <.   **4.** a) 5.3.

b) -6.4. c) -7.7. d) -2.7. e) 13. f) -2.4.
g) -3.2. h) -1.12. i) -7. j) -3.6. k) -1.728.
l) -2.5.   **5.** a) -5.7. b) 0. c) -2. d) 4.6.
e) 12.4. f) 12.

6. a)

3.5	-8.1	-0.6
-7	16.2	1.2
0	23.2	8.2
0	11.6	4.1
-4.5	7.1	-0.4
-1	-1	-1

b) -1.   **7.** a) 6. b) -2.74. c) 8.0. d) -5.
8. a) 1.71. b) 17.14. c) 0.58. d) 0.06.
9. a)

x	-6	-5	-4	-3	-2		
y	-3.5	-2.5	-1.5	-0.5	0.5		
(x con'd)	-1	0	1	2	3	4	5
(y con'd)	1.5	2.5	3.5	4.5	5.5	6.5	7.5

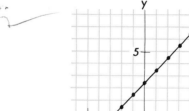

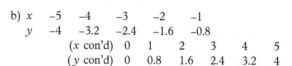

b)

x	-5	-4	-3	-2	-1	
y	-4	-3.2	-2.4	-1.6	-0.8	
(x con'd)	0	1	2	3	4	5
(y con'd)	0	0.8	1.6	2.4	3.2	4

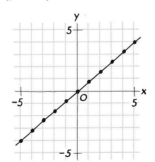

c)

x	−4	−3	−2	−1
y	1	1.3	2	4

(x con'd)	0	1	2	3	4
(y con'd)	—	−4	−2	−1.3	−1

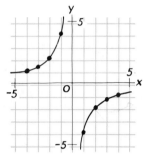

d)

x	−2	−1	0	1	2	3
y	6	2	0	0	2	6

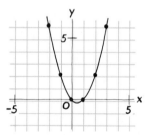

10. a) $y = x + 2.5$ and $y = 0.8x$. b) $y = 0.8x$.

c) $y = \dfrac{-4}{x}$ and $y = x^2 - x$. d) $y = \dfrac{-4}{x}$.

Chapter 5, Lesson 1

4. a) If a certain number is multiplied by 2, the result is 8. b) If 3 is added to a certain number, the result is 11. c) If a certain number is divided by 6, the result is 5. d) If a certain number is cubed, the result is 1. e) If 5 is raised to a certain power, the result is 25. **5.** a) True. b) False. c) Neither. d) True. e) False. f) Neither. **6.** a) 95. b) 20. c) 9. d) 4. e) 36. f) 8. g) 4 and −4. h) 0. i) No number, because a number cannot be 2 more than itself. j) 0 and 1. k) −24. l) −7. m) −10. n) 0. o) 0. p) 1. q) True for any number. r) 2. s) −1. t) No number, because the square of a number is not negative. u) −1. v) 2. w) 3. x) No number, because every power of 1 is 1. **7.** a) 1 and 3. b) 0, −2, and 5. c) −3 and −4. d) 1, 3, 8, −2, and −9.

Chapter 5, Lesson 2

4. a) Subtraction. b) Division. c) Addition. d) Multiplication. **5.** a) Divide by 5. b) Subtract 5. c) Add 2. d) Multiply by 2. e) Subtract −8 (or add 8). f) Divide by −8. g) Multiply by −6. h) Add −6 (or subtract 6). i) Add x. j) Multiply by x. **6.** a) 7. b) $x − 3$. c) 12. d) $x + 5$. e) 4. f) $\dfrac{x}{8}$. g) 18. h) $2x$. **7.** a) $2x + 6$.

b) $2(x + 6)$. c) $\dfrac{x}{5} - 1$. d) $\dfrac{x - 1}{5}$.

e) $(x + 3) - y$. f) $(x − y) + 3$. g) $\dfrac{4x}{y}$. h) $4\left(\dfrac{x}{y}\right)$.

8. a) Subtract 6 and divide the result by 2. b) Divide 2 and subtract 6 from the result. c) Add 1 and then multiply the result by 5. d) Multiply by 5 and then add 1 to the result. e) Add y and then subtract 3 from the result. f) Subtract 3 and then add y to the result. g) Multiply by y and divide the result by 4. h) Divide by 4 and multiply the result by y.

9. a)

				e)			
−3	−2	−1	0	−4	0	4	8
−8	−7	−6	−5	−1	0	1	2
−1	0	1	2	7	8	9	10

b)

				f)			
−3	−2	−1	0	−4	0	4	8
4	5	6	7	4	8	12	16
−1	0	1	2	1	2	3	4

c)

				g)			
−2	−1	0	1	−6	−4	−2	0
−6	−3	0	3	−36	−24	−12	0
−8	−5	−2	1	−18	−12	−6	0

d)

				h)			
−2	−1	0	1	−6	−4	−2	0
−4	−3	−2	−1	−3	−2	−1	0
−12	−9	−6	−3	−18	−12	−6	0

i) Tables a and b; tables g and h. **10.** a) False. b) False. c) True.

Chapter 5, Lesson 3

4. a) Four 1-pound weights were removed from each pan. b) The contents of each pan was divided by 3. c) $3x + 4 = 10$. d) $3x = 6$. e) $x = 2$. f) Subtract 4 from each side of the equation. g) Divide each side of the equation by 3. h) The weight of one brick is 2 pounds. **5.** a) $2x = 5$. b) $x = 0$. c) $7x = 14$. d) $x = −72$. e) $3x = −6$. f) $x = −6$. **6.** a) Divide each side of the equation by 4. b) Multiply each side of the equation by 4.

c) Add 5 to each side of the equation. d) Subtract 5 from each side of the equation. e) Subtract 9 from each side of the equation. f) Multiply each side of the equation by 6. g) Subtract 21 from each side of the equation. h) Divide each side of the equation by 3. i) Add 1 to each side of the equation. j) Multiply each side of the equation by 4. k) Subtract 6 from each side of the equation. l) Multiply each side of the equation by –1.
7. a) –9. b) 0.25. c) 4. d) 5. e) –22. f) –40. g) –10. h) –7. **8.** a) 7.8. b) 7.5. c) 10.2. d) 10.8. e) –7.5. f) 4. g) –7.3. h) –2.1. **9.** a) –2. b) 8. c) 48. d) –12. e) 3. f) –9. **10.** a) –6. b) –2. c) 0. d) 18. e) 6. f) –12.

Chapter 5, Lesson 4

4. a) Yes. b) Yes. c) No. d) Yes. e) Yes. f) Yes. g) No. h) Yes. i) No. j) Yes. k) Yes. l) Yes. **5.** a) $1 + 8x$. b) $(x + 3)x$. c) $x^2 + 3x$. d) $(5 + 2) + y$ or $7 + y$. e) $(5 \cdot 2)y$ or $10y$. f) $24x - 6$. g) $x + 1$. h) $7(xx)$ or $7x^2$.
6. a) Commutative property of multiplication. b) Distributive property of multiplication over addition. c) Associative property of addition. d) Distributive property of division over subtraction. e) Associative property of multiplication. f) Commutative property of addition. **7.** a) $10x$. b) $x + 10$. c) $8x$. d) $4x^2$. e) $8x$. f) 0. g) $16x^2$. h) $10x$. i) $8x$. j) x. k) $-x$. **8.** a) $7x$. b) $8x^3$. c) $3x + 4$. d) $8x^2$. e) $x - 7$. f) $x - 9$. g) $4x - 5$. h) $10x$.
9. a) Think of a number: x
　　　Subtract 3:　　　　$x - 3$
　　　Multiply by 2:　　$2(x - 3) = 2x - 6$
　　　Add 8:　　　　　$(2x - 6) + 8 = 2x + 2$
　　　Divide by 2:　　　$\dfrac{2x + 2}{2} = x + 1$
　　　Subtract the number
　　　　first thought of:　$(x + 1) - x = 1$
b) 1.

Chapter 5, Lesson 5

4. a) One box was removed from each pan. b) Two circles were removed from each pan. c) The contents of each pan was divided by 5 (because 5 boxes are balanced by 5 colored circles, one box would be balanced by 1 colored circle).

d) $6x + 2 = x - 3$. e) $5x + 2 = -3$. f) $5x = -5$. g) $x = -1$. h) Subtract x from each side of the equation. i) Subtract 2 from each side of the equation. j) Divide each side of the equation by 5. k) $x = -1$.
5.

a)
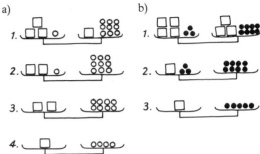
b)

6. a) 4. b) 2.2. c) –1. d) 5. e) 15. f) –7. g) 6. h) 1.2. **7.** a) 8. b) –4. c) 7. d) –2.5. e) –18. f) 3.6. g) 9. h) –89. **8.** a) 1.1. b) –8.5. c) 5.7. d) 3.3. e) –82.

Chapter 5, Lesson 6

4. a) $x + 8$. b) $4x$. c) $2x + 20$. d) $5x + 10$.
5. a) $7x$. b) $4x^2$. c) $8x - 32$.
6. $x + (x + 2) + 5 = 19$; 6, 8, and 5. **7.** a) 5, 8, 5, and 8. b) 4.5, 4.5, and 7. c) 4, 11, 4, and 11. d) 7, 4, 3, 1, and 2. **8.** a) 9 and 4. b) 6 and 5. c) 5 and 7. **9.** a) AE = 35; CD = 39. b) AE = 24.5; EB = 13.5. c) AB = 9; CE = 2; ED = 7. d) AE = 1.6; EB = 21.6; CD = 23.2.

Chapter 5, Lesson 7

4.　　a)

b)

c)

5. a)

t	0	1	2	3	4
d	0	45	90	135	180

b) Directly. c) $d = 45t$.
6. a)

r	400	450	500	550
t	9.9	8.8	7.92	7.2

b) $t = \dfrac{3{,}960}{r}$. c) Inversely. **7.** a) 15. b) 50. c) 0.5. d) $65x$. e) $\dfrac{25}{y}$. f) $\dfrac{z}{75}$.

8. a) $80x = 24{,}900$.   b) 311.25 miles per day.
9. a) 2.04 meters per second; 1.82 meters per second; 1.72 meters per second; 1.66 meters per second.
b) 122.45 meters per minute.   c) $7{,}346.94$ meters per hour.   d) 7.35 kilometers per hour.

Chapter 5, Lesson 8

4. a) $x + 2$.
b)
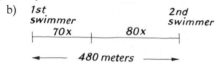

c) $660(x + 2) = 720x$.   d) 22 minutes.   e) $15{,}840$ meters.   **5.** a) $70x$ meters and $80x$ meters, respectively.

b)

1st swimmer   2nd swimmer

| $70x$ | $80x$ |

←———— 480 meters ————→

c) $70x + 80x = 480$.   d) 3.2 minutes.   e) 224 meters and 256 meters.   **6.** a) $2 - x$ hours.

b)

| $110x$ | With wind |
| $90(2-x)$ | Against wind |

c) $110x = 90(2 - x)$.   d) 0.9.   e) 198 miles.

Chapter 5, Review

1. a) True.   b) False.   c) True.   d) Neither.
2. a) -11.   b) 13.   c) No number, because no number equals itself increased by 3.   d) 48.   e) 5 or -5.   f) 3.   **3.** a) 2 and -4.   b) 6 and -1.   c) 3, 4, 5, 6, and 7.   **4.** a) Divide by 6 and subtract 2.
b) Add 5 and multiply by 10.   c) Subtract y and divide by 3.   d) Multiply by 7 and add y.
5. a) -7.   b) 7.   c) -0.75.   d) -1.   e) 72.   f) -2.
6. a) -5.   b) 39.   c) 2.   d) -4.   e) -13.
7. a) $x + (9 + 3)$ or $x + 12$.   b) $3(x + 9)$.
c) $2x + 1$.   d) $2(xx)$ or $2x^2$.   e) $x^2 - 10x$.
8. a) $9x$.   b) $9 + x$.   c) $20x$.   d) $7x^2$.   e) $7 + 2x$.
f) $9x$.   **9.** a) 0.5.   b) 4.   c) -3.   d) 8.   e) -6.
f) 1.4.   **10.** a) 5.   b) -2.5.   c) 4.   d) -14.5.
11. a) Perimeter, $2(x + 2)$ or $2x + 4$; area, $2x$.
b) Perimeter, $4(5x)$ or $20x$; area, $(5x)^2$ or $25x^2$.
c) Perimeter, $2x + 22$; area, $4(x + 7)$ or $4x + 28$.
12. a) 5, 12, and 8.   b) 6.5, 9, 19.5, and 13.   c) 5, 7,

5, and 7.   d) 8, 12, 8, and 12.   **13.** a) $CE = 24$; $ED = 20$.   b) $AE = 7$; $CD = 36$.   c) $AB = 46$; $CE = 6$; $ED = 40$.
14. a)
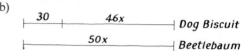

| r | 350 | 400 | 500 |
| t | 2 | 1.75 | 1.4 |

b) $t = \dfrac{700}{r}$.   c) Inversely.   **15.** a) 27 miles per day.   b) $27x$ miles.   **16.** a) Dog Biscuit, $46x$ feet; Beetlebaum, $50x$ feet.
b)

| 30 | $46x$ | Dog Biscuit |
| $50x$ | | Beetlebaum |

c) $30 + 46x = 50x$.   d) 7.5.   e) Dog Biscuit, 345 feet; Beetlebaum, 375 feet.

Chapter 6, Lesson 1

4. a) 15.   b) 27.   c) 15.   d) 69.   e) 2.   f) -22.
g) -10.   h) 2.   i) 14.   j) 34.   k) 34.   l) 34.   m) 4.
n) 40.   o) 49.   p) 40.   **5.** a) Yes.   b) No.
c) Yes.   d) No.   e) Yes.   f) No.   g) No.   h) Yes.
i) No.   j) Yes.   k) Yes.   l) Yes.   m) No.   n) Yes.
o) No.   p) Yes.   **6.** a) $2x + 6y$.
b) $2x + 6y = 30$.   c) Yes.   d) Yes.   e) No.
f) $3xy$.   g) $3xy = 36$.   h) Yes.   i) No.   j) Yes.
7. a)

| x | 1 | 0 | -1 | -2 | -3 |
| y | 13 | 9 | 5 | 1 | -3 |

b) y is decreased by 4.

c)
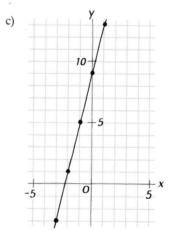

d) They lie on a straight line.

8. a) $(1, 5)$, $(2, 4)$, $(3, 3)$, $(4, 2)$, and $(5, 1)$.   b) $(1, 5)$, $(2, 3)$, and $(3, 1)$.   c) Unlimited number.   d) $(1, 10)$,

(2, 5), (5, 2), and (10, 1).   e) (1, 2) and (6, 1).
f) None.   g) Unlimited number.   h) (2, 1).

Chapter 6, Lesson 2

4. a) $y = x + 3$.   b) $y = x - 8$.   c) $y = 4x$.

d) $y = \dfrac{x}{2}$.   e) $y = 10 - x$.   f) $y = 5x + 1$.

5. a) $x = y - 3$.   b) $x = y + 8$.   c) $x = \dfrac{y}{4}$.

d) $x = 2y$.   e) $x = 10 - y$.   f) $x = \dfrac{y - 1}{5}$.

6. a) $x = 3 - y$.   b) $y = 3 - x$.   c) $x = \dfrac{6}{y}$.

d) $y = \dfrac{6}{x}$.   e) $x = \dfrac{5y}{2}$.   f) $y = \dfrac{2x}{5}$.   g) $x = \dfrac{y + 8}{4}$.

h) $y = 4x - 8$.   **7.** a) $a = bh$.   b) $b = \dfrac{a}{h}$.

c) $h = \dfrac{a}{b}$.   **8.** a) $x = 2a - y$.   b) $y = 2a - x$.

c) 7.   d) $3 = 2(7) - 11 = 14 - 11$.   e) –3.
f) $-8 = 2(-3) - 2 = -6 - 2$.   **9.** a) 85 meters.

b) $a = h - \dfrac{r^2}{20}$.   c) 70 meters.   **10.** a) 37 cents.

b) $n = \dfrac{c - 5}{4}$.   c) $8 = \dfrac{37 - 5}{4} = \dfrac{32}{4} = 8$.

11. a) 120.   b) $m = \dfrac{ci}{100}$.   c) $18 = \dfrac{15(120)}{100} = 18$.

d) $c = \dfrac{100m}{i}$.

Chapter 6, Lesson 3

4. a) $6x + 1y = 2$; $a = 6$, $b = 1$, $c = 2$.
b) $2x + 3y = -7$; $a = 2$, $b = 3$, $c = -7$.
c) $2x + 8y = 9$; $a = 2$, $b = 8$, $c = 9$.
d) $4x + -5y = 1$; $a = 4$, $b = -5$, $c = 1$.
e) $3x + 0y = 11$; $a = 3$, $b = 0$, $c = 11$.
f) $1x + 2y = 8$; $a = 1$, $b = 2$, $c = 8$.
g) $1x + -6y = -10$; $a = 1$, $b = -6$, $c = -10$.
h) $-0.5x + 7y = 0$; $a = -0.5$, $b = 7$, $c = 0$.

5. a) $y = 4x + 3$.   b) $y = 9x - 1$.   c) $y = \dfrac{x - 10}{5}$.

d) $y = \dfrac{1 - 8x}{2}$.   e) $y = -5x$.   f) $y = 6x - 7$.

6. a) $y = \dfrac{15 - 5x}{2}$.

b) Example table:

x	0	1	2	3
y	7.5	5	2.5	0

c)

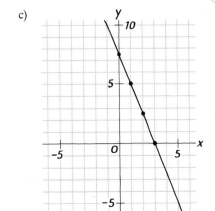

d) Yes.   e) Yes.   f) No.   g) No.

7. a) Example table:

x	–1	0	1	2
y	2	2	2	2

b)

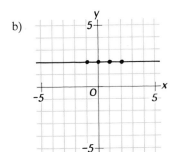

c) No.   d) No.   e) Yes.   f) Yes.

8.

a)

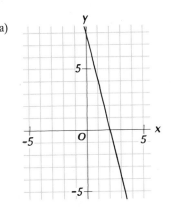

b)

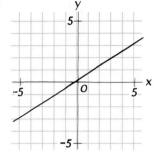

c)

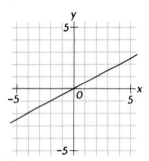

d)

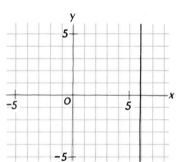

9. a–d)

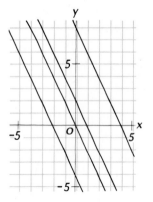

e) All the lines are parallel.

10. a–d)

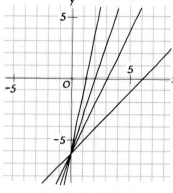

e) All the lines pass through the point (0, –6)

Chapter 6, Lesson 4

4. a) *x*-intercept, 3; *y*-intercept, 4. b) *x*-intercept, –3; *y*-intercept, 1. c) *x*-intercept, –2; *y*-intercept, –5. d) *x*-intercept, 0; *y*-intercept, 0. e) *x*-intercept, 4; No *y*-intercept. **5.** a) 10 and 4. b) 1 and 8. c) 12 and –4. d) 0 and 0. e) –4.5 and 2. f) –3 and 6. g) 2.8; no *y*-intercept. h) No *x*-intercept; 5.5. **6.** a) –6 and 2. b) –6 and 2. c) –6 and 2. d) –6 and 2. e) All the graphs are the same line (because they have the same *x*- and *y*-intercepts). f) The equations are equivalent. **7.** a) 6 and 12.

b)

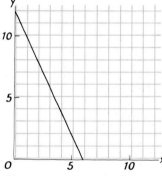

c) Yes. d) Yes. e) No. f) No.
8.
a)

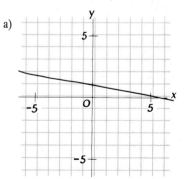

b)

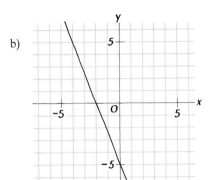

c)

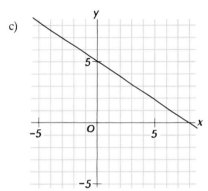

d)

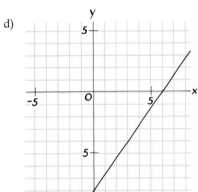

e)

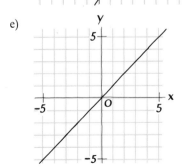

(The line goes through the origin; so, to plot it, a point other than the intercepts must be found.)

f)

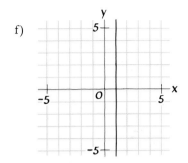

9. a–c)

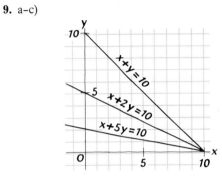

d) All have *x*-intercept 10.

10. a–c)

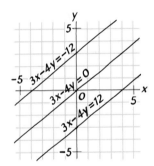

d) All the lines are parallel.

Chapter 6, Lesson 5

4. a) 4. b) 0.5. c) –3. d) 0.

5. a)

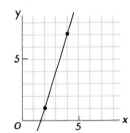

Slope = 3.

b)

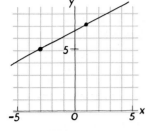

Slope = 0.5.

c)

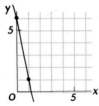

Slope = –5.

d)

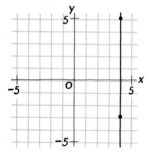

Slope is not defined.

6. a)

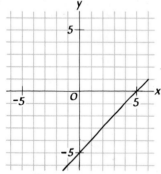

Slope = 1.

b)

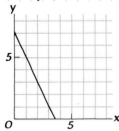

Slope = –2.

c)

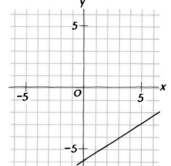

Slope = 0.6.

d)

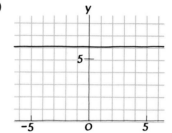

Slope = 0.

7. a and b)

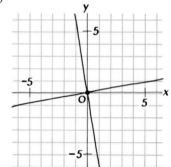

c and d)

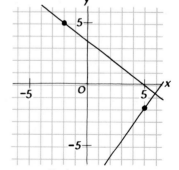

e) They are perpendicular.

8.

a)

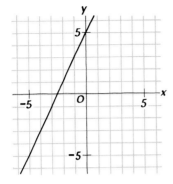

b)

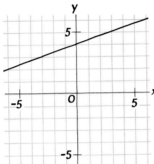

c)

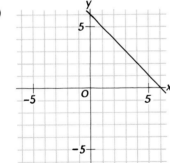

d)

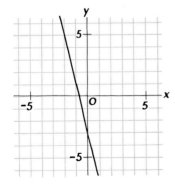

e)

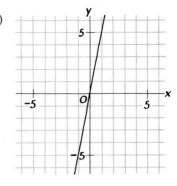

9. a) Slope, 2; y-intercept, 5.   b) Slope, $\frac{1}{3}$; y-intercept, 4.   c) Slope, -1; y-intercept, 6.
d) Slope, -4; y-intercept, -3.   e) Slope, 5; y-intercept, 0.   f) Slope, 7; y-intercept, 2.   g) Slope, -3; y-intercept, -8.

Chapter 6, Lesson 6

4. a) Linear.   b) Straight lines.   c) The slope.
d) The y-intercept.   **5.** a) Slope, 8; y-intercept, 3.
b) Slope, $\frac{1}{2}$; y-intercept, 5.   c) Slope, 1; y-intercept, -7.   d) Slope, 6; y-intercept, -12.   e) Slope, 2; y-intercept, 0.   f) Slope, -1; y-intercept, 0.   g) Slope, -3; y-intercept, 10.   h) Slope, 0; y-intercept, 4.
6. a) $y = 9x + 1$.   b) $y = \frac{2}{3}x - 6$.
c) $y = -4x + 7$.   d) $y = 2.5x$.   e) $y = -8$.
f) $y = 0$.   **7.** a) $y = 2x + 3$.   b) $y = \frac{1}{3}x - 4$.
c) $y = -x + 5$.   d) $y = -1$.

8.

a)

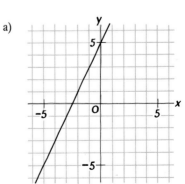

b)

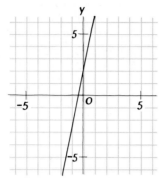

c)

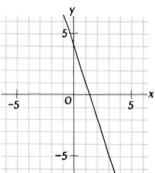

d)

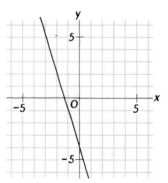

e)

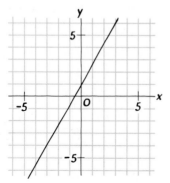

f)

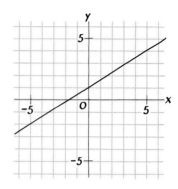

9. a) $y = x + 5$.

b)

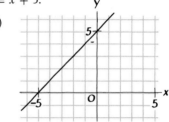

c) $(-5, 0)$. d) $(0, 5)$. e) 1. **10.** a) $y = 2x + 4$.

b)

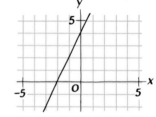

c) $(-2, 0)$. d) $(0, 4)$. e) 2.

Chapter 6, Review

1. a) 14. b) –8. c) 14. d) 32. e) 9. f) 9.
2. a) Yes. b) No. c) Yes. d) No. e) Yes.
f) Yes. **3.** a) (13, 1) and (6, 2). b) Unlimited
number. c) (1, 12), (2, 6), (3, 4), (4, 3), (6, 2), and
(12, 1). d) None. **4.** a) $x = \dfrac{y + 7}{3}$.

b) $y = 3x - 7$. c) $x = 9 - 4y$. d) $y = \dfrac{9 - x}{4}$.

5. a) $m = dv$. b) $v = \dfrac{m}{d}$. c) = 3 kilograms per

liter. d) $7.5 = 3(2.5) = 7.5$ e) $2.5 = \dfrac{7.5}{3} = 2.5$.

6. a) 3. b) 126 pounds. c) $y = \dfrac{x - w}{3}$. d) 20.

7. a) $6x + -1y = 12$; $a = 6$, $b = -1$, $c = 12$.
b) $6x + 2y = 5$; $a = 6$, $b = 2$, $c = 5$.
c) $4x + -1y = -1$; $a = 4$, $b = -1$, $c = -1$.
d) $0x + 7y = 9$; $a = 0$, $b = 7$, $c = 9$.

8. a) $y = 5x + 10$. b) $y = \dfrac{x + 3}{4}$. c) $y = \dfrac{9 - 6x}{2}$.

d) $y = 8x - 1$.   **9.** a) x-intercept, 14; y-intercept, 4. b) x-intercept, 1.8; y-intercept, –9.
c) x-intercept, 8; y-intercept, 3. d) x-intercept, 3; no y-intercept.

10. a)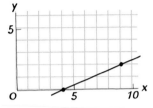

Slope $= \dfrac{2}{5}$.

b)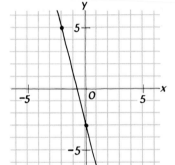

Slope $= -4$.

c)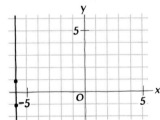

Slope is not defined.

11.
a)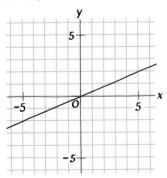

b)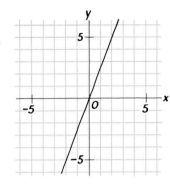

c)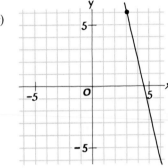

d)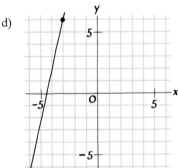

12. a) Slope, 4; y-intercept, –5. b) Slope, $\dfrac{1}{2}$; y-intercept, 0. c) Slope, –1; y-intercept, 11.
d) Slope, 0; y-intercept, 8.   **13.** a) $y = 3x + 2$.
b) $y = -\dfrac{5}{2}x + 5$. c) $y = x$. d) $y = 0$.

14.
a)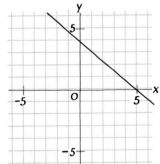

b)

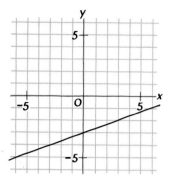

c)

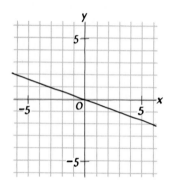

d)

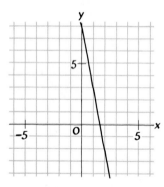

e)

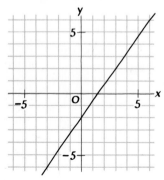

f)

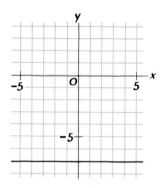

Chapter 7, Lesson 1

4. a) Yes. b) No. c) No. d) No. e) Yes.
f) No. g) Yes. h) Yes. i) Yes. **5.** a) $2x = 14$.
b) $x = 7$. c) $y = 2$. d) $(7, 2)$. e) $7 - 2(2) = 3$.
6. a) $3x + y = 22$. b) $3x - y = 14$. c) $x = 6$;
$y = 4$. d) $3(6) + 4 = 22$ and $3(6) - 4 = 14$.
7. $x = 11$; $y = 8$. **8.** $x = 8$; $y = 10.5$.
9. $(15, 6)$. **10.** $(2, -1)$. **11.** $(-11, 14)$.
12. $(5, 27)$. **13.** $(2.5, 1)$. **14.** $(11, 0)$.

Chapter 7, Lesson 2

4. a) 5. b) 7. c) $2x + y = 17$ and $x + y = 12$.
d) $(5, 7)$. **5.** $4x + 3y = 57$, $4x + y = 43$; $x = 9$,
$y = 7$. **6.** $5x - 2y = 36$, $x - 2y = 2$; $x = 8.5$,
$y = 3.25$. **7.** a) $6x = 30$; $x = 5$, $y = 6$.
b) $2y = 12$; $y = 6$, $x = 5$. c) $-2y = -12$; $y = 6$,
$x = 5$. **8.** $(4, 1)$. **9.** $(7, -5)$. **10.** $(8, -3)$.
11. $(2.5, 0)$. **12.** $(-2, -9)$. **13.** $(-3, 40)$.
14. $(4.5, -1.5)$. **15.** $(25, 0)$. **16.** $(7, 2.25)$.

Chapter 7, Lesson 3

3. $x + 2y = 16$, $2x + 4y = 32$; multiply each side by
2. **4.** $2x - y = 5$, $6x - 3y = 15$; multiply each
side by 3. **5.** $4x + 4y = 28$, $x + y = 7$;
divide each side by 4. **6.** a) $3x + 12y = 21$.
b) $16x - 40y = 8$. c) $4x + y = 10$.
d) $-6x + y = 3$. e) $5x - 15y = 0$.
f) $3x - 2y = 9$. **7.** a) $12x + 3y = 90$.
b) $13x = 91$. c) $x = 7$. d) $y = 2$.
e) $4(7) + 2 = 30$, $28 + 2 = 30$, $30 = 30$;
$7 - 3(2) = 1$, $7 - 6 = 1$, $1 = 1$. f) $4x - 12y = 4$.
g) $13y = 26$. h) $y = 2$. i) $x = 7$.
8. $(11, -4)$. **9.** $(7, 3)$. **10.** $(4, 0)$.
11. $(-1, 5)$. **12.** $(8, -2)$. **13.** $(-6, 10)$.

Chapter 7, Lesson 4

4. $(4, -1)$.   **5.** $(-2, 3)$.   **6.** $(-3.5, -5)$.

7. a, c, and e)

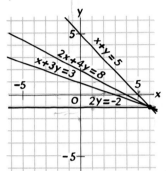

b) $2x + 4y = 8$.   d) $2y = -2$.   f) All four lines go through $(6, -1)$.

8. a, c, and e)

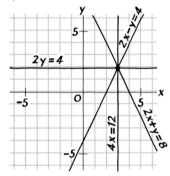

b) $4x = 12$.   d) $2y = 4$.   f) All four lines go through $(3, 2)$.

9.

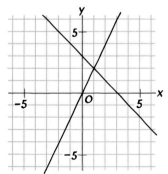

Solution: $(1, 2)$.

10.

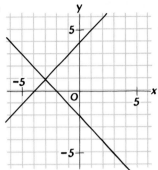

Solution: $(-3, 1)$.

11.

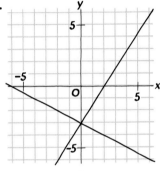

Solution: $(0, -3)$.

12.

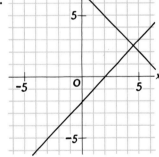

Solution: $(4.5, 2.5)$.

Chapter 7, Lesson 5

4. a) There is one solution: $(-3, 2)$.   b) There are no solutions (the equations are inconsistent).   c) There are infinitely many solutions (the equations are equivalent).   **5.** a) $2x + 2y = 25$; $x + y = 12$.
c) Inconsistent.   **6.** a) $x + 2y = 14$; $3x + 6y = 42$.   c) Equivalent.
7. a) $2x - 2y = 16$; $5x - 5y = 40$.
c) Equivalent.   **8.** a) $3x - y = 7$; $6x - 2y = 10$.
c) Inconsistent.

9. a)

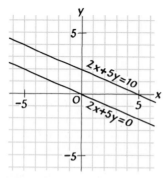

b) The lines are parallel. c) None. d) Inconsistent.

10. a)

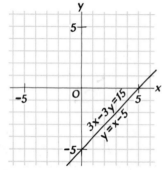

b) Each graph is the same line. c) Infinitely many.
d) Equivalent.

11.

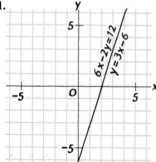

The equations have infinitely many solutions.

12.

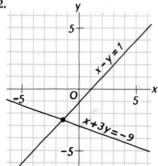

The equations have one solution: $(-1.5, -2.5)$.

13.

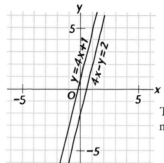

The equations have no solutions.

Chapter 7, Lesson 6

4. $2x + y = 35$; $y = 3x$. Solution: $(7, 21)$.
5. $x + 3y = 20$; $x = y + 8$. Solution: $(11, 3)$.
6. $2x + 3 = y$; $4x + 2y = 42$. Solution: $(4.5, 12)$.
7. $2x + 14 = 4y$; $9 + 3x = 4y$. Solution: $(5, 6)$.
8. $(8, 8)$.   **9.** $(11, 9)$.   **10.** $(-1, 3)$.
11. $(5, 14)$.   **12.** a) $2(3y + 7) - 6y = 11$,
$6y + 14 - 6y = 11$, $14 = 11$. b) The equations are
inconsistent. c) Parallel lines.
13. a) $4(2x + 1) - 8x = 4$, $8x + 4 - 8x = 4$,
$4 = 4$. b) The equations are equivalent. c) A single
line.   **14.** a) $x + y = 35$; $3x = 12y$. b) $(28, 7)$.
c) $28 + 7 = 35$, $35 = 35$; $3(28) = 12(7)$, $84 = 84$.
15. a) $x + y = 20$; $6x = 4y$. b) $(8, 12)$.
c) $8 + 12 = 20$, $20 = 20$; $6(8) = 4(12)$, $48 = 48$.
16. a) $x + y = 24$; $3x = 5y$. b) $(15, 9)$.
c) $15 + 9 = 24$, $24 = 24$; $3(15) = 5(9)$, $45 = 45$.

Chapter 7, Lesson 7

4. a) $x + y = 52$. b) $5x$. c) $10y$.
d) $5x + 10y = 450$. e) $x = 14$, $y = 38$. f) 14
nickels and 38 dimes.   **5.** a) $x + y = 20$. b) $45x$.
c) $60y$. d) $45x + 60y = 960$. e) $x = 16$, $y = 4$.
f) 16 liters of apple juice and 4 liters of cranberry
juice.   **6.** a) $x + y = 95$; $3x + 5y = 353$.
b) $x = 61$, $y = 34$. c) 61 plants at \$3 and 34 plants
at \$5.   **7.** a) $x + y = 10$; $7.2x + 11.0y = 83.4$.
b) $x = 7$, $y = 3$. c) 50.4 grams of tin and 33 grams
of lead.

Chapter 7, Review

1. a) Yes. b) No. c) Yes. d) Yes. e) No.
2. a) $7x = 14$. b) $5x - 6y = 10$. c) $2x - y = 4$.
d) $6x + 18y = 12$.   **3.** $(25.5, 17.5)$.
4. $(3, -11)$.   **5.** $(-1, 7)$.   **6.** $(-4, -5)$.
7. $(9, 2)$.   **8.** $(5, -1)$.

9.

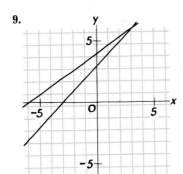

The equations have one solution: (3, 6).

10.

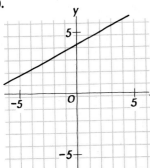

The equations have infinitely many solutions.

11.

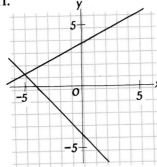

The equations have one solution: (–5, 1).

12.

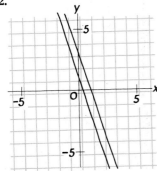

The equations have no solutions.

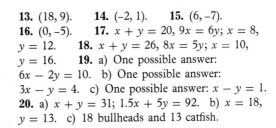

13. (18, 9).　　**14.** (–2, 1).　　**15.** (6, –7).
16. (0, –5).　　**17.** $x + y = 20$, $9x = 6y$; $x = 8$, $y = 12$.　　**18.** $x + y = 26$, $8x = 5y$; $x = 10$, $y = 16$.　　**19.** a) One possible answer: $6x - 2y = 10$.　b) One possible answer: $3x - y = 4$.　c) One possible answer: $x - y = 1$.
20. a) $x + y = 31$; $1.5x + 5y = 92$.　b) $x = 18$, $y = 13$.　c) 18 bullheads and 13 catfish.

Chapter 8, Lesson 1

4. a) One hundred.　b) One thousand.　c) Ten thousand.　d) One hundred thousand.　e) One million.　**5.** a) 10^8.　b) 10^{11}.　c) 10^6.　d) 10^4.
6. a) 34,000.　b) 57.2.　c) 9.　d) 0.016.　e) 10^{16}.
f) 8×10^8.　g) 3×10^{11} or 0.3×10^{12}.　　**7.** a) 75.
b) 0.28.　c) 0.0001.　d) 10^5.　e) 4.1×10^9 or 41×10^8.　**8.** a) 2,000.　b) 1,000,000,000.
c) 75,000.　d) 60.2.　e) 300,000,000,000.
f) 84,000.　　**9.** a) 42,000,000.　b) 4.2×10^7.
10. a) 3×10^4.　b) 8×10^1.　c) 7.2×10^2.
d) 2.01×10^7.　e) 1.984×10^3.　f) 6.0005×10^5.
11. a) 1×10^{12}.　b) 5×10^6.　c) 9.4×10^{10}.
d) 3×10^{19}.　e) 7.6×10^1.　　**12.** a) 40,000.
b) 4×10^4.　　**13.** a) Eleven million.
b) 1.1×10^7.　　**14.** a) 330,000.　b) Three hundred thirty thousand.　　**15.** a) 6,227,000,000.
b) Six billion two hundred twenty-seven million.

Chapter 8, Lesson 2

4. a) x^8.　b) y^5.　c) x^2y^7.　d) x^6y.　e) x^4y^2.
5. a) x^5.　b) y^{10}.　c) x^8.　d) y^6.　e) x^9y^9.　f) x^3y^{21}.
g) $16x^{10}$.　h) $15x^8$.　　**6.** a) 11.　b) 7.　c) 10.
d) 8.　e) 7.　f) 3.　g) 20.　h) 8.
7. a) $3^7 = 2{,}187$.　b) $3^{11} = 177{,}147$.
c) $3^5 \cdot 3^9 = 3^{14} = 4{,}782{,}969$.
d) $3^8 \cdot 3^8 = 3^{16} = 43{,}046{,}721$.
e) $3^3 \cdot 3^4 \cdot 3^5 = 3^{12} = 531{,}441$.
f) $3^2 \cdot 3^2 \cdot 3^2 \cdot 3^2 \cdot 3^2 = 3^{10} = 59{,}049$.　g) False.
h) True.　i) False.　j) True.　　**8.** a) 8×10^8.
b) 6×10^7.　c) 4×10^{12}.　d) 2.1×10^{15}.
e) 9×10^5.　f) 8×10^{16}.　　**9.** a) 3×10^{24}.
b) 3,000,000,000,000,000,000,000,000.
10. a) 5.84×10^8 and 4.6×10^9.　b) 2.7×10^{18}.

Chapter 8, Lesson 3

4. a) $4^5 \cdot 4^2 = (4 \cdot 4 \cdot 4 \cdot 4 \cdot 4)(4 \cdot 4) = 4 \cdot 4 \cdot 4 \cdot 4 \cdot 4 \cdot 4 \cdot 4 = 4^7$. b) $\dfrac{4^5}{4^2} = \dfrac{4 \cdot 4 \cdot 4 \cdot 4 \cdot 4}{4 \cdot 4} = 4 \cdot 4 \cdot 4 = 4^3$.

c) $(4^5)^2 = (4^5)(4^5) = 4^{5+5} = 4^{10}$.   **5.** a) x^9. b) x.
c) x^7. d) x^{14}. e) x^{14}. f) x^{36}.   **6.** a) 6. b) 5.
c) 1. d) 17. e) 8. f) 8. g) 11. h) 28. i) 10.
j) 16. k) 9. l) 4. m) 18. n) 6.
7. a) $4^5 = 1{,}024$. b) $4^{14} = 268{,}435{,}456$.

c) $\dfrac{4^{11}}{4^3} = 4^8 = 65{,}536$. d) $\dfrac{4^{14}}{4^7} = 4^7 = 16{,}384$.

e) $4^{15} = 1{,}073{,}741{,}824$. f) $4^8 = 65{,}536$.
g) $(4^2)^6 = 4^{12} = 16{,}777{,}216$.
h) $(4^6)^2 = 4^{12} = 16{,}777{,}216$. i) False. j) True.
k) True. l) False.   **8.** a) 3×10^6. b) 2×10^{10}.
c) 5×10^9. d) 7×10^2. e) 3×10^{12}.
f) 7.5×10^8.   **9.** a) 133,000,000 and
4,000,000,000. b) 1.33×10^8 and 4×10^9.
c) 30.   **10.** 500 seconds.

Chapter 8, Lesson 4

4. a) 1. b) 10. c) 0. d) 1. e) 1. f) $\dfrac{1}{6}$. g) -32.

h) $\dfrac{1}{25}$. i) 64. j) $\dfrac{1}{64}$. k) -64. l) $-\dfrac{1}{64}$. m) $-\dfrac{1}{8}$.

n) 1.   **5.** a) $5^{-1} = 0.2$. b) $5^{-6} = 0.000064$.
c) $5^5 = 3{,}125$. d) $5^{-5} = 0.00032$.

e) $5^{-4} \cdot 5^6 = 5^2 = 25$. f) $\dfrac{5^1}{5^{-5}} = 5^6 = 15{,}625$.

g) $5^7 \cdot 5^{-7} = 5^0 = 1$. h) $5^8 = 390{,}625$.
i) $(5^5)^{-1} = 5^{-5} = 0.00032$. j) 1.
6. a)

x	4	3	2	1	0	-1	-2	-3
y	16	8	4	2	1	0.5	0.25	0.125

7. a) $<$. b) $=$. c) $>$. d) $>$. e) $>$. f) $<$.
g) $<$. h) $<$.   **8.** a) x^2. b) x^{-15}. c) x^{10}. d) x^0.
e) x^0. f) x^{-16}. g) x^{-7}. h) x^6. i) x^{-8}. j) x^{-1}.
k) x^{-5}. l) x^0. m) x^7. n) x^{-12}. o) x^0. p) x^{-11}.
q) x^7. r) x^{10}.

Chapter 8, Lesson 5

4. a) One. b) One-tenth. c) One-one hundredth.
d) One-one thousandth. e) One-ten thousandth.
5. a) 0.7. b) 123.45. c) 10^4. d) 10^{-2}.
e) 90×10^{-11} or 9×10^{-10}. f) 2×10^{-6} or
0.2×10^{-5}.   **6.** a) 0.0004. b) 3.33. c) 10^7.
d) 10^{-9}. e) 5×10^{-6} or 0.5×10^{-5}.
7. a) 2×10^3. b) 2×10^{-4}. c) 7.5×10^1.

d) 7.5×10^{-1}. e) 3.14×10^{-2}. f) 3.14×10^{-10}.
g) 8×10^0. h) 1.00×10^{-6}.   **8.** a) 600.
b) 0.06. c) 0.0003. d) 0.00033. e) 105,000,000.
f) 0.0000000105.   **9.** a) 0.00006. b) 6×10^{-5}.
10. a) 8×10^4. b) 8×10^{-2}. c) 4×10^5.
d) 4×10^{-7}.   **11.** a) $<$. b) $>$. c) $<$. d) $>$.
e) $=$. f) $<$.   **12.** a) 6×10^{-5}. b) 4×10^1.
c) 6.3×10^{-1}. d) 2×10^7. e) 5×10^{-8}.
f) 4×10^{-1}.   **13.** a) 0.00002. b) Two-one
hundred thousandths.   **14.** a) 4×10^6 and
1×10^{-5}. b) 4×10^{11}. c) Four hundred billion.

Chapter 8, Lesson 6

4. a) x^6y^6. b) $x^{-1}y^{-1}$. c) $8x^3$. d) $625y^4$. e) x^{14}.

f) x^5y^{30}. g) $16x^{-10}$. h) $-27y^{12}$. i) $\dfrac{x^8}{y^8}$. j) $\dfrac{y^{10}}{32}$.

k) $\dfrac{1}{x^{18}}$. l) $\dfrac{49x^2}{y^{14}}$.   **5.** a) $<$. b) $=$. c) $>$.

d) $<$. e) $=$. f) $>$. g) $=$. h) $>$. i) $>$.
j) $=$.   **6.** a) $2^3 \cdot 4^3 = 8^3 = 512$. b) $8^{-1} = 0.125$.
c) $8^4 = 4{,}096$. d) $8^5 = 32{,}768$.
e) $8^2 \cdot 8^4 = 8^6 = 262{,}144$. f) $8^2 = 64$.
g) $8^{-2} = 0.015625$.

h) $\left(\dfrac{1}{8}\right)^3 = (8^{-1})^3 = 8^{-3} = 0.001953125$.

7. a) 8×10^{12}. b) 4.9×10^5. c) 2.43×10^{-3}.
d) 2×10^{-4}. e) 1×10^{-16}. f) 2.5×10^{11}.
8. a) 36. b) 4. c) 9. d) 4. e) 12. f) 6. g) 12.
h) 4. i) 4. j) 3. k) 12. l) 8.

Chapter 8, Lesson 7

4. a) $8 \cdot 10^3$. b) $5 \cdot 3^7$. c) $11 \cdot 4^2$. d) $6 \cdot 10^{-4}$.
e) $2 \cdot 7^{-5}$.   **5.** a) 500. b) 200,000. c) 9. d) 4.
e) 8. f) 0.032.
6. a)

x	0	1	2	3	4
y	1	4	16	64	256

b) y is multiplied by 4. c) y is multiplied by 16.
d) y is multiplied by 64.
7. a)

x	-2	-1	0	1	2
y	-6	-3	0	3	6

b)

x	-2	-1	0	1	2
y	-8	-1	0	1	8

c)

x	-2	-1	0	1	2
y	$\dfrac{1}{9}$	$\dfrac{1}{3}$	1	3	9

d)

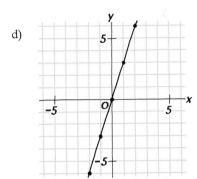

e)

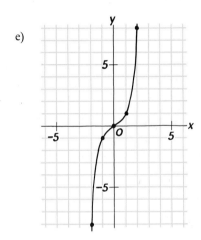

f)

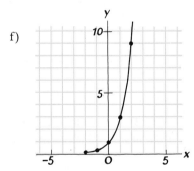

8. a) $4,500. b) $3,375. c) $2,531.25. d) $337.88.

9. a)

x	0	3	6	9	12	15	18	21	24
y	1	2	4	8	16	32	64	128	256

b) 16. c) 256. d) 65,536.

Chapter 8, Review

1. a) 10^{13}. b) 0.36. c) 10^6. d) 8.21.
e) 7×10^{-5}.   **2.** a) 8,000. b) 0.008. c) 0.45.

d) 720.   **3.** a) 5×10^7. b) 8.1×10^{10}.
c) 6×10^{-3}. d) 6×10^{-6}.   **4.** a) 3×10^{-9}.
b) Three-billionths.   **5.** a) 1. b) 1. c) 81.
d) $\frac{1}{81}$. e) $\frac{1}{64}$. f) $-\frac{1}{7}$.   **6.** a) x^5. b) x^{-36}.
c) x^{-3}. d) x^8. e) x^{-6}. f) x^{-4}.   **7.** a) $<$.
b) $>$. c) $=$. d) $>$.   **8.** a) $32x^5$. b) $x^{-1}y^{-1}$.
c) $81y^{12}$. d) $\frac{x^6}{y^3}$.
9. a) 3.2×10^{13}. b) 2.7×10^{-4}. c) 2.5×10^4.
d) 3.5×10^{-12}. e) 6.4×10^{10}. f) 4×10^{10}.
10. a) $>$. b) $<$. c) $>$. d) $=$. e) $<$. f) $=$.
g) $>$.   **11.** a) 6. b) 8. c) 9. d) 4. e) 8.
f) 5.   **12.** a) 33,000,000,000. b) 3.3×10^{10}.
c) 9×10^7; 90 million.
13. a)

x	0	1	2	3	4
y	0.5	1	2	4	8

b) y is doubled.

c)

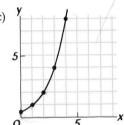

14. a) $5,600. b) $4,480. c) No; the first year it decreases by $1,400 and the second year it decreases by $1,120.

Midterm Review

1.

```
o o o    o o o o
o o o    o o o o
o o o    o o o o
         o o o o
```

2. $x = \dfrac{y - 7}{3}$.   **3.** $-6x^4$.   **4.** 3.

5.

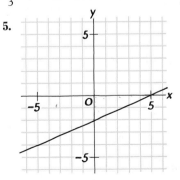

6. 3.   **7.** $4x - y = 1$.   **8.** –3.5.   **9.** 212.

10.

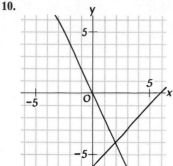

Solution: $(2, -4)$.

11. 8.

12.

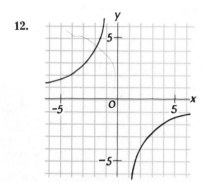

13. 1.5×10^7.   **14.** –3.5.   **15.** 3^5.
16. Subtract x from each side of the equation.

17. $x^3 > \dfrac{x}{2}$.   **18.** –7.   **19.** $(8, 1)$ and $(1, 2)$.

20. –5.   **21.** $y = 15 - x$.

22.

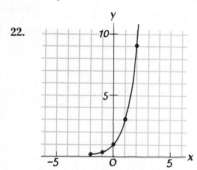

23. 4.2×10^9.   **24.** $(7, -1)$.   **25.** $<$.
26. 1.6.   **27.** 314,000.   **28.** Odd.   **29.** –19.
30. 73.   **31.** 0.36.   **32.** 18^5.
33. $y = 1 - 3x$.   **34.** 30.   **35.** $y = 3x + 4$.
36. $40(5 - x)$.   **37.** $50x + 40(5 - x) = 215$.
38. 1.5 hours.   **39.** –132.   **40.** $(2, -1)$.

Chapter 9, Lesson 1

4. a) Coefficient, 7; degree, 2.   b) Coefficient, 1;
degree, 6.   c) Coefficient, –4; degree, 1.
d) Coefficient, 5; degree, 0.   **5.** a) Yes.   b) No.
c) Yes.   d) Yes.   e) No.   f) No.   **6.** a) $9x$.
b) Not possible.   c) Not possible.   d) $5x^3$.   e) $10x$.
f) Not possible.   g) $15x$.   h) $3x^5$.   i) $9x^2$.   j) Not
possible.   k) $2x^7$.   l) Not possible.   m) $-4x$.   n) Not
possible.   o) $-4x^4$.   **7.** a) $14x^2$.   b) $7x^3$.   c) x^9.
d) $4x^6$.   e) $-11x^2$.   f) $-121x$.   g) $125x^3$.   h) x^{15}.
i) $24x^6$.   j) $8x^9$.   k) $24x^{11}$.   l) $-5x^6$.   m) $-6x^7$.
n) $16x^{10}$.   o) $-125x^{12}$.   **8.** a) 25.   b) 100.   c) 3.
d) 6.   **9.** a) $15x^2$.   b) $26x^4$.   c) Not possible.
d) $26x^3$.   e) $7x^5$.   f) $-8x^{10}$.   g) Not possible.
h) $-8x^6$.   i) $12x^4$.   j) $64x^{12}$.   k) $10x^{10}$.   l) $32x^{50}$.
10. a) Perimeter, $16x$; area, $15x^2$.   b) Perimeter, $4x^3$;
area, x^6.   c) Perimeter, $12x^2 + 2x$; area, $6x^3$.
d) Perimeter, $2x^4 + 8x$; area, $4x^5$.

Chapter 9, Lesson 2

4. a) Yes.   b) Yes.   c) No.   d) Yes.   e) Yes.
f) No.   **5.** a) $x^3 + -4$.   b) $1 + 2x + -x^2$.
c) $-5x + -10$.   **6.** a) $x^5 + x$.   b) $5y^2 - 7y + 3$.
c) $2x^4 + 3x^3 + 4x^2 - x$.   d) $-y^6 + 1$.
e) $-2x^8 - x^6 + 5x^4 + 20$.   f) $y + 2^5$.
7. a) $6x + 16$.   b) $-5y^2 + 5$.   c) $x^4 + 7x^3 - 6x^2$.
d) $3x^3 - 6x^2 + 15x$.   e) $x^5 - 2x^4 + 5x^3$.
f) $-4y^6 + 8y^2$.   **8.** a) $-x^2 - 5x - 10$.
b) $-y^5 + 7y$.   c) $6x^2 - 11$.
d) $-1 + 2y - 3y^2 + 4y^3$.   **9.** a) –1.   b) 7.
c) 17.   d) –7.   e) –11.   f) 15.   g) 12.   h) 12.
i) –60.   j) –15.   k) –12.   l) –12.   m) 60.   n) 1.
o) 5.   p) 1.   q) 31.   r) 11.   s) 11,111.   t) 9,091.
10. a) $6x^2 + 6x$.   b) $2x^3 + 3x^2 - 5x$.   c) $6x^2 - 7x$.
d) $x^3 + 7x - 7$.   e) $x^4 - 8x^2 + 4$.   f) $2x^7 + 14x$.

Chapter 9, Lesson 3

4. a) $6x + 1$.   b) $2x - 4y$.   c) $4x^2 + 8$.
d) $3x^2 - x$.   e) $4x^3 - 4x^2 + 4x + 4$.
5. a) $7x + 4$.   b) x.   c) $-x^2 - 6x$.
d) $2x^2 + 10x - 1$.   e) $x^4 - 9x^2 + 9x + 2$.
6. a) $4x^2 - 3x + 1$.   b) $2x^2 + 5x - 5$.

c)

x	4	5
polynomial A	50	78
polynomial B	3	8
polynomial C	53	86
polynomial D	47	70

d) The values of polynomials C and D are the sums and differences, respectively, of the corresponding values of polynomials A and B.
7. a) $2x - 1$. b) $20x - 13$. c) $x^3 + x^2 + x$.
d) $9x - 4y - z$.　**8.** a) $6x^2 + 6y$. b) $-3x - 24$.
c) $6x^2 - 2x - 4$. d) $x^3 - x^2 - x + 1$.
9. a) $9x - 3y$. b) $8x^2 - x + 10$.
c) $x^3 - x^2 - x + 1$.

Chapter 9, Lesson 4

4. a) The 24 on the fourth line should have been written one digit to the left.
b)

	20	4
10	200	40
5	100	20

$$
\begin{array}{r}
20 + 4 \\
\times\ 10 + 5 \\
\hline
100 + 20 \quad 120 \\
200 + 40 \quad 240 \\
\hline
360
\end{array}
$$

5. a)
$$
\begin{array}{r}
37 \\
\times\ 52 \\
\hline
74 \\
185 \\
\hline
1924
\end{array}
$$

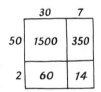

	30	7
50	1500	350
2	60	14

b)
$$
\begin{array}{r}
8 \\
\times\ 419 \\
\hline
72 \\
8 \\
32 \\
\hline
3352
\end{array}
$$

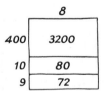

	8
400	3200
10	80
9	72

c)
$$
\begin{array}{r}
206 \\
\times\ 93 \\
\hline
618 \\
1854 \\
\hline
19158
\end{array}
$$

	200	6
90	18000	540
3	600	18

6. a) $(2x + 1)(x + 5)$; $2x^2 + 11x + 5$.
b) $6y(3y - 7)$; $18y^2 - 42y$.
c) $(a^2 + 6a - 2)(4a + 3)$; $4a^3 + 27a^2 + 10a - 6$.
d) $(a - b)(a + b)$; $a^2 - b^2$.

7. a)

	x	11
x	x^2	$11x$
4	$4x$	44

$x^2 + 15x + 44$.

b)

	$3y$	-1
$5y$	$15y^2$	$-5y$
2	$6y$	-2

$15y^2 + y - 2$.

c)

	a^2	4
a^3	a^5	$4a^3$

$a^5 + 4a^3$.

d)

	b^2	$7b$	-1
b	b^3	$7b^2$	$-b$
7	$7b^2$	$49b$	-7

$b^3 + 14b^2 + 48b - 7$.

8. a) $x^2 + 14x + 24$. b) $x^2 + x - 72$.
c) $6x^2 - 7x - 20$. d) $36x^2 + 12x + 1$.
e) $36x^2 - 1$. f) $21x^2 - 58x + 21$. g) $16 - x^2$.
h) $x^6 + 7x^3 - 8$. i) $2x^3 + x^2 - 44x + 5$.
j) $x^4 - 1$. k) $x^4 - 4x^3 + 5x^2 - 28x + 32$.
l) $x^4 + 2x^2 + 9$.

Chapter 9, Lesson 5

4. a) $x^2 + 8x + 12$. b) $x^2 + 4x - 12$.
c) $x^2 - 11x + 24$. d) $x^2 + x - 20$.
e) $x^2 + 14x + 49$. f) $x^2 - 49$. g) $6x^2 + 26x + 8$.
h) $6x^2 - 26x + 8$. i) $30x^2 + 13x - 10$.
j) $4x^2 - 31x - 90$. k) $x^2 - 2xy + y^2$.
l) $x^2 - y^2$.　**5.** a) $10x^2 - x - 3$.
b)

x	1	3	10
polynomial A	2	12	47
polynomial B	3	7	21
polynomial C	6	84	987

c) The values of polynomial C are the products of the corresponding values of polynomials A and B.
6. a) $x^4 + 4x^2 - 5$. b) $x^4 - 9x^3 + x^2 - 9x$.
c) $x^3 + 8$. d) $x^4 + 2x^3 - x^2 + 10x - 3$.
e) $2x^3 + 4x^2 - 48x$. f) $x^3 - 6x^2 + 11x - 6$.
7. a) $x^2 - 1$. b) $x^3 - 1$. c) $x^4 - 1$. d) $x^5 - 1$.
e) $x^{11} - 1$.

Chapter 9, Lesson 6

4. a) $(x + 5)^2 = x^2 + 10x + 25$.
b) $(3y - 1)(3y + 1) = 9y^2 - 1$.
c) $(a - 6)^2 = a^2 - 12a + 36$.
d) $(2 + 7b)(2 - 7b) = 4 - 49b^2$.

5.

a)
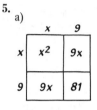

	x	9
x	x^2	$9x$
9	$9x$	81

$x^2 + 18x + 81$.

c)
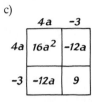

	$4a$	-3
$4a$	$16a^2$	$-12a$
-3	$-12a$	9

$16a^2 - 24a + 9$.

b)

	$5x$	$2y$
$5x$	$25x^2$	$10xy$
$2y$	$10xy$	$4y^2$

$25x^2 + 20xy + 4y^2$.

d)
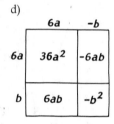

	$6a$	$-b$
$6a$	$36a^2$	$-6ab$
b	$6ab$	$-b^2$

$36a^2 - b^2$.

6. a) By squaring the second term of the binomial.
b) By taking twice the product of the first term and the second term of the binomial.
7. a) $x^2 + 16x + 64$. b) $x^2 - 64$.
c) $144a^2 + 24a + 1$. d) $9b^2 - 60b + 100$.
e) $49x^2 - y^2$. f) $4x^2 - 12xy + 9y^2$.
8. a) 1,849. b) 1,591. c) 1,849. d) 2,451.
9. a) $x^4 + 6x^2 + 9$. b) $9x^4$. c) $16y^{10} - 8y^5 + 1$.
d) $16y^{10} - 1$. e) a^6b^6. f) $a^6 - 2a^3b^3 + b^6$.
10. a) $x^2 + 8x + 16$; $x + 4$. b) $x^2 + 2x + 1$;
$x + 1$. c) $x^2 - 18x + 81$; $x - 9$.
d) $x^2 - 10x + 25$; $x - 5$.

Chapter 9, Lesson 7

3. a)

	$4x$	5
$3x$	$12x^2$	$15x$
8	$32x$	40

b)
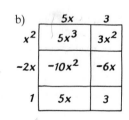

	$5x$	3
x^2	$5x^3$	$3x^2$
$-2x$	$-10x^2$	$-6x$
1	$5x$	3

c)

	$3x^2$	$7x$	-2
$6x$	$18x^3$	$42x^2$	$-12x$
-4	$-12x^2$	$-28x$	8

4. a) $3x + 8$. b) $x^2 - 2x + 1$. c) $6x - 4$.
5.

a)

	$7x$	1
$2x$	$14x^2$	$2x$
3	$21x$	3

$2x + 3$.

b)

	$4x$	9
x^2	$4x^3$	$9x^2$
$-3x$	$-12x^2$	$-27x$
1	$4x$	9

$x^2 - 3x + 1$.

6. a) $2x + 3$. b) $x^2 - 3x + 1$. **7.** a) $5x + 8$.
b) $x^2 + 4x - 6$. c) $3x - 7$. d) $x^2 - 2x + 4$.
e) $2x - 3$. f) $x^2 + 5$. **8.** a) $x^2 + 0x + 6$.
b) $8x^3 + 0x^2 + 0x + 1$.
c) $x^4 + 0x^3 + 6x^2 + 2x - 3$.
d) $-x^5 + 0x^4 + 0x^3 + 0x^2 + 10x + 0$.
9. a) $x^2 - 3x + 8$. b) $5x^2 + 4x + 5$.
c) $8x^3 + 20x^2 + 50x + 125$.

Chapter 9, Review

1. a) True. b) True. c) False. d) True.
e) False. **2.** a) 57. b) −57. c) 16. d) 448.
e) 8. f) 1,000. **3.** a) Not possible. b) $3x^5$.
c) Not possible. d) $-8x^5$. e) $16x^{10}$. f) $-2,048x^5$.
g) Not possible. h) $-4x^6$.
4.

	1	7
1	1^2	$1 \cdot 7$
7	$1 \cdot 7$	7^2

$(1 + 7)^2 = 1^2 + 2(1 \cdot 7) + 7^2$.

	2	6
2	2^2	$2 \cdot 6$
6	$2 \cdot 6$	6^2

$(2 + 6)^2 = 2^2 + 2(2 \cdot 6) + 6^2$.

	4	4
4	4^2	$4 \cdot 4$
4	$4 \cdot 4$	4^2

$(4 + 4)^2 = 4^2 + 2(4 \cdot 4) + 4^2$.

Every equation checks out because each side gives 64, the number of squares on the checkerboard.

5. a)

$4x^2 + 28x + 49.$

b)

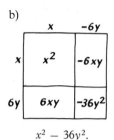

$x^2 - 36y^2.$

c)

	9x	7
4x	36x²	28x
-3	-27x	-21

$36x^2 + x - 21.$

d)

	2y	1
3x	6xy	3x
-1	-2y	-1

$6xy + 3x - 2y - 1.$

6. a) $-x^5 + 7x^3 + 10.$ b) $x^8 - x^2 + 6x.$
c) $-x^4 + 2.$ d) $4x^3 - 2x^2.$ **7.** a) $x^2 + 5x.$
b) $x^2 + 7x + 10.$ c) $6x^2 - 18x.$
d) $6x^2 - 19x + 3.$ e) $x^2 + 3x - 28.$
f) $28x^2 + 3x - 1.$ g) $x^2 + 16x + 64.$ h) $64x^2.$
i) $4x^2 - 81.$ j) $9x^2 - 6x + 1.$ k) $x^8 + 8x^4 + 16.$
l) $x^8 - 16.$ **8.** a) $9y.$ b) $7x^3 + x - 1.$
c) $x^4 + 10x - 10.$ d) $x - 8y - 3.$
e) $24x^2 + 19x - 120.$ f) $4x^3 + 19x^2 - 7x - 10.$
g) $2x^2 + 5x - 4.$ h) $8x^3 + 4x^2 + 2x + 1.$
9. a) Perimeter, $12x + 10$; area, $24x + 4.$
b) Perimeter, $8x^3$; area, $4x^6.$ c) Perimeter, $12x + 18$;
area, $5x^2 + 49x - 10.$ d) Perimeter, $16x$; area,
$16x^2 - 49.$ **10.** a) $20x.$ b) $625.$ c) $8xy.$
d) $16.$ **11.** a) $10x - 10.$ b) $21x^2 - 62x + 16.$
c) $x^3 - x^2 - 31x + 55.$ d) $x^2 + 4x - 10.$
e) $4x^2 + 2x - 6.$ f) $4x^2 - 2x - 12.$
g) $8x^3 + 12x^2 - 18x - 27.$ h) $2x - 3.$

Chapter 10, Lesson 1

4. 21 $3 \cdot 7$
22 $2 \cdot 11$
23 prime
24 $2^3 \cdot 3$
25 5^2
26 $2 \cdot 13$
27 3^3
28 $2^2 \cdot 7$
29 prime
30 $2 \cdot 3 \cdot 5$

5. a) $2^2 \cdot 5^2.$ b) $2^2 \cdot 3^2 \cdot 5.$ c) $3 \cdot 7 \cdot 11.$
d) $2 \cdot 3 \cdot 5 \cdot 7 \cdot 11.$ e) $2^4 \cdot 3 \cdot 17.$ f) $2^6 \cdot 3 \cdot 17.$
g) $3^2 \cdot 5^2.$ h) $3^6 \cdot 5^6.$ i) $2^6 \cdot 7^3.$ j) $3^{16}.$
6. a) Yes. b) Yes. c) No. d) No. e) Yes.
f) Yes. g) No. h) Yes. i) No. j) Yes. k) Yes.
l) No. **7.** a) 1, 5, 11, and 55. b) 1, 2, 4, 8, 16,
and 32. c) 1 and 71. d) 1, 2, 3, 4, 6, 8, 9, 12, 18,
24, 36, and 72. e) 1, 5, 5^2, and $5^3.$ f) 1, 3, 3^2, 3^3,
3^4, and $3^5.$ **8.** a) 5. b) 8. c) 1. d) 12. e) 18.
f) 2. g) 1. h) $3^4.$ i) 64. j) 1.

Chapter 10, Lesson 2

4. a) $56x^2.$ b) $56x^2.$ c) $15x^8.$ d) $-x^{20}.$ e) $36x^8.$
f) $64x^{24}.$ g) $abx^{a+b}.$ h) $a^2x^{2a}.$ **5.** a) (1)(18),
(2)(9), (3)(6). b) (1)(34x), (2)(17x), (17)(2x), (34)(x).
c) $(1)(25x^2), (5)(5x^2), (25)(x^2), (x)(25x), (5x)(5x).$
d) $(1)(2x^3), (2)(x^3), (x)(2x^2), (2x)(x^2).$ **6.** a) $4^2.$
b) $(x^8)^2.$ c) $(3x^5)^2.$ d) $(5xy)^2.$ **7.** a) 1, 2, 3, 4, 6,
8, 12, and 24. b) 1, 3, 5, 15, x, $3x$, $5x$, and $15x.$
c) 1, 7, 49, x, $7x$, $49x$, x^2, $7x^2$, and $49x^2.$ d) 1, x, x^2,
x^3, x^4, x^5, and $x^6.$ e) 1, 3, x, $3x$, x^2, $3x^2$, x^3, and
$3x^3.$ **8.** a) $16x^4.$ b) $5x^8.$ c) $x^3.$ d) $-3y.$
e) $x^7y^3.$ f) $2x^7.$ **9.** a) 3. b) $5x.$ c) 1. d) $x^7.$
e) 1. f) 11. g) $8xy.$ h) $13xy^2.$ i) 1. j) 3. k) 1.
l) $2x.$

Chapter 10, Lesson 3

4. a) $16x - 24y.$ b) $5x^2 + 20x.$ c) $x^5 - 2x^4.$
d) $4x^2y - 4xy^2 + 4xy.$ **5.** a) $2x.$ b) $x^3.$ c) 1.
d) 8.
6.
a)

	x	2
4	4x	8

$4x + 8 = 4(x + 2).$

b)

	x	5
x	x²	5x

$x^2 + 5x = x(x + 5).$

c)

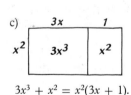

$3x^3 + x^2 = x^2(3x + 1).$

d)

	x³	6
2x	2x⁴	12x

$2x^4 + 12x = 2x(x^3 + 6).$

7. a) $3(x + 4)$. b) $5(x + 1)$. c) $2(x + y)$.
d) $1(2x + 3y)$. e) $4(x - 1)$. f) $5(2x - 3)$.
g) $8(4x - y)$. h) $x(x + 6)$. i) $2(x^3 + 1)$.
j) $1(3x^2 + 2)$. k) $x^2(x^2 - 8)$. l) $1(x^4 + y^4)$.
8. a) $x(7 + 3) = 10x$. b) $x(5 - 1) = 4x$.
c) $x^2(1 + 8) = 9x^2$. d) $x^3(2 - 5) = -3x^3$.
e) $2xy(2 + 1) = 6xy$. f) $xy^2(1 - 9) = -8xy^2$.
9. a) $23(6 + 4) = 23(10) = 230$.
b) $12(31 - 1) = 12(30) = 360$.
c) $45(36 - 16) = 45(20) = 900$.
d) $99(99 + 1) = 99(100) = 9,900$.   **10.** a) 3.
b) $x^3 - 6x + 1$. c) $2x^4 + x^2 - 3$.
d) $5 + x - y$.   **11.** a) $4x(3x^2 - 4x + 2)$.
b) $1(6x^2 + xy + 6y^2)$. c) $x^5(10x^5 - 1)$.
d) $2x^2(1 + 3x^4 + 5x^8)$.

Chapter 10, Lesson 4

4. a) $x^2 + 16x + 15$. b) $x^2 - 16x + 15$.
c) $x^2 + 8x + 15$. d) $x^2 - 8x + 15$.
e) $x^2 + 14x - 15$. f) $x^2 - 14x - 15$.
g) $x^2 + 2x - 15$. h) $x^2 - 2x - 15$.
5. a) $x^2 + 8x + 12 = (x + 6)(x + 2)$.
b) $x^2 - 13x + 30 = (x - 3)(x - 10)$.
c) $x^2 + 3x - 28 = (x - 4)(x + 7)$.
d) $x^2 - x - 72 = (x + 8)(x - 9)$.
6. a) $10 = (2)(5)$ and $2 + 5 = 7$.
b) $10 = (-2)(-5)$ and $-2 + -5 = -7$.
c) $-33 = (-3)(11)$ and $-3 + 11 = 8$.
d) $-33 = (3)(-11)$ and $3 + -11 = -8$.
e) $42 = (6)(7)$ and $6 + 7 = 13$. f) $42 = (3)(14)$
and $3 + 14 = 17$. g) $-36 = (-2)(18)$ and
$-2 + 18 = 16$. h) $-36 = (-6)(6)$ and
$-6 + 6 = 0$.   **7.** a) $x + 9$. b) $x - 4$. c) $x + 7$.
d) $x - 8$. e) $x - 12$. f) $x + 5$.
8. a) $(x + 1)(x + 5)$. b) $(x - 1)(x - 5)$.
c) $(x + 3)(x + 9)$. d) $(x + 4)(x + 8)$.
e) $(x - 2)(x + 11)$. f) $(x + 2)(x - 11)$.
g) $(x + 7)(x + 7)$. h) $(x + 7)(x - 7)$.
i) $(x - 5)(x - 12)$. j) $(x - 3)(x - 20)$.
k) $(x - 4)(x + 15)$. l) $(x + 6)(x - 10)$.
9. a) $(x + 2y)(x + y)$. b) $(x + 8y)(x - 5y)$.
c) $(x + 6)(y + 4)$. d) $(x^2 - 10)(x^2 - 7)$.
10. a) $(x + 3y)(x + 5y)$. b) $(x - 2y)(x + 9y)$.
c) $(x + 4)(y + 4)$. d) $(x^2 + 1)(x^2 - 10)$.

Chapter 10, Lesson 5

4. a) $x^2 - 400$. b) $16x^2 - 1$. c) $25x^2 - y^2$.

d) $x^6 - 9$. e) $x^2 - y^4$. f) $1 - x^2y^2$.   **5.** a) 8^2.
b) $(x^4)^2$. c) Not possible. d) $(x^{18})^2$. e) $(3x^2)^2$.
f) Not possible. g) $(4xy)^2$. h) $(10x^{50})^2$.
6. a) $x^2 - 9 = (x + 3)(x - 3)$.
b) $25x^2 - 49 = (5x - 7)(5x + 7)$.
c) $x^2 - 16y^2 = (x + 4y)(x - 4y)$.
d) $x^6 - y^4 = (x^3 - y^2)(x^3 + y^2)$.   **7.** a) 0.
b) 40. c) 91. d) -9. e) -5. f) 16. g) 0. h) 40.
i) 91. j) -9. k) -5. l) 16.
8. a) $(x - 8)(x + 8)$. b) Not possible.
c) $(4x + 3y)(4x - 3y)$. d) $(9 + x)(9 - x)$.
e) $8(x + 1)(x - 1)$. f) $4(2x^2 - 1)$. g) $5(x^2 + 4)$.
h) $(x^2 - 7)(x^2 + 7)$. i) $(6x^5 - 1)(6x^5 + 1)$.
j) $2(x^3 - 4)(x^3 + 4)$.
9. a) $(x + y + 3)(x + y - 3)$.
b) $(4x + 13)(4x + 1)$. c) $(6 - x)(4 + x)$.
d) $x(16 - x)$.

Chapter 10, Lesson 6

4. a) $x^2 + 6x + 9 = (x + 3)^2$.
b) $x^2 - 20x + 100 = (x - 10)^2$.
c) $16x^2 + 8xy + y^2 = (4x + y)^2$.
d) $x^8 - 2x^4y + y^2 = (x^4 - y)^2$.
5. a) $x^2 + 14x + 49$. b) $x^2 - 14x + 49$.
c) $16x^2 - 8x + 1$. d) $25x^2 + 20xy + 4y^2$.
e) $4x^2 + 20xy + 25y^2$. f) $x^4 + 2x^2y^5 + y^{10}$.
6. a) $(3x)^2$. b) $(x^8)^2$. c) 20^2. d) Not possible.
e) $(5x^{18})^2$. f) Not possible. g) $(8x^3y^2)^2$. h) Not
possible.   **7.** a) $16x$. b) 121. c) $24x$. d) 1.
8. a) $(x + 6)^2$. b) $(x - y)^2$. c) $(2x + 1)^2$. d) Not
possible. e) $(3x - 10)^2$. f) Not possible.
g) $(7x + y)^2$. h) $(5x - 8y)^2$.   **9.** a) 36. b) 64.
c) 225. d) 0. e) 16. f) 36. g) 64. h) 225.
i) 0. j) 16.   **10.** a) $(x^2 + 11)^2$. b) $(x^3 - 4)^2$.
c) $(x^5 + y^5)^2$. d) $2(x - 3)^2$. e) $5(x^2 - 5)^2$.
f) $(4x - y^8)^2$.

Chapter 10, Lesson 7

4. a) $2x^2 + 17x + 36 = (x + 4)(2x + 9)$.
b) $5x^2 - 8x + 3 = (5x - 3)(x - 1)$.
c) $12x^2 + 7x - 12 = (3x + 4)(4x - 3)$.
d) $36x^2 - 25 = (6x - 5)(6x + 5)$.
5. a) $3x^2 + 37x + 12$. b) $3x^2 + 35x - 12$.
c) $3x^2 + 20x + 12$. d) $3x^2 + 16x - 12$.
e) $3x^2 + 15x + 12$. f) $3x^2 + 9x - 12$.
g) $3x^2 + 12x + 12$. h) $3x^2 - 12$.
6. a) $(2x + 1)(x + 7)$. b) $(2x - 1)(x + 7)$.

c) $(5x - 1)^2$. d) $(5x + 1)(5x - 1)$.
e) $(3x - 2)(x + 3)$. f) $3(x - 1)(x - 2)$.
g) $(4x + 5)(2x + 3)$. h) $(4x - 5)(2x - 3)$.
i) $(3x + 11)(2x - 1)$. j) $(3x - 11)(2x + 1)$.
7. a) $(2x - y)(3x + y)$. b) $(x - y)(8x - 3y)$.
c) $(4x + 1)(y - 5)$. d) $(5x^2 + 6)(x^2 + 2)$.
8. a) $(5x + y)(x + 3y)$. b) $(4x - y)(2x - 7y)$.
c) $(x + 5)(2y - 1)$. d) $(3x^2 + 2)(x^2 + 10)$.

Chapter 10, Lesson 8

4. a) $10x^2 - 5x - 5$. b) $10x^2 - 5x - 5$.
c) $10x^2 - 5x - 5$. d) $10x^2 - 5x - 5$. **5.** a) $6x^3$.
b) $2x^3 + 6x^2$. c) $x^3 + 5x^2 + 6x$. d) All have
degree 3. **6.** a) $x(x^2 + 10)$. b) $4x(x + 2)(x - 2)$.
c) $x^2(x + 4)(x + 6)$. d) $3x(x - 1)(x - 5)$.
e) $6x^2(x^2 + x + 3)$. f) $2x^3(x + 7)(x - 5)$.
g) $5x^2(2x + 3)(2x - 3)$. h) $x(5x^2 + 10x + 1)$.
i) $(x + 1)(x - 1)(x^2 + 1)$. j) $x^3(3x + 1)(x + 3)$.
7. a) 11. b) 47. c) $4x + 1$. d) $x - 9$.
e) $3x + 1$. f) $x^2 + x + 1$. **8.** a) $x(x^2 + 2y)$.
b) $xy(x + y)(x - y)$. c) $2(x - y)^2$.
d) $x(x + y)(x + 3y)$. e) $(3x + 15)(y - 2)$.
f) $(x + 3y)(x - 3y)(x^2 + y^2)$.

Chapter 10, Review

1. a) $11 \cdot 13$. b) $3^3 \cdot 5^2$. c) $2 \cdot 7^2 \cdot 19$. d) $3^2 \cdot 5^2$.
e) $2^{12} \cdot 5^6$. f) $2^{12} \cdot 3^{12}$. **2.** a) 7. b) 5. c) 1.
d) 12. e) 49. f) 1. **3.** a) 1, 2, 3, 4, 6, 8, 12, 16,
24, and 48. b) 1, 3, 11, 33, x, $3x$, $11x$, and $33x$.
c) 1, x, x^2, x^3, x^4, and x^5. d) 1, 3, 9, x, $3x$, $9x$, x^2,
$3x^2$, and $9x^2$.
4. a) Example: 1,234

$$\frac{12,341,234}{73} = 169,058$$

$$\frac{12,341,234}{137} = 90,082$$

b) Writing a four-digit number twice to form an
eight-digit number is equivalent to multiplying the
number by 10,001. $10,001 = 73 \cdot 137$. **5.** a) 2.
b) x. c) x^6. d) $2xy$. e) 1. f) $6x$.
6. a) $56(39 + 61) = 56(100) = 5,600$.
b) $90^2 - 2^2 = 8,100 - 4 = 8,096$.
c) $25(81 - 1) = 25(80) = 2,000$.
d) $(127 + 27)(127 - 27) = (154)(100) = 15,400$.
7. a) 81. b) $60x$. **8.** a) $3x^2$. b) $-2x^5$. c) 8.
d) $10 - x^2$. e) $x - 2$. f) $x + 6$. g) $3x - 7$.
h) $5x + 2$. **9.** a) $7(2x + 5)$. b) $x(x - 8)$.

c) $(x + 4)(x - 4)$. d) $x^3(4x + 3)$. e) $(x - 10)^2$.
f) $(x + 2)(x + 11)$. g) $(x + 6)(x - 5)$.
h) $x(x - 1)^2$. i) $x^2 + 9$. j) $(7x + 2)^2$.
k) $(3x + 1)(x + 5)$. l) $(2x - 7)(x + 4)$.
m) $3(x - 3)^2$. n) $x(x + 8)(x - 5)$. o) $4x(x^2 + 4)$.
p) $2x^2(x - 1)(x - 2)$. **10.** a) $y(3x + y^2)$.
b) $4(2x - y^2)$. c) $(x - 9y)(x - y)$.
d) $(x - 1)(y + 6)$. e) $(5x + 2y)(5x - 2y)$.
f) $(x^2 + y)^2$.

Chapter 11, Lesson 1

4. A, $\dfrac{3}{4}$; B, $\dfrac{9}{4}$; C, $-\dfrac{5}{4}$; D, $\dfrac{14}{4}$ or $\dfrac{7}{2}$.
5. a–c)

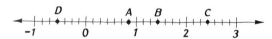

d) 1. e) $\dfrac{10}{7}$. f) $1\dfrac{1}{7}$ or $\dfrac{8}{7}$. g) $\dfrac{1}{7}$. **6.** a) 0.75.

b) 0.75. c) 0.17. d) 0.17. e) 0.18. f) 5.50.
g) 0.62. h) 2.43. i) 0.55. j) 0.50. **7.** a) 0 and
1. b) 2 and 3. c) –1 and 0. d) –2 and –1. e.) 9
and 10. f) 90 and 91. **8.** a) $\dfrac{1}{2}$. b) $\dfrac{2}{3}$. c) $\dfrac{5}{17}$.

d) $\dfrac{17}{5}$. e) $\dfrac{11}{101}$. f) $\dfrac{1}{91}$. g) $\dfrac{7}{9}$. h) $\dfrac{43}{81}$. **9.** a) $\dfrac{8}{10}$

and $\dfrac{9}{10}$. b) $\dfrac{9}{21}$ and $\dfrac{49}{21}$. c) $\dfrac{3}{72}$ and $\dfrac{2}{72}$. d) $\dfrac{20}{24}$

and $\dfrac{15}{24}$. e) $\dfrac{20}{60}$, $\dfrac{15}{60}$, and $\dfrac{12}{60}$. f) $\dfrac{12}{16}$, $\dfrac{14}{16}$, and $\dfrac{15}{16}$.

10. a) $\dfrac{2}{7}$, $\dfrac{3}{7}$, $\dfrac{4}{7}$. b) $\dfrac{7}{4}$, $\dfrac{7}{3}$, $\dfrac{7}{2}$. c) $\dfrac{3}{10}$, $\dfrac{1}{3}$, $\dfrac{11}{30}$.

d) $\dfrac{19}{8}$, $\dfrac{5}{2}$, $\dfrac{11}{4}$. **11.** a) $-\dfrac{3}{7}$. b) $\dfrac{-8}{-5}$. c) $\dfrac{-1}{-2}$.

d) $\dfrac{4}{-9}$. **12.** a) 3. b) 0. c) –2. d) Not possible.

13. a) 3. b) 2. c) 5. d) 9. e) –7. f) 5. g) 5.
h) Because $\dfrac{0}{0}$ is not defined. i) 5. j) 5.

14. a) 12. b) 35. c) 14. d) 15. e) 41. f) 28.
g) 3. h) 4.

Chapter 11, Lesson 2

4. a) 0. b) 1. c) –5. d) No value. e) 2 and –2.
f) No value. **5.** a) $\dfrac{3x}{8}$. b) $\dfrac{x}{y}$. c) $\dfrac{x}{4}$. d) $\dfrac{1}{x^2}$.

e) $\frac{y}{x}$. f) $\frac{x+1}{2}$. g) $\frac{x}{x-2}$. h) $\frac{3}{x+3}$. **6.** a) $\frac{2x}{10}$ and $\frac{x}{10}$. b) $\frac{2}{3x}$ and $\frac{18}{3x}$. c) $\frac{8x}{x^2}$ and $\frac{2}{x^2}$. d) $\frac{x^2}{9x}$ and $\frac{27}{9x}$. e) $\frac{y}{xy}$ and $\frac{x}{xy}$. f) $\frac{2x+8}{(x+2)(x+4)}$ and $\frac{4x+8}{(x+2)(x+4)}$. g) $\frac{4x}{10}, \frac{25x}{10}$, and $\frac{x}{10}$. h) $\frac{3x^2}{x^3}, \frac{4x}{x^3}$, and $\frac{5}{x^3}$. **7.** a) 10. b) 2. c) 4. d) 1. e) –1. f) Not possible. **8.** a) 3x. b) 5. c) 2x + 6. d) 10x. e) 4x – 32. f) x + 5. **9.** a) $\frac{1}{3}$. b) Not possible. c) $\frac{4x}{x-2}$. d) Not possible. e) $\frac{4}{5}$. f) $\frac{1}{3}$. g) $\frac{x}{x+7}$. h) $\frac{2}{x-1}$. i) x. j) $\frac{3}{x}$. k) $\frac{1}{x-4}$. l) $\frac{5}{x-5}$. **10.** a) 1. b) 1.6. c) 3. d) 1. e) 1.6. f) 3. g) Yes, for all values of x except 2 because $\frac{x^2+3x-10}{5x-10} = \frac{(x+5)(x-2)}{5(x-2)} = \frac{x+5}{5}$.

Chapter 11, Lesson 3

4. a) 1. b) $\frac{2}{5}$. c) $\frac{x}{3}$. d) $\frac{12}{x}$. e) 2x. f) $\frac{10+2x}{5}$ or $\frac{2(5+x)}{5}$. g) x. h) $\frac{5}{4}$. i) $\frac{5x}{4}$. j) 10. k) $\frac{x-10}{5}$. l) $\frac{20}{x-10}$. m) $\frac{x+10}{x-10}$. n) 1. o) $\frac{2x}{x+2}$. p) $\frac{-3}{x+2}$. q) 1. r) x. **5.** a) $\frac{1}{2}$. b) $\frac{7}{12}$. c) $\frac{9}{4x}$. d) $\frac{4x}{77}$. e) $\frac{2x^3-3x^2}{6}$. f) $\frac{3-2x}{x^3}$. g) $\frac{x^2+25}{5x}$. h) $\frac{3x}{4}$. i) $\frac{1}{2x+6}$. j) $\frac{x+3}{10}$. k) $\frac{2x^2}{x^2-1}$. l) $\frac{16x}{x^2-16}$. **6.** a) $\frac{y+x}{xy}$. b) $\frac{2x}{y}$. c) $\frac{2xy}{x^2-y^2}$. d) –1. e) $\frac{x^2-2xy+y^2}{xy}$ or $\frac{(x-y)^2}{xy}$.

Chapter 11, Lesson 4

4. a) $\frac{41}{8}$. b) $\frac{41}{5}$. c) $\frac{17}{7}$. d) $\frac{23}{7}$. e) $\frac{35}{6}$. f) $\frac{11}{15}$. **5.** a) $\frac{4x+1}{x}$. b) $\frac{xy-x}{y}$. c) $\frac{3x+x^2}{3}$. d) $\frac{9x}{2}$. e) $\frac{2x-1}{x}$. f) $\frac{2x+9}{x+6}$. g) $\frac{x^2+x}{x+7}$. h) $\frac{5x^2-10}{4}$.

i) $\frac{x^2-8}{x-3}$. j) $\frac{x^2+y^2}{x+y}$. **6.** a) 16. b) 12. c) 0. d) 16. e) 12. f) 0. g) Yes, for all values of x except 3 because

$$x + 2 + \frac{6+x}{x-3} = \frac{x+2}{1} + \frac{6+x}{x-3}$$
$$= \frac{(x+2)(x-3)}{x-3} + \frac{6+x}{x-3}$$
$$= \frac{x^2-x-6+6+x}{x-3} = \frac{x^2}{x-3}$$

7. a) $\frac{6}{x}$. b) $\frac{1}{2}$. c) $\frac{1}{9}$. d) $\frac{8x}{x+1}$. e) $\frac{2}{x^3}$. f) $\frac{2x^3}{x^3-6}$. g) $\frac{7}{x-7}$. h) 1. **8.** a) $\frac{x}{5}+\frac{y}{5}$. b) $\frac{x}{3}-\frac{1}{6}$. c) $\frac{x}{x-3}+\frac{3}{x-3}$. d) $8+\frac{7}{x}$. e) $1+\frac{z}{x-y}$. f) $\frac{2x^2}{x+3}-1$. **9.** a) 2x + 1. b) 1. c) x + y. d) x^2.

Chapter 11, Lesson 5

4. a) $\frac{8}{15}$. b) $\frac{1}{12}$. c) $\frac{25}{49}$. d) 1. e) 6. f) $\frac{15}{14}$. **5.** a) $\frac{3}{4}$. b) $\frac{5}{64}$. c) $\frac{11}{x}$. d) $\frac{28}{x^2}$. e) $\frac{5x}{6}$. f) $\frac{x^2}{9}$. g) 1. h) $\frac{x^2+100}{10x}$. i) $\frac{2}{x+1}$. j) $\frac{1}{(x+1)^2}$. k) $\frac{2x+3}{x^3}$. l) $\frac{6}{x^5}$. m) $\frac{18}{x}$. n) $\frac{x}{2}$. o) $\frac{216}{x^3}$. p) $\frac{x^3}{216}$. **6.** a) $\frac{21}{x}$. b) $\frac{x}{30}$. c) $\frac{x^2}{x^2-4}$. d) $\frac{x^6}{4}$. e) 1. f) $\frac{(x+y)^2}{(x-y)^2}$ or $\frac{x^2+2xy+y^2}{x^2-2xy+y^2}$. g) $\frac{x}{x+3}$. h) $\frac{1}{x(x-1)}$. **7.** a) $\frac{x^2}{2}$. b) x. c) $\frac{3}{4}$. d) $\frac{x+4}{x+2}$.

Chapter 11, Lesson 6

4. a) $\frac{12}{5}$. b) 11. c) $-\frac{5}{2}$. d) 20. e) $\frac{x}{2}$. f) –8. g) $\frac{1}{5}$. h) $\frac{3}{4x}$. **5.** a) 4x. b) $\frac{x+36}{3}$. c) 10. d) $\frac{2x^2-5}{x}$. e) $\frac{6x}{x-6}$. f) $\frac{7x-36}{x-6}$. g) $\frac{x^2}{3}$. h) $\frac{1-3x^4}{3x}$. i) $\frac{x(x-1)}{4}$ or $\frac{x^2-x}{4}$. j) $\frac{5x-4}{4}$. k) 1. l) $\frac{x^2+14x+50}{x+7}$. **6.** a) $\frac{x(x-5)}{5}$ or

$\dfrac{x^2 - 5x}{5}$. b) $8(x + 1)$ or $8x + 8$. c) $\dfrac{3x^4}{3 + x}$.

d) $4(6x^3 - 1)$ or $24x^3 - 4$. e) $\dfrac{x^2 - 49}{x^2 + 49}$.

f) $\dfrac{(x + 10)(2x - 1)}{2x}$ or $\dfrac{2x^2 + 19x - 10}{2x}$.

7. a) $x^2 + x + \dfrac{6}{49}$. b) $\dfrac{4x^2}{9} - \dfrac{1}{4}$.

c) $x^2 - 12 + \dfrac{35}{x^2}$. d) $12x^2 - x - \dfrac{2}{25}$.

8. a) $\dfrac{49x^2 + 49x + 6}{49}$. b) $\dfrac{16x^2 - 9}{36}$.

c) $\dfrac{x^4 - 12x^2 + 35}{x^2}$. d) $\dfrac{300x^2 - 25x - 2}{25}$.

Chapter 11, Lesson 7

4. a) $\dfrac{1}{4}$. b) 3. c) $\dfrac{7}{5}$. d) x. e) $\dfrac{1}{2x + 1}$.

f) $\dfrac{x - 3}{x^2}$. **5.** a) $\dfrac{8}{7}$. b) 36. c) 20. d) $\dfrac{1}{90}$.

e) 8. f) $\dfrac{1}{2}$. **6.** a) $\dfrac{x^3}{8}$. b) $2x$. c) $\dfrac{2x^2 - x}{4}$.

d) 45. e) $5x^2$. f) $\dfrac{10}{x + 1}$. g) $\dfrac{24}{(x + 1)^2}$. h) $\dfrac{3}{2}$.

i) $\dfrac{x - 6}{x}$. j) $\dfrac{x}{x - 6}$. **7.** a) $\dfrac{1}{x^2}$. b) $\dfrac{x^3}{y^3}$. c) $\dfrac{x^3}{2}$.

d) $\dfrac{2(x + 2)}{x}$. e) $\dfrac{3}{4(x - 1)}$. f) -1. **8.** a) $\dfrac{5}{16}$.

b) 12. c) $\dfrac{(x + 3)(x + 4)}{(x - 3)(x - 4)}$ or $\dfrac{x^2 + 7x + 12}{x^2 - 7x + 12}$.

d) $\dfrac{1}{x + y}$. e) $\dfrac{x - 1}{x + 1}$. f) $\dfrac{x^2 + 4}{x^2}$.

Chapter 11, Lesson 8

4. a) $\dfrac{2}{5}$. b) $\dfrac{8}{11}$. c) $\dfrac{28}{1}$. d) $\dfrac{1}{9}$. e) $\dfrac{37}{9}$. f) $\dfrac{7}{25}$.

5. a) $\dfrac{2(x + 2)}{3(x + 3)}$ or $\dfrac{2x + 4}{3x + 9}$. b) $\dfrac{8x - 3}{8x + 5}$. c) $\dfrac{x}{x - 1}$.

d) $\dfrac{10}{3}$. e) $\dfrac{7}{x}$. f) $\dfrac{1}{4x(x - 4)}$. g) $x + 1$.

h) $\dfrac{x - 6}{x - 3}$. **6.** a) $\dfrac{1}{xy}$. b) $\dfrac{x^2}{y^2}$. c) $\dfrac{y - x}{1}$ or

$y - x$. d) $\dfrac{x^2 + x + 1}{x^2 + 1}$.

Chapter 11, Review

1. a) The second fraction was divided by the first.

b) $\dfrac{5}{9} \div \dfrac{1}{8} = \dfrac{40}{9} = 4\dfrac{4}{9}$; $\dfrac{8}{9} \div \dfrac{1}{3} = \dfrac{24}{9} = 2\dfrac{2}{3}$.

2. a) Not true; example: $x = 6$, $y = 3$, $\dfrac{6 - 2}{3 - 2} \neq \dfrac{6}{3}$.

b) Not true; example: $x = 1$, $y = 0$, $\dfrac{1 + 0}{1 - 0} \neq -1$.

c) True. d) True. e) True. f) Not true; example:

$x = 4$, $\dfrac{4^2 + 4}{4 + 2} \neq 4 + 2$. **3.** a) 9. b) 0. c) -1

and 1. **4.** a) $\dfrac{16}{25}$. b) $\dfrac{7}{x}$. c) $\dfrac{x - 3}{x}$. d) $\dfrac{x}{5}$.

e) Not possible. f) $x + 2$. **5.** a) $\dfrac{3}{14}$. b) $\dfrac{15}{3}$.

c) $\dfrac{4}{5}$. d) $\dfrac{16}{61}$. **6.** a) $2x$. b) $x + 1$. c) $\dfrac{4x + 4}{x^2 + 2x}$.

d) $\dfrac{6y - 6x}{xy}$. e) $\dfrac{x^4 - 1}{x^2}$. f) $\dfrac{x^2 + 9}{x - 3}$. **7.** a) $\dfrac{x}{3} + \dfrac{1}{5}$.

b) $\dfrac{x}{x - 4y} - \dfrac{2y}{x - 4y}$. c) $\dfrac{x^3}{x - 6} + 1$.

d) $3x - \dfrac{21}{x}$. **8.** a) $3x^4$. b) $\dfrac{x + 5}{x - 5}$. c) 1.

d) $\dfrac{x(3 - x)}{3}$ or $\dfrac{3x - x^2}{3}$. e) $\dfrac{(x + y)^2}{2xy}$ or

$\dfrac{x^2 + 2xy + y^2}{2xy}$. f) $\dfrac{9}{16}$. **9.** a) $x^2 - \dfrac{16}{81}$.

b) $6 - \dfrac{13}{x} + \dfrac{6}{x^2}$. **10.** a) $\dfrac{81x^2 - 16}{81}$.

b) $\dfrac{6x^2 - 13x + 6}{x^2}$. **11.** a) $\dfrac{3x}{4}$. b) $\dfrac{x - 10}{2(x - 5)}$ or

$\dfrac{x - 10}{2x - 10}$. c) $4x - 16$. d) $\dfrac{2}{5x^4}$. e) $\dfrac{x^6}{2}$. f) -1.

12. a) $\dfrac{x^2}{3}$. b) $\dfrac{x + 8}{x - 8}$. c) $\dfrac{xy}{y + x}$. d) $\dfrac{-1}{x - 1}$.

Chapter 12, Lesson 1

4. a) 7 and -7. b) 2 and -2. c) None. d) 30 and
-30. e) 0. f) None. **5.** a) 2,025. b) 2,916.
c) 6.71. d) Not possible. e) 78. f) 37. g) 51.84.
h) 7.3. **6.** a) 3 and 4. b) 4 and 5. c) -2 and
-1. d) -6 and -5. e) 11 and 12. f) 35 and 36.
7. a) 64. b) 6,400. c) 640,000. d) 64,000,000.
e) 6. f) 19. g) 60. h) 190.
8. a) $(11.2)^2 = 125.44$; $(11.22)^2 = 125.8884$;
$(11.225)^2 = 126.000625$. b) 11.2 is smaller than
$\sqrt{126}$; 11.22 is smaller than $\sqrt{126}$; 11.225 is larger
than $\sqrt{126}$. **9.** a) 12. b) -9. c) 16. d) 16.
e) 7. f) 7. g) 7. **10.** a) $=$. b) $=$. c) $=$.

d) =. e) <. f) >. g) =. h) >. i) <. j) >.
k) <. l) =. **11.** a) 8. b) 4. c) 6. d) 6.
e) 13. f) 17.

Chapter 12, Lesson 2

4. a) $\sqrt{10}$. b) $\sqrt{6x}$. c) $6\sqrt{x}$. d) $\sqrt{x^3}$.
e) $3y\sqrt{x}$. f) $3x\sqrt{y}$. g) $3\sqrt{xy}$. h) $8^{\frac{1}{2}}$. i) $5x^{\frac{1}{2}}$.
j) $(5x)^{\frac{1}{2}}$. k) $(xy)^{\frac{1}{2}}$. l) $x^{\frac{1}{2}}y$. m) $x(7y)^{\frac{1}{2}}$. n) $10xy^{\frac{1}{2}}$.
5. a) 3. b) 5. c) 24. d) 81. e) 2. f) 5. g) 20.
h) 93. **6.** a) $2\sqrt{3}$. b) $3\sqrt{7}$. c) $4\sqrt{2}$.
d) $3\sqrt{10}$. e) $5\sqrt{5}$. f) $6\sqrt{11}$. g) $3\sqrt{83}$.
h) $30\sqrt{6}$. **7.** a) 14.388. b) 14.424. c) 26.46.
d) 83.67. **8.** a) 7. b) x. c) $2x^3$. d) x^4. e) x^2.
f) $2x^{18}$. **9.** a) $3x$. b) x^{18}. c) Not possible.
d) $36x$. **10.** a) $8\sqrt{x}$. b) x^{32}. c) $x^3\sqrt{x}$.
d) $2x\sqrt{10}$. e) $11x\sqrt{x}$. f) $3x^9\sqrt{2}$.

Chapter 12, Lesson 3

4. a) $\dfrac{5}{25}$. b) $\dfrac{6}{9}$. c) $\dfrac{22}{36}$. d) $\dfrac{6x}{36}$. e) $\dfrac{4x}{x^4}$. f) $\dfrac{10x}{16x^2}$.
5. a) 4. b) $\dfrac{2}{5}$. c) 7. d) $\sqrt{2}$. **6.** a) $\dfrac{\sqrt{5}}{3}$.
b) $\dfrac{\sqrt{22}}{2}$. c) $\dfrac{\sqrt{42}}{7}$. d) $\dfrac{\sqrt{2}}{4}$. e) $\dfrac{\sqrt{21}}{6}$. f) $\dfrac{\sqrt{26}}{10}$.
7. a) 0.866. b) 0.3162. c) 1.291. d) 0.5916.
8. a) 8. b) $\sqrt{3}$. c) $\sqrt{14x}$. d) $\dfrac{x^8}{4}$. e) x^3.
f) 10. **9.** a) $\dfrac{\sqrt{x}}{9}$. b) $\dfrac{\sqrt{3}}{x^2}$. c) $\dfrac{x^3}{5}$. d) $\dfrac{\sqrt{10x}}{x}$.
e) $\dfrac{\sqrt{3x-6}}{3}$. f) $\dfrac{\sqrt{x}}{2x^2}$. **10.** a) 10 and 10. b) 30
and 30. c) 0 and 0. d) 5 and –5. e) 20 and –20.
f) 1 and 1. g) 2 and 2. h) 1 and –1. i) Yes.
j) No. k) Yes. l) No.

Chapter 12, Lesson 4

4. a) $3\sqrt{5}$. b) $5\sqrt{3}$. c) $7\sqrt{x}$. d) $4\sqrt{11x}$.
e) $\sqrt{10}+\sqrt{10}$.
f) $\sqrt{x}+\sqrt{x}+\sqrt{x}+\sqrt{x}+\sqrt{x}$. **5.** a) $11\sqrt{2}$.
b) $\sqrt{2}$. c) $7\sqrt{2}$. d) $5\sqrt{2}$. e) Not possible.
f) Not possible. g) $5\sqrt{3}$. h) $3\sqrt{5}$. i) $6\sqrt{2}$.
j) $4\sqrt{11}$. k) $\sqrt{6}$. l) $-\sqrt{6}$. **6.** a) 8. b) 6.3.
c) 5.6. d) 4. e) 0.2. f) 1. g) 3.5. h) 4.5. i) 4.
j) 2.8. **7.** a) $8+5\sqrt{5}$. b) $9+\sqrt{3}+\sqrt{6}$.
c) $20+4\sqrt{2}$. d) 2. e) $1+\sqrt{2}$. f) $48+2\sqrt{3}$.

8. a) $20\sqrt{x}$. b) $6x$. c) $x+\sqrt{x}$. d) $4\sqrt{x}$.
e) $72\sqrt{xy}$. f) Not possible. g) $3\sqrt{7x}$.
h) $2\sqrt{2x}$. **9.** a) $\sqrt{2}$. b) $\sqrt{19}-3$. c) $11\sqrt{11}$.
d) –4. e) $2\sqrt{2}$. f) 8.

Chapter 12, Lesson 5

4. a) 9. b) $2\sqrt{10}$. c) 33. d) 60. e) $10\sqrt{15}$.
f) 360. g) 600. h) 42. i) $16+2\sqrt{8}$.
j) $4+8\sqrt{2}$. k) $7-\sqrt{7}$. l) 12. **5.** a) $9x$.
b) x^5. c) $12x$. d) $35x^6$. e) $16+4\sqrt{x}$.
f) $4\sqrt{x}+x$. **6.** a) 8. b) 15. c) 45. d) 75.
e) $14+6\sqrt{5}$. f) $28+10\sqrt{3}$. g) $8+2\sqrt{15}$.
h) 8. **7.** a) 9. b) 167. c) 11. d) $x-16$.
e) x^2-10. f) $xy-6$. **8.** a) $11+16\sqrt{2}$.
b) $20-14\sqrt{2}$. c) $1+\sqrt{5}+\sqrt{6}+\sqrt{30}$.
d) $30+21\sqrt{3}$.

Chapter 12, Lesson 6

4. a) 5. b) 4. c) $\dfrac{1}{2}$. d) $\sqrt{11}$. e) $2\sqrt{3}$. f) $\dfrac{\sqrt{5}}{5}$.
5. a) 2.83. b) 0.84. c) 0.14. d) 1.58. **6.** a) 5.
b) x^{12}. c) $\dfrac{x}{2}$. d) $2x^3$. **7.** a) $8-\sqrt{2}$.
b) $\sqrt{15}+6$. c) $x+\sqrt{y}$. d) $\sqrt{x}-\sqrt{y}$.
8. a) $\dfrac{3-\sqrt{5}}{4}$. b) $3+\sqrt{6}$. c) $\sqrt{11}-\sqrt{2}$.
d) $\dfrac{4+\sqrt{7}}{3}$. **9.** a) 2.73. b) 2.13. c) 1.46.
d) 1.69. **10.** a) $\dfrac{\sqrt{x}}{x}$. b) $\dfrac{\sqrt{x}-1}{x-1}$. c) $x+\sqrt{y}$.
d) $\dfrac{x-2\sqrt{xy}+y}{x-y}$. **11.** a) $\sqrt{5}$. b) $\sqrt{11}$.
c) $5\sqrt{15}$. d) $5\sqrt{3}$. e) $\sqrt{2}$. f) $6\sqrt{3}$.

Chapter 12, Lesson 7

4. a) 6. b) $\dfrac{1}{2}$. c) 4. d) 14. e) 5. f) 7. g) Not
possible. h) –7. **5.** a) 3. b) 7. c) 10. d) 12.
e) It is 2 more than x.
f) $\sqrt{x^2+4x+4}=\sqrt{(x+2)^2}=x+2$. g) No;
$\sqrt{(-10)^2+4(-10)+4}=8$, 8 is not 2 more than
–10. **6.** a) Yes. b) No. c) Yes. d) Yes.
7. a) 45. b) 9. c) 220. d) 100. e) No solution.
f) –4. g) No solution. h) –2. i) 16. j) No

solution. k) 10. l) 2.25. m) –115. n) No
solution. **8.** a) $10 + \sqrt{10}$. b) $10\sqrt{10}$. c) $2\sqrt{3}$.
d) 3. e) $\sqrt{2}$. f) $2\sqrt{2}$.

Chapter 12, Review

1. a) 8 and –8. b) None. c) 0. d) $\frac{1}{3}$ and $-\frac{1}{3}$.

2. a) 25. b) 25. c) 10. d) 48. e) 18. f) 30.
3. a) $=$. b) $=$. c) $<$. d) $=$. e) $>$. f) $=$.
4. a) $3\sqrt{5}$. b) $7\sqrt{2}$. c) $2\sqrt{26}$. d) $100\sqrt{6}$.
5. a) $6\sqrt{x}$. b) x^{18}. c) $x^5\sqrt{10}$. d) $3x\sqrt{x}$.
6. a) $\dfrac{\sqrt{6}}{5}$. b) $\dfrac{\sqrt{2}}{6}$. c) $\dfrac{\sqrt{7x}}{x}$. d) $\dfrac{\sqrt{6x}}{2}$.

7. a) 5.5. b) 7.7. c) 3.9. d) 4.5. **8.** a) $8\sqrt{5}$.
b) $8\sqrt{2}$. c) $9\sqrt{3}$. d) Not possible. e) Not
possible. f) $9\sqrt{x}$. **9.** a) $7 + 7\sqrt{7}$.
b) $34 + 5\sqrt{2}$. c) $-5\sqrt{3}$. **10.** a) 48.
b) $19 + 8\sqrt{3}$. c) $7 - 4\sqrt{3}$. d) 18. **11.** a) 10.
b) 33. c) $24 - 4\sqrt{6}$. d) x^4. e) $x + 7\sqrt{x}$.
f) $10 + 7\sqrt{2}$. g) $x - 144$. h) $4 - 5\sqrt{x} + x$.
12. a) 4. b) $2\sqrt{3}$. c) $12(2 - \sqrt{3})$. d) $\dfrac{\sqrt{66} + 6}{5}$.

13. a) 0.77. b) 0.26. c) 3.74. **14.** a) $2\sqrt{3}$.
b) $2\sqrt{7}$. c) 160. d) 1,590. e) 900. f) 2. g) No
solution. h) 64.

Chapter 13, Lesson 1

4. a) $4x - 11 = 0$; degree 1. b) $x^3 - 6x^2 + 8 = 0$;
degree 3. c) $x^2 + 2x - 5 = 0$; degree 2.
d) $2x^5 - 1 = 0$; degree 5. e) $7x - 20 = 0$;
degree 1. f) $2x^4 + 9x^2 - x = 0$; degree 4.
5. a) Yes. b) Yes. c) Yes. d) No. e) Yes.
f) Yes. g) Yes. h) Yes. i) Yes. j) Yes. k) Yes.
l) Yes. **6.** a) Yes. b) No. c) Yes. d) No.
e) Zero is a solution if x is a factor of every term on
the left side. **7.** a) –10. b) 19. c) 4. d) 0.
e) –0.6. f) –8. g) 7. h) –25.

Chapter 13, Lesson 2

4. a) 1. b) Linear. c) A straight line.
d)

x	–3	–2	–1	0	1	2	3
y	–1	1	3	5	7	9	11

e)

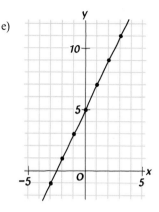

5. a) 2. b) Quadratic. c) A curve.
d)

x	–4	–3	–2	–1	0	1	2	3	4
y	20	12	6	2	0	0	2	6	12

e)

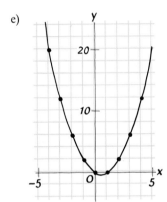

6. a–c)

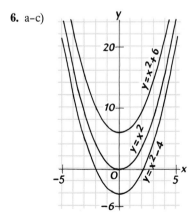

d) All have the same shape.

7. a)

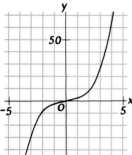

b)

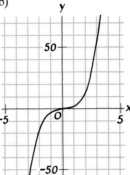

c)

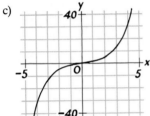

Chapter 13, Lesson 3

4. a) −1.   b) 0 and 4.5.   c) 2.

5.

a)

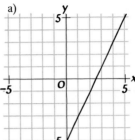

b)

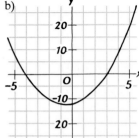

c)

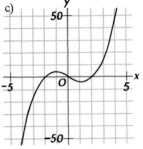

d)

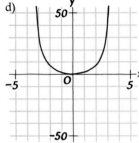

6. a) 2.5.   b) −4 and 3.   c) −2, 0, and 2.   d) 0.

7.

a)

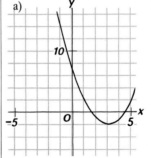

b)

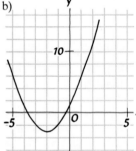

c)

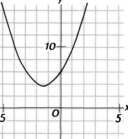

d)

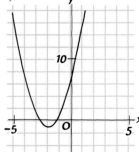

8.

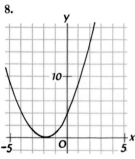

9.

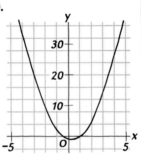

10.

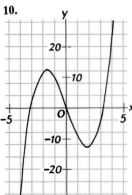

11.

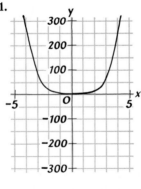

8. a) Between 1 and 2; between 4 and 5.
b) Between –4 and –3; between –1 and 0.   **c)** No
solution.   **d)** Between –3 and –2; between –2 and –1.

Chapter 13, Lesson 4

4. a) 4 and 5.   **b)** 0 and –8.   **c)** 0.5.   **d)** 0 and 10.

e) –1 and 3.   **f)** 4 and –0.25.   **g)** $-a$ and $-b$.   **h)** $\dfrac{1}{a}$

and $\dfrac{c}{b}$.   **5. a)** –3 and 3.   **b)** 0 and 9.   **c)** 0 and

–1.   **d)** –3 and –4.   **e)** 5.   **f)** –7 and 2.   **g)** 0 and

–12.   **h)** 8 and –8.   **i)** –16 and 4.   **j)** $\dfrac{1}{3}$ and $-\dfrac{1}{3}$.

k) 0 and $\dfrac{8}{9}$.   **l)** $-\dfrac{1}{9}$ and 1.   **6. a)** $x^2 - 25 = 0$; 5

and –5.   **b)** $x^2 + 9x - 22 = 0$; 2 and –11.
c) $x^2 - 10x = 0$; 0 and 10.   **d)** $x^2 + 8x + 7 = 0$; –1
and –7.   **e)** $4x^2 + 6x - 4 = 0$; –2 and 0.5.
f) $2x^2 - 2x - 24 = 0$; 4 and –3.

Chapter 13, Lesson 5

4. a) 12 and –12.   **b)** $\sqrt{30}$ and $-\sqrt{30}$.   **c)** $2\sqrt{3}$ and
$-2\sqrt{3}$.   **d)** $5\sqrt{6}$ and $-5\sqrt{6}$.   **5. a)** 9, –3.   **b)** –4,
–18.   **c)** 2, $\dfrac{6}{5}$.   **d)** $10 + \sqrt{3}$, $10 - \sqrt{3}$.

e) $1 + 4\sqrt{5}$, $1 - 4\sqrt{5}$.   **f)** $2 + \sqrt{2}$, $2 - \sqrt{2}$.
6. a) $x^2 - 6x - 27 = 0$; 9 and –3.   **b)** $x - 3 = \pm6$;
9 and –3.   **c)** The square-root method.
7. a) $-4 + \sqrt{15}$ and $-4 - \sqrt{15}$.
b) $x^2 + 8x - 1 = 0$; $x^2 + 8x - 1$ doesn't factor.
8. a) $3x - 6 = \pm\sqrt{45}$; $2 + \sqrt{5}$ and $2 - \sqrt{5}$.
b) 4.236 and –0.236.   **c)** $[3(4.236) - 6]^2 = 45$,
$(12.708 - 6)^2 = 45$, $(6.708)^2 = 45$, $44.997264 \approx 45$;
$[3(-0.236) - 6]^2 = 45$, $(-0.708 - 6)^2 = 45$,
$(-6.708)^2 = 45$, $44.997264 \approx 45$.   **9. a)** $4\sqrt{3}$ and
$-4\sqrt{3}$.   **b)** 7 and 3.   **c)** 3 and –7.   **d)** $-1 + \sqrt{10}$ and
$-1 - \sqrt{10}$.   **e)** 6 and –3.   **f)** 1 and $-\dfrac{11}{3}$.

10. a) 7 and –1.   **b)** $-8 + 3\sqrt{5}$ and $-8 - 3\sqrt{5}$.

c) 0 and –5.   **d)** $\dfrac{1 + \sqrt{5}}{5}$ and $\dfrac{1 - \sqrt{5}}{5}$.

11. a) 9.236 and 4.764.   **b)** 2.317 and –4.317.
c) 0.833.   **d)** 0.020 and –0.770.

Chapter 13, Lesson 6

4. a) $x^2 + 8x + 16 = (x + 4)^2$.
b) $x^2 - 30x + 225 = (x - 15)^2$.
c) $4x^2 - 12x + 9 = (2x - 3)^2$.
d) $25x^2 + 70x + 49 = (5x + 7)^2$.
e) $x^2 + 2ax + a^2 = (x + a)^2$.
f) $a^2x^2 - 4ax + 4 = (ax - 2)^2$.   **5. a)** Seven was
added to each side of the equation.   **b)** Four was
added to each side of the equation and the left-hand
side was factored as a binomial square.   **c)** The
square root of each side of the equation was taken.
d) Two was subtracted from each side of the
equation.   **e)** Each side of the equation was divided
by 3.   **6. a)** $x^2 + 6x - 40 = 0$,
$(x + 10)(x - 4) = 0$; –10 and 4.
b) $x^2 + 6x + 9 = 40 + 9$; 4 and –10.
7. a) $x^2 - 10x + 25 = 36 + 25$; $5 + \sqrt{61}$ and
$5 - \sqrt{61}$.   **b)** 12.81 and –2.81.
c) $(12.81)^2 - 10(12.81) = 36$, $164.0961 - 128.1 = 36$,
$35.9961 \approx 36$; $(-2.81)^2 - 10(-2.81) = 36$,
$7.8961 + 28.1 = 36$, $35.9961 \approx 36$.

8. a)

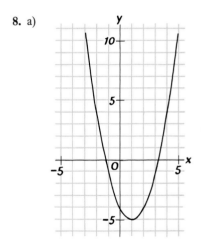

b) Between –2 and –1 and between 3 and 4.
c) $1 + \sqrt{5}$ and $1 - \sqrt{5}$.   **d)** 3.24 and –1.24.
9. a) 7 and –11.   **b)** 7 and 5.   **c)** –9.   **d)** $1 + \sqrt{11}$
and $1 - \sqrt{11}$.   **e)** $-8 + \sqrt{65}$ and $-8 - \sqrt{65}$.

f) $3 + 3\sqrt{2}$ and $3 - 3\sqrt{2}$.   **g)** $\dfrac{7}{3}$ and $-\dfrac{5}{3}$.

h) $\dfrac{-5 + \sqrt{14}}{4}$ and $\dfrac{-5 - \sqrt{14}}{4}$.   **10. a)** $-2 + \sqrt{26}$

and $-2 - \sqrt{26}$.   **b)** $2\sqrt{6}$ and $-2\sqrt{6}$.   **c)** $4 + \sqrt{22}$
and $4 - \sqrt{22}$.   **d)** $-6 + \sqrt{34}$ and $-6 - \sqrt{34}$.

Chapter 13, Lesson 7

4. a) $x^2 - 5x + 3 = 0$; $a = 1$, $b = -5$, $c = 3$.
b) $4x^2 - x - 9 = 0$; $a = 4$, $b = -1$, $c = -9$.
c) $6x^2 + 3x = 0$; $a = 6$, $b = 3$, $c = 0$.
d) $2x^2 + 14x - 11 = 0$; $a = 2$, $b = 14$, $c = -11$.
5. a) $3 + \sqrt{21}$ and $3 - \sqrt{21}$. b) $3 + \sqrt{21}$ and $3 - \sqrt{21}$. c) 7.58 and –1.58.
d) $(7.58)^2 - 6(7.58) = 12$, $57.4564 - 45.48 = 12$, $11.9764 \approx 12$; $(-1.58)^2 - 6(-1.58) = 12$, $2.4964 + 9.48 = 12$, $11.9764 \approx 12$.
6. a) $\dfrac{-3 + \sqrt{5}}{2}$ and $\dfrac{-3 - \sqrt{5}}{2}$. b) $\dfrac{-3 + \sqrt{5}}{2}$ and $\dfrac{-3 - \sqrt{5}}{2}$. c) The quadratic formula. **7.** a) 7 and –2. b) 7.1 and –2.1. c) 7.2 and –2.2. d) 6.6 and 0.5. e) 7.6 and 0.4. f) 8.7 and 0.4. **8.** a) 8 and –2. b) $-\dfrac{1}{4}$ and –2. c) $\dfrac{5 + \sqrt{17}}{4}$ and $\dfrac{5 - \sqrt{17}}{4}$.
d) $\dfrac{\sqrt{10}}{4}$ and $\dfrac{-\sqrt{10}}{4}$. e) 0 and $-\dfrac{1}{3}$. f) –5.
9. a) $\dfrac{-1 + \sqrt{61}}{2}$ and $\dfrac{-1 - \sqrt{61}}{2}$. b) $\dfrac{1 + \sqrt{129}}{8}$ and $\dfrac{1 - \sqrt{129}}{8}$. c) –1 and –8. d) $2 + 2\sqrt{5}$ and $2 - 2\sqrt{5}$. **10.** a) $\dfrac{-b + \sqrt{b^2 - 4c}}{2}$ and $\dfrac{-b - \sqrt{b^2 - 4c}}{2}$. b) $\dfrac{1 + \sqrt{1 + 4de}}{2d}$ and $\dfrac{1 - \sqrt{1 + 4de}}{2d}$. c) $\dfrac{\sqrt{a}}{a}$ and $-\dfrac{\sqrt{a}}{a}$. d) 0 and $-\dfrac{g}{f}$.

Chapter 13, Lesson 8

4. a) 4; two solutions. b) 0; one solution. c) –8; no solutions.

5. a)

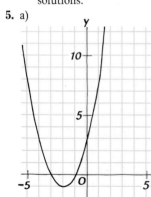

b)

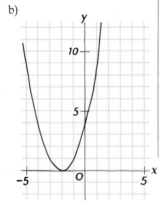

c)

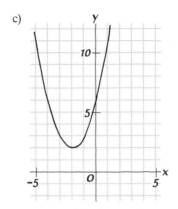

d) –3 and –1. e) –2. f) No solutions. **6.** a) –1 and –3. b) –2. c) No solutions. **7.** a) –1 and –3. b) –2. c) No solutions. **8.** a) 8; two solutions. b) –15; no solutions. c) 0; one solution. d) 20; two solutions. **9.** a) 1.5 and 4.5. b) No solutions. c) –0.5. d) 2.2 and –2.2. **10.** a) 4.41 and 1.59. b) No solutions. c) –0.5. d) 2.24 and –2.24.

Chapter 13, Lesson 9

3. a) $x^5 - 8x = 0$; 5. b) $x^3 + 4x - 10 = 0$; 3. c) $3x + 1 = 0$; 1. d) $x^4 - 7x^3 + 2x - 14 = 0$; 4. **4.** a) Yes. b) No. c) Yes. d) Yes. **5.** a) –4, –1.3, and 4. b) –5, –2, 0, 3, and 5.

6. a)

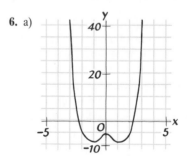

b) –2.2 and 2.2. **7.** a) 0, –1, and 2. b) 0, $\dfrac{1 + \sqrt{13}}{2}$, and $\dfrac{1 - \sqrt{13}}{2}$. c) –1, 1, –3, and 3. d) $\sqrt{5}$ and $-\sqrt{5}$. **8.** a) 0, –3, and 0.5. b) $3 + \sqrt{5}$ and $3 - \sqrt{5}$. c) –2 and 2.

Chapter 13, Review

1. a) $6x^5 - 2x + 3 = 0$; quintic; 5. b) $7x - 44 = 0$;

linear; 1. c) $x^3 - 81x = 0$; cubic; 3.   **2.** a) No.
b) Yes.   **3.** a) $x^4 - 2x^3 - 12 = 0$. b) 2. c) –1.6
and 2.6.

4. a)

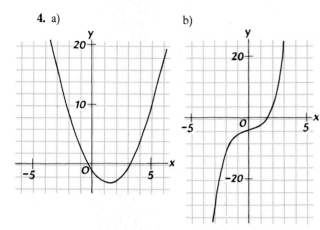

b)

5. a) –0.3 and 3.3. b) 1.6.   **6.** a) 8 and –8.
b) –7 and 5. c) 1 and $-\frac{1}{3}$. d) 2 and 9.
7. a) $3\sqrt{2}$ and $-3\sqrt{2}$. b) $-1 + \sqrt{10}$ and
$-1 - \sqrt{10}$. c) 3 and $-\frac{5}{3}$. d) 4 and –1.   **8.** a) 10
and –8. b) $-4 + \sqrt{22}$ and $-4 - \sqrt{22}$.   **9.** a) –71;
no solutions. b) 4; two solutions. c) 0; one
solution.   **10.** a) 1 and –0.2. b) $4 + \sqrt{11}$ and
$4 - \sqrt{11}$. c) $\frac{1}{3}$. d) $\frac{-1 + \sqrt{13}}{3}$ and $\frac{-1 - \sqrt{13}}{3}$.
11. a) 0, $\sqrt{6}$, and $-\sqrt{6}$. b) 2 and –2.   **12.** a) 2
and 5. b) 10 and –4. c) $-1 + \sqrt{6}$ and $-1 - \sqrt{6}$.
d) 7 and –3. e) $\frac{1 + \sqrt{29}}{2}$ and $\frac{1 - \sqrt{29}}{2}$.
f) 0, 5, and –5.

Chapter 14, Lesson 1

4.

x	$\dfrac{1}{x}$	$\dfrac{1}{x}$
1	1.00000	1
2	0.50000	0.5
3	0.33333	$0.\overline{3}$
4	0.25000	0.25
5	0.20000	0.2
6	0.16666	$0.1\overline{6}$
7	0.14285	$0.\overline{142857}$
8	0.12500	0.125

5. a) 0.2. b) 0.04. c) 0.008. d) 0.0016. e) It
ends. f) Even. g) n. h) 0.00032. i) 0.00000256.
j) Yes.   **6.** a) $0.\overline{09}$. b) $0.\overline{009}$. c) $0.\overline{0009}$. d) It
repeats. e) 10. f) 9.   **7.** a) $\frac{33}{10}$. b) $\frac{68}{25}$.
c) $\frac{1}{250}$. d) $\frac{7}{16}$.   **8.** a) $\frac{2}{3}$. b) $\frac{23}{9}$. c) $\frac{5}{11}$.
d) $\frac{1}{12}$. e) $\frac{13}{22}$. f) $\frac{11}{54}$.   **9.** a) 19. b) It would
stop. c) 18.   **10.** a) $\frac{1}{3}$; $0.\overline{3}$. b) $\frac{2}{5}$; 0.4. c) $\frac{5}{11}$;
$0.\overline{45}$. d) $\frac{50}{101}$; $0.\overline{4950}$. e) It increases.

Chapter 14, Lesson 2

4. a–j)

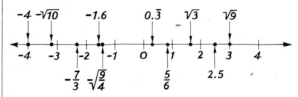

k) $\sqrt{3}$ and $-\sqrt{10}$.   **5.** a) Rational. b) Irrational.
c) Rational. d) Irrational. e) Irrational.
f) Rational. g) Irrational. h) Rational.
6. a) Not possible; irrational. b) 4; rational. c) Not
possible; irrational. d) 3.5; rational. e) Not
possible; irrational. f) 6; rational. g) Not possible;
irrational. h) 0.5; rational. i) $0.2\overline{6}$; rational.
j) $0.1\overline{3}$; rational.   **7.** a) 1.4. b) Smaller. c) $\frac{49}{25}$.
d) Smaller. e) $1.41\overline{6}$. f) Larger. g) $\frac{289}{144}$.
h) Larger.   **8.** a) 36.000. b) 34.810. c) 35.046.
d) 34.999. e) 35.000. f) Each number has a
decimal form that ends. g) 5.9 and 5.916 are smaller
than $\sqrt{35}$; 6, 5.92, and 5.9161 are larger than $\sqrt{35}$.
h) No, because $\sqrt{35}$ is irrational.   **9.** a) 1 and $\frac{1}{4}$.
b) $\frac{3 + \sqrt{5}}{4}$ and $\frac{3 - \sqrt{5}}{4}$. c) $\frac{1}{2}$. d) $\frac{2 + \sqrt{3}}{2}$ and
$\frac{2 - \sqrt{3}}{2}$. e) It must be the square of an integer.

Chapter 14, Lesson 3

4.

x	x^2	x^3	x^4	x^5
1	1	1	1	1
2	4	8	16	32
3	9	27	81	243
4	16	64	256	1,024
5	25	125	625	3,125

5. a) Irrational. b) Rational; 2. c) Rational; 5. d) Irrational. e) Rational; 9. f) Rational; 3. g) Rational; $\frac{1}{3}$. h) Rational; $\frac{2}{3}$. i) Irrational. j) Rational; $\frac{2}{5}$. **6.** a) It gets smaller. b) 1. c) It gets larger. d) 1. **7.** a) The y-coordinate is the square of the x-coordinate. b) The x-coordinate is the square root of the y-coordinate. c)

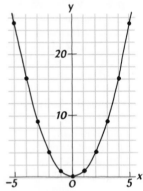

d) –3.2 and 3.2.
e) –4.2 and 4.2.
f) –4.7 and 4.7.
g) Two. h) None. **8.** a)

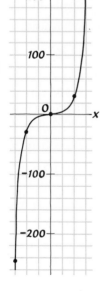

b) 1.8. c) 2.4. d) –1.6. e) One. f) One.
9. a) 10. b) –2. c) –5 and 5. d) None exists. e) 1. f) –3. **10.** a) $>$. b) $>$. c) $=$. d) $<$. e) $<$. f) $>$. g) $>$. h) $>$.

Chapter 14, Lesson 4

4. a) $\frac{256}{81}$. b) 3.1605. c) Larger. d) $\left(\frac{4}{3}\right)^4$ is rational and π is irrational.
5. a)

Radius	1	2	3	4	5
Circumference	2π	4π	6π	8π	10π

b) Yes. c) Directly. **6.** a) Circumference, π cm; area, 0.25π square cm. b) Circumference, $2\pi(\sqrt{2})$ cm; area, 2π square cm. c) Circumference, $2\pi^2$ cm; area, π^3 square cm. d) Circumference, 6 cm; area, $\frac{9}{\pi}$ square cm. **7.** a) 4.5. b) $\frac{7}{\pi}$. c) 3. d) $2\sqrt{5}$.
8. a) 1,600. b) $16x^2$. c) 400π. d) $4\pi x^2$.
9. a) 27π. b) $3\pi x^2$. c) $\frac{3}{4}$. **10.** a) 113 inches.
b) 2 inches. c) 53 square centimeters. d) 40 centimeters.

Chapter 14, Lesson 5

4. a) 8, $\sqrt[3]{8} = 2$, and $8^2 = 64$. b) $0^8 = 0$ and –8. c) 0.8, $0.\overline{8}$, and $\frac{1}{8}$. d) $\sqrt{8}$ and 8π. **5.** a) No; example: $1 - 2 = -1$. b) Yes. c) No; example: $\frac{1}{2} = 0.5$. d) Yes. e) No; example: $\sqrt{2}$ is irrational. **6.** a) Yes. b) Yes. c) Yes. d) No; example: $\frac{2}{4} = 0.5$. e) Yes. **7.** a) Yes. b) Yes. c) Yes. d) Yes. e) Yes. f) No; example: $\sqrt{2}$ is irrational. **8.** a) 0.0008, 0.009, 0.07, 0.6. b) 0.183, $0.\overline{183}$, $0.18\overline{3}$, $0.1\overline{83}$. c) $\sqrt[4]{10}$, $\sqrt[3]{10}$, $\sqrt{10}$. d) $\sqrt[3]{-2}$, $\sqrt[3]{-1}$, $\sqrt[3]{1}$, $\sqrt[3]{2}$. e) $-\pi$, $-\frac{\pi}{2}$, $\frac{\pi}{2}$, π. **9.** a) 1.82.
b) 0.43. c) –0.04. d) 0.32. e) –2.21. **10.** a) $\frac{22}{7}$; rational. b) 1; counting number. c) $\sqrt{2}$; irrational. d) $\sqrt{5}$ and $-\sqrt{5}$; irrational. e) 5 and –1; integers. f) 49; counting number.

Chapter 14, Review

1. a) True. b) True. c) True. d) False.
e) True. **2.** a) 2.5. b) 0.28. c) 4.$\overline{8}$.

d) 0.$\overline{153846}$. **3.** a) $\frac{3}{50}$. b) $\frac{1}{15}$. c) $\frac{9}{8}$.

d) $\frac{1124}{999}$. **4.** a) 17. b) It repeats zeros.

c) 16. **5.** a) –1. b) Not possible. c) $\frac{5}{3}$. d) $\frac{5}{4}$.

6. 0.027, 0.0$\overline{27}$, 0.0$\overline{27}$, 0.02$\overline{7}$. **7.** a) –4. b) 3
and –3. c) 0. d) 2. **8.** a) $<$. b) $=$. c) $<$.
d) $>$. **9.** a) 5 and 1. b) $3 + \sqrt{3}$ and $3 - \sqrt{3}$.
c) $3 + \sqrt{2}$ and $3 - \sqrt{2}$. d) 3. e) It must be the
square of an integer. **10.** a) 3.141$\overline{8}$. b) Larger.
c) π is not rational. **11.** a) $2\pi\sqrt{5}$. b) 0.09π.

c) $\frac{7}{\pi}$. d) $10\sqrt{2}$. **12.** a) 75π. b) $3\pi x^2$. c) $\frac{3}{4}$.

13. a) 38 inches. b) 7,850 square centimeters.
14. a) 9 and $\sqrt{9} = 3$. b) $0^9 = 0$ and –9. c) 0.9

and $\frac{1}{9}$. d) $\sqrt[3]{9}$ and 9π. **15.** a) No; example:

$3 + 1 = 4$. b) No; example: $3 - 1 = 2$. c) Yes.

d) No; example: $1 \div 3 = \frac{1}{3}$. e) Yes. **16.** a) 2.4;

rational. b) $3\sqrt{6}$; irrational. c) $\sqrt{6}$ and $-\sqrt{6}$;
irrational. d) 121; counting number.

Chapter 15, Lesson 1

4. a) $\frac{9}{19}$. b) $\frac{1}{19}$. c) 9. d) 1. **5.** a) $\frac{2}{3}$. b) Not

possible. c) $\frac{2x + 1}{3x + 1}$. d) $\frac{2}{3}$. e) Not possible.

f) $\frac{1}{x^2}$. g) $\frac{4}{x^3}$. h) Not possible. **6.** a) 6. b) 13.2.

c) $\frac{5}{3}$. d) 41. e) True for all x except –1. f) 10

and –10. g) No solution. h) –1.3. i) 8 and –8.
j) 1.
7. a)

Lucy's age	1	2	3	6	10	20	30
Ethel's age	7	8	9	12	16	26	36

b) 7, 4, 3, 2, 1.6, 1.3, 1.2. c) $r = \frac{n + 6}{n}$. d) It gets

smaller. e) No; if the ratio of their ages became
equal to 1, Ethel and Lucy would be the same age.
8. a)

Time in minutes	1	5	6.25	12.5
Amount of gas in ounces	1.6	8	10	20

b) 1.6. c) Directly. d) $g = 1.6t$.

Chapter 15, Lesson 2

4. a) $20 + x$. b) $3x - 2$. c) $8 + 2x$. d) $x^2 - 15$.
e) $4x + 26$. f) $3x^2 + 4x - 2$.

5. a) $x - 12 = \frac{3x}{4}$. b) $\frac{4x}{3} - 16 = x$.

c) $4x - 48 = 3x$. e) It contains no fractions.
6. a) 0.2. b) 2.5. c) –3. d) 42. e) 15. f) No
solution. g) 0. h) 8. i) –3.5. j) 5.
7. a) $(15, 8)$. b) $(6, -11)$.

Chapter 15, Lesson 3

4. a) 5 and –5. b) 2 and –0.5. c) 8 and –5.
d) –1. **5.** a) 8.48 and –0.48. b) 3.65 and –1.65.

c) 0.17. d) –0.38 and –2.62. **6.** a) Ollie, $\frac{1}{80}$;

Alice, $\frac{1}{60}$. b) Ollie, $\frac{x}{80}$; Alice, $\frac{x}{60}$.

c) $\frac{x}{80} + \frac{x}{60} = 1$. d) Approximately 34.3

minutes. **7.** a) Time going, $\frac{x}{15}$; time returning, $\frac{x}{10}$.

b) $\frac{x}{15} + \frac{x}{10}$. c) 12 kilometers per hour.

Chapter 15, Lesson 4

4. a) $h = ab$. b) About 4,184 hits. **5.** a) $x = \frac{-b}{a}$.

b) $\frac{-6}{2} = -3$. c) 0.6. **6.** a) –10. b) –21.

c) $-ab$. d) 4. e) 9. f) a. **7.** a) b. b) $\frac{1 + b}{a}$.

c) $\frac{a(a + b)}{b}$. d) 0. e) $\frac{a + b}{c}$. f) $a + b$.

g) $\pm\sqrt{a^2 - 1}$. h) $\frac{a^2 + b^2}{c^2}$. **8.** a) 55 degrees.

b) $n = 4(F - 40)$. c) 128. d) 40.

Chapter 15, Lesson 5

4. a) 7. b) 13. c) $a + b$. d) $\frac{1}{3}$. e) $\frac{1}{7}$.

f) $\frac{1}{a - 1}$. **5.** a) $\frac{1}{a + 1}$. b) $\frac{a}{1 - b}$. c) $\frac{a + b}{a - b}$.

d) $\frac{abc}{b - a}$. e) $\frac{a}{1 + ab}$. f) $\pm\sqrt{a^2 + b^2}$.

6. a) 39. b) $a = \frac{(b + c)h}{2}$. c) $h = \frac{2a}{b + c}$. d) 4.8.

7. a)

A	20	35	50	65
p	112	121	134.5	152.5

b) It gets higher. c) No. **8.** a) 4.8 centimeters.

b) $a = \dfrac{bf}{b - f}$. c) 15 centimeters.

Chapter 15, Review

1. a) $\dfrac{5}{7}$. b) $\dfrac{5}{2}$. c) $\dfrac{7}{2}$. **2.** a) $\dfrac{1}{3x - 1}$. b) Not

possible. c) $\dfrac{1}{3}$. d) Not possible. **3.** a) $45 + x$.

b) $7x - 2$. c) $4 + 6x$. d) $x^2 - x - 24$.
4. a) 4.9. b) 1.25. c) True for all numbers except
–5. d) 5. e) –36. f) –15. g) 1.2. h) $\sqrt{10}$ and

$-\sqrt{10}$. i) $\dfrac{-5 + 2\sqrt{5}}{5}$ and $\dfrac{-5 - 2\sqrt{5}}{5}$. j) –4 and

2.5. **5.** a) 16π cubic centimeters. b) $b = \dfrac{3v}{4\pi a^2}$.

c) 8 centimeters. **6.** a) $\dfrac{a^2 - b^2}{b^2}$. b) $a + 1$.

c) $\sqrt{a}$ and $-\sqrt{a}$. d) $\dfrac{b}{a - b}$. e) $\dfrac{a}{ab + 1}$.

7. a) 273.$\overline{3}$ degrees. b) $F = \dfrac{9K - 2300}{5}$.

c) 1,475 degrees. **8.** a) $\dfrac{4}{3}$ hours.

b) $x = \dfrac{abc}{bc + ac + ab}$.

c) $\dfrac{3 \cdot 4 \cdot 6}{4 \cdot 6 + 3 \cdot 6 + 3 \cdot 4} = \dfrac{72}{24 + 18 + 12} = \dfrac{72}{54} = \dfrac{4}{3}$.

Chapter 16, Lesson 1

4. a) $>$. b) $<$. c) $<$. d) $>$. e) $>$. f) $<$.
g) $=$. h) $<$. i) $>$. j) $>$. k) $<$. l) $>$.
m) $<$. n) $=$. o) $>$. **5.** a) $2 < 9$ and $9 > 2$.
b) $-4 < 4$ and $4 > -4$. c) $0 < x$ and $x > 0$.
d) $y < -3$ and $-3 > y$. **6.** a) E. b) A. c) F.
d) B. e) C. f) D. **7.** a) $<$. b) $<$. c) $=$.
d) $>$. e) $>$. f) $<$. g) $>$. h) $<$. i) $<$. j) $<$.
k) $<$. l) $<$. m) $<$. n) $<$. **8.** a) $<$. b) $<$.
c) $<$. d) $<$. e) $<$. f) $<$. g) $<$. h) $<$.
i) $<$. j) Not defined. k) $>$. l) $>$. m) $<$.
n) $>$. **9.** a) $<$. b) None. c) $\geq$. d) None.
e) None. f) $>$. **10.** a) $>$. b) $<$. c) None.
d) $>$. e) $>$. f) None.

Chapter 16, Lesson 2

4.
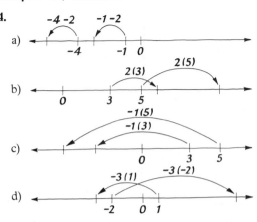

5. a) The direction is the same. b) The direction is
reversed. c) The direction is the same.
6. a) Yes. b) Yes. c) No. d) Yes. e) Yes.
f) Yes. g) Yes. h) No. i) Yes. j) Yes. k) No.
l) Yes. m) Yes. n) No. o) No. **7.** a) Seven was
added to both sides. b) Both sides were divided by
three. c) Ten was subtracted from both sides.
d) Both sides were divided by –6. e) Both sides
were multiplied (or divided) by –1. f) Both sides
were multiplied by –4. **8.** a) $x < -7$. b) $x < \dfrac{1}{8}$.

c) $x > -10$. d) $x < 3$. e) $x \leq -18$. f) $x \geq -18$.
g) $7 < x < 14$. h) $-8 \leq x \leq 2$. **9.** a) $x > -3$.

b) $x \geq -2$. c) $x \geq 15$. d) $x > \dfrac{3}{4}$. e) $5 < x < 10$.

f) $x < 5$. g) True for all numbers. h) $40 \geq x \geq 0$.

Chapter 16, Lesson 3

4. a) $x > 2$; Figure C. b) $x > 5$; Figure A.
c) $x \leq 5$; Figure D. d) $x > 2$; Figure C.
e) $2 < x < 5$; Figure B. f) $5 \geq x$; Figure D.
5.

a) $x < 3$.

b) $x > -8$.

c) $2 \geq x$.

d) $x \leq 0$.

e) $-1 > x$.

f) $3 < x$.

g) $-24 \geq x$.

h) $x > 4$.

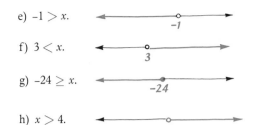

6. a) $x > \dfrac{b}{a}$. b) $x > \dfrac{a}{b}$. c) $0 \leq x \leq b - a$.

d) $x > \dfrac{-1}{a - b}$. **7.** a) $x < 3 + 5$, $x < 8$;

$3 < x + 5$, $-2 < x$; $5 < x + 3$, $2 < x$.

b) $x < 12 + x + 6$, $0 < 18$; $12 < x + x + 6$, $3 < x$;

$x + 6 < x + 12$, $6 < 12$. c) $x < 2x + 6$, $-6 < x$;

$6 < x + 2x$, $2 < x$; $2x < x + 6$, $x < 6$.

d) $4 < (2 + x) + (8 - x)$, $4 < 10$;

$2 + x < 4 + (8 - x)$, $x < 5$; $8 - x < 4 + (2 + x)$,

$1 < x$.

Chapter 16, Lesson 4

4. a) 15 and -15. b) 17 and 5. c) 17 and -5. d) 6 and -10. e) $x + 3$ and $x - 3$. f) $b + a$ and $b - a$. **5.** a) C. b) A. c) D. d) B.

6. a) Yes. b) No. c) Yes. d) No. e) Yes.

f) Yes. g) Yes. h) No. i) Yes. j) Yes. k) No.

l) Yes. **7.** a) 8 and -8. b) 0. c) No solution.

d) $-3 < x < 3$. e) $-5 \leq x \leq 5$. f) $x \geq 4$ or

$x \leq -4$. g) True for all numbers. h) $x > 30$ or

$x < -30$.

8. a)
b)
c)
d)
e)
f)
g)
h)
i)

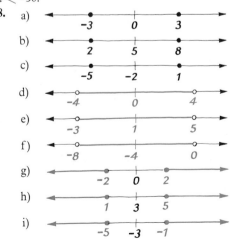

9. a) 9 and -9. b) 9 and 5. c) $-13 < x < 13$.
d) $-5 < x < 13$. e) $x \geq 12$ or $x \leq -12$.
f) $x \geq 12$ or $x \leq -10$. g) $-13 \leq x \leq 3$. h) True
for all numbers. **10.** a) $171 \leq x \leq 173$.
b) $|x - 172| \leq 1$.

Chapter 16, Review

1. a) $<$. b) $>$. c) $>$. d) $<$. e) $>$. f) $=$.
g) $>$. h) $=$.   **2.** a) $x \leq 5$. b) $0 \leq x \leq 8$.
c) $x > -2$. d) $-6 < x < 4$.   **3.** a) $>$. b) None.
c) $\geq$. d) None.   **4.** a) $>$. b) None. c) $<$.
d) $>$.   **5.** a) Yes. b) No. c) No. d) No.
e) Yes. f) Yes. g) No. h) Yes. i) Yes.

6. a) $x < -3$. b) $x < \dfrac{2}{5}$. c) $x < -10$. d) $x \leq 3$.

e) $6 < x < 7$. f) $x > -6$. g) $x \leq 20$. h) $x < \dfrac{1}{2}$.

i) $x > 12$. j) $x > -7$. k) $x < -7$. l) $x \leq 12$.

7. a) $x > -\dfrac{b}{a}$. b) $x > \dfrac{ab}{c}$. c) $x \geq \dfrac{b}{a - c}$.

d) $-a > x > -b$.   **8.** a) 8 and -8. b) 18 and -12.
c) 2 and -12. d) $y + x$ and $y - x$.

9.

a)
b)
c)
d)
e)
f)

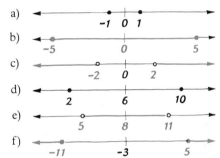

10. a) -8 and 8. b) $-14 < x < 14$. c) -10 and 10.
d) 10 and 8. e) True for all numbers. f) $x > -5$
or $x < -11$. g) -2. h) True for all numbers.

Chapter 17, Lesson 1

4. a) Arithmetic; common difference, 7.
b) Geometric; common ratio, 2. c) Neither.
d) Arithmetic; common difference, -2. e) Geometric;
common ratio, 10. f) Neither. g) Both; common
difference, 0; common ratio, 1. h) Geometric;
common ratio, 1.5. i) Arithmetic; common
difference, 3.5. j) Neither. k) Geometric; common
ratio, -3. l) Neither. m) Geometric; common

ratio, $\dfrac{1}{2}$. n) Arithmetic; common difference, $\dfrac{1}{6}$.

o) Geometric; common ratio, $\frac{2}{3}$.　**5.** a) 22.

b) 324.　c) 17.　d) –23.　e) 6.　f) 125.　g) $\frac{6}{5}$.　h) $\frac{5}{4}$.
i) 0.025.　j) 1,875.　k) $\sqrt{11}$.　l) –0.3.　**6.** a) 7, 9,
11, 13, 15.　b) By adding 2.　c) Arithmetic.
7. a) 10, 20, 40, 80, 160.　b) By multiplying by 2.
c) Geometric.　**8.** a) –30, 90.　b) 2, 250.　c) 10,
17.　d) 6, 12.　**9.** a) $t_n = n + 3$.　b) $t_n = n^2$.
c) $t_n = 5n$.　d) $t_n = -8$.　e) $t_n = 4^n$.
f) $t_n = n - 6$.　g) $t_n = \frac{7}{n}$.　h) $t_n = \sqrt{n}$.

Chapter 17, Lesson 2

4. a) 3, 5, 7, 9, 11.　b) 2.

c)　　　　　d)　　　　　e)

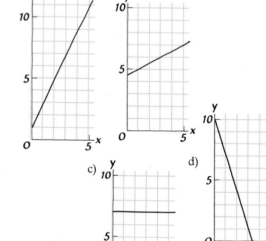

5.

a)　　　　　b)

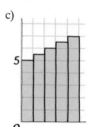
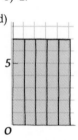
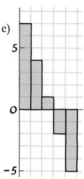

c)　　　　　d)

e) Linear.　**6.** a) 8, 11, 14.　b) 3.　c) 35.
d) 305.　e) 215.　f) 15,650.　**7.** a) 6, –4, –14.
b) –10.　c) –84.　d) –984.　e) –390.　f) –48,900.
8. a) 23.　b) 20.　c) 17, 31.　d) 19, 4.　e) $6\frac{1}{3}, 6\frac{2}{3}$.

f) 7.5, –7.5.　g) 1, 4, 7.　h) $\frac{7}{12}$.

9. a) $t_n = 2 + (n - 1)5$; 27.　b) $t_n = 16 + (n - 1)9$;
97.　c) $t_n = 92 - (n - 1)3$; 32.
d) $t_n = 20 - (n - 1)11$; –101.　e) $t_n = -6$; –6.
f) $t_n = -70 + (n - 1)12$; 1,130.　**10.** a) 126.
b) 306.　c) 465.　d) 1,275.　e) 1,800.　f) –2,450.
11. a) 280.　b) 780.　c) –253.　d) 1,387.5.
12. a) 21.　b) 78.　**13.** a) 21.5 miles.
b) 935 miles.

Chapter 17, Lesson 3

4. a) 1, 3, 9, 27.　b) 3.

c)　　　　　d)　　　　　e)

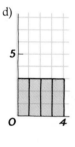
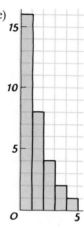

5.

a)　　　　　b)　　　　　c)

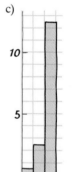

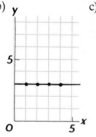

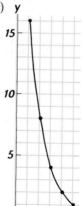

d) Exponential.　**6.** a) 0.5, 2, 8.　b) 4.　c) $\frac{1}{2} \cdot 4^9$.

d) $\frac{1}{2} \cdot 4^{99}$.　**7.** a) 15, 5, $\frac{5}{3}$.　b) $\frac{1}{3}$.　c) $15 \cdot \left(\frac{1}{3}\right)^9$.

d) $15 \cdot \left(\frac{1}{3}\right)^{99}$.　**8.** a) 2,000.　b) 12.　c) 4, 16.

d) 5, –135.　e) 7^0, 7^1.　f) $\frac{1}{10}$.

9. a) $t_n = 6 \cdot 2^{n-1}$.　b) $t_n = 64 \cdot \left(\frac{1}{8}\right)^{n-1}$.

c) $t_n = 5 \cdot (-1)^{n-1}$.　d) $t_n = 4 \cdot \left(\frac{7}{2}\right)^{n-1}$.

e) $t_n = 2 \cdot (\sqrt{6})^{n-1}$.　f) $t_n = 300 \cdot \left(-\frac{1}{10}\right)^{n-1}$.

10. a) 728.　b) 390,624.　c) –341.　d) 65,532.
11. a) 2, 4, 8, 16, 32, 64, 128, 256, 512, 1,024.
b) 2,046.　c) 2^n.　d) $S_n = 2(2^n - 1)$.　**12.** a) 1, 3,
9, 27, 81, 243, 729, 2,187.　b) 3,280.　c) 3^{63}.

d) $\frac{3^{64} - 1}{2}$.

Chapter 17, Lesson 4

4. a) 1.2.　b) 0.9.　c) 1.　d) –0.5.　e) –1.　f) –1.4.
g) 0.3.　h) 1.01.　**5.** a) Sequences a, f, and h; their
absolute values are greater than 1.　b) Sequences b,
d, and g; their absolute values are less than 1.

6. a) 4, 4.8, 5.8, 6.9, 8.3, 10.

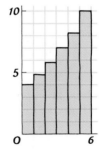

c) 5, –5.5, 6.1,
　–6.7, 7.3, –8.1.

b) 4, 3.2, 2.6, 2, 1.6, 1.3.

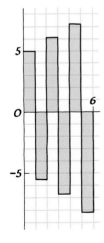

d) 5, –4.5, 4.1, –3.6, 3.3, –3.

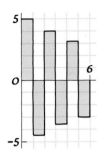

7. a) 81.　b) $\frac{3}{4}$.　c) 192.　d) $\frac{5}{18}$.　**8.** a) $\frac{1}{3}$.

b) $\frac{5}{11}$.　c) $\frac{8}{37}$.　d) $\frac{7}{90}$.　**9.** a) 16, 8, 4, 2, 1.

b) 31.　c) 32.　**10.** a) 62.5 meters.

Chapter 17, Review

1. a) Arithmetic; common difference, 6.　b) Neither.
c) Geometric; common ratio, 2.5.　d) Neither.
e) Both; common difference, 0; common ratio, 1.

f) Geometric; common ratio, $\frac{1}{9}$.　**2.** a) 80.

b) 33.　c) $2\sqrt{5}$.　d) $\frac{3}{10}$.　e) 0.081.　f) 50.

3. a) 24.　b) 27.　c) 25.　d) 20.　**4.** a) 1, 8, 27,
64.　b) 3, 9, 27, 81.　c) 11, 16, 21, 26.　d) 7, 6, 5,

4.　e) $\frac{1}{2}, \frac{2}{3}, \frac{3}{4}, \frac{4}{5}$.　f) 40, 400, 4,000, 40,000.

g) c and d.　h) b and f.　**5.** a) $t_n = n^2$; 121.

b) $t_n = n - 7$; 43.　c) $t_n = \frac{1}{n}$; $\frac{1}{35}$.　d) $t_n = -4n$;

–80.　e) $t_n = 3^n$; 2,187.　f) $t_n = \sqrt{n + 2}$; 10.

g) $t_n = \frac{n}{8}$; 3.　h) $t_n = 3 \cdot 6^{n-1}$; 23,328.

6. a) 287.　b) –24.　c) 5,700.　d) 1,560.
7. a) 2,186.　b) 23,325.　c) –2,046.　**8.** a) 0.1.

b) 1.2.　c) –0.45.　**9.** a) 50.　b) 243.　c) $\frac{3}{7}$.

10. a) $\frac{1}{9}$.　b) $\frac{6}{11}$.　c) $\frac{41}{333}$.　**11.** a) 829, 1,039,
1,249, 1,459, 1,669.　b) $t_n = 199 + (n - 1)210$.
c) 2,299.　d) Composite; 2,299 = 11 · 209.
12. a) 256, 128, 64, 32, 16, 8, 4, 2, 1.　b) 9.
c) 511.　**13.** a) 12 centimeters.　b) 42
centimeters.　c) Shortly after end of second month;

second month: $42 + \frac{2}{5}(42) = 58.8$, third month:

$58.8 + \frac{2}{5}(58.8) = 82.32$.

Final Review

1. False. **2.** $\frac{9}{x}$. **3.** Slope, 1; y-intercept, 7.

4. $\frac{1}{5x}$. **5.** $3\sqrt{3}$. **6.** -15 and 1. **7.** 1.125.

8. Quadratic. **9.** -3. **10.** $-8 \le x \le 8$.

11. $x\sqrt{x}$. **12.** $0.35, 0.3\overline{5}, 0.3\overline{5}$. **13.** $<$.

14. $(5, 2)$. **15.** 3.6×10^9. **16.** $\frac{3x}{10}$. **17.** $\frac{1}{8}$.

18. 8 and -8. **19.** $y = 2^x$. **20.** $\frac{4}{3}$.

21. acx^{b+d}. **22.** 3.7×10^5. **23.** 42.

24. 4 and -4. **25.** $t_n = 3^n$. **26.** $\frac{3}{2}$.

27. $-4 < x < 10$. **28.** $2 \cdot 5^2 \cdot 7$.

29. $x + 11y$. **30.** Rational. **31.** $2 + \frac{5}{x}$.

32. 1.5. **33.** -0.5. **34.** 13.5. **35.** $x^3 - 1$.

36. $\frac{x}{x+2}$. **37.** $3 + \sqrt{7}$ and $3 - \sqrt{7}$.

38. $\frac{1}{6}$. **39.** $\frac{x-3}{x}$. **40.** True.

41. $x = ab + 1$. **42.** True. **43.** $x^2 + x - 3$.
44. 100.
45.

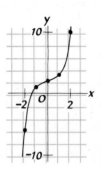

46. $x^5 + 4x^2 - 5x$. **47.** 0, 2, and -5.
48. $2\sqrt{3}$. **49.** $a < c$. **50.** $3y(5x + 3)$.

51. $x < \frac{1-a}{b}$. **52.** 3. **53.** It is reversed.

54. $6x^2 + 5x + 1$. **55.** 180.

56. $(x^3 + 2)(x^3 - 2)$. **57.** $(3x^8)^2$. **58.** $\frac{1}{x+y}$.

59. $x \ge 4$ or $4 \le x$. **60.** $x < 9$.
61. $(x - 3)(x - 7)$. **62.** $3\sqrt{2}$. **63.** $3x + 30$.
64. 3 and 4. **65.** -3, 0, and 2.
66. $x(x + y)^2$. **67.** -5. **68.** $x^2 - x + 2$.
69. 5 and -2. **70.** $x^6 + 2x^3 + 1$. **71.** It gets

smaller (closer and closer to 1). **72.** $\frac{x}{5}$.

73. 2 and -2. **74.** $p = 2a + b$. **75.** $A = \frac{1}{2}hb$.

76. $x + y = 15$; $20x + 30y = 340$. **77.** 11 packets

at 20 cents, 4 packets at 30 cents. **78.** $y = \frac{x}{7}$ or

$y = \frac{1}{7}x$. **79.** Directly. **80.** It would be a

straight line through the origin.

Index